Names, Formulas, and Charges of Common I[ons]

Positive Ions (Cations)			Negative Ions (Anions)		
1+	Ammonium	NH_4^+	Acetate	$C_2H_3O_2^-$	
	Copper(I)	Cu^+	Bromate	BrO_3^-	
	Potassium	K^+	Bromide	Br^-	
	Silver	Ag^+	Chlorate	ClO_3^-	
	Sodium	Na^+	Chloride	Cl^-	
2+	Barium	Ba^{2+}	Chlorite	ClO_2^-	
	Cadmium	Cd^{2+}	Cyanide	CN^-	
	Calcium	Ca^{2+}	Fluoride	F^-	
	Cobalt(II)	Co^{2+}	Hydride	H^-	
	Copper(II)	Cu^{2+}	Hydrogen carbonate (Bicarbonate)	HCO_3^-	
	Iron(II)	Fe^{2+}	Hydrogen sulfate (Bisulfate)	HSO_4^-	
	Lead(II)	Pb^{2+}	Hydrogen sulfite (Bisulfite)	HSO_3^-	
	Magnesium	Mg^{2+}	Hydroxide	OH^-	1-
	Manganese(II)	Mn^{2+}	Hypochlorite	ClO^-	
	Mercury(II)	Hg^{2+}	Iodate	IO_3^-	
	Nickel(II)	Ni^{2+}	Iodide	I^-	
	Tin(II)	Sn^{2+}	Nitrate	NO_3^-	
	Zinc	Zn^{2+}	Nitrite	NO_2^-	
3+	Aluminum	Al^{3+}	Perchlorate	ClO_4^-	
	Antimony(III)	Sb^{3+}	Permanganate	MnO_4^-	
	Arsenic(III)	As^{3+}	Thiocyanate	SCN^-	
	Bismuth(III)	Bi^{3+}	Carbonate	CO_3^{2-}	
	Chromium(III)	Cr^{3+}	Chromate	CrO_4^{2-}	
	Iron(III)	Fe^{3+}	Dichromate	$Cr_2O_7^{2-}$	
	Titanium(III)	Ti^{3+}	Oxalate	$C_2O_4^{2-}$	
4+	Manganese (IV)	Mn^{4+}	Oxide	O^{2-}	2-
	Tin(IV)	Sn^{4+}	Peroxide	O_2^{2-}	
	Titanium(IV)	Ti^{4+}	Silicate	SiO_3^{2-}	
5+	Antimony(V)	Sb^{5+}	Sulfate	SO_4^{2-}	
	Arsenic(V)	As^{5+}	Sulfide	S^{2-}	
			Sulfite	SO_3^{2-}	
			Arsenate	AsO_4^{3-}	
			Borate	BO_3^{3-}	
			Phosphate	PO_4^{3-}	3-
			Phosphide	P^{3-}	
			Phosphite	PO_3^{3-}	

BASIC CONCEPTS OF CHEMISTRY

NINTH EDITION

Leo J. Malone
Saint Louis University

Theodore O. Dolter
Southwestern Illinois College

In Collaboration With

Steven Gentemann
Southwestern Illinois College

John Wiley & Sons, Inc.

VICE PRESIDENT AND EXECUTIVE PUBLISHER	Kaye Pace
ASSOCIATE PUBLISHER	Petra Recter
ACQUISITIONS EDITOR	Nick Ferrari
PROJECT EDITOR	Jennifer Yee
MARKETING MANAGER	Kristine Ruff
PRODUCT DESIGNER	Geraldine Osnato
MEDIA SPECIALIST	Evelyn Brigandi
CREATIVE DIRECTOR	Harry Nolan
SENIOR DESIGNER	James O'Shea
SENIOR PHOTO EDITOR	Lisa Gee
EDITORIAL ASSISTANT	Lauren Stauber
PRODUCTION SERVICES	Ingrao Associates
COVER IMAGE	© Sylvain Grandadam/age fotostock

This volume contains selected illustrations from the following texts, reprinted by permission from John Wiley and Sons, Inc.

- Brady, James E.; Senese, Fred, *Chemistry: Matter and Its Changes, Fourth Edition*, ©2004.
- Brady, James E.; Senese, Fred, *Chemistry: Matter and Its Changes, Fifth Edition*, ©2009.
- Hein, Morris; Arena, Susan, *Foundations of College Chemistry, Twelfth Edition*, ©2007.
- Murck, Barbara; Skinner, Brian, MacKenzie, *Visualizing Geology, First Edition*, ©2008.
- Olmsted III, John; Williams, Gregory M., *Chemistry, Fourth Edition*, ©2006.
- Pratt, Charlotte W.; Cornely, Kathleen, *Essential Biochemistry, First Edition*, ©2004.
- Raymond, Kenneth, *General, Organic, and Biological Chemistry: An Integrated Approach, Second Edition*, ©2008.

This book was set in ITC New Baskerville 10/12 by Preparé, and printed and bound by Courier/Kendallville. The cover was printed by Courier/Kendallville.

This book is printed on acid-free paper.

Founded in 1807, John Wiley & Sons, Inc. has been a valued source of knowledge and understanding for more than 200 years, helping people around the world meet their needs and fulfill their aspirations. Our company is built on a foundation of principles that include responsibility to the communities we serve and where we live and work. In 2008, we launched a Corporate Citizenship Initiative, a global effort to address the environmental, social, economic, and ethical challenges we face in our business. Among the issues we are addressing are carbon impact, paper specifications and procurement, ethical conduct within our business and among our vendors, and community and charitable support. For more information, please visit our website: *www.wiley.com/go/citizenship*.

Library of Congress Cataloging-in-Publication Data

Malone, Leo J., 1938-
  Basic concepts of chemistry / Leo J. Malone, Theodore O. Dolter, Steve Gentemann. – 9th ed.
    p. cm.
  Includes index.
  ISBN 978-0-470-93845-4 (hardback)
  1. Chemistry.   I. Dolter, Theodore O., 1965-   II. Gentemann, Steve.   III. Title.
    QD31.3.M344 2012
    540–dc23

                            2011019567

ISBN 978-0-470-93845-4

Printed in the United States of America
10 9 8 7 6 5 4 3 2

▶ LEO J. MALONE

Leo Malone is a native of Kansas where he received his B.S. in Chemistry from Wichita State University in 1960 and M.S. in Chemistry in 1962. At WSU he worked under the direction of Dr. Robert Christian. He moved on to the University of Michigan where he received his Ph.D. in 1964 under the direction of Dr. Robert Parry. Dr. Malone began his teaching career at Saint Louis University in 1965 where he remained until his retirement as Professor Emeritus in 2005. Although his early research at SLU involved boron hydride chemistry, he eventually concentrated his efforts on the teaching of basic chemistry and in the field of chemical education.

▶ THEODORE (TED) O. DOLTER

Ted Dolter received his B.S. in Chemistry from St. Louis University in 1987, where he was a student of Dr. Malone's. He went on to the University of Illinois where he received a Masters of Chemical Education in 1990. He concurrently earned a secondary teaching certificate, and it was there that he received most of his training in modern educational theory. After six years of teaching high school chemistry and evening courses at Southwestern Illinois College, he joined the faculty at SWIC full time. He served as department chair of the Physical Science Department from 2004 to 2010, during which time he was in charge of the department's outcomes assessment activities.

Professor Dolter's background in educational training and his knowledge of the needs of the growing numbers of community college students compliment Dr. Malone's years of traditional university educational experience. Together, they have produced a text that remains flexible and applicable to the rapidly changing face of today's post-secondary student population.

BRIEF CONTENTS

Contents

v

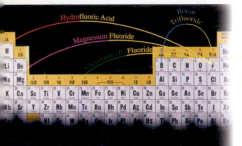

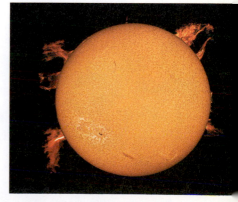

Why we wrote this book

Basic Concepts of Chemistry was written over thirty years ago to address the needs of general chemistry students with little or no background in chemistry. Over time, the text evolved beyond purposes solely aimed at a preparatory chemistry course. For some preparatory chemistry students, a main sequence in general chemistry may follow, but for others, a semester of organic and biochemistry may follow. Other students enroll to simply satisfy a basic science or chemistry requirement. The text was written with a level and functionality designed to accommodate the needs of each of these varied groups of students. *Basic Concepts of Chemistry* was designed with a flexibility that allowed instructors to emphasize or omit certain clearly delineated sections. Over the next eight editions, the mission of the text has evolved in response to the increased diversity of students and the emphasis on outcomes assessment. Leo Malone, Professor Emeritus in the Chemistry Department of St. Louis University, was the sole author of the first seven editions of *Basic Concepts of Chemistry*. Ted Dolter, Professor of Chemistry at SWIC (Southwestern Illinois Community College) joined with Leo Malone for the eighth edition. Ted brought a rich talent in Chemical Education with expertise in outcomes assessment as an integral part of the eighth, and now ninth, edition. Professor Steve Gentemann, a colleague of Ted's at SWIC, has collaborated with the team for the ninth edition. Steve, a St. Louis native, received his B.S. in 1987 and M.S. in 1989, both from the Department of Chemistry of St. Louis University. In 1994, Steve received his Ph.D. from Washington University where he studied the ultrafast photophysical behavior of porphyrins and authored several publications. Steve joined the faculty at Southwestern Illinois College in 1998 and has taught introductory, general, and organic chemistry.

Basic Concepts of Chemistry Today

Today, more students are entering post-secondary education with a diversity of learning styles and varied stages of preparedness to further their studies. Recent data on student populations indicate that many of them are visual and kinesthetic learners. It is critical therefore, that textbook authors both recognize and integrate this pedagogy into their books and overall learning solution. Achievement of this goal has been of paramount importance to us. Furthermore, the students that enroll in preparatory chemistry have diverse mathematical backgrounds and so we have provided essential resources to assist students struggling with the mathematics encountered in studying chemistry. To accommodate visual and kinesthetic learners, and to support students with weaker math backgrounds, we continue to emphasize outcomes assessment, which provides timely feedback. This emphasis on learning outcomes is evident both within the text and in our online teaching and learning solution, *WileyPLUS*.

The Role of Outcomes Assessment in the Ninth Edition

Surveys of programs across the country show that chemistry is being taught in numerous ways. From packed lecture halls to intimate classrooms, from Ph.D.s to adjunct instructors, to teaching assistants, chemistry education is being delivered in multiple formats, often within the same institution. Outcomes Assessment is an attempt to insure consistency in evaluating student achievement across these multiple formats. By delineating the expected outcomes or objectives for each chapter, and then devising assessment tools, such as homework and exam questions, lab experiments, group work, and the like, that are designed to target those specific objectives, schools can insure that all students in all sections are being served. By incorporating outcomes assessment into the curriculum, students can receive the same topical instruction and be evaluated against the same standard.

Implementing and justifying an outcomes assessment program can be time consuming. To that end, the authors, having already been through an outcomes assessment program, have designed an overarching solution with objectives and assessments already in place, along with the required information needed to show how all of it ties together.

Learning Objectives

In the Ninth Edition, each chapter section includes a list of the relevant objectives for that chapter.

These are measurable outcomes that the student should master by the completion of that part of the chapter. Assessments of varying complexity follow each section so that the student, upon completion, can evaluate to what degree the material has been internalized. These assessments are also available within *WileyPLUS* so that students can engage in the material and get immediate answer feedback, along with some question assistance. *Level 1* tests basic definitions and concepts. *Level 2* requires application of section concepts and *Level 3* requires a more thorough understanding and may require extension to more advanced concepts, or incorporation of previously learned material.

▶ **ASSESSING THE OBJECTIVE FOR SECTION 3-1**

EXERCISE 3-1(a) LEVEL 1: Identify the following as an example of a physical or chemical change:

(a) boiling	**(c)** evaporating	**(e)** rotting
(b) burning	**(d)** rusting	**(f)** dissolving

EXERCISE 3-1(b) LEVEL 2: Calcium, an element, is a dull, gray solid that melts at 839°C. When it is placed in water, bubbles form, as the solid calcium slowly disappears in the water. When the water is evaporated, a white powder remains, but elemental calcium is not recovered. Which are the physical properties of calcium? Which is a chemical property?

EXERCISE 3-1(c) LEVEL 2: A beaker of an unknown clear and colorless liquid has a volume of 100.0 mL and a mass of 78.9 g. Initially, its temperature is 25°C. When heated, it boils at 78.5°C. If ignited, it burns completely with a blue flame, leaving no residue behind. Which of these pieces of information will be helpful in identifying the liquid?

EXERCISE 3-1(d) LEVEL 3: In lab you are handed a metallic object and charged with determining its identity. What types of things can you do to determine what the object is made from?

For additional practice, work chapter problems 3-4, 3-6, and 3-10.

Each chapter concludes with an objectives grid which correlates the objectives to examples within the sections, assessment exercises at the end of each section, and relevant chapter problems at the end of each chapter. This correlation is emulated within the *WileyPLUS* course. The organization of our content around objectives holds utility for both the student and instructor, allowing them to easily locate tools within the text and in our *WileyPLUS* course to help address a specific student need.

OBJECTIVES

SECTION	YOU SHOULD BE ABLE TO...	EXAMPLES	EXERCISES	CHAPTER PROBLEMS
3-1	List several properties of matter and distinguish them as physical or chemical.		1a, 1b, 1c	4, 5, 6, 9, 11
3-2	Perform calculations involving the density of liquids and solids.	3-1, 3-2, 3-3, 3-4	2a, 2b, 2c	12, 14, 16, 20, 22, 24 28, 29
3-3	Perform calculations involving the percent of a pure substance in a mixture.	3-5, 3-6	3a, 3b, 3c, 3d, 3e, 3f, 3g	41, 42, 48, 49, 50, 52, 53, 55
3-4	Define terms associated with energy exchanges in physical and chemical processes.		4a, 4b, 4c	56, 57, 59, 60
3-5	Perform calculations involving the specific heat of a substance.	3-7, 3-8, 3-9, 3-10	5a, 5b, 5c, 5d, 5e, 5f	63, 64, 67, 68, 69, 71, 79, 81

Additional Math Support For Students

- **Math Check** - The chemical concepts mentioned in Chapter 1 are again presented in Chapters 2 and 3 but with more discussion and detail. When a math concept is to be required in an example problem in these two chapters a new feature called **Math Check** is presented, which allows a student to quickly assess the needed basic skill. If the student is not mathematically ready for the problem, he or she is referred back to Chapter 1 or the math appendices for further review before tackling the problem. With this new feature, it is now a viable option for instructors who wish to avoid the heavy early emphasis on mathematics to begin the text with Chapter 2 and use the Math Checks to alert the student to skills as needed, along with the references to the relevant discussions.

✔ MATH CHECK:

Math Operations and Significant Figures

In the example problem that follows, you will be asked to divide two numbers expressed in different degrees of precision. Correctly expressing the results of mathematical operations using numbers with different numbers of decimal places or significant figures is a necessary skill in chemistry. Unfortunately, the calculator can't help us here. We have to know the rules. The following questions test your understanding of these principles.

    a. What is the sum of 11.841 mL, 0.009 mL, and 5.1 mL?

    b. What is the result and unit of the following calculation: 98.020 cm^3 divided by 32.0 cm?

Answers: a. 17.0 mL **b.** 3.06 cm^2

Refer to Section 1-2.1 for help with significant figures related to addition and subtraction.

Refer to Section 1-2.2 for help with significant figures related to multiplication and division.

Facilitating Problem Solving

- **Chapter Synthesis Problem** - At the end of each chapter we have included a multi-faceted problem that brings the key concepts of the chapter into one encompassing problem. It is a challenging but entirely workable problem that tests the overall understanding of what may seem like diverse concepts. The worked-out solutions to the problem immediately follow. We then present a mirror problem for the student entitled **Your Turn** with only the answers provided at the end of the chapter.

CHAPTER 3 SYNTHESIS PROBLEM

In this chapter, we learned that density and specific heat are intensive physical properties that can be used to identify pure substances. These properties can also be used to convert among a number of measurements. Density can be used to convert between volume and mass and specific heat can be used to convert among heat, mass, and temperature change. Solutions are homogeneous mixtures where the concentration can be expressed as percent by mass, which allows us to convert among mass of solute, mass of solvent, and mass of solution. At a specified concentration, solutions also have a certain density and specific heat. Consider the following solution of ethanol (ethyl alcohol) in water.

PROBLEM	SOLUTION
a. The density of water at 25.0°C is 0.9971 g/mL. What is the mass of 1255 mL of water?	**a.** $1255 \text{ mL} \times \dfrac{0.9971 \text{ g}}{\text{mL}} = \underline{1251 \text{ g water}}$
b. What mass of ethanol must be mixed with this amount of water to form a 5.45% by mass ethanol solution?	**b.** For a 5.45% solution, in 100 g solution, there are 5.45 g ethanol and $100.00 - 5.45 = 94.55$ g water $1251 \text{ g water} \times \dfrac{5.45 \text{ g ethanol}}{94.55 \text{ g water}} = \underline{72.1 \text{ g ethanol}}$
c. What volume of ethanol is needed for this solution? (See Table 3-1.)	**c.** $72.1 \text{ g ethanol} \times \dfrac{1 \text{ mL}}{0.790 \text{ g ethanol}} = \underline{91.3 \text{ mL}}$
d. What is the total mass of the solution?	**d.** $72.1 \text{ g ethanol} + 1251 \text{ g water} = \underline{1323 \text{ g (solution)}}$
e. Determine the amount of heat (in kJ) needed to warm the solution from 25.0°C to 35.0°C. The specific heat of the solution is 3.97 J/g · °C.	**e.** $\text{specific heat} = \dfrac{\text{joules}}{\text{g} \cdot °\text{C}}$ $\text{joules} = \text{specific heat} \times \text{g} \times °\text{C} = \dfrac{3.97 \text{ J}}{\text{g} \cdot °\text{C}} \times 10.0 °\text{C} \times 1323 \text{ g}$ $= 52500 \text{ J} \times \dfrac{1 \text{ kJ}}{10^3 \text{ J}} = \underline{52.5 \text{ kJ}}$

YOUR TURN

Windshield-washing solvent is a solution of methanol and water (and a small amount of detergent which we will neglect in this problem).

    a. Determine the volume of water (density = 0.9971 g/mL) that must be combined with 375 mL of methanol (density = 0.7914 g/mL) to make a 45.0% by mass methanol solution.

    b. Determine the total mass of the solution.

    c. Determine the amount of heat (in kJ) that is required to warm the solution from 25.0°C to 55.0°C. The specific heat of the solution is 3.52 J/g · °C

Answers on p. 111.

Organization

The **Prologue** is a unique feature which introduces the origin of science in general and chemistry in particular. There are no quizzes, exercises, or problems; rather, it is meant to be a relaxing, historical glimpse at the origin of this fascinating subject and how it now affects our lives. Our intent is build interest and engage the student in further study. Also included in the prologue is a discussion of the scientific method and tips on studying chemistry and effectively using this textbook.

We recognize the changing needs of students and balance that with the requirements to successfully study chemistry. As such, we continue to provide the necessary support for students continuing on in the study of chemistry. **Chapter 1, Measurements in Chemistry,** provides the necessary math tools in a non-threatening way. This chapter now includes brief introductions to several fundamental chemical concepts such as elements, compounds, atoms, molecules and atomic weight. The intention is to bring relevance to the math concepts used in chemistry, which are introduced in the first chapter. Based upon feedback from reviewers, the authors agreed that a math introduction would be more meaningful if the student first understood why the skills are necessary.

In **Chapter 2, Elements and Compounds,** the elements are introduced starting from what we see and sense about us (the macroscopic) to the atoms of which they are composed, and finally into the structure within the atom (the microscopic and submicroscopic.) We do the same for compounds in the second part of this chapter. **Chapter 3, The Properties of Energy and Matter,** continues the discussion of matter and its properties. Some additional yet relevant math concepts such as density, percent composition, and specific heat are introduced in this chapter as properties of matter. **Chapter 4, The Periodic Table and Chemical Nomenclature,** allows us to draw in one of the primary tools of the chemist, the periodic table. We see its functionality and organization, and begin using it in a thorough discussion of how to name most common chemicals, whose structure was discussed in Chapter 2.

In this edition, **Quantities in Chemistry** (now **Chapter 5**) has been switched with **Chemical Reactions** (now **Chapter 6**). It seems that more instructors prefer this order although the two chapters can still be switched without any disadvantage. **Chapter 7** continues with **Quantitative Relationships in Chemical Reactions**. **Chapter 8, Modern Atomic Theory** and **Chapter 9, The Chemical Bond** follow. Still, the two latter chapters can be moved ahead of Chapter 5 without prejudice, depending on the preferences of the instructor, and the ease with which the material can be incorporated into the overall curriculum. Many still prefer to cover the more abstract concepts of the atom before the quantitative aspects. **Chapter 10, The Gaseous State,** begins a three chapter in-depth study of the states of matter by examining the unique and predictable behaviors of gases. This is followed by similar discussions of the condensed states of matter in **Chapter 11, The Liquid and Solid States,** which includes a thorough but understandable discussion of intermolecular forces and how those affect the properties of matter. This discussion continues in **Chapter 12, Aqueous Solutions,** where we discuss both the qualitative and quantitative aspects of solute-solvent interactions. This chapter serves as a wrap-up of the quantitative relationships introduced in Chapters 5 and 7.

Chapter 13, Acids, Bases, and Salts, provides an in-depth discussion of how these important classes of compounds interact with water and each other. A second class of reaction is explored in depth in **Chapter 14, Oxidation-Reduction Reactions.** The relationship of these types of reactions to the creation of batteries and electrical currents is explained.

The remaining chapters offer a survey of topics that are of general chemical interest, and are appropriate for those looking to expose their students to a broad set of topics. **Chapter 15, Reaction Rate and Equilibrium,** introduces the concepts of Kinetics and Equilibrium and gives a taste of the sophisticated mathematical treatment these topics receive. **Chapter 16, Nuclear Chemistry,** explores how radioactivity is a phenomenon associated with certain elements and isotopes. **Chapter 17, Organic Chemistry,** gives the briefest introduction to organic functional groups, structure, and bonding. **Chapter 18, Biochemistry,** gives a similar treatment to common biochemical structures like carbohydrates, proteins, lipids, and nucleic acids. The latter two chapters are available on the Web.

Hallmark Features of the Ninth Edition

- Every concept in the text is clearly illustrated with one or more **step by step examples**. Most examples, in addition to a *Procedure* and *Solution* step, are followed by two steps: *Analysis* and *Synthesis*. The *Analysis* step discusses the problem in light of the reasonableness of the answer, or perhaps suggests an alternate way to solve the problem involving different learning modes. The concluding *Synthesis* step gives the student the opportunity to delve deeper, asking the student to extend their knowledge. These added steps promote critical thinking and facilitate deeper conceptual understanding.
- **Making it Real** essays have been updated to present timely and engaging real-world applications, emphasizing to the student the relevance of the material they are learning. For example, in the high-interest field of forensics, we describe how glass shards from crime scenes can be identified by their density in Chapter 3, and then in Chapter 7, explore how the refractive index of glass is also used as important evidence in solving crimes. In Chapter 7, we take a look at the chemical reactions involved in the breathalyzer and in Chapter 13, how salts are used to analyze fingerprints. Other essays emphasize useful types of energy.
- This edition continues the end of chapter **Student Workshop** activities. These are intended to cater to the many different student learning styles and to engage students in the practical aspect of the material discussed in the chapter. Each "Student Workshop" includes a statement of purpose and an estimated time for completion. These activities work well as dry labs, and for those students involved in recitation sessions and small group work.

Teaching and Learning Resources

WileyPLUS is an online teaching and learning environment that integrates the **entire digital textbook** with the most effective instructor and student resources to fit every learning style. It contains a variety of rich repositories of assessment, much of which are algorithmic. The diverse problem types are designed to enable and support problem-solving skills development and conceptual understanding. *WileyPLUS* offers three unique repositories of questions which provide breadth, depth and flexibility in instructional and assessment content.

- **End of chapter questions** are available, featuring immediate answer feedback. A subset of these end of chapter questions are linked to **Guided Online Tutorials** which are stepped out problem-solving tutorials that walk the student through the problem, offering individualized feedback at each step. The **testbank** is also offered as assignable questions for homework. In addition to the test bank and end of chapter questions, *WileyPLUS* offers an assignment type called **CATALYST**, which are **prebuilt concept mastery assignments**, organized by topic and concept, allowing for iterative drill and skill practice. For more information on CATALYST, visit: **www.wiley.com/college/catalyst**.

For Students *WileyPLUS* addresses different learning styles, different levels of proficiency, and different levels of preparation—each of your students is unique. *WileyPLUS* empowers them to take advantage of their individual strengths:

- Students receive timely access to resources that address their demonstrated needs, and get immediate feedback and remediation when needed.
- Integrated, multimedia resources—including audio and visual exhibits, demonstration problems, and much more—provide multiple study-paths to fit each student's learning preferences and encourage more active learning.
- *WileyPLUS* includes many opportunities for self-assessment linked to the relevant portions of the text. Students can take control of their own learning and practice until they master the material.

For Instructors *WileyPLUS* empowers you with the tools and resources you need to make your teaching even more effective:

- You can customize your classroom presentation with a wealth of resources and functionality from PowerPoint slides to a database of rich visuals. You can even add your own materials to your *WileyPLUS* course.
- With *WileyPLUS* you can identify those students who are falling behind and intervene accordingly, without having to wait for them to come to office hours.
- *WileyPLUS* simplifies and automates such tasks as student performance assessment, marking assignments, scoring student work, keeping grades, and more.

WileyPLUS can be used in conjunction with your textbook or it can replace the printed text altogether, as a complete eBook comes standard.

How Do I Access *WileyPLUS*?

To access *WileyPLUS*, students need a *WileyPLUS* registration code. This can be purchased stand alone, or the code can be bundled with a textbook. For more information and/or to request a WileyPLUS demonstration, contact your local Wiley sales representative or visit **www.wileyplus.com**.

Additional Instructor Resources

All of these resources can be accessed within WileyPLUS or by contacting your local Wiley sales representative.

Instructor's Manual and Test Bank by Leo J. Malone, St Louis University; Ted Dolter and Steve Gentemann, Southwestern Illinois College; and Kyle Beran, University of Texas-Permian Basin. The Instructor's Manual consists of two parts; the first part includes an overview and comments for each chapter by Ted Dolter, followed by daily lesson plans by Steve Gentemann. The second part of the manual contains answers and worked-out solutions to all chapter-end problems in the text by Leo Malone. The test bank, by Kyle Beran, consists of multiple choice, short answer, and fill in the blank questions. PC and Macintosh compatible versions of the entire test bank are available with full editing features to help the instructor to customize tests.

Instructor's Manual for Experiments in Basic Chemistry, written by Steven Murov and Brian Stedjee, Modesto Junior College, contains answers to post-lab questions, lists of chemicals needed, suggestions for other experiments, as well as suggestions for experimental set-ups.

Power Point Lecture Slides, created by Wyatt Murphy, of Seton Hall University, contain key topics from each chapter of the text, along with supporting artwork and figures from the text. The slides also contain assessment questions that can be used to facilitate discussions during lecture.

PowerPoint Art Slides, PPT slides containing images, tables, and figures from the text.

Digital Image Archive. The text web site includes downloadable files of text images in JPEG format.

Additional Student Resources

Study Guide/Solutions Manual by Leo J. Malone is available to accompany this text. In the Study Guide/Solutions Manual, the same topics in a specific section are also grouped in the same manner for review, discussion, and testing. In this manner, the Study Guide/Solutions Manual can be put to use before the chapter is completed. The Study Guide/Solutions Manual contains answers and worked-out solutions to all problems in green lettering in the text. [ISBN: 978-1-118-15643-8]

Experiments in Basic Chemistry, Seventh Edition by Steven Murov and Brian Stedjee, Modesto Junior College. Taking an exploratory approach to chemistry, this hands-on lab manual for preparatory chemistry encourages critical thinking and allows students to make discoveries as they experiment. The manual contains 26 experiments that parallel text organization and provides learning objectives, discussion sections outlining each experiment, easy-to-follow procedures, post-lab questions, and additional exercises.   [ISBN: 978-0-470-42373-8]

Acknowledgments

A revision of this magnitude involves efforts spanning several years and requiring the input of many people. In particular, Dr. Malone thanks his colleagues at Saint Louis University for their helpful comments in previous editions of this text. He's grateful to his wife Meg, who demonstrated patience and put up with occasional crabbiness during the new text's preparation. Dr. Malone also appreciates the support of his children and their spouses: Lisa and Chris, Mary and Brian, Katie and Rob, and Bill. They and their eleven children were both a source of great inspiration and a large amount of noise.

Professor Dolter continues to thank his general chemistry instructor, who provided a sound understanding of the basic principles needed in forming his craft. Although his professor (LJM) was a taskmaster, the end result was worth it. He'd also like to thank his colleagues at SWIC, who happily allowed him to bounce ideas off of them, and delighted in giving their advice. Professor Dolter is thankful for his wife Peggy, who endured the single-parent lifestyle many weekends during the writing of this text. A nod also goes to, Isabel and Zachary, who wondered if their Dad was ever going to finish this book; they're owed some Dad-time.

The authors also wish to thank the many people at John Wiley who helped and encouraged this project, including Nick Ferrari, our editor, for his support of this revision; Jennifer Yee, our Project Editor; Lauren Stauber, Editorial Assistant; Elizabeth Swain, Production Editor; Lisa Gee, Photo Editor; James O'Shea, Designer, Janet Foxman, Production Editor, and Suzanne Ingrao of Ingrao Associates, who kept an eye on every production detail, and kindly made sure we kept on schedule. Thank you all for remaining wonderfully patient in the face of missed deadlines and family conflicts.

Finally, the following people offered many useful comments and suggestions for the development of the Ninth Edition:

Dale Arrington *North Idaho College*
Satinder Bains *Paradise Valley Community College*
Ruth Birch *Saint Louis University*
Joseph Caddell *Modesto Junior College*
Jeff Cavalier *SUNY Dutchess*
Bertrand Chiasson *College of Southern Nevada*
Claire Cohen *The University of Toledo*
Michael Cross *Northern Essex Community College*
Paul Edwards *Edinboro University of Pennsylvania*
Eugenio Jaramillo *Texas A&M International University*
James Falender *Central Michigan University*
Rick Fletcher *University of Idaho*
Nancy Foote *Chandler Gilbert Community College*
Joy Frazier-Earhart *James Madison University*
Darlene Gandolfi *Manhattanville College*
Gregory Hanson *Otterbein University*
Theresa Hill *Rochester Community & Technical College*

Rebecca Hoenigman *Community College of Aurora*
Byron Howell *Tyler Junior College*
Jason Jadin *Rochester Community & Technical College*
Jodi Kreiling *University of Nebraska at Omaha*
Carla Kegley-Owen *University of Nebraska at Kearney*
Michael Lewis *Saint Louis University*
Richard Lomneth *University of Nebraska at Omaha*
Hussein Samha *Southern Utah University*
John Seeley *Oakland University*
Kris Slowinski *California State University, Long Branch*
Duane Smith *Nicholls State University*
Mackay Steffensen *Southern Utah University*

Leo J. Malone
Saint Louis University
Theodore O. Dolter
Southwestern Illinois College

BASIC CONCEPTS OF CHEMISTRY

Science and the Magnificent Human Mind

A campfire on a chilly night provides not only warmth but also serenity. Fire is an awesome force of nature which our ancient ancestors tamed and put to use.

SETTING THE STAGE

Among the animal kingdom, only humans have the ability to take their minds beyond simple survival. We also analyze, ponder, and predict the future based on observations. This has led us to a remarkable understanding of all that we see and otherwise sense about us. So this prologue is dedicated to how the wonderful workings of our minds have allowed us to establish the realm of modern science. In Section A, we present an abbreviated history that begins with the first chemical process that occurred billions of years ago and proceed to how the human race eventually put these processes to work starting with the taming of fire. In Section B, we take note of how science in general and chemistry in particular has progressed from random discoveries and serendipity to the complex technical world in which we exist today. This section emphasizes what we call the "scientific method." Finally, in Section C, we discuss how an individual can come to understand and appreciate this wonderful branch of science we know as "chemistry." After proceeding through the prologue, we will build this science from the most basic substances of the universe to the complex world of chemistry that serves us in so many ways today.

A Brief History of Chemistry

The history of chemistry can be developed on three levels. First, what was the first chemical process and when did it occur? Second, when and how did the human race put chemical processes to practical use? And finally, when did we arrive at reasonable explanations of these chemical processes?

A-1 The First Chemical Process

Chemistry has been happening for a very long time, indeed. Billions of years ago, when the universe was still very young, matter was composed of very minuscule, individual particles called *atoms*. Most of these were the simple atoms of the most fundamental element that we know as *hydrogen* (symbol H). At the time the cosmos was extremely hot, which meant that the atoms moved so fast that when atoms collided, they simply bounced off of each other. But as time went on, the cosmos cooled and then something very significant happened. When two atoms of hydrogen, now moving somewhat slower, collided they stuck or bonded to each other to form a *molecule* (symbol H_2). (See Figure P-1.) If two atoms were not more stable bonded together rather than apart, we could not be here. The trillions of trillions of atoms that make up the complex molecules of our bodies (and everything else) would simply exist as individual atoms scattered throughout space. So chemical processes began in the early universe when atoms came together to form chemical bonds. In fact, current astronomers have discovered hundreds of more complex molecules that have formed in outer space.

A-2 The Application of Chemical Processes

Chemistry has been going on since almost the beginning of time, but when did chemical processes become useful to the human race? Our second aspect of the history of chemistry lies with fire. This phenomenon first made its appearance on Earth about 400 million years ago when carbon in the form of vegetation, oxygen in the atmosphere, and lightning all came together. Fire is a chemical process where chemical bonds are changing and heat and light energy are being given off. So fire certainly falls into the realm of chemistry.

It is difficult to imagine how our ancient ancestors could have managed without fire. Humans do not have sharp night vision like the raccoon, but fire brought light to the long, dark night. We have no protective fur like the deer, but fire lessened the chill of winter. We do not have sharp teeth or powerful jaws like the lion, but fire rendered meat tender. Humans are not as strong or as powerful as the other large animals, but fire repels even the most ferocious of beasts. It seems reasonable to suggest that the taming of fire was one of the most monumental events in the history of the human race. The use of fire made our species dominant over all others.

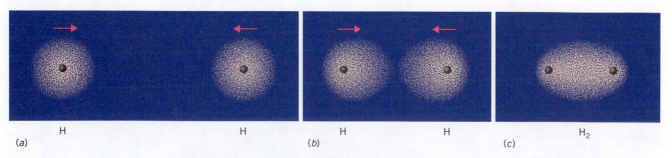

| H | H | H | H | H_2 |
| (a) | (b) | | (c) | |

FIGURE P-1 A Hydrogen Molecule Chemistry first occurred when two hydrogen atoms bonded together to form a hydrogen molecule.

Let's fast-forward in time to near the end of the Stone Age, about 10,000 years ago, when fire was purposely applied to effect a unique chemical process. In the Stone Age, weapons and utensils were fashioned from rocks and a few chunks of copper metal (an element) that were found in nature. Copper was superior to stone because pounding could easily shape it into fine points and sharp blades. Unfortunately, native copper was quite rare. But about 7000 years ago this changed. Some anthropologists speculate that some resident of ancient Persia found copper metal in the ashes remaining from a hot charcoal fire. The free copper had not been there before, so it must have come from a green stone called *malachite* (see Figure P-2), which probably lined the fire pit. Imagine the commotion that this discovery must have caused. Hot coals could transform a particular but plentiful stone into a valuable metal. Fire was the key that launched the human population into the age of metals. The recovery of metals from their ores is now a branch of chemical science called *metallurgy*. The ancient persians must have considered this discovery a dramatic example of the magic of fire.

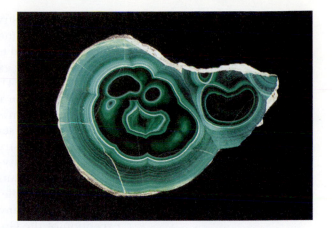

FIGURE P-2 Malachite Malachite is a copper ore. When heated with charcoal, it forms metallic copper.

Other civilizations used chemistry in various ways. About 3000 B.C., the Egyptians learned how to dye cloth and embalm their dead through the use of certain chemicals found in nature. They were very good at what they did. In fact, we can still determine from ancient mummies the cause of death and even diseases the person may have had. The Egyptians were good chemists, but they had no idea why any of these procedures worked. Every chemical process they used was discovered by accident.

The early centuries of the Middle Ages (A.D. 500–1600) in Europe are sometimes referred to as the Dark Ages because of the lack of art and literature and the decline of central governments. The civilizations that Egypt, Greece, and Rome had previously built began to decline. Chemistry, however, began to grow during this period, especially in the area of experimentation. Chemistry was then considered a combination of magic and art rather than a science. Many of those who practiced chemistry in Europe were known as *alchemists*. Some of these alchemists were simply con artists who tried to convince greedy kings that they could transform cheaper metals such as lead and zinc into gold. Gold was thought to be the perfect metal. Such a task was impossible, of course; so many of these alchemists met a drastic fate for their lack of success. However, all was not lost. Many important laboratory procedures such as distillation and crystallization were developed. Alchemists also discovered or prepared many previously unknown chemicals that we now know as elements and compounds.

A-3 The Understanding of Chemistry

Now we step back a little in time where we find the first theoretical explanations of chemical phenomena. Around 400 B.C., while some Greeks were speculating about their various gods, other Greek philosophers were trying to understand and describe nature. These great thinkers argued about why things occurred in the world around them, but they were not inclined (or able) to check out their ideas by experimentation or to put them to practical use. At the time, however, people believed that there were four basic elements of nature—earth, air, water, and fire. Everything else was simply a specific combination of these basic elements. Of the original four elements, fire was obviously the most mysterious. It was the transforming element; that is, it had the capacity to change one substance into another (e.g., certain rocks into metals). We now call such transformations "chemistry." Fire itself consists of the hot, glowing gases associated with certain chemical changes. If fire is a result of an ongoing chemical transformation, then it is reasonable to suggest that chemistry and many significant advances in the human race are very much related.

Modern chemistry has its foundation in the late 1700s when the use of the *analytical balance* became widespread. Chemistry then became a quantitative science in which theories had to be correlated with the results of direct laboratory experimentation. From these experiments and observations came the *modern atomic theory*, first proposed by John Dalton around 1803. This theory, in a slightly modified form, is still the basis of our understanding of nature today. Dalton's theory gave chemistry the solid base from which it could serve humanity on an impressive scale. Actually, most of our understanding of chemistry has evolved in the past 130 years. In a way, this makes chemistry a very young science. However, if we mark the beginning of the use of chemistry with fire, it is also the oldest science.

From the Persians of five millennia ago, to the ancient Egyptians, to the alchemists of the Middle Ages, various cultures have stumbled on assorted chemical procedures. In many cases, these were used to improve the quality of life. With the exception of the Greek philosophers, there was little attention given to why a certain process worked. The *why* is very important. In fact, the tremendous explosion of scientific knowledge and applications in the past 200 years can be attributed to how science is now approached. This is called the "scientific method," which we will discuss next.

B The Scientific Method

B-1 The Development of the Scientific Method

In ancient times, scientific advances were discovered by accident. This still occurs to some extent, but we have made great strides in how we approach science such that most modern advances occur by design. Later we will discuss how our modern approach to science produces so many new and wonderful things such as drugs that cure certain cancers or other illnesses. The modern approach is known as the *scientific method*.

The first step in the scientific method is a long way from producing a useful drug. It simply involves *making observations and gathering data*. As an example, imagine that we are the first to make a simple observation about nature: "The sun rises in the East and sets in the West." This never seems to vary and, as far as we can tell from history, it has always been so. In other words, our scientific observation is strictly *reproducible*. So now we ask, "Why?" We are ready for a hypothesis. A **hypothesis** *is a tentative explanation of observations.* The first plausible hypothesis to explain our observations was advanced by Claudius Ptolemy, a Greek philosopher, in A.D. 150. He suggested that the sun, as well as the rest of the universe, revolves around the Earth from east to west. That made sense. It certainly explained the observation. In fact, this concept became an article of religious faith in much of the Western world. However, Ptolemy's hypothesis did not explain other observations known at the time, which included the movement of the planets across the sky and the phases of the moon.

Sometimes new or contradictory evidence means a hypothesis, just like a broken-down old car, must either receive a major overhaul or be discarded entirely. In 1543, a new hypothesis was proposed. Nicolaus Copernicus explained all of the observations about the sun, moon, and planets by suggesting that Earth and the other planets orbit around the sun instead of vice versa. Even though this hypothesis explained the mysteries of the heavenly bodies, it was considered extremely radical and even heretical at the time. (It was believed that God made Earth the center of the universe.) In 1609, a Venetian scientist by the name of Galileo Galilei built a telescope, which had just been invented, to view ships still far out at sea. When he turned the telescope up to the sky, he eventually produced almost unquestionable proof that Copernicus was correct. Galileo is sometimes credited with the beginning of the modern scientific method because he provided direct experimental data in support of a concept. The hypothesis had withstood the challenge of experiments and thus could be considered a theory. A **theory** *is a well-established hypothesis.* A theory should not only explain known phenomena, but also predict the results of future experiments or observations. A scientific theory must explain or incorporate all the known relevant facts. It cannot be selective.

The next part of this story comes in 1684, when an English scientist named Sir Isaac Newton stated a law that governs the motion of planets around the sun. A **law** *is a concise scientific statement of fact to which no exceptions are known.* Newton's law of universal gravitation states that planets are held by gravity in stationary orbits around the sun. (See Figure P-3.)

In summary, these were the steps that led to a law of nature:

FIGURE P-3 The Solar System The Copernican theory became the basis of a natural law of the universe.

1. Reproducible observations (e.g., the sun rises in the East)
2. A hypothesis advanced by Ptolemy and then a better one by Copernicus
3. Experimental data gathered by Galileo in support of the Copernican hypothesis and eventual acceptance of the hypothesis as a theory
4. The statement by Newton of a universal law based on the theory

Variations on the scientific method serve us well today as we pursue an urgent search for cures of diseases. An example follows.

B-2 The Scientific Method in Action

The healing power of plants and plant extracts has been known for thousands of years. For example, ancient Sumerians and Egyptians used willow leaves to relieve the pain of arthritis. We now know that extracts of the common willow contain a drug very closely related to aspirin. This is the observation that starts us on our journey to new drugs. An obvious hypothesis comes from this observation, namely, that there are many other useful drugs among the plants and soils of the world. We should be able to find them. There are several modern discoveries that support this hypothesis. For example, the rosy periwinkle is a common tropical plant not too different from thousands of other tropical plants except that this one saves lives. The innocent-looking plant contains a powerful chemical called *vincristine*, which can cure childhood leukemia. Another relatively new drug called Taxol has been extracted from the pacific yew tree. (See Figure P-4.) Taxol was originally used exclusively in treating ovarian cancer but is now used to treat breast cancer. Others include *cyclosporine*, isolated from a fungus in 1957, which made organ transplants possible, and *digoxin*, isolated from the foxglove plant, used for treatment of heart failure. In fact, many of the best-selling medicines in the United States originated from plants and other natural sources. Besides those mentioned, other drugs treat conditions such as high blood pressure, cancer, glaucoma, and malaria. The search for effective drugs from natural sources and newly synthesized compounds is very active today. These chemicals are screened for potential anticancer, antiarthritic, and anti-AIDS activity. Since the greatest variety of plants, molds, and fungi are found in tropical forests, these species are receiving considerable attention. Lately,

FIGURE P-4 **The Pacific Yew**
The bark of this tree is a source of an anticancer drug known as Taxol.

scientists are focusing on plants and algae in the ocean for possible drugs. The introduction of a new medicine from a natural source involves the following steps.

1. *Collection of materials.* "Chemical prospectors" scour the backwoods of the United States and the tropical forests such as those in Costa Rica, collecting and labeling samples of leaves, barks, and roots. Soil samples containing fungi and molds are also collected and carefully labeled.

2. *Testing of activity.* Scientists at several large chemical and pharmaceutical companies make extracts of the sample in the laboratory. These extracts are run through a series of chemical tests to determine whether there is any antidisease activity among the chemicals in the extract. New methods such as high-throughput-screening (HTS) allow certain chemical companies to screen thousands of chemicals in a single day. If there is antidisease activity, it is considered a "hit" and the extract is taken to the next step.

3. *Isolation and identification of the active ingredient.* The next painstaking task is to separate the one chemical that has the desired activity from among the soup of chemicals present. Once that's done, the particular structure of the active chemical must be determined. A hypothesis is then advanced about what part of the structure is important and how the chemical works. The hypothesis is tested by attempting to make other more effective drugs (or ones with fewer side effects) based on the chemical's structure. It is then determined whether the chemical or a modified version of it is worth further testing.

4. *Testing on animals.* If the chemical is considered promising, it is now ready to be tested on animals. This is usually done in government and university labs under strictly controlled conditions. Scientists study toxicity, side effects, and the chemical's activity against the particular disease for which it is being tested. If, after careful study, the chemical is considered both effective and safe, it is ready for the next step.

5. *Testing on humans.* Perhaps the most important step regards clinical trials. This involves careful testing on humans, which is carefully monitored by agencies such as the FDA. Effectiveness, dosage, and long-term side effects are carefully recorded and evaluated. All told, it currently takes from 10 to 15 years for a new drug to make it all the way from research and development to market. Sometimes the process is speeded up if the drug is especially effective or treats a rare disease.

When chemicals with the desired activity are randomly discovered, only about 1 in 1000 may actually find its way into general use. Still, the process works. Many chemicals active against cancer and even AIDS are now in the pipeline for testing. There is some urgency in all of this. Not only are we anxious to cure specific diseases, but the tropical forests that contain the most diverse plants are disappearing at an alarming rate. In any case, nature is certainly our most important chemical laboratory.

At this time, most of the drugs that originated from synthetic or natural sources are considered cases of "serendipity." We are now moving more toward the concept of "rational drug design." Here, the goal is to identify the active sites on the molecules of diseases such as viruses, tumors, or bacteria. The next step is to deliberately synthesize a drug that attaches to that active site and either destroys or otherwise alters the disease molecules. This sounds easy, but it is not. It requires that we know more about the structure and geometry of these disease molecules. We can then advance hypotheses as to how designed molecules would interact. This is the direction in which pharmaceutical chemists are heading, however.

C The Study of Chemistry and Using This Textbook

Chemistry is the *fundamental natural science.* This is not just an idle boast. Chemistry is concerned with the basic structure and properties of all matter, be it a huge star or a microscopic virus. Biology, physics, and geology, as well as all branches of engineering and medicine, are based on an understanding of the chemical substances of which nature is composed. Chemistry is the beginning point in the course of studies that eventually produce all scientists, engineers, and physicians. But it is also important for all responsible citizens. Our environment is very fragile—more fragile than we realized just a few years ago. Many of the chemicals that make life easier also affect our surroundings. Control of air, water, and land pollution needs as much attention from citizens and scientists as the invention of new materials did in the twentieth century. (See Figure P-5.) We all have a big stake in the future, so it is reasonable that chemistry is a prerequisite not only for courses of study but also for life, especially in these complex times. What follows is a brief look at academic self-disciplines and skills needed in the study of chemistry.

FIGURE P-5 Our Environment Is a Fragile System Pollution of ocean waters is a serious problem as seen from the Gulf oil spill in 2010.

C-1 Time Management and Study Skills

Establish a Schedule Chemistry is not unlike basketball or piano—it requires lots of practice. It is not only a question of putting in the time but *how* and *when* you put in the time. One does not wait until the night before the big game to first practice jump shots or the night before the recital to first practice the sonata. A master schedule should be prepared, with chemistry (as well as all other subjects) receiving a regularly scheduled study time. The study period should follow as soon as possible after the lecture. In setting up your schedule, it is wise to select short but frequent study periods. Two separate one-hour sessions are more effective than one two-hour period. After a length of time, the mind tends to wander into areas not directly related to chemistry (e.g., baseball—what else?).

Organize Problems and Exercises Now that we have a good study schedule, we need to plan an efficient approach to the tasks at hand. Many of the problems that are required in chemistry require a step-by-step approach. It is a lot like planning an extended trip. Usually, we don't just get in the car and drive. We plan our journey so that we can take the shortest or the easiest route. We may also plan ahead on what we want to accomplish and when we want to do it. The secret is advance planning. This is certainly true with chemistry problems. We don't just start doing a calculation—we should take some time to plan the journey through the problem from beginning to end. This requires that we first write down the steps that we will take.

Take Good Notes The most useful approach takes some work but is *extremely* effective. As soon as you can after a lecture, recopy your class notes. As you do this, imagine that you are explaining the material to someone else (out loud, if you can). If you do this, you will become aware of concepts that are still hazy to you. Also, the logical progression of problems may now seem more obvious. As you recopy your notes, leave about one-third of the paper as a blank margin so that you can add thoughts, emphasis, or questions for later review. All this assumes that one is going to the lectures. No electronic presentation can substitute for being present at a lecture.

Be Able to Memorize Most of us wish that we could just grasp a concept and, with little effort, have it permanently imprinted in our brains. Unfortunately, it rarely works that way. In many cases, we need to memorize a definition or rule so as to be able to categorize facts as fitting the concept or not. We certainly could not begin to play a sport such as soccer without first knowing the object of the game and some basic rules. Likewise, in chemistry we wouldn't be able to write a molecular structure until we first know the steps involved. Even in this age, the use of 3×5 note cards can help especially if you are expected to know the formulas and names of ions. An old tried-and-true method of memorizing still is the most effective.

Read Ahead This one is hard to do but it does help. Recall that even in high school, football and basketball opponents are scouted before the game. Every team likes to know what they are up against and what lies ahead. In studying science, the equivalent to scouting is reading ahead. Even if you do not grasp the concepts, reading ahead will give you a feeling for some of the material that you will be discussing in the next lecture or two. If you know something is coming that seems confusing, you will be more alert when the concept is discussed. Reading ahead is also a time saver. When you know that certain definitions or tables are in the book, you can save note-taking time.

Attendance and Participation Despite all of the support media from various web sites, almost all instructors still give face-to-face lectures in front of real people. The importance of regular class attendance cannot be overemphasized. You really need to be there. This text is a great ally, but what material your instructor covers and in what depth can be discovered only in class. Nor can you sense your instructor's emphasis or benefit from his or her problem-solving hints by reading someone else's notes. The key to what will be on the exams is found in the lecture. You need to be there.

If you hesitate to ask questions in class or the class is too big, take advantage of help sessions or your instructor's office hours. In many cases, instructors are a significantly underutilized resource during their office hours. Most teachers would still rather talk to you in person rather than answer an e-mail or text message. You may discover that, in the office, the instructor is really a very nice person and very helpful. The bottom line is that you owe yourself answers and understanding. Do what you have to do. Somebody is paying big bucks for you to take this course.

Perseverance Everyone hopes to start off with an A on the first test or quiz. However, few do that, including many who are subsequently very successful in the study of chemistry. If you are disappointed with an exam grade, reanalyze your study habits, make adjustments, and try again. Don't expect better results by doing the same thing. Remember to use your instructor as your primary resource for advice in this matter. Perhaps your problem is not the material but test anxiety. This is a very real phenomenon and has to be acknowledged. Most colleges have a counseling center that can help you overcome this problem. Sometimes your instructor can give you helpful hints on taking chemistry tests so that you won't have the fear of freezing up. If you are ultimately not successful in the study of chemistry, at least you can say that you gave it your best try.

C-2 Using This Textbook

Finally, make good use of this textbook. Most textbooks have many study aids within the chapters. This text is no exception and, in fact, has many unique features that can help you understand the concepts as well as assess and evaluate your progress. The following are special aids and features to help you through this course.

Setting a Goal and Section Objectives Each chapter is divided into at least two distinct parts. At the beginning of the chapter, the parts are listed along with the topics within each part. A brief discussion titled *Setting the Stage* gives an introduction to the purpose of the chapter. *Part A* then begins with a statement titled *Setting a Goal* and a list of *Objectives* that relate to this overall goal.

Continuity Between Sections Each section begins with a short paragraph entitled *Looking Ahead*. The intention is to give an overview of what the section will cover and perhaps how it relates to the previous topic.

Topics and Subtopics This text helps you appreciate the fact that chemistry builds slowly, one concept at a time. The sections are divided into bite-sized subheadings, each covering a separate concept. For example, consider the following heading with three subheadings:

2-3 The Composition of the Atom
2-3.1 The Electron and Electrostatic Forces

> ### 2-3.2 The Nuclear Model of the Atom
> ### 2-3.3 The Particles in the Nucleus

The authors have put together what they feel is a logical sequence of topics for the introduction of chemistry. There are other logical ways to introduce this science, however. The text has been written with this in mind, so it is quite flexible in order to accommodate alternative sequences.

Example Problems   This text has many in-chapter practice problems. These are worked through in a step-by-step manner. Most of the example problems start with a *Procedure* that analyzes the problem and establishes a path to the answer. The *Solution* follows, which works through the problem. The *Analysis* has us look at whether the answer is reasonable and may suggest alternative ways to solve the problem using different learning modes. Finally, the *Synthesis* takes the discussion a little deeper and may project how this problem relates to future concepts. The analysis and synthesis steps, which are unique to this text, promote critical thinking and a deeper understanding.

Assessing Progress   At the end of each section is a collection of exercises titled *Assessing the Objectives*. This provides you with a chance to evaluate your understanding and mastery of the objectives for that section. The problems are presented in three levels: *Level 1*, *Level 2*, and *Level 3*. Each level is progressively more sophisticated. It is expected that the student should be able to answer the Level 1 and Level 2 questions correctly. The Level 1 exercises would assure basic understanding, Level 2 tests a slightly higher level of understanding, and Level 3 indicates mastery of the topic with possible applications to related concepts. The answers to these exercises are at the end of the chapter.

Summarizing Concepts   At the end of each part you will find a list of *Key Terms* introduced in the relevant sections. The key term is shown in bold and is part of a definition. Also at the end of each part is at least one *Summary Chart* that can help you visualize a key concept. At the end of the chapter there is a *Chapter Summary*, which is an easy way to quickly check or review your understanding of the key concepts of the chapter. There is also an *Objectives Grid*, which can help you refer to the Examples, Exercises, and Chapter Problems relevant to each objective.

Chapter Synthesis Problem   This problem is unique to this text and is presented at the end of the chapter. It is a multiconcept problem that requires the application of several different topics presented in the chapter. It contains a brief discussion and two problems. The first problem is immediately followed by worked-out solutions. The second problem, titled "Your Turn," presents the students with a similar comprehensive problem to work on their own. The answers to the second problem are included at the end of the chapter.

End-of-Chapter Problems   A large number of *Chapter Problems* are provided to establish and practice your skills for each topic listed. There are a wide variety of problems, some quite basic, some that test your knowledge at a higher level, and some that are quite challenging. The latter are shown with an asterisk (*). The problems with numbers shown in color have answers at the end of the text.

Making it Real   Finally, we have included at least two *Making It Real* essays in each chapter. We hope you find these short discussions fun to read and can then see the connections between the chapter topics being discussed and the real world.

Good luck in the study of chemistry. The authors and the publisher hope you enjoy the course and find this textbook a very real aid in your understanding and appreciation of this science.

KEY TERMS

B-1   A **hypothesis** is a tentative explanation of observations. p. 6
B-1   A **theory** is a well-established hypothesis. p. 7
B-1   A **law** is a concise scientific statement of fact to which no exceptions are known. p. 7

Chemistry and Measurements

Students of chemistry are intimately involved in making careful measurements. The numbers expressed in such measurements are of extreme importance to the chemist. In this chapter, we will use a few basic chemical concepts that will serve as an entry to discussing the quality of the numbers in meaningful measurements. The chemical concepts themselves will get more detailed discussion in the next two chapters.

SETTING THE STAGE

Chemistry is a science that requires us to deal with all the stuff that we see around us and even some that we can't see. This stuff, which we call matter, is subject to the laws of nature. Many of these laws originate from reproducible quantitative measurements. Measurements naturally contain numbers. The quality, meaning, magnitude, and manipulation of the numbers we use in chemistry form the beginning point of our study.

It can be confusing and maybe distracting to start the study of chemistry with what seems more like a math course. So, in this chapter we will present some basic chemical principles such as elements, compounds, atoms, and molecules to bring context to the math required. We will see that math is an integral part of chemistry. In the next chapter, we will discuss the chemistry concepts in much more detail.

It is likely that most students need at least a bit of a mathematical workout to get into shape. For some, a simple review will do, and that is provided in the text. At appropriate points in this chapter and other chapters, you will also be referred to a specific appendix in the back of the book for more extensive review. Review appendixes include basic arithmetical operations (Appendix A), basic algebra operations (Appendix B), and scientific notation (Appendix C). If you are worried about the math, you are not alone. Just remember that this text was written with your concerns in mind. We hope to make the math as painless as possible so that it doesn't interfere with your understanding of the fascinating and exciting world of chemistry

The measurements we use in chemistry consist of two parts—a number and a specific unit. We will discuss these two parts separately. In Part A, we will address questions about the numerical quantity of a measurement. The units that are used in the measurement are then discussed in Part B.

Part A

The Numbers Used in Chemistry

SETTING A GOAL

- You will learn how to apply and manipulate measurements to produce scientifically meaningful outcomes.

OBJECTIVES

1-1 Determine the number of significant figures and the degree of uncertainty in a measurement.

1-2 Perform arithmetic operations and percent error calculations, rounding the answer to the appropriate number of significant figures.

1-3 Perform arithmetic operations involving scientific notation.

▶ **OBJECTIVE FOR SECTION 1-1**
Determine the number of significant figures and the degree of uncertainty in a measurement.

1-1 The Numerical Value of a Measurement

LOOKING AHEAD! Chemistry, as well as all other sciences, involves the interpretation of numbers. How reliable are the numbers and what do they really tell us? In this section, we will evaluate the quality and reliability of the numbers that are part of the measurements in chemistry. First, however, we will define the basic components of matter and then proceed to the numbers involved in their measurements. ■

1-1.1 Chemistry—Elements, Compounds, Atoms, and Molecules

Chemistry *can be defined as the study of matter and the changes it undergoes.*

First, consider **matter**, *which is defined as anything that has mass and occupies space.* We can classify matter in several ways, but the most fundamental classification of matter is as either *elements* or *compounds.*

Next, consider the *elements.* There are only 90 or so elements that exist in the earth. We are familiar with some elements that exist as free elements in nature uncombined with other elements. The gold and diamond in a ring, the oxygen and the nitrogen in the air are all examples of free elements. (See Figure 1-1a and also Figure 2-1.) In the next chapter we will describe the elements in more detail including their names, symbols, and distribution.

Most of the elements on earth are not present in their free form but are combined with other elements to form unique types of matter known as *compounds.* For example, the elements hydrogen and oxygen are combined to form the water in rain. Carbon and oxygen are combined to form carbon dioxide in the atmosphere or carbon monoxide, a deadly gas. There are literally millions of known compounds. (See Figure 1-1b and also Figure 2-8.)

Now consider the composition of elements. Elements, as well as all matter, are composed of extremely tiny basic particles called *atoms.* Most compounds are composed of *molecules,* which are combinations of the

FIGURE 1-1
Elements and Compounds Gold and diamond (carbon) at the top are two elements found as free elements in nature. Water and dry ice (carbon dioxide) are common compounds.

atoms of two or more elements. (Some elements such as oxygen are also composed of molecules.) This is illustrated in Figure 1-2 with one atom of the element hydrogen and one atom of the element bromine, which, when bonded together, form a molecule of the compound known as hydrogen bromide. (See also Figure 2-9.)

Hydrogen   Bromine   Hydrogen bromide
atom   atom   molecule

FIGURE 1-2 The Molecule of a Compound The molecule of a typical compound is formed from one atom of hydrogen and one atom of bromine.

1-1.2 Measurement of the Mass of an Element and Significant Figures

The elements are listed in alphabetical order inside the front cover. The atoms of each element have specific masses (i.e., the atomic masses). First, we will examine the masses of the atoms of two elements individually. The mass of a hydrogen atom is listed as 1.00794 and a bromine atom is 79.904. The units of these masses are known as *atomic mass units* (*amu*). This is a very specific measure of mass, which is not important to our current discussion but will be examined later.

The mass of a hydrogen bromide molecule is naturally, the sum of the masses of its two component atoms. So, do we just add them together? In fact, it is not that easy. Notice that there is a subtle but important difference in how the two masses are expressed. The mass of hydrogen is expressed to five decimal places and the bromine to just three. (The difference arises from the details of how the masses are measured.) So, before we do the math, let's step back and formally define a measurement and discuss the meaning of the number that is expressed in the measurement. After we lay this important groundwork, we will return to the actual addition calculation in the next section so we can understand the quality of the answer and how to express it correctly.

A **measurement** *determines the quantity, dimensions, or extent of something, usually in comparison to a specific unit.* A **unit** *is a definite quantity adopted as a standard of measurement.* Thus, a measurement (e.g., 15.8 grams) consists of two parts: a numerical quantity (15.8) followed by a specific unit (grams). The units common in chemistry will be discussed later in this chapter. In this discussion we will first focus on the numerical part of the measurement.

Our first mathematical discussion centers on the question of how much confidence we can place in a measurement such as the 15.8 grams. This question takes us to the concept of significant figures. To answer this we will consider the analogy of attendance at a rock concert. Assume that the evening news informs us that 12,000 people gathered at the concert. Did they mean exactly 12,000? Not really—actually, it was just an estimate. Two other experts might have estimated the same crowd at 13,000 and 11,000, respectively. This means that the original estimate had an *uncertainty* of ±1000 (i.e. plus or minus 1000). Thus only the 1 and 2 are considered significant in this estimate. The three zeros simply tell us the magnitude of the number. In a measurement, a **significant figure** *is a digit that is either reliably known or closely estimated.* In the number 12,000, we can assume the 1 is reliable and reproducible from any number of estimates, but the 2 is estimated. The zeros are not significant since they actually have no specific numerical meaning. Thus, in our example, there are two significant figures: the 1 and the 2. The number of significant figures or digits in a measurement is simply the number of measured digits and refers to the precision of the measurement. **Precision** *relates to the degree of reproducibility or uncertainty of the measurement.* Indeed, all measured values have an uncertainty that is expressed in the last significant figure to the right.

Now assume the same crowd at the concert is seated in the bleachers of a stadium instead of milling about. In this case, a more precise estimate is possible since the exact capacity of the stadium is known. The crowd can now be estimated at between 12,400 and 12,600, or an average of 12,500. This is a measurement with three significant figures. The 1 and 2 are now reliable, but the third significant figure, the 5, is estimated. The extra significant figure means that the uncertainty is now reduced to ±100. The more significant figures in a measurement, the more precise it is and the less uncertainty.

Consider another example. Participants in the sport of riflery (target shooting) are judged on two points: how close the bullet holes are to each other (the pattern) and how close the pattern is to the center of the target known as the bull's-eye. The precision of the contestant's shooting is measured by the tightness of the pattern. How close the pattern is to the bull's-eye is a measure of the shooter's accuracy. **Accuracy** *in a measurement refers to how close the measurement is to the true value.* Usually, the more precise the measurement, the more accurate it is—but not always. In our example, if a certain competitor has a faulty sight on the rifle, the shots may be close together (precise) but off center (inaccurate). (See Figure 1-3.)

Accuracy in measurements depends on how carefully the instrument of measurement has been calibrated (compared to a reliable standard). For example, what if we attempted to measure length with a plastic ruler that became warped after being left in the hot sun? We obviously would not obtain accurate readings. We would need to recalibrate the ruler by comparing its length divisions to a reliable standard.

1-1.3 Zero as a Significant Figure

It would be easy to determine the number of significant figures in a measurement if it were not for the number zero. All nonzero numbers in a measurement are considered significant. Zero, however, serves two functions: as a reliable or estimated digit, or simply as a marker to locate the decimal point (such as the three zeros in the estimated crowd of 12,000 people). Since the zeros look alike in both cases, it is important for us to know whether a zero is significant or is there simply to locate the decimal point. The following rules can be used to tell us about zero. Digits that are underlined are significant.

1. When a zero is between other nonzero digits, it is significant: 7<u>0</u>9 has three significant figures. Zeros between two nonzero digits are always significant.
2. Zeros to the right of a nonzero digit and to the right of the decimal point are significant: 8.<u>0</u> has two significant figures, just as 7.9 and 8.1 do; 7.9<u>00</u> has four significant figures. There is no value difference between 8.0 and just 8. Why, then, is the zero added? The answer is simply because it is an estimated digit and should be included. For the number 8, the uncertainty is ±1. For the number 8.0, the uncertainty is ±0.1.
3. Zeros to the left of the first nonzero digit are not significant: 0.00<u>78</u> and 0.<u>45</u> both have two significant figures; 0.0<u>4060</u> has four significant figures. In this case, the zeros to the left of the digit are simply showing the decimal place and are not measured or estimated digits. Therefore, they are not significant.
4. Zeros to the left of an implied decimal point may or may not be significant. In most cases, they are not. The crowd of <u>12</u>,000 has two significant figures,

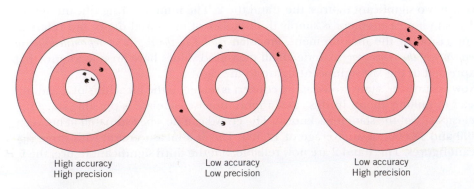

FIGURE 1-3 Precision and Accuracy High precision does not necessarily mean high accuracy.

High accuracy
High precision

Low accuracy
Low precision

Low accuracy
High precision

as does 6600. Just as in the previous case, the zeros to the left are decimal place holders, not actual measurements. What if the zero in a number such as 890 is actually an estimated digit and thus significant? This is a tough question. Some texts use a line over the zero to indicate that it is significant (e.g., 89$\overline{0}$), and others simply place a decimal point after the zero (e.g., 890.). As we will see in Section 1-3.2, there is a solution to this dilemma.

In most problems used in this text, measurements are expressed to three significant figures. Therefore, in calculations where other numbers have three significant figures, we will assume that numbers such as 890 also have three significant figures.

EXAMPLE 1-1   Evaluating Zero as a Significant Figure

How many significant figures are in the following measurements? What is the uncertainty in each of the measurements?

(a) 1508 cm

(b) 300.0 ft

(c) 20.003 lb

(d) 0.00705 gal

PROCEDURE

To determine the number of significant figures, refer to the rules regarding zero that were listed. Since these numbers all involve measurements, the last significant figure to the right is estimated. This digit indicates the uncertainty of the measurement.

SOLUTION

(a) The zero between the two nonzero digits is significant. There are four significant figures. The uncertainty is in the last digit to the right, so is ± 1 cm.

(b) The zero to the right of the decimal point is significant because it is displayed. Therefore, the other two zeros are also significant because they lie between significant figures. There are four significant figures. The uncertainty is in the last zero to the right, so is ± 0.1 ft.

(c) All the zeros are between nonzero digits, so they are all significant. There are five significant figures. The uncertainty is in the last digit to the right, so is ± 0.001 lb.

(d) The three zeros to the left of the first nonzero digit are not significant. The zero between the two nonzero digits is significant. Thus, there are three significant figures. The uncertainty is in the last digit to the right, so is ± 0.00001 gal.

ANALYSIS

Be certain not to confuse the concept of "significant" with the idea of "important." In the measurement 0.00705 gallons, the two zeros immediately following the decimal are most assuredly important to the value of the number. Without them, the value changes. But *significant* refers to a measurement. In that particular example, the measurement doesn't start until you reach the 7. Therefore, the zeros, while necessary, are not significant.

SYNTHESIS

The more significant figures there are in a measurement, the more precise it is. The closest the Earth and sun ever get to one another is 147,098,000 km. The uncertainty is ± 1000 km. The distance between New York and Philadelphia is 127 km, known with a precision to the nearest kilometer. One would assume that knowing a distance to a kilometer would be more precise than to the nearest 1000 kilometers. And yet, because the measured distance to the sun has six significant figures in it, it is 1000 times more precise than the measured driving distance between the two cities. Which of the four measurements in the problem is the most precise? 20.003 lb. It has more significant figures than any of the others.

▶ **ASSESSING THE OBJECTIVE FOR SECTION 1-1**

EXERCISE 1-1(a) **LEVEL 1:** Students in different class sections attempted to measure the mass of a 100.0-gram object. Describe the series of measurements as precise, precise and accurate, or neither.

(a) class 1: 94.2 g, 93.8 g, 94.4 g, 94.0 g *Precise*

(b) class 2: 94.3 g, 89.7 g, 102.4 g, 97.8 g *Neither*

(c) class 3: 100.2 g, 100.0 g, 99.8 g, 99.8 g *Accurate*

EXERCISE 1-1(b) **LEVEL 3:** You are determining the boiling point of an unknown compound in degrees Celsius. On three successive attempts, you record 145.5°C, 145.8°C, and 146.0°C. Are your data precise? Are your data accurate? Is there extra information needed to make either of those claims?

EXERCISE 1-1(c) **LEVEL 2:** Determine the number of significant figures in the following measurements:

(a) 45.00 oz   **(b)** 0.045 lb   **(c)** 45000 s   **(d)** 0.450 in.

EXERCISE 1-1(d) **LEVEL 3:** Two models of analytical balance are available for purchase. The first measures to ±0.01 gram. The second is more expensive, but measures to ±0.0001 gram. How many significant figures will each give when measuring the mass of an object of about 25 grams? What considerations would you include in determining which model to buy?

(Throughout the text, answers to all Assessing the Objectives exercises can be found at the end of each chapter.)

For additional practice, work chapter problems 1-1, 1-3, 1-4, and 1-6.

▶ **OBJECTIVE FOR SECTION 1-2**
Perform arithmetic operations, and the percent error calculations, rounding the answer to the appropriate number of siginificant figures.

1-2 Significant Figures in Chemical Calculations

LOOKING AHEAD! In chemistry we are continually adding, multiplying, or performing other calculations involving various measurements. Often these measurements involve different degrees of precision, so we need to understand how to report the results of such calculations. ∎

1-2.1 The Mass of a Compound

The compound hydrogen bromide was mentioned in the previous section. The mass of the compound is naturally the sum of its two parts, that is, the mass of one hydrogen atom (1.00794 amu) plus one bromine atom (79.904 amu). This is known as the compound's *molecular weight*.

For under $10 we can own a small calculator that can almost instantly carry out this and any other calculation that we may encounter in general chemistry. This device is truly phenomenal, especially to older scientists who had to do without this modern marvel (i.e., think slide-rule). The use of the calculator does have one serious drawback, however. It does not necessarily report the answer to a calculation involving measurements to the proper precision or number of significant figures. For example, if we add the numbers given above (1.00794 amu + 79.904 amu), the calculator actually gives us a scientifically misleading result (i.e., 80.91194 amu). Since the calculator does not know how to deal with significant figures in calculations, we need to know how many significant figures should be expressed in the answer. It is very important in chemistry that the results of calculations be properly expressed. Otherwise, systematic errors are introduced into our results. What follows are guidelines on how we express the results of calculations in chemistry.

1-2.2 Molecular Weight and the Rules for Addition and Subtraction and Rounding Off

There are two sets of rules for properly expressing the result of a mathematical operation: one applies to addition and subtraction and the other applies to multiplication and division. We will discuss the rule for addition and subtraction first.

The calculator does not give the answer to the proper number of decimal places or significant figures. For the calculation described in the text, the answer should be rounded off to 80.912.

When numbers are added or subtracted, the answer is expressed to the same number of decimal places as the measurement with the fewest decimal places. In other words, the summation must have the same degree of **uncertainty** as the measurement with the most uncertainty (e.g., ±100 has more uncertainty than ±10 and ±0.1 more than ±0.01). Consider the following addition problem encountered in chemistry.

Let's again consider our example of the molecular weight of hydrogen bromide. The summation can be illustrated as follows.

$$
\begin{array}{ll}
1.00794 \text{ amu} & \pm0.00001 \text{ (five decimal places)} \\
+79.904 \text{ amu} & \pm0.001 \text{ (three decimal places)} \\
\hline
80.91194 = 80.912 & \pm0.001 \text{ (three decimal places)}
\end{array}
$$

The numbers to the far right (e.g., ±0.0001) represent the degree of uncertainty in the particular measurement. The summation should be expressed to the same degree of uncertainty as the measurement with the least uncertainty. Thus, the answer is 80.912, which has the same uncertainty as the mass of bromine (i.e., ±0.001). Expressing the "94" in the summation has no meaning and cannot be included except for rounding-off purposes.

The rules for rounding off a number are as follows (the examples are all to be rounded off to three significant figures).

1. If the digit to be dropped is less than 5, simply drop that digit (e.g., 12.44 is rounded down to 12.4).
2. If the digit to be dropped is 5 or greater, increase the preceding digit by one (e.g., 0.3568 is rounded up to 0.357, and 13.65 is rounded up to 13.7). Note that the number 12.448 is rounded up to 12.45 if it is to be expressed to four significant figures or two decimal places, and down to 12.4 if it is to be expressed to three significant figures or one decimal place.

Recall our example of a concert crowd estimated at 12,000 people. In cases like this where a decimal point is not shown, we must be careful how we add or subtract numbers from zeros that are not significant. The answer must still show the same precision as before (e.g., ±1000). For example, if we tried to subtract 8 or 80 from 12,000, it would not change our estimate since the resulting numbers must still be rounded off to 12,000. Subtracting 800, however, would affect the result since we can now round off the number to 11,000. These three operations are illustrated as follows.

$$
\begin{array}{lll}
12,000 & 12,000 & 12,000 \\
-\ \ \ \ 8 & -\ \ \ 80 & -\ \ 800 \\
\hline
12,992 = \underline{12,000} & 12,920 = \underline{12,000} & 11,200 = \underline{11,000}
\end{array}
$$

EXAMPLE 1-2   **Expressing Summations and Subtractions to the Proper Decimal Place**

Carry out the following calculations, rounding off the answer to the proper decimal place.

(a)   7.56
       0.375
    +14.2203
    _____

(b) 14,000
       580
      +75
    _____

(c)   0.0327
    −0.00068
    _____

PROCEDURE

Refer to the rules on addition. In (a) the answer should be rounded off to two decimal places since the uncertainty is ±0.01, as in 7.56. In (b), the zeros are all to be considered nonsignificant, so the answer should be rounded off to ±1000, which is the implied uncertainty in 14,000. In (c), the answer should have an uncertainty of ±0.0001, as in 0.0327.

SOLUTION

(a) 22.16 (Round off 22.1553 to two decimal places.)

(b) 15,000 (Round off 14,655 to thousands.)

(c) 0.0320 (Round off 0.03202 to four decimal places.)

ANALYSIS

Make certain to distinguish between the answer to a calculation and a measurement made in a lab setting. Calculations need to be rounded according to the rules in order to make sense. Measurements themselves should be made as precisely as the measuring device allows; they are never rounded.

SYNTHESIS

These are not just rules to be applied, but common sense ways of dealing with measurements. If you had two friends who lived in a town 450 miles away, and one lived a mile farther down the road than the other, you wouldn't say that the one was 450 miles from you and the other was 451 miles. You'd describe both friends as living 450 miles from you. Compared to the uncertainty in the larger measurement, which may be several miles, that extra mile really makes no difference to the total distance.

1-2.3 The Rules for Multiplication and Division

In multiplication and division, we consider the number of significant figures in the answer rather than the number of decimal places. *The answer is expressed with the same number of significant figures as the multiplier, dividend, or divisor with the least number of significant figures.* In other words, the answer can be only as precise as the least precise part of the calculation (i.e., the chain is only as strong as its weakest link). We must be careful with this rule, however. If one of the multipliers represents a whole-number count, rather than a measurement, then the multiplication is essentially a shortcut to addition. In this case, the rules for addition are appropriate. For example, if one cheeseburger weighs 352.4 grams, then three cheeseburgers weigh 1057.2 grams [i.e., 3 (exact) × 352.4 = 1057.2 grams]. In this case, because the multiplication of 352.4 by 3 is the same calculation as 352.4 + 352.4 + 352.4, the answer should be expressed to one decimal place. This can be a confusing wrinkle to that rule, but we will remind you of this exception when it shows up in some future exercises. Just remember that in addition and subtraction we are concerned with the decimal point, but in multiplication and division we are concerned with the number of significant figures.

EXAMPLE 1-3   Expressing Multiplications and Divisions to the Proper Number of Significant Figures

Carry out the following calculations. Assume the numbers represent measurements; so the answer should be rounded off to the proper number of significant figures or decimal places.

(a) 2.34×3.225 **(b)** $\dfrac{11.688}{4.0}$ **(c)** $(0.56 \times 11.73) + 22.34$

PROCEDURE

Refer to the rules on multiplication and division. The answer should have the same number of significant figures as the number in the calculation with the least.

SOLUTION

(a) The answer on the calculator reads 7.5465. Since the first multiplier has three significant figures and the second has four, the answer should be expressed to three significant figures. The answer is rounded off to 7.55.

(b) The answer shown on the calculator is 2.922 but should be rounded off to two significant figures because 4.0 has two significant figures. The answer is 2.9.

(c) In cases where we must use both rules, carry out the exercise in parentheses first, round off to the proper number of significant figures, and then add and finally round off to the proper decimal place.

$$0.56 \times 11.73 = 6.6 \qquad 6.6 + 22.34 = 28.9$$

ANALYSIS

Numbers in chemistry are fundamentally different from numbers in math. In math, a 6 is a 6 and nothing else. In chemistry, if you measure something to be 6°C, it may actually be only 5.8°C. Every measurement has a degree of uncertainty in it. As such, any time calculations are done with measured values, there is uncertainty in the calculations. When numbers are multiplied, the uncertainties are multiplied, too. Small errors in several measurements that are to be multiplied can lead to very large uncertainties in the final values. Therefore, we always strive to make our measurements as accurate as we possibly can.

Consider another example of the difference between math and science: in math $6 \times 6 = 36$, but in science, 6 in. $\times$ 6 in. = 40 in.^{2}. Do you understand why?

SYNTHESIS

Smaller measuring devices like graduated cylinders are generally more accurate than larger ones. Large devices hold more. What's a good rule of thumb when using measuring equipment as regards which one to choose? Pick the smallest cylinder that will hold what it is that you're trying to measure. Make only one measurement. The smallest device will give you the most precise reading, but it defeats the purpose if you have to refill it several times.

1-2.4 Calculating Percent Error

An excellent example of incorporating the principles of accuracy and precision with the rules for dealing with significant figures in measurements is the determination of the percent error of a measurement. In many cases when experimental data is being collected, the actual, or accepted value is already known. By doing an experiment, we are merely verifying the data, or determining the validity of a technique or accuracy of an instrument. *The difference between what we expect and what we measure is the* error *in the measurement.*

Not all errors are the same. Your bathroom scale might measure objects to within the nearest pound or kilogram, whereas an electronic balance in a laboratory can determine masses accurately to hundredths of a gram. The electronic balance is inherently more accurate, but are all measurements done with the electronic balance automatically closer to their accepted value? Not necessarily; a better sense of how accurately a measurement has been made is given by a value called the percent error. **Percent error** *is a comparison of the experimental value of* *a measurement to its accepted value, reported as a percent.* The formula for percent error is:

$$\% \text{ error } = \frac{(\text{known value } - \text{ measured value})}{\text{known value}} \times 100\%$$

The absolute value bars around the equation ensure that the percent error is always reported as a positive value. For additional discussion of percent refer to Appendix A-6.

EXAMPLE 1-4 **Determining the Percent Error in an Experimental Measurement**

Calculate the percent error when

(a) A known value of 5.65 is measured to be 5.80.   **(b)** A known value of 1245 is measured to be 1241.

PROCEDURE

Insert the numbers for the experimental (i.e., measured) and the accepted (i.e., known) value into the equation for % error. Follow the rules for significant figures, first for the subtraction of two numbers, and then the division of two values.

SOLUTION

(a) % error $= \dfrac{5.80 - 5.65}{5.80} \times 100\% = 2.7\%*$

(b) % error $= \dfrac{1245 - 1241}{1245} \times 100\% = 0.3\%$

ANALYSIS

Notice that the absolute error in the second measurement (4) is larger than the first measurement (0.15). However, because the accepted value is so much larger as well, the percent error is actually smaller. This establishes an important principle in science—the larger the sample that can be measured, the more accurate the measurement that can be made, all other things being equal.

SYNTHESIS

As a rule of thumb, percent errors less than 5% are considered acceptable within the scientific community. If you are trying to verify a result, or prove a hypothesis, and your calculations consistently give you results of 5% or less, you can comfortably assert your ideas with the backing of your data. As your error increases above 5%, the validity of your data becomes more and more suspect. Generally, scientists would expect you to "go back to the drawing board" and refine your technique or procedure until you could produce results with percent errors under that 5% threshold.

*Many scientists believe that since error is a factor that exists only in the last digit of a measurement, all percent errors should be reported to only one significant figure. It would not be unreasonable to say that the error in this measurement is 3%.

▶ **ASSESSING THE OBJECTIVE FOR SECTION 1-2**

EXERCISE 1-2(a) LEVEL 1: To how many significant figures should the following calculations be rounded, assuming that the numbers are all measurements?

(a) 45.0×1.43 **(b)** 0.0034×143.7 **(c)** 50×138

EXERCISE 1-2(b) **LEVEL 1:** To what decimal place should the following calculations be rounded, assuming that the numbers are all measurements?

(a) $15.9 + 0.27$ **(b)** $420 + 38.34$ **(c)** $0.04 + 0.26$

EXERCISE 1-2(c) **LEVEL 2:** Perform the following calculations and round your answer to the appropriate number of significant figures or decimal places.

(a) $19.63 + 0.366$ **(b)** 0.200×12.765 **(c)** $(12.45 - 11.65) \times 2.68$

EXERCISE 1-2(d) **LEVEL 2:** Determine the percent error when a school with 772 students is estimated to have 750 enrolled.

EXERCISE 1-2(e) **LEVEL 3:** Which calculation is more precise, one made from many very precise measurements and a single less precise one, or one made from several modestly precise measurements?

EXERCISE 1-2(f) **LEVEL 3:** What experimental modifications can you make in lab to reduce the percent error of a procedure?

For additional practice, work chapter problems 1-8, 1-12, 1-14, 1-22, and 1-28.

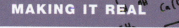

MAKING IT REAL

Ted Williams and Significant Figures

Ted Williams, a baseball legend.

Ted Williams was possibly the best hitter that baseball has ever known. He played from 1939 to 1960, interrupted by service as a fighter pilot in World War II and again in the Korean War. He was a true American hero. The "Splendid Splinter," as he was known, passed away on July 5, 2002, at age 83.

So what does Ted Williams have to do with chemistry? Not much, actually. But his most important record has a lot to do with significant figures and how numbers are rounded off. Among many other records that he set, Williams was the last major leaguer to hit the magical "400" in a season. In 1941, Williams, of the Boston Red Sox, had 185 hits in 456 at-bats. Baseball averages are computed by dividing the number of hits by the number of at-bats,

expressed as a decimal fraction rounded off to three significant figures. Thus his official final average was 406 (i.e., 0.406).

But there is more to this story. With one final day of baseball remaining, Williams had 179 hits in 448 at-bats for an average of 0.39955, which would officially round off to 0.400. If he had had only one less hit for the entire season, his average would have been 0.397—great but still not 400. The Boston manager, Joe Cronin, offered to let Ted sit out the final day's games (a double-header) and preserve his 400 average for the record books. Ted was a true professional, however, so he put his 400 average at risk and played both games. If he got only three hits out of eight at-bats for the day, his average would drop to 0.39912, which now rounds off to 0.399. He would have to go four for eight or at least three for seven to preserve his average. Williams ended the day with a home run, a double, and four singles out of eight at-bats, which brought his average up to 0.4057 for an official average of 0.406.

The great Ted Williams did not want to settle for a record that had to be "rounded off."

1-3 Expressing the Large and Small Numbers Used in Chemistry

LOOKING AHEAD! The study of chemistry requires that we deal with some extremely large or small numbers. Keeping track of five or more nonsignificant zeros in these numbers is close to impossible. So how can we express the numbers in a more compact and readable form? That is the subject we review in this section, with further information and exercises in Appendix C. ∎

1-3.1 Chemistry and Scientific Notation

The atom is such a tiny thing that it takes a unimaginably large number of them to even become visible let alone measure in a laboratory. An example of such a huge number of atoms that we will deal with later is 602,200,000,000,000,000,000,000. The magnitude of that number is almost impossible to determine without tediously counting zeros. A much more convenient and readable way to express this number is as follows

$$6.022 \times 10^{23}$$

When measurements require the use of many nonsignificant zeros, we express these numbers in a form of exponential notation known as scientific notation. *In* **scientific notation,** *a given value is expressed as a number written with one nonzero digit to the left of the decimal point and all other significant digits to the right of it* (*known as the* **coefficient**). The coefficient of 6.022 shown above indicates a precision of four significant figures. This number is then multiplied by 10 raised to a given power, called the exponent. The **exponent** *indicates the magnitude of the number.* Following are some powers of 10 and their equivalent numbers.

$$10^0 = 1$$

$$10^1 = 10$$

$$10^2 = 10 \times 10 = 100$$

$$10^3 = 10 \times 10 \times 10 = 1000$$

$$10^4 = 10 \times 10 \times 10 \times 10 = 10,000$$

etc.

$$10^{-1} = \frac{1}{10^1} = 0.1$$

$$10^{-2} = \frac{1}{10^2} = 0.01$$

$$10^{-3} = \frac{1}{10^3} = 0.001$$

$$10^{-4} = \frac{1}{10^4} = 0.0001$$

etc.

1-3.2 Mathematical Manipulation of Scientific Notation

In the following exercises, we will give examples of how numbers are expressed in scientific notation and how scientific notation is handled in multiplication and division. These examples can serve as a brief review, but if further practice is needed, see Appendix C for additional discussion on adding, squaring, and taking square roots of numbers in scientific notation. Before we consider the examples, we can see how scientific notation can remove the ambiguity of numbers such as 12,000, where the zeros may or may not be significant. Notice that by expressing the number in scientific notation, we can make it clear whether one or more of the zeros are actually significant.

$$1.2 \times 10^4 \text{ has two significant figures}$$
$$1.20 \times 10^4 \text{ has three significant figures}$$
$$1.200 \times 10^4 \text{ has four significant figures}$$

EXAMPLE 1-5   Changing Ordinary Numbers to Scientific Notation

Express each of the following numbers in scientific notation.

(a) 47,500   (b) 5,030,000   (c) 0.0023   (d) 0.0000470

PROCEDURE

A practical way to convert ordinary numbers to scientific notation is to count places from the original decimal point to where you wish to move the decimal point. In scientific notation we seek a number between 1 and 10 multiplied by 10^x. For numbers greater than one, the decimal (which may not be shown) is on the right of the number. Count in from the original decimal point to the point where the new one should be. The number of places the decimal is moved to the left will be the positive exponent of 10. For numbers less than one, count in from the original decimal on the left to where the new one should be. The number of places that the decimal is moved to the right will be the negative exponent of 10.

SOLUTION

(a) The number 47,500 can be factored as $4.75 \times 10,000$. Since $10,000 = 10^4$ the number can be expressed as

$$4.75 \times 10^4$$

We can also obtain this result by moving the decimal point four places to the left.

(b) The decimal will be moved six places to the left so the exponent will be six.

$$\overset{6\ \ 5\ \ 4\ \ 3\ \ 2\ \ 1}{5,0\ 3\ 0,0\ 0\ 0\ 0} = 5.03 \times 10^6$$

(c) The number 0.0023 can be factored into 2.3×0.001. Since $0.001 = 10^{-3}$, the number can be expressed as

$$2.3 \times 10^{-3}$$

Or, we can move the decimal point three spaces to the right.

$$\overset{1\ \ 2\ \ 3}{0.0\ 0\ 2\ 3} = 2.3 \times 10^{-3}$$

(d) The decimal point is moved five places to the right.

$$\overset{1\ \ 2\ \ 3\ \ 4\ \ 5}{0.0\ 0\ 0\ 0\ 4\ 7\ 0} = 4.70 \times 10^{-5}$$

ANALYSIS

You may find a need to go back and forth between scientific notation and a number's long form. Don't get lost when considering how many places to move the decimal point and in which direction. Quite simply, the exponent tells you the number of spaces to move the decimal point from where it is in the number to where it needs to be for the correct notation format, or vice versa. In regard to whether to move it to the right or to the left, simply realize that positive exponents are for numbers larger than 1 and negative exponents are for numbers smaller than 1, and adjust accordingly. Some may find it is easier to imagine moving the exponent up or down rather than right or left. Consider Example C-2 in Appendix C.

SYNTHESIS

When should a number be written in scientific notation? When it has become so large or so small that its magnitude is difficult to see at a glance—alternatively, when it has so many zeros in it that it becomes more convenient to use scientific notation. What is 60.4 written in scientific notation? 6.04×10^1. Clearly, scientific notation is not the appropriate choice here. On the other hand, 1,240,000,000 is much more conveniently stated as 1.24×10^9. What's the cutoff? That's generally an issue of personal preference.

EXAMPLE 1-6   Multiplying and Dividing Numbers Expressed in Scientific Notation

(Consult Appendix E, Section 3, on how to express exponents of 10 on a calculator.)

Carry out the following operations. Express the answer to the proper number of significant figures.

(a) $(8.25 \times 10^{-5}) \times (5.442 \times 10^{-3})$

(b) $(4.68 \times 10^{16}) \div (9.1 \times 10^{-5})$

PROCEDURE

In both (a) and (b), group the digits and the powers of 10. In (a), carry out the multiplication of the digits and round off the answer to three significant figures. Add the exponents. Change the number to scientific notation. In (b), group the digits and carry out the division; express the answer to two significant figures. Change the number to scientific notation.

SOLUTION

(a) $(8.25 \times 5.442) \times (10^{-5} \times 10^{-3}) = 44.9 \times 10^{-8}$

$$= 4.49 \times 10^{-7} \text{ (three significant figures)}$$

(b) $\dfrac{4.68 \times 10^{16}}{9.1 \times 10^{-5}} = \dfrac{4.68}{9.1} \times \dfrac{10^{16}}{10^{-5}}$

$$= 0.51 \times 10^{16-(-5)}$$
$$= 0.51 \times 10^{21}$$
$$= 5.1 \times 10^{20} \text{ (two significant figures)}$$

ANALYSIS

One of the advantages to scientific notation is that, because of the format, you can tell instantly how many significant figures are in a given measurement. It's always just the number of digits in the coefficient. By rule 2 in Section 1–3.2, any zeros found in correctly written scientific notation are automatically significant.

SYNTHESIS

Scientific notation can be used when there is no other convenient way to express the answer. Consider this tricky little problem: $15.6 \times 64.1 = ????$. When entered into a calculator, 999.96 is the answer returned. By the rules of significant figures, the answer should have three digits, but nines round up, so then the answer rounds to 1000. As written, though, there's no way to know the number of significant figures. Only scientific notation can clearly illustrate the three significant digits. What, then, is the best expression of the answer? 1.00×10^{3}.

▶ **ASSESSING THE OBJECTIVE FOR SECTION 1-3**

EXERCISE 1-3(a) LEVEL 1: Write the following numbers in scientific notation:

(a) 34,500   **(b)** 0.00000540   **(c)** 0.2

EXERCISE 1-3(b) LEVEL 1: Convert the following numbers to decimal form:

(a) 9.854×10^{8} **(b)** 1.400×10^{-3} **(c)** 5.4×10^{0}

EXERCISE 1-3(c) LEVEL 2: With the aid of a calculator, perform the following calculations, rounding your answer as appropriate:

(a) $(9.41 \times 10^{12}) \times (2.7722 \times 10^{-5})$ **(b)** $(4.856 \times 10^{10}) \div (2.0 \times 10^{4})$

For additional practice, work chapter problems 1-34, 1-36, 1-38, and 1-40.

KEY TERMS

1-1.1	**Chemistry** is the study of matter and the changes it undergoes. **Matter** is anything that has mass and occupies space p. 14
1-1.2	A **measurement** contains a numerical value and a specific **unit**. p. 15
1-2.1	The number of **significant figures** in a measurement is the number of measured or estimated digits. p. 15
1-2.1	The **precision** of a measurement relates to the number of significant figures, while the **accuracy** of a measurement relates to how close it is to the actual value. pp. 15–16
1-2.2	The precision of a measurement relates to its **uncertainty**. p. 19
1-2.4	The **percent error** is calculated by comparing the measured value to the actual value. p. 21
1-3.1	Very large and very small numbers are conveniently expressed in **scientific notation**. p. 24
1-3.1	In scientific notation, the power of 10 is known as the **exponent**, while the number with one digit to the left of the decimal point is the **coefficient**. p. 24

SUMMARY CHART

Measurement Qualities

Precision

Refers to the degree of reproducibility or the number of significant figures.

Accuracy

Refers to how close the measurement is to the true value.

Calculations

Addition or Subtraction

Answer expressed to the proper decimal place or uncertainty.

Multiplication or Division

Answer expressed to the proper number of significant figures.

Part B

The Measurements used in Chemistry

SETTING A GOAL

- You will learn how to manipulate the units of measurement for the purposes of problem solving.

OBJECTIVES

1-4 List several fundamental and derived units of measurement in the metric (SI) system.

1-5 Interconvert measurements by dimensional analysis.

1-6 Interconvert several key points on the Fahrenheit, Celsius, and Kelvin temperature scales.

1-4   The Units Used in Chemistry

LOOKING AHEAD! Now that we have examined the numerical part of a measurement, we are ready to examine the other part—the units that are expressed in the measurement. We will consider some of the units used in chemistry and how they are defined. ∎

▶ **OBJECTIVE FOR SECTION 1-4**

List several fundamental and derived units of measurement in the metric (SI) system.

1-4.1 The Metric System

There are several systems of measurements that have been developed throughout time. You may be most familiar with the English system of feet, gallons, and pounds. Most of the world and the sciences, however, use a different system known as the **metric system** of measurement for length, volume, and mass. *The principal metric unit for length is the* **meter**. **Volume** *is the space that a sample of matter occupies and its principal metric unit is the* **liter**. **Mass** *is the quantity of matter that a sample contains. The principal metric unit for mass is the* **gram**. (See Figure 1-4.) Since 1975, there have been plans to convert to the metric system in the United States, but there has been little progress toward this goal. However, most citizens are becoming more familiar with this system, such as purchasing soda in a 2-L plastic bottle or an automobile engine now being reported in liters rather than cubic inches. (See Figure 1-5.) Metric units also form the basis of the SI system, after the French Système International (International System). The fundamental SI units that will concern us are listed in Table 1-1. (There are other SI units for measurements that are not used in this text.) The fundamental units of the SI system have very precisely defined standards based on certain known properties of matter and light. For example, the unit of one meter is defined as the distance light travels in a vacuum in $1/299{,}792{,}458^{th}$ of a second. (Originally, it was one-millionth of the distance between the North Pole and the equator on a meridian passing through Paris.) Obviously, the current standard is extremely precise, but when we are aiming a spaceship at a planet billions of miles away (e.g., Neptune), we need a great deal of precision in our units.

TABLE 1-1

Fundamental SI Units

MEASUREMENT	UNIT	SYMBOL
Mass	kilogram	kg
Length	meter	m
Time	second	s
Temperature	kelvin	K
Quantity	mole	mol

1-4.2 Fundamental versus Derived Units

Of the numerous possible measurements that can be made, only seven are considered **fundamental**, which means that they cannot be described in terms of anything else. The five fundamental units listed in Table 1-1 can be combined in various useful and meaningful ways to make many other **derived** units. For example, the fundamental unit of distance can be divided by the fundamental unit of time to produce a derived unit called speed. In the English system, speed is

FIGURE 1-4 Length, Volume, and Mass These properties of a quantity of matter are measured with common laboratory equipment: (a) metric ruler; (b) graduated cylinders, burets, volumetric flasks, and pipet; (c) an electric balance.

(a)

(b)

(c)

FIGURE 1-5 Metric Units The use of metric units in the United States is becoming common.

TABLE 1-2

Some Derived SI Units

MEASUREMENT	UNIT	SYMBOL	DERIVATION
Volume	liter	L	Length^3
Energy	joule	J	$\text{Mass} \times \text{Distance}^2 / \text{Time}^2$
Pressure	pascal	Pa	$\text{Mass}/(\text{Time}^2 \times \text{Distance})$

measured in the familiar distance and time units of miles per hour. In the metric system, the unit for speed is the meter per second, or m/s. By combining two or more fundamental units through multiplication and division, we can produce all of the other measurements needed by scientists. In Table 1-2. three examples of common derived units are listed. Only volume will concern us at this point. Energy and pressure will be applied in later chapters.

1-4.3 The Prefixes Used in the Metric System

The metric system uses prefixes, which are exact multiples of 10 of the basic unit (e.g., meters, liters, or grams). Some of the prefixes refer to smaller portions of the basic unit (e.g., *centi-, milli-, nano-*) and others to larger portions (e.g., *deca-, kilo-, mega-*). (See Table 1-3.)

The most common prefixes used in chemistry are *milli-* and *kilo-. Centi-* is also commonly used in length (i.e., *cm*). In Making It Real, "Worlds from the Small to the Distant—Picometers to Terameters," we encounter the use of some of these prefixes as well as those expressing smaller and larger dimensions.

There is one other convenient feature of the metric system. There is an exact relationship between length and volume. The SI unit for volume is the cubic meter (m^3).

TABLE 1-3

Prefixes Used in the Metric System

PREFIX	SYMBOL	RELATION TO BASIC UNIT	PREFIX	SYMBOL	RELATION TO BASIC UNIT
tera-	T	10^{12}	deci-	d	10^{-1}
giga-	G	10^{9}	centi-	c	10^{-2}
mega-	M	10^{6}	milli-	m	10^{-3}
kilo-	k	10^{3}	micro-	μ[a]	10^{-6}
hecto-	h	10^{2}	nano-	n	10^{-9}
deca-	da	10^{1}	pico-	p	10^{-12}

[a]A Greek letter, mu.

Since this is a rather large volume for typical laboratory situations, the liter is used. One liter is defined as the exact volume of 1 cubic decimeter (i.e., 1 L = 1 dm^3). On a smaller scale, 1 milliliter is the exact volume of 1 cubic centimeter (1 mL = 1 cm^3 = cc). Thus the units milliliter and cubic centimeter can be used interchangeably when expressing volume. (See Figure 1-6.)

The basic metric unit of mass is the gram, but the SI unit is the kilogram (kg), which is equal to 1000 grams (g). The terms *mass* and *weight* are often used interchangeably, but they actually refer to different concepts. **Weight** *is a measure of the attraction of gravity for the sample.* An astronaut has the same mass on the moon as on Earth. Mass is the same anywhere in the universe. An astronaut who weighs 170 lb on Earth, however, weighs only about 29 lb on the moon. In earth orbit, where the effect of gravity is counteracted, the astronaut is "weightless" and floats free. You can lose all your weight by going into orbit, but obviously your body (mass) is still there. However, since our relevant universe is confined mostly to the surface of Earth, we often use weight as a measure of mass. In this text, we will use the term *mass*, as it is the more scientific term.

1-4.4 The English System of Measurements

In the United States, we use a measurement of length called the *foot*. The standard for this unit was not particularly scientific, as it referred to the length of a certain English king's foot (Henry VIII). Another disadvantage of the English system is that it can be awkward. For example, a trophy bass out of a Midwestern lake tips the scale at 9 lb 6 oz. An eagerly recruited basketball player for a men's college team tops out at 6 ft 11 in. Thus, we often need to use two units (e.g., feet and inches) to report only one measurement. A third major problem with these units is that they lack any systematic relationship between units (see Table 1-4). The monetary system in the United States is an exception because it is based on the decimal system. Therefore, only one unit (dollars and decimal fractions of dollars) is needed to show a typical student's dismal financial condition (e.g., having about $11.98). The British system was decimalized in 1971. Before that time, it had 12 pence to the shilling and 20 shillings to the pound.

Although some English units were formerly based on strange standards such as a king's anatomy, in modern times many have been redefined more precisely based on a corresponding metric unit. For example, one inch is now defined as exactly equal to 2.54 centimeters. The relationship between the familiar mile and

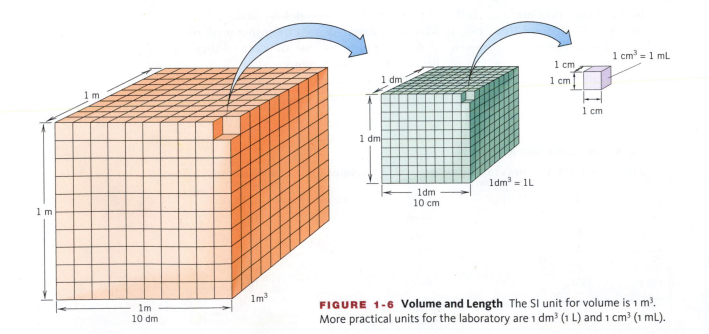

FIGURE 1-6 Volume and Length The SI unit for volume is 1 m^3. More practical units for the laboratory are 1 dm^3 (1 L) and 1 cm^3 (1 mL).

TABLE 1-4

Relationships Among English Units

LENGTH	VOLUME	MASS
12 inches (in.) = 1 foot (ft)	2 pints (pt) = 1 quart (qt)	16 ounces (oz) = 1 pound (lb)
3 ft = 1 yard (yd)	4 qt = 1 gallon (gal)	2000 lb = 1 ton
1760 yd = 1 mile (mi)	42 gal = 1 barrel (bbl)	

the kilometer originated from a measurement, but 1 mile is now defined as exactly equal to 1.609344 km. For simplicity in our calculations, however, the relationship is rounded off to four significant figures. That is, 1.000 mi = 1.609 km.

1-4.5 Relationships Between the Metric and English Systems

Several relationships between the metric and English systems are listed in Table 1-5. (See also Figure 1-7.) The relationships within systems (e.g., Tables 1-3 and 1-4) are always defined and exact numbers. Because most of the relationships between systems, however, are measured, they are not necessarily exact. Thus, they can be expressed to various degrees of precision. Most of the relationships shown in Table 1-5 are known to many more than the four significant figures given.

TABLE 1-5

The Relationship Between English and Metric Units

	ENGLISH	METRIC EQUIVALENT
Length	1.000 in.	2.540 cm (exact)
	1.000 mi	1.609 km
Mass	1.000 lb	453.6 g
	2.205 lb	1.000 kg
Volume	1.057 qt	1.000 L
	1.000 gal	3.785 L
Time	1.000 s	1.000 s

1 yd

1 m

(a)

1 L   1 Qt

(b)

1 kg   1 lb

(c)

FIGURE 1-7 Comparison of Metric and English Units
(a) 1 meter and 1 yard,
(b) 1 quart and 1 liter, and
(c) 1 kilogram and 1 pound.

EXERCISE 1-4(a) LEVEL 1: List the metric units for the following.

(a) mass   **(b)** volume   **(c)** energy   **(d)** temperature

EXERCISE 1-4(b) LEVEL 1: What are the powers of 10 for the following metric prefixes?

(a) kilo- 10^3 **(b)** centi- 10^{-2} **(c)** milli- 10^{-3} **(d)** Mega- 10^6

EXERCISE 1-4(c) LEVEL 3: What are some of the advantages of using the metric system as compared to the English system? Are there any drawbacks?

For additional practice, work chapter problems 1-51 and 1-52.

▶ **OBJECTIVE FOR SECTION 1-5**

Interconvert measurements by dimensional analysis .

1-5 Conversion of Units by Dimensional Analysis

LOOKING AHEAD! When one learns a new language, the first goal is to be able to translate sentences from the new to the familiar. Likewise, to become comfortable with a different system of measurement, we need to convert from one to the other. Converting between systems can be accomplished with an important tool that we will introduce in this section. This will give us a chance to practice solving many of the types of problems that we will eventually encounter. ■

1-5.1 Relationships as Conversion Factors

In Chapter 5, we will encounter a strange but extremely important unit known as "the mole." This unit is used to measure a large number of atoms or molecules that represents a specific mass of those atoms or molecules. Although we will not solve problems relating to the mole at this time, we can see how conversions between units are very important in the study of chemistry. The mole relates to a number and a mass as illustrated for the element carbon.

1.000 mole of carbon = 6.022×10^{23} atoms of carbon = 12.011 g of carbon

(the mass in grams listed inside the front cover)

Eventually, it will become necessary that we *interconvert* among the three units—moles, number of atoms, and the mass. Dimensional analysis of conversions will be employed to guide us through these and other similar calculations that come later in the text.

Making conversions in science can be extremely important. In 1999, a rocket destined to land softly on the planet Mars crashed into the surface instead. Apparently, someone did not convert between English and metric units in programming the onboard computers that controlled the landing. This mistake cost about $1 billion and disappointed a lot of scientists. In fact, we convert between one thing and another all the time. For example, electrical converters change alternating current (AC) to direct current (DC) for devices such as electrical razors or to charge a cell phone battery. What we need is a mathematical converter to change measurements in one unit to those in another. In chemistry this mathematical converter is known as a conversion factor. *A* **conversion factor** *is a relationship between two units or quantities expressed in fractional form.* **Dimensional analysis** *converts a measurement in one unit to another by the use of conversion factors.*

For example, a conversion factor can be constructed from the exact relationship within the metric system.

$$10^3 \text{ m} = 1 \text{ km}$$

This equality can be expressed as either of two fractions as shown here.

$$\textbf{(1)} \ \frac{1 \text{ km}}{10^3 \text{ m}} \quad \text{and} \quad \textbf{(2)} \ \frac{10^3 \text{ m}}{1 \text{ km}} \ (\text{or simply} \ \frac{10^3 \text{ m}}{\text{km}})$$

MAKING IT REAL

Worlds from the Small to the Distant—Picometers to Terameters

Saturn is about 1Tm from Earth

2 meters tall

Blood cells are about 1μm in diameter

The meter is just about right for the world of human beings. For example, our height may be a little less than 2 meters and the room may be about 7 square meters. If people were the size of ants, however, the meter would not be a very practical unit of length. Fortunately, the metric system allows us to venture into the far reaches of other worlds, either small or distant. All we need is a few prefixes to adjust the unit. First, let's shrink ourselves into the world of the small. Our first stop is the "milli" world (10^{-3} m), where normally small specks now appear large. Those little bugs called mites are of this size; if we lived in this world, it would look a lot more hostile than our world of the meter. When we go to the "micro" world (10^{-6} m), we are well into the invisible world of the tiny. Blood cells that circulate in our veins have these dimensions (i.e., 1 μm). Next, we descend to the "nano" world (10^{-9} m), where we encounter one of the tiniest forms of life, called viruses. Some of these nasty creatures measure 10 nm across. Our final stop is the "pico" region (10^{-12} m), where we notice that the basic components of matter, atoms, appear as large spheres. The aluminum atom in a can of soda measures 125 pm in diameter.

Now we take ourselves out away from the surface of Earth and view our surroundings as we ascend into the sky. One km (10^3 m) from the surface we are not that far up. We can still see cars and buildings. But if we ascend to one megameter (10^6 m), we are certainly in outer space. In this region the artificial satellites drift silently by. Our planet still looms large in our vision, with a diameter of about 6 Mm. Now let's ascend to one gigameter (10^9 m) from the surface. We find ourselves well past the moon, which would lie about halfway between us and Earth. Venus is still 32 Gm away. Finally, at one terameter (10^{12} m), we are nearly to the planet Saturn, which lies about 1.2 Tm from Earth at its closest approach.

The metric system obviously provides convenient units from the smallest to the largest or farthest. Even more units exist that take us to even smaller and farther regions.

These fractional relationships read as "1 kilometer per 10^3 meters" and "10^3 meters per 1 kilometer." The latter fraction is usually simplified to "10^3 meters per kilometer." When just a unit (no number) is read or written in the denominator,

To recharge this cell phone, AC current must be converted into DC current.

it is assumed to be 1 of that unit. *Factors that relate a quantity in a certain unit to 1 of another unit are sometimes referred to as* **unit factors**. Unit factors are written without the 1 in the denominator in this text. It should be understood, however, that the 1 in the numerator or implied in the denominator can be considered exact in calculations.

Two other conversion factors can be constructed from an equality between English and metric units. For example, the exact relationship between inches (in.) and centimeters (cm) is

$$1 \text{ in.} = 2.54 \text{ cm}$$

which can be expressed in fractional form as

$$\textbf{(3)} \; \frac{1 \text{ in.}}{2.54 \text{ cm}} \quad \text{and} \quad \textbf{(4)} \; \frac{2.54 \text{ cm}}{\text{in.}}$$

Each of the four fractions can be used to convert one unit into the other. The general procedure for a one-step conversion by dimensional analysis is

$$(\text{what's given}) \times (\text{conversion factor}) = (\text{what's requested})$$

In most conversions, "what's given" and "what's requested" each have one unit. The conversion factor has two: the requested or new unit in the numerator and the given or old unit in the denominator. (In this scheme we are assuming that the new or requested unit is in the numerator.)

$$A \; (\text{given unit}) \times \frac{B \; (\text{requested unit})}{C \; (\text{given unit})} = D \; (\text{requested unit})$$

When combined in the calculation, the given units will appear in both the numerator and the denominator and thus cancel, just like any numerical quantity. The requested unit survives. This is illustrated as follows, where A, B, C, and D represent the numerical parts of the measurements or definitions.

$$\frac{A \; (\text{given unit}) \times B \; (\text{requested unit})}{C \; (\text{given unit})} = D \; (\text{requested unit})$$

Doing problems by this method is like taking a trip. The first two things we know about a trip are where we start and where we want to end up. The conversion factor is *how* we get from here to there. The key to using the dimensional analysis is to select the right conversion factor that relates the given and requested units. Recall that the units are to be treated like numerical quantities; that is, they are multiplied, divided, or canceled in the course of the calculation.

1-5.2 One-Step Conversions and Unit Maps

We will now give some examples of how dimensional analysis is used in some one-step conversions. The first example (1-7) uses familiar units. The second example (1-8) will put to use one of the two conversion factors that we constructed within the metric system, and the next (Example 1-9) will use one of the two conversion factors between the English and metric systems. In the examples, we proceed carefully through four steps.

1. From what is given and what is requested, decide what the conversion factor must do (e.g., convert in. to cm). You can express the procedure in a shorthand method that we will refer to as the **unit map** in this text (e.g., $\boxed{\text{in.}} \rightarrow \boxed{\text{cm}}$).
2. Find the proper relationship between units from a table (if necessary). Express the relationship as a conversion factor so that the new unit is in the numerator.
3. Put the problem together, making sure that the proper unit cancels and your answer has the requested unit or units.
4. Express your answer to the proper number of significant figures.

Be aware of conversion factors that are not the result of measurements but instead are exact definitions. The following relationships are examples of exact definitions.

$$12 \text{ in.} = 1 \text{ ft} \qquad 4 \text{ qt} = 1 \text{ gal} \qquad 1 \text{ km} = 10^3 \text{ m}$$

That is, there are exactly 12 in. (12.0000, etc.) in 1 ft (1.0000, etc.) and so forth. Since exact relationships are considered to have unlimited precision in a calculation, they can be ignored in determining the number of significant figures in an answer. Exact definitions are generally relationships *within* a measurement system (e.g., in. and ft, km and m).

Measurements are never exact, so the number of significant figures shown reflects the precision of the instrument used to make the measurement. The use of exact relationships in calculations can be somewhat confusing at first, so we will remind you when one is being used in calculations in this chapter.

EXAMPLE 1-7   Converting Hours to Days

Convert 129 hours to an equivalent number of days.

PROCEDURE

1. The unit map for the conversion is

2. The well-known relationship is that there are exactly 24 hours in one day. This exact relationship can be written as two conversion factors

$$\frac{24 \text{ hr}}{\text{day}} \quad \text{and} \quad \frac{1 \text{ day}}{24 \text{ hr}}$$

The latter is the appropriate factor since hr (the given unit) is in the denominator and day (the new requested unit) is in the numerator.

SOLUTION

$$129 \text{ hr} \times \frac{1 \text{ day}}{24 \text{ hr}} = \underline{5.38 \text{ day}}$$

ANALYSIS

What would have happened if we had used the first conversion factor (i.e., 24 hr/day) by mistake? In that case, the strange units that result (i.e., hr^2/day) would alert us that we made a mistake. Paying close attention to the units that cancel and those that don't will be a big help in correctly solving equivalent chemistry problems. Notice also that if we multiplied, the numerical answer would be 3096, which is not a reasonable value.

SYNTHESIS

Students sometimes have problems writing a proper conversion factor. For instance, if the conversion is between liters (L) and milliliters (mL), they recognize that there is a relationship of 1000 to 1 between the two. However, is it

$$\frac{1000 \text{ mL}}{\text{L}} \quad \text{or} \quad \frac{1000 \text{ L}}{\text{mL}} \text{?}$$

Rarely would someone make that mistake with units that they're very familiar with, like feet and inches. We all know that it's

$$\frac{12 \text{ in.}}{\text{ft}} \quad \text{and not} \quad \frac{12 \text{ ft}}{\text{in.}}$$

However, the concept here is worth noting. Notice that the larger numerical value goes with the smaller-sized unit. That is because the two units in a conversion factor are meant to be equivalent. Keeping that in mind will help prevent you from writing and using incorrect conversions like 100 m/cm.

EXAMPLE 1-8   Converting Meters to Kilometers

Convert 0.468 m to **(a)** km and **(b)** mm.

PROCEDURE (a)

1. The unit map for the conversion in (a) is

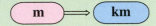

2. From Table 1-3, we find the relationship is 10^3 m = 1 km. When expressed as the proper conversion factor, km is in the numerator (requested) and m is in the denominator (given). The conversion factor is

$$\frac{1\ km}{10^3\ m}$$

SOLUTION (a)

$$0.468\ m \times \frac{1\ km}{10^3\ m} = 0.468 \times 10^{-3}\ km = \underline{\underline{4.68 \times 10^{-4}\ km}}$$

PROCEDURE (b)

1. The unit map for the conversion in (b) is

m ⟹ mm

2. From Table 1-3, we find the proper relationship is 1 mm = 10^{-3} m. When shown as the proper conversion factor, mm is in the numerator (requested) and m is in the denominator (given). The conversion factor is

$$\frac{1\ mm}{10^{-3}\ m}$$

SOLUTION (b)

$$0.468\ m \times \frac{1\ mm}{10^{-3}\ m} = 0.468 \times 10^3\ mm = \underline{\underline{468\ mm}}$$

ANALYSIS

Notice that both conversion factors are exact definitions, so do not limit the number of significant figures in the answers. We would expect (a) to be a smaller number than the given number, since the new unit is larger, and (b) to be larger, since the new unit is smaller.

SYNTHESIS

How might you convert from mm and km? This could be done as a two-step conversion, as discussed in the next section: millimeters → meters → kilometers, with each step having its own conversion factor.

$$mm \times \frac{10^{-3}\ m}{mm} \times \frac{1\ km}{10^3\ m}$$

Alternatively, you can realize that it is six powers of 10 from milli to kilo, and write a single conversion factor:

$$\frac{1\ km}{10^6\ mm}$$

Use the strategy that makes the most sense to you.

EXAMPLE 1-9   Converting Centimeters to Inches

Convert 825 cm to in.

PROCEDURE

1. The unit map for the conversion is

2. From Table 1-5, we find that the proper relationship is 1 in. = 2.54 cm. Expressed as a conversion factor, in. is in the numerator (requested) and cm is in the denominator (given). The conversion factor is

$$\frac{1\text{ in.}}{2.54\text{ cm}}$$

SOLUTION

$$825\text{ cm} \times \frac{1\text{ in.}}{2.54\text{ cm}} = \underline{325\text{ in.}}$$

ANALYSIS

Notice that the answer is reasonable, since we expect there to be fewer inches than cm.

SYNTHESIS

When it comes to having conversion factors at your fingertips, we assume that you are familiar with most English system conversions, like 3 ft per yd, or 16 oz per lb. Conversions within the metric system are all based on factors of 10: 100, 1000, 0.01, etc. Learn these as quickly as possible. It would also be helpful to memorize one conversion from the English to the metric system for mass, length, and volume. Then you will be able to convert from any given measurement to any other without need for reference, by employing the multistep conversions discussed in the next section.

In conversions between the metric and English systems, we will use a conversion factor with four significant figures. (Recall that the relationship between inches and centimeters, used in Example 1-8, is exact, however.) Since most of the measurements are given to three significant figures, this means that the conversion factor does not limit the precision of the answer. The answer should then be expressed to the same number of significant figures as the original measurement.

1-5.3 Multistep Conversions

A one-step conversion like those that have been worked so far is analogous to a direct, nonstop airline flight between your home city and your destination. A multistep conversion is analogous to the situation in which a nonstop flight is not available and you have to make two or more intervening stops before reaching your destination. Each stop in the flight is a separate journey, but it gets you closer to your ultimate destination. When one plans such a journey, one must carefully plan each step, perhaps with the help of a map. Many of our conversion problems also require more than one step and must be carefully planned. In multistep conversion problems, each step along the way requires a separate conversion factor, which will be illustrated by the unit map.

For example, let's consider a problem requiring a conversion between two possible units of quantity—number of apples and boxes of apples. Let's assume we know that exactly six apples can be placed in each sack and exactly four sacks make up one box. The problem is, "How many boxes of apples can be prepared with 254 apples?" Since we don't have a direct relationship between apples and boxes, we need a "game plan," which is our unit map. First, we can convert number of apples to number of sacks (conversion a) and then number of sacks to number of boxes (conversion b). The problem and the unit map are set up and completed as follows.

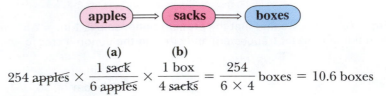

$$254 \text{ apples} \times \underset{\text{(a)}}{\frac{1 \text{ sack}}{6 \text{ apples}}} \times \underset{\text{(b)}}{\frac{1 \text{ box}}{4 \text{ sacks}}} = \frac{254}{6 \times 4} \text{ boxes} = 10.6 \text{ boxes}$$

Notice that the two conversion factors represent exact relationships, so they do not affect the number of significant figures in the answer.

We are now ready to work through some real multistep conversions between units of measurement.

EXAMPLE 1-10   Converting Dollars to Doughnuts

How many doughnuts can one purchase for \$123 if doughnuts cost \$3.25 per dozen?

PROCEDURE

1. A conversion factor between dollars and individual doughnuts is not directly given. We can solve this problem by making a two-step conversion by (a) converting dollars (\$) to dozen (doz) and then (b) converting dozen to number of doughnuts. The unit map for the conversions is shown as

$$\$ \xrightarrow{\text{(a)}} \text{doz} \xrightarrow{\text{(b)}} \text{doughnuts}$$

2. The conversion factors arise from the two relationships: \$3.25 per dozen and 12 doughnuts per dozen (an exact relationship). In conversion factor (a), \$ must be in the denominator to convert to doz. In conversion factor (b), doz [the new unit from the conversion in (a)] must be in the denominator to convert to doughnuts. The two conversion factors are

$$\text{(a) } \frac{1 \text{ doz}}{\$3.25} \quad \text{and} \quad \text{(b) } \frac{12 \text{ doughnuts}}{\text{doz}}$$

SOLUTION

$$\$123 \times \frac{1 \text{ doz}}{\$3.25} \times \frac{12 \text{ doughnuts}}{\text{doz}} = \underline{454 \text{ doughnuts}}$$

ANALYSIS

As you are writing conversion factors for a problem, realize that it's not critical to put it in the exact form needed in the problem immediately. Once you have all the factors that you need, you can tell what unit needs to go into the numerator and what needs to be in the denominator, based on the units you are converting. You can enter them correctly when you're setting up the problem.

SYNTHESIS

It's always wise to estimate your answer to see if what's calculated is at least reasonable. A quick estimate tells us that at around \$3.00 a dozen, about \$120 will buy us approximately 40 dozen: $40 \times 12 = 480$, which means that our answer is reasonable.

EXAMPLE 1-11 Converting Liters to Gallons

Convert 9.85 L to gal.

PROCEDURE

1. A two-step conversion is needed. L can be converted into qt (a) and qt then converted into gal (b). The unit map for the conversions is expressed as

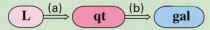

2. The two relationships needed are 1.057 qt = 1.000 L (from Table 1-5) for conversion (a) and 4 qt = 1 gal (from Table 1-4) for conversion (b). The two relationships properly expressed as conversion factors are

$$\text{(a)} \ \frac{1.057 \ \text{qt}}{\text{L}} \quad \text{and} \quad \text{(b)} \ \frac{1 \ \text{gal}}{4 \ \text{qt}}$$

SOLUTION

$$9.85 \ \cancel{\text{L}} \times \overset{\text{(a)}}{\frac{1.057 \ \cancel{\text{qt}}}{\cancel{\text{L}}}} \times \overset{\text{(b)}}{\frac{1 \ \text{gal*}}{4 \ \cancel{\text{qt}}}} = \underline{2.60 \ \text{gal}}$$

ANALYSIS

Remember that the relationship between quarts and gallons is exact, since it is within the English system. We start with about 10 L, which is about 10 qt. Ten quarts is 2.5 gal, so our answer is reasonable.

SYNTHESIS

There is not always just one way to do a multistep problem. Suppose you knew that there were 3785 mL in 1 gallon. What conversion factors would be useful in solving the problem with this information? The unit map would be L → mL → gal, and the conversion factors necessary would be:

$$\frac{1000 \ \text{mL}}{1 \ \text{L}} \quad \text{and} \quad \frac{1 \ \text{gal}}{3785 \ \text{mL}}$$

*This represents an exact relationship.

EXAMPLE 1-12 Converting Miles per Hour to Meters per Minute

Convert 55 mi/hr to m/min.

PROCEDURE

1. In this case, we can convert mi/hr to km/hr (a) and then km/hr to m/hr (b). It is then necessary to change the units of the denominator from hr to min (c). The unit map for the conversions is expressed as

$$\boxed{\text{mi/hr}} \overset{\text{(a)}}{\Longrightarrow} \boxed{\text{km/hr}} \overset{\text{(b)}}{\Longrightarrow} \boxed{\text{m/hr}} \overset{\text{(c)}}{\Longrightarrow} \boxed{\text{m/min}}$$

2. The needed relationships are 1.000 mi = 1.609 km (from Table 1-5), 10^3 m = 1 km (from Table 1-3), and 60 min = 1 hr. The relationships expressed as conversion factors are

$$\text{(a)} \ \frac{1.609 \ \text{km}}{\text{mi}} \qquad \text{(b)} \ \frac{10^3 \ \text{m}}{\text{km}} \qquad \text{(c)} \ \frac{1 \ \text{hr}}{60 \ \text{min}}$$

Notice that factor (c) has the requested unit (minutes) in the denominator and the given unit (hours) in the numerator. This is different from all the previous examples because we are changing a denominator in the original unit.

SOLUTION

$$55 \,\frac{\text{mi}}{\text{hr}} \times \overset{\text{(a)}}{\frac{1.609 \text{ km}}{\text{mi}}} \times \overset{\text{(b)}}{\frac{10^3 \text{ m}}{\text{km}}} \times \overset{\text{(c)}}{\frac{1 \text{ hr}}{60 \text{ min}}} = \underline{\underline{1.5 \times 10^3 \text{ m/min}}}$$

ANALYSIS

Recall that the relationships between meters and kilometers and between hours and minutes are exact. Thus the answer should be expressed to two significant figures because of the two digits in the original measurement (55).

SYNTHESIS

Though this text uses a single continuous calculation to solve most problems, many students prefer to break problems like this down into two individual problems. First, you can convert the units of distance, miles into meters, and then you can convert the units of time, hours into minutes. Finally, you can take those answers and divide the converted distance by the converted time to arrive at the same answer.

▶ **ASSESSING THE OBJECTIVE FOR SECTION 1-5**

EXERCISE 1-5(a) LEVEL 1: What are the conversion factors necessary to perform the following conversions?

(a) g into kg   **(b)** hr into s   **(c)** in. into yd   **(d)** m into cm

EXERCISE 1-5(b) LEVEL 1: What are the two conversion factors necessary to perform each of the following conversions?

(a) oz into kg   **(b)** cm into ft   **(c)** mL into qt

EXERCISE 1-5(c) LEVEL 2: How many miles are in 25.0 km?

EXERCISE 1-5(d) LEVEL 2: A certain size of nail costs $1.25/lb. What is the cost of 3.25 kg of these nails?

EXERCISE 1-5(e) LEVEL 3: Estimate how many times your heart beats in one day. In one year.

EXERCISE 1-5(f) LEVEL 3: A 2.0-L bottle of cola costs $1.09. A six-pack of 12-oz cans costs $1.89. There are 355 mL in 12 oz. What is the cost per ounce of the 2.0-L bottle and the cost per ounce of the cans? Which is the better deal?

For additional practice, work chapter problems 1-54, 1-58, 1-61, 1-62, 1-68, 1-73, and 1-82.

▶ **OBJECTIVE FOR SECTION 1-6**

Interconvert several key points on the Fahrenheit, Celsius, and Kelvin temperature scales.

1-6 Measurement of Temperature

LOOKING AHEAD! The final measurement we will consider in this chapter is temperature. Once again, those who grew up in the United States are at a disadvantage. The scale we are most familiar with is not the same as that used by science and most of the rest of the world. We will become more familiar with two other temperature scales in this section. ■

1-6.1 Thermometer Scales

When a log burns, along with a chemical effect that converts the wood to ash and gas there is the liberation of a large amount of heat. One of the effects of that heat that we can measure is a rise in the temperature around the burning log. **Temperature** *is a* *measure of the heat energy of a substance.* (We will have more to say about the meaning of heat energy in Chapter 3.) When someone says an item is "hot," we can interpret this in several ways. It could mean that this item is something we should have, that it looks great, that it is stolen, or that it has a somewhat higher temperature than "warm." In this section, we will be concerned with the last concept. "Hot" in science refers to a high temperature. *Temperature can be measured with a device called a* **thermometer**.

The thermometer scale with which we are most familiar in the United States is the **Fahrenheit** scale (°F), but the **Celsius** scale (°C) is commonly used elsewhere and in science. Many U.S. television news shows once broadcast the temperature in both scales, but, unfortunately, this practice has been mostly discontinued. Thermometer scales are established by reference to the freezing point and boiling point of pure water. These two temperatures are constant and unchanging (under constant air pressure). Also, when pure water is freezing or boiling, the temperature remains constant. We can take advantage of these facts to compare the two temperature scales and establish a relationship between them. In Figure 1-8, the temperature of an ice-and-water mixture is shown to be exactly 0°C. This temperature was originally established by definition and corresponds to exactly 32°F on the Fahrenheit thermometer. The boiling point of pure water is exactly 100°C, which corresponds to 212°F.

1-6.2 Relationships Between Scales

On the Celsius scale, there are 100 equal divisions (div.) between these two temperatures, whereas on the Fahrenheit scale, there are $212 - 32 = 180$ equal divisions (div.) between the two temperatures. Thus, we have the following relationship between the scale divisions:

$$100 \text{ C div.} = 180 \text{ F div.}$$

This relationship can be used to construct conversion factors between an equivalent number of Celsius and Fahrenheit degrees. The relationship results from exact definitions, so it does not affect the number of significant figures in the calculation.

$$\frac{100 \text{ C div.}}{180 \text{ F div.}} = \frac{1 \text{ C div.}}{1.8 \text{ F div.}} \qquad \text{The inverse relationship is} \qquad \frac{1.8 \text{ F div.}}{\text{C div.}}$$

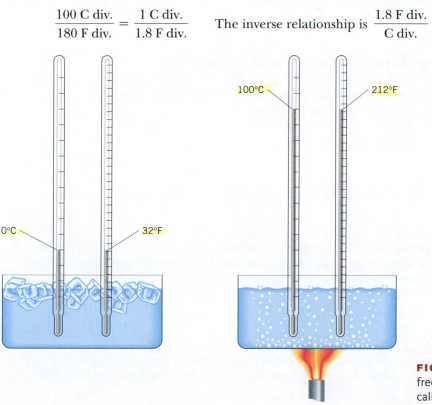

FIGURE 1-8 The Temperature Scales The freezing and boiling points of water are used to calibrate the temperature scales.

To convert the Celsius temperature [$T(°C)$] to Fahrenheit temperature [$T(°F)$]:

1. Multiply the Celsius temperature by the proper conversion factor (1.8°F/1°C) to get the equivalent number of Fahrenheit degrees.
2. Add 32°F to this number so that both scales start at the same point (the freezing point of water)

$$T(°F) = \left[T(°C) \times \frac{1.8°F}{1°C} \right] + 32°F = [T(°C) \times 1.8] + 32$$

To convert the Fahrenheit temperature to the Celsius temperature,

1. Subtract 32°F from the Fahrenheit temperature so that both scales start at the same point.
2. Multiply the number by the proper conversion factor (1°C/1.8°F) to convert to the equivalent number of Celsius degrees.

$$T(°C) = [T(°F) - 32°F] \times \frac{1°C}{1.8°F} = \frac{[T(°F) - 32]}{1.8}$$

EXAMPLE 1-13   Converting Between Fahrenheit and Celsius

A person with a cold has a fever of 102°F. What would be the reading on a Celsius thermometer?

PROCEDURE

Use the equation that converts °F to °C: $T(°C) = \dfrac{[T(°F) - 32]}{1.8}$

SOLUTION

$$T(°C) = \frac{[102 - 32]}{1.8} = \underline{\underline{39°C}}$$

EXAMPLE 1-14   Converting Between Celsius and Fahrenheit

On a cold winter day the temperature is −10.0°C. What is the reading on the Fahrenheit scale?

PROCEDURE

Use the equation that converts °C to °F: $T(°F) = (1.8 \times °C) + 32$

SOLUTION

$$T(°F) = (-10.0 \times 1.8) + 32 = \underline{14.0°F}$$

ANALYSIS

These two calculations represent one of the few times in this text that memorization of an equation is necessary. For most of the problems you'll be doing, dimensional analysis is useful and strongly encouraged. In this way, relatively few things need to be committed to memory. In this case, however, because the two scales are offset by 32 degrees, the specific formula is the only way to convert them.

SYNTHESIS

Since both scales, Fahrenheit and Celsius, represent straight-line relationships, there is a point where both lines intersect. Said another way, there is a temperature that is the same value on both scales. Put an x into the equation for both °F and °C, and solve. What value do you calculate for x? (See chapter problem 1–107.)

The SI temperature unit is called the kelvin (K). Notice that a degree symbol (°) is not shown with a kelvin temperature. The zero on the **Kelvin scale** is theoretically the lowest possible temperature (the temperature at which the heat energy is zero). This corresponds to $-273°C$ (or, more precisely, $-273.15°C$). Since the Kelvin scale also has exactly 100 divisions between the freezing point and the boiling point of water, the magnitude of a kelvin and a Celsius degree is the same. Thus we have the following simple relationship between the two scales. The temperature in kelvins is represented by $T(K)$, and $T(°C)$ represents the Celsius temperature.

$$T(K) = T(°C) + 273$$

Thus, the freezing point of water is 0°C or 273 K, and the boiling point is 100°C or 373 K. We will use the Kelvin scale more in later chapters.

▶ **ASSESSING THE OBJECTIVE FOR SECTION 1-6**

EXERCISE 1-6(a) LEVEL 1: Fill in the following table with appropriate temperatures:

	FAHRENHEIT	CELSIUS	KELVIN
Water boiling			
Water freezing			
Absolute zero			

EXERCISE 1-6(b) LEVEL 2: The temperature of cooking oil reaches 248°F. What is this on the Celsius and Kelvin scales?

EXERCISE 1-6(c) LEVEL 2: The record cold temperature in the continental U.S. is $-70°F$ on January 20, 1954 in Rogers Pass, Montana. What is this temperature in Celsius and Kelvin?

EXERCISE 1-6(d) LEVEL 3: Despite U.S. insistence on using the Fahrenheit scale, Celsius seems to make much more sense. It is divided into 100 equal parts and has obvious set points at 100°C and 0°C. Why do you suppose Fahrenheit uses such odd numbers for boiling and freezing?

For additional practice, work chapter problems 1-98, 1-99, 1-104, and 1-105.

PART B SUMMARY

KEY TERMS

1-4.1	Most measurements in chemistry are expressed using the **metric system**. p. 28
1-4.1	**Volume** is the space that a given sample of matter occupies and **mass** refers to the quantity of matter that the sample contains. p. 28
1-4.1	In the metric system, the principal unit for length is the **meter**, for volume it is the **liter**, and for mass it is the **gram**. p. 28
1-4.2	Five **fundamental** units can produce many **derived** units. p. 28
1-4.3	**Weight** refers to the attraction of gravity for a quantity of matter. p. 30
1-5.1	Converting from one unit of measurement to another requires the use of **conversion factors**. If the factor relates a quantity in one system to 1 in another system, is called a **unit factor**. pp. 32, 34
1-5.1	The procedure to convert from one measurement to another is **dimensional analysis**. p. 32
1-5.2	An outline of a conversion procedure is expressed as a **unit map**. p. 34
1-6	The **temperature** is measured by a device called a **thermometer**. p. 41
1-6.1	The **Fahrenheit** scale is used in the United States, but the **Celsius** scale is used in science. p. 41
1-6.2	The **Kelvin scale** begins at absolute zero. p. 43

SUMMARY CHART

Sample Conversion

Problem: Convert 10.0 km to miles.

Unit map: km $\Longrightarrow$ mi

Relationship: 1.609 km = 1.000 mi

Conversion factor: $\dfrac{1 \text{ mi (requested)}}{1.609 \text{ km (given)}}$

Solution: $10.0 \text{ km} \times \dfrac{1 \text{ mi}}{1.609 \text{ km}} = 6.22 \text{ mi}$

CHAPTER 1 SYNTHESIS PROBLEM

We all have some idea about the meaning of *density*. Wood floats because it is less dense than water. Iron sinks because it is denser than water. Helium balloons rise because helium is less dense or lighter than air. We will discuss density in detail in Chapter 3, but we will use this property here to review many of the concepts presented in this chapter. Density is calculated from two measurements, the mass of an object (usually in grams) and the volume of the object (usually in mL). Consider the calculation of the density and other related problems of a certain familiar metal in the following problems. The worked-out solutions are to the right.

PROBLEM	SOLUTION
a. The mass of a sample of a metal is 28.738 g. The volume of this same sample is 4.10 mL. What is the number of significant figures and the degree of uncertainty in each of the two measurements?	**a.** 28.738 g has five significant figures and an uncertainty of ± 0.001. 4.10 mL has three significant figures and an uncertainty of ± 0.01. Recall that the trailing zero in 4.10 is significant.
b. What is the density of the metal in g/mL?	**b.** 28.738 g/4.10 mL = 7.01 g/mL (Express to three significant figures because of the denominator.)
c. The metal turns out to be *tin*. The actual density of tin is 7.31 g/mL. What is the percent error in the calculation?	**c.** $(7.31 - 7.01)/7.31 \times 100\% = 4.1\%$ (two significant figures in the answer since the subtraction in the numerator is done first. In fact, percent error is usually reported to a whole number so an accepted answer is 4% error)
d. Density can be used as a conversion factor to convert between mass and volume of a sample using dimensional analysis. Using the actual density of tin, what is the mass of a sample of the metal that occupies 2.37×10^4 mL expressed in scientific notation?	**d.** $2.37 \times 10^4 \text{ mL} \times \dfrac{7.31 \text{ g}}{\text{mL}} = 1.73 \times 10^5 \text{ g}$. (express one digit to the left of the decimal point.)
e. What is the mass expressed in kg?	**e.** $1.73 \times 10^5 \text{ g} \times \dfrac{1 \text{ kg}}{10^3 \text{ g}} = 173 \text{ kg}$
f. What is the volume of a sample of tin that has a mass of 16.02 g?	**f.** $16.02 \text{ g} \times \dfrac{1 \text{ mL}}{7.31 \text{ g}} = 2.19 \text{ mL}$

YOUR TURN

A sample of a metal has a mass of 198.2 grams and has a volume of 23.0 mL. The metal is actually nickel with a density of 8.909 g/mL.

a. How many significant figures are in the two measurements and the uncertainty in each?

b. What is the calculated density of the metal?

c. What is the percent error compared to the actual density?

d. A sample of nickel has a volume of 8.78×10^{-2} L. Convert this volume to mL.

e. Using the actual density of nickel, convert the volume in part d to the mass of nickel.

f. What is the volume of a sample of nickel that has a mass of 628 g?

Answers are on p. 46.

CHAPTER SUMMARY

Chemistry, probably more than any other science, is a science of **measurements**. Certainly, that is not all there is—many other concepts are also central to the science—but handling measurements properly is of the highest priority. That is why we needed to address this subject very early in the text. A measurement consists of a number and a **unit**. Our first item of business was to examine the numerical part of a measurement as to the significance of the numbers that are reported. The numerical value has two qualities: **precision** (number of **significant figures** or **uncertainty**) and **accuracy**. The **percent error** expresses the relative accuracy of a measurement.

Perhaps a greater challenge than just understanding the precision of one measurement is to handle measurements of varying precision in mathematical operations. Addition and subtraction focus on the precision, or decimal place, in the answer, while multiplication and division are concerned with the number of significant figures.

We won't get too far into the text before we encounter some extremely small and extremely large numbers. Writing large numbers of zeros is cumbersome. **Scientific notation**, however, allows us to write a number in an easy-to-read manner, that is, as a number (the **coefficient**) between 1 and 10 multiplied by 10 raised to a power indicated by the **exponent**. Some of the ambiguities of zero as a significant figure are removed when the number is expressed in scientific notation. Remember, help is available in Appendix C if you are uneasy with scientific notation and using it in calculations.

Next, we turned our attention to units of measurements. Most of us have to become familiar with the SI system and the **metric system** on which it is based, specifically with regard to units of length (**meters**), **mass** or **weight** (**grams**), and volume (**liters**). Compared to the rather unsystematic English system, the metric system has the advantage of decimal relationships between units. The most common prefixes are *milli*, *centi*, and *kilo*.

Because we need to convert between systems of measurement, we then introduced our main problem-solving technique—**dimensional analysis**. In this, we change between units of measurement using **conversion factors** (usually **unit factors**), which are constructed from relationships between units. When a direct relationship is available between what's given and what's requested, conversion can be accomplished in a one-step calculation as shown in a **unit map**. If not, the calculation is more involved and requires a game plan that outlines step-by-step conversions from given to requested. This problem-solving method is the language of this text. If you do not follow this method, it is hard to take full advantage of the worked-out examples in this and future chapters.

The final topic of this chapter concerned the measurement of **temperature** using **thermometers**. By comparing the number of divisions between the freezing point and boiling point of water, we can set up conversion factors between the two temperature scales, **Fahrenheit** (°F) and **Celsius** (°C). A third system, closely related to the Celsius scale, is the **Kelvin scale**. Using an algebraic approach, we are able to convert between Fahrenheit and Celsius as well as between Celsius and Kelvin temperatures.

OBJECTIVES

SECTION	YOU SHOULD BE ABLE TO...	EXAMPLES	EXERCISES	CHAPTER PROBLEMS
1-1	Determine the number of significant figures and the degree of uncertainty in a measurement.	1-1	1a, 1b, 1c, 1d	1, 2, 3, 4, 5
1-2	Perform arithmetic operations and percent error calculations, rounding the answer to the appropriate number of significant figures.	1-2, 1-3, 1-4	2a, 2b, 2c, 2d	8, 9, 10, 14, 16, 18, 20, 24, 28, 32
1-3	Perform arithmetic operations involving scientific notation.	1-5, 1-6	3a, 3b, 3c	32, 34, 36, 38, 42, 44, 46, 48
1-4	List several fundamental and derived units of measurement in the metric (SI) system.		4a, 4b, 4c	51, 52, 53
1-5	Interconvert measurements by dimensional analysis.	1-7, 1-8, 1-9, 1-10, 1-11. 1-12	5a, 5b, 5c, 5d, 5e, 5f	58, 60, 62, 64, 67, 70, 74, 78, 82, 87
1-6	Interconvert several key points on the Fahrenheit, Celsius and Kelvin temperature scales.	1-13, 1-14	6a, 6b, 6c	104, 105, 107

▶ANSWERS TO ASSESSING THE OBJECTIVES

Part A

EXERCISES

1-1(a) class 1—precise; class 2—neither; class 3—precise and accurate

1-1(b) The data are precise, based on the consistency of values. Without knowing what the chemical is and knowing its exact boiling point, it is impossible to say whether the data are accurate.

1-1(c) (a) four significant figures (b) two (c) two (d) three

1-1(d) 3%

1-1(e) The first balance would give four significant figures. The second balance would give six significant figures. Which one to buy really depends on the need for the extra precision versus the cost of the balance. In a teaching lab, it is probably reasonable to purchase the cheaper balance. In a research lab, the extra added precision is the more valuable commodity.

1-2(a) (a) three significant figures (b) two (c) just one

1-2(b) (a) rounded to the tenths place (b) tens place (c) hundredths place

1-2(c) (a) 20.00 (b) 2.55 (c) 2.1

1-2(d) 2.8% (or 3%)

1-2(e) One very large error destroys the accuracy of many precisely made measurements. The calculation with several modestly precise measurements will probably be more accurate.

1-2(f) Error can be reduced in two ways. First is by the use of more accurate equipment. Measure volume with a micropipette instead of a graduated cylinder. Use a milligram balance instead of a centigram balance. The second way is by using a larger sample size. With the same magnitude of error, the percent error is reduced.

1-3(a) (a) 3.45×10^4 (b) 5.40×10^{-6} (c) 2×10^{-1}

1-3(b) (a) 985,400,000 (b) 0.001400 (be sure to include the trailing zeros to show the precision in the number) (c) 5.4 ($10^0 = 1$)

1-3(c) (a) 2.61×10^8 (b) 2.4×10^6

Part B

EXERCISES

1-4(a) (a) kilogram (b) liter (c) joule (d) kelvin

1-4(b) (a) 10^3 (b) 10^{-2} (c) 10^{-3} (d) 10^6

1-4(c) The metric system is accepted worldwide. Each prefix is 10 times larger or smaller than the one before it, so changing from one unit to another is as easy as moving a decimal. Each prefix can be applied to any unit, so less memorization is required. It can easily be scaled up or down as far as we need to go. The only real drawback might be our lack of familiarity with it. Full immersion would take care of this problem in a very short period of time.

1-5(a) (a) 1 kg/1000 g (b) 3600 s/hr (c) 1 yd/36 in. (d) 100 cm/m

1-5(b) (a) 1 lb/16 oz and 1 kg/2.205 lb (b) 1 in./2.54 cm and 1 ft/12 in. (c) 1 L/1000 mL and 1.057 qt/1 L

1-5(c) 15.5 mi

1-5(d) $8.96

1-5(e) At 75 beats per minute; 1.1×10^5 beats per day; 3.9×10^7 beats per year. Adjust up or down for your particular heart rate.

1-5(f) For the 2.0-L bottle, the cost is 1.6¢/oz. For the six-pack, the cost is 2.6¢/oz. Buying soda in 2.0-L bottles is about 40% cheaper. Of course, the downside is that it goes flat faster, and cans are more convenient. Generally, as you know, people buy 2.0-L bottles for parties and cans for individual consumption.

1-6(a)

	FAHRENHEIT	CELSIUS	KELVIN
Water boiling	212	100	373
Water freezing	32	0	273
Absolute zero	−460	−273	0

1-6(b) 120°C; 393 K

1-6(c) −57°C; 216 K

1-6(d) The reason Fahrenheit has such odd numbers for boiling and freezing is that it wasn't based on those measurements. Instead, it was based on two other specific points. Is it a coincidence that body temperature is approximately 100°F? Consider checking Wikipedia for a number of possible sources.

ANSWERS TO CHAPTER SYNTHSIS PROBLEM

a. mass—four significant figures, uncertainty = ±0.1, volume—three significant figures, uncertainty = ± 0.1
b. 8.62 g/mL **c.** 3.3% or 3% error **d.** 87.8 mL **e.** 782 mL **f.** 7.05 mL

CHAPTER PROBLEMS

Throughout the text, answers to all exercises in color are given in Appendix E. The more difficult exercises are marked with an asterisk.

Significant Figures (SECTION 1-1)

1-1. Which of the following measurements is the most precise?
(a) 75.2 gal
(b) 74.212 gal
(c) 75.22 gal
(d) 75 gal

1-2. How can a measurement be precise but not accurate?

1-3. The actual length of a certain plank is 26.782 in. Which of the following measurements is the most precise and which is the most accurate?

(a) 26.5 in. (c) 26.202 in.

(b) 26.8 in. (d) 26.98 in.

1-4. How many significant figures are in each of the following measurements?

(a) 7030 g (d) 0.01 ft (g) 8200 km

(b) 4.0 kg (e) 4002 m (h) 0.00705 yd

(c) 4.01 lb (f) 0.060 hr

1-5. How many significant figures are in each of the following measurements?

(a) 0.045 in. (d) 21.0 m (g) 0.0080 in.

(b) 405 ft (e) 7.060 qt (h) 2200 lb

(c) 0.340 cm (f) 2.0010 yd

1-6. What is the uncertainty in each of the measurements in problem 1-4? Recall that the uncertainty is ±1 expressed in the last significant figure to the right (e.g., 5670 = ±10).

1-7. What is the uncertainty in each of the measurements in problem 1-5? Recall that the uncertainty is ±1 expressed in the last significant figure to the right.

Significant Figures in Mathematical Operations
(SECTION 1-2)

1-8. Round off each of the following numbers to three significant figures.

(a) 15.9994 (d) 488.5 (g) 301.4

(b) 1.0080 (e) 87,550

(c) 0.6654 (f) 0.027225

1-9. Round off each of the following numbers to two significant figures.

(a) 115 (d) 0.47322 (g) 1,557,000

(b) 27.678 (e) 55.6 (h) 321

(c) 37,500 (f) 0.0396

1-10. Express the following fractions in decimal form to three significant figures.

(a) $\dfrac{1}{4}$ (b) $\dfrac{4}{5}$ (c) $\dfrac{5}{3}$ (d) $\dfrac{7}{6}$

1-11. Express the following fractions in decimal form to three significant figures.

(a) $\dfrac{2}{3}$ (b) $\dfrac{2}{5}$ (c) $\dfrac{5}{8}$ (d) $\dfrac{13}{4}$

1-12. Without doing the calculation, determine the uncertainty in the answer (e.g., if the answer is to two decimal places, uncertainty = ±0.01).

(a) 12.34 + 0.003 + 1.2 (c) 45.66 + 12 + 0.002

(b) 26,000 + 450 + 132,500 (d) 0.055 + 13.43 + 1.202

1-13. Without doing the calculation, determine the uncertainty in the answer.

(a) 13,330 + 0.8 + 1554 (c) 14,230 + 34 + 1932

(b) 0.0002 + 0.164 + 0.00005 (d) 567 + 7 + 47

1-14. Carry out each of the following operations. Assume that the numbers represent measurements, and express the answer to the proper decimal place.

(a) 14.72 + 0.611 + 173 (d) 47 + 0.91 − 0.286

(b) 0.062 + 11.38 + 1.4578 (e) 0.125 + 0.71

(c) 1600 − 4 + 700

1-15. Carry out each of the following operations. Assume that the numbers represent measurements and express the answer to the proper decimal place.

(a) 0.013 + 0.7217 + 0.04 (c) 35.48 − 4 + 0.04

(b) 15.3 + 1.12 − 3.377 (d) 337 + 0.8 − 12.0

1-16. A container holds 32.8 qt of water. The following portions of water are then added to the container: 0.12 qt, 3.7 qt, and 1.266 qt. What is the new volume of water?

1-17. A container holds 3760 lb (three significant figures) of sand. The following portions are added to the container: 1.8 lb, 32 lb, and 13.55 lb. What is the final mass of the sand?

1-18. Supply the missing measurements.

(a)
$$\begin{array}{r} 6.03 \\ + (\quad\quad) \\ \hline 13.0 \end{array}$$
(d)
$$\begin{array}{r} 0.5668 \\ - (\quad\quad) \\ \hline 0.122 \end{array}$$

(b)
$$\begin{array}{r} (\quad\quad) \\ + 0.9 \\ \hline 138 \end{array}$$
(e)
$$\begin{array}{r} 0.0468 \\ + (\quad\quad) \\ \hline 3.25 \end{array}$$

(c)
$$\begin{array}{r} (\quad\quad) \\ + 0.48 \\ \hline 192 \end{array}$$
(f)
$$\begin{array}{r} 47.9 \\ - (\quad\quad) \\ \hline 45.0 \end{array}$$

1-19. Supply the missing measurements.

(a)
$$\begin{array}{r} 98.732 \\ + (\quad\quad) \\ \hline 98.780 \end{array}$$
(d)
$$\begin{array}{r} (\quad\quad) \\ + 0.0468 \\ \hline 2.227 \end{array}$$

(b)
$$\begin{array}{r} 98.732 \\ + (\quad\quad) \\ \hline 98.90 \end{array}$$
(e)
$$\begin{array}{r} 9.05 \\ + (\quad\quad) \\ \hline 11.6 \end{array}$$

(c)
$$\begin{array}{r} 98.732 \\ + (\quad\quad) \\ \hline 99.7 \end{array}$$
(f)
$$\begin{array}{r} 0.0377 \\ + (\quad\quad) \\ \hline 0.16 \end{array}$$

1-20. Without doing the calculation, determine the number of significant figures in the answer from the following.

(a) $0.59 \times 87 \times 23.0$ (c) $\dfrac{176 \times 0.20}{33.45}$

(b) $\dfrac{17.0}{2.334}$ (d) $\dfrac{0.30 \times 22.42}{0.03}$

1-21. Without doing the calculation, determine the number of significant figures in the answer from the following.

(a) $135,200 \times 0.330$ (c) $\dfrac{0.4005 \times 1.2235}{0.0821 \times 298.5}$

(b) $\dfrac{0.0303}{6.022}$ (d) $\dfrac{23.44 \times 0.050}{0.006 \times 75}$

1-22. Supply the missing measurements.

(a) $22.4 \times (\quad) = 136$ (e) $878.8/(\quad) = 221$

(b) $22.400 \times (\quad) = 135.5$ (f) $878.8/(\quad) = 221.2$

(c) $6.482 \times (\quad) = 55.9$ (g) $0.7820/(\quad) = 1.534$

(d) $7 \times (\quad) = 50$

1-23. Supply the missing measurements.

(a) $0.272 \times (\quad) = 1.65$ (d) $(\quad)/0.22 = 110$

(b) $0.517 \times (\quad) = 0.04$ (e) $25.1/(\quad) = 114$

(c) $6.482 \times (\quad) = 55.90$ (f) $0.782/(\quad) = 1.5$

1-24. Carry out the following calculations. Express your answer to the proper number of significant figures. Specify units.

(a) $40.0 \text{ cm} \times 3.0 \text{ cm}$

(c) $\dfrac{4.386 \text{ cm}^2}{2 \text{ cm}}$

(b) $179 \text{ ft} \times 2.20 \text{ ft}$

(d) $\dfrac{14.65 \text{ in.} \times 0.32 \text{ in.}}{2.00 \text{ in.}}$

1-25. Carry out the following calculations. Express your answer to the proper number of significant figures. Specify units.

(a) $\dfrac{243 \text{ m}^2}{0.05 \text{ m}}$

(c) $0.0575 \text{ in.} \times 21.0 \text{ in.}$

(b) $3.0 \text{ ft} \times 472 \text{ ft}$

(d) $\dfrac{1.84 \text{ yd} \times 42.8 \text{ yd}}{0.8 \text{ yd}}$

*1-26. Carry out the following calculations. Express your answer to the proper number of significant figures or decimal places.

(a) $\dfrac{146}{2.3} + 75.0$ (c) $(12.688 - 10.0) \times (7.85 + 2.666)$

(b) $(157 - 112) \times 25.6$

1-27. Carry out the following calculations. Express your answer to the proper number of significant figures or decimal places.

(a) $(67.43 \times 0.44) - 23.456$

(b) $(0.22 + 12.451 + 1.782) \times 0.876$

(c) $(1.20 \times 0.8842) + (7.332 \times 0.0580)$

1-28. A chemical reaction is expected to produce 13.5 g of product, but only produces 11.7 g. What is the percent of product lost?

1-29. The boiling point of water at sea level is measured in the lab to be 99°C. What is the error in this measurement? (Temperature errors must be calculated in kelvins.)

1-30. Mount Everest is currently measured to be 29,035 ft in height above sea level. Its height was first estimated in 1856 to be 29,002 ft. What was the percent error in that measurement?

1-31. The height of a 55-ft tower is measured to the nearest foot. The distance between two cities is 150 miles, measured to the nearest mile. Which measurement has the lower percentage error?

Scientific Notation (SECTION 1-3)

1-32. Express the following numbers in scientific notation (one digit to the left of the decimal point).

(a) 157

(b) 0.157

(c) 0.0300

(d) 40,000,000
(two significant figures)

(e) 0.0349

(f) 32,000

(g) 32 billion

(h) 0.000771

(i) 2340

1-33. Express the following numbers in scientific notation (one digit to the left of the decimal point).

(a) 423,000

(b) 433.8

(c) 0.0020

(d) 880

(e) 0.00008

(f) 82,000,000
(three significant figures)

(g) 75 trillion

(h) 0.00000106

1-34. Using scientific notation, express the number 87,000,000 to (a) one significant figure, (b) two significant figures, and (c) three significant figures.

1-35. Using scientific notation, express the number 23,600 to (a) one significant figure, (b) two significant figures, (c) three significant figures, and (d) four significant figures.

1-36. Express the following as ordinary decimal numbers.

(a) 4.76×10^{-4}

(b) 6.55×10^3

(c) 78×10^{-4}

(d) 0.489×10^5

(e) 475×10^{-2}

(f) 0.0034×10^{-3}

1-37. Express the following as ordinary decimal numbers.

(a) 64×10^{-3}

(b) 8.34×10^3

(c) 0.022×10^4

(d) 0.342×10^{-2}

1-38. Change the following numbers to scientific notation (one digit to the left of the decimal point).

(a) 489×10^{-6}

(b) 0.456×10^{-4}

(c) 0.0078×10^6

(d) 571×10^{-4}

(e) 4975×10^5

(f) 0.030×10^{-2}

1-39. Change the following numbers to scientific notation.

(a) 0.078×10^{-8}

(b) $72,000 \times 10^{-5}$

(c) 3450×10^{16}

(d) 280.0×10^8

(e) 0.000690×10^{-10}

(f) 0.0023×10^6

1-40. Order the following numbers from the smallest to the largest.

(a) 12

(b) 0.042×10^{-3}

(c) 48×10^5

(d) 0.084×10^2

(e) 3.7×10^6

(f) 8.6×10^{-5}

(g) 0.0022

1-41. Order the following numbers from smallest to largest.

(a) 0.40×10^2

(b) 40×10^4

(c) 0.077×10^{-2}

(d) 4.8×10^5

(e) 510×10^2

(f) 8.9

(g) 2.7×10^{-4}

1-42. Carry out each of the following operations. Assume that the numbers represent measurements, so the answer should be expressed to the proper decimal place.

(a) $(1.82 \times 10^{-4}) + (0.037 \times 10^{-4}) + (14.11 \times 10^{-4})$

(b) $(13.7 \times 10^6) - (2.31 \times 10^6) + (116.28 \times 10^5)$

(c) $(0.61 \times 10^{-6}) + (0.11 \times 10^{-4}) + (0.0232 \times 10^{-3})$

(d) $(372 \times 10^{12}) + (1200 \times 10^{10}) - (0.18 \times 10^{15})$

1-43. Carry out each of the following operations. Assume that the numbers represent measurements, so the answer should be expressed to the proper decimal place.

(a) $(1.42 \times 10^{-10}) + (0.17 \times 10^{-10}) - (0.009 \times 10^{-10})$

(b) $(146 \times 10^8) + (0.723 \times 10^{10}) + (11 \times 10^8)$

(c) $(1.48 \times 10^{-7}) + (2911 \times 10^{-9}) + (0.6318 \times 10^{-6})$

(d) $(299 \times 10^{10}) + (823 \times 10^8) + (0.75 \times 10^{11})$

1-44. Carry out the following calculations without the aid of a calculator.

(a) $10^3 \times 10^4$

(b) $10^6 \times 10^{-6}$

(c) $\dfrac{10^{26}}{10^{-3}}$

(d) $\dfrac{10^4 \times 10^{-8}}{10^{-13}}$

1-45. Carry out the following calculations without the aid of a calculator.

(a) $\dfrac{10^8}{10^{-8}}$

(c) $\dfrac{10^{16} \times 10^{-12}}{10^4}$

(b) $\dfrac{10^{22} \times 10^{-4}}{10^{17} \times 10^8}$

(d) $10^{21} \times 10^{-28}$

1-46. Carry out each of the following operations. Assume that the numbers represent measurements, so the answer should be expressed to the proper number of significant figures.

(a) $(149 \times 10^6) \times (0.21 \times 10^3)$

(b) $\dfrac{0.371 \times 10^{14}}{2 \times 10^4}$

(c) $(6 \times 10^6) \times (6 \times 10^6)$

(d) $(0.1186 \times 10^6) \times (12 \times 10^{-5})$

(e) $\dfrac{18.21 \times 10^{-10}}{0.0712 \times 10^6}$

1-47. Carry out each of the following operations. Assume that the numbers represent measurements, so the answer should be expressed to the proper number of significant figures.

(a) $(76.0 \times 10^7) \times (0.6 \times 10^8)$

(b) $(7 \times 10^{-5}) \times (7.0 \times 10^{-5})$

(c) $\dfrac{0.786 \times 10^{-7}}{0.47 \times 10^7}$

(d) $\dfrac{3798 \times 10^{18}}{0.00301 \times 10^{12}}$

(e) $(0.06000 \times 10^{18}) \times (84{,}921 \times 10^{-9})$

1-48. Supply the missing measurement. Express in scientific notation.

(a) $(4.0 \times 10^{12})/(\quad) = 2.0$

(b) $(\quad) \times (5.18 \times 10^{-8}) = 1.9 \times 10^9$

(c) $(6.0 \times 10^4) \times (\quad) = 3.6 \times 10^7$

(d) $(4.0 \times 10^{12})/(\quad) = 2 \times 10^{24}$

(e) $(8.520 \times 10^{-8}) \times (\quad) = 16$

1-49. Supply the missing measurement. Express in scientific notation.

(a) $(\quad)/(7.890 \times 10^6) = 1.552$

(b) $(\quad)/(7.50 \times 10^4) = 1.20 \times 10^{-16}$

(c) $(9.00 \times 10^{-12})/(\quad) = 3.0 \times 10^{12}$

(d) $(9.0 \times 10^4) \times (\quad) = 8 \times 10^9$

(e) $(8.002 \times 10^{15})/(\quad) = 8 \times 10^5$

1-50. Supply the missing measurement. Express in scientific notation.

(a) $(\quad)/(5.32 \times 10^{10}) = 3.4 \times 10^8$

(b) $(8.55 \times 10^{12})/(\quad) = 2 \times 10^8$

(c) $(8.55 \times 10^{12})/(\quad) = 4.13$

(d) $(9.00 \times 10^4) \times (\quad) = 8.1 \times 10^8$

Length, Volume, and Mass in the Metric System
(SECTION 1-4)

1-51. Write the proper prefix, unit, and symbol for the following. Refer to Tables 1-1, 1-2, 1-3.

(a) 10^{-3} L

(c) 10^{-9} J

(e) 10^{-6} g

(b) 10^2 g

(d) $1/100$ m

(f) 10^{-1} Pa

1-52. Write the proper prefix, unit, and symbol for the following. Refer to Tables 1-1, 1-2, and 1-3.

(a) 10^3 m

(c) $1/1000$ L

(e) 10^{-9} m

(b) 10^{-3} g

(d) 10^3 s

(f) $1/1000$ mol

1-53. Complete the following table.

	mm	cm	m	km
Example	108	10.8	0.108	1.08×10^{-4}
(a)	7.2×10^3	_____	_____	_____
(b)	_____	_____	56.4	_____
(c)	_____	_____	_____	0.250

1-54. Complete the following table.

	mg	g	kg
(a)	8.9×10^3	_____	_____
(b)	_____	25.7	_____
(c)	_____	_____	1.25

1-55. Complete the following table.

	mL	L	kL
(a)	_____	_____	6.8
(b)	_____	0.786	_____
(c)	4452	_____	_____

Conversions Between Units of Measurement
(SECTION 1-5)

1-56. Which of the following are "exact" relationships?

(a) 12 = 1 doz

(d) 1.06 qt = 1 L

(b) 1 gal = 3.78 L

(e) 10^3 m = 1 km

(c) 3 ft = 1 yd

(f) 454 g = 1 lb

1-57. How many significant figures are in each of the following relationships? (If exact, the answer is "infinite.")

(a) 10^3 m = 1 km

(d) 1.609 km = 1 mi

(b) 4 qt = 1 gal

(e) 1 gal = 3.8 L

(c) 28.38 g = 1 oz

(f) 2 pt = 1 qt

1-58. Write a relationship in factor form that would be used in making the following conversions. Refer to Table 1-3.

(a) mg to g

(c) cL to L

(b) m to km

(d) mm to km (two factors)

1-59. Write a relationship in factor form that would be used in making the following conversions. Refer to Table 1-3.

(a) kL to L

(c) kg to Mg (two factors)

(b) Mg to g

(d) cg to hg (two factors)

1-60. Write a relationship in factor form that would be used in making the following conversions. Refer to Tables 1-4 and 1-5.

(a) in. to ft

(d) L to qt

(b) in. to cm

(e) pt to L (two steps)

(c) mi to ft

1-61. Write a relationship in factor form that would be used in making the following conversions. Refer to Tables 1-4 and 1-5.

(a) qt to gal

(c) gal to L

(b) kg to lb

(d) ft to km (two steps)

1-62. Convert each of these measurement to a unit in the same system that will produce a number between 1 and 100 (e.g., 150 cm converts to 1.5 m).

(a) 4.7×10^4 mL =   (e) 9.2×10^{10} nm =

(b) 98×10^{-5} km =   (f) 1725 qt =

(c) 9780 ft =   (g) 32×10^{12} mg =

(d) 1856 in. =

1-63. Convert each of these measurements to a unit in the same system that will produce a number between 1 and 100.

(a) 9.5×10^4 oz =   (e) 49×10^5 cm =

(b) 1548 in. =   (f) 2.52×10^2 pt =

(c) 8.22×10^6 mm =   (g) 1.5×10^{-12} Mg =

(d) 652×10^{-7} kL =

1-64. Complete the following table.

	mi	ft	m	km
(a)	_____	_____	7.8×10^3	_____
(b)	0.450	_____	_____	_____
(c)	_____	8.98×10^3	_____	_____
(d)	_____	_____	_____	6.78

1-65. Complete the following table.

	gal	qt	L
(a)	6.78	_____	_____
(b)	_____	670	_____
(c)	_____	_____	7.68×10^3

1-66. Complete the following table.

	lb	g	kg
(a)	_____	_____	0.780
(b)	_____	985	_____
(c)	16.0	_____	_____

1-67. If a person has a mass of 122 lb, what is her mass in kilograms?

1-68. The moon is 238,700 miles from Earth. What is this distance in kilometers?

1-69. A punter on a professional football team averaged 28.0 m per kick. What is his average in yards? Should he be kept on the team?

1-70. A can of soda has a volume of 355 mL. What is this volume in quarts?

1-71. If a student drinks a 12-oz (0.375-qt) can of soda, what volume did she drink in liters?

1-72. The meat in a "quarter-pounder" should weigh 4.00 oz. What is its mass in grams?

1-73. A prospective basketball player is 6 ft $10\frac{1}{2}$ in. tall and weighs 212 lb. What are his height in meters and his weight in kilograms?

1-74. Gasoline is sold by the liter in Europe. How many gallons does a 55.0-L gas tank hold?

1-75. If the length of a football field is changed from 100.0 yd to 100.0 m, will the field be longer or shorter than the current field? How many yards would a "first and ten" be on the metric field?

1-76. Bourbon used to be sold by the "fifth" (one-fifth of a gallon). A bottle now contains 750 mL. Which is greater?

1-77. A marathon runner must cover 26 mi 385 yd. How far is this in kilometers?

1-78. If the speed limit is 65.0 mi/hr, what is the speed limit in km/hr?

1-79. Mount Everest is 29,028 ft in elevation. How high is this in kilometers?

1-80. It is 525 mi from St. Louis to Detroit. How far is this in kilometers?

1-81. A small pizza has a diameter of 9.00 in. What is this length in millimeters?

1-82. Gasoline sold for as low as $0.899 per gallon in 2001. (That's hard to believe.) What was the cost per liter? What did it cost to fill an 80.0-L tank in the good old days? In 2010, the price was $2.759 per gallon. What did it cost to fill the 80.0-L tank in 2010?

1-83. At the price of gas in the preceding problem (in 2010), how much does it cost to drive 551 mi if your car averages 21.0 mi/gal? How much does it cost to drive 482 km?

1-84. Using the information from the two preceding problems, how many kilometers can you drive for $75.00?

1-85. An aspirin tablet contains 0.324 g (5.00 grains) of aspirin. How many pounds of aspirin are in a 500-tablet bottle?

1-86. A hamburger in Canada sold for $4.55 (Canadian dollars) in 2002. The exchange rate at that time was $1.56 Canadian per one U.S. dollar. (In 2011, they were about even.) What was the cost in U.S. dollars?

1-87. A certain type of nail costs $0.95/lb. If there are 145 nails per pound, how many nails can you purchase for $2.50?

1-88. Another type of nail costs $0.92/lb, and there are 185 nails per pound. What is the cost of 5670 nails?

1-89. If a hybrid automobile gets 38.5 mi/gal of gasoline and gasoline costs $2.759/gal, what would it cost to drive 858 km? How much would it cost if the car is an SUV that gets only 18.5 mi/gal?

1-90. If a train travels at a speed of 85 mi/hr, how many hours does it take to travel 17,000 ft? How many yards can it travel in 37 min?

1-91. If grapes sell for $2.15/lb and there are 255 grapes per pound, how many grapes can you buy for $8.15?

1-92. A high-speed train in Europe travels 215 km/hr. How long would it take for this train to travel nonstop from Boston to Washington, D.C., which is 442 miles?

1-93. The exchange rates among currencies vary from day to day and are usually expressed with up to five significant figures. The new currency in Europe since the beginning of 2002 is the eurodollar (or simply the "euro"). It began trading at about 1.12 euros per U.S. dollar. (a) What was the cost in euros of a sandwich that would cost $6.50 in the United States? (b) What was the cost in U.S. dollars of 1 liter of wine in France that sold for 12.65 euros? The exchange rate in 2010 was $1.427/euro. (c) What is the cost in dollars for 1 liter of wine that now sells for 15.58 euros?

1-94. The traditional pound as well as the euro is used in Britain. The British pound traded for 0.694 pound per U.S. dollar in 2010. What is the current conversion rate between the pound and the euro using the conversion rate between the dollar and euro? If an automobile sells for 25,600 euros in Germany, about how much would it cost in pounds?

1-95. In Mexico, the peso exchanged for 13.51 pesos per U.S. dollar in 2010. What is the cost in dollars for a Corona beer that sells for 35.0 pesos in Cancun? What would it cost in euros and in British pounds? (See the previous two problems.)

*1-96. At a speed of 35 mi/hr, how many centimeters do you travel per second?

*1-97. The planet Jupiter is about 4.0×10^8 mi from Earth. If radio signals travel at the speed of light, which is 3.0×10^{10} cm/s, how long would it take a radio command from Earth to reach a spacecraft passing Jupiter?

Temperature (SECTION 1-6)

1-98. The temperature of the water around a nuclear reactor core is about 300°C. What is this temperature in degrees Fahrenheit?

1-99. The temperature on a comfortable day is 76°F. What is this temperature in degrees Celsius?

1-100. The lowest possible temperature is −273°C and is referred to as absolute zero. What is this temperature in degrees Fahrenheit?

1-101. Mercury thermometers cannot be used in cold arctic climates because mercury freezes at −39°C. What is this temperature in degrees Fahrenheit?

1-102. The coldest temperature recorded on Earth was −110°F. What is this temperature in degrees Celsius?

1-103. A hot day in the U.S. Midwest is 35.0°C. What is this in degrees Fahrenheit?

1-104. Convert the following Kelvin temperatures to degrees Celsius.

(a) 175 K   (d) 225 K
(b) 295 K   (e) 873 K
(c) 300 K

1-105. Convert the following temperatures to the Kelvin scale.

(a) 47°C   (d) −12°C
(b) 23°C   (e) 65°F
(c) −73°C   (f) −20°F

1-106. Make the following temperature conversions.

(a) 37°C to K   (d) 127 K to °F
(b) 135°C to K   (e) 100°F to K
(c) 205 K to °C   (f) −25°C to K

*1-107. At what temperature are the Celsius and Fahrenheit scales numerically equal?

General Problems

1-108. Carry out the following calculations. Express the answer to the proper number of significant figures.

(a) $\dfrac{12.61 + 0.22 + 0.037}{0.04}$

(b) $0.333 \text{ g} \times (23.60 + 1.2) \text{ cm}$

(c) $\dfrac{6.286 \text{ g}}{(13.68 - 12.48) \text{ mL}}$

(d) $\dfrac{44.35 + 0.03 + 0.057}{22.35 - 20.018}$

1-109. Write in factor form the two relationships needed to convert the following.

(a) mg to lb   (c) hm to mi
(b) L to pt   (d) cm to ft

1-110. Convert 5.34×10^{10} ng to pounds.

1-111. Convert 7.88×10^{-4} mL to gallons.

1-112. If gold costs $1,249/oz in 2010, what is the cost of 1.00 kg of gold? (Metals are traded as troy ounces. There are exactly 12 troy ounces per troy pound, and one troy pound is equal to 373 g, to three significant figures.)

1-113. Construct unit factors from the following information: A 82.3-doz. quantity of oranges weighs 247 lb.

(a) What is the mass per dozen oranges?

(b) How many dozen oranges are there per pound?

1-114. A unit of length in horse racing is the furlong. The height of a horse is measured in hands. There are exactly 8 furlongs per mile, and 1 hand is exactly 4 inches. How many hands are there in 12.0 furlongs? (Express the answer in standard scientific notation.)

1-115. The unit price of groceries is sometimes listed in cost per ounce. Which has the smaller cost per ounce: 16 oz of baked beans costing $1.45 or 26 oz costing $2.10?

1-116. A cigarette contains 11.0 mg of tar. How many packages of cigarettes (20 cigarettes per package) would have to be smoked to produce 0.500 lb of tar? If a person smoked two packs per day, how many years would it take to accumulate 0.500 lb of tar?

1-117. An automobile engine has a volume of 306 in.^{3}. What is this volume in liters?

1-118. A U.S. quarter has a mass of 5.70 g. How many dollars is 1 pound of quarters worth?

1-119. The surface of the sun is at a temperature of about 3.0×10^7°C What is this temperature in degrees Fahrenheit? In kelvins?

1-120. In Saudi Arabia, gasoline costs 30.0 haliala per liter. If there are exactly 100 halialas in 1 royal and 1 royal exchanges for about 25.0 cents (U.S.), what is the cost in cents/gal?

*1-121. If snow were piled up by 1.00 ft on a roof 30.0 ft × 50.0 ft, what would be the mass of snow on the roof in pounds and in tons? Assume that 1 ft^3 of snow is equivalent to 0.100 ft^3 of water and that 1 ft^3 of water has a mass of 62.0 lb.

Elements and Compounds

This small robot explorer found convincing evidence that our neighbor planet once had water. This common compound on Earth is essential for life, at least as we know it.

SETTING THE STAGE

In early 2004, two small rovers named *Spirit* and *Opportunity* began their historic exploration on opposite sides of the dry, harsh surface of the planet Mars. Their primary mission was to explore this barren landscape for signs of the familiar substance, *water*. Apparently their mission was successful; photos and analysis of nearby rock formations did indeed indicate that water had at one time been present in the form of a lake or shallow ocean. In fact, spacecraft sent to Mars decades ago relayed photos showing channels that must have been formed from a flowing liquid. Since then, we have wondered whether, at one time, conditions on Mars may have allowed liquid water to exist on its surface much like it does on Earth today. If so, then there is a reasonable chance that some elementary form of life may have existed on Mars and perhaps still does below the surface. Scientists are convinced that water is a necessary component for the formation of living creatures. It is the medium in which other substances can rearrange and combine into the most basic forms of living creatures. In 2010, substantial water in the form of ice was also discovered in the polar region of our moon. This is an important discovery although scientists do not suggest the possibility that life does or ever existed on the moon.

Water is just one example of matter. Mars, the moon, and all of the flickering stars in the night sky are all composed of the same kinds of matter that we find on Earth. The forms of matter that we see on Earth are often changing. Plants grow, die, and decay; rocks weather, crumble, and become part of the fertile soil of the plains or deposits in the oceans. These changes are also the domain of chemistry.

We will first describe the fundamental composition of matter of which there are two types—elements and compounds. In Part A in this chapter we will discuss elements, which are the most basic form of visible matter. In Part B we will explore the more complex form of matter, compounds. In both parts we begin our discussion with the matter that we can see in front of us (the macroscopic) and then delve into the unseen world within the matter (the microscopic).

The Elements and their Composition

SETTING A GOAL

- You will become familiar with the basic components of matter and the properties that make each type of matter unique.

OBJECTIVES

2-1 List the names and symbols of common elements.

2-2 List the postulates of the atomic theory.

2-3 List the components of an atom and their relative masses, charges, and location in the atom.

2-4 From the percent abundance of specific isotopes of an element, determine the atomic mass of the element and the number of protons, neutrons and electrons in each isotope.

▶ **OBJECTIVE FOR SECTION 2-1**

List the names and symbols of common elements.

2-1 The Elements

LOOKING AHEAD! Many of the most basic forms of matter—the elements—are familiar to us, but some are not. The names and symbolic representation of the elements are the first topics that we will discuss. ■

At first glance, the world around us seems so complex. However, we can simplify our understanding by organizing it into general classifications. Many sciences group their disciplines into categories. For example, biology is divided into the study of plants (the flora) and animals (the fauna.) Geology divides its study into the continents and the oceans. The matter we study in chemistry can also be placed into one of two categories.

Chemistry is defined as the study of matter and the changes it undergoes. Matter has mass and occupies space. All of the matter that we see around us from the contents of your room to the farthest stars in space is essentially composed of fewer than 90 unique substances called *elements. Because it cannot be broken down into simpler substances, an* **element** *is the most basic form of matter that exists under ordinary conditions.* The more complex forms of matter are known as compounds. *A* **compound** *is a unique substance that is composed of two or more elements that are chemically combined.* When we say chemically combined we do not mean merely a mixture of elements. Rather, the elements are intimately joined together into a unique form of matter that is distinct from the elements that compose the compound. We will return to a discussion of compounds in Part B.

2-1.1 Free Elements in Nature

Only a few elements are found around us in their free state; that is, they are not combined with any other element. The shiny gold in a ring, the life-supporting oxygen in air, and the carbon in a sparkling diamond are examples of free elements. (See Figure 2-1.)

Other free elements that we have put to use include the aluminum in a can and the iron in a bridge support. These elements were not originally found in the free state in nature but in compounds containing other elements. (A small amount of iron is found in the free state in certain meteorites.) In most cases, elements are extracted from their compounds by rigorous chemical processes.

2-1.2 The Names of Elements

The names of elements come from many sources. Some are derived from Greek, Latin, or German words for colors—for example, bismuth (white mass), iridium (rainbow), rubidium (deep red), and chlorine (greenish-yellow). Some relate to

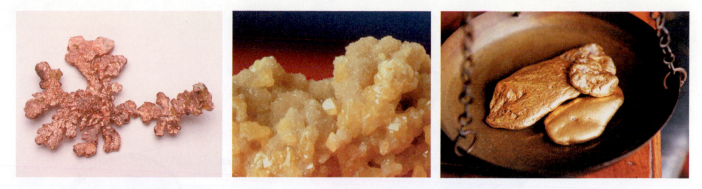

FIGURE 2-1 Some Elements Found in Nature Three elements found in their free state are copper (left), sulfur, and gold.

the locality where the element was discovered (e.g., germanium, francium, and californium). Four elements (yttrium, erbium, terbium, and ytterbium) are all named after a town in Sweden (Ytterby) near where they were discovered. Other elements honor noted scientists (e.g., seaborgium, einsteinium, fermium, and curium) or mythological figures (e.g., plutonium, uranium, titanium, and mercury.) Many of the oldest known elements have names with obscure origins.

2-1.3 The Distribution of the Elements

In the earliest stage of the universe, about 14 billion years ago, only three elements existed: hydrogen (90%), helium (10%), and just a trace of lithium. Since that time, all the other elements have been produced from the original hydrogen in the cores of billions of stars and from the supernova explosions at the end of the stars' lives. Still, since the beginning of time, the abundances of the elements have changed little. In fact, all of this solar activity has converted only about 0.25% of the mass of the universe into elements heavier than helium. Fortunately, that was enough to form the solid earth on which we exist.

Earth and the other planets in our system were formed 4.5 billion years ago from the original elements and the debris of earlier stars. There is comparatively little hydrogen and helium on Earth, although these elements are predominant in the universe as a whole. Figure 2-2a shows the relative abundances of the elements present in Earth's crust in percent by weight. The crust is the outer few miles of the solid surface plus the oceans and atmosphere. Since the core of this planet is mostly iron and nickel, these two elements are more plentiful for Earth as a whole. However, the crust is the region from which we can most easily acquire all our natural resources. It is therefore more meaningful to us to evaluate the distribution of elements found there. Notice that of all the elements, just the top 10 constitute 99% of the mass of the crust. Similarly, consider the human body. Over 96% of the mass of our bodies is composed of only four elements: oxygen, carbon, nitrogen, and hydrogen. This is shown in Figure 2-2b.

2-1.4 The Symbols of the Elements

An element can be conveniently identified by a symbol. *A **symbol** is usually the first one or two letters of the element's English or Latin name*. When an element has a two-letter symbol, the first is capitalized but the second is not. The table on the inside cover of the text includes a complete list of elements along with their symbols. Some common elements and their symbols are listed in Table 2-1. The symbols of some common elements are derived from their original Latin names (or, in the case of tungsten, its German name—Wolfram). The symbols of these elements are listed in Table 2-2.

TABLE 2-1	
Some Common Elements	
ELEMENT	**SYMBOL**
Aluminum	Al
Bromine	Br
Calcium	Ca
Carbon	C
Chlorine	Cl
Chromium	Cr
Fluorine	F
Helium	He
Hydrogen	H
Iodine	I
Magnesium	Mg
Nickel	Ni
Nitrogen	N
Oxygen	O
Phosphorus	P
Silicon	Si
Sulfur	S
Zinc	Zn

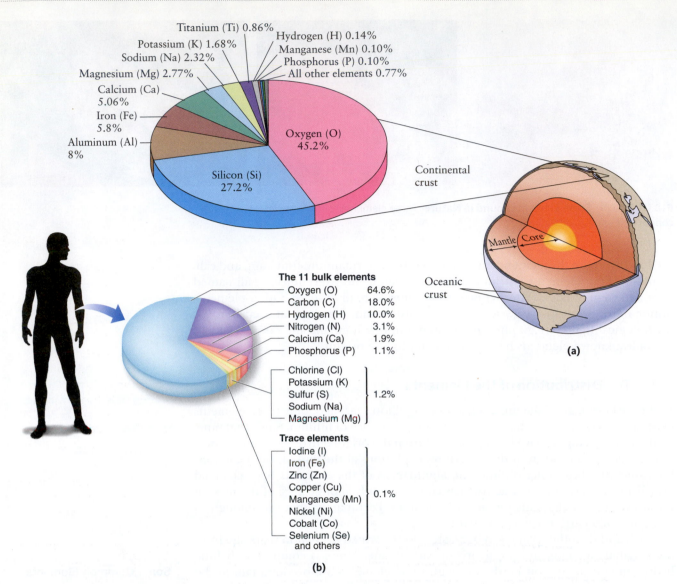

Titanium (Ti) 0.86%
Potassium (K) 1.68%
Sodium (Na) 2.32%
Magnesium (Mg) 2.77%
Calcium (Ca) 5.06%
Iron (Fe) 5.8%
Aluminum (Al) 8%
Silicon (Si) 27.2%
Hydrogen (H) 0.14%
Manganese (Mn) 0.10%
Phosphorus (P) 0.10%
All other elements 0.77%
Oxygen (O) 45.2%
Continental crust
Mantle Core
Oceanic crust

(a)

The 11 bulk elements
Oxygen (O)   64.6%
Carbon (C)   18.0%
Hydrogen (H)   10.0%
Nitrogen (N)   3.1%
Calcium (Ca)   1.9%
Phosphorus (P)   1.1%

Chlorine (Cl)
Potassium (K)
Sulfur (S)   } 1.2%
Sodium (Na)
Magnesium (Mg)

Trace elements
Iodine (I)
Iron (Fe)
Zinc (Zn)
Copper (Cu)   } 0.1%
Manganese (Mn)
Nickel (Ni)
Cobalt (Co)
Selenium (Se)
and others

(b)

FIGURE 2-2 **The Distribution of the Elements** Most of Earth's crust is composed of surprisingly few elements (a). The human body is composed mostly of just three elements (b).

TABLE 2-2		
Elements with Symbols from Earlier or Alternative Names		
ELEMENT	**SYMBOL**	**FORMER LATIN NAME**
Antimony	Sb	Stibium
Copper	Cu	Cuprum
Gold	Au	Aurum
Iron	Fe	Ferrum
Lead	Pb	Plumbum
Mercury	Hg	Hydragyrum
Potassium	K	Kalium
Silver	Ag	Argentum
Sodium	Na	Natrium
Tin	Sn	Stannum
Tungsten	W	Wolfram

MAKING IT REAL

Iridium, the Missing Dinosaurs, and the Scientific Method

The dinosaurs, such as the T-Rex, probably had a catastrophic end.

"It's elementary, my dear Watson." If Sherlock Holmes had been on the case of the missing dinosaurs, that statement would have been brilliant. The first major clue was indeed "elementary" or, more specifically, "elemental." Iridium is a very rare element found on Earth's surface. Four billion years ago the Earth was molten, and most iridium (which is very dense) sank deep into the interior. However, matter from space including meteors, asteroids, and comets also contain comparatively high amounts of this element. The following application of the scientific method, as discussed in the Prologue, provides us with a theory of how the element iridium is connected to the extinction of the dinosaurs.

In 1979, American scientists discovered a thin layer of sediment in various locations around the world that was deposited about 65 million years ago, coincidentally the same timeframe in which the dinosaurs became extinct. Indeed, there were dinosaur fossils below that layer but none above. Interestingly, that layer contained comparatively high amounts of iridium. Scientists proposed that this layer contained the dust and debris from a collision of a huge asteroid or comet (about 6 miles in diameter) with Earth. They concluded that a large cloud of dust must have formed, encircling Earth and completely shutting out the sunlight. A bitter cold wave followed, and most animals and plants quickly died. A *hypothesis* (a tentative explanation of facts) was proposed that the dinosaurs must have been among the casualties. After many months the dust settled, forming a thin layer of sediment. Scientists further proposed that small mammals and some reptiles had survived and inherited the planet.

More information has since been discovered to support the original hypothesis. Perhaps most important was the discovery in 1991 of a huge impact crater near the Yucatan Peninsula in Mexico. The crater, 110 miles wide, is buried miles under the ocean surface and was formed about 65 million years ago, evidence that surely a huge asteroid or comet had collided with Earth. Since it landed in the ocean, huge tidal waves must have formed. Evidence of waves over 1 mile high has been found in North and Central America.

Based on all the evidence, most scientists have now embraced the originally proposed hypothesis as a plausible *theory* (a well-established hypothesis). Despite its acceptance, there are still unanswered questions regarding the death of the dinosaurs. So, the issue may not be completely closed. However, it looks likely that the end many of the dinosaurs was sudden and dramatic. Now, instead of worrying about a nasty Tyrannosaurus Rex lurking in the forest, we can instead worry about collisions with asteroids!

▶ ASSESSING THE OBJECTIVE FOR SECTION 2-1

EXERCISE 2-1(a) LEVEL 1: The following chemicals are either elements or compounds. Which are elements and which are compounds?

(a) tin   **(b)** baking soda   **(c)** water   **(d)** quartz   **(e)** mercury

EXERCISE 2-1(b) LEVEL 1: Provide the symbols for the following elements.

(a) copper   **(b)** sulfur   **(c)** calcium

EXERCISE 2-1(c) LEVEL 1: Provide the name of the element from its symbol.
(a) Pb   **(b)** P   **(c)** Na

EXERCISE 2-1(d) LEVEL 3: There are nearly 120 different elements. Would you expect the number of compounds to be roughly the same, slightly more, or significantly more than that?

EXERCISE 2-1(f) LEVEL 3: How could you tell if an unknown substance was an element or a compound?

For additional practice, work chapter problems 2-1, 2-5, 2-7, and 2-32.

▶ **OBJECTIVE FOR SECTION 2-2**

List the postulates of the atomic theory.

2-2 The Composition of Elements: Atomic Theory

LOOKING AHEAD! We are now ready to look deeper into the composition of the elements. We will see that they are composed of basic particles called atoms. ■

Our modern understanding of the particulate nature of matter actually had its beginning over 2000 years ago. A Greek philosopher named Democritus suggested that all matter is like grains of sand on a beach. In other words, he proposed that matter is composed of tiny indivisible particles that he called *atoms.* However, as recently as two centuries ago, the idea of matter being composed of atoms was not accepted. Most knowledgeable scientists thought that a sample of an element such as copper could be divided (theoretically) into infinitely smaller pieces without changing its nature. In other words, they believed that matter was continuous.

2-2.1 The Atomic Theory

In 1803, an English scientist named John Dalton (1766–1844) proposed a theory of matter based on the original thoughts of Democritus. His ideas are now known as **atomic theory.** The major conclusions of atomic theory are as follows:

- Matter is composed of small, indivisible particles called atoms.
- Atoms of the same element are identical and have the same properties.
- Chemical compounds are composed of atoms of different elements combined in small whole-number ratios.
- Chemical reactions are merely the rearrangement of atoms into different combinations.

The atomic theory is now universally accepted as our current view of matter. Thus, we may define an **atom** *as the smallest fundamental particle of an element that has the properties of that element.*

Democritus' proposal 2000 years ago was simply the product of his own rational thought. Dalton's theory was a brilliant and logical explanation of many experimental observations and laws that were known at the time but had not been explained. One of these laws, the law of conservation of mass, will be discussed in the next chapter.

Why are we so sure that Dalton was right? Besides the overwhelming amount of indirect experimental evidence, we now have direct proof. In recent years, a highly sophisticated instrument called the scanning tunneling microscope (STM) has produced images of atoms of several elements. Although these images are

FIGURE 2-3 STM of the Atoms of an Element The atoms are shown in an orderly pattern.

somewhat fuzzy, they indicate that an element such as gold is composed of spherical atoms packed closely together, just as you would find in a container of marbles all of the same size. (See Figure 2-3.)

2-2.2 The Size of an Atom

When we look at a small piece of copper wire, it is hard to imagine that it is not continuous. This is because it is so difficult to comprehend the small size of the atom. Since the diameter of a typical atom is on the order of 0.00000001 cm (1×10^{-8} cm), it would take about 10 quadrillion atoms to appear as a tiny speck. The piece of copper wire is like a brick wall: from a distance it looks completely featureless, but up close we notice that it is actually composed of closely packed basic units.

MATH CHECK:

Exponential Notation and the Metric System

The atom is extremely small. Exponential notation (i.e., 1×10^{-8}) can be used to conveniently represent its diameter. Notice also that the unit of diameter is expressed in the metric system of centimeters (cm). The following questions are meant to check your understanding in these two areas.

1. **Exponential and Scientific Notation**

   a. Convert the number 89,260,000 into scientific notation. 8.926×10^7

   b. Express the number 0.00825×10^8 in scientific notation. 8.25×10^5

2. **The Metric System**

   c. What are the principal units and the symbols in the metric system for mass, volume, and length?   Gram, liter, meters

   d. Give the weight in grams of 0.078 kg. Give the volume in L of 345 mL. Give the length in meters of 45.6 cm. $345 \times \dfrac{10^{-?}}{1\,mL} \times \underline{\quad}$

Answers: **1. a.** 8.926×10^7 **b.** 8.25×10^5 **2. c.** mass−grams (g), volume−liters (L) length−meters (m) **d.** 78 g, 0.345 L, 0.456 m

Refer to Sections 1-3.1, 1.3.2, and Appendix C for help with scientific notation.

Refer to Sections 1-4.1 and 1-4.2 for help with the metric system.

▶ **ASSESSING THE OBJECTIVE FOR SECTION 2-2**

EXERCISE 2-2(a) LEVEL 1: Fill in the blanks.

An _____ is composed of small, indivisible particles called _____.

Atoms of the same _____ are identical and have the same _____.

Chemical _____ are composed of atoms of different elements combined in small _____ ratios. Chemical _____ are rearrangements of _____ into different combinations.

EXERCISE 2-2(b) LEVEL 2: Are each of the following statements supported by any of Dalton's postulates?

(a) Any chemical experiment on an element should always yield the same result.

(b) Any two elements will always form the same compound.

(c) There is a limit to how far a substance can be divided.

(d) All atoms involved before a chemical reaction are there after the chemical reaction.

EXERCISE 2-2(c) **LEVEL 3:** The ideas of Democritus and Dalton are very similar. Why, then, is Dalton considered the father of modern chemistry? Why is Dalton's statement of his ideas superior to that of Democritus?

For additional practice, work chapter problem 2-8.

> ▶ **OBJECTIVE**
> **FOR SECTION 2-3**
> List the components of an atom and their relative masses, charges, and location in the atom.

2-3 Composition of the Atom

LOOKING AHEAD! We now look even deeper into the nature of matter—the structure of the atom itself. In this section, we will describe the contents of the atom as if we could peer into its tiny confines. ■

Just over 130 years ago, scientists perceived the atom to be a hard, featureless sphere. However, beginning in the late 1880s and continuing today, the mysteries and complexities of the atom have been slowly discovered and understood. Ingenious experiments of brilliant scientists such as Thomson, Rutherford, Becquerel, Curie, and Roentgen, among others, contributed to the current model of the atom.

2-3.1 The Electron and Electrostatic Forces

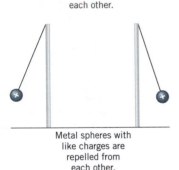

Metal spheres with opposite charges are attracted to each other.

Metal spheres with like charges are repelled from each other.

FIGURE 2-4 Electrostatic Forces Opposite charges attract; like charges repel.

If we had a magical microscope that would allow us to see inside the atom, we would find that the atom itself is composed of basic particles. The first particle that we would notice is relatively small (compared to the other particles in the atom) and is called an **electron.** The electron was the first subatomic particle to be identified. In 1897, J. J. Thomson characterized the electron by proving that it has a negative electrical charge (assigned a value of −1) and is common to the atoms of all elements. The identification of electrons indicated that matter is electrical in nature and that electrostatic forces are at work within the confines of the atom. **Electrostatic forces** *consist of forces of attraction between unlike charges and forces of repulsion between like charges.* (See Figure 2-4.) Atoms themselves have no net electrical charge, so they must contain positive charges that counterbalance the negative charge of the electrons.

The first model of the atom based on this information was proposed by Thomson and was known as the *plum pudding* model of the atom (plum pudding was a popular English dessert). It was suggested that the positive charge would be diffuse and evenly distributed throughout the volume of the atom (analogous to pudding). This is logical—the like positive charges would tend to spread out as much as possible because of their mutual repulsion. The negative particles (electrons) would be embedded throughout the atom like raisins in the pudding. There would be enough electrons to balance the positive charge. Since few of us are familiar with plum pudding, a better analogy would be to picture the atom as a ball of cotton with tiny seeds (representing electrons) distributed throughout the cotton. (See Figure 2-5.)

FIGURE 2-5 The Plum Pudding Model Electrons were thought to be like tiny particles distributed in a positively charged medium.

2-3.2 The Nuclear Model of the Atom

The next major development in understanding the atom occurred in 1911. Ernest Rutherford in England conducted experiments that he fully expected would support the accepted model of Thomson. His results, however, suggested a radically different model. Radioactivity had recently been discovered, one form of which is called *alpha* radiation. It is composed of positively charged helium atoms that are spontaneously ejected at high velocities from certain heavy elements. When students in his laboratories bombarded a thin foil of gold with the small, fast-moving alpha particles, they expected that the alpha particles would pass right through the large atoms of gold with very little effect. The small, hard alpha particles should easily

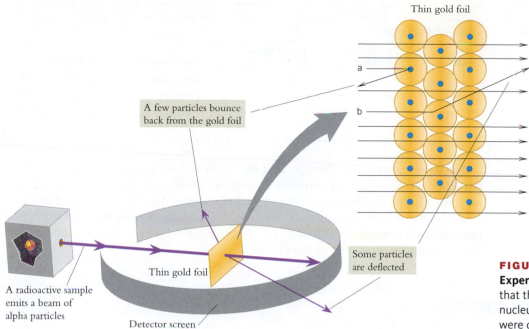

FIGURE 2-6 Rutherford's Experiment Rutherford suggested that the atom contains a small nucleus, since a few alpha particles were deflected or reflected back.

push the tiny electrons within the gold atoms aside and pass unaffected through the diffuse positive charge, similar to bullets through a bale of cotton. Instead, a small number of the alpha particles were deflected significantly from their path; a few even came straight back. Imagine shooting a volley of bullets into a thin bale of cotton and finding one or two bullets being ricocheted at a large angle. The conclusion would have to be that something hard, like a rock, was embedded in the soft cotton. Likewise, Rutherford was forced to conclude that the gold atoms had a small, hard core containing most of the mass of the atom and all of the positive charge.

Rutherford's new model was needed to explain the experimental results. The close encounters between the alpha particles and core, or **nucleus**, would cause the alpha particles to be deflected because of the repulsion of like positive charges. Occasional "direct hits" onto the nucleus would reflect the alpha particle back toward the source. (See Figure 2-6.) The alpha particles that were unaffected indicated that most of the volume of the atom is actually empty space containing the very small electrons. At first scientists were skeptical because they wondered why the electrons would not be pulled into the positive nucleus. However, an explanation of this puzzle would eventually be advanced through modern theories. To get an idea about proportions of the atom, imagine a nucleus expanded to the size of a softball. In this case, the radius of the atom would extend for about 1 mile.

2-3.3 The Particles in the Nucleus

Later experiments showed that the nucleus is composed of particles called **nucleons**. There are two types of nucleons: **protons**, which have a positive charge (assigned a value of +1, equal and opposite to that of an electron), and **neutrons**, which do not carry a charge. (See Figure 2-7.) Data on these three particles in the atom are summarized in Table 2-3. The proton and neutron have roughly the same mass, which is about "1 amu" (1.67×10^{-24} g). The amu (atomic mass unit) is a convenient unit for the masses of individual atoms and subatomic particles. This unit will be defined more precisely in the next section.

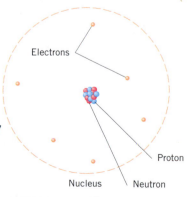

FIGURE 2-7 The Composition of the Atom A diamond is composed of carbon atoms. The carbon atoms are composed of electrons, protons, and neutrons.

TABLE 2-3

Atomic Particles

NAME	SYMBOL	ELECTRICAL CHARGE	MASS (amu)	MASS (g)
Electron	e	-1	0.000549	9.110×10^{-28}
Proton	p	$+1$	1.00728	1.673×10^{-24}
Neutron	n	0	1.00867	1.675×10^{-24}

For some time, it was thought that the atom was composed of just these three particles. As experimental procedures became more elaborate and sophisticated, however, this model became outdated. It now appears that the proton and the neutron are themselves composed of various combinations of even more fundamental particles called *quarks.* Fortunately for us, the original three-particle model of the atom still meets the needs of the chemist.

MATH CHECK:

Significant Figures

Consider the measurements shown in Table 2-3. The masses are expressed in various degrees of precision such as 0.000549 and 1.00728. This relates to how we express measurements. Questions a and b check your knowledge of significant figures.

   a. How many significant figures are in the numbers 9.110 and 1.00728? Which is the more precise?

   b. Round off the following number to three significant figures: 7.8956

   c. What is the degree of uncertainty in the number 9.110?

Answers: **a.** 9.110—four, 1.00728—six.   1.00728 is more precise.   **b.** 7.90   **c.** ±0.001

Refer to Sections 1-1.1 and 1-1.2 for help with significant figures.

▶ **ASSESSING THE OBJECTIVE FOR SECTION 2-3**

EXERCISE 2-3(a) LEVEL 1: Identify the specific subatomic particle(s).
(a) smallest of the three
(b) found in the nucleus
(c) most massive
(d) positive in charge

EXERCISE 2-3(b) LEVEL 1: Complete the following table.

PARTICLE	MASS (amu)	CHARGE	LOCATION
Proton			
		Neutral	
			Outside of nucleus

EXERCISE 2-3(c) LEVEL 2: Answer the following questions.
(a) Where is virtually all of the mass of an atom located?
(b) What takes up most of the volume of an atom?
(c) Historically, which was the first particle to be identified?

EXERCISE 2-3(d) LEVEL 3: It has been proposed that neutrons are simply particles formed when protons and electrons are combined into one particle. What facts about neutrons suggest this might be the case?

For additional practice, work chapter problems 2-9 and 2-10.

2-4   Atomic Number, Mass Number, and Atomic Mass

▶ **OBJECTIVE FOR SECTION 2-4**
From the percent abundance of specific isotopes of an element, determine the atomic mass of the element and the number of protons, neutrons and electrons in each isotope.

LOOKING AHEAD! We are now ready to look into how protons, neutrons, and electrons define the atoms of a particular element. We will see in this section why not all atoms of an element are exactly the same and how the atoms of one element differ from another. ■

2-4.1 Atomic Number, Mass Number, and Isotopes

In Dalton's original atomic theory, he suggested that all atoms of an element are identical. But if we look at a number of atoms of most elements, we find that this statement is not exactly true. For example, consider the atoms of the element copper. Most atoms are composed of a nucleus containing a total of 63 nucleons, of which 29 are protons and 34 are neutrons. The atom also contains 29 electrons that exactly balance the positive charge of the protons, resulting in a neutral atom. *The number of protons in the nucleus (which is equal to the total positive charge) is referred to as the atom's* **atomic number**. *The total number of nucleons (protons and neutrons) is called the* **mass number**. Therefore, this particular copper atom has an atomic number of 29 and a mass number of 63. There are other copper atoms that are not exactly the same, however. These atoms of copper have a mass number of 65 rather than 63. This means that these atoms have 36 neutrons as well as 29 protons. *An atom of a specific element with a specific mass number is known as an* **isotope**. Isotopes of an element have the same atomic number but different mass numbers.

Most elements that are present in nature exist as a mixture of isotopes. *It is the atomic number, however, that distinguishes one element from another.* Any atom with an atomic number of 29, regardless of any other consideration, is an atom of copper. If the atomic number is 28, the element is nickel; if it is 30, the element is zinc.

Specific isotopes are written in a form known as *isotopic notation*. In isotopic notation, the mass number is written as a superscript to the left of the element. Sometimes, the atomic number is written as a subscript, also on the left. The indication of the atomic number is strictly a convenience. Since the atomic number determines the identity of the element, it can therefore be determined from the symbol. The isotopic notations for the two isotopes of copper are written as follows.

Mass number (number of nucleons)
(29 protons and 34 neutrons)

(29 protons and 36 neutrons)

$$^{63}_{29}\text{Cu} \qquad ^{65}_{29}\text{Cu}$$

Atomic number (number of protons)
(29 protons)

The convention for verbally naming specific isotopes is to use the element's name followed by its mass number. For example, ^{63}Cu is called copper-63, and ^{65}Cu is called copper-65. From either the written isotopic notation or the isotope name, we can determine the number of each type of particle in an isotope, as we will see in Example 2-1.

EXAMPLE 2-1 **Calculating the Number of Particles in an Isotope**

How many protons, neutrons, and electrons are present in $^{90}_{38}$Sr?

PROCEDURE

The subscript (and symbol) provide us information on protons. Since this is a neutral atom, the number of protons and electrons is the same. The superscript gives us the total number of protons and neutrons. The number of neutrons alone is the difference.

SOLUTION

$$\text{number of protons} = \text{atomic number} = \underline{38}$$

$$\text{number of neutrons} = \text{mass number} - \text{number of protons } 90 - 38 = \underline{52}$$

$$\text{number of electrons} = \text{number of protons} = \underline{38}$$

ANALYSIS

If you knew only the name of an element, which, if any, of the fundamental particles could you determine? Protons and electrons can be determined directly from the identity of the element, but the number of neutrons must be determined from the mass number. Knowing the specific isotope of an element allows you to determine the exact numbers of protons, neutrons, and electrons.

SYNTHESIS

What would happen in an atom if the number of protons and electrons were not exactly balanced? What if there was an extra electron? Or one too few? Would the positive and negative charges cancel out? Clearly not. If there was an extra electron, the atom would have an overall negative charge. If there was one too few, there would be an overall positive charge. This situation occurs and is explored later in this chapter.

In Chapter 5, we'll consider how the mass of one element compares to another. The mass of the electrons is extremely small compared to the masses of the protons and neutrons, so it is not included in the mass of an isotope. Thus, the mass number of an isotope is a convenient but rather imprecise measure of its mass. It is imprecise because electrons are not included and protons and neutrons do not have exactly the same mass.

2-4.2 Isotopic Mass and Atomic Mass

A more precise measure of the mass of one isotope relative to another is known as the isotopic mass. **Isotopic mass** *is determined by comparison to a standard,* ^{12}C*, which is defined as having a mass of exactly 12 atomic mass units. Therefore, one* **atomic mass unit** (*amu*) *is a mass of exactly 1/12 of the mass of* ^{12}C. For example, precise measurements show that the mass of ^{10}B is 0.83442 times the mass of ^{12}C, which means it has an isotopic mass of 10.013 amu. From similar calculations, we find that the atomic mass of ^{11}B is 11.009 amu. Since boron, as well as most other naturally occurring elements, is found in nature as a mixture of isotopes, the atomic mass of the element reflects this mixture. *The* **atomic mass** *of an element is obtained from the weighted average of the atomic masses of all isotopes present in nature.* A weighted average relates the isotopic mass of each isotope present to its percent abundance. It can be considered as the isotopic mass of an "average atom," although an average atom does not itself exist. Example 2-2 illustrates how atomic mass relates to the distribution of isotopes.

MAKING IT REAL

Isotopes and the History of Earth's Weather

Greenland.

Lately it seems like we are always hearing of some great storm or heat record. Is something strange going on, or are the media just doing a better job of reporting these events? Actually, we are all anxious to know whether the climate is permanently changing and, if so, whether it is going to get worse. The key to predicting the future, however, may lie in understanding the past. An ingenious key to past climates is found with the naturally occurring isotopes of oxygen and hydrogen. We will consider here how the isotopes of oxygen are put to use.

Oxygen is composed of 99.76% ^{16}O, 0.04% ^{17}O, and 0.20% ^{18}O, which produces a weighted average of 15.9994. This means that only 24 out of 10,000 oxygen atoms are "heavy" oxygen atoms (i.e., ^{17}O or ^{18}O). When ^{17}O or ^{18}O is part of a water molecule, the properties are very slightly different from normal water (H_2 ^{16}O). For example, the heavier water evaporates just a little more slowly than normal water. The colder the temperature of the oceans, the less heavy oxygen ends up in the clouds. This means that the amount of water with the heavier oxygen is slightly less in fresh water (which comes from precipitation of the evaporated water) than in the ocean. But even this can vary ever so slightly. When the climate is colder than normal, less water evaporates from the ocean and so the amount of heavy water in precipitation is less than normal.

Greenland and Antarctica are our natural weather-history laboratories. Greenland has been covered with ice for more than 100,000 years and Antarctica for almost 1 million years. In some areas of Greenland, the ice is now 2 miles deep. Year by year, the snow has been accumulating and has been pressed down into layers that resemble tree rings. Scientists have bored through the ice and analyzed the layers for their content of dust, trapped gases, and the ratio of oxygen isotopes. Results of isotope studies indicate that the Northern Hemisphere was a whopping 20°C (or 36°F) colder during the last ice age (about 30,000 years ago) than it is now. More important, variations in the weather from then to now have also been determined. These studies will help us understand and perhaps predict what's going to happen to our weather in the future. Is our current warming a natural occurrence or a new phenomenon? Most evidence indicates that humans have a hand in the warming weather.

MATH CHECK:

Percent and Decimal Fractions

In the following example, you will be asked to convert percent to a decimal fraction. To check your ability with percent, answer the following two questions.

a. A crowd of spectators is composed of 42 women and 74 men. What is the decimal fraction of men and what is the percent of women present? Express both answers to two significant figures.

b. Another crowd of 611 people had the same percent of women and men. Using the results from **a** above, calculate how many women were in this crowd?

Answers: a. 0.64 men, 36% women   **b.** 220 women

Refer to Appendix A-6 for help with percent and decimal fractions.

EXAMPLE 2-2 Calculating the Atomic Mass of an Element from Percent Distribution of Isotopes

In nature, the element boron occurs as 19.9% ^{10}B and 80.1% ^{11}B. If the isotopic mass of ^{10}B is 10.013 and that of ^{11}B is 11.009 amu, what is the atomic mass of boron?

PROCEDURE

Find the contribution of each isotope toward the atomic mass by multiplying the percent in decimal form by the isotopic mass. Recall that percent can be expressed as a normal fraction or a decimal fraction (e.g., 25% = 1/4 = 0.25).

SOLUTION

$$^{10}B \ 0.199 \times 10.013 = \ \ 1.99 \ \text{amu}$$
$$^{11}B \ 0.801 \times 11.009 = \ \underline{8.82 \ \text{amu}}$$
$$\text{atomic mass of boron} = \underline{10.81}$$

ANALYSIS

Based on the percentages of boron-10 and boron-11, does this answer seem reasonable? Since the atomic mass is a weighted average of the two isotopes, and they have mass numbers of 10 and 11, the answer should be between those values. It should be closer to the isotope with the higher percent abundance. The answer of 10.81 is between the two numbers and closer to the isotope of 80.1% abundance.

SYNTHESIS

The atomic mass of boron is approximately 80% of the way between the mass numbers of boron-10 and boron-11. What can you conclude about the percentages of various isotopes of an element whose atomic mass is very close to an integer? Consider carbon as an example. Its atomic mass is 12.011, just a little higher than 12 exactly. We would most likely conclude that the isotope with mass 12 is by far the most common (actually, more than 99%).

▶ **ASSESSING THE OBJECTIVE FOR SECTION 2-4**

EXERCISE 2-4(a) LEVEL 1: Identify the following as atomic number, mass number, or isotopic mass.

(a) always an integer value

(b) the superscript in isotopic notation

(c) the total mass of the atom

(d) the number of protons

(e) determines the number of protons and neutrons combined.

EXERCISE 2-4(b) LEVEL 2: Determine the number of protons, neutrons, and electrons in the following isotopes.

(a) $^{13}_{6}C$ (b) $^{1}_{1}H$ (c) $^{238}_{92}U$

EXERCISE 2-4(c) LEVEL 2: Write the isotopic notation for the following three species.

(a) protons = 9, neutrons = 10, electrons = 9

(b) protons = 35, neutrons = 44, electrons = 35

(c) protons = 20, neutrons = 20, electrons = 20

EXERCISE 2-4(d) LEVEL 2: Antimony (Sb) has two naturally occurring isotopes: ^{121}Sb, with a mass of 120.903 amu and a 57.3% abundance, and ^{123}Sb, with a mass of 122.904 amu and a 42.7% abundance. Determine antimony's atomic mass.

EXERCISE 2-4(e) LEVEL 2: Lithium has two naturally occurring isotopes: ^{6}Li has a mass of 6.015 and is 7.42% abundant; ^{7}Li has a mass of 7.016 and is 92.58% abundant. Calculate the atomic mass of lithium.

EXERCISE 2-4(f) LEVEL 3: All through the Middle Ages, alchemists searched for ways to turn base metals, such as lead, into gold. Why was this a chemically fruitless endeavor?

EXERCISE 2-4(g) LEVEL 3: Bromine has two naturally occurring isotopes, ^{79}Br and ^{81}Br. From its atomic weight, estimate the percentage abundances of the two isotopes.

For additional practice, work chapter problems 2-11, 2-13, 2-15, 2-26, and 2-28.

PART A SUMMARY

KEY TERMS

2-1	All substances can be classified as either **elements** or **compounds**. p. 54
2-1.4	Each element is designated by a unique **symbol**. p. 55
2-2.1	According to **atomic theory**, the basic particle of an element is an **atom**. p. 58
2-3.1	The discovery of the **electron** proved the presence of **electrostatic forces** in the atom. p. 60
2-3.2	Experiments indicated that the atom contained a central positively charged core called a **nucleus**. p. 61
2-3.3	The nucleus consists of particles called **nucleons**, which are either **neutrons** or **protons**. p. 61
2-4.1	The number of protons in a nucleus is known as its **atomic number**, and the number of nucleons is known as its **mass number**. p. 63
2-4.1	Naturally occurring elements are composed of one or more **isotopes**. p. 63
2-4.2	The standard of mass is ^{12}C, which is defined as exactly 12 **atomic mass units** (amu). The **isotopic mass** of an isotope is determined experimentally by comparison to this standard. p. 64
2-4.2	The **atomic mass** of an element is determined by its naturally occurring isotopes and their percent abundance. p. 64

SUMMARY CHART

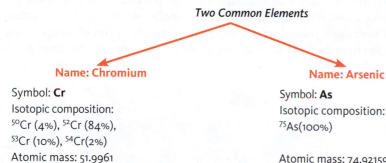

Two Common Elements

Name: Chromium

Symbol: **Cr**
Isotopic composition:
^{50}Cr (4%), ^{52}Cr (84%),
^{53}Cr (10%), ^{54}Cr (2%)
Atomic mass: 51.9961

Name: Arsenic

Symbol: **As**
Isotopic composition:
^{75}As (100%)

Atomic mass: 74.92159

Part B

Compounds and Their Composition

SETTING A GOAL

- You will learn to distinguish between ionic and molecular compounds based on their chemical structure and general properties.

OBJECTIVES

2-5 List the characteristics of compounds composed of molecules.

2-6 Write the formulas of simple ionic compounds given the charges on the ions.

▶ **OBJECTIVE
FOR SECTION 2.5**
List the characteristics of compounds composed of molecules.

2-5 Molecular Compounds

LOOKING AHEAD! The atoms of elements are usually joined together with other atoms of the same element or other elements into a more complex form of particulate matter. In this section, we will examine one of the two principal forms of matter formed by the combined atoms. ■

Every element and compound has a unique set of properties. Properties *describe the particular characteristics or traits of a substance.* We will have more to say about specific properties in the next chapter. Elements and compounds can be referred to as pure substances. **Pure substances** *have definite compositions and definite, unchanging properties.* (See Figure 2-8.)

2-5.1 Recognizing the Names of Compounds

The names of the simplest compounds are usually based on the elements from which they are composed, and most contain two words. Carbon dioxide, sodium sulfite, and silver nitrate all refer to specific compounds. Notice that the second word ends in *-ide, -ite,* or *-ate.* A few compounds have three words, such as sodium hydrogen carbonate. A number of compounds have common names of one word, such as water, ammonia, lye, and methane. We will talk in more detail about how we determine the names of compounds in Chapter 4.

2-5.2 Molecules, Molecular Compounds, and Covalent Bonds

About three centuries ago, water was thought to be an element. When scientists were able to decompose water into hydrogen and oxygen, it became apparent that water is a compound. Just as the basic particles of most elements are atoms, the basic particles of a particular type of compound are known as molecules. *A* **molecule** *is formed by the chemical combination of two or more atoms. Molecules composed of different atoms are the basic particles of* **molecular compounds**. *The atoms in a molecule are joined and held together by a force called the* **covalent bond**. (The nature of the covalent bond will be examined in more detail in Chapter 9.) If it were possible to magnify a droplet of water and visualize its basic particles, we would see that it is an example of a molecular compound. Each molecule of water is composed of two atoms of hydrogen joined by covalent bonds to one atom of oxygen. (See Figure 2-9.) Molecules can contain as few as two atoms or, in the case of the

FIGURE 2-8 Some Familiar Compounds These products all contain one compound. Notice their common names do not indicate the elements in the compounds.

complex molecules on which life is based, millions of atoms. We will concentrate on molecular compounds in the remainder of this section. In the next section we will discuss a second category of compounds known as ionic compounds.

2-5.3 The Formulas of Molecular Compounds

A compound is represented by the symbols of the elements of which it is composed. This is called the **formula** *of the compound.* The familiar formula for water is therefore H_2O. Note that the 2 is written as a subscript, indicating that the molecule has two hydrogen atoms. When there is only one atom of a given element present (e.g., oxygen), a subscript of "1" is assumed but not shown.

What makes one molecular compound different from another? The answer is that each chemical compound has a unique formula or arrangement of atoms in its molecules. For example, there is another compound composed of just hydrogen and oxygen, but it has the formula H_2O_2. Its name is hydrogen peroxide, and its properties are distinctly different from those of water (H_2O). Figure 2-10 illustrates how the *atoms* of hydrogen and oxygen combine to form the *molecules* of two different compounds. The formulas of other well-known compounds are $C_{12}H_{22}O_{11}$ (sucrose, which we know as table sugar), $C_9H_8O_4$ (aspirin), NH_3 (ammonia), and CH_4 (methane).

Sometimes two or more compounds may share the same chemical formula. What, then, makes them unique? In this case, their difference stems from the sequence of the atoms within the molecule. For example, ethyl alcohol and dimethyl ether are two distinct compounds, but both have the formula C_2H_6O. The difference in the two compounds lies in the order of the bonded atoms.

Notice that in ethyl alcohol the order of bonds is $C—C—O$ and in ether the order is $C—O—C$. (The dashes between atoms represent covalent chemical bonds, which hold the atoms together.) The difference in the arrangement has a profound effect on the properties of these two compounds. Ingestion of alcohol causes intoxication, while a similar amount of ether may cause death. *Formulas that show the order and arrangement of specific atoms are known as* **structural formulas**.

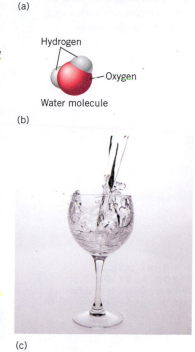
(a)

Hydrogen atom   Oxygen atom

(b)
Hydrogen
Oxygen
Water molecule

(c)

FIGURE 2-9 Molecules of Water A molecule of water is composed of one atom of oxygen and two atoms of hydrogen.

ethyl alcohol

dimethyl ether

2-5.4 Molecular Elements

Each breath of fresh air that we inhale is primarily composed of just three elements—nitrogen (78%), oxygen (21%), and argon (less than 1%), although there are traces of other gases as well. What would we see if we could magnify a sample of air so that the atoms of these three elements could become visible? The most noticeable difference among these elements is that argon exists as solitary atoms, but atoms of nitrogen and oxygen are joined together in pairs to form molecules. (See Figure 2-11.)

Hydrogen, fluorine, chlorine, bromine, and iodine in their elemental form also exist as diatomic (two-atom) molecules under normal temperature conditions. A form of elemental phosphorus consists of molecules composed of four atoms, and a form of sulfur consists of molecules composed of eight atoms. There is also a

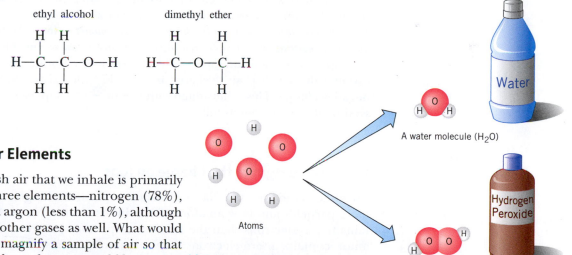

Atoms

A water molecule (H_2O)

Water

Hydrogen Peroxide

A hydrogen peroxide molecule (H_2O_2)

FIGURE 2-10 Atoms and Molecules Atoms of hydrogen and atoms of oxygen can combine to form molecules of two different compounds.

FIGURE 2-11 The **Composition of the Atmosphere** Our atmosphere is composed mostly of nitrogen and oxygen molecules with a small amount of argon atoms.

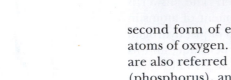

second form of elemental oxygen, known as ozone, which is composed of three atoms of oxygen. Molecules composed of two or more atoms of the same element are also referred to by formulas such as I_2 (iodine), O_2 (oxygen), O_3 (ozone), P_4 (phosphorus), and S_8 (sulfur).

2-6 Ionic Compounds

LOOKING AHEAD! There is a second classification of compounds known as ionic compounds. You may have noticed that molecular compounds are often found as gases or liquids. Ionic compounds are always solids at normal temperatures. We will discuss ionic compounds next. ∎

▶ **OBJECTIVE FOR SECTION 2.6**
Write the formulas of simple ionic compounds given the charges on the ions.

2-6.1 Cations and Anions

When you scrape your stocking feet along a carpet in a dry room, you often pick up a charge of static electricity that discharges when you touch something metallic, resulting in an unpleasant shock. Atoms can also achieve an electrostatic charge. *When atoms have an electrostatic charge, they are known as* **ions**. *Positively charged ions are known as* **cations**, *and negatively charged ions are called* **anions**. Ordinary table salt is a compound named sodium chloride. In sodium chloride, the sodium exists as a cation with a single positive charge and the chlorine exists as an anion with a single negative charge. This is illustrated with a + or − as a superscript to the right of the symbol of the element as follows.

$$Na^+ \quad and \quad Cl^-$$

2-6.2 The Origin of the Charge on Ions

To understand why the Na^+ cation has a positive charge, we must look at its basic particles and how an atom of Na differs from a Na^+ cation. A cation contains fewer electrons than the number of protons found in a neutral atom. An anion contains more electrons than there are protons in a neutral atom. *In both cases, it is the electrons that are out of balance, not the protons.* A +1 charge on a cation indicates that it has one *less* electron than its number of protons (its atomic number), and a +2 charge indicates that the cation has two *fewer* electrons than its atomic number. For example, the Na^+ cation has 11 protons (the atomic number of Na) and 10 electrons. The +1 charge arises from this imbalance [i.e., $(11p \times +1) + (10e \times -1) = +1$]. An anion with a −1 charge has one *more* electron than the atomic number of the element, and a −2 charge indicates two *more* electrons than its atomic number. For example, the S^{2-} ion has 16 protons in its nucleus and 18 electrons. The -2 charge arises from the two extra electrons [i.e., $(16p \times +1) + (18e \times -1) = -2$].

Why does Na specifically form a $+1$ charge and Cl a -1 charge and not vice versa or some other charge? In fact, there are valid reasons for this, which we will discuss in detail in a later chapter. For now, however, we can make some simple observations that provide us with some solid clues. It concerns a group of elements, sometimes called a *family*, known as the *noble gases*. These elements exist in nature as free atoms forming neither cations nor anions. The number of electrons that these elements possess lends them particular stability so they have little or (in most cases) no tendency to lose or gain electrons. As a result they all exist in nature as gaseous, monatomic elements (e.g., Ar in the air). The noble gases are the elements helium, neon, argon, krypton, and xenon. A neutral helium atom (He) has 2 electrons, neon (Ne) has 10, argon (Ar) has 18, krypton (Kr) has 36, and Xenon (Xe) has 54. Elements with atomic numbers that are near these elements tend to add or lose electrons so as to have the same number of electrons as their neighboring noble gas. For example, chlorine has 17 electrons. By adding 1 electron it has 18 (like argon) but now has a -1 charge. Sodium has 11 electrons. By losing 1 electron it has 10 (like neon) but acquires a $+1$ charge. In general, elements with 1 to 3 extra electrons beyond those of a noble gas lose electrons to form cations, while elements with 1 to 3 fewer electrons than a noble gas gain electrons to form anions. These generalizations do not work for all elements, however. Determining the charge on many of these ions will be made much simpler when we employ the *periodic table*, which we will do in Chapter 4. In Chapter 8, we will take up the topic again to present the theoretical bases of ion formation.

2-6.3 The Formulas of Ionic Compounds

Since the sodium cations and the chlorine anions are oppositely charged, the ions are held together by electrostatic forces of attraction. In Figure 2-12 the ions in sodium chloride are shown as they would appear if sufficient magnification were possible. Note that each cation (one of the smaller spheres) is attached to more than one anion. In fact, each ion is surrounded by six oppositely charged ions. The reason that cations are usually small compared to anions will be discussed in Chapter 8. Bonding in these compounds is much different from that of the previously discussed molecular compounds, in which atoms bond together to form discrete entities (molecules.)

Compounds consisting of ions are known as **ionic compounds.** *The electrostatic forces holding the ions together are known as* **ionic bonds.** The ions in ionic compounds are locked tightly in their positions by the strong electrostatic attractions. This results in solid compounds that are almost all hard and rigid. They are the material of most rocks and minerals.

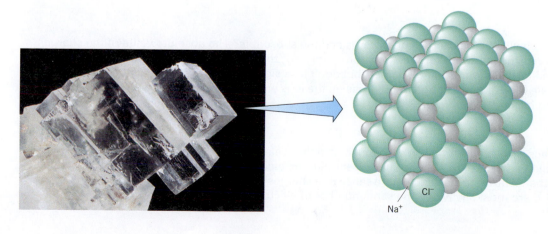

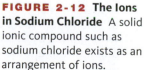

FIGURE 2-12 The Ions in Sodium Chloride A solid ionic compound such as sodium chloride exists as an arrangement of ions.

The formula of sodium chloride is

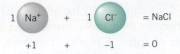

$$+1 \quad + \quad -1 \quad = 0$$

This represents the simplest ratio of cations to anions present (in this case, one-to-one). This ratio reflects the fact that the two ions have equal and opposite charges and that *in any ionic compound, the anions and cations exist together in a ratio such that the negative charge balances the positive charge* [e.g., $+1 + (-1) = 0$]. Notice that the charges are not displayed in the formula. *The simplest whole-number ratio of ions in an ionic compound is referred to as a* **formula unit**.

Ions may also have charges greater than 1. In these cases, the ratio of ions in the formula may not be simply one-to-one.

Calcium chloride is a compound composed of Ca^{2+} cations and Cl^- anions. Two chlorine anions are needed to balance the $+2$ charge on the calcium [$+2 + (\mathbf{2} \times -1) = 0$]. Thus the formula is

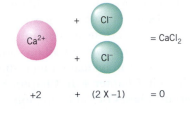

$$+2 \quad + \quad (2 \times -1) \quad = 0$$

EXAMPLE 2-3 Determining the Formula of an Ionic Compound

What is the formula unit of an ionic compound formed from the Al^{3+} cation and the S^{2-} anion?

PROCEDURE
In an ionic compound, the total positive charge of all cations and the total negative charge of all anions must be the same absolute value. We need to determine how many $+3$'s will cancel with how many -2's. Look for the least common multiple of the charges.

SOLUTION
The least common multiple of 2 and 3 is 6:

$$\mathbf{2} \times (+3) = +6$$
$$\mathbf{3} \times (-2) = -6$$

$(+6) + (-6) = 0$, a neutral compound. It will take **2** Al^{3+} and **3** S^{2-} to form an ionic compound. The formula is written $\mathbf{Al_2S_3}$.

ANALYSIS
Do you notice a pattern between the charges on the ions and the subscripts? The absolute value of the charge on the sulfur, 2, becomes the subscript for the aluminum. Similarly, the charge on the aluminum, 3, becomes the subscript for the sulfur. This works in most cases, but don't lose sight of the fact that we do this to *balance the charge*. Do not write Mg_2S_2, even though the charges on the ions are $+2$ and -2 respectively. It is sufficient to have one of each to balance the charge: MgS.

SYNTHESIS
A positive charge means a loss of electrons. What happened to these electrons? A negative charge means a gain of electrons. Where did they come from? It becomes clear that the electrons lost by one atom in a compound were gained by the other. It now makes all the more sense that charges have to balance. As electrons shift from one atom to another, there must always be the same number of electrons lost as gained.

2-6.4 The Formulas of Compounds Containing Polyatomic Ions

Groups of atoms that are covalently bonded to each other may as a whole also be cations or anions. They are known as **polyatomic ions**. Examples include the nitrate anion (NO_3^-), the perchlorate anion (ClO_4^-), and the ammonium cation (NH_4^+). While properly classified as ionic, compounds containing polyatomic ions can be viewed as a combination of the two types of substances we've discussed in this chapter. The elements involved in the polyatomic ion are joined by covalent bonds, the same as in neutral molecular compounds. Then, when the cation is added, the entirety is ionic. When more than one polyatomic ion is in a formula unit, parentheses are placed around the polyatomic ion and a subscript used. When there is only one polyatomic ion, no parentheses are used. Barium perchlorate (a compound containing one Ba^{2+} ion and two ClO_4^- ions) is represented in the margin in (a), and calcium carbonate (a compound containing one Ca^{2+} ion and one CO_3^{2-} ion) is represented in (b).

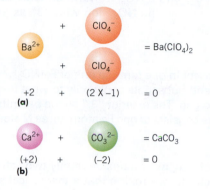

$$Ba^{2+} + ClO_4^- + ClO_4^- = Ba(ClO_4)_2$$

$$+2 + (2 \times -1) = 0$$

(a)

$$Ca^{2+} + CO_3^{2-} = CaCO_3$$

$$(+2) + (-2) = 0$$

(b)

Sometimes the formulas of molecular compounds are confused with those of ions. (See Figure 2-13.) For example, NO_2 is the formula of a molecular compound known as nitrogen dioxide. The NO_2 molecules are neutral entities. This compound is a brownish gas responsible for some air pollution. Notice that no charge is shown by the formula. The NO_2^- species (known as the nitrite ion) contains a negative charge, which means that it is a polyatomic ion. It exists only with a cation as part of an ionic compound (e.g., $NaNO_2$, sodium nitrite). This compound is a solid. Remember that cations and anions do not normally exist alone but only as the two oppositely charged parts of an ionic compound. We will spend more time writing and naming ionic compounds in Chapter 4.

The charge on polyatomic ions also arises from an imbalance of electrons. For example, the CO_3^{2-} ion has a total of 30 protons in the four nuclei [i.e., $6(C) + (3 \times 8)$ (O) = 30]. The presence of a -2 charge indicates that 32 negatively charged electrons are present in the ion.

In the next chapter we will see how the properties of molecular and ionic compounds differ.

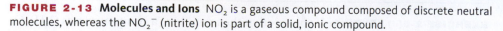

FIGURE 2-13 Molecules and Ions NO_2 is a gaseous compound composed of discrete neutral molecules, whereas the NO_2^- (nitrite) ion is part of a solid, ionic compound.

EXAMPLE 2-4 **Determining the Formula of an Ionic Compound Containing a Polyatomic Ion**

What is the formula unit of an ionic compound formed from the Fe^{3+} cation and the NO_3^- anion?

PROCEDURE

Again, the total positive charge of all cations and the total negative charge of all anions must be the same absolute value. Determine the least common multiple of the two charges, and then use the necessary numbers as subscripts to get a neutral entity.

SOLUTION

The least common multiple of 3 and 1 is 3:

$$1 \times (+3) = +3$$
$$3 \times (-1) = -3$$

$(+3) + (-3) = 0$, a neutral compound. It will take **1** Fe^{3+} and **3** NO_3^- to form an ionic compound with no net charge (neutral). Remember to include parentheses around the NO_3^- ion with "3" as a subscript to indicate three NO_3^- ions along with the one Fe^{3+} ion: **$Fe(NO_3)_3$**.

ANALYSIS

How many atoms of each type—Fe, N, and O—are present in one formula unit of $Fe(NO_3)_3$? When the subscript is "1," as in the case of iron, it is understood to be present. Subscripts apply only to what they immediately follow. The interior "3," next to oxygen, applies only to the oxygen. The exterior "3," by the parenthesis, applies to everything within the parenthesis. The formula unit therefore consists of one Fe atom, three N atoms, and nine O atoms.

SYNTHESIS

Is it possible to form an ionic compound composed of only two anions or of only two cations? If we again think of charges as resulting from the gain or loss of electrons, we realize that a theoretical substance made from two anions would have a total number of electrons greater than the number of protons. The substance would not be neutral and therefore not stable. The same argument applies to theoretical substances made entirely of cations. You must always match a cation with an anion to generate a neutral compound.

▶ **ASSESSING THE OBJECTIVES FOR SECTIONS 2-5 and 2-6**

EXERCISE 2-6(a) LEVEL 1: Determine whether the following statements apply to molecular compounds, ionic compounds, or both.
(a) composed of charged particles
(b) have properties different from their constituent elements
(c) consist of individual molecules
(d) have an overall neutral charge
(e) formula written as the smallest whole-number ratio of elements
(f) held together by a covalent bond

EXERCISE 2-6(b) LEVEL 2: What is the formula of the ionic compounds formed by combining the following?

(a) Mg^{2+} and I^- (b) Li^+ and SO_4^{2-} (c) Al^{3+} and Se^{2-} (d) Na^+ and HCO_3^-

EXERCISE 2-6(c) LEVEL 2: Complete the table by writing the chemical formula for ionic compounds formed from the corresponding cations and anions:

	Cs^+	Zn^{2+}	Fe^{3+}
ClO_3^-			
O^{2-}			
PO_4^{3-}			

EXERCISE 2-6(d) LEVEL 3: If you saw that the formula of a compound was $C_6H_{12}O_6$ and were asked whether the compound was ionic or molecular, which would you answer, and what evidence would you cite?

EXERCISE 2-6(e) LEVEL 3: You are a research scientist analyzing athletes' blood for a particular illegal chemical whose formula is $C_8H_{12}O_2N$. You found a chemical with that exact formula in one of the athelete's blood samples. Is that enough evidence to indict the athlete?

For additional practice, work chapter problems 2-36, 2-42, 2-45, 2-46, and 2-56.

MAKING IT REAL

Ionic Compounds and Essential Elements

As our bodies age, it becomes more important that we include adequate amounts of trace elements in our diets. Some of these elements include boron, calcium, chromium, copper, iodine, iron, magnesium, phosphorus, potassium, and zinc. This is not an all-inclusive list—there are several others. In fact, we are learning more all the time about the role of other trace elements in our body chemistry. As a result, the list will only grow. Deficiencies of any of these elements can cause serious health effects. For example, one of the most widespread maladies is *anemia*. Anemia is caused by a shortage of red blood cells, which are involved in the transport of oxygen from the lungs to the tissues. Iron plays a key role in the action of the hemoglobin in red blood cells. If there is too little iron in the body, anemia results; on the other hand, too much iron in the body can also cause serious problems. Genetic problems can cause some individuals to accumulate excess iron in certain organs, resulting in a condition known as *hemochromatosis*. The victims of such "iron overload" can be very sick, indeed. This condition is usually treated by periodically removing blood from the individual. However, iron deficiency is the more common malady.

Of course, we all hear on television of the need for older individuals to take plenty of calcium supplements. As we age, bones may become brittle, especially among women. This condition is known as *osteoporosis*. It is important to take calcium supplements so as to slow or even stop this degenerative process.

The best way to make sure our bodies get all the necessary elements is to eat a balanced diet. This includes green leafy vegetables as well as the usual meats, vegetables, dairy products, and carbohydrates. If that isn't enough, we may include a multivitamin on a daily basis. A typical multivitamin includes (besides vitamins) all of the trace elements, usually referred to as minerals, that we need. Actually, the minerals are not present as free elements but rather as components of a compound. For example, the most common calcium supplement is actually calcium carbonate ($CaCO_3$). In nature, $CaCO_3$ is known as limestone, chalk, or marble.

The essential element, the name of the compound containing the element, and its formula are shown in the accompanying table. Except for boric acid, all of the compounds shown are ionic. All are solid compounds that can be included in a solid pill.

Supplement Facts

Serving Size 2 tablets
Servings Per Container 180

Amount Per Serving		% Daily Value
Vitamin A (as beta carotene)	5,000IU	100%
Vitamin C (as ascorbic acid)	200mg	333%
Vitamin D3 (as cholecalciferol)	400IU	100%
Vitamin E (as d-alpha tocopheryl succinate)	100IU	333%
Thiamine (as thiamine hydrochloride)	30mg	2000%
Riboflavin	30mg	1765%
Niacin (as niacinamide)	100mg	500%
Vitamin B6 (as pyridoxine hydrochloride)	30mg	1500%
Folic Acid	400mcg	100%
Vitamin B12 (as cyanocobalamin)	100mcg	1667%
Biotin (as d-biotin)	25mcg	8%
Pantothenic Acid (as calcium pantothenate)	30mg	300%
Calcium (as calcium carbonate and dicalcium phosphate)	200mg	20%
Iron (as ferrous fumarate)	18mg	100%
Phosphorous (as dicalcium phosphate)	60mg	6%
Iodine (from kelp)	150mcg	100%
Magnesium (as magnesium oxide)	20mg	5%
Zinc (as zinc oxide)	5mg	33%
Copper (as copper gluconate)	2mg	100%
Manganese (as manganese carbonate)	7mg	350%
Proprietary Blend	148mg	*

Citrus Bioflavonoids, Inositol, Choline Bitartrate, Rutin, Alfalfa Leaf Powder, Lecithin (Soy), Parsley Leaf Powder, Watercress Powder.

* Daily Value not established

ELEMENT	NAME OF COMPOUND	FORMULA OF COMPOUND
Boron (B)	boric acid	H_3BO_3
Calcium (Ca)	calcium carbonate	$CaCO_3$
	calcium citrate	$Ca_3(C_6H_5O_7)_2$
Chromium (Cr)	chromic chloride	$CrCl_3$
Copper (Cu)	cupric sulfate	$CuSO_4$
	cupric gluconate	$Cu(C_6H_{11}O_7)_2$
Iodine (I)	sodium iodide	NaI
Iron (Fe)	ferrous sulfate	$FeSO_4$
	ferrous fumarate	$FeC_4H_2O_4$
	ferrous gluconate	$Fe(C_6H_{11}O_7)_2$
Magnesium (Mg)	magnesium oxide	MgO
Phosphorus (P)	calcium phosphate	$Ca_3(PO_4)_2$

Refer to the Student Workshop at the end of the chapter.

PART B SUMMARY

KEY TERMS

2-5	A sample of a particular element or compound is known as a **pure substance**. p. 68
2-5.2	**Molecular compounds** are composed of discrete **molecules**. p. 68
2-5.2	The atoms in molecules are held together by **covalent bonds**. p. 68
2-5.3	The **formula** of a molecular compound indicates the number of atoms of each element in one molecule. p. 69
2-5.3	The **structural formula** of a compound shows the position of the atoms relative to each other. p. 69
2-6.1	**Ions** are charged species that are either **cations** (positive) or **anions** (negative). p. 70
2-6.3	**Ionic compounds** are held together by **ionic bonds**, which are electrostatic interactions between positive and negative ions. p. 71
2-6.3	A **formula unit** of an ionic compound contains the simplest whole-number ratio of ions that balances the positive and negative charge. p. 72
2-6.4	Ionic compounds may contain **polyatomic ions**, which are groups of atoms that are held together by covalent bonds but that have a positive or negative charge. p. 73

SUMMARY CHART

Compounds

Molecular

Composition: discrete neutral molecules
Example formula: NH_3 (ammonia)
One formula unit:

$$H—N—H \text{ molecules}$$
$$\quad\ \ |$$
$$\quad\ \ H$$

Binding force: covalent bonds between atoms

Ionic

Composition: cations (+) and anions (−)
Example formula: Na_2S (sodium sulfide)
One formula unit:

$$2Na^+S^{2-} \text{ ions}$$

Binding force: ionic bonds between cations and anions

CHAPTER 2 SYNTHESIS PROBLEM

This chapter presents a wealth of information about the basic forms of matter—elements and compounds—and their composition. The purpose of this problem is to apply the information from Chapter 2 in some detective work to eventually establish the formula of an ionic compound and some additional information about the cation. Here are the facts. An ion of an element has 12 protons and 10 electrons. This cation forms a compound with a polyatomic anion that has a −3 charge and a total of 68 electrons. The anion is composed of four oxygen atoms and one atom of another element. An isotope of the cation has a mass of 4.34×10^{-23} g and has an 11.01% abundance in nature.

PROBLEM	SOLUTION
a. Write the symbol of the cation and its charge.	**a.** The cation has 12 protons, which is its atomic number. From inside the front cover (or the periodic table), we find that this is the element magnesium (symbol Mg). Since it has two fewer electrons than protons, it has a +2 charge. The ion is Mg^{2+}.
b. Write the formula of the anion including its charge.	**b.** (See Section 2-6.4.) Four neutral oxygen atoms have a total of $4 \times 8 = 32$ electrons. If the ion has a −3 charge, it has three additional electrons plus those of the neutral oxygen atoms (i.e., 32 + 3). The other element thus has $68 − 35 = 33$ electrons. This is the element arsenic (As). The anion is $AsO_4{}^{3-}$.

c. What is the formula of the compound formed from these two ions?	**c.** Three cations ($3 \times +2 = +6$) are needed to balance two anions ($2 \times -3 = -6$). Parentheses are needed for the polyatomic anion. The formula of the compound is $$Mg_3(AsO_4)_2$$
d. How many neutrons are in the isotope of the cation described above? (Refer to Table 2-3.)	**d.** From Table 2-3, notice that neutrons and protons each weigh about 1.67×10^{-24} g/am. Therefore, the mass of the isotope in amu is about $$4.34 \times 10^{-23} \text{ g}/1.67 \times 10^{-24} \text{ g/amu} = 26.0 \text{ amu.}$$ The total of neutrons and protons is therefore 26. There are $26 - 12 = \underline{14 \text{ neutrons.}}$
e. What is the approximate mass in amu that the isotope described above contributes to the atomic mass of the element?	**e.** The mass contribution in amu to three significant figures is this mass times its abundance in decimal fraction form. $$26.0 \text{ amu} \times 0.1101 = \underline{2.86 \text{ amu}}$$

YOUR TURN

A common element forms a monatomic cation with a charge of +3. This cation has a total of 23 electrons. It forms a compound with a polyatomic anion that is composed of one chromium atom and four oxygen atoms with a total of 58 electrons. The cation has an isotope with a mass of 9.35×10^{-23} g and has an abundance of 91.8% in nature.

a. What is the symbol of the cation and its charge?

b. Write the formula of the anion and its charge.

c. What is the formula of the compound formed from these two ions?

d. How many neutrons are in the isotope of the cation described above?

e. What is the approximate mass in amu that the isotope described above contributes to the atomic mass of the element?

Answers are on p. 79.

CHAPTER SUMMARY

All of the various forms of nature around us, from the simple elements in the air to the complex **compounds** of living systems, are composed of only a few basic forms of matter called **elements**. Each element has a name and a unique one- or two-letter **symbol**.

A little more than 200 years ago, John Dalton's **atomic theory** introduced the concept that elements are composed of fundamental particles called **atoms**. It has been about 30 years since we have been able to produce images of these atoms with a special microscope.

The atom is the smallest unique particle that characterizes an element. It is composed of more basic particles called **electrons** and **nucleons**. There are two types of nucleons, **protons** and **neutrons**. The relative charges and masses of these three particles are summarized in Table 2-3. The proton and electron are attracted to each other by **electrostatic forces**.

Rather than being a hard sphere, an atom is mostly empty space containing the negatively charged electrons. The protons and neutrons are located in a small dense core called the **nucleus**. The number of protons in an atom is known as its **atomic number**, which distinguishes the atoms of one element from those of another. The total number of nucleons in an atom is known as its **mass number**. Atoms of the same element may have different mass numbers and are known as **isotopes** of that element. An atom is neutral because it has the same number of electrons as protons.

Since protons and neutrons do not have exactly the same mass, the mass number is not an exact measure of the comparative masses of isotopes. A more precise measure of mass is the **isotopic mass**. This is obtained by comparing the mass of the particular isotope with the mass of ^{12}C, which is defined as having a mass of exactly 12 **atomic mass units** (amu). The **atomic mass** of an element is the weighted average of all of the naturally occurring isotopes found in nature.

Most elements are present in nature as aggregates of individual atoms. In some elements, however, two or more atoms are combined by **covalent bonds** to produce basic units called **molecules. Molecular compounds** are also composed of molecules, although, in this case, the atoms of at least two different elements are involved. The **formula** of a molecular compound represents the actual number of atoms of each element contained in a molecular unit. Each compound has a unique arrangement of atoms in the molecular unit. These are sometimes conveniently represented by **structural formulas.**

There is another type of compound, however. Atoms can become electrically charged to form **ions.** An atom can have a net positive charge if there are fewer electrons than protons in the nucleus or a net negative charge when there are more electrons than protons. Groups of atoms that are covalently bonded together can also have a net charge and are known as **polyatomic ions.** An **ionic compound** is composed of **cations** (positive ions) and **anions** (negative ions). The interactions of cations and anions are known as **ionic bonds.** The formula of an ionic compound shows the type and number of ions

in a **formula unit**. A formula unit represents the smallest whole-number ratio of cations and anions, which reflects the fact that the positive charge is balanced by the negative charge. The four most common ways that we find atoms in nature are summarized as follows:

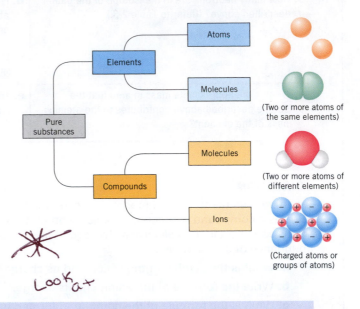

(Two or more atoms of the same elements)

(Two or more atoms of different elements)

(Charged atoms or groups of atoms)

OBJECTIVES

SECTION	YOU SHOULD BE ABLE TO...	EXAMPLES	EXERCISES	CHAPTER PROBLEMS
2-1	List the names and symbols of common elements.		1a, 1b, 1c, 1d, 1e, 1f, 1g	1, 5, 6, 7, 31, 32, 34
2-2	List the postulates of the atomic theory.		2a, 2b, 2c	8
2-3	List the components of an atom and their relative masses, charges, and location in the atom.		3a, 3b, 3c, 3d	9, 10
2-4	From the percent abundance of specific isotopes of an element, determine the atomic mass of the element and the number of protons, neutrons and electrons in each isotope.	2-1, 2-2	4a, 4b, 4c, 4d, 4e, 4f, 4g	11, 13, 14, 18, 19, 22, 26, 27, 28
2-5	List the characteristics of compounds composed of molecules.			44, 45, 56, 63
2-6	Write the formulas of simple ionic compounds given the charges on the ions.	2-3, 2-4	6a, 6d, 6e	39, 41, 43, 46, 50

▶ ANSWERS TO ASSESSING THE OBJECTIVES

Part A

EXERCISES

2-1 (a) Look at the table of elements inside the front cover. If the substance is present, it is an element. **(a)** element **(b)** compound (made from Na, C, H, and O) **(c)** compound (made from H and O) **(d)** compound (made from Si and O) **(e)** element

2-1 (b) **(a)** Cu **(b)** S **(c)** Ca

2-1 (c) **(a)** lead **(b)** phosphorus **(c)** sodium

2-1 (d) Compounds are combinations of elements. How many different ways can 120 objects be combined? Theoretically, there are an infinite number of combinations. The actual number is in the tens of millions, with new ones created every day.

2-1 (e) The elements with Latin roots for their names and symbols were likely discovered in ancient times (like gold). Since technology was not as advanced back then, it stands to reason that those elements must be the most stable and the most easily found in their free state.

2-1 (f) You would have to run some sort of chemical test on it to see if it could be broken down any further. If test after test failed to break the substance down any further, you might begin to assume that the unknown substance was an element.

2-2 (a) *An element* is composed of small, indivisible particles called *atoms*. Atoms of the same *element* are identical and have the same *properties*. Chemical *compounds* are composed of atoms of different elements combined in small *whole-number* ratios. Chemical *reactions* are rearrangements of *atoms* into different combinations.

2-2 (b) (a) True. If all atoms of an element are identical, then an experiment should always yield the same result. **(b)** False. Even though compounds have specific whole number ratios of atoms, nothing prohibits them from combining in different numbers to form a different compound. **(c)** True. Indivisible means indivisible, at least as far as chemical properties are concerned. **(d)** True. The rearrangement that occurs in a chemical reaction ensures that no atoms disappear, nor are any atoms formed, in the reaction.

2-2 (c) Democritus, the philosopher, based his ideas on nothing more than rational thought. Dalton used experimental evidence to support his conclusions. These conclusions have stood up to 200 years of scrutiny.

2-3 (a) (a) electron **(b)** proton and neutron **(c)** neutron (just barely) **(d)** proton

2-3 (b)

PARTICLE	MASS (AMU)	CHARGE	LOCATION
Proton	1	+1	Nucleus
Neutron	1	Neutral	Nucleus
Electron	0	−1	Outside nucleus

2-3 (c) (a) Virtually all of the mass is in the nucleus with the protons and neutrons. **(b)** The electrons take up most of the volume. **(c)** The electron was first to be identified by Thompson in 1897. Its existence implied the proton. The neutron was the last identified.

2-3 (d) Neutrons are neutrally charged, which is what would happen if a positive proton and negative electron combined into one particle. Also, the mass of the neutron is just a little more than the combined mass of the proton and the electron.

2-4 (a) (a) atomic number and mass number **(b)** mass number **(c)** isotopic mass **(d)** atomic number **(e)** mass number

2-4 (b) The protons are determined by matching an elemental symbol to its atomic number. The numbers of electrons in these neutral species are the same. The number of neutrons is equal to the difference between the atomic number and the mass number. **(a)** protons = 6, neutrons = 7, electrons = 6 **(b)** protons = 1, neutrons = 0, electrons = 1 **(c)** protons = 92, neutrons = 146, electrons = 92

2-4 (c) (a) ^{19}F **(b)** ^{79}Br **(c)** ^{40}Ca

2-4 (d) $(120.903 \times 0.573) + (122.904 \times 0.427) = 122$

2-4 (e) $(6.015 \times 0.0742) + (7.016 \times 0.9258) = 6.94$. Compare this answer to lithium's atomic mass on the periodic table. It's a match.

2-4 (f) Lead and gold are both elements with their own sets of protons, neutrons, and electrons. To convert one element to another, these three particles, and specifically the number of protons, would have to change. There is no chemical way to add or subtract protons from the nucleus of an atom.

2-4 (g) The weighted average (atomic mass) of the two isotopes is 79.9, or almost 80. This is halfway between the two isotopes. It stands to reason that each isotope contributes roughly evenly to the mass number. Therefore, the percentage abundance of each is nearly 50%.

Part B

EXERCISES

2-6 (a) (a) ionic **(b)** both **(c)** molecular **(d)** both **(e)** ionic **(f)** molecular

2-6 (b) (a) MgI_2 **(b)** Li_2SO_4 **(c)** Al_2Se_3 **(d)** $NaHCO_3$

2-6 (c)

	Cs^+	Zn^{2+}	Fe^{3+}
ClO_3^-	$CsClO_3$	$Zn(ClO_3)_2$	$Fe(ClO_3)_3$
O^{2-}	Cs_2O	ZnO	Fe_2O_3
PO_4^{3-}	Cs_3PO_4	$Zn_3(PO_4)_2$	$FePO_4$

2-6 (d) In ionic compounds, atoms are present in the smallest whole-number ratio. If this were an ionic compound, the formula would be CH_2O. Since the formula is six times bigger, it must be molecular.

2-6 (e) No, not nearly. Two or more compounds can share the same formula, but the order of their bonds might be quite different. Further tests would be required to establish the identity of the substance beyond a reasonable doubt.

ANSWERS TO CHAPTER SYNTHESIS PROBLEM

a. Fe^{3+} **b.** CrO_4^{2-} **c.** $Fe_2(CrO_4)_3$ **d.** 30 neutrons **e.** 51.4 amu

CHAPTER PROBLEMS

Throughout the text, answers to all exercises in color are given in Appendix E. The more difficult exercises are marked with an asterisk.

Names and Symbols of the Elements (SECTION 2-1)

2-1. Write the symbols of the following elements. Try to do this without referring to a table of elements.

(a) bromine   (c) lead   (e) sodium

(b) oxygen   (d) tin   (f) sulfur

2-2. The following elements all have symbols that begin with the letter C: cadmium, calcium, californium, carbon, cerium, cesium, chlorine, chromium, cobalt, copper, and curium. The symbols are C, Ca, Cd, Ce, Cf, Cl, Cm, Co, Cr, Cs, and Cu. Match each symbol with an element and then check with the table of elements inside the front cover.

2-3. The names of seven elements begin with the letter B. What are their names and symbols?

2-4. The names of nine elements begin with the letter S. What are their names and symbols?

2-5. Using the table inside the front cover, write the symbols for the following elements.

(a) barium   (c) cesium   (e) manganese

(b) neon   (d) platinum   (f) tungsten

2-6. Name the elements corresponding to the following symbols. Try to do this without reference to a table of the elements.

(a) S   (c) Fe   (e) Mg

(b) K   (d) N   (f) Al

2-7. Using the table, name the elements corresponding to the following symbols.

(a) B   (c) Ge   (e) Co   (g) Be

(b) Bi   (d) U   (f) Hg   (h) As

Composition of the Atom (SECTIONS 2-2 AND 2-3)

2-8. Which of the following were not part of Dalton's atomic theory?

(a) Atoms are the basic building blocks of nature.

(b) Atoms are composed of electrons, neutrons, and protons.

√(c) Atoms are reshuffled in chemical reactions.

√(d) The atoms of an element are identical.

(e) Different isotopes can exist for the same element.

2-9. Which of the following describes a neutron?

(a) +1 charge, mass 1 amu   (c) 0 charge, mass 1 amu

(b) +1 charge, mass 0 amu   (d) −1 charge, mass 0 amu

2-10. Which of the following describes an electron?

(a) +1 charge, mass 1 amu   (c) −1 charge, mass 1 amu

(b) +1 charge, mass 0 amu   (d) −1 charge, mass 0 amu

2-11. Give the mass numbers and atomic numbers of the following isotopes. Refer to the table of the elements inside the front cover.

(a) ^{193}Au (b) ^{132}Te (c) ^{118}I (d) ^{39}Cl

2-12. Give the numbers of protons, neutrons, and electrons in each of the following isotopes. Refer to the table of the elements inside the front cover.

(a) ^{45}Sc (b) ^{232}Th (c) ^{223}Fr (d) ^{90}Sr

2-13. Three isotopes of uranium are ^{234}U, ^{235}U, and ^{238}U. How many protons, neutrons, and electrons are in each isotope?

2-14. Using the table of elements inside the front cover, complete the following table for neutral isotopes.

Isotope	Isotopic Notation	Atomic Number	Mass Number	Subatomic Particles Protons	Neutrons	Electrons
molybdenum-96	$^{96}_{42}Mo$	42	96	42	54	42
(a)	$^{?}_{?}Ag$				61	
(b)		14			14	
(c)			39		20	
(d) cerium-140						
(e)				26	30	
(f)		50	110			
(g)	$^{118}_{?}I$					
(h) mercury-?						116

2-15. Using the table of elements inside the front cover, complete the following table for neutral isotopes.

Isotope	Isotopic Notation	Atomic Number	Mass Number	Subatomic Particles Protons	Neutrons	Electrons
(a) tungsten-?			184			
(b)					12	11
(c)	$^{200}_{?}Au$					
(d)	$^{?}_{?}Pm$				87	
(e)			109	46		
(f)			48			23
(g)		21			29	

2-16. Write the isotopic notation for an isotope of cobalt that has the same number of neutrons as ^{60}Ni.

2-17. Write the isotopic notation for an isotope of uranium that has the same number of neutrons as ^{240}Pu.

Atomic Number and Mass (SECTION 2-4)

2-18. How do the following concepts relate and differ?

(a) element and atomic number

(b) atomic mass and atomic number

(c) mass number and atomic mass

(d) isotopes and number of protons

(e) isotopes and number of neutrons

2-19. Determine the atomic number and the atomic mass of each of the following elements. Use the table inside the front cover.

(a) Re   (b) Co   (c) Br   (d) Si

2-20. About 75% of a U.S. "nickel" is an element with an atomic mass of 63.546 amu. What is the element?

2-21. White gold is a mixture of gold containing an element with an atomic mass of 106.4 amu. What is the element?

2-22. The elements O, N, Si, and Ca are among several that are composed *primarily* of one isotope. Using the table inside the front cover, write the atomic number and mass number of the principal isotope of each of these elements.

2-23. The atomic mass of hydrogen is given inside the front cover as 1.00794. The three isotopes of hydrogen are ^1H, ^2H, and ^3H. What does the atomic mass tell us about the relative abundances of the three isotopes?

2-24. A given element has a mass 5.81 times that of ^{12}C. What is the atomic mass of the element? What is the element?

2-25. The atomic mass of a given element is about 3.33 times that of ^{12}C. Give the atomic mass, the name, and the symbol of the element.

2-26. Bromine is composed of 50.5% ^{79}Br and 49.5% ^{81}Br. The isotopic mass of ^{79}Br is 78.92 amu and that of ^{81}Br is 80.92 amu. What is the atomic mass of the element?

2-27. Silicon occurs in nature as a mixture of three isotopes: ^{28}Si (27.98 amu), ^{29}Si (28.98 amu), and ^{30}Si (29.97 amu). The mixture is 92.21% ^{28}Si, 4.70% ^{29}Si, and 3.09% ^{30}Si. Calculate the atomic mass of naturally occurring silicon.

2-28. Naturally occurring Cu is 69.09% ^{63}Cu (62.96 amu). The only other isotope is ^{65}Cu (64.96 amu). What is the atomic mass of copper?

*2-29. Chlorine occurs in nature as a mixture of ^{35}Cl and ^{37}Cl. If the isotopic mass of ^{35}Cl is approximately 35.0 amu and that of ^{37}Cl is 37.0 amu, and the atomic mass of the mixture as it occurs in nature is 35.5 amu, what is the proportion of the two isotopes?

*2-30. The atomic mass of the element gallium is 69.72 amu. If it is composed of two isotopes, ^{69}Ga (68.926 amu) and ^{71}Ga (70.925 amu), what is the percent of ^{69}Ga?

Molecular Compounds and Formulas (SECTION 2-5)

2-31. How do the following concepts relate and differ?

(a) a molecule and an atom

(b) a molecule and a compound

(c) an element and a compound

(d) a molecular element and a monatomic element

2-32. Which of the following are formulas of elements rather than compounds?

(a) P_4O_{10} (c) F_2O (e) MgO

(b) Br_2 (d) S_8 (f) P_4

2-33. Name all of the elements shown in the previous problem.

2-34. What is the difference between Hf and HF?

2-35. What is the difference between NO and No?

2-36. Which of the following is the formula of a diatomic element? Which is the formula of a diatomic compound?

(a) NO_2 (c) K_2O (e) N_2

(b) CO   (d) $(NH_4)_2S$ (f) CO_2

2-37. Give the name and number of atoms of each element in the formulas of the following compounds.

(a) H_2SeO_3 (c) NI_3 (e) $Ba(BrO_3)_2$

(b) Na_4SiO_4 (d) NiI_2 (f) $B_3N_3(CH_3)_6$

2-38. What is the total number of atoms in each formula unit for the compounds in problem 2-37?

2-39. Determine the number of atoms of each element in the formulas of the following compounds.

(a) $C_6H_4Cl_2$

(b) C_2H_5OH (ethyl alcohol)

(c) $CuSO_4 \cdot 9H_2O$ (H_2O's are part of a single formula unit)

(d) $C_9H_8O_4$ (aspirin)

(e) $Al_2(SO_4)_3$

(f) $(NH_4)_2CO_3$

2-40. What is the total number of atoms in each molecule or formula unit for the compounds listed in problem 2-39?

2-41. How many carbon atoms are in each molecule or formula unit of the following compounds?

(a) C_8H_{18} (octane in gasoline)   (c) $Fe(C_2O_4)_2$

(b) $NaC_7H_4O_3NS$ (saccharin)   (d) $Al_2(CO_3)_3$

2-42. Write the formulas of the following molecular compounds.

(a) sulfur dioxide (one sulfur and two oxygen atoms)

(b) carbon dioxide (one carbon and two oxygen atoms)

(c) sulfuric acid (two hydrogens, one sulfur, and four oxygen atoms)

(d) acetylene (two carbons and two hydrogens)

2-43. Write the formulas of the following molecular compounds.

(a) phosphorus trichloride (one phosphorus and three chlorines)

(b) naphthalene (ten carbons and eight hydrogens)

(c) dibromine trioxide (two bromines and three oxygens)

Ions and Ionic Compounds (SECTION 2-6)

2-44. How do the following concepts relate and differ?

(a) an atom and an ion

(b) a molecule and a polyatomic ion

(c) a cation and an anion

(d) a molecular and an ionic compound

(e) a molecular unit and an ionic formula unit

2-45. The gaseous compound HF contains covalent bonds, and the compound KF contains ionic bonds. Sketch how the basic particles of these two compounds appear.

2-46. Write the formulas of the following ionic compounds.

(a) calcium perchlorate (one Ca^{2+} and two ClO_4^- ions)

(b) ammonium phosphate (three NH_4^+ ions and one PO_4^{3-} ion)

(c) iron(II) sulfate (one Fe^{2+} and one SO_4^{2-} ion)

2-47. What is the number of atoms of each element present in the compounds in problem 2-46?

2-48. Write the formulas of the following ionic compounds.

(a) calcium hypochlorite (one Ca^{2+} ion and two ClO^- ions)

(b) magnesium phosphate (three Mg^{2+} ions and two PO_4^{3-} ions)

(c) chromium(III) oxalate (two Cr^{3+} ions and three $C_2O_4^{2-}$ ions)

2-49. What is the number of atoms of each element present in the compounds in problem 2-48?

2-50. The formula of an ionic compound indicates one Fe^{2+} ion combined with one anion. Which of the following could be the other ion?

(a) F^- **(b)** Ca^{2+} **(c)** S^{2-} **(d)** N^{3-}

2-51. An ionic compound is composed of two ClO_4^- ions and one cation. Which of the following could be the other ion?

(a) SO_4^{2-} **(b)** Ni^{2+} **(c)** Al^{3+} **(d)** Na^+

2-52. An ionic compound is composed of one SO_3^{2-} and two cations. Which of the following could be the cations?

(a) I^- **(b)** Ba^{2+} **(c)** Fe^{3+} **(d)** Li^+

2-53. An ionic compound is composed of two Al^{3+} ions and three anions. Which of the following could be the anions?

(a) S^{2-} **(b)** Cl^- **(c)** Sr^{2+} **(d)** N^{3-}

2-54. Write the formulas of the compounds in problems 2-50 and 2-52.

2-55. Write the formulas of the compounds in problems 2-51 and 2-53.

2-56. Explain the difference between SO_3 and SO_3^{2-}. Which one would be a gas?

2-57. What are the total number of protons and the total number of electrons in each of the following ions?

(a) K^+ **(c)** S^{2-} **(e)** Al^{3+}

(b) Br^- **(d)** NO_2^- **(f)** NH_4^+

2-58. What are the total number of protons and the total number of electrons in each of the following ions?

(a) Sr^{2+} **(c)** V^{3+} **(e)** SO_3^{2-}

(b) P^{3-} **(d)** NO^+

2-59. Write the element symbol or symbols and the charge for the following ions.

(a) 20 protons and 18 electrons

(b) 52 protons and 54 electrons

(c) one phosphorus and three oxygens with a total of 42 electrons

(d) one nitrogen and two oxygens with a total of 22 electrons

2-60. Write the element symbol or symbols and the charge for the following ions.

(a) 50 protons and 48 electrons

(b) 53 protons and 54 electrons

(c) one aluminum and two oxygens with a total of 30 electrons

(d) one chlorine and three fluorines with a total of 43 electrons

2-61. A monatomic bromine species has 36 electrons. Does it exist independently?

2-62. A species is composed of one chlorine atom chemically bonded to two oxygen atoms. It has a total of 33 electrons. Is this species most likely a gaseous molecular compound or part of an ionic compound?

General Problems

2-63. Describe the difference between a molecular and an ionic compound. Of the two types of compounds discussed, is a stone more likely to be a molecular or an ionic compound? Is a liquid more likely to be a molecular or an ionic compound?

2-64. Write the symbol, mass number, atomic number, and electrical charge of the element given the following information. Refer to the table of the elements.

(a) An ion of Sr contains 36 electrons and 52 neutrons.

(b) An ion contains 24 protons, 28 neutrons, and 21 electrons.

(c) An ion contains 36 electrons and 45 neutrons and has a −2 charge.

(d) An ion of nitrogen contains 7 neutrons and 10 electrons.

(e) An ion contains 54 electrons and 139 nucleons and has a +3 charge.

2-65. Write the symbol, mass number, atomic number, and electrical charge of the element given the following information. Refer to the table of the elements.

(a) An ion of Sn contains 68 neutrons and 48 electrons.

(b) An ion contains 204 nucleons and 78 electrons and has a +3 charge.

(c) An ion contains 45 neutrons and 36 electrons and has a −1 charge.

(d) An ion of aluminum has 14 neutrons and a +3 charge.

2-66. Give the number of protons, electrons, and neutrons represented by the following species. These elements are composed almost entirely of one isotope, which is implied by the atomic mass.

(a) Na and Na$^+$ **(c)** F and F$^-$

(b) Ca and Ca^{2+} **(d)** Sc and Sc^{3+}

2-67. Give the number of protons, electrons, and neutrons represented by the following species. These elements are composed almost entirely of one isotope, which is implied by the atomic mass.

(a) Cr and Cr^{2+} **(c)** I and I$^-$

(b) Au and Au^{3+} **(d)** P and P^{3-}

*2-68. An isotope of iodine has a mass number that is 10 amu less than two-thirds the mass number of an isotope of thallium. The total mass number of the two isotopes is 340 amu. What is the mass number of each isotope? (*Hint:* There are two equations and two unknowns.)

*2-69. An isotope of gallium has a mass number that is 22 amu more than one-fourth the mass number of an isotope of osmium. The osmium isotope is 122 amu heavier than the gallium isotope. What is the mass number of each isotope? (*Hint:* There are two equations and two unknowns.)

*2-70. A given element is composed of 57.5% of an isotope with an isotopic mass of 120.90 amu. The remaining percentage of isotope has an isotopic mass of 122.90 amu. What is the atomic mass of the element? What is the element? How many electrons are in a cation of this element if it has a charge of 3? How many neutrons are in each of the two isotopes of this element? What percent of the isotopic mass of each isotope is due to neutrons?

2-71. A given isotope has a mass number of 196, and 60.2% of the nucleons are neutrons. How many electrons are in a cation of this element if it has a charge of 2?

2-72. A given isotope has a mass number of 206. The isotope has 51.2% more neutrons than protons. What is the element?

2-73. A given molecular compound is composed of one atom of nitrogen and one atom of another element. The mass of nitrogen accounts for 46.7% of the mass of one molecule. What is the other element? What is the formula of the compound? This molecule can lose one electron to form a polyatomic ion. How many electrons are in this ion?

2-74. A given molecular compound is composed of one atom of carbon and two atoms of another element. The mass of carbon accounts for 15.8% of the mass of one molecule. What is the other element? What is the formula of the compound?

2-75. If the isotopic mass of ^{12}C were defined as exactly 8 instead of 12, what would be the atomic mass of the following elements to three significant figures? Assume that the elements have the same masses relative to each other as before: that is, hydrogen still has a mass of one-twelfth that of carbon.

(a) H   **(b)** N   **(c)** Na   **(d)** Ca

2-76. Assume that the isotopic mass of ^{12}C is defined as exactly 10 and that the atomic mass of an element is 43.3 amu on this basis. What is the element?

2-77. Assume that the isotopic mass of ^{12}C is defined as exactly 20 instead of 12 and that the atomic mass of an element is 212.7 amu on this basis. What is the element?

STUDENT WORKSHOP

Chemical Formulas

Purpose: To evaluate chemical compounds for their composition, and to create new compounds from their constituent parts. (Work in groups of three or four. Estimated time: 25 min.)

Divide up the compounds listed in Making It Real, "Ionic Compounds and Essential Elements," so that each person has three or four. Using a periodic table, do the following:

- Name the elements found in each compound.
- Determine how many atoms of each element are contained within each compound.

- Calculate the total number of protons in each compound.
- Do the same for the number of neutrons. (You should assume that the isotope present is the one with the closest integer value to the mass number found on the periodic table.)
- Determine whether the compound contains a polyatomic ion.

Now, using the following ions (all found in the compounds above), construct three or four new compounds each, with the necessary numbers to make their formulas neutral.

Cations: H$^+$ Cr^{2+} Cu^{2+} Na$^+$ Fe^{2+} Mg^{2+} Ca^{2+}

Anions: BO$_3$$^{3-}$ CO$_3$$^{2-}$ I$^-$ SO$_4$$^{2-}$ O^{2-} PO$_4$$^{3-}$

The Properties of Matter and Energy

A copper dome and statue quickly tarnish due to the formation of a blue-green coating of copper(II) carbonate. The chemical and physical characteristics of substances such as copper are among the subjects of this

SETTING THE STAGE

In the previous chapter, we discussed recent findings on our neighboring planet, Mars. Now we will focus on recent amazing discoveries from farther out in space that involve Titan, the mysterious moon of Saturn. Titan's surface is perpetually hidden from view by thick clouds. Scientists were not able to see through these clouds, however, until a Saturn orbiter named *Cassini* was launched from Earth in 1997. It finished its long journey to orbit Saturn in 2004. In 2005, Cassini came near Titan and launched a probe, named *Huygens*, into its atmosphere and which then parachuted to the surface. An unbelievable scene was revealed. There were huge lakes, valleys eroded by liquids, and obvious signs of precipitation. In other words, it looked somewhat like parts of Earth. But it was actually very different. The temperature was much too low (i.e., $-178°C$) for this liquid to be ordinary water. More likely, the bodies of liquid and the rain that formed the valleys would be composed of ethane (C_2H_6) and methane (CH_4). How did the scientists determine all this? The answer lies in the known characteristics or properties of the compounds involved, like water, ethane, and methane. For example, under such cold conditions methane, which is a gas (i.e., natural gas) on Earth, would be a liquid on Titan. In fact, the liquid methane could evaporate and condense into cold clouds and fall as precipitation just like water does on our planet. Titan is a strange world somewhat like Earth but in fact very different.

Now we turn our focus to our home planet, with its abundant life. Life is dependent not only on the presence of liquid water but also on the constant supply of energy from the sun. *Energy, which has no mass, is a second component of the universe in addition to matter.* When a log burns in the fireplace, it is obvious that a change in matter has occurred. The log is transformed into a small pile of ashes and hot gases. But there is more involved than simply a change in matter. The burning of the log warms us—it has given off heat. The heat and light liberated by the burning process are forms of energy. Energy is a more abstract concept than matter. It can't be weighed, and it doesn't have shape, form, or dimensions. But energy can be measured, and it does interact with matter (e.g., it warms us, starts our car, and makes trees grow). Since energy is involved in the changes that matter undergoes, it is also important in the study of the properties of matter.

How we describe the properties of matter is the topic of Part A of this chapter. In Part B, we will discuss the properties and measurement of energy.

Part A

The Properties of Matter

SETTING A GOAL

■ You will learn how a sample of matter can be described by its properties and how they can be quantitatively expressed.

OBJECTIVES

3-1 List several properties of matter and distinguish them as physical or chemical.

3-2 Perform calculations involving the density of liquids and solids.

3-3 Perform calculations involving the percent of a pure substance in a mixture.

▶ **OBJECTIVE FOR SECTION 3-1**

List several properties of matter and distinguish them as physical or chemical.

3-1 The Physical and Chemical Properties of Matter

LOOKING AHEAD! People have a physical description, such as that found on a driver's license, and a personality description. Your physical description tells how you are observed, and your personality description tells how you interact with others. Similarly, there are two ways to describe an element or compound, depending on whether it is an observable description (a physical property) or on how it interacts with other substances (a chemical property). We will first consider physical properties and then proceed to chemical properties. ■

"The suspect in the robbery was a six-foot-two-inch male with a heavy build, short hair, and a thin mustache." These are a few of the physical properties of a person that we may hear about in a news bulletin. These properties can be observed without interacting with this individual or getting too close. Like an individual, *the physical properties of a substance are those that can be observed or measured without changing the substance into another substance.* Some physical properties can be simply observed. Physical state, color, and sometimes odor are such properties that a substance may display. First we will consider physical state.

3-1.1 The Physical States of Matter

In the previous two chapters, you learned that all matter is categorized as either elements or compounds. Recall that elements and compounds are composed of extremely tiny particles (atoms, molecules, or ions). Sometimes a sample of matter, such as in a container of argon, is composed of single atoms, but most matter is made of molecules or ions. The distance between these particles and their relative motion is what determines the **physical state** of a sample of matter: solid, liquid, or gas.

A **solid** is composed of matter where the particles are close together and remain in relatively fixed positions. Movement of the particles is very restricted and confined mostly to vibrations about these positions. Because of the fixed positions of the particles, *solids have a definite shape and a definite volume.*

A **liquid** is composed of matter where the particles are close together but are able to move past one another. Because of the movement of the particles, liquids flow and take the shape of the lower part of a container. *Liquids have a definite volume but not a definite shape.*

A **gas** is made up of atoms or molecules that are not all close to one another and move independently in all directions with random motion. The particles in gases fill a container uniformly. *Gases have neither a definite volume nor a definite shape.* (See Figure 3-1.)

We are already familiar with many examples of all three physical states. Ice, rock, salt, and steel are substances that exist as solids; water, gasoline, and alcohol

(a)   **(b)**   **(c)**

FIGURE 3-1 The Three States of Matter (a) Solids have a definite shape and volume. (b) Liquids have a definite volume but an indefinite shape. (c) Gases have an indefinite shape and volume.

are liquids; ammonia, natural gas, and the components of air are present as gases. However, whether a particular element or compound is a solid, liquid, or gas depends not only on the nature of the substance but also on the temperature. For example, at low temperatures (i.e., below 0°C), liquid water freezes to form a solid (ice), and at high temperatures (i.e., above 100°C), liquid water boils to form a gas (vapor or steam). At very low temperatures (below −196°C), even the gases that form our atmosphere condense to liquid.

3-1.2 Changes in Physical State

The temperature at which a pure substance changes from one physical state to another is a fundamental and constant physical property. A substance *melts* when it changes from the solid to the liquid state and *freezes* when it changes from the liquid to the solid state. *The **melting point** is the temperature at which a particular element or compound changes from the solid state to the liquid state.* For example, ice begins to melt when the temperature is 0°C. In the reverse process, liquid water begins to change to the solid state when it is cooled to 0°C. This is known as the **freezing point**.

At a higher temperature, the liquid begins to boil. Boiling occurs when bubbles of vapor form in the liquid and rise to the surface. *The **boiling point is** the temperature at which boiling begins.* In the reverse process, *the change from the gaseous state to the liquid state is known as* **condensation**. The formation of dew on the grass on a summer morning is a result of condensation. The boiling point of a liquid is also a constant, but only at specific atmospheric pressure. For example, water boils at 100°C at average sea-level pressure but boils at 69°C on the top of Mt. Everest, the world's highest mountain. Boiling-point temperatures are usually listed as the boiling point of the liquid at average sea-level atmospheric pressure. These phase changes are summarized as follows:

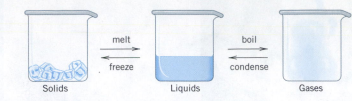

When a liquid freezes or boils, it undergoes a change to another physical state. However, it is still the same substance. *A* **physical change** *in a substance does not involve a change in the composition of the substance but is simply a change in physical state or dimensions.* Liquid water, ice, and steam are all physical states of the same compound. (See Figure 3-2.)

3-1.3 Types of Physical Properties

Intensive properties, such as clarity, color, melting point, and density, *are those that do not depend on the amount of material present.* **Extensive properties**, such as mass and volume, *are those that do depend on the amount of material present.* Both types of properties are used to identify a particular unknown substance, but intensive properties are more definitive. For example, an unidentified clear, colorless liquid (an extensive property) that freezes at 0°C and boils at 100°C (two intensive properties) is most likely water.

3-1.4 Chemical Changes and Chemical Properties

When water is cooled, it solidifies into ice. When it is allowed to warm, the ice melts back to liquid water. Ice and liquid water are two different physical states of the same compound. However, when we heat a raw egg, it solidifies. When we cool the egg, it stays solid. Obviously, the contents of the egg are not the same—they have undergone profound changes into other compounds. When iron rusts, vegetation decays, and wood burns, the original substances have been transformed into one or more other substances. These processes all describe chemical changes.

Chemical properties *of a pure substance refer to its tendency to undergo chemical changes.* **Chemical changes** *transform one substance into one or more other substances.* A chemical property of the element iron is its tendency to react with oxygen from the air in the presence of water to form rust (a compound composed of iron and oxygen). The conversion of rust back into iron and oxygen is an involved and difficult chemical process. In some cases, chemical properties relate to the absence of specific changes. For example, a chemical property of the element gold is that it maintains its lustrous appearance because it resists rusting or tarnishing. Chemical properties of three substances are shown in Figure 3-3.

3-1.5 Chemical Change and Conservation of Mass

What happens when a chemical change occurs? In fact, the total mass of the elements and compounds involved does not change—only the identity of the substances involved. *The* **law of conservation of mass** *states that matter is neither created nor destroyed in a chemical reaction.* Only three centuries ago, scientists were still puzzled over the apparent disappearance of mass when wood burned, since only a small portion of the original mass remained in the form of ashes. At that time, however, the involvement of gases in chemical reactions was not understood. We now know that most of the solid compounds of the wood have been simply transformed in the combustion process into gaseous compounds and smoke that drift away in the atmosphere. The mass of the wood plus the mass of the oxygen from the air equals the mass of the ashes plus the mass of the gaseous combustion products, as stated by the law of conservation of mass.

FIGURE 3-2 A Physical Change The ice is undergoing a physical change to liquid water at its melting point (32°F, or 0°C).

(a)   (b)   (c)

FIGURE 3-3 Chemical Properties and Changes Zinc reacts with acid (a), sulfur burns in air (b), and iron rusts (c).

▶ **ASSESSING THE OBJECTIVE FOR SECTION 3-1**

EXERCISE 3-1(a) LEVEL 1: Identify the following as an example of a physical or chemical change:

(a) boiling   (c) evaporating   (e) rotting
(b) burning   (d) rusting   (f) dissolving

EXERCISE 3-1(b) LEVEL 2: Calcium, an element, is a dull, gray solid that melts at 839°C. When it is placed in water, bubbles form, as the solid calcium slowly disappears in the water. When the water is evaporated, a white powder remains, but elemental calcium is not recovered. Which are the physical properties of calcium? Which is a chemical property?

EXERCISE 3-1(c) LEVEL 2: A beaker of an unknown clear and colorless liquid has a volume of 100.0 mL and a mass of 78.9 g. Initially, its temperature is 25°C. When heated, it boils at 78.5°C. If ignited, it burns completely with a blue flame, leaving no residue behind. Which of these pieces of information will be helpful in identifying the liquid?

EXERCISE 3-1(d) LEVEL 3: In lab you are handed a metallic object and charged with determining its identity. What types of things can you do to determine what the object is made from?

For additional practice, work chapter problems 3-4, 3-6, and 3-10.

3-2 Density—A Physical Property

▶ **OBJECTIVE FOR SECTION 3-2**
Perform calculations involving the density of liquids and solids.

LOOKING AHEAD! Some physical properties such as odor, color, and physical state can be determined by observation. Others, such as melting or boiling point, must be determined by measurements. Another important intensive physical property that is obtained from measurements is density. ∎

A Styrofoam coffee cup is "light," but a lead car battery is "heavy." Actually, by themselves these terms, *light* and *heavy*, are not very useful because a truckload of Styrofoam would be quite heavy. The volume and mass of a substance are extensive properties

size of the sample. The physical property of density is an inten- ...llows us to compare the mass of substances for a specific volume. *f the mass (usually in grams) to the volume (usually in milliliters for a solid* *gas). That is, density = mass/volume.* Like all extensive properties, the ...nce does not depend on the amount present. The density of lead ...ter than the density of Styrofoam, no matter how much is present.

as a Physical Property

pure substance is a property that can be used to identify a particu- ...ompound. The densities of several liquids and solids are listed in ...use the volume of liquids and solids expands slightly as the tempera- ...ties are usually given at a specific temperature. In this case, 20°C is ...mperature.) Because 1 mL is the same as 1 cm^3 (see Section 1-4.3), ...xpressed as g/cm^3. The densities of gases are discussed in Chapter 10. ...e that the density of a substance does not depend on sample size with ...nstration. If you had a beaker of water, you could measure its volume and ...from those two measurements calculate its density. If your lab partners were evaluating some water in a smaller beaker, they would measure an appropriately smaller mass and smaller volume. If you compared the results, you would find that the density of water is the same regardless of the original amount present.

TABLE 3-1

Density (at 20°C)			
SUBSTANCE (LIQUID)	**DENSITY (g/mL)**	**SUBSTANCE (SOLID)**	**DENSITY (g/mL)**
Ethyl alcohol	0.790	Aluminum	2.70
Gasoline (a mixture)	~ 0.67 (variable)	Gold	19.3
Carbon tetrachloride	1.60	Ice	0.92 (0°C)
Kerosene (a mixture)	0.82	Lead	11.3
Water	1.00	Lithium	0.53
Mercury	13.6	Magnesium	1.74
Table salt	2.16		

✔ MATH CHECK:

Math Operations and Significant Figures

In the example problem that follows, you will be asked to divide two numbers expressed in different degrees of precision. Correctly expressing the results of mathematical operations using numbers with different numbers of decimal places or significant figures is a necessary skill in chemistry. Unfortunately, the calculator can't help us here. We have to know the rules. The following questions test your understanding of these principles.

    a. What is the sum of 11.841 mL, 0.009 mL, and 5.1 mL?

    b. What is the result and unit of the following calculation: 98.020 cm^3 divided by 32.0 cm?

Answers: a. 17.0 mL   **b.** 3.06 cm^2

Refer to Section 1-2.1 for help with significant figures related to addition and subtraction.

Refer to Section 1-2.2 for help with significant figures related to multiplication and division.

The calculation of density from the two measurements is discussed in the following two examples. As we will see, the volume of an unknown sample is often measured by the displacement of water. When a substance is added to water it will either sink or float depending on its density (assuming that it doesn't dissolve in or react with water). If the unknown is less dense than water, it will float and it is said to be *buoyant* in water. If it is denser, it sinks. In the example that follows, the unknown sinks, so we can conclude that it is denser than water and we can measure its volume by the volume of water that is displaced.

EXAMPLE 3-1   Determining Density

A sample of a pure substance is found to have a mass of 52.11 g. As shown below, a measured quantity of water has a volume of 12.5 mL. When the substance is placed in the water, it sinks and the volume of the water and substance now read 31.8 mL. What is the density?

PROCEDURE

The density equals the mass measured in grams divided by the volume measured in milliliters. The volume of the sample is the volume of water displaced when the substance is placed in the water. Therefore, the volume is calculated as the difference in the volume of the water before and after the sample is added.

SOLUTION

As shown below, the mass is obtained by placing the substance on an electronic balance, while the volume is measured from the volume differences of the water.

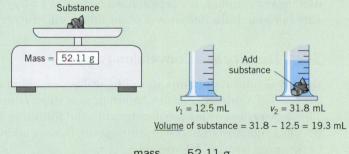

Substance

Mass = 52.11 g

Add substance

$v_1 = 12.5$ mL $v_2 = 31.8$ mL

Volume of substance = 31.8 − 12.5 = 19.3 mL

$$\text{Density} = \frac{\text{mass}}{\text{volume}} = \frac{52.11 \text{ g}}{19.3 \text{ mL}} = \underline{2.70 \text{ g/mL}}$$

ANALYSIS

The problem stated that the sample was a pure substance. With the information given to you in Table 3-1, can you determine the substance? (Yes. Table 3-1 indicates that aluminum has a density of 2.70 g/mL. Therefore, the sample is most likely aluminum.)

SYNTHESIS

The volume of a symmetrical, uniform solid, like a cube, is easy to measure if you know its dimensions. You just witnessed an example of how to measure the volume of an asymmetrical, or nonuniform, solid using water displacement. Can you measure the volume of all asymmetrical solids using this method? (No. Not if they float above the surface, like Styrofoam, or dissolve, like salt.)

EXAMPLE 3-2   Identifying a Substance from Density

A person was interested in purchasing a ring of pure gold having a mass of 89.9 g. Being wise, she wished to confirm that it was actually gold before she paid for it. With a quick test using a graduated cylinder like that shown in the previous example, she found that the ring had a volume of 7.96 mL. Was it made of gold?

PROCEDURE

By calculating the density of the ring, we can provide evidence as to its identity. Density is a constant and unchanging property of a pure element or compound (at a specific temperature). From the volume and the mass, the density can be calculated by dividing the mass by the volume. Compare this value to that of gold in Table 3-1.

SOLUTION

$$\text{density of the ring} = 89.9 \text{ g} /7.96 \text{ mL} = 11.3 \text{ g/mL}$$

ANALYSIS

Is the ring pure gold? Should she buy it? Comparing the result to the values in Table 3-1, we see that it's not pure gold. Pure gold would have a density of 19.3 g/mL

SYNTHESIS

What do you think the ring is made of? What might be going on here? On reexamining Table 3-1, you will discover that the ring has the same density as lead. It's probably a lead ring plated with a thin layer of gold to fool the unsuspecting. Lead is cheaper than gold, and so it would seem that the seller is trying to pull a fast one.

Density also has medical applications. The density of a person can be a measure of his or her amount of body fat. The density of fat is 0.900 g/mL, fat-free muscle is 1.066 g/mL, and normal bone is 3.317 g/mL. The more fat on a person, the lower the person's density and the less he or she will weigh when submerged in water. By consulting a chart, a specialist can give an accurate estimate of body fat from the person's density. For example, if a person has a density of 1.07 g/mL, the individual has about 12% body fat. If the density is down to 1.03 g/mL, the person has about 28% body fat.

3-2.2 Density as a Conversion Factor

Density is not only an important physical property but can also be used to convert the mass of a substance with a known density to an equivalent volume, or vice versa. These two conversions are illustrated in the following examples. Notice that since density originates from two measurements, it is not an exact factor such as the factor between meters and kilometers.

✔ MATH CHECK:

Making Conversions by Dimensional Analysis

In the following section, you will be asked to convert a mass of a substance to its volume (or vice versa) using density as a conversion factor. Although density problems can be solved algebraically by substitution, we will use dimensional analysis where the cancellation of the same unit leads us to the proper answer. This method will serve us well later on where substitution is not practical. The following questions test your ability to solve problems using this method.

a. There are exactly 42 gallons in one barrel and exactly four quarts in one gallon. A large Gulf oil leak in 2010 spewed out 1.5×10^6 quarts a day. Using dimensional analysis exclusively, convert this to number of barrels of oil.

b. In 2010, the exchange rate for currency was 12.8 pesos per U.S. dollar and 1.23 euros per dollar. In Cancun, Mexico, a silver bracelet sold for 116 pesos. Using dimensional analysis exclusively, convert pesos to the equivalent number of euros.

Answers: a. $1.5 \times 10^6 \, \text{qts} \times \dfrac{1 \, \text{gal}}{4 \, \text{qts}} \times \dfrac{1 \, \text{barrel}}{42 \, \text{gal}} = 6.8 \times 10^3 = \underline{8900 \text{ barrels/day}}$;

b. $116 \, \text{pesos} \times \dfrac{1\$}{12.8 \, \text{pesos}} \times \dfrac{1.23 \, \text{euros}}{\$} = \underline{11.1 \text{ euros}}$

Refer to Section 1-5 for help with solving problems by dimensional analysis.

EXAMPLE 3-3   Determining the Volume from the Mass Using Density

What is the volume in milliliters occupied by 485 g of table salt?

PROCEDURE

Use the density of table salt as a conversion factor from mass to volume.

1. The unit map for the conversion is

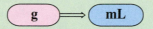

2. The density of table salt is given in Table 3-1 as 2.16 g/mL. We need to invert this relationship so that g is in the denominator and mL is in the numerator. This factor then converts mass to the equivalent volume. The factor for this conversion is

$$\dfrac{1 \, \text{mL}}{2.16 \, \text{g}}$$

SOLUTION

$$485 \, \text{g} \times \dfrac{1 \, \text{mL}}{2.16 \, \text{g}} = \underline{225 \text{ mL}}$$

ANALYSIS

To check your answer, you can plug it back into the density formula.

$$\text{density} = \text{mass} \div \text{volume}$$
$$\text{density} = 485 \, \text{g} \div 225 \, \text{mL} = 2.16 \, \text{g/ml}$$

This is the original reported density. The answer checks out.

SYNTHESIS

Does this answer make sense? If 1 mL has a mass of about 2 g, there would be about 240 mL (i.e., 485 g ÷ 2 g/mL = 242 mL) in the given sample. 225 mL is very close to the estimated value. The answer makes sense. Could you measure the volume by adding the salt to water as in the previous examples? No. Salt dissolves in water, so the use of density as a conversion factor is appropriate.

EXAMPLE 3-4   Determining the Mass from the Volume Using Density

What is the mass in grams of 1.52 L of kerosene?

PROCEDURE

Using the density of kerosene from Table 3-1 as a conversion factor, convert volume to mass. Notice that the volume is not given in milliliters, however. A two-step conversion is necessary.

1. In the first step (a), L is converted into mL, and in the second step (b), mL is converted into g. The unit map for the conversion is

2. The proper relationships are 1 mL = 10^{-3} L (from Table 1-4) and 0.82 g/mL (from Table 3-1). Expressed as proper conversion factors, the relationships are

$$\text{(a) } \frac{1 \text{ mL}}{10^{-3} \text{ L}} \quad \text{and} \quad \text{(b) } \frac{0.82 \text{ g}}{\text{mL}}$$

SOLUTION

$$1.52 \ \cancel{\text{L}} \times \frac{1 \ \cancel{\text{mL}}}{10^{-3} \cancel{\text{L}}} \times \frac{0.82 \text{ g}}{\cancel{\text{mL}}} = \underline{1.2 \times 10^3 \text{ g}}$$

ANALYSIS

The same value for the density can also be expressed as kg/L as well as g/mL (i.e., 0.82 kg/L and 0.82 g/mL). If the density is a little less than 1 kg per liter, we would expect the mass of 1 L to be around 1 kg, which it is. What does the unit map for the conversion using kg/L look like?

The necessary conversions are 0.82 kg/L and 1000 g/kg.

SYNTHESIS

When solving a problem involving conversion of units, is there a single correct way to work it out? Clearly, there isn't. We've supplied two different pathways for this single problem. Interestingly, though the pathways differ, the exact same numbers end up being manipulated. We multiply by 0.82 and then either multiply by 10^3 or divide by 10^{-3}, which works out the same.

3-2.3 Specific Gravity

In place of density, certain applications, especially in the medical field, use the term *specific gravity*. **Specific gravity** *is the ratio of the mass of a substance to the mass of an equal volume of water under the same conditions.* Since the mass 1 mL of water is 1.00 g, specific gravity has the same value as density, only expressed without units. For example, the density of aluminum is 2.70 g/mL, so its specific gravity is simply 2.70.

▶ **ASSESSING THE OBJECTIVE FOR SECTION 3-2**

EXERCISE 3-2(a) LEVEL 1: Refer to Table 3-1. If you were given exactly 100 grams of the following compounds, list them in order from smallest to largest volume: water, ethyl alcohol, carbon tetrachloride, gasoline, kerosene.

EXERCISE 3-2(b) LEVEL 2: What is the volume in milliliters occupied by 285 g of iron? The density of iron is 7.87 g/mL.

EXERCISE 3-2(c) LEVEL 2: The volume of water in a graduated cylinder measures 21.95 mL. After addition of a 53.5-g sample of cadmium metal, the volume of the water and metal reads 32.72 mL. What is the density of the metal?

EXERCISE 3-2(d) LEVEL 2: A sample of a given pure liquid has a mass of 254 g and a volume of 159 mL. What might the liquid be? Refer to Table 3-1. Does this liquid float or sink when mixed in a beaker with water?

EXERCISE 3-2(e) LEVEL 3: Antifreeze is made up of equal portions of ethylene glycol and water. Ethylene glycol is denser than water, but the two liquids mix together. How can density be used to determine the quality of the antifreeze in a radiator?

For additional practice, work chapter problems 3-12, 3-17, and 3-21.

MAKING IT REAL

Forensic Chemistry: Identifying a Glass Shard from a Crime Scene by Density

A woman is struck down by a hit-and-run driver, but there are glass fragments from the car in the street. A burglar breaks through a glass display window but doesn't realize that some small glass shards stick to his clothes. Identification of the origin of a tiny piece of glass by matching it to a specific automobile or to a certain plate-glass window provides strong forensic evidence that can help connect the car or a suspect to a crime scene.

Glass may all look the same, but it is not. Even glass from different glass windows in the same building may have very small variations in properties depending on the type, the manufacturer, or even the time when it was made. However, these small differences can be used effectively by the crime scene investigator. Glass fragments can be matched with other glass by the use of density and refractive index. When two pieces of glass have the same or very close values for these two measurements, it can be assumed that they have the same origin. We will see how density is used at the crime scene in this discussion. Refractive index relates to the nature of light, so it will be discussed at the appropriate time in Chapter 5.

The density of a piece of glass can be determined by its buoyancy. Two liquids are used for the determination of glass density, bromoform ($CHBr_3$, density 2.8899 g/mL) and bromobenzene (C_6H_5Br, density 1.4950 g/mL). Window glass has a density range of 2.47 to 2.54 g/mL. Thus, all glass sinks in pure bromobenzene but floats in pure bromoform. By mixing the two liquids, however, we can produce a mixture that has a precise range of densities intermediate between the two pure liquids. By adding the less dense liquid to the more dense, the mixture of the two liquids gradually becomes less dense. Eventually, a mixture is produced whereby the glass shard is no longer buoyant and becomes suspended in the liquid. At that point, the density of the glass is the same as that of the liquid. The density is then determined from a table that relates the density of the mixture to the measured proportion of the two liquids. Under ideal conditions, this procedure has a precision of around ±0.0001 g/mL. It is a very simple laboratory procedure but can provide valuable information. If two fragments of glass have the same density, they most likely originated from the same source. Refractive index, which is described later, should confirm this evidence.

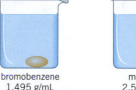

| bromoform 2.890 g/mL (glass shard floats) | bromobenzene 1.495 g/mL (glass shard sinks) | mixture 2.51 g/mL (glass shard suspended) |

Refer to the Student Workshop at the end of the chapter.

3-3 The Properties of Mixtures

▶ OBJECTIVE
FOR SECTION 3-3
Perform calculations involving the percent of a pure substance in a mixture.

LOOKING AHEAD! It is not often in the world around us that we encounter an element or compound in such a highly concentrated state that we may judge it as "pure." Even the water we drink contains other compounds (e.g., salt) and elements (e.g., oxygen). How we describe mixtures is the subject of this discussion. ■

3-3.1 Heterogeneous Mixtures

You probably know that you can't mix oil and water. Oil and water form a heterogeneous mixture. *A* **heterogeneous mixture** *is a nonuniform mixture containing two or more phases with definite boundaries or interfaces between the phases. A* **phase** *is one physical state (solid, liquid, or gas) with distinct boundaries and uniform properties.*

Besides oil and water (two liquid phases), an obvious heterogeneous mixture is a handful of soil from the backyard. If we look closely, we see bits of sand, some black matter, and perhaps pieces of vegetation. One can easily detect several solid phases with the naked eye. Other examples of heterogeneous mixtures are carbonated beverages (liquid and gas) and muddy water (liquid and solid). Heterogeneous mixtures can often be separated into their components by simple laboratory procedures. For example, suspended solid matter can be removed from water by *filtration*. (See Figure 3-4.) When the turbid water is passed through the filter, the suspended matter remains on the filter paper and the clear liquid phase passes through. In the purification of water for drinking purposes, the first step is the removal of suspended particulate matter.

Sometimes heterogeneous mixtures cannot be detected with the naked eye. For example, creamy salad dressing and smoke both appear uniform at first glance. However, if we were to magnify each, it becomes apparent that these are actually heterogeneous mixtures. The salad dressing has little droplets of oil suspended in the vinegar (two liquid phases), and the smoke has tiny solid and liquid particles suspended in the air (solid, liquid, and gas phases).

FIGURE 3-4 **Filtration** A heterogeneous mixture of a solid and liquid can be separated by filtration.

3-3.2 Homogeneous Mixtures and Solutions

Oil and water don't mix, but when you pour alcohol into water, both liquids disperse into each other and no boundary between the two liquids is apparent. *A* **homogeneous mixture** *is the same throughout and contains only one phase.* In Figure 3-5, a heterogeneous mixture, on the left, is compared to a homogeneous mixture, on the right. Notice on the left that the oil floats on top of the water because the density of oil (≈ 0.90 g/mL) is less than that of water (1.00 g/mL).

In heterogeneous mixtures, portions of each component are large enough to be detected, although some magnification may be necessary. In homogeneous mixtures, the components disperse uniformly into each other. As mentioned earlier, matter is composed of fundamental particles. In a typical homogeneous mixture, the mixing extends all the way to the molecular or ionic level. Thus, there is no detectable boundary between components. No amount of magnification would reveal pieces of solid salt when it is dissolved in the water. When table salt is added to water, it forms a solution. *A* **solution** *usually refers to homogeneous mixtures with one liquid phase.* As such, one of the two components must be dissolved in the other. Saying that something is *dissolved* means that *its structure has been broken down to the molecular or ionic level.* Thus components of a solution cannot be separated by filtration. However, the two components can be separated by a laboratory procedure called *distillation.* (See Figure 3-6.) In the distillation of a solution containing a dissolved red substance, the water is boiled away from the solution and then retrieved by condensation through a water-cooled tube. When all the water has boiled away, the solid substance remains behind in the distilling flask.

(a)   (b)   (c)   (d)

FIGURE 3-5 **Mixtures** Water and oil (a, b) form a heterogeneous mixture with two liquid phases. Water and an alcohol (c) form a homogeneous mixture with one liquid phase (d).

A glass of salt water and a glass of pure water look exactly the same. They do taste different, however. In fact, *since both solutions and pure substances (elements and compounds) are homogeneous matter, one must examine the physical properties to distinguish between the two.* Mixtures have properties that vary with the proportion of the components. Elements and compounds have definite and unchanging properties. A simple example of a variable property is the taste and color of a cup of coffee. The more coffee that is dissolved in the water, the stronger the taste and the darker the solution.

Now consider the properties of two compounds alone: table salt and water. Solid table salt (sodium chloride) melts at 801°C and water ice melts at 0°C. A solution of salt in water begins to freeze anywhere from −18°C to just under 0°C, depending on the amount of salt dissolved. Also, a particular saltwater solution does not have a sharp, unchanging boiling point or freezing point, as does pure water.

Density is another physical property that is different for a solution compared to the pure liquid component. For example, battery acid is a solution of a compound, sulfuric acid, in water. Its density is greater than that of pure water. The more sulfuric acid present, the denser is the solution. In a fully charged battery, the density is about 1.30 g/mL; if it is mostly discharged, the density is about 1.15 g/mL.

The classification of matter from the most complex, a sample of heterogeneous matter (at the upper left), to the most basic form of homogeneous matter, a pure element (at the lower left), is illustrated in Figure 3-7.

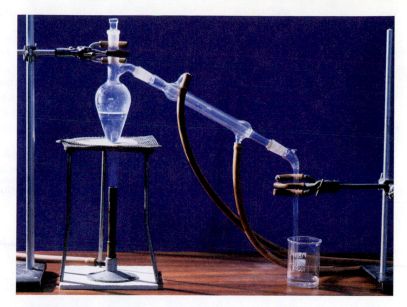

FIGURE 3-6 Distillation A homogeneous mixture of a solid in a liquid or two liquids can be separated by distillation.

3-3.3 Percent Composition of Solutions

On the label of a typical bottle of wine, we see that it is composed of 13% alcohol. This certainly gives us an idea of how potent the wine may be. A solution is composed of the **solvent**, *which is the dissolving medium* (in this case water), and the **solute**, *the substance dissolved in the solvent* (in this case alcohol). In the case of wine, there are several different solutes present besides alcohol. *The* **concentration** *of the solution refers to how much of a specific solute is present in a specific amount of solvent or solution.*

There are several ways that we can express concentration of solutions. Each of these has a particular application. We will delay discussion of the various concentration units until Chapter 12 where we can provide appropriate applications. One unit that we will cover at this time is useful in many laboratory situations. This unit expresses concentration as percent by mass. **Percent by mass** *is the ratio of the mass of the solute to the mass of solution multiplied by 100 to express it as percent.*

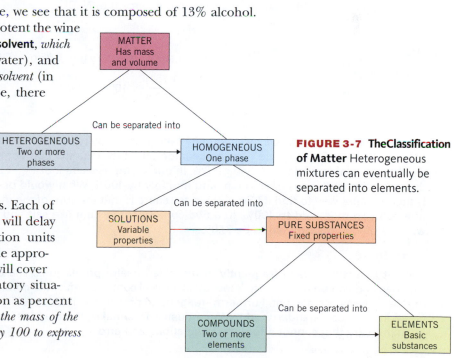

FIGURE 3-7 The Classification of Matter Heterogeneous mixtures can eventually be separated into elements.

✔

MATH CHECK:

Using Percent as a Conversion Factor

In Example 2-2, we used percent as a decimal fraction to convert from the total mass to the component mass. We can also solve problems involving percent by dimensional analysis. Recalling that *percent* simply means "parts per 100," we can also set up percent as a conversion factor. Use percent as a conversion factor to solve the following two problems.

a. Gasoline is produced from crude oil in an oil refinery. If 42% of the oil can be converted into gasoline, how many barrels of gasoline can be obtained from 195 barrels of oil?

b. How many barrels of oil are needed to produce 885 L of gasoline? One barrel contains 159 L.

Answers: a. $195 \text{ bbls oil} \times \dfrac{42 \text{ bbls gasoline}}{100 \text{ bbls oil}} = 82 \text{ barrels gasoline};$

b. $885 \text{ L gasoline} \times \dfrac{1 \text{ bbl gasoline}}{159 \text{ L gasoline}} \times \dfrac{100 \text{ bbl oil}}{42 \text{ bbl gasoline}} = 13 \text{ bbl oil}$

Refer to Appendix A-6 for review of decimal fractions and percent as a conversion factor.

EXAMPLE 3-5   Calculating Mass Percent

What is the percent by mass of a solution of NaCl if 1.75 g of NaCl is dissolved in 5.85 g of H_2O?

PROCEDURE

Find the total mass of the solution and then the percent of NaCl.

SOLUTION

$$\text{Total mass of solution} = 1.75 \text{ g} + 5.85 \text{ g} = 7.60 \text{ g}$$

$$\frac{1.75 \text{ g}}{7.60 \text{ g}} \times 100\% = \underline{23.0\%}$$

ANALYSIS

In general, percent is always calculated by taking what you're looking for, dividing by the total, and then multiplying by 100%. To calculate your percentage on a test, take the number correct, divide by the total number of points available, and multiply by 100%. In our example, we're looking at the % salt. So take the mass of salt, divide by the total mass of solution, and multiply by 100%. What would be the percent by mass of the water? Is there a quick way to find it? Something to realize is that the sum of the mass percents of all the components of a solution must add to 100%. In a two-component system like salt and water, the percent by mass of the water is quickly determined to be 77.0%.

SYNTHESIS

Percent by mass is used frequently in the allied health professions. The calculations involved are relatively friendly, and you do not need to know the chemical composition of the components involved. Simply determining the masses is sufficient. Furthermore, general practices call for specific concentrations of solutions, whose creation has been standardized by the industry. This makes the creation of these solutions a bit more failsafe in an industry that cannot afford miscalculations and errors.

EXAMPLE 3-6 Calculating Amount of Solution from Mass Percent

A solution is 14.0% by mass H_2SO_4. What mass of solution contains 365 g of H_2SO_4?

PROCEDURE

Find the mass of the solution by multiplying the mass of compound by the percent in the proper fraction form. In this case, the percent factor should be inverted.

SOLUTION

$$365 \text{ g } H_2SO_4 \times \frac{100 \text{ g solution}}{14.0 \text{ g } H_2SO_4} = \underline{2610 \text{ g solution}}$$

ANALYSIS

Any problem using mass percent will use one of three conversion factors. When we say that a solution is 14.0% solute, we can use any of these relationships: 14.0 g solute/100 g solution, 14.0 g solute/86.0 g solvent, or 86.0 g solvent/100 g solution. Which to use depends on the question that is being asked. Look for references to the amount of solute, solvent, or solution in the problem to find the appropriate conversion factor.

SYNTHESIS

Along with percent by mass, which is often designated (w/w) or (weight per weight) to avoid confusion, we can create a short list of concentration calculations that all depend on measuring the amount of solute and solvent by either mass or volume. For example, percent of mass per unit volume, or (w/v), is used frequently in measurements of human bodily fluids. The concentration of solutes in blood is often expressed in terms of w/v percent. For example, a 0.89% (w/v) solution of a saline solution (NaCl) has the same concentration of ions as human blood. Percent by volume (v/v) is useful when both solute and solvent are liquids, such as the concentration of alcohol in alcoholic beverages. A wine that lists its alcohol content at 11.5% means that there are 11.5 mL of pure ethyl alcohol present in each 100 mL of solution. In all cases, knowing the chemical formulas of the components is unnecessary, since the units of mass and volume can be measured directly. This has the advantage of easy preparation of solutions. We will revisit these units of concentration in Chapter 12.

▶ ASSESSING THE OBJECTIVE FOR SECTION 3-3

EXERCISE 3-3(a) LEVEL 1: Is the listed property a description of a heterogeneous or homogenous mixture?

(a) has distinct boundaries   (c) contains only one phase

(b) visibly shows several components   (d) has a single concentration throughout

EXERCISE 3-3(b) LEVEL 2: Carbon tetrachloride and water are two liquids that do not mix homogeneously with one another. What type of mixture will they form? Describe the appearance of the mixture.

EXERCISE 3-3(c) LEVEL 2: In a fishbowl, there appears to be only water. A fish placed in there, however, can retrieve oxygen from the water with its gills. What type of mixture do the water and oxygen form?

EXERCISE 3-3(d) LEVEL 3: The separation technique you use is dependent on the type of mixture you have. Could you use distillation to separate the heterogeneous mixture of water and sand? Could you use filtration to separate the homogeneous mixture of water and salt?

EXERCISE 3-3(e) LEVEL 3: An aqueous saline solution is 11.2% NaCl. What mass of water is needed to make 500 g of solution?

EXERCISE 3-3(f) LEVEL 3: The gold in a ring is actually a homogeneous mixture of metals known as an *alloy*. For example, 14-K gold is 58.0% gold. What is the weight of pure gold in 4.00 oz of 14-K gold?

EXERCISE 3-3(g) LEVEL 3: What mass of 14-K gold can be made from 128 g of pure gold?

For additional practice, work chapter problems 3-41, 3-48, 3-50, and 3-54.

PART A SUMMARY

KEY TERMS

3-1   A **physical property** can be observed or measured without changing the substance. p. 86

3-1.1   **Physical state** describes whether the substance is a **solid**, **liquid**, or **gas** at a specific temperature. p. 86

3-1.2   **Melting** and **freezing point** is the temperature at which a substance changes between the liquid and solid states. p. 87

3-1.2   **Boiling** and **condensation** refer to the changes between the gaseous and liquid states. The **boiling point** of a pure liquid is a constant at a specific pressure. p. 87

3-1.2   A **physical change** involves a change in phase or dimensions. p. 88

3-1.3   An **extensive property** depends on the amount present, while an **intensive property** is independent of the amount. p. 88

3-1.4   A **chemical property** refers to the tendency to undergo a chemical change from one substance into another substance. p. 88

3-1.5   Chemical reactions illustrate the **law of conservation of mass**. p. 88

3-2   **Density** is the mass per unit volume. It is an intensive physical property that can be used to help identify a substance and convert between mass and volume. p.90

3-2.3   **Specific gravity** is used in certain medical applications and has the same numerical value as density. p. 94

3-3.1   **Heterogeneous mixtures** exhibit more than one identifiable **phase**. p. 95

3-3.2   **Homogeneous mixtures** and **solutions** involve an intimate mixture with one phase. p. 96

3-3.3   Solutions contain a **solute** dissolved in a **solvent**. The **concentration** of solute can be measured by **percent by mass**. p. 100

SUMMARY CHART

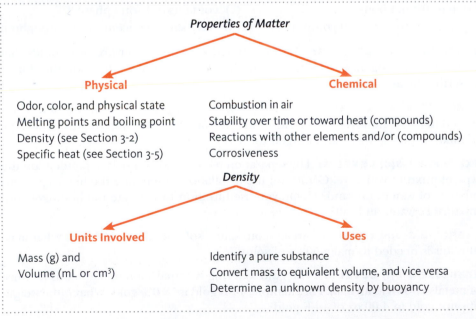

Properties of Matter

Physical

Odor, color, and physical state
Melting points and boiling point
Density (see Section 3-2)
Specific heat (see Section 3-5)

Chemical

Combustion in air
Stability over time or toward heat (compounds)
Reactions with other elements and/or (compounds)
Corrosiveness

Density

Units Involved

Mass (g) and
Volume (mL or cm³)

Uses

Identify a pure substance
Convert mass to equivalent volume, and vice versa
Determine an unknown density by buoyancy

Part B

The Properties of Energy

SETTING A GOAL

■ You will be able to qualitatively and quantitatively describe processes in terms of the forms and types of energy associated with them.

OBJECTIVES

3-4 Define terms associated with energy exchanges in physical and chemical processes.

3-5 Perform calculations involving the specific heat of a substance.

3-4 The Forms and Types of Energy

LOOKING AHEAD! We bask in the heat from the sun, but energy can take other forms as well. How we describe and measure heat energy is the focus of this section. ■

▶ **OBJECTIVE FOR SECTION 3-4**
Define terms associated with energy exchanges in physical and chemical processes.

When we don't feel like we have much energy, we sure don't feel like working. Actually, that is the definition of energy. **Energy** *is the capacity or the ability to do work.* Work involves the transfer of energy when an object is moved a certain distance.

3-4.1 Forms of Energy

Just as matter has more than one physical state, energy has more than one form. Most of the energy on Earth originates from the sun. Deep in the interior of our ordinary star, transformations of elements occur that liberate a form of energy called *nuclear energy.* Some of this energy, however, changes in the sun into *light* or *radiant energy* that then travels through space to illuminate Earth. Light energy from the sun bathes our planet and shines on the surface vegetation, where some of it is converted into *chemical energy* by a process called photosynthesis. Chemical energy is stored in the energy-rich compounds that make up the bulk of the vegetation. When logs from a tree are burned, the chemical energy is released in the form of *heat energy.* In a similar process, the metabolism of food in our bodies releases energy to keep us alive. In the burning and metabolism processes energy-poor compounds are produced and recycle to the environment.

In the production of electrical power, the heat energy is used to produce steam that turns a turbine. The movement of the turbine is *mechanical energy.* The mechanical energy powers a generator that converts the mechanical energy into *electrical energy.* Other conversions between energy forms are possible. For example, in the chemical change that occurs in a car battery, chemical energy is converted directly into electrical energy. Solar cells directly convert light energy into electrical energy. (See Figure 3-8.) Energy changes are subject to the same law as matter changes in chemical reactions.

The **law of conservation of energy** *states that energy cannot be created or destroyed but only transformed from one form to another.* In the production of electrical energy in a fossil fuel plant, only about 35% of the chemical energy is eventually transformed into electrical energy. The rest of the energy is lost as heat energy in the various transformations. However, the total energy remains constant.

FIGURE 3-8 Solar Energy
These solar cells directly convert light energy into electrical energy.

3-4.2 Exothermic and Endothermic Changes

Chemical or physical changes may be accompanied by either the release or the absorption of heat energy. *When a change releases heat, it is said to be* **exothermic**. *When a change absorbs heat, it is said to be* **endothermic**. Combustion (burning) is a common example of an exothermic chemical reaction. An "instant cold pack" is an example of an endothermic process. When the compounds ammonium nitrate (a solid) and water are brought together in a plastic bag, a solution is formed. The endothermic solution process causes enough cooling to make an ice pack useful for treating sprains and minor aches. (See Figure 3-9.) The advantage of this is that cold can be produced without the need for a refrigerator. Melting ice is a physical change but is also an endothermic process. That is, to melt ice cubes, we supply heat by taking a tray of cubes out of the freezer and letting it sit in the warmer room. Boiling water is another example of an endothermic process: when water absorbs enough heat from the stove, it boils.

FIGURE 3-9 Instant Cold Pack When capsules of ammonium nitrate are broken and mixed with water, a cooling effect results.

3-4.3 Kinetic and Potential Energy

In addition to the *forms* of energy, there are two *types* of energy. These depend on whether the energy is available but not being used or is actually in use. **Kinetic energy** *is the energy produced from motion.* A moving baseball, a speeding train, and water flowing down a spillway from a dam (see Figure 3-10) all have kinetic energy. **Potential energy** *is energy that is available because of position or composition.* For example, a weight suspended above the ground has energy available because of its position and the attraction of gravity for the weight. Water stored behind a dam (Figure 3-10), a compressed spring, and a stretched rubber band all have potential energy. The chemical energy stored in the compounds of a tree log is also classified as potential energy.

▶ **ASSESSING THE OBJECTIVE FOR SECTION 3-4**

EXERCISE 3-4(a) LEVEL 1: Identify the following as endothermic or exothermic.
(a) sweat evaporating   **(c)** wax melting   **(e)** a lake icing over
(b) natural gas combusting   **(d)** leaves decaying

EXERCISE 3-4(b) LEVEL 1: Identify the principal type of energy, kinetic or potential, exhibited by each of the following.
(a) a car parked on a hill   **(d)** an uncoiling spring in an alarm clock
(b) a train traveling at 60 mph   **(e)** a falling brick
(c) chemical energy of a nutrition bar

EXERCISE 3-4(c) LEVEL 3: Sunlight causes corn to grow. A substance in corn can be converted to ethanol. The ethanol is used as automobile fuel. What forms of energy are involved in each of these changes? Is a particular form of energy potential or kinetic energy?

For additional practice, work chapter problems 3-56, 3-59, and 3-62.

FIGURE 3-10 Potential and Kinetic Energy Water stored behind a dam has potential energy. The water flowing over the spillway has kinetic energy.

▶ **OBJECTIVE FOR SECTION 3-5**
Perform calculations involving the specific heat of a substance.

3-5 Energy Measurement and Specific Heat

LOOKING AHEAD! Heating a substance may cause it to melt or boil. If adding heat does not cause a physical or chemical change, it just raises the temperature. However, each compound or element is affected to a different extent by the same amount of heat. We will now consider how heat affects the temperature of a given substance. ∎

The most obvious thing that happens when heat is applied to a pan of water is that the temperature of the pan and water increases. The amount of temperature change for a specific mass of substance is an intensive physical property known as specific heat capacity, or simply specific heat. **Specific heat** *is defined as the amount of heat required to raise the temperature of 1 gram of a substance 1 degree Celsius (or Kelvin).*

MATH CHECK:

Temperature Scales

If you are accustomed to the Fahrenheit scale to express temperature, you also need to be familiar with the Celsius scale. The Celsius and the Kelvin scales are used exclusively in science (and most other countries except the United States). Answer the following questions.

  a. What are the freezing point and the boiling point of water on the Fahrenheit, Celsius, and Kelvin scales?

  b. Express 50°C on the Fahrenheit scale and 50°F on the Celsius and Kelvin scales.

Answers: a. freezing point – 32°F, 0°C, 273 K; boiling point – 212°F, 100°C, 373 K
  b. 122°F, 10°C, and 283 K

If you are not familiar with the Celsius and Kelvin scales and how to convert between them, refer to Section 1-6.

3-5.1 Units of Heat Energy

Units of heat energy are based on the specific heat of water. *A* **calorie** *is the amount of heat energy required to raise the temperature of 1 gram of water from 14.5°C to 15.5°C.* The unit of heat energy most often used in chemistry is the SI unit called the **joule**. *The calorie is now defined in terms of the joule.*

$$1 \text{ cal} = 4.184 \text{ joule}(J) \text{ (exactly)}$$

The definition of the calorie provides us with the specific heat of water in joules:

$$1.000 \text{ cal}/(g \cdot °C) = 4.184 \text{ J}/(g \cdot °C)$$

Notice that there are approximately 4 joules per calorie. The units read: calories (or joules) per gram per degree Celsius. The degree Celsius unit represents a *temperature change*, not a specific temperature reading. (Change is represented by the Greek letter Δ, pronounced *delta*.) Thus, a change in temperature is represented as ΔT. The Kelvin scale can also be used in place of the Celsius scale since the Kelvin and Celsius degrees are the same. The formula used to calculate the specific heat is

$$\text{Specific heat} = \frac{\text{heat energy (J)}}{\text{mass (g)} \times \Delta T \text{ (°C)}}$$

The specific heats of several pure substances are listed in Table 3-2.

Examples 3.7 and 3.8 illustrate the calculation of specific heat and the use of specific heat to calculate temperature change.

TABLE 3-2		
Specific Heats		
SUBSTANCE	SPECIFIC HEAT [cal/(g · °C)]	SPECIFIC HEAT [J/(g · °C)]
Water	1.000	4.184
Ice	0.492	2.06
Aluminum (Al)	0.214	0.895
Gold (Au)	0.031	0.129
Copper (Cu)	0.092	0.385
Zinc (Zn)	0.093	0.388
Iron (Fe)	0.106	0.444

EXAMPLE 3-7 **Determining Specific Heat**

It takes 628 J to raise the temperature of a 125-g quantity of silver from 25.00°C to 32.14°C. What is the specific heat of silver in joules?

PROCEDURE

Determine a specific heat from the heat required to change a given mass and a given temperature change.
(a) Calculate the actual temperature change $[\Delta T = T(\text{final}) - T(\text{initial})]$.
(b) The specific heat is calculated by substituting the appropriate quantities in the formula.

SOLUTION

$\Delta T = 32.14°C - 25.00°C = 7.14°C$

$$\text{Specific Heat} = \frac{628 \text{ J}}{125 \text{ g} \times 7.14°C} = 0.704 \frac{\text{J}}{\text{g} \cdot °C}$$

ANALYSIS

What would this value be in units of cal/g · °C? Use the relationship that 1 cal = 4.184 J. Then

$$0.704 \frac{\cancel{J}}{\text{g} \cdot °C} \times \frac{1 \text{ cal}}{4.184 \cancel{J}} = 0.168 \frac{\text{cal}}{\text{g} \cdot °C}$$

SYNTHESIS

From an evaluation of Table 3-2, what seems to be the relationship between the atomic mass of a metal and its specific heat? The higher the atomic mass, the lower the specific heat of an element. That is referred to as an *inverse relationship*. It results from there being fewer atoms in a gram of a heavier metal compared to a lighter metal. Therefore, there are fewer particles to absorb energy.

EXAMPLE 3-8 **Determining Temperature Change Using Specific Heat**

If 1.22 kJ of heat is added to 50.0 g of water at 25.0°C, what is the final temperature of the water? [The specific heat of water is 4.184 J/(g · °C).]

PROCEDURE

Calculate how much the temperature increases (from 25°C) by addition of a specific amount of heat to a specific mass of water.

(a) Convert kJ to J using the relationship 1 kJ = 10^3 J.
(b) Solve the specific heat formula for ΔT and substitute in the known values.

$$\text{Specific heat} = \frac{\text{heat energy (J)}}{\text{Mass (g)} \times \Delta T (°C)}$$

$$\Delta T = \frac{\text{heat energy (J)}}{\text{Mass (g)} \times \text{specific heat}}$$

(c) Find the final temperature of the water by adding the temperature change to the original temperature.

SOLUTION

(a) $1.22 \text{ kJ} \times \dfrac{10^3 \text{J}}{\text{kJ}} = 1.22 \times 10^3 \text{J}$ **(b)** $\Delta T = \dfrac{1.22 \times 10^3 \text{J}}{50.0 \cancel{g} \times 4.184 \dfrac{\cancel{J}}{\cancel{g} \cdot °C}} = 5.83°C$

The temperature change of the water is 5.83°C. Make sure you read on.

ANALYSIS

Is 5.83°C the answer to the question "what is the final temperature"? It can't be. Heat was added, and yet the answer is less than the original temperature of 25.0°C. The 5.83 is the *change* in temperature, and since heat was added, it makes sense that the new temperature is 5.83 degrees *higher* than the initial temperature.
(c) $T_{\text{final}}(°C) = 25.0 + 5.83 = \underline{30.8°C}$

SYNTHESIS

Is our answer reasonable now? Adding heat to the water will cause the temperature to go up. Notice from the specific heat that about 4 joules raises 1 gram of water 1 degree, so it would take about 200 J (50×4) to raise 50 g of water 1 degree. 1220 J divided by 200 J indicates a temperature rise of about 6 degrees. The answer is reasonable.

You may have noticed how fast an iron skillet heats up compared to an equivalent amount of water. It takes longer to heat water because it has a comparatively high specific heat. Notice in Table 3-2 that the specific heat of water is almost 10 times higher than that of the iron in the skillet. Thus, in a calculation similar to the one in Example 3-8b, the same amount of heat will raise the temperature of the iron almost 10°C for every 1°C for the same weight of water. (Iron also conducts heat rapidly from the fire to the handle.)

3-5.2 The Nutritional Calorie and Heat Exchange

The amount of energy that we obtain from food can also be measured and expressed in calories. The nutritional calorie is actually one *kilocalorie* (10^3 cal) as defined above. To distinguish between the two calories, the *c* in *calorie* is capitalized in the nutritional Calorie (1 Cal).

The Calorie content of a portion of food is measured by burning the dried food. The heat released by the combustion process is then used to heat a known amount of water at a known original temperature. The principle of heat exchange states that *heat lost equals heat gained*, assuming no heat is lost to the surroundings. Heat exchange is a practical example of the conservation of energy. In this case, the water gains the heat lost by the portion of food. This is illustrated in Example 3-9.

The application of the principle of heat exchange can also be used to measure the specific heat of a substance. When two substances at different temperatures are mixed, the hotter item will lose heat energy and the cooler one will gain heat energy. Thus, the temperature of the hotter item comes down as the temperature of the other increases. Eventually, the two substances come to the same temperature, which is somewhere in between the two original temperatures. In the following example, we have a known mass of water and a metal with an unknown specific heat. The metal starts at a high temperature and the water at a lower temperature. When the two are mixed, the temperature settles between the two original temperatures.

Nutrition Facts

Serving Size 1 piece (34 g)
Servings Per Container 6

Amount Per Serving	
Calories 180	Calories from Fat 90

% Daily Value*

Total Fat 10 g	15%
Saturated Fat 3.5 g	18%
Cholesterol 0 mg	0%
Sodium 130 mg	5%
Total Carbohydrate 19 g	6%
Dietary Fiber 1 g	4%
Sugars 16 g	
Protein 4 g	

Vitamin A 0%	•	Vitamin C 0%
Calcium 0%		

Many people are very conscious of the energy content (Calories) of the food they eat.

EXAMPLE 3-9 **Counting Calories**

A piece of cake is dried and burned so that all the heat energy released heats some water. If 3.15 L of water is heated a total of 75.0°C, how many Calories does the cake contain?

PROCEDURE

Use the equation for specific heat to solve for the amount of heat in calories. The heat required to heat the specific amount of water is the heat content of the piece of cake.

(a) Determine the mass. First, convert the volume of water in liters to milliliters using 1 L = 10^3 mL. Next, convert the volume of water in milliliters to the mass of water using the density, which is 1.00 g/mL.

(b) Use the formula for specific heat (in cal) to solve for the amount of heat energy.

$$\text{amount of heat energy} = \text{mass} \times \Delta T \times \text{specific heat}$$

(c) Convert calories to Calories using the relationship 1 Cal = 10^3 cal.

SOLUTION

(a) $3.15 \text{ L} \times \dfrac{10^3 \text{ mL}}{\text{L}} \times \dfrac{1.00 \text{ g}}{\text{mL}} = 3.15 \times 10^3 \text{ g}$

(c) $236 \times 10^3 \text{ cal} \times \dfrac{1 \text{ Cal}}{10^3 \text{ cal}} = \underline{236 \text{ Cal}}$

(b) $3.15 \times 10^3 \text{ g} \times 75.0°C \times \dfrac{1.00 \text{ cal}}{\text{g} \cdot °C} = 236 \times 10^3 \text{ cal}$

ANALYSIS

We could have saved a step if we had noted that specific heat could be expressed as 1.00 kcal (Cal)/kg (water) $\times \Delta T$. If we use those units, what does the second calculation look like? 315 kg $\times$ 75°C $\cdot$ 1.00 Cal/kg $\cdot$ °C = 236 Cal.

SYNTHESIS

Does this answer make sense? If we were interested in weight control, we'd have to exercise off that amount of energy. Typical exercise machines like stair climbers, rowing machines, and bikes that allow you to count calories can burn about 500 Calories each hour, depending on the activity. Is eating the cake worth it? 236 Calories for a piece of cake is a very reasonable answer. At 500 Calories each hour, it would take half that time to burn 236 Calories, or about half an hour of constant exercise. Only you can decide whether the enjoyment of a slice of cake is worth that amount of exertion.

MAKING IT REAL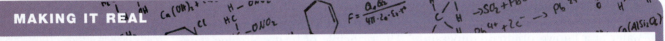

Body Solutions—Lose Weight (actually money) While You Sleep

There are two ways to lose weight. Either reduce the total amount of Calories that you consume to a level lower than you burn, or increase your activity so that the amount of Calories you burn rises above the amount you consume. Ideally, someone interested in weight loss would do both at the same time. Back in 1999, however, there appeared to be a third option that was receiving a lot of attention.

A company called Mark Nutritionals Inc. began marketing a product called Body Solutions Evening Weight Loss Formula. Its claim was that by using its product, you could lose weight while you slept. By taking a teaspoon of the product with a glass of water before you went to bed and by refraining from eating for three hours before going to bed, you would begin to lose weight. The marketing strategy was to have well-known and respected DJs in major markets use and promote the product. The airwaves were busy with testimonials touting the product's benefits. Many people who tried the product initially experienced weight loss, which they were happy to talk about,

adding to the excitement about the product. By May 2002, Body Solutions was becoming available in reputable drugstores throughout the country.

By October, though, the Federal Trade Commission (FTC) was actively pursuing Mark Nutritionals Inc. for advertising fraud. The company had paid the radio personalities to make claims without merit and given them a percentage of the profits that their market generated. Scientific studies found no evidence supporting Body Solutions as a weight-loss supplement, and there were even reports of adverse health effects from some of Mark Nutritionals' supplementary products. In the settlement of the case, Body Solutions was put out of business. Why, then, did so many people experience an initial weight loss?

If it is part of your normal lifestyle to eat throughout the evening, but instead you stopped eating for three hours before you went to bed, it is possible that you would eat 400 or 500 fewer Calories each day than you otherwise would. A pound of fat in the body stores 3500 Calories, so by refraining from eating for three hours before going to bed, and not consuming more during the other hours, you could expect to lose 1 pound per week by doing nothing else. It really comes down to just reducing Calories; it had nothing to do with the diet supplement, for which consumers had paid a total of $155 million over three years.

EXAMPLE 3-10   Determining a Specific Heat by Heat Exchange

A 440-g quantity of a certain metal is heated to 100.0°C. It is immediately thrust into 258 g of water that is initially at 25.0°C. The temperature of the metal–water mixture eventually settles at 36.5°C. What is the metal? (Refer to Table 3-2.)

PROCEDURE

To establish its specific heat, we need the metal's mass, the temperature change, and the heat released by cooling to cause the change. The heat released by the metal is the same as the heat gained by the water.

(a) The heat gained by the water is given by the equation

$$\text{heat gained} = \Delta°C(\text{water}) \times \text{mass (g water)} \times \text{specific heat (water)}$$
$$\Delta°C = T_{final} - T_{ininitial}$$

(b) The heat gained by the water is equal to the heat lost by the metal. Heat lost is assigned a negative sign by convention. The heat lost by the metal is

$$\text{heat lost} = -\Delta°C(\text{metal}) \times \text{mass (g metal)} \times \text{specific heat(metal)}$$

(c) Set the heat gained by the water equal to the heat lost by the metal. Solve for the specific heat of the metal.

$$-\text{heat lost} = \text{heat gained}$$
$$-\Delta°C(\text{water}) \times \text{mass(water)} \times \text{specific heat(water)} = \Delta°C(\text{metal}) \times \text{mass(metal)} \times \text{specific heat(metal)}$$

SOLUTION

Substituting all of the given values we have

$$-(25.0 - 36.5)°C \times 258 \text{ g} \times 4.184 \text{ J/°C} \cdot \text{g} = (100 - 36.5)°C \times 440 \text{ g} \times \text{specific heat (metal)}$$

Solving for specific heat (metal),

$$\text{specific heat (metal)} = \frac{-(25.0 - 36.5)°C \times 258 \text{ g} \times 4.184 \text{ J/g} \cdot °C}{(100 - 36.5)°C \times 440 \text{ g}}$$

$$\text{specific heat (metal)} = \underline{0.444 \text{ J/g} \cdot °C}$$

ANALYSIS

What is the metal? Matching the calculated value with those in Table 3-2, we find that the metal is *iron*.

SYNTHESIS

Notice that the temperature of the water increases by only 11.5°C, while the temperature of the metal decreases by 63.5°C. This is mostly due to the comparatively low specific heat of a metal. The same amount of heat causes more temperature change in the metal compared to water. If the metal in question were gold, instead of iron, how would that have affected the overall temperature change of the water and the metal? Since gold has an even lower specific heat than iron, there is less heat to transfer into the water. The overall temperature change of the water would be even less than before, while the temperature change of the gold would be greater.

▶ **ASSESSING THE OBJECTIVE FOR SECTION 3-5**

EXERCISE 3-5(a) LEVEL 1: Compare the specific heats of water and copper. If a hot piece of copper was placed in a similar mass of cool water, would the final temperature of the two be closer to the original temperature of the water or the copper?

EXERCISE 3-5(b) LEVEL 2: A 44.0-g sample of an element absorbs 1870 J of energy and increases in temperature from 25.0°C to 72.5°C, What is the specific heat of the element? What is the element?

EXERCISE 3-5(c) **LEVEL 2:** How much energy (in joules) is released when 18.5 g of copper cools from 285°C down to 45°C?

EXERCISE 3-5(d) **LEVEL 2:** A 20.0-g piece of hot iron at 225°C is placed in 51.0 g of water. No heat escapes to the surroundings. The water and iron equilibrate at 45.0°C. How much did the temperature of the water increase?

EXERCISE 3-5(e) **LEVEL 2:** A 0.0625-lb sample of fat (1 oz) is burned. The heat from the combustion raises the temperature of 10.0 L of water from 25.0°C to 45.5°C. How many Calories are in the sample of fat?

EXERCISE 3-5(f) **LEVEL 3:** If you've ever been camping and used aluminum foil to cook with, you know you can reach right into the fire and pull out the aluminum foil without burning yourself. Why doesn't the aluminum foil feel hot?

For additional practice, work chapter problems 3-63, 3-66, and 3-74.

MAKING IT REAL

Dark Matter and Energy

In this chapter we have discussed the matter and energy components of the universe that we can see or measure. However, there is apparently much more out there, and it borders on the bizarre. In recent years, scientists have come to understand that there may be other forms of matter and energy that reveal their presence only indirectly.

First consider the possibility of invisible matter. A Swiss astronomer, Fritz Zwicky, first proposed its existence in 1933 by observation of the stars. Galaxies are huge groupings of stars rotating around a central core. (Our home galaxy is known as the Milky Way.) Zwicky noted that the visible matter in the galaxy could not control the motion of the stars around the core. It was as if some unseen, exotic part of nature was at work controlling the motions of stars. Most scientists now accept the existence of this invisible stuff, known as dark matter. In fact, as much as 95% of the matter in the universe may actually be dark matter. But it is hard to prove its existence if it is invisible. We have only indirect evidence (gravity) of its existence.

Quite recently, in 1998, more science fiction became reality. While dark matter accounts for most of the motion of the stars within a galaxy, it now appears that the galaxies of the universe are moving more and more rapidly away from one another. This is a reversal of thinking of just a few years ago, when we thought the expansion of the universe was slowing because of gravity. It has been proposed that an exotic form of energy also exists that causes matter to move apart. This "antigravity" energy has been called dark energy. The effect of dark energy would be opposite that of normal gravity.

Science fiction has long suggested an invisible world and devices that defy gravity. Maybe they are not as far out as we think. There seems to be an invisible mass around us that leaves only gravity as the marker of its existence. And now we also have antigravity—a force that would make us "fall up" rather than "fall down."

The study of dark matter and dark energy are rapidly developing and exciting areas of astronomy. We will know more in the next few years as we look for ways to prove their existence.

PART B   SUMMARY

KEY TERMS

3-4   **Energy** has several different forms, including heat, electrical, mechanical, chemical, and nuclear energy. p. 101

3-4.1   The **law of conservation of energy** states that energy cannot be created or destroyed. p. 101

3-4.2   An **exothermic** process refers to the production of heat, while **endothermic** refers to the absorption of heat. p. 102

3-4.3   Energy may be either used as **kinetic energy** or stored as **potential energy**. p. 102

3-5   **Specific heat** is a physical property that interrelates heat, mass, and temperature change. p. 102

3-5.1   Heat is measured in **joules** or **calories**. p. 103

SUMMARY CHART

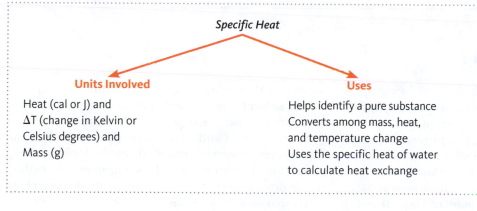

Specific Heat

Units Involved

Heat (cal or J) and
ΔT (change in Kelvin or
Celsius degrees) and
Mass (g)

Uses

Helps identify a pure substance
Converts among mass, heat,
and temperature change
Uses the specific heat of water
to calculate heat exchange

CHAPTER 3 SYNTHESIS PROBLEM

In this chapter, we learned that density and specific heat are intensive physical properties that can be used to identify pure substances. These properties can also be used to convert among a number of measurements. Density can be used to convert between volume and mass and specific heat can be used to convert among heat, mass, and temperature change. Solutions are homogeneous mixtures where the concentration can be expressed as percent by mass, which allows us to convert among mass of solute, mass of solvent, and mass of solution. At a specified concentration, solutions also have a certain density and specific heat. Consider the following solution of ethanol (ethyl alcohol) in water.

PROBLEM	SOLUTION
a. The density of water at 25.0°C is 0.9971 g/mL. What is the mass of 1255 mL of water?	**a.** $1255 \text{ mL} \times \dfrac{0.9971 \text{ g}}{\text{mL}} = \underline{1251 \text{ g water}}$
b. What mass of ethanol must be mixed with this amount of water to form a 5.45% by mass ethanol solution?	**b.** For a 5.45% solution, in 100 g solution, there are 5.45 g ethanol and $100.00 - 5.45 = 94.55$ g water $1251 \text{ g water} \times \dfrac{5.45 \text{ g ethanol}}{94.55 \text{ g water}} = \underline{72.1 \text{ g ethanol}}$
c. What volume of ethanol is needed for this solution? (See Table 3-1.)	**c.** $72.1 \text{ g ethanol} \times \dfrac{1 \text{ mL}}{0.790 \text{ g ethanol}} = \underline{91.3 \text{ mL}}$
d. What is the total mass of the solution?	**d.** $72.1 \text{ g ethanol} + 1251 \text{ g water} = \underline{1323 \text{ g (solution)}}$
e. Determine the amount of heat (in kJ) needed to warm the solution from 25.0°C to 35.0°C. The specific heat of the solution is 3.97 J/g · °C	**e.** $\text{specific heat} = \dfrac{\text{joules}}{\text{g} \cdot °\text{C}}$ $\text{joules} = \text{specific heat} \times \text{g} \times °\text{C} = \dfrac{3.97 \text{ J}}{\text{g} \cdot °\text{C}} \times 10.0 \, °\text{C} \times 1323 \text{ g}$ $= 52500 \text{ J} \times \dfrac{1 \text{ kJ}}{10^3 \text{ J}} = \underline{52.5 \text{ kJ}}$

30.48

YOUR TURN

Windshield-washing solvent is a solution of methanol and water (and a small amount of detergent which we will neglect in this problem).

a. Determine the volume of water (density = 0.9971 g/mL) that must be combined with 375 mL of methanol (density = 0.7914 g/mL) to make a 45.0% by mass methanol solution.

b. Determine the total mass of the solution.

c. Determine the amount of heat (in kJ) that is required to warm the solution from 25.0°C to 55.0°C. The specific heat of the solution is 3.52 J/g · °C

Answers are on p. 111.

CHAPTER SUMMARY

Elements and compounds are distinguished by their properties. They display **physical properties** and undergo **physical changes**. An important physical property of a substance is its **physical state: solid, liquid**, or **gas**. A physical change takes place when the substance changes to a different physical state. The temperatures at which an element or compound changes to a different **phase** are known as the **melting point** of a solid (or **freezing point** of a liquid) and **boiling point** of a liquid (or **condensation** point of a gas). A pure substance also has **chemical properties** that relate to the **chemical changes** it undergoes. In chemical changes the **law of conservation of mass** is observed.

An important physical property of a substance is its **density**. Density relates the mass of a substance to its volume; thus, it is independent of the size of the sample. Mass and volume are **extensive properties**, which may vary. Density is independent of the amount present, so it is an **intensive property**. Density can be used as an identifying property as well as a conversion factor between the mass and volume of a sample. When density is expressed in units of g/mL or g/cm^3, **specific gravity** has the same numerical quantity but is expressed without units.

Elements and compounds may form either **homogeneous** or **heterogeneous mixtures**. Heterogeneous mixtures of solids and liquids can be separated by filtration. A **solution** is an example of a homogeneous mixture with one liquid phase. A solution is composed of the **solvent** and the **solute**. The components of a solution can be separated by distillation. Aqueous (water) solutions and pure water often look identical. Pure water, however, has definite and unchanging properties. The properties of a solution vary according to the proportions of the mixture. Among those properties that vary for a solution is density. The **concentration** of a solution is often expressed as **percent by mass**.

The other component of the universe, **energy**, is also intimately involved in physical and chemical processes. Besides heat, there are other *forms* of energy such as chemical, mechanical, and electrical energy. There are also two *types* of energy: **potential energy** and **kinetic energy**. Forms and types of energy can be converted into each other, but in all cases the **law of conservation of energy** is observed. When a chemical or physical process liberates heat energy, the chemical change is said to be **exothermic**. When heat energy is absorbed, the reaction is said to be **endothermic**.

How much the temperature of 1 gram of a substance is affected by a certain amount of heat is a physical property known as **specific heat**. Specific heat is a measure of the amount of heat in **calories** or **joules** (or kcal or kJ) that will raise the temperature of 1 gram of the substance 1 degree Celsius.

OBJECTIVES

SECTION	YOU SHOULD BE ABLE TO...	EXAMPLES	EXERCISES	CHAPTER PROBLEMS
3-1	List several properties of matter and distinguish them as physical or chemical.		1a, 1b, 1c	4, 5, 6, 9, 11
3-2	Perform calculations involving the density of liquids and solids.	3-1, 3-2, 3-3, 3-4	2a, 2b, 2c	12, 14, 16, 20, 22, 24 28, 29
3-3	Perform calculations involving the percent of a pure substance in a mixture.	3-5, 3-6	3a, 3b, 3c, 3d, 3e, 3f, 3g	41, 42, 48, 49, 50, 52, 53, 55
3-4	Define terms associated with energy exchanges in physical and chemical processes.		4a, 4b, 4c	56, 57, 59, 60
3-5	Perform calculations involving the specific heat of a substance.	3-7, 3-8, 3-9, 3-10	5a, 5b, 5c, 5d, 5e, 5f	63, 64, 67, 68, 69, 71, 79, 81

▶ **ANSWERS TO ASSESSING THE OBJECTIVES**

Part A
EXERCISE

3-1(a) **(a)** physical   **(b)** chemical   **(c)** physical
(d) chemical   **(e)** chemical   **(f)** physical

3-1(b) The physical properties of calcium are its gray color, its solid state, its dull sheen, and its reported melting point. The chemical property is its reactivity with water to form a gas and a chemical soluble in water.

3-1(c) The fact that it's a liquid at room temperature tells us its melting point is lower than 25°C. Clear and colorless are also important properties. The fact that it's flammable gives us a good idea of the type of chemical it is. The most important piece of information is its exact boiling point.

3-1(d) Make note of the intensive properties of the substance. Is it silver or gold? Heavy or light? Hard or soft? At what temperature does it melt? Consult a handbook or Internet site to find a list of physical properties of metals, and see if you can find a match.

3.2(a) The larger the density, the smaller the volume for substances of the same mass. The items should be listed in order of decreasing density: carbon tetrachloride < water < kerosene < ethyl alcohol < gasoline.

3-2(b) The volume the sample occupies is 36.2 mL.

3-2(c) The volume of the cadmium is 10.77 mL. Its density is 4.97 g/mL (three significant figures).

3-2(d) Density = 1.60 g/mL. We can now scan the list of densities in Table 3-1 and note that carbon tetrachloride also has a density of 1.60 g/mL. It is very likely that the pure liquid is carbon tetrachloride. Since the liquid does not dissolve in water and its density is greater than 1.0 g/mL, carbon tetrachloride sinks to the bottom of a beaker of water.

3-2(e) The density of antifreeze should be halfway between the densities of water and ethylene glycol. If you measured the density of a small sample of antifreeze, and it was closer to the density of water, then clearly some of the glycol has broken down and more needs to be added.

3-3(a) **(a)** heterogeneous **(b)** heterogeneous **(c)** homogeneous **(d)** homogeneous

3-3(b) Since they do not mix, one will not dissolve in the other. Therefore, they will form two individual phases separated by a visible interface (or boundary). This is a heterogeneous mixture. A quick check of densities reveals the water layer to be the less dense, which means that it will be found on top.

3-3(c) The oxygen is completely dissolved (no bubbles). Furthermore, the concentration of the oxygen is the same throughout the bowl. This is a *solution* of a gas dissolved in a liquid.

3-3(d) Distillation will separate water and sand, but filtration would be easier. The water boils away, leaving the sand behind. Filtration will not separate water and salt. Since the salt is dissolved, it will pass through the filter along with the water. Distillation is necessary to separate salt and water.

3-3(e) 444 g

3-3(f) 2.32 oz

3-3(g) 221 g of 14-K gold

Part B
EXERCISE

3-4(a) Does the process require energy, in the form of heat? Or does the process release heat energy when it occurs? **(a)** endothermic **(b)** exothermic **(c)** endothermic **(d)** exothermic **(e)** exothermic (The lake must lose energy to freeze.)

3-4(b) Is the energy stored? Or does it result from motion? **(a)** potential **(b)** kinetic **(c)** potential **(d)** kinetic **(e)** kinetic

3-4(c) The sunlight is solar energy that becomes stored potential chemical energy in the corn. The energy in the corn is converted into a useful source by its fermentation to ethanol. As the ethanol combusts in the car's engine, both the form and type of energy change. Potential energy stored in the ethanol is converted to kinetic energy as the car moves. This is also a conversion of chemical into mechanical energy.

3-5(a) The water. With the higher specific heat, it takes more energy to change water's temperature.

3-5(b) The temperature change is 47.5°C. The specific heat is $0.895 \, J/(g \cdot °C)$. The element is most likely aluminum.

3-5(c) $1.71 \times 10^3 \, J$

3-5(d) The hot piece of iron lost $1.60 \times 10^3 \, J$ of energy. The water absorbed the same. The water's temperature would have increased 7.5°C. (Its original temperature before heating was therefore 37.5°C.)

3-5(e) 205 Cal ($8.58 \times 10^5 \, J$)

3-5(f) There are two reasons. The aluminum foil is very thin, so its mass is not great. Further, metal (in this case, aluminum) has a much lower specific heat than water (a major component of your fingers). Both these factors combine to produce very little heat transfer to your fingertips, which then do not heat up very much at all.

ANSWERS TO CHAPTER SYNTHESIS PROBLEM

a. The volume of water is 364 mL.   **b.** The mass of the solution is 660 g.   **c.** The heat required is 69.7 kJ.

CHAPTER PROBLEMS

Throughout the text, answers to all exercises in color are given in Appendix E. The more difficult exercises are marked with an asterisk.

Physical and Chemical Properties and Changes (SECTION 3-1)

3-1. Which of the following describes the liquid phase?

(a) It has a definite shape and a definite volume.

(b) It has a definite shape but not a definite volume.

(c) It has a definite volume but not a definite shape.

(d) It has neither a definite shape nor a definite volume.

3-2. Which state of matter is compressible? Why?

3-3. A fluid is a substance that flows and can be poured. Which state or states can be classified as fluids?

3-4. Identify the following as either a physical or a chemical property.

(a) Diamond is one of the hardest known substances.

(b) Carbon monoxide is a poisonous gas.

(c) Soap is slippery.

(d) Silver tarnishes.

(e) Gold does not rust.

(f) Carbon dioxide freezes at −78°C.

(g) Tin is a shiny, gray metal.

(h) Sulfur burns in air.

(i) Aluminum has a low density.

3-5. Identify the following as either a physical or a chemical property.

(a) Sodium burns in the presence of chlorine gas.

(b) Mercury is a liquid at room temperature.

(c) Water boils at 100°C at average sea-level pressure.

(d) Limestone gives off carbon dioxide when heated.

(e) Hydrogen sulfide has a pungent odor.

3-6. Identify the following as a physical or a chemical change.

(a) the frying of an egg (d) the burning of gasoline

(b) the vaporization of dry ice (e) the breaking of glass

(c) the boiling of water

3-7. Identify the following as a physical or a chemical change.

(a) the souring of milk

(b) the fermentation of apple cider

(c) the compression of a spring

(d) the grinding of a stone

3-8. When table sugar is heated to its melting point, it bubbles and turns black. When it cools, it remains a black solid. Describe the change as chemical or physical.

3-9. Aluminum metal melts at 660°C and burns in oxygen to form aluminum oxide. Identify a physical property and change and a chemical property and change.

3-10. A pure substance is a green solid. When heated, it gives off a colorless gas and leaves a brown, shiny solid that melts at 1083°C. The shiny solid cannot be decomposed to simpler substances, but the gas can. List all the properties given and tell whether they are chemical or physical. Tell whether each substance is a compound or an element.

3-11. A pure substance is a greenish-yellow, pungent gas that condenses to a liquid at −35°C. It undergoes a chemical reaction with a given substance to form a white solid that melts at 801°C and a brown, corrosive liquid with a density of 3.12 g/mL. The white solid can be decomposed to simpler substances, but the gas and the liquid cannot. List all the properties given and tell whether they are chemical or physical. Tell whether each substance is a compound or an element.

Density (SECTION 3-2)

3-12. A handful of sand has a mass of 208 g and displaces a volume of 80.0 mL. What is its density?

3-13. A 125-g quantity of iron has a volume of 15.8 mL. What is its density?

3-14. A given liquid has a volume of 0.657 L and a mass of 1064 g. What might the liquid be? (Refer to Table 3-1.)

3-15. A 25.0-g quantity of magnesium has a volume of 14.4 mL. What is its density? What is the volume of 1.00 kg of magnesium?

3-16. What is the volume in milliliters occupied by 285 g of mercury? (See Table 3-1.)

3-17. What is the mass of 671 mL of table salt? (See Table 3-1.)

3-18. What is the mass of 1.00 L of gasoline? (See Table 3-1.)

3-19. What is the mass in pounds of 1.00 gal of gasoline? (See Tables 3-1 and 1-5.)

3-20. What is the mass of 1.50 L of gold? (See Table 3-1.)

3-21. What is the volume in milliliters occupied by 1.00 kg of carbon tetrachloride?

3-22. The volume of liquid in a graduated cylinder is 14.00 mL. When a certain solid is added, the volume reads 92.5 mL. The mass of the solid is 136.5 g. What might the solid be? (Refer to Table 3-1.)

3-23. A 1.05-lb quantity of a solid has a volume of 1.00 L. What is its density? Does this material float on water?

3-24. Pumice is a volcanic rock that contains many trapped air bubbles. A 155-g sample is found to have a volume of 163 mL. What is the density of pumice? What is the volume of a 4.56-kg sample? Will pumice float or sink in water? In ethyl alcohol? (Refer to Table 3-1.)

3-25. A bottle weighs 44.75 g. When 13.0 mL of a liquid is added to the bottle, it weighs 65.55 g. What might the liquid be? (Refer to Table 3-1.) Does it form a layer below water or above water?

3-26. The density of diamond is 3.51 g/mL. What is the volume of the Hope diamond if it has a mass of 44.0 carats? (1 carat = 0.200 g)

3-27. A small box is filled with liquid mercury. The dimensions of the box are 3.00 cm wide, 8.50 cm long, and 6.00 cm high. What is the mass of the mercury in the box? (1.00 mL = 1.00 cm^3)

3-28. A 125-mL flask has a mass of 32.5 g when empty. When the flask is completely filled with a certain liquid, it weighs 143.5 g. What is the density of the liquid?

3-29. Which has a greater volume, 1 kg of lead or 1 kg of gold?

3-30. Which has the greater mass, 1 L of gasoline or 1 L of water?

3-31. A large nugget of a shiny metal is found in a mountain stream. It weighs 5.65 oz (16 oz/lb). 25.0 mL of water is placed in a graduated cylinder. When the nugget is placed in the cylinder, the water level reads 33.3 mL. Did we find a nugget of gold? (Refer to Tables 3-1 and 1-5.)

3-32. What is the difference between density and specific gravity?

*3-33. What is the mass of 1 gal of carbon tetrachloride in grams? In pounds?

*3-34. Calculate the density of water in pounds per cubic foot (lb/ft^3).

*3-35. In certain stars, matter is tremendously compressed. In some cases the density is as high as 2.0×10^7 g/mL. A tablespoon full of this matter is about 4.5 mL. What is the mass of this tablespoon of star matter in pounds?

Mixtures and Pure Substances (SECTION 3-3)

3-36. Carbon dioxide is not a mixture of carbon and oxygen. Explain.

3-37. Which of the following is a mixture: water, sulfur dioxide, pewter, or ammonia?

3-38. List the following waters in order of increasing purity: ocean water, rainwater, and drinking water. Explain.

3-39. Iron is attracted to a magnet, but iron compounds are not. How could you use this information to tell whether a mixture of iron and sulfur forms a compound when heated?

3-40. When a teaspoon of solid sugar is dissolved in a glass of liquid water, what phase or phases are present after mixing?

(a) liquid only   (c) solid only   (b) still solid and liquid

3-41. Identify the following as homogeneous or heterogeneous matter.

(a) gasoline   (d) alcohol   (g) aerosol spray
(b) dirt   (e) a new nail   (h) air
(c) smog   (f) vinegar

3-42. Identify the following as homogeneous or heterogeneous matter.

(a) a cloud   (c) whipped cream   (e) natural gas
(b) dry ice   (d) bourbon   (f) a grapefruit

3-43. In which physical state or states does each of the substances listed in problem 3-41 exist?

3-44. In which physical state or states does each of the substances listed in problem 3-42 exist?

3-45. Carbon tetrachloride and kerosene mix with each other, but neither mixes with water. How can water be used to keep the carbon tetrachloride and kerosene apart? Which liquid is on the top? (Refer to Table 3-1.)

3-46. How could liquid mercury be used to tell whether a certain sample of metal is lead or gold? (Refer to Table 3-1.)

3-47. Why does ice float on water? (Refer to Table 3-1.) Is ice floating in water homogeneous or heterogeneous matter? Pure or a mixture?

3-48. Tell whether each of the following properties describes a heterogeneous mixture, a solution (homogeneous mixture), a compound, or an element.

(a) a homogeneous liquid that, when boiled away, leaves a solid residue.

(b) a cloudy liquid that, after a time, seems more cloudy toward the bottom

(c) a uniform red solid that has a definite, sharp melting point and cannot be decomposed into simpler substances

(d) a colorless liquid that boils at one unchanging temperature and can be decomposed into simpler substances

(e) a liquid that first boils at one temperature but, as the heating continues, boils at slowly increasing temperatures (there is only one liquid phase)

3-49. Tell whether each of the following properties describes a heterogeneous mixture, a solution (homogeneous mixture), a compound, or an element.

(a) a nonuniform powder that, when heated, first turns mushy and then continues to melt as the temperature rises

(b) a colored gas that can be decomposed into a solid and another gas (the entire sample of the new gas seems to have the same chemical properties)

(c) a sample of colorless gas, only part of which reacts with hot copper

3-50. A solution is prepared by dissolving 22.6 g of NaCl in 855 g of water. What is the percent by mass NaCl in the solution? What is the mass percent of water?

3-51. An aqueous solution weighs 22.8 lb. It contains 20.8 lb of water, and the rest is KCl. What is the mass percent KCl in this solution?

3-52. A solution contains 25% by mass of solute. What mass of the solute is in 255 kg of the solution?

3-53. A certain ring is a homogeneous solid made of 10-K gold, which is only 42% gold. What mass of gold is in 186 g of 10-K gold?

3-54. Duriron is a homogeneous solid used to make pipes and kettles. It contains 14% silicon, with the remainder being iron. What mass of duriron can be made from 122 lb of iron?

3-55. A certain solution is 8.45% by mass glucose sugar. This solution is made with 175 mL of water. What is the mass of glucose and the mass of the solution?

Energy (SECTION 3-4)

3-56. From your own experiences, tell whether the following processes are exothermic or endothermic.

(a) decay of grass clippings

(b) melting of ice

(c) change in an egg when it is fried

(d) condensation of steam

(e) curing of freshly poured cement

3-57. A car battery can be recharged after the engine starts. Trace the different energy conversions from the burning of gasoline to energy stored in the battery.

3-58. Windmills are used to generate electricity. What are all of the different forms of energy involved in the generation of electricity by this method?

3-59. Identify the principal type of energy (kinetic or potential) exhibited by each of the following.

(a) a book on top of a table   (d) the wind

(b) a mud slide   (e) waves on a beach

(c) methane gas (CH_4)

3-60. Identify the following as having either potential or kinetic energy or both.

(a) an arrow in a fully extended bow

(b) a baseball traveling high in the air

(c) two magnets that are separated from each other

(d) a chair on the fourth floor of a building

3-61. When you apply your brakes to a moving car, the car loses kinetic energy. What happens to the lost energy?

3-62. When a person rides on a swing, at what point in the movement is kinetic energy the greatest? At what point is potential energy the greatest? Assume that once started, the person will swing to the same height each time without an additional push. At what point is the total of the kinetic energy and the potential energy greatest?

Specific Heat (SECTION 3-5)

3-63. It took 73.2 J of heat to raise the temperature of 10.0 g of a substance 8.58°C. What is the specific heat of the substance?

3-64. When 365 g of a certain pure metal cooled from 100°C to 95°C, it liberated 56.6 cal. Identify the metal from among those listed in Table 3-2.

3-65. A 10.0-g sample of a metal requires 22.4 J of heat to raise the temperature from 37.0°C to 39.5°C. Identify the metal from those listed in Table 3-2.

3-66. If 150 cal of heat energy is added to 50.0 g of copper at 25°C, what is the final temperature of the copper? Compare this temperature rise with that of 50.0 g of water initially at 25°C. (Refer to Table 3-2.)

3-67. A large cube of ice weighing 558 g is cooled in a freezer to −15.0°C. It is removed and allowed to warm to 0.0°C but does not melt. What is the specific heat of ice if 17.2 kJ of heat is required in the process?

3-68. How many joules are evolved if 43.5 g of aluminum is cooled by 13°C? (Refer to Table 3-2.)

3-69. What mass of iron is needed to absorb 16.0 cal if the temperature of the sample rises from 25°C to 58°C?

3-70. If one has a copper and an iron skillet of the same weight, which would fry an egg faster? Explain.

3-71. When 486 g of zinc absorbs 265 J of heat energy, what is the rise in temperature of the metal?

3-72. Given 12.0-g samples of iron, gold, and water, calculate the temperature rise that would occur when 50.0 J of heat is added to each.

3-73. A 10.0-g sample of water cools 2.00°C. What mass of aluminum is required to undergo the same temperature change?

3-74. A can of diet soda contains 1.00 Cal (1.00 kcal) of heat energy. If this energy was transferred to 50.0 g of water at 25°C, what would be the final temperature?

3-75. The specific heat of platinum is one of the lowest among the metals. What is its specific heat if a 5.44-g quantity evolves 7.36 cal when it cools from 55.0°C to 12.3°C?

3-76. If 50.0 g of aluminum at 100.0°C is allowed to cool to 35.0°C, how many joules are evolved?

3-77. A 22.5-J quantity of heat raises the temperature of a piece of zinc 2.0°C. How many degrees Celsius would the same amount of heat raise the temperature of the same amount of aluminum?

3-78. In the preceding problem, assume that the hot aluminum was added to water originally at 30.0°C. What mass of water was present if the final temperature of the water was 35.0°C?

3-79. If 50.0 g of water at 75.0°C is added to 75.0 g of water at 42.0°C, what is the final temperature?

*3-80. A 100.0-mL volume of water is originally at 25.0°C. When a chunk of lead that is originally at 42.8°C is added to the water, the temperature of the water increases to 28.7°C. If the specific heat of lead is 0.128 J/g · °C, what was the weight of the lead?

*3-81. If 100.0 g of a metal at 100.0°C is added to 100.0 g of water at 25.0°C, the final temperature is 31.3°C. What is the specific heat of the metal? Identify the metal from Table 3-2.

General Problems

3-82. A 22-mL quantity of liquid A has a mass of 19 g. A 35-mL quantity of liquid B has a mass of 31 g. If they form a heterogeneous mixture, explain what happens.

3-83. The volume of water in a graduated cylinder reads 25.5 mL. What does the volume read when a 25.0-g quantity of pure nickel is added to the cylinder? (The density of nickel is 8.91 g/mL.)

3-84. A solution of table sugar in water is 14% by mass sugar. The density of the solution is 1.06 g/mL. What is the mass of sugar in 100 mL of the solution?

3-85. A solid metal weighing 62.485 g was introduced into a small flask having a total volume of 24.96 mL. To completely fill the flask, 18.22 mL of water was required. What is the density of the metal expressed to the proper number of significant figures?

3-86. What is the volume occupied by 22.175 g of the metal in problem 3-85?

3-87. A 10.0-g quantity of table salt was added to 305 mL of water. What is the percent table salt in the solution?

3-88. Battery acid is 35% sulfuric acid in water and has a density of 1.29 g/mL. What is the mass of sulfuric acid in 1.00 L of battery acid?

3-89. An alloy is prepared by mixing 50.0 mL of gold with 50.0 mL of aluminum. What is the mass percent gold in the alloy?

3-90. How many kilocalories are required to raise 1.25 L of ice from $-45°C$ to its melting point? (Refer to Tables 3-1 and 3-2.)

3-91. When 215 J of heat is added to a 25.0-g sample of a given substance, the temperature increases from 25°C to 91 °C. What is the volume of the sample? (Refer to Tables 3-1 and 3-2.)

3-92. A 21.8-mL quantity of magnesium is at an initial temperature of 25.0°C. When 18.4 cal of heat energy is added to the sample, the temperature increases to 27.0°C. What is the specific heat of magnesium? (Refer to Table 3-1.)

3-93. 25.0 mL of a given pure substance has a mass of 67.5 g. How much heat in kilojoules (kJ) is required to heat this sample from 15°C to 88°C? (Refer to Tables 3-1 and 3-2.)

3-94. When a log burns, the ashes have less mass than the log. When zinc reacts with sulfur, the zinc sulfide has the same mass as the combined mass of zinc and sulfur. When iron burns in air, the compound formed has more mass than the original iron. Explain each reaction in terms of the law of conservation of mass.

STUDENT WORKSHOP

Graphing Physical Properties

Purpose: To depict a physical property using graphical analysis, and evaluate an unknown by comparison to a standard. (Work in groups of three or four. Estimated time: 20 min.)

This exercise is based on Making It Real, "Identifying a Glass Shard from a Crime Scene by Density." Assuming that the volumes of solutions of bromoform and bromobenzene are additive, we can construct a graph that shows what the density of each exact mixture of the two chemicals would be.

1. Obtain a piece of graph paper. Draw your x-axis 20 spaces across. This axis will be your solution percentage, and each space will represent a 5% increase in the amount of bromoform (0% on the left, 100% on the right). Label every other line.

2. Draw your y-axis 28 spaces high. This axis is your density and will start at 1.50 g/mL, the approximate density of bromobenzene. It will continue to 2.90 g/mL, the density of bromoform. Each line, then, is 0.05 g/mL. Label every fourth line.

3. The density of the different percent solutions should be a straight-line relationship. Plot a point at 0% and 1.50 g/mL for pure bromobenzene, and at 100% and 2.90 g/mL for bromoform. Connect the two points with a straight line.

You now have a graph on which you can read the density for any percentage of bromoform and bromobenzene or can find a percentage for any given density. Now proceed to analyze some data. Several glass samples were obtained, and they were experimentally analyzed in the bromobenzene/bromoform mixtures. The percentage of bromoform indicates the mixture that suspended the particles.

GLASS SAMPLES	PERCENTAGE OF BROMOFORM	ESTIMATED DENSITY
A	71%	???
B	75%	???
C	63%	???
D	78%	???
E	67%	???

Note that the glass samples were all suspended over a very small range of mixtures. That's because the densities of glass do not vary much, one from another. In reality, the range is even smaller.

Using your graph, determine the approximate densities for the five samples. A sixth sample of glass found on a suspect had a density of 2.60 g/ml. Which of the five evidence samples matches up with the suspect's? If that was the sample taken from the crime scene, our suspect will have some explaining to do.

4

The Periodic Table and Chemical Nomenclature

The periodic table allows the chemist to systematize and categorize the elements. It is also used for the naming of compounds formed by the elements. The vast amount of information presented in this table and its use in naming compounds are the subjects of this chapter.

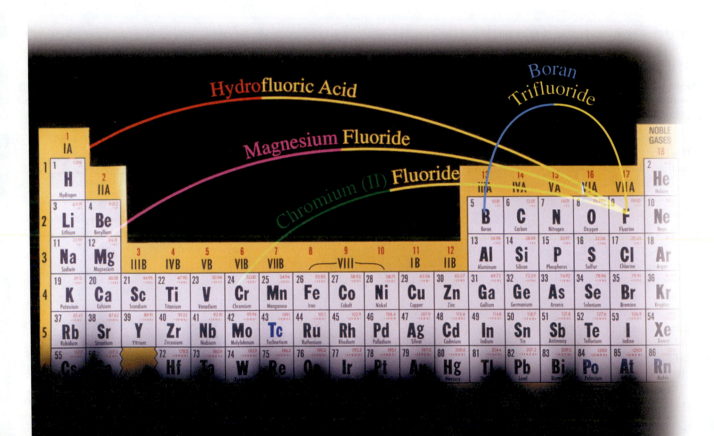

SETTING THE STAGE

In any geography course we would certainly find a globe of the earth. Let's consider the location of two cities on the globe that are in the United States: Duluth, Minnesota, and Miami, Florida. We can make some fairly good observations just from their respective locations. On one hand, we would predict that in December, Duluth would have short days and a ton of snow. On the other hand, Miami is located far to the south and so we would predict it to have longer days and maybe warm, beach-like weather at this time of the year. Is there anything for chemistry students that would be as useful as a globe is to geography students? Indeed there is. It is known as the *Periodic Table of the Elements*. By locating an element on the table, we can predict many of its physical and chemical properties. It brings much order to the huge number of facts inherent to this science. It is no wonder that we find this magnificent table in almost any chemistry lab or lecture hall.

Along with the periodic table, chemists are also known for their own language and vocabulary for the over 10 million known compounds. The vocabulary of chemistry is known as *chemical nomenclature*. Like any other language, learning chemical nomenclature requires organization and, of course, some memorization.

These two topics, the periodic table and chemical nomenclature, may at first seem unrelated. In fact, to properly name many compounds we need access to some information displayed by a periodic table.

In Part A of this chapter we will present the periodic table and much of the information it displays. We will put the periodic table to use in Part B by writing formulas and naming compounds.

Part A

Relationships Among the Elements and the Periodic Table

SETTING A GOAL

■ You will be able to explain the significance of the periodic table, its origins, and how different properties of an element can be predicted by its location on the table.

OBJECTIVES

4-1 Describe the origins of the periodic table and locate the metals and nonmetals.

4-2 Using the periodic table, identify a specific element with its group number and name, period, and physical state.

▶ **OBJECTIVE FOR SECTION 4-1**
Describe the origins of the periodic table and locate the metals and nonmetals.

4-1 The Origin of the Periodic Table

LOOKING AHEAD! It had been known for centuries that there were chemical similarities among certain elements. How these similarities led to a chart called the periodic table is the subject of this section. ■

4-1.1 Construction of the Periodic Table

Ever since ancient times, it has been known that the properties of the elements were not all completely random. There were strong similarities among certain elements so that they could be grouped into several *families*. For example, one family of elements is composed of copper, silver, and gold. These were known as the *noble metals* and were found in the free state in nature. They are chemically unreactive so they have been used in coins and jewelry for thousands of years. (See Figure 4-1.) Three other elements—lithium, sodium, and potassium — were members of a different family and are known as the *active metals*. Their properties were very distinct from the noble metals. They are chemically reactive and as a result are found only as parts of compounds in nature.

In the middle of the 19th century, two scientists realized that elements in families were related in a periodic fashion that could be systematized by a table. The earliest version of this table was introduced in 1869 by Dmitri Mendeleev of Russia. Lothar Meyer of Germany independently presented a similar table in 1870. When these two scientists arranged the elements in order of increasing atomic masses (atomic numbers were still unknown), they observed that elements in families appeared at regular (periodic) intervals. The **periodic table** was constructed so that elements in the same families (e.g., Li, Na, and K) fell into vertical columns.

At first this did not always happen. Sometimes the next-heaviest element did not seem to chemically fit in a certain family. For example, the next-heaviest known element after zinc (Zn) was arsenic (As). Arsenic seemed to belong under phosphorus (P), not aluminum (Al). Mendeleev solved this problem by leaving two blank spaces. For example, a space was left under silicon (Si) and above tin (Sn) for what Mendeleev suggested was a yet-undiscovered element. Mendeleev called the missing element "ekasilicon" (meaning one place away from silicon). Later, an element was discovered that had properties intermediate between silicon and tin, as predicted by the location of the blank space. The element was later named germanium (Ge). Gallium (Ga) was eventually discovered and fit comfortably under aluminum.

A second problem involved some misfits when the elements were ordered according to atomic mass. For example, notice that tellurium (number 52, Te) is heavier than iodine (number 53, I). But Mendeleev realized that iodine clearly belonged under bromine and tellurium under selenium, not vice versa. Mendeleev simply reversed the order, suggesting that perhaps the atomic masses reported were in error. (That was known to happen sometimes.) We now know that this problem does not occur when the elements are listed in order of increasing atomic number

FIGURE 4-1 Silver and Gold With a little polish, these gold and silver coins regained their original luster after three centuries at the bottom of the ocean.

The Modern Periodic Table of the Elements

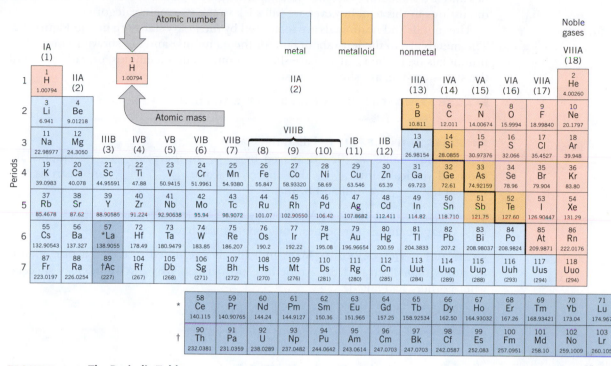

FIGURE 4-2 The Periodic Table

instead of atomic mass. This method of ordering conveniently displays the **periodic law**, which states that *the properties of elements are periodic functions of their atomic numbers.*

The modern periodic table is shown in Figure 4-2 and inside the back cover. Before we put it to use, we should be aware of the vast amount of information presented in this table. In a later chapter, we will examine the theoretical foundation for the ordering of the elements in the table.

4-1.2 Metals and Nonmetals

It seems like many sciences divide their subject into two categories. Biology studies flora (plants) and fauna (animals). Geology studies the continents and the oceans. Before going forward, we will use the periodic table to locate the two main categories of elements, which are *metals* and *nonmetals*. But first, we will discuss the difference between these two categories. **Metals** *are generally hard, lustrous elements that are malleable* (can be pounded into thin sheets) *and ductile* (can be drawn into wires). (See Figure 4-3.) We also know they readily conduct electricity and heat. Many metals such as iron, copper, and aluminum form the strong framework on which our modern society is built. Recall from the Prologue of this text that the discovery and use of metals over 5000 years ago moved civilization beyond the Stone Age.

The second type of element is noted by its *lack* of metallic properties. These are the nonmetals. **Nonmetals** *are generally gases or soft solids that do not conduct electricity.* (See Figure 4-4.) Notice that the lightest element, hydrogen, is a nonmetal.

There are notable exceptions to the general properties of metals and nonmetals. Still, almost everyone has a general idea of what a metal is like. In addition to these physical properties, there are some very important chemical differences between metals and nonmetals, which we will explore in Chapter 9. The division between metallic

FIGURE 4-3 Two Properties of Metals (a) Metals can be pressed into thin sheets (malleable) or (b) drawn into wires (ductile).

and nonmetallic properties is not distinct. Some elements have intermediate properties and are sometimes classified as a separate group called *metalloids*. In most of the discussions in this text, however, we will stick to the two main categories.

The metals and nonmetals are separated by the heavy stair-step line in Figure 4-2. The metals are to the left (about 80% of the elements) and are shown in blue. The nonmetals lie to the right and are shown in pink. The metalloids, on either side of the stair-step line, are shown in orange.

▶ **ASSESSING THE OBJECTIVE FOR SECTION 4-1**

EXERCISE 4-1(a) LEVEL 1: Fill in the blanks.
The periodic table was first displayed by _____ in the year _____. Elements are grouped vertically into _____ that share the same _____. _____ are lustrous and malleable. _____ are dull and brittle.

EXERCISE 4-1(b) LEVEL 2: Metals conduct electricity. Nonmetals serve as insulators. What electronic properties would metalloids have?

EXERCISE 4-1(c) LEVEL 3: Many individuals attempted to create arrangements of atoms that illustrated their repetitive properties. Why did Mendeleev succeed where others had failed?

For additional practice, work chapter problems 4-1 and 4-2.

▶ **OBJECTIVE
FOR SECTION 4-2**
Using the periodic table, identify a specific element with its group number and name, period, and physical state.

4-2 Using the Periodic Table

LOOKING AHEAD! There is a good deal of information that can be determined about a specific element from its position in the periodic table. We will discuss some of the categories and properties of elements that are predictable from their positions in the table. ∎

4-2.1 Periods

The periodic table allows us to locate families of elements in vertical columns. In fact, there are common characteristics in the horizontal rows as well. *Horizontal rows of elements in the table are called* **periods**. Each period ends with a member of the family of elements called the *noble gases*. The first period contains only 2 elements, hydrogen and helium. The second and third contain 8 each (Figure 4-4); the fourth and fifth contain 18 each; the sixth and seventh contain 32. The last element synthesized is at the end of the seventh period. Should another element be synthesized in the future it would be in the eighth period.

4-2.2 Groups

Families of elements fall into vertical columns called **groups**. Each group is designated by a number at the top of the group. The most commonly used label employs Roman numerals followed by an A or a B. Another method, which is becoming more common, numbers the groups 1 through 18. The periodic tables used in the text display both numbering systems. For instructional reasons, however, we will use the traditional method involving Roman numerals along with the letters A and B.

The groups of elements can be classified even further into four main categories of elements.

1. **The Main Group or Representative Elements (Groups IA–VIIA)** Most of the familiar elements that we will discuss and use as examples in this text are **main group** or **representative elements**. These are the elements labeled A in the periodic table. For example, the four main elements of life—carbon, oxygen, nitrogen,

FIGURE 4-4 Nonmetals The bottle on the left contains liquid bromine and its vapor. The flask in the back contains pale-green chlorine gas. Solid iodine is in the flask on the right. Powdered red phosphorus is in the dish in the middle, and black powdered graphite (carbon) in the watch glass in front. Lumps of yellow sulfur are shown in the front.

and hydrogen—are in this category. One group of the representative elements includes the highly reactive metals that we discussed earlier. Notice that lithium, sodium, and potassium are found in Group IA. This family of elements (not including hydrogen) is known as the **alkali metals**. Another set of metals that are also somewhat chemically reactive is found in Group IIA and is known as the **alkaline earth metals**. Group VIIA elements are all nonmetals and are known as the **halogens**. The other representative element groups (IIIA, IVA, VA, and VIA) are not generally referred to by a family name but are instead identified by the element at the top of the column. Hence, any element in column VA is part of the nitrogen family. In Chapter 9, we will discuss in more detail some of the physical and chemical properties of the representative elements.

2. **The Noble or Inert Gases (Group VIIIA)** These elements form few chemical compounds. In fact, helium, neon, and argon do not form any compounds. The **noble gases** all exist as individual atoms in nature.

3. **The Transition Metals (Group B Elements)** Transition metals include many of the familiar structural metals, such as iron and chromium, as well as the noble metals—copper, silver, and gold (Group IB)—which we discussed earlier.

4. **The Inner Transition Metals** The 14 **inner transition metals** between lanthanum (number 57) and hafnium (number 72) are known as the **lanthanides**. These metals are also known as the **rare earths** and several of them are essential for uses in sophisticated electronic equipment. Deposits are found mainly in China. The 14 metals between actinium (number 89) and rutherfordium (number 104) are known as the **actinides**. All of the elements are radioactive.

4-2.3 Physical States and the Periodic Table

Next we will consider what the periodic table can tell us about the physical state of an element. We must be cautious, however, as to the temperature conditions we define. If the temperature is low enough, all elements exist as solids (except helium); if it is high enough, all elements are in the gaseous state. On Triton, a moon of the planet Neptune, the temperature is $-236°C$ (37 K). The atmosphere is very thin because most substances that are gases under Earth conditions are solids or liquids under Triton conditions. At the outer part of the sun, however, the temperature is 50,000°C, so only gases exist. Thus, we must come to some agreement as to a reference temperature to define the physical state of an element. **Room temperature**, which is defined as exactly *25°C, is the standard reference temperature* used to describe physical state. At this temperature, all three physical states are found among the elements on the periodic table. Fortunately, except for hydrogen, the gaseous elements are all found at the extreme right top of the table (nitrogen, oxygen, fluorine, and chlorine) and in the right-hand vertical column (the noble gases). There are only two liquids: a metal, mercury, and a nonmetal, bromine. All other elements are solids. (See Figure 4-5.) (Two solid metals, gallium and cesium, melt to become liquids at approximately 29°C, slightly above the reference temperature. Since body temperature is 37°C, both of these elements would melt in your hand.)

Many of the nonmetals exist as diatomic molecules rather than individual atoms. The periodic table can help us locate these elements. All the gaseous elements except for the noble gases, which exist as individual atoms, are composed of diatomic molecules (e.g., N_2, O_2, and H_2). Or one might notice that all the naturally occurring halogens (Group VIIA) are diatomic. Not all molecular nonmetals are diatomic, however. The formula for the most common form of phosphorus is P_4, which indicates the presence of molecules composed of four atoms. The most common form of sulfur is S_8, indicating eight-atom molecules. The forms of carbon will be discussed in Chapter 11.

FIGURE 4-5 The Physical States of the Elements The elements are found in all three physical states at room temperature.

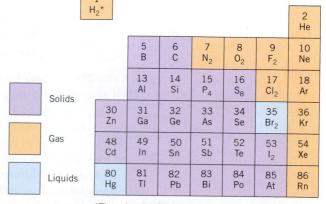

*The subscript 2 indicates diatomic molecules at 25°C

MAKING IT REAL

The Discovery of a Group VIIA Element, Iodine

It was a cold day in a small seaside town on the Atlantic coast of France in 1811. A few dozen seamen were extracting potassium salts with acid from the sludge of seaweed. Bernard Courtois, the employer of these seamen, was a chemist by training and a graduate of the Polytechnical Institute in Paris. His factory prepared saltpeter (potassium nitrate) to be used in ammunition for Napoleon's armies. Today, however, the workers' efforts turned fruitless. One of the workers decided to use a more concentrated form of acid. At that point, a huge volume of violet fumes rose from the tanks and dark crystals started depositing on every cold surface that was nearby. Their observations would lead to the discovery of a very important element.

Courtois collected those unique crystals for examination and found out they would combine with hydrogen and phosphorus, but not with oxygen. He also discovered they would form an explosive compound with ammonia. He later gave samples to two of his Paris Polytechnical Institute friends, C. Desormes and N. Clement, who published the discoveries two years later. Soon thereafter, Frenchman Joseph Louis Gay-Lussac and Englishman Sir Humphry Davy announced the discovery of a new element, which was first named *iode* (from the Greek word for "violet") by Gay-Lussac and finally *iodine*, to give it the same ending as *chlorine*, an element with similar properties. (Because of its chemical properties, iodine would later be classified as a *halogen* along with chlorine and bromine in Group VIIA.)

What happened on that day in 1811? Seaweed concentrates several ionic compounds other than sodium chloride (table salt), but no one had attached very much importance to them. That day, however, the iodine compounds must have become considerably concentrated after the extraction of the sodium chloride. The concentrated acid converted the iodine compounds to elemental iodine. The iodine vaporized (sublimed) but quickly condensed on the cool surfaces.

The practical applications of the new element had an immediate impact on patients with goiter, the enlargement of the thyroid gland. In 1820, Jean Francois Coinder associated the lack of goiter among seamen with the presence of iodine compounds in their working environment. The thyroid gland, he concluded, needs iodine to function properly and this could be achieved by adding small amounts of sodium iodide to table salt (i.e., iodized salt).

▶ **ASSESSING THE OBJECTIVE FOR SECTION 4-2**

(Answer these questions using a periodic table.)

EXERCISE 4-2(a) LEVEL 1: Provide the term being defined.
   i. _____ A horizontal column on the periodic table.
  ii. _____ Any element found in a B column.
 iii. _____ Any entire row of the periodic table.
  iv. _____ The nonreactive elements in the VIIIA column.
   v. _____ Any element in an A column.

EXERCISE 4-2(b) LEVEL 1: Fill in the blanks.
Of the two general classifications of elements, nickel is a _____ and sulfur is a _____. Some borderline elements such as germanium are sometimes referred to as _____. In the periodic table, elements are ordered according to increasing _____ so that _____ of elements fall into vertical columns. Of the four general categories of elements, calcium is a _____ element, nickel is a _____, and xenon is a _____. An element in Group IIA, such as calcium, is also known as an _____ metal. A solid nonmetal that is composed of diatomic molecules is in the group known as the _____. Metals are all solids except for _____.

EXERCISE 4-2(c) LEVEL 2: Assign each of the following elements to one of the four major categories, give its symbol and physical state, tell whether it's a metal or nonmetal, and give its period and group number.

(a) copper   **(b)** argon   **(c)** barium   **(d)** bromine   **(e)** uranium

EXERCISE 4-2(d) LEVEL 2: Identify the element that fits each description.

(a) the fourth-period   **(c)** the IIIA nonmetal   **(e)** the gaseous element
     alkaline earth metal   **(d)** the last transition metal   in Group VA.
(b) the liquid halogen   of the fifth period

EXERCISE 4-2(e) LEVEL 3: A few elements that have been created in recent years have never existed outside of a laboratory. Yet we can still use the periodic table to predict what these elements would be like if we could somehow isolate a measurable quantity. Evaluate element 118 and the hypothetical 119 for category, physical state, period, group number, and whether they are metals or nonmetals.

For additional practice, work chapter problems 4-4, 4-10, 4-12, and 4-14.

PART A SUMMARY

KEY TERMS

4-1.1   The **periodic table** is a chart that displays the relationships among the elements. p. 118

4-1.1   The **periodic law** describes the ordering of the elements based on atomic numbers. p. 119

4-1.2   **Metals** are generally hard, lustrous elements that are malleable and ductile. p. 119
          Nonmetals are generally gases or soft solids that do not conduct electricity. p. 119

4-2.1   All elements can be assigned to a specific **period**. p. 120

4-2.2   The main classifications of **groups** of elements are the **main groups** or **representative elements**, the **noble gases**, the **transition metals**, and the **inner transition metals**. pp. 120–121

4-2.2   The named groups of representative elements are the **alkali metals**, the **alkaline earth metals**, and the **halogens**. p. 121

4-2.2   The inner transition elements are known as either the **lanthanides** (**rare earths**) or **actinides**. p. 121

4-2.3   The physical state of an element at **room temperature** can be determined from the periodic table. p. 121

SUMMARY CHART

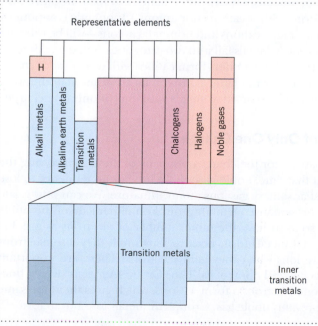

Part B

The Formulas and Names of Compounds

SETTING A GOAL

■ You will learn how to systematically name various types of molecular and ionic compounds and determine the formulas of compounds from the names.

OBJECTIVES

4-3 Write the names and formulas of ionic compounds between metals and nonmetals using the IUPAC conventions.

4-4 Write the name and formulas of compounds containing polyatomic ions.

4-5 Name binary molecular (nonmetal–nonmetal) compounds using proper Greek prefixes or common names.

4-6 Write the names and formulas of acids.

▶ **OBJECTIVE FOR SECTION 4-3**
Write the names and formulas of ionic compounds between metals and nonmetals using the IUPAC conventions.

4-3 Naming and Writing Formulas of Metal–Nonmetal Binary Compounds

LOOKING AHEAD! There are now over 10 million recorded compounds. With such a number, systematic methods of naming the compounds must be employed. We will need a periodic table, since we need to know whether an element is a metal or nonmetal and the charge on its monatomic ion. Our first task is to look at binary (i.e., two-element) compounds made from a metal and a nonmetal. For the most part, these are ionic compounds, which is one of the two major classes of compounds described in Chapter 2. ■

Chemical compounds can be roughly divided into two groups: organic and inorganic. Organic compounds are composed principally of carbon, hydrogen, and oxygen. These are the compounds of life. All other compounds are called inorganic compounds, and we will focus on these. As mentioned, many inorganic compounds were given common names, and some of these names have survived the years. For example, H_2O is certainly known as *water* rather than the more exact systematic name of *hydrogen oxide*. Ammonia (NH_3) and methane (CH_4) are two other compounds that are known exclusively by their common names. Other compounds have ancient common names but are best described by their more modern systematic names, which we will use here.

One of the most important chemical properties of a metal is its tendency to form positive ions (cations). Most nonmetals, on the other hand, can form negative ions (anions). (The noble gases form neither cations nor anions.) Inorganic compounds composed of just two elements—metal cations and nonmetal anions—will be referred to as *metal–nonmetal binary compounds*. Metals fall into two categories. Some metals form cations with only one charge (e.g., Ca^{2+}). Others form cations with two or more charges (e.g., Fe^{2+} and Fe^{3+}). The rules for the latter compounds are somewhat different from those for the first. So we initially consider metals with ions of only one charge.

4-3.1 Metals with Ions of Only One Charge

In Chapter 2, we noticed a pattern for the charges on many of the ions involving the noble gases. As mentioned at that time, noble gases have little or no tendency to lose or gain electrons and so exist as solitary gaseous atoms in nature. We can now relate this to the periodic table. Representative element groups on either side of the noble gases gain or lose electrons so as to have the same number of electrons as a noble gas. For this reason, we see that metals form positive ions when they lose electrons and nonmetals form negative ions when they gain electrons. These are important chemical characteristics of metals and nonmetals. Thus Group IA metals (which does not include hydrogen) lose one electron to form +1 ions, which gives them the same number of electrons as the previous noble gas. Group IIA metals lose two to form +2 ions, and Group IIIA *metals* lose three to form +3 ions for the same reason. Except for Al, the other Group IIIA metals also form a +1 ion. (Since formation of positive ions is a metal characteristic, boron is excluded.) On the other side of the periodic table, notice Groups VIIA nonmetals (and hydrogen) form −1 ions by gaining one electron to have the same number as the next noble gas. Group VIA gain two to form −2 ions, and Group VA nonmetals, N and P, gain three to form −3 ions. (See Figure 4-6.) The noble gas stability does not explain all of the representative ions or the transition metal ions. We will explore the reasons for formation of these ions in Chapter 8.

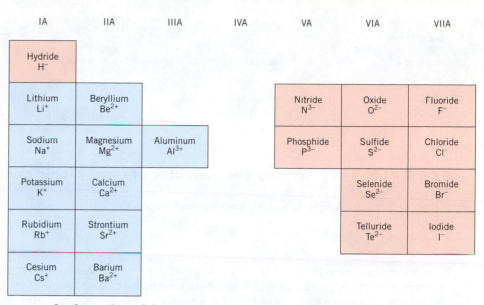

FIGURE 4-6 Monatomic Ions of the Representative Elements Metals form specific cations, and nonmetals form specific anions.

Note that metal and nonmetal ions are distinctly different from their neutral atoms. When we monitor our dietary intake of sodium or calcium we are in fact referring to the intake of sodium or calcium ions. The metal ions are parts of ionic compounds such as sodium chloride or calcium carbonate. We would not wish to ingest the neutral atoms of the metal, which would be in the form of a solid metal. (See Making It Real in this chapter.)

Note also that positive and negative ions do not exist separately. They are always found together in ordinary matter. In both naming and writing the formula for a binary ionic compound, the metal comes first and the nonmetal second. The unchanged English name of the metal is used. (If a metal cation is named alone, the word *ion* is also included to distinguish it from the free metal.) The name of the anion includes only the English root plus *-ide*. For example, *chlorine* as an anion is named *chloride* and *oxygen* as an anion is named *oxide*. So the names for NaCl and CaO are sodium chloride and calcium oxide.

Some other examples of writing names from formulas are shown in Example 4-1. Writing formulas from names can be a somewhat more challenging task, since we must determine the number of each element present in the formula. In Chapter 2, we practiced writing formulas given specific ions with specific charges. In this chapter, we are now able to determine the charge on the ion by reference to the periodic table. Recall also from Chapter 2 that the formulas represent neutral compounds where the positive and negative charges add to zero. In other words, the total positive charge is canceled by the total negative charge. Thus, NaCl is neutral because one Na^+ is balanced by one Cl^- [i.e., $+1 + (−1) = 0$] and CaO is neutral because one Ca^{2+} is balanced by one O^{2-} [i.e., $+2 + (−2) = 0$]. The formula for magnesium chloride, however, requires two Cl^- ions to balance the one Mg^{2+} ion, so it is written as $MgCl_2$ [i.e., $+2 + (2 \times −1) = 0$]. We will practice writing formulas from names in Example 4-2.

EXAMPLE 4-1   Naming Binary Ionic Compounds

Name the following binary ionic compounds: KI, Li_2S, and Mg_3N_2.

PROCEDURE

A quick glance at the periodic table reveals that the first element in each compound is a representative metal and the second is a representative nonmetal. Use the metal name unchanged, and change the ending of the nonmetals to *-ide*.

SOLUTION

FORMULA	METAL	NONMETAL	COMPOUND NAME
KI	potassium	iodine	potassium iodide
Li_2S	lithium	sulfur	lithium sulfide
Mg_3N_2	magnesium	nitrogen	magnesium nitride

EXAMPLE 4-2   Writing the formulas of binary ionic compounds

Write the formulas for the following binary metal–nonmetal compounds: **(a)** aluminum fluoride and **(b)** calcium selenide.

PROCEDURE

From the names, recognize that these compounds are representative elements. Identify their charges based on their column on the periodic table. Determine how many of each element are necessary to produce a neutral compound. Use these as subscripts in the formula.

SOLUTION

(a) Aluminum is in Group IIIA, so it forms a cation with a +3 charge exclusively. Fluorine is in Group VIIA, so it forms an anion with a −1 charge. Since the positive charge must be balanced by the negative charge, we need three F^{1-} anions to balance one Al^{3+} ion [e.g., $+3 + (3 \times -1) = 0$]. Therefore, the formula is written as

$$Al^{3+} + 3(F^-) = \underline{AlF_3}$$

(b) Calcium is in Group IIA, so it forms a +2 cation. Selenium is in Group VIA, so it forms a −2 anion. Together the charges add to zero, so one of each atom is sufficient in the formula:

$$Ca^{2+} + Se^{2-} = \underline{CaSe}$$

ANALYSIS

Another convenient way to establish the formula is to write the ions with their appropriate charges side by side. The numerical value of the charge on the cation becomes the subscript on the anion, and vice versa. The number 1 is understood instead of written as a subscript. This is known as the cross-charge method. Notice that, in (b), by exchanging values of the charge, we first indicate a formula of Ca_2Se_2. This is not a correct representation, however. Ionic compounds should be expressed with the simplest whole numbers for subscripts. Therefore, the proper formula is written as CaSe.

SYNTHESIS

What is the formula of a compound made from a metal (M) with a +3 charge and a nonmetal (X) with a −2 charge? Using the cross-charge method, we get a formula of M_2X_3. Charges are rarely higher than +3 or −3, so this is typically as complicated as balancing the charges in ionic compounds ever gets.

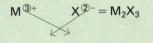

4-3.2 Metals with Ions of More than One Charge

Except for Groups IA, IIA, and aluminum, other representative metals and most transition metals can form more than one cation. Therefore, a name such as iron chloride would be ambiguous since there are two iron chlorides, $FeCl_2$ and $FeCl_3$. An even more dramatic case is that of the oxides of manganese—there are five different compounds (MnO, Mn_3O_4, Mn_2O_3, MnO_2, and Mn_2O_7). To distinguish among these compounds, we use the Stock method. *In the **Stock method**, the charge on the metal ion follows the name of the metal in Roman numerals and in parentheses.* By this convention, the two chlorides of iron are named iron(II) chloride ($FeCl_2$) and iron(III) chloride ($FeCl_3$).

In order to use the Stock method, it is sometimes necessary to determine the charge on the metal cation by working backward from the known charge on the anion. For example, in a compound with the formula FeS, we can establish from Figure 4-6 that the charge on the S is -2. Therefore, the charge on the one Fe must be $+2$. The compound is named *iron(II) sulfide*. In the following examples, we will use a simple algebra equation that will help us determine the charge on the metal and thus the proper Roman numeral to use for more complex compounds. The equation is

$$(\text{number of metal cations}) \times (+ \text{ charge on metal}) +$$
$$(\text{number of nonmetal anions}) \times (- \text{ charge on nonmetal}) = 0$$

4-3.3 The Classical Method for Metals with More than One Charge (Optional)

Another method of naming compounds is known as the **classical method**. Though no longer widely used in chemistry, it is still found in many pharmaceuticals and drug labels. In this system, *the name of the metal ion that has the lower charge ends in -ous and the higher ends in -ic.*

If the symbol of the element is derived from a Latin word, the Latin root is generally used rather than an English root. Thus the two chlorides of iron are named ferrous chloride ($FeCl_2$) and ferric chloride ($FeCl_3$). The disadvantage of this method is that you have to remember the possible charges of a specific metal, whereas the Stock method tells you explicitly. The classical method will not be included in examples or problems in this text. However, for reference purposes, several common examples of names of ions of metals that form more than one cation are shown in Table 4-1.

TABLE 4-1

Metals that Form More than One Ion

ION	STOCK NAME	CLASSICAL NAME
Cr^{2+}	chromium(II)	chromous
Cr^{3+}	chromium(III)	chromic
Fe^{2+}	iron(II)	ferrous
Fe^{3+}	iron(III)	ferric
Pb^{2+}	lead(II)	plumbous
Pb^{4+}	lead(IV)	plumbic
Au^+	gold(I)	aurous
Au^{3+}	gold(III)	auric
Cu^+	copper(I)	cuprous
Cu^{2+}	copper(II)	cupric
Mn^{2+}	manganese(II)	manganous
Mn^{3+}	manganese(III)	manganic
Sn^{2+}	tin(II)	stannous
Sn^{4+}	tin(IV)	stannic
Co^{2+}	cobalt(II)	cobaltous
Co^{3+}	cobalt(III)	cobaltic

EXAMPLE 4-3   Naming Binary Compounds with Metals with More than One Charge

Name the following compounds: SnO_2 and Co_2S_3.

PROCEDURE

To determine the charge on the metal cation, we can work backward from the known charge on the anion. Recall that Group VIA nonmetals form a -2 charge in binary compounds. We can use the general algebra equation introduced in the preceding discussion to determine the proper Roman numeral to use for the metal.

SOLUTION

The equation applied to the given compound is

$$[1 \times (\text{Sn charge})] + [2 \times (\text{O charge})] = 0$$

Substitute the known charge on the oxygen and solve for the charge on the tin.

$$\text{Sn} + (2 \times -2) = 0$$
$$\text{Sn} - 4 = 0$$
$$\text{Sn} = +4(\text{IV})$$

The name of the compound is, therefore, <u>tin (IV) oxide</u>.

For Co_2S_3, we can construct the following equation.

$$[2 \times (\text{Co charge})] + [3 \times (\text{S charge})] = 0$$

The charge on a Group VIA nonmetal such as S is −2.

$$2Co + (3 \times -2) = 0$$
$$2Co = +6$$
$$Co = +3 (III)$$

The name of the compound is <u>cobalt(III) sulfide</u>.

EXAMPLE 4-4 Writing the Formulas of Compounds with Metals that Form More than One Charge

Write the formulas for lead(IV) oxide, nickel(II) bromide, and chromium(III) sulfide.

PROCEDURE

These types of compounds are actually easier than those with normal representative elements. The charge on the metal is already stated explicitly for you. Determine the charge on the nonmetal, balance the charges, and write the formula.

SOLUTION

Lead(IV) oxide: If lead has a +4 charge, two O^{2-} ions are needed to form a neutral compound [i.e., $(1 \times +4) - (2 \times -2) = 0$].

$$\underline{PbO_2}$$

Nickel(II) bromide: Two bromines are needed to balance the +2 nickel [i.e., $\mathbf{1} \times +2) - (\mathbf{2} \times -1) = 0$].

$$\underline{NiBr_2}$$

Chromium(III) sulfide: Two Cr^{3+} ions are needed to balance three S^{2-} ions [i.e., $(\mathbf{2} \times +3) - (3 \times -2) = 0$].

$$\underline{Cr_2S_3}$$

ANALYSIS

The key to successful nomenclature is recognizing the type of compound you are dealing with, and then naming it according to the rules set down for that type of compound. A periodic table is indispensable in this regard. With it, you can determine whether the metal in question is in column I or II (or Al) and therefore does not require a charge in the name, or whether it is any of the other metals available to us and therefore does.

SYNTHESIS

Students often become confused over the meaning of the number in parenthesis. Remember that it is the *charge* on the metal, not the number of metal atoms in the compound. So what are the formulas of copper(I) sulfide and copper(II) sulfide? Cu_2S and CuS, respectively. Notice that copper(I) requires two ions of copper and copper(II) requires one ion of copper. Can you see from the charges why this is so?

▶ **ASSESSING THE OBJECTIVE FOR SECTION 4-3**

EXERCISE 4-3(a) LEVEL 1: For which of the following metals must a charge be placed in parenthesis when naming one of its compounds? Co, Li, Sn, Al, Ba

EXERCISE 4-3(b) LEVEL 2: Name the following compounds.
(a) Li_2O **(b)** CrI_3 **(c)** PbS **(d)** Mg_3N_2 **(e)** Ni_3P_2

EXERCISE 4-3(c) LEVEL 3: Provide the formula for the following compounds.
(a) aluminum iodide   **(c)** tin(IV) bromide   **(e)** sodium sulfide
(b) iron(III) oxide   **(d)** calcium nitride   **(f)** copper(I) phosphide

EXERCISE 4-3(d) LEVEL 3: Using an M to represent the metal and an X to represent the nonmetal, write theoretical formulas for all possible combinations of M and X with charges of +1, +2, +3 and -1, -2, and -3.

For additional practice, work chapter problems 4-18, 4-20, 4-22, and 4-24.

4-4 Naming and Writing Formulas of Compounds with Polyatomic ions

▶ OBJECTIVE
FOR SECTION 4-4
Write the names and formulas of compounds containing polyatomic ions.

LOOKING AHEAD! Two or more atoms that are chemically combined with covalent bonds may have an imbalance of electrons and protons just like monatomic cations and anions. This results in a charged species called a polyatomic ion. How compounds containing these ions are named is our next topic. Again, you should recognize these as ionic compounds, so we find them as solids under normal conditions. ■

We use bicarbonates and carbonates for indigestion, as well as sulfites and nitrites to preserve foods. So we are probably familiar with some of these names, which are commonly used. A list of some common polyatomic ions is given in Table 4-2. Notice that all but one (NH_4^+ ammonium) are anions.

4-4.1 Oxyanions

There is some systematization possible that will help in understanding Table 4-2. In most cases, *the anions are composed of oxygen and one other element.* Thus these anions are called **oxyanions**. When there are two oxyanions of the same element (e.g., SO_3^{2-} and SO_4^{2-}), they, of course, have different names. The anion with the smaller number of oxygens uses the root of the element plus *-ite*. The one with the higher number uses the root plus *-ate*.

$$SO_3^{2-} \text{ sulf}ite \qquad SO_4^{2-} \text{ sulf}ate$$

There are four oxyanions containing Cl. The middle two are named as before (i.e., with *-ite* and *-ate*). The one with one less oxygen than the chlorite has a prefix of *hypo-*, which means "under." (as in *hypo*glycemic, meaning "low blood sugar"). The one with one more oxygen than chlorate has a prefix of *per-*, which in this usage means "highest" (a shortening of *hyper*, as in hy*per*active, meaning "overactive").

$$ClO^- \text{ hypochlorite} \qquad ClO_3^- \text{ chlorate}$$
$$ClO_2^- \text{ chlorite} \qquad ClO_4^- \text{ perchlorate}$$

Certain anions are composed of more than one atom but behave similarly to monatomic anions in many of their chemical reactions. Two such examples in Table 4-2 are the CN^- ion and the OH^- ion. Both of these have *-ide* endings similar to the monatomic anions. Thus, the CN^- anion is known as the cyanide ion and the OH^- as the hydroxide ion, just as the Cl^- ion is named the chloride ion.

Most of the ionic compounds that we have just named are also referred to as salts. *A* **salt** *is an ionic compound formed by the combination of a cation with an anion.* (Cations combined with hydroxide or oxide form a class of compounds that are not considered salts and are discussed in the next chapter.) For example, potassium nitrate is a salt composed of K^+ and NO_3^- ions, and calcium sulfate is a salt composed of Ca^{2+} and SO_4^{2-} ions. Ordinary table salt is NaCl, composed of Na^+ and Cl^- ions.

4-4.2 Naming and Writing the Formulas of Salts with Polyatomic Ions

In naming and writing the formulas of compounds with polyatomic ions, as in Example 4-5, we follow the same procedures as with metal–nonmetal compounds. The metal is written first with its charge (if it is not Al or in Group IA or IIA) followed by the name of the polyatomic ion. To calculate the charge on the cation, if necessary, we

TABLE 4-2	
Polyatomic Ions	
ION	**NAME**
$C_2H_3O_2^-$	acetate
NH_4^+	ammonium
CO_3^{2-}	carbonate
ClO_3^-	chlorate
ClO_2^-	chlorite
CrO_4^{2-}	chromate
CN^-	cyanide
$Cr_2O_7^{2-}$	dichromate
$H_2PO_4^-$	dihydrogen phosphate
HCO_3^-	hydrogen carbonate or bicarbonate
HPO_4^{2-}	hydrogen phosphate
HSO_4^-	hydrogen sulfate or bisulfate
HSO_3^-	hydrogen sulfite or bisulfite
OH^-	hydroxide
ClO^-	hypochlorite
NO_3^-	nitrate
NO_2^-	nitrite
$C_2O_4^{2-}$	oxalate
ClO_4^-	perchlorate
MnO_4^-	permanganate
PO_4^{3-}	phosphate
SO_4^{2-}	sulfate
SO_3^{2-}	sulfite

can use the same simple algebra equation as before. For example, consider $Cr_2(SO_4)_3$. Since the sulfate ion has a -2 charge, the charge on the chromium is

$$(2 \times Cr) + (3 \times -2) = 0$$
$$2\ Cr = +6$$
$$Cr = +3$$

The name of the compound is *chromium(III) sulfate.*

When writing formulas from names, as in Example 4-6, we recall from Chapter 2 that when more than one polyatomic ion is present in the compound, parentheses enclose the polyatomic ion. If only one polyatomic ion is present, parentheses are not used (e.g., $CaCO_3$).

EXAMPLE 4-5   Naming Compounds Containing Polyatomic Ions

Name the following compounds: K_2CO_3 and $Fe_2(SO_4)_3$.

PROCEDURE
Evaluate the metal as in Section 4-3. Recognize that after the metal in each compound is a grouping of atoms that form polyatomic ions. Name the anion with the appropriate ending: *-ate*, *-ite*, or *-ide*.

SOLUTION
K_2CO_3: The cation is K^+ (Group IA). The charge is not included in the name because it forms a +1 ion only. The anion is CO_3^{2-} (the carbonate ion).

<u>potassium carbonate</u>

$Fe_2(SO_4)_3$: The charge on the Fe cation can be determined from the charge on the SO_4^{2-} (sulfate) ion.

$$2Fe + 3SO_4^{2-} = 0$$
$$2Fe + 3(-2) = 0$$
$$2Fe = +6$$
$$Fe = +3$$

<u>Iron (III) sulfate</u>

EXAMPLE 4-6   Writing the Formulas for Ionic Compounds Containing Polyatomic Ions

Give the formulas for barium acetate, ammonium sulfate, and manganese(II) phosphate.

PROCEDURE
Write the formulas for the metal ion and the polyatomic ion, including their charges. Determine how many of each are needed to balance the charges. Write the formulas with those appropriate subscripts.

SOLUTION
Barium acetate: Barium is in Group IIA, so it has a +2 charge. Acetate is the $C_2H_3O_2^-$ ion. Two acetates are required to balance one Ba^{2+} ion. The formula is

<u>$Ba(C_2H_3O_2)_2$</u>

Ammonium sulfate: From Table 4–2, ammonium = NH_4^+, sulfate = SO_4^{2-}. Two ammoniums are needed to balance one sulfate. The formula is

<u>$(NH_4)_2SO_4$</u>

Manganese(II) phosphate: The stated charge on manganese is +2. Phosphate is the PO_4^{3-} ion. Using the cross-charge method (Example 4-2b), we will need three manganese ions to balance two phosphate ions. Since there is more than one polyatomic ion, the whole ion is placed in parentheses.

<u>$Mn_3(PO_4)_2$</u>

ANALYSIS

While it is not a major mistake to use parentheses when there is only one polyatomic ion, as in $Mg(SO_3)$, parentheses do have an important purpose. Subscripts affect only what they immediately follow. In order to demonstrate two or three of the group of atoms that make up the polyatomic ion, parentheses are placed around the entire group so that the subscript refers to all. What is the formula of calcium hydroxide? It is $Ca(OH)_2$. Some students may write $CaOH_2$, assuming that because hydroxide has no subscript itself, no parentheses are necessary. However, notice the difference in the two formulas. The first shows one calcium, two oxygens, and two hydrogens. The second incorrectly shows one calcium, one oxygen, and two hydrogens. The parentheses are necessary to indicate the correct number of elements.

SYNTHESIS

There are clear patterns in the names of the polyatomic ions that can be used to decipher the names of compounds with ions that are not in Table 4-2. What would we name $NaBrO_4$ and $Cu(IO)_2$? When Mendeleev organized the periodic table, he placed elements with similar properties in the same column. Notice that Br and I are in the same column as Cl. It is not unreasonable, then, to assume that the ions BrO_4^- and IO^- might be named similarly to ClO_4^- and ClO^-. Perchlorate, then, becomes the model for perbromate and hypochlorite suggests hypoiodite. The charges, too, should remain constant. $NaBrO_4$ then becomes sodium perbromate. $Cu(IO)_2$ requires two -1's to balance the single Cu ion. Copper in this compound, then, must be $+2$, and the compound is named copper(II) hypoiodite.

MAKING IT REAL

Ionic Compounds in the Treatment of Disease

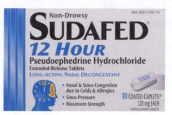

Since antiquity, it has been believed that drinking mineral waters from certain spas can have a soothing effect on the mood as well as the body. Some of these magical minerals may have been compounds containing the lithium ion. Lithium (in the form of the Li^+ ion) is found in certain rocks, in seawater, and in some freshwater springs. It is not known for sure that lithium was present in these waters, but it may have been an ancient treatment for manic-depressive disorder, which is now known more commonly as *bipolar disorder*. This is a debilitating disease that affects about 2.5 million Americans alone. Its symptoms include exaggerated mood swings from exhilarating highs to extremely depressive lows. In the manic mood, a person has boundless energy but acts impulsively and makes poor judgments. In the depressive state, the same person goes in the other direction and the world feels hopelessly glum. If untreated, about one in five commit suicide when in the depressive stage. The disease is known to be hereditary, with a neurological basis in certain chromosomes. It is certainly not due to weakness or any character flaw.

In 1949, John Cade, an Australian physician, found that lithium compounds had a calming effect on small animals. He then tested the toxicity of the compound on himself. Since he seemed okay and suffered no

adverse effects, he began treating manic patients with lithium. Unfortunately, it was not known at the time that lithium can build up to toxic levels in the blood. In other uses, such as for cardiac patients, lithium caused several deaths, so its use in treating bipolar disorder did not gain much ground at that time in the United States. In the 1960s, however, it was reintroduced along with careful monitoring of the levels of lithium in the patient's blood. Under controlled conditions, studies quickly found that lithium was extremely effective. It was approved by the Food and Drug Administration (FDA) in 1970 to treat mania. In 1974, it was approved for use as a preventive (i.e., prophylactic) treatment for bipolar disease.

Treatment for bipolar disorder is most effective when lithium is combined with "talk therapy" with a trained professional. It is taken in the form of lithium carbonate (Li_2CO_3) or lithium citrate ($Li_3C_6H_5O_7$). In any case, many of those that suffer from this illness can now lead normal, productive lives.

There are many other ionic compounds that we ingest for their curative value. Magnesium hydroxide [$Mg(OH)_2$] and calcium carbonate ($CaCO_3$) treat indigestion or "heartburn." Many antihistamines and decongestants are also ionic compounds. If the drug name ends in "hydrochloride" or "HCl," it is an ionic compound. For example, pseudoephedrine hydrochloride ($C_{10}H_{14}NOH_2^+Cl^-$) is an ionic compound used as a nasal decongestant in products such as Sudafed. Like other ionic compounds, it is a solid and can be taken in pill form.

▶ **ASSESSING THE OBJECTIVE FOR SECTION 4-4**

EXERCISE 4-4(a) LEVEL 1: What are the names of the following ions?
(a) NO_2^- **(b)** OH^- **(c)** $C_2H_3O_2^-$ **(d)** PO_4^{3-} **(e)** HSO_4^-

EXERCISE 4-4(b) LEVEL 2: What are the formulas for the following compounds?
(a) nickel(II) cyanide   **(c)** magnesium bicarbonate   **(e)** ammonium phosphate
(b) aluminum sulfite   **(d)** iron(III) perchlorate   **(f)** sodium dichromate

EXERCISE 4-4(c) LEVEL 3: What are the names of the following compounds?
(a) $Ba(OH)_2$ **(b)** LiH_2PO_4 **(c)** $AgClO$ **(d)** K_2CrO_4 **(e)** $Co_2(CO_3)_3$

EXERCISE 4-4(d) LEVEL 3: The prefix *thio-* means to replace one of the oxygen atoms with a sulfur atom. If OCN^- is cyanate, what is iron(III) thiocyanate? What is sodium thiosulfate?

For additional practice, work chapter problems 4-29, 4-32, 4-34, and 4-35.

▶ **OBJECTIVE FOR SECTION 4-5**
Name binary molecular (nonmetal–nonmetal) compounds using proper Greek prefixes or common names.

4-5 Naming Nonmetal-Nonmetal Binary Compounds

LOOKING AHEAD! The compounds that we have named so far are generally ionic compounds. They constitute much of the hard, solid part of nature. When nonmetals bond to other nonmetals, however, molecular compounds are formed. These are very likely to be gases or liquids. How we name molecular compounds is the topic of this section. ■

4-5.1 Writing the Formulas of Binary Molecular Compounds

When a metal is combined with a nonmetal, it is simple to decide which element to name and write first. But which do we write first if neither is a metal? In these cases, we generally write the one closer to being a metal first—that is, the nonmetal closer to the metal–nonmetal border in the periodic table (farther down or farther to the left). Thus, we write CO_2 rather than O_2C but OF_2 rather than F_2O. In cases where both elements are equidistant from the border, Cl is written before O (e.g., Cl_2O) and the others in the order S, N, then Br (e.g., S_4N_4 and NBr_3). When hydrogen is one of the nonmetals and is combined with nonmetals in Groups VIA and VIIA (e.g., H_2O and HF), hydrogen is written first. When combined with other nonmetals (Groups IIIA, IVA, and VA) and metals, however, it is written second (e.g., NH_3 and CH_4).*

Chemical names are familiar to us in many common drugs and cleansers.

TABLE 4-3

Greek Prefixes

NUMBER	PREFIX
1	mono-
2	di-
3	tri-
4	tetra-
5	penta-
6	hexa-
7	hepta-
8	octa-
9	nona-
10	deca-

4-5.2 Naming Binary Molecular Compounds

The nonmetal closer to the metal borderline is also named first, using its English name. The less metallic is named second, using its English root plus *-ide*, as discussed before. These are not ionic compounds, so there are no charges that require balancing. However, this also means that more than one combination of the two elements is possible. Because more than one compound of the same two nonmetals could exist, *the number of atoms of each element present in the compound is indicated by the use of* **Greek prefixes**. (See Table 4-3.) Table 4-4 illustrates the nomenclature of nonmetal–nonmetal compounds with the six oxides of nitrogen. Notice that if there is only one atom of the nonmetal written first, *mono-* is not used. However, if there is only one of the second nonmetal, *mono-* is used. (Notice that the second *o* in *mono* is dropped in *monoxide* for ease in pronunciation.) The Stock method is

---

*With few exceptions, in organic compounds containing C, H, and other elements, the C is written first followed by H and then other elements that are present.

TABLE 4-4

The Oxides of Nitrogen

FORMULA	NAME
N_2O	dinitrogen monoxide (sometimes referred to as nitrous oxide)
NO	nitrogen monoxide (sometimes referred to as nitric oxide)
N_2O_3	dinitrogen trioxide
NO_2	nitrogen dioxide
N_2O_4	dinitrogen tetroxide[a]
N_2O_5	dinitrogen pentoxide[a]

[a]*The a is often omitted from* tetra *and* penta *with oxides for ease in pronunciation.*

Nitrogen dioxide contributes to the brownish haze in polluted air.

rarely applied to the naming of nonmetal–nonmetal compounds because it can be ambiguous in some cases. For example, both NO_2 and N_2O_4 could be named nitrogen(IV) oxide.

Several of these compounds are known only by their common names, such as water (H_2O), methane (CH_4) and related carbon–hydrogen compounds, and ammonia (NH_3).

According to the rules, compounds such as TiO_2 and UF_6 should be named by the Stock method—for example, titanium(IV) oxide and uranium(VI) fluoride, respectively. Sometimes, however, we hear them named in the same manner as nonmetal–nonmetal binary compounds (i.e., titanium dioxide and uranium hexafluoride). The rationale for the latter names is that when the charge on the metal exceeds +3, the compound has properties more typical of a molecular nonmetal–nonmetal binary compound than an ionic one. For example, UF_6 is a liquid at room temperature, whereas true ionic compounds are all solids under these conditions. In any case, in this text we will identify all metal–nonmetal binary compounds by the Stock method regardless of their properties and confine the use of Greek prefixes to the nonmetal–nonmetal compounds.

EXAMPLE 4-7 Naming Binary Molecular Compounds

What are the names of **(a)** $SeBr_4$ and **(b)** B_2O_3? What are the formulas of **(c)** dichlorine trioxide and **(d)** sulfur hexafluoride?

PROCEDURE

Check in the periodic table for the location of all the elements involved. Note that they are all nonmetals, meaning that these are examples of binary molecular compounds. Use the Greek prefixes and the ending -*ide* in the names.

SOLUTION

(a) selenium tetrabromide   **(b)** diboron trioxide   **(c)** Cl_2O_3 **(d)** SF_6

ANALYSIS

These are some of the easiest types of compounds to deal with. The prefixes essentially do all the work for you. The real chore comes in identifying these as binary molecular compounds in the first place.

SYNTHESIS

Sometimes formulas can be very confusing. In Chapter 2 (Figure 2-13) we discussed the difference between NO_2 (nitrogen dioxide) and NO_2^- (the nitrite ion). Another case involves SO_3 and SO_3^{2-}. How would you expect these two to differ from each other? SO_3 is a binary molecular compound named sulfur trioxide. It is a molecular compound made from one sulfur and three oxygen atoms. It is a gas at room temperature. SO_3^{2-} is the sulfite anion. It is not a compound in and of itself. It cannot stand alone; instead, it needs to be combined with a charge-balancing cation, such as Na^+. It would then form the solid ionic compound, sodium sulfite (Na_2SO_3).

TABLE 4-5

The Straight-Chained Alkanes

FORMULA	NAME
CH_4	methane
C_2H_6	ethane
C_3H_8	propane
C_4H_{10}	butane
C_5H_{12}	pentane
C_6H_{14}	hexane
C_7H_{16}	heptane
C_8H_{18}	octane
C_9H_{20}	nonane
$C_{10}H_{22}$	decane

4-5.3 Naming Alkanes

There is a special class of binary molecular compounds generally referred to as **hydrocarbons**, *which contain only carbon and hydrogen.* These compounds are of immense importance in our society. They serve as the fuels that power our cars, heat our homes, and cook our food. One specific class of hydrocarbons is distinguished by its relative number of carbons to hydrogens. **Alkanes** have the general formula C_nH_{2n+2}. In other words, for the number of carbon atoms in the molecule, there are double that number of hydrogen atoms plus two. Alkanes make up some of the best-known fuels that we use. Methane is natural gas used in furnaces and gas stoves. Propane is used as a portable fuel because it remains as a liquid under moderate pressure. Butane is found in lighters because it is a liquid under normal atmospheric pressures. It also serves as a propellant in aerosol cans. Other hydrocarbons will be dealt with in Chapter 17. But for now, it will be helpful to know the names of the first 10 alkanes based on the number of carbon atoms. (See Table 4-5.) After the first four on the list, the names are derived from the prefix for the number of carbons, and the ending *-ane*, indicating the carbon-to-hydrogen relationship.

▶ **ASSESSING THE OBJECTIVE FOR SECTION 4-5**

EXERCISE 4-5(a) LEVEL 1: What are the Greek prefixes for the following numbers?
(a) one   **(b)** four   **(c)** six   **(d)** ten

EXERCISE 4-5(b) LEVEL 2: What are the formulas of the following compounds?
(a) carbon tetrachloride
(b) phosphorus pentabromide
(c) sulfur dioxide
(d) carbon monoxide
(e) pentane
(f) octane

EXERCISE 4-5(c) LEVEL 3: What are the names of the following compounds?
(a) SiS_2 **(b)** P_2O_5 **(c)** BF_3 **(d)** AsH_3

For additional practice, work chapter problems 4-43, 4-44, 4-45, and 4-46.

▶ **OBJECTIVE FOR SECTION 4-6**
Write the names and formulas of acids.

4-6 Naming Acids

LOOKING AHEAD! We have one last category of compounds. These involve most of the anions listed in Figure 4-6 and Table 4-2 when combined with hydrogen. Since hydrogen is not a metal, these are molecular compounds in the pure state. However, many of these compounds have an important property when present in water. This special property allows us to give them special names, and we will do so in this section. ■

When hydrogen is combined with an anion such as Cl^-, the formula of the resulting compound is HCl. The fact that HCl is a gas, not a hard solid, at room temperature indicates that HCl is molecular rather than ionic. When dissolved in water, however, the HCl is ionized by the water molecules to form H^+ ions and Cl^- ions. This ionization is illustrated as

$$HCl \xrightarrow{H_2O} H^+ + Cl^-$$

Most of the hydrogen compounds formed from the anions in Figure 4-6 and Table 4-2 behave in a similar manner, at least to some extent. *This common property of forming H^+ in aqueous solution is a property of a class of compounds called* **acids**. Acids are important enough to earn their own nomenclature. The chemical nature of acids is discussed in more detail in Chapters 6 and 13.

4-6.1 Binary Acids

The acids formed from the anions listed in Figure 4-6 *are composed of hydrogen plus one other element, so they are called* **binary acids**. These compounds can be named in two ways. In the pure state, the hydrogen is named like a metal with only one charge (+1). That is, HCl is named *hydrogen chloride*, and H_2S is named *hydrogen sulfide*. When dissolved in water, however, these compounds are generally referred to by their acid names. The acid name is obtained by dropping the word *hydrogen*, adding the prefix *hydro-* to the anion root, and changing the *-ide* ending to *-ic* followed by the word *acid*. Both types of names are illustrated in Table 4-6. Polyatomic anions that have an *-ide* ending are also named in the same manner as the binary acids. For example, the acid formed by the cyanide ion (CN^-) has the formula HCN and is named hydrocyanic acid.

The following hydrogen compounds of anions listed in Figure 4-6 are not generally considered to be binary acids: H_2O, NH_3, CH_4, and PH_3.

4-6.2 Oxyacids

The acids formed by combination of hydrogen with most of the polyatomic anions in Table 4-2 are known as **oxyacids** *because they are formed from oxyanions.* To name an oxyacid, we use the root of the anion to form the name of the acid. If the name of the oxyanion ends in *-ate*, it is changed to *-ic* followed by the word *-acid*. If the name of the anion ends in *-ite*, it is changed to *-ous* plus the word *-acid*. Most hydrogen compounds of oxyanions do not exist in the pure state as do the binary acids. Generally, only the acid name is used in the naming of these compounds. For example, HNO_3 is called nitric acid, not hydrogen nitrate. Development of the acid name from the anion name is shown for some anions in Table 4-7.

In summary, acids -are named as follows:

ENDING ON ANION	CHANGE	ANION EXAMPLE	ACID NAME
-ide	add *hydro-* and change ending to *-ic*	brom-*ide*	*hydro-*brom*ic* acid
-ite	change ending to *-ous*	hypochlor*ite*	hypochlor*ous* acid
-ate	change ending to *-ic*	perchlor*ate*	perchlor*ic* acid

TABLE 4-6

Binary Acids

ANION	FORMULA OF ACID	COMPOUND NAME	ACID NAME
Cl^-	HCl	hydrogen chloride	hydrochloric acid
F^-	HF	hydrogen fluoride	hydrofluoric acid
I^-	HI	hydrogen iodide	hydroiodic acid
S^{2-}	H_2S	hydrogen sulfide	hydrosulfuric acid

TABLE 4-7

Oxyacids

ANION	NAME OF ANION	FORMULA OF ACID	NAME OF ACID
$C_2H_3O_2^-$	acetate	$HC_2H_3O_2$	acetic acid
CO_3^{2-}	carbonate	H_2CO_3	carbonic acid
NO_3^-	nitrate	HNO_3	nitric acid
PO_4^{3-}	phosphate	H_3PO_4	phosphoric acid
ClO_2^-	chlorite	$HClO_2$	chlorous acid
ClO_4^-	perchlorate	$HClO_4$	perchloric acid
SO_3^{2-}	sulfite	H_2SO_3	sulfurous acid
SO_4^{2-}	sulfate	H_2SO_4	sulfuric acid

EXAMPLE 4-8   Naming Acids

Name the following acids: H_2Se, $H_2C_2O_4$, and $HClO$.

PROCEDURE

Identify the type of acid and the typical ending of the anion (*-ide*, *-ate*, or *-ite*). Change the ending as appropriate and add the word *acid* to the name.

SOLUTION

ACID	ANION	ANION NAME	ACID NAME
H_2Se	Se^{2-}	selenide	hydroselenic acid
$H_2C_2O_4$	$C_2O_4^{2-}$	oxalate	oxalic acid
$HClO$	ClO^-	hypochlorite	hypochlorous acid

EXAMPLE 4-9   Writing Formulas of Acids

Give formulas for the following: permanganic acid, dichromic acid, and acetic acid.

PROCEDURE

Pay attention to the ending for each acid. Decide from what anion the acid is derived. Add enough hydrogens to that anion to balance the charge.

SOLUTION

ACID NAME	ANION NAME	ANION	ACID
permanganic acid	permanganate	MnO_4^-	$HMnO_4$
dichromic acid	dichromate	$Cr_2O_7^{2-}$	$H_2Cr_2O_7$
acetic acid	acetate	$C_2H_3O_2^-$	$HC_2H_3O_2$

ANALYSIS

It helps to have a few models on which to base the naming patterns. Many people are aware that HCl is hydrochloric acid and that H_2SO_4 is sulfuric acid. These are among the most common of all acids. Use the principles of similarity to name other acids that resemble these in much the same fashion. HBr then becomes hydrobromic acid and $HBrO_3$ becomes bromic acid.

SYNTHESIS

The hydrogen in a pure acid is not present as a positive ion. It forms a covalent bond to the anion. Since hydrogen is nearer the metal–nonmetal borderline, it is more metallic and is written first. When acids are present in aqueous (water) solutions, however, they do form positive hydrogen ions. This is what makes them distinctive compounds. Hydrogen can be present in compounds as a negative ion, however. What is it called when the hydrogen is negatively charged? A hydride (just as Cl^- would be called a chloride). In a molecule like acetic acid, $HC_2H_3O_2$, hydrogen is written in two separate locations. The first hydrogen is an acidic one. The next three are neutral, part of the covalent polyatomic anion. Consider the following three compounds: NaH, HI, and CH_4. These represent three different situations for hydrogen. NaH is a typical ionic compound where the hydrogen exists as a −1 anion. HI is an acid that forms H^+ and I^- ions in aqueous solution. CH_4 is a typical molecular compound where the hydrogen is covalently bonded to the carbon. The hydrogens do not ionize in water, nor are they present as anions. We will learn more about this in Chapter 9.

▶ ASSESSING THE OBJECTIVE FOR SECTION 4-6

EXERCISE 4-6(a) LEVEL 1: What changes are made to the name of an acidic compound whose anion would normally end in the following?
(a) -ide   **(b)** -ite   **(c)** -ate

EXERCISE 4-6(b) LEVEL 2: What are the formulas of the following compounds?
(a) acetic acid   **(c)** hydroiodic acid   **(e)** hydrofluoric acid
(b) nitrous acid   **(d)** perchloric acid

EXERCISE 4-6(c) LEVEL 2: What are the names of the following compounds?
(a) H_3PO_4 **(b)** HClO   **(c)** HCN   **(d)** $HClO_3$ **(e)** H_2SO_3

EXERCISE 4-6(d) LEVEL 3: Classify each compound as being an acid, a binary molecular compound, a binary ionic compound with either a representative or a variable cation, or an ionic compound with a polyatomic ion.
(a) $SeBr_4$ **(c)** HNO_3 **(e)** CS_2 **(g)** K_3N
(b) PbO   **(d)** $Mg(OH)_2$ **(f)** $AgNO_3$ **(h)** HF

EXERCISE 4-6(e) LEVEL 3: Water is a unique chemical that can be viewed as fitting in many of the classes of compounds we've outlined. How would you scientifically name water according to the rules of:
(a) binary molecular compounds
(b) acids
(c) ionic compounds with a polyatomic anion (OH^-)
(d) ionic compound of representative elements

For additional practice, work chapter problems 4-47 and 4-48.

PART B SUMMARY

KEY TERMS

4-3.2	Metals that have more than one charge are named by the **Stock method**. p. 127
4-3.3	Metals that have more than one charge can also be named by the **classical method**. p. 127
4-4.1	Most polyatomic anions are **oxyanions**. A **salt** is an ionic compound. p. 129
4-5.2	Binary molecular compounds are named with the use of **Greek prefixes**. p. 132
4-5.3	**Alkanes** have the general formula C_nH_{2n+2}. p. 134
4-6	**Acids** are compounds that produce in ion in aqueous solution. p. 135
4-6.1	**Binary acids** contain hydrogen combined with a monatomic anion. p. 135
4-6.2	**Oxyacids** contain hydrogen combined with an oxyanion. p. 135

SUMMARY CHART

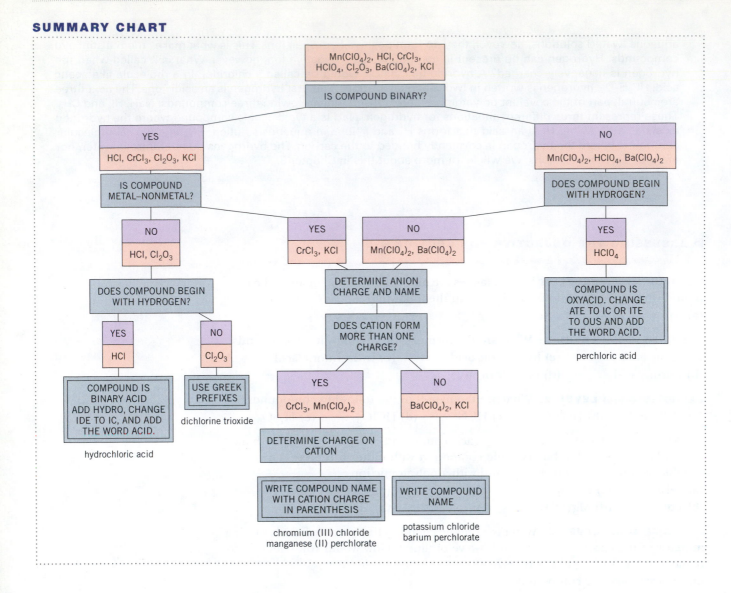

CHAPTER 4 SYNTHESIS PROBLEM

The periodic table is as essential to a chemist as a road map is to a traveler (at least for those who do not totally trust a GPS system). Consider the element fluorine. It is the most reactive element, forming compounds with all other elements except He, Ne, and Ar. Answer the following questions about fluorine and some of the compounds it forms. The last question is a more challenging problem.

PROBLEM	SOLUTION
a. By its location on the periodic table, give as much information about the element as possible including its group number, group name, physical state, charge on its ions, and its formula.	**a.** Locate fluorine in the upper-right side of the periodic table. Fluorine is a nonmetal and a member of group VIIA known as the halogens. It is a gas with the formula F_2. In binary compounds with metals it forms a -1 ion.

PROBLEM	SOLUTION
b. Give names for the following fluorine compounds. **i.** BF_3 **ii.** AlF_3 **iii.** IF_3 **iv.** TlF_3 **v.** IrF_3 **vi.** HF	**b. i.** Boron is a nonmetal in group IIIA, so this is a nonmetal-nonmetal binary compound. Use Greek prefixes. The name is <u>boron trifluoride</u>. **ii.** Aluminum is a metal in group IIIA that forms only one charge of +3. Its name is <u>aluminum fluoride</u>. **iii.** Iodine is a halogen in group VIIA. This is also a nonmetal–nonmetal compound. Its name is <u>iodine trifluoride</u>. **iv.** This is a metal–nonmetal binary compound where the metal has more than one charge. Since the fluorine has a −1 charge the thallium must have a +3 charge. The name is <u>thallium(III) fluoride</u>. **v.** Iridium is a transition metal and would also have a +3 charge. Its name is <u>iridium(III) fluoride</u>. **vi.** This is a binary acid with two names. The pure compound is named <u>hydrogen fluoride</u>. As an acid it is named <u>hydrofluoric acid</u>.
c. Give the formulas of the following compounds formed by fluorine: **i.** xenon tetrafluoride **ii.** chromium(III) fluoride **iii.** barium fluoride **iv.** ammonium fluoride	**c. i.** XeF_4 **ii.** CrF_3 **iii.** BaF_2 **iv.** NH_4F
d. Classify each of the compounds in **b** and **c** above as molecular or ionic compounds.	**d. b.** BF_3, IF_3, and HF are molecular compounds. AlF_3, TlF_3, and IrF_3 are ionic compounds. **c.** xenon tetrafluoride is a molecular compound. The other three are all ionic compounds.
e. Give the name and formula of the anion and the name and formula of the compound from the following information. The metal is in the sixth period and in group VIB. The anion is composed of one sulfur and several oxygens and has a total of 50 electrons. Two of the anions and one cation are in the formula of the compound.	**e.** The metal is tungsten (W). If the anion has 50 electrons and 16 are from the sulfur, that leaves 34 electrons. Four oxygens would provide 32 electrons. Two are left over from one sulfur and four oxygens, forming a −2 charge. This is the sulfate ion, SO_4^{2-}. The formula is $W(SO_4)_2$ and the name is tungsten(IV) sulfate.

YOUR TURN

Oxygen is the second-most-reactive element. It also forms compounds with almost all other elements. Answer the following questions concerning oxygen.

   a. Locate oxygen in the periodic table and give as much information as you can from its location.

   b. Give names for the following compounds; **i.** SrO   **ii.** SeO_2 **iii.** H_2O **iv.** OF_2 **v.** Ti_2O_3 **vi.** Cs_2O

   c. Write formulas for **i.** aluminum oxide **ii.** dichlorine monoxide **iii.** tin(iv) oxide

   d. Describe the compounds in part b and c as either ionic or molecular.

   e. Give the name and formula of the anion and the name and formula of the compound from the following information. The metal is in the fifth period and is in group IIIA. The anion has one silicon and four oxygens and has a total of 48 electrons. Three anions are present for two cations.

Answers are on p. 141.

CHAPTER SUMMARY

There is no more important time-saving device for the chemist than the **periodic table**, which demonstrates in table form the **periodic law** for the elements. One important function of the periodic table is that it shows a clear boundary between elements that are classified as **metals** and **nonmetals**. Some metals and nonmetals have intermediate characteristics and may be referred to as **metalloids**.

Horizontal rows are known as **periods**, and vertical columns are known as **groups**. Although there are four categories of elements in the table, the category that we will emphasize in the text is the **main group** or **representative elements**. This category includes some named groups such as the **alkali metals**, the **alkaline earth metals**, and the **halogens**. The other three categories are the **noble gases**, the **transition metals**, and the **inner transition metals**. The last category includes the **lanthanides** and the **actinides**.

All three physical states are found among the elements at the reference temperature of 25°C, or **room temperature**. Some of the nonmetals also exist as molecules rather than individual atoms at the reference temperature.

Our first important use of the periodic table is to aid us in **chemical nomenclature** by determining the type of chemical compound we are attempting to name. For example, binary compounds containing a metal and a nonmetal are named differently from those composed of two nonmetals. Also, all metals form cations, but some form only one charge and others form more than one. In using the **Stock method** of nomenclature, we need the periodic table to tell us which metals are in the latter group. The **classical method** is still used in the medical field. In addition to the binary compounds, we discussed the naming of **salts** containing **oxyanions**. Binary molecular compounds are named using **Greek prefixes** rather than the Stock method. A category of carbon-hydrogen compounds, known as **alkanes**, are identified by common names. Finally, a class of hydrogen compounds called **acids** (both **binary** and **oxyacids**) was discussed as a special group of molecular compounds.

OBJECTIVES

SECTION	YOU SHOULD BE ABLE TO ...	EXAMPLES	EXERCISES	CHAPTER PROBLEMS
4-1	Describe the origins of the periodic table and identify the locations of the metals and nonmetals.		1a, 1b	
4-2	Using the periodic table, identify a specific element with its group number and name, period, and physical state.		2a, 2b, 2c, 2d	3, 4, 6, 7, 9, 11, 14, 15
4-3	Write the names and formulas of ionic compounds between metals and nonmetals using the IUPAC conventions.	4-1, 4-2, 4-3, 4-4	3a, 3b, 3c, 3d	18, 19, 20, 21, 22, 23, 24, 25
4-4	Write the names and formulas of containing polyatomic ions.	4-5, 4-6	4a, 4b, 4c	32, 33, 34, 35, 36, 38, 40
4-5	Name binary molecular (nonmetal-nonmetal) compounds using proper Greek prefixes, or common names.	4-7	5a, 5b, 5c	41, 43, 44, 45, 46
4-6	Write the names and formulas of acids.	4-8, 4-9	6a, 6b, 6c, 6d	47, 48, 49, 50

▶ ANSWERS TO ASSESSING THE OBJECTIVE

Part A

EXERCISES

4-1(a) The periodic table was first displayed by <u>Dimitri Mendeleev</u> in the year <u>1869</u>. Elements are grouped vertically into <u>families</u> that share the same <u>chemical properties</u>. <u>Metals</u> are lustrous and malleable. <u>Nonmetals</u> are dull and brittle.

4-1(b) Metalloids have properties halfway between metals and nonmetals. They conduct electricity to a small extent. They are *semiconductors*.

4-1(c) Mendeleev chose to focus on chemical properties rather than physical properties, and he was not afraid to shift things around when those properties didn't line up. He also correctly assumed that not all elements had been discovered at that time, so he left spaces in his table to be filled in later.

4-2(a) i. group or family   ii. transition metal   iii. period   iv. noble gas   v. representative or main group element

4-2(b) Of the two general classifications of elements, nickel is a <u>metal</u> and sulfur is a <u>nonmetal</u>. Some borderline elements such as germanium are sometimes referred to as <u>metalloids.</u> In the periodic table, elements are ordered according to increasing <u>atomic number</u> so that <u>families</u> of elements fall into vertical columns. Of the four general categories of elements, calcium is a <u>representative</u> element, nickel is a <u>transition metal</u>, and xenon is a <u>noble gas</u>. An element in Group IIA, such as calcium, is also known as an <u>alkaline earth</u> metal. A solid nonmetal that is composed of diatomic molecules is in the group known as the <u>halogens</u>. Metals are all solids except for <u>mercury</u>.

4-2(c) (a) transition metal, Cu, solid, metal, fourth period, Group IB (b) noble gas, Ar, gas, nonmetal, third period, Group VIIIA (c) representative element, Ba, solid, metal, sixth period, Group IIA (d) representative element, Br, liquid, nonmetal, fourth period, Group VIIA (e) inner transition metal, U, solid, metal, seventh period, no group number

4-2(d) (a) calcium (b) bromine (c) boron (d) cadmium (e) oxygen

4-2(e) 118: noble gas, gas, seventh period, VIIIA, nonmetal 119: representative element, solid, eighth period, IA, metal

Part B
EXERCISES

4-3(a) Co and Sn

4-3(b) (a) lithium oxide (b) chromium(III) iodide (c) lead(II) sulfide (d) magnesium nitride (e) nickel(II) phosphide

4-3(c) (a) AlI_3 (b) Fe_2O_3 (c) $SnBr_4$ (d) Ca_3N_2 (e) Na_2S (f) Cu_3P

4-3(d) MX, MX_2, MX_3, M_2X, M_2X_3, M_3X, M_3X_2

4-4(a) (a) nitrite (b) hydroxide (c) acetate (d) phosphate (e) bisulfate

4-4(b) (a) $Ni(CN)_2$ (b) $Al_2(SO_3)_3$ (c) $Mg(HCO_3)_2$ (d) $Fe(ClO_4)_3$ (e) $(NH_4)_3PO_4$ (f) $Na_2Cr_2O_7$

4-4(c) (a) barium hydroxide (b) lithium dihydrogen phosphate (c) silver(I) hypochlorite (d) potassium chromate (e) cobalt(III) carbonate

4-4(d) $Fe(SCN)_3$; $Na_2S_2O_3$

4-5(a) (a) mono (b) tetra (c) hexa (d) deca

4-5(b) (a) CCl_4 (b) PBr_5 (c) SO_2 (d) CO (e) C_5H_{12} (f) C_8H_{18}

4-5(c) (a) silicon disulfide (b) diphosphorus pentoxide (c) boron trifluoride (d) arsenic trihydride (e) ethane (f) decane

4-6(a) (a) -ide $\rightarrow$ hydro- -ic acid (b) -ite $\rightarrow$ -ous acid (c) -ate $\rightarrow$ -ic acid

4-6(b) (a) $HC_2H_3O_2$ (b) HNO_2 (c) HI (d) $HClO_4$ (e) HF

4-6(c) (a) phosphoric acid (b) hypochlorous acid (c) hydrocyanic acid (d) chloric acid (e) sulfurous acid

4-6(d) (a) binary molecular compound (b) ionic compound with a variable cation (c) acid (with an oxyanion) (d) ionic compound with a representative cation and a polyatomic anion (e) binary molecular compound (f) ionic compound with a variable cation and a polyatomic anion (g) ionic compound with a representative cation (h) acid (binary)

4-6(e) (a) dihydrogen monoxide (b) hydroxic acid (c) hydrogen hydroxide (d) hydrogen oxide

---

ANSWERS TO CHAPTER SYNTHESIS PROBLEM

a. Oxygen is a nonmetal in group VIA. It is a gas that forms diatomic molecules (O_2) and, under special circumstances, triatomic molecules, (O_3) known as ozone. It forms a -2 ion with metals in binary compounds.

b. i. strontium oxide **ii.** selenium dioxide **iii.** water **iv.** oxygen difluoride **v.** titanium(III) oxide **vi.** cesium oxide

c. i. Al_2O_3 **ii.** Cl_2O **iii.** SnO_2

d. SeO_2, H_2O, OF_2, and Cl_2O are molecular compounds. SrO, Ti_2O_3, Cs_2O, Al_2O_3, and SnO_2 are ionic. (SnO_2 is actually on the border.)

e. The metal is indium (In). The anion is SiO_3^{2-} known as the silicate ion. (The CO_3^{2-} ion in the period above is the carbonate ion.) The formula of the compound is $In_2(SiO_4)_3$ and its name is indium(III) silicate.

CHAPTER PROBLEMS

Throughout the text, answers to all exercises in color are given in Appendix E. The more difficult exercises are marked with an asterisk.

The Periodic Table (SECTION 4-2)

4-1. How is an active metal different from a noble metal?

4-2. How many elements are in the recently completed seventh period?

4-3. Which of the following elements are halogens?

(a) O_2 (c) I_2 (e) Li   (g) Br_2
(b) P_4 (d) N_2 (f) H_2

4-4. Which of the following elements are alkaline earth metals?

(a) Sr   (c) B   (e) Na
(b) C   (d) Be   (f) K

4-5. Classify the following elements into one of the four main categories of elements.

(a) Fe   (c) Pm   (e) Xe   (g) In
(b) Te   (d) La   (f) H

4-6. Classify the following elements into one of the four main categories of elements.

(a) Se   (c) Ni   (e) Zn   (g) Er

(b) Ti   (d) Sr   (f) I

4-7. Which of the following elements are transition metals?

(a) In   (c) Ca   (e) Pd   (g) Ag

(b) Ti   (d) Xe   (f) Tl

Physical States of the Elements (SECTION 4-2)

4-8. What is the most common physical state of the elements at room temperature? Which are more common, metals or nonmetals?

4-9. Which metals, if any, are gases at room temperature? Which metals, if any, are liquids? Which nonmetals, if any, are liquids at room temperature?

4-10. Referring to Figure 4-5, tell which of the following are gases at room temperature.

(a) Ne   (c) B   (e) Br   (g) Na

(b) S   (d) Cl   (f) N

4-11. Referring to Figure 4-5, tell which of the following elements exist as diatomic molecules under normal conditions.

(a) N   (c) Ar   (e) H   (g) Xe

(b) C   (d) F   (f) B   (h) Hg

4-12. Referring to Figure 4-4, tell which of the following elements are metals.

(a) Ru   (c) Hf   (e) Ar   (g) Se

(b) Sn   (d) Te   (f) B   (h) W

4-13. Which, if any, of the elements in problem 4-12 can be classified as a metalloid?

4-14. Identify the following elements using the information in Figures 4-4 and 4-5.

(a) a nonmetal, monatomic gas in the third period

(b) a transition metal that is a liquid

(c) a diatomic gas in Group VA

(d) the second metal in the second period

4-15. Identify the following elements using the information in Figures 4-4 and 4-5.

(a) a nonmetal, diatomic liquid

(b) the last element in the third period

(c) a nonmetal, diatomic solid

(d) the only member of a group that is a nonmetal

4-16. What properties would you expect for element 118 if enough atoms were produced to actually study? (Assume that the border between metals and nonmetals continues as before.)

4-17. What is the atomic number of the alkaline earth metal that would appear after element 118?

Metal-Nonmetal Binary Compounds (SECTION 4-3)

4-18. Name the following compounds.

(a) LiF   (c) Sr_3N_2 (e) $AlCl_3$

(b) BaTe   (d) BaH_2

4-19. Name the following compounds.

(a) CaI_2 (c) BeSe   (e) RaS

(b) FrF   (d) Mg_3P_2

4-20. Give formulas for the following compounds.

(a) rubidium selenide   (d) aluminum telluride

(b) strontium hydride   (e) beryllium fluoride

(c) radium oxide

4-21. Give formulas for the following compounds.

(a) potassium hydride   (c) potassium phosphide

(b) cesium sulfide   (d) barium telluride

4-22. Name the following compounds using the Stock method.

(a) Bi_2O_5 (c) SnS_2 (e) TiO_2

(b) SnS   (d) Cu_2Te

4-23. Name the following compounds using the Stock method.

(a) CrI_3 (c) IrO_4 (e) $NiCl_2$

(b) $TiCl_4$ (d) MnH_2

4-24. Give formulas for the following compounds.

(a) copper(I) sulfide   (d) nickel(II) phosphide

(b) vanadium(III) oxide   (e) chromium(VI) oxide

(c) gold(I) bromide

4-25. Give formulas for the following compounds.

(a) yttrium(III) hydride   (c) bismuth(V) fluoride

(b) lead(IV) chloride   (d) palladium(II) selenide

4-26. From the magnitude of the charges on the metals, predict which of the compounds in problems 4-22 and 4-24 may be molecular compounds.

4-27. From the magnitude of the charges on the metals, predict which of the compounds in problems 4-23 and 4-25 may be molecular compounds.

Compounds with Polyatomic Ions (SECTION 4-4)

4-28. Which of the following is the chlorate ion?

(a) ClO_2^- (c) ClO_3^- (e) ClO_3^+

(b) ClO_4^- (d) Cl_3O^-

4-29. Which of the following ions have a -2 charge?

(a) sulfate   (c) chlorite   (e) sulfite

(b) nitrite   (d) carbonate   (f) phosphate

4-30. What are the name and formula of the most common polyatomic cation?

4-31. Which of the following oxyanions contain four oxygen atoms?

(a) nitrate   (d) sulfite   (g) carbonate

(b) permanganate   (e) phosphate

(c) perchlorate   (f) oxalate

4-32. Name the following compounds. Use the Stock method where appropriate.

(a) $CrSO_4$ (d) $RbHCO_3$ (g) $Bi(OH)_3$

(b) $Al_2(SO_3)_3$ (e) $(NH_4)_2CO_3$

(c) $Fe(CN)_2$ (f) NH_4NO_3

4-33. Name the following compounds. Use the Stock method where appropriate.

(a) $Na_2C_2O_4$ (c) $Fe_2(CO_3)_3$

(b) $CaCrO_4$ (d) $Cu(OH)_2$

4-34. Give formulas for the following compounds.

(a) magnesium permanganate

(b) cobalt(II) cyanide

(c) strontium hydroxide

(d) thallium(I) sulfite

(e) iron(III) oxalate

(f) ammonium dichromate

(g) mercury(I) acetate [The mercury(I) ion exists as Hg_2^{2+}.]

4-35. Give formulas for the following compounds.

(a) zirconium(IV) phosphate

(b) sodium cyanide

(c) thallium(I) nitrite

(d) nickel(II) hydroxide

(e) radium hydrogen sulfate

(f) beryllium phosphate

(g) chromium(III) hypochlorite

4-36. Complete the following table. Write the appropriate anion at the top and the appropriate cation to the left. Write the formulas and names in other blanks as is done in the upper-left-hand box.

Cation/ Anion	HSO_3^-	_____	_____
NH_4^+	NH_4HSO_3 Ammonium bisulfite	_____	_____
_____	_____	CoTe _____ (name)	_____
_____	_____	_____	_____ (formula) aluminum phosphate

4-37. Complete the following table. Write the appropriate anion at the top and the appropriate cation to the left. Write the formulas and names in other blanks as is done in the upper-left-hand box.

Cation/ Anion	_____	$C_2O_4^{2-}$	_____
	_____ (formula) thallium (I) hydroxide	_____	_____
Sr^{2+}	_____	_____	_____
_____	_____	_____	TiN (name)

4-38. Give the systematic name for each of the following.

	Common Name	Formula
(a)	table salt	$NaCl$
(b)	baking soda	$NaHCO_3$
(c)	marble or limestone	$CaCO_3$
(d)	lye	$NaOH$
(e)	Chile saltpeter	$NaNO_3$
(f)	sal ammoniac	$NH4Cl$
(g)	alumina	Al_2O_3
(h)	slaked lime	$Ca(OH)_2$
(i)	caustic potash	KOH

4-39. The perxenate ion has the formula XeO_6^{4-}. Write formulas of compounds of perxenate with the following.

(a) calcium   (b) potassium   (c) aluminum

*4-40. Name the following compounds. In these compounds, an ion is involved that is not in Table 4-2. However, the name can be determined by reference to other ions of the central element or from ions in Table 4-2 in which the central atom is in the same group.

(a) PH_4F (c) $Co(IO_3)_3$ (e) $AlPO_3$

(b) $KBrO$ (d) $CaSiO_3$ (f) $CrMoO_4$

Nonmetal-Nonmetal Binary Compounds (SECTION 4-5)

4-41. The following pairs of elements combine to form binary compounds. Which element should be written and named first?

(a) Si and S   (c) H and Se   (e) H and F

(b) F and I   (d) Kr and F   (f) H and As

4-42. The following pairs of elements combine to make binary compounds. Which element should be written and named first?

(a) S and P   (c) O and Br

(b) O and S   (d) As and Cl

4-43. Name the following.

(a) CS_2 (c) P_4O_{10} (e) CH_4 (g) PCl_5

(b) BF_3 (d) Br_2O_3 (f) Cl_2O (h) SF_6

4-44. Name the following.

(a) PF_3 (c) ClO_2 (e) $SeCl_4$

(b) I_2O_3 (d) AsF_5 (f) SiH_4

4-45. Write the formulas for the following.

(a) tetraphosphorus hexoxide   (d) hexane

(b) carbon tetrachloride   (e) sulfur hexafluoride

(c) iodine trifluoride   (f) xenon dioxide

4-46. Write formulas for the following.

(a) xenon trioxide

(b) sulfur dichloride

(c) dibromine monoxide

(d) pentane

(e) diboron hexahydride (also known as diborane)

Naming Acids (SECTION 4-6)

4-47. Name the following acids.

(a) HCl   (c) HClO   (e) HIO_4

(b) HNO_3 (d) $HMnO_4$ (f) HBr

4-48. Write formulas for the following acids.

(a) hydrocyanic acid   (d) carbonic acid

(b) hydroselenic acid   (e) hydroiodic acid

(c) chlorous acid   (f) acetic acid

4-49. Write formulas for the following acids.

(a) oxalic acid   (c) dichromic acid

(b) nitrous acid   (d) phosphoric acid

***4-50.** Refer to the ions in problems 4-39 and 4-40. Write the acid names for the following.

(a) HBrO   (c) H_3PO_3 (e) H_4XeO_6

(b) HIO_3 (d) $HMoO_4$

***4-51.** Write the formulas and the names of the acids formed from the arsenite (AsO_3^{3-}) ion and the arsenate (AsO_4^{3-}) ion.

General Problems

4-52. A gaseous compound is composed of two oxygens and one chlorine. It has been used to kill anthrax spores in contaminated buildings. Write the formula of the compound and give its name.

4-53. The halogen (A_2) with the lowest atomic number forms a compound with another halogen (X_2) that is a liquid at room temperature. The compound has the formula A_5X or XA_5. Write the correct formula with the actual elemental symbols and the name.

4-54. A metal that has only a +2 ion and is the third member of the group forms a compound with a nonmetal that has a −2 ion and is in the same period. What are the formula and name of the compound?

4-55. The only gas in a certain group forms a compound with a metal that has only a +3 ion. The compound contains one ion of each element. What are the formula and name of the compound? What are the formula and name of the compound the gas forms with a Ti^{2+} ion?

4-56. An alkali metal in the fourth period forms a compound with the phosphide ion. What are the formula and name of the compound?

4-57. A transition metal ion with a charge of +2 has 25 electrons. It forms a compound with a nonmetal that has only a −1 ion. The anion has 36 electrons. What are the formula and name of the compound?

4-58. The lightest element forms a compound with a certain metal in the third period that has a +2 ion and with a nonmetal in the same period that has a −2 ion. What are the formulas and names of the two compounds?

4-59. The thiosulfate ion has the formula. $S_2O_3^{2-}$. What are the formula and name of the compound formed between the thiosulfate ion and an Rb ion; an Al ion; an Ni^{2+} ion; and a Ti^{4+} ion? What are the formula and name of the acid formed from the thiosulfate ion?

4-60. Name the following compounds: NiI_2, H_3PO_4, $Sr(ClO_3)_2$, H_2Te, As_2O_3, Sb_2O_3, and SnC_2O_4.

4-61. Name the following compounds: SiO_2, SnO_2, MgO, $Pb_3(PO_4)_2$, $HClO_2$, $BaSO_4$, and HI.

4-62. Give formulas for the following compounds: tin(II) hypochlorite, chromic acid, xenon hexafluoride, barium nitride, hydrofluoric acid, iron(III) telluride and lithium phosphate.

4-63. Which of the following is composed of a metal that can have one charge and a polyatomic ion?

(a) H_2CO_3 (c) B_2O_3 (e) $Rb_2C_2O_4$

(b) CaH_2 (d) $V(NO_3)_3$

4-64. Which of the following is composed of a metal that can have more than one charge and a monatomic anion?

(a) $Ti(ClO_4)_2$ (c) Cu_2Se (e) $MgCrO_4$

(b) Mg_2S (d) H_2Se

4-65. The peroxide ion has the formula O_2^{2-}. What are the formulas of compounds formed with Rb, Mg, Al, and Ti^{4+}? What is the formula of the acid for this anion? What is the name of this compound as a pure compound and as an acid?

4-66. The cyanamide ion has the formula CN_2^{2-} What are the formulas of compounds formed with Li, Ba, Sc^{3+}, and Sn^{4+}? What is its formula as an acid? What is the name of this compound as a pure compound and as an acid?

4-67. Give the formulas of the following common compounds.

(a) sodium carbonate   (d) aluminum nitrate

(b) calcium chloride   (e) calcium hydroxide

(c) potassium perchlorate   (f) ammonium chloride

4-68. Give the names of the following common compounds.

(a) $TiCl_4$ (d) $Mg(OH)_2$ (g) $NaClO_4$

(b) NH_4NO_3 (e) HNO_3

(c) LiH   (f) H_2SO_4

4-69. Nitrogen is found in five ions mentioned in this chapter. Write the formulas and names of these ions.

4-70. Carbon is found in five ions mentioned in this chapter. Write the formulas and names of these ions.

4-71. Which of the following is the correct name for $Cr_2(CO_3)_3$?

(a) dichromium tricarbonate

(b) chromium carbonate

(c) chromium(II) carbonate

(d) chromium(III) tricarbonate

(e) chromium(III) carbonate

4-72. Which of the following is the correct name for $SiCl_4$?

(a) sulfur tetrachloride   (d) sulfur chloride

(b) silicon tetrachloride   (e) silicon chloride

(c) silicon(IV) chloride

4-73. Which of the following is the correct name for $Ba(ClO_2)_2$?

(a) barium dichlorite   (d) barium chlorite(II)

(b) barium(II) chlorite   (e) barium chlorate

(c) barium chlorite

4-74. Which of the following is the correct name for H_2CrO_4?

(a) hydrogen(I) chromate   (d) dichromic acid

(b) hydrogen chromate   (e) dihydrogen chromate

(c) chromic acid   (f) chromous acid

STUDENT WORKSHOP

A New Periodic Table

Purpose: To create a table that illustrates the relationships in properties between elements. (Work in groups of three or four. Estimated time: 40 min.)

1. Cut heavy cardstock into 18 equal-sized squares, approximately 2 in. per side. Onto each of them, transfer the data from one line in the following table.

2. Arrange the cards into rows and columns using whatever information seems relevant. Try to use at least one property for your rows and one for your columns to establish a pattern.

3. Be prepared to discuss your arrangement and give justifications for it. Have any other patterns emerged other than the two that you used to establish rows and columns?

MELTING POINT	BOILING POINT	DENSITY	MASS	FORMULA WITH Cl	FORMULA WITH O
−272	−269	0.18	4	NONE	NONE
−259	−253	0.09	1	XCL	X_2O
−249	−246	0.90	20	NONE	NONE
−220	−188	1.70	19	NONE	X_2O
−218	−183	1.43	16	NONE	NONE
−210	−196	1.25	14	XCL_3	MULTIPLE
−189	−186	1.78	40	NONE	NONE
−101	−35	3.21	36	NONE	MULTIPLE
44	280	1.82	31	XCL_3	MULTIPLE
98	553	0.97	23	XCL	X_2O
113	445	2.07	32	MULTIPLE	MULTIPLE
180	1347	0.53	7	XCL	X_2O
650	1107	1.74	24	XCL_2	XO
660	2467	2.70	27	XCL_3	X_2O_3
1278	2970	1.85	9	XCL_2	XO
1410	2355	2.33	28	XCL_4	XO_2
2300	2550	2.34	11	XCL_3	X_2O_3
3500	4827	2.62	12	XCL_4	XO, XO_2

Chapter 5

Quantities in Chemistry

Modern farming techniques require the addition of nitrogen to the soil. The composition of ammonium nitrate makes it suitable for delivering a large amount of nitrogen. The composition of compounds is a topic of this chapter.

AMMONIUM NITRATE
BASED FERTILIZER
34,4% N

TOTAL NITROGEN (N)	34,4%
NITRIC NITROGEN (N)	17,2%
AMMONIACAL NITROGEN (N)	17,2%

500 KG NET

5.1

UN 2067

SETTING THE STAGE

Consider the molecules of ordinary water (H_2O). The smallness of these molecules defies the ability of our human minds to comprehend. Because of their size, it takes a huge number to become visible to the naked eye. Even the tiniest droplet of water that we can barely see contains between 10^{16} and 10^{17} molecules (over 10 quadrillion). To form this tiny droplet, we would need to combine enormous numbers of hydrogen and oxygen atoms. From the formula of water notice that we would need two atoms of hydrogen for each atom of oxygen. For this tiny sample of water it would certainly be impossible to count the number of atoms needed. Fortunately, in chemistry we count atoms by weighing rather than counting. This works because, on average, all atoms of an element weigh the same. However, we now need a new counting unit that describes a huge number appropriate for large amounts of atoms.

The counting unit of chemistry will provide us with an important tool for specific calculations. For example, consider a very practical application in the use of fertilizers to provide the nitrogen that is essential for crops. Without fertilizers, it is estimated that two out of five of us would not be alive today. Nature just does not supply enough fixed nitrogen to produce the crops to feed nearly 7 billion people. Two manufactured compounds that are widely used as a source of nitrogen for soil are ammonia (NH_3) and ammonium nitrate (NH_4NO_3). The advantage of ammonia is that 100 kg of ammonia supplies 82 kg of nitrogen to the soil. Its disadvantage is that it is a gas and must be injected into the soil. Ammonium nitrate, an ionic compound, is a solid and can be spread on the surface. Its disadvantage is that it supplies only 35 kg of nitrogen for each 100 kg of fertilizer applied. The decision about which to use depends on cost, availability, and soil conditions. The mass of nitrogen in a given amount of each of these two compounds relates to its formula and to the relative masses of the elements in each compound, as we will discuss in this chapter.

In Part A, we will examine the main unit of mass and number as it relates to elements and compounds. We will use this information in Part B to examine the elemental composition of compounds.

Part A

The Measurement of Masses of Elements and Compounds

SETTING A GOAL

- You will become proficient at working with the units of moles, mass, and numbers of atoms and molecules, and at converting between each of these.

OBJECTIVES

5-1 Calculate the masses of equivalent numbers of atoms of different elements.

5-2 Define the mole in terms of numbers of atoms and atomic masses of elements.

5-3 Perform calculations involving masses, moles and number of molecules or formula units for a compound.

▶ **OBJECTIVE FOR SECTION 5-1**
Calculate the masses of equivalent numbers of atoms of different elements.

5-1 Relative Masses of Elements

LOOKING AHEAD! A good place to start is by looking at how the mass of an average atom of one element relates to the average mass of an atom of another element. From this we relate the masses of elements containing any equivalent numbers of atoms. ■

5-1.1 Counting by Weighing

Say a carpenter needs a large number of bolts with the same number of nuts. Counting them out one by one is certainly not the easy way to do it. However, if we know the average weight of a nut and a bolt we can obtain equivalent numbers by use of a straightforward calculation and a scale. So we buy each by the pound or kilogram, knowing that these weights contain a specific number of nuts and bolts. Like nuts and bolts, atoms of a specific element all weigh the same on average, so we can use a scale to obtain specific numbers of atoms. We will illustrate these calculations for nuts and bolts as well as for hydrogen and oxygen atoms.

We will assume that our carpenter needed 175 bolts and the same number of nuts. If an average bolt weighed 10.5 g, the following calculation tells the carpenter the weight of bolts needed.

$$175 \ \text{bolts} \times \frac{10.5 \ \text{g}}{\text{bolt}} \times \frac{1 \ \text{kg}}{10^3 \ \text{g}} = \underline{1.84 \ \text{kg bolts}}$$

A nut weighs 2.25 g. What weight of nuts is needed to provide a nut for each bolt?

$$175 \ \text{nuts} \times \frac{2.25 \ \text{g}}{\text{nut}} \times \frac{1 \ \text{kg}}{10^3 \ \text{g}} = \underline{0.394 \ \text{kg of nuts}}$$

Actually, if we just want to have the same number of nuts as bolts but don't care what the number is, we can use a shortcut. Once we know the mass ratio of bolts to nuts, we can easily convert an equivalent number of nuts to bolts, and vice versa. The ratio of the masses can be expressed as follows.

$$\frac{10.5 \ \text{g bolt}}{2.25 \ \text{g nut}} \quad \text{and} \quad \frac{2.25 \ \text{g nut}}{10.5 \ \text{g bolt}}$$

The mass ratio holds for any units of mass. For example, there will be the same number of nuts as bolts in 1.84 pounds of bolts and 2.25 pounds of nuts. Using the appropriate mass ratio expressed in kilograms rather than grams, we can make the previous calculation easier. The following calculation allows us to measure the same number of nuts as there are in 1.84 kg of bolts.

Each bolt requires one nut.

$$1.84 \text{ kg bolts} \times \frac{2.25 \text{ kg nuts}}{10.5 \text{ kg bolts}} = 0.394 \text{ kg nuts}$$

5-1.2 The Mass of an Average Atom

Before we apply the same principles to the atoms of an element, consider how their mass units are expressed. Each carbon atom in a tiny smudge of graphite is a specific isotope of carbon. Most of the atoms are ^{12}C, which is used as the standard and is assigned a mass of exactly 12 amu. The masses of other carbon isotopes and the isotopes of any other element are calculated by comparison to ^{12}C. Most elements occur in nature as mixtures of isotopes, so the atomic mass in the periodic table represents the weighted average mass of all isotopes of that element. In this chapter, we will treat the atoms of an element as if they were all identical, with the atomic mass representing the mass of this hypothetical "average" atom.

In most of the calculations that follow in this chapter, we do not need to know the *actual* number of atoms involved in a certain weighed amount, but we do need to know the *relative* numbers of atoms of different elements present. However, if we know the relative masses of the individual atoms, this is no problem. For example, if one hydrogen atom has a mass of 1.00 amu and one oxygen atom has a mass of 16.0 amu, their masses are in the following ratio.

$$\frac{1.00 \text{ amu H}}{16.0 \text{ amu O}}$$

In fact, any time hydrogen and oxygen are present in a 1.00:16.0 mass ratio *regardless of the units of mass*, we can conclude that the same number of atoms of each element is present. (See Table 5-1.) We can generalize this statement to all the elements. *When the masses of samples of any two elements are in the same ratio as that of their atomic masses, the samples have the same number of atoms.* Thus, if we wanted the same number of hydrogen atoms as the number of atoms present in 45.0 g of oxygen, we do not have to count or even know what the number is. The following calculation tells us what we want.

$$45.0 \text{ g O} \times \frac{1.00 \text{ g H}}{16.0 \text{ g O}} = 2.81 \text{ g H}$$

TABLE 5-1

Mass Relation of O and H

O	H	NUMBER OF ATOMS OF EACH ELEMENT PRESENT
16.0 amu	1.00 amu	1
32.0 amu	2.00 amu	2
16.0 g	1.00 g	6.02×10^{23}
32.0 g	2.00 g	1.20×10^{24}
16.0 lb	1.00 lb	2.73×10^{26}
32.0 ton	2.00 ton	1.09×10^{30}

EXAMPLE 5-1   Calculating the Relative Masses of Elements in a Compound

The formula of the compound magnesium sulfide (MgS) indicates that there is one atom of Mg for every atom of S. What mass of sulfur is combined with 46.0 lb of magnesium?

PROCEDURE

From the atomic masses in the periodic table, note that one atom of magnesium has a mass of 24.31 amu and one atom of sulfur has a mass of 32.07 amu. Likewise, 24.31 lb of magnesium and 32.07 lb of sulfur have an equal number of atoms. This statement can be represented by two conversion factors, which we can use to change a mass of one element to an equivalent mass of the other.

$$(1) \; \frac{24.31 \text{ lb Mg}}{32.07 \text{ lb S}} \qquad (2) \; \frac{32.07 \text{ lb S}}{24.31 \text{ lb Mg}}$$

Use factor **(2)** to convert the mass of Mg to an equivalent mass of S.

SOLUTION

$$46.0 \text{ lb Mg} \times \frac{32.07 \text{ lb S}}{24.31 \text{ lb Mg}} = 60.7 \text{ lb S}$$

Thus, 60.7 lb of sulfur has the same number of atoms as 46.0 lb of magnesium.

ANALYSIS

Again, let us stress that we are saying nothing about the actual number of atoms present. But we don't need to. It is not really important to know whether there are a million atoms or a million and ten. Just knowing that there is the same number of atoms is the most important thing chemically as compound formulas are made up of specific numbers of atoms.

SYNTHESIS

Can this be done with compounds made up of atoms of three different elements? Consider sodium carbonate, Na_2CO_3. What would the ratio of the masses look like for this molecule? Since there are two sodium atoms, the total relative mass of sodium would be $23 \times 2 = 46$. The relative mass of oxygen would be $16 \times 3 = 48$. So any of these ratios or any of their inverses could be used to solve problems.

$$\frac{46 \text{ amu Na}}{12 \text{ amu C}} \qquad \frac{46 \text{ amu Na}}{48 \text{ amu O}} \qquad \frac{48 \text{ amu O}}{12 \text{ amu C}}$$

Also, any unit of mass could be put into the equation: pounds, grams, tons. Typically grams will be the most convenient.

▶ **ASSESSING THE OBJECTIVE FOR SECTION 5-1**

EXERCISE 5-1(a) LEVEL 1: What mass of each of these elements has the same number of atoms as does 12.00 g of ^{12}C?
(a) Cl   **(b)** Na   **(c)** Pb

EXERCISE 5-1(b) LEVEL 1: What is the approximate ratio of masses of elements in the following compounds?
(a) SiC   **(b)** HF   **(c)** SO_2

EXERCISE 5-1(c) LEVEL 2: A compound has the formula NO. What mass of oxygen is present in the compound for each 25.0 g of nitrogen?

EXERCISE 5-1(d) LEVEL 2: What mass of fluorine would combine with 8.50 g of Si to make SiF_4?

EXERCISE 5-1(e) LEVEL 3: If 15.0 g of Fe and 8.00 g of S combined to form FeS, would you use all of the reactants? Would anything be left over? How much?

For additional practice, work chapter problems 5-4, 5-6, and 5-9.

▶ **OBJECTIVE FOR SECTION 5-2**

Define the mole in terms of the numbers of atoms and atomic masses of elements.

5-2 The Mole and the Molar Mass of Elements

LOOKING AHEAD! From the previous discussion we can conclude that the atomic masses of the elements, regardless of the mass units, represent equal numbers of atoms. But what is this number and just how big is it? That will be discussed in this section. ■

5-2.1 The Definition of the Mole

The mass of individual atoms has, so far, been expressed in terms of the mass unit *amu* (atomic mass unit). This is valuable when comparing the masses of individual atoms, but it has no practical value in a laboratory situation. The mass unit used in the laboratory is the gram. The mass of an "average" carbon atom (12.01 amu) expressed in grams (1.994×10^{-23} g) is obviously too small to measure. Since even the best laboratory balance can detect nothing less than 10^{-3} g, it is reasonable that we need many atoms at a time to register on our scales. Thus, we must "scale-up" the numbers of atoms so that the amounts are detectable with our laboratory instruments. In order to scale-up our measurements, we need an appropriate counting unit for a huge number of atoms.

The number of atoms represented by the atomic mass of an element expressed in grams is a unit known as a **mole**. (The SI symbol is **mol**.) The following equality expresses the number that one mole represents.

$$1.000 \text{ mol} = 6.022 \times 10^{23} \text{ objects or particles}$$

This number, 6.022×10^{23}, is referred to as **Avogadro's number** (named in honor of Amedeo Avogadro, 1776–1856, a pioneer investigator of the quantitative aspects of chemistry). Avogadro's number has been determined experimentally by various methods. The formal definition of one mole concerns an isotope of carbon, ^{12}C. *One mole is defined as the number of atoms in exactly 12 grams of* ^{12}C. This number, of course, is Avogadro's number. *Thus the atomic mass of one mole of any element, expressed in grams, contains the same number of basic particles as there are in exactly 12 grams of* ^{12}C. Avogadro's number is not an exact, defined number such as 12 in 1 dozen or 144 in 1 gross, but it is known to many more significant figures than the four (i.e., 6.022) that are shown and used in this text.

Many common counting units represent a number consistent with their use. On one hand, a baker sells a dozen doughnuts at a time because 12 is a practical number for that purpose. On the other hand, we buy a ream of computer paper, which contains 500 sheets. A ream of doughnuts and a dozen sheets of computer paper are not practical amounts to purchase for most purposes. Counting units of a dozen or a ream are likewise of little use to a chemist because they don't include enough individual objects. For example, grouping 10^{20} atoms into about 10^{19} dozen atoms does us little good. Because atoms are so small, the counting unit used by chemists needs to include an extremely large number of individual units in order to be practical. (See Figure 5-1.)

Although Avogadro's number is valuable to a chemist, its size defies the ability of the human mind to comprehend. For example, if an atom were the size of a marble and one mole of marbles were spread over the surface of Earth, our planet would be covered by a 50-mi-thick layer. Or, if the marbles were laid end to end and extended into outer space, they would reach past the farthest planets almost to the center of the galaxy. It takes light moving at 186,000 miles per second over 30,000 years to travel from Earth to the center of the galaxy. A new supercomputer can count all the people in the United States in one-quarter of a second, but it would take almost 2 million years for it to count one mole of people at the same rate. A glass of water, which is about 10 moles of water, contains more water molecules than there are grains of sand in the Sahara desert. That's difficult to imagine.

FIGURE 5-1 Counting Units One dozen, one ream, and one mole all have applications dependent on the amount needed.

5-2.2 The Meaning of a Mole

Note that one mole of a certain element implies two things.

1. *The atomic mass expressed in grams,* which is *different* for each element. This mass is known as the **molar mass of the element**. (Most of the measurements that we will use in the examples and exercises that follow are expressed to three significant figures. We will express the molar mass and Avogadro's number to four significant figures so that these factors will not limit the precision of the answer.) Thus, the mass of one mole of oxygen atoms is 16.00 g, the mass of one mole of helium atoms is 4.002 g, and the mass of one mole of uranium atoms is 238.0 g.
2. *Avogadro's number of atoms,* which is the *same* for all elements. (See Figure 5-2.)

FIGURE 5-2 Moles of Elements One mole of iron (the paper clips), copper, liquid mercury, and sulfur are shown. Each sample contains the same number of atoms (Avogadro's number) but has a different mass.

At this point, notice that we have established relationships among a unit (mole), a number (6.022×10^{23}), and the mass of an element expressed in grams. It is now necessary for us to be able to convert among these three quantities. The mechanics of the conversions are not complex, however. So before we do some conversions involving atoms and so on, let's consider similar conversions involving an everyday object like a dozen oranges. In the following two calculations, assume that we are dealing with identical oranges that are weighed and sold by the dozen instead of individually. One dozen of these oranges has a mass of 5.960 lb, so we can consider this to be the "dozen mass" of oranges just as there is a "molar mass" for atoms of elements. The number in 1 dozen, however, is a much more manageable 12 rather than the 6.022×10^{23} in one mole. The relationships between mass of oranges and the number in 1 dozen can be summarized as follows. Notice that the four ratios that result can be used as conversion factors in calculations using dimensional analysis (introduced in Chapter 1).

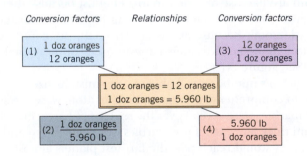

5-2.3 Relationships Among Dozens (the Unit), Number, and Mass

If we are asked to calculate how many dozens of oranges there are in 17.9 lb, we would need to convert from a mass (lb) to a unit (dozen). Factor (2) is used for this conversion, as follows.

$$17.9 \; \cancel{lb} \times \frac{1 \; doz \; oranges}{5.960 \; \cancel{lb}} = \underline{3.00 \; doz \; oranges}$$

In a second problem, if we are asked to calculate the mass of 142 oranges, we need to convert between mass in pounds and number of oranges. Since a direct conversion is not given, the calculation can be done in two steps. We first convert the number of oranges to the dozen unit [factor (1)] and, in the second step, convert the dozen unit to mass [factor (4)]. The unit map appears as

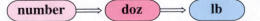

The calculation is as follows.

$$142 \; \cancel{oranges} \times \frac{1 \; \cancel{doz \; oranges}}{12 \; \cancel{oranges}} \times \frac{5.960 \; lb}{\cancel{doz \; oranges}} = \underline{70.5 \; lb}$$

5-2.4 Relationships among Moles (the Unit), Number, and Mass

We are now ready to work very similar problems using moles, grams, and 6.022×10^{23} rather than dozens, pounds, and 12. If you're still a little uneasy with scientific notation, see Chapter 1 or Appendix C for a review.

In these problems, we will assume that the atoms are all identical, with the same size, shape, and mass. From the periodic table we find that the molar mass (the mass of 6.022×10^{23}) of sodium atoms is 22.99 g. The two relationships just given can be either written as equalities or set up as ratios that can be eventually used as conversion factors.

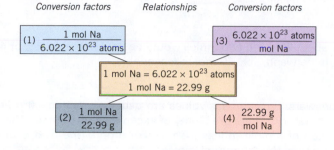

Conversion factors   Relationships   Conversion factors

$(1) \dfrac{1 \text{ mol Na}}{6.022 \times 10^{23} \text{ atoms}}$ $(3) \dfrac{6.022 \times 10^{23} \text{ atoms}}{\text{mol Na}}$

$1 \text{ mol Na} = 6.022 \times 10^{23} \text{ atoms}$
$1 \text{ mol Na} = 22.99 \text{ g}$

$(2) \dfrac{1 \text{ mol Na}}{22.99 \text{ g}}$ $(4) \dfrac{22.99 \text{ g}}{\text{mol Na}}$

EXAMPLE 5-2   Converting Moles to Mass

What is the mass of 3.00 mol of Na?

PROCEDURE

We need a conversion factor that converts units of moles to units of mass, or g. Thus, the conversion factor would have mole in the denominator and grams in the numerator. This is factor (4).

$$\times \frac{g}{mol}$$

$$\text{mol} \Longrightarrow \text{mass}$$

SOLUTION

$$3.00 \text{ mol Na} \times \frac{22.99 \text{ g}}{\text{mol Na}} = \underline{69.0 \text{ g}}$$

ANALYSIS

It's worth repeating that whenever you're solving a problem, the unit of the answer is the unit that belongs in the numerator of your conversion factor. If the unit in the denominator is not the one for which given information is provided, then a second conversion factor is needed, again with this new unit in the numerator. This process continues until you reach the unit with which you're starting.

SYNTHESIS

What would be the mass of the same amount of chlorine? 3.00 mol Cl × 35.45 g/mol = 106 g. So there is the same number of moles, and therefore the same number of atoms, in 69.0 g of Na as there is in 106 g of Cl. If you were trying to make NaCl from its elements, you wouldn't use the same mass of each element. That would be a waste of sodium. You'd want to use about one-and-a-half times as much chlorine. The larger the mass of each atom, the larger the mass needed to give you the same number of moles.

EXAMPLE 5-3   Converting Mass to Moles

How many moles are present in 34.5 g of Na?

PROCEDURE

This is the reverse of the previous problem, so we need a conversion factor that has grams in the denominator and mole in the numerator. This is factor (2).

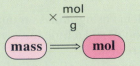

$$34.5 \ \cancel{g} \times \frac{1 \ mol \ Na}{22.99 \ \cancel{g}} = \underline{1.50 \ mol \ Na}$$

ANALYSIS

Does this answer make sense? One mole is about 23 g and two moles would be double that mass, or 46 grams. Clearly the amount given is about halfway in between, so 1.50 mol is reasonable.

SYNTHESIS

Consider what is reasonable for all your answers. Typical mass values are usually between 1.0 g and 100 g, masses that are easy to manipulate in the laboratory. (There are plenty of exceptions, of course.) The numbers of moles are smaller than that, from a fraction of a mole to less than 10. For atoms (the next example), the numbers are around 20 powers of 10 or more than the number of moles.

EXAMPLE 5-4   Converting Number to Mass

What is the mass of 1.20×10^{24} atoms of Na?

PROCEDURE

In this conversion, a direct relationship between mass and number is not given. This requires a two-step conversion. In the first step, we convert the number of atoms to moles [factor (1)]; in the second step, we convert moles to mass [factor (4)].

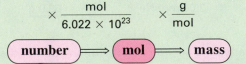

SOLUTION

$$1.20 \times 10^{24} \ \cancel{atoms \ Na} \times \frac{1 \ \cancel{mol \ Na}}{6.022 \times 10^{23} \ \cancel{atoms \ Na}} \times \frac{22.99 \ g}{\cancel{mol \ Na}} = \underline{45.8 \ g}$$

ANALYSIS

Does this answer make sense? One mole (6.02×10^{23} atoms) is about 23 g, and two moles (1.20×10^{24} atoms) would be double that mass, or 46 grams. Clearly the answer should be around 46 g.

SYNTHESIS

What if we simplified the question and asked how many moles are in 8.75×10^{22} atoms of aluminum? The solution would be

$$8.75 \times 10^{22} \ \cancel{atoms \ Al} \times \frac{1 \ mol \ Al}{6.022 \times 10^{23} \ \cancel{atoms \ Al}} = 0.145 \ mol$$

So, then, how many moles are present in 8.75×10^{22} atoms of iron? The same number. Since we are not considering grams, the different molar masses of aluminum and iron are not part of the calculation. It would be analogous to asking how many dozen roses are 24? How many dozen eggs are 24? There are 2 dozen in each case. There's 12 anything in 1 dozen and there's 6.022×10^{23} anything in one mole.

EXAMPLE 5-5   Converting Mass to Number

How many individual atoms are present in 11.5 g of Na?

PROCEDURE

This is the reverse of the previous problem and also requires a two-step conversion. We will convert grams to moles using factor (2) and then moles to number using factor (3).

$$\times \frac{mol}{g} \qquad \times \frac{6.022 \times 10^{23}}{mol}$$

$$mass \Longrightarrow mol \Longrightarrow number$$

SOLUTION

$$11.5 \ g \times \frac{1 \ mol \ Na}{22.99 \ g} \times \frac{6.022 \times 10^{23} \ atoms \ Na}{mol \ Na} = \underline{3.01 \times 10^{23} \ atoms \ Na}$$

ANALYSIS

Always check to see that your answer makes sense. If you were asked to determine grams and the answer you calculated was 3×10^{23} g, is this likely? That is more than the mass of several moons in the solar system. However, if you were calculating the number of atoms and the calculator said 10.25, again, you would want to recheck how you set up the solution. Do the units cancel? Are the conversion factors correct? Was scientific notation entered into the calculator correctly? Any of these could lead to an answer that should send up red flags.

SYNTHESIS

Can you take these two-step analysis problems one more step? What if the element in question was a liquid, like mercury, and we wanted to know the number of atoms in 8.50 mL of Hg. What extra conversion factor is now needed? How would the problem be set up? What we need is a conversion between grams and mL. This is the density of mercury, as we discussed in Chapter 3. In the case of mercury, the factor is 13.5 g/ml. So the problem sets up:

$$mL \times \left(\frac{g}{mL}\right) \times \left(\frac{mol}{g}\right) \times \left(\frac{atoms}{mol}\right) = atoms$$

▶ **ASSESSING THE OBJECTIVE FOR SECTION 5-2**

EXERCISE 5-2(a) LEVEL 1: What is needed for a conversion factor between grams and moles? Between molecules and moles?

EXERCISE 5-2(b) LEVEL 1: Answer the following without the use of a calculator:
(a) How many moles are in 36 grams of carbon?
(b) How many grams are in 0.50 moles of calcium?
(c) How many atoms are in 2.0 moles of sodium?
(d) How many moles are in 6.022×10^{25} atoms of tin?
(e) How many moles are in 10.0 grams of argon?

EXERCISE 5-2(c) LEVEL 2: How many moles of vanadium atoms are represented by 4.82×10^{24} atoms? What is the mass of this number of atoms of vanadium?

EXERCISE 5-2(d) LEVEL 3: Order these items by increasing mass.
(a) 4.55×10^{22} atoms of Pb   **(c)** 0.280 moles of Ca
(b) 8.50 g of C   **(d)** 7.14×10^{22} atoms of Zn

EXERCISE 5-2(e) LEVEL 3: What element has 1.74×10^{22} atoms in 3.12 g of material?

For additional practice, work chapter problems 5-14, 5-16, 5-18, and 5-20.

▶ **OBJECTIVE FOR SECTION 5-3**
Perform calculations involving masses, moles, and number of molecules or formula units for compounds.

5-3 The Molar Mass of Compounds

LOOKING AHEAD! Using *moles* of atoms rather than individual atoms allows us to scale-up our measurements to convenient laboratory quantities. Most of the substances around us, however, are composed of compounds rather than free elements. In this section we will extend the concept of the mole from elements to compounds. ∎

5-3.1 The Formula Weight of a Compound

Just as the mass of an automobile is the sum of the masses of all its component parts, the mass of a molecule is the sum of the masses of its component atoms. First, we will examine the masses of single molecules. *The **formula weight** of a compound is determined from the number of atoms and the atomic mass (in amu) of each element indicated by a chemical formula.* Recall that chemical formulas represent two types of compounds: molecular and ionic. The formula of a molecular compound represents a discrete molecular unit, whereas the formula of an ionic compound represents a formula unit, which is the whole-number ratio of cations to anions (e.g., one formula unit of K_2SO_4 contains two K^+ ions and one SO_4^{2-} ion). The following examples illustrate the calculation of the formula weight of a molecular compound (Example 5-6) and an ionic compound (Example 5-7).

EXAMPLE 5-6 **Calculating the Formula Weight of a Molecular Compound**

What is the formula weight of CO_2?

PROCEDURE

Determine the number of atoms of each type. Multiply the number of atoms by their atomic masses, and add the totals together.

SOLUTION

ATOM	NUMBER OF ATOMS IN MOLECULE	ATOMIC MASS		TOTAL MASS OF ATOMS IN MOLECULE
C	1	× 12.01 amu	=	12.01 amu
O	2	× 16.00 amu	=	32.00 amu
				44.01 amu

The formula weight of CO_2 is

$$44.01 \text{ amu}$$

EXAMPLE 5-7 **Calculating the Formula Weight of an Ionic Compound**

What is the formula weight of $Fe_2(SO_4)_3$?

PROCEDURE

The same process holds for both molecular and ionic compounds.

SOLUTION

ATOM	NUMBER OF ATOMS IN FORMULA UNIT	ATOMIC MASS		TOTAL MASS OF ATOMS IN FORMULA UNIT
Fe	2	× 55.85 amu	=	111.70 amu
S	3	× 32.07 amu	=	96.21 amu
O	12	× 16.00 amu	=	192.00 amu
				399.91 amu

The formula weight of $Fe_2(SO_4)_3$ (to four significant figures) is

$$399.9 \text{ amu}$$

ANALYSIS

In these multiplications we observe the rules for addition and subtraction for significant figures because the numbers of atoms are exact and the multiplication could be considered the same as addition. Therefore, the decimal place takes precedence over the number of significant figures.

SYNTHESIS

When we had single atoms, we were able to scale-up their individual masses from amu to grams to accumulate a more usable quantity of matter. We will shortly see how we can do the same thing with compounds.

5-3.2 The Formula Weight of Hydrates

Certain ionic compounds can have water molecules attached to their cations and/or anions. These compounds are known as *hydrates* and have some distinctive properties compared to their dehydrated (no-water) forms. For example, $CuSO_4$ [copper(II) sulfate] is a pale-green solid, whereas $CuSO_4 \cdot 5H_2O$ [copper(II) sulfate pentahydrate] is a dark-blue solid. The waters of hydration usually can be removed by heating the solid. The formula weight of a hydrate includes the mass of the water molecules. For example, the molar mass of $CuSO_4 \cdot 5H_2O$ is calculated by summing the atomic masses of copper, sulfur, four oxygens, and five waters. That is, 63.55 amu (Cu) + 32.07 amu (S) + $(4 \times 16.00 \text{ amu})$ (O) + $(5 \times 18.02 \text{ amu})$ (2H + O) = 249.7 amu.

5-3.3 The Molar Mass of a Compound

The formula weights that we have calculated represent one molecule or one formula unit. Once again, we need to scale-up our numbers so that we have a workable amount that can be measured on a laboratory balance. Thus, we extend the definition of the mole to include compounds. *The mass of one mole $(6.022 \times 10^{23}$ molecules or formula units) is referred to as the* **molar mass of the compound** *and is the formula weight expressed in grams.* As was the case with atoms of elements, the molar masses of various compounds differ, but the number of molecules or formula units remains the same. (See Figure 5-3.)

FIGURE 5-3 Moles of Compounds One mole of white sodium chloride (NaCl), blue copper sulfate hydrate ($CuSO_4 \cdot 5H_2O$), yellow sodium chromate (Na_2CrO_4), and water are shown. Each sample contains Avogadro's number of molecules or formula units.

EXAMPLE 5-8   Converting Moles to Mass and Number of Formula Units

(a) What is the mass of 0.345 mol of $Al_2(CO_3)_3$?

(b) How many individual ionic formula units does this amount represent?

PROCEDURE

The problem is worked much like the examples in which we were dealing with moles of atoms rather than compounds. In this case, however, we need to find the molar mass of the compound, which is the formula weight expressed in grams. The units of molar mass are g/mol. The molar mass equals

$$2Al + 3C + 9O = [(2 \times 26.98) + (3 \times 12.01) + (9 \times 16.00)] = 234.0 \text{ amu}$$

Thus, the molar mass is 234.0 g/mol.
The conversions are

$$\times \frac{g}{mol}$$

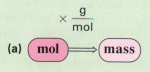

(a)   mol ⟹ mass

and

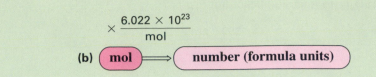

$$\times \frac{6.022 \times 10^{23}}{mol}$$

(b) mol $\Longrightarrow$ number (formula units)

SOLUTION

(a) $0.345 \text{ mol Al}_2(CO_3)_3 \times \dfrac{234.0 \text{ g}}{\text{mol Al}_2(CO_3)_3} = \underline{80.7 \text{ g}}$

(b) $0.345 \text{ mol Al}_2(CO_3)_3 \times \dfrac{6.022 \times 10^{23} \text{ formula units}}{\text{mol Al}_2(CO_3)_3} = \underline{2.08 \times 10^{23} \text{ formula units}}$

ANALYSIS

Most quantitative calculations use moles as the basis for their conversion. As a rule, if mass is mentioned in the problem, either in the given information or as the quantity to be calculated, that should be a tip-off that the molar mass of that compound or element is a necessary conversion for solving the problem. Any mention of molecule or atom indicates a sure bet that Avogadro's number, 6.022×10^{23} molecules/mol, is needed.

SYNTHESIS

Looking ahead, then, are there other quantities that convert directly to moles? Volumes of gases and solutions both do. These will be explored in Chapters 10 and 12, respectively. In addition, reactants in an equation relate to each other by moles as well, as will be explored in Chapter 7. The mole is a central concept for working quantitatively in chemistry, and the first step in many problems is to convert the given amount of material into moles.

EXAMPLE 5-9   Converting Mass to Number of Molecules

How many individual molecules are present in 25.0 g of N_2O_5?

PROCEDURE

The formula weight of $N_2O_5 = [(2 \times 14.01) + (5 \times 16.00)] = 108.0$ amu. The molar mass is 108.0 g/mol. This is a two-step conversion as follows.

$$\times \frac{mol}{g} \qquad \times \frac{6.022 \times 10^{23}}{mol}$$

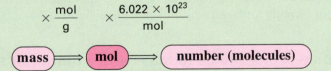

mass $\Longrightarrow$ mol $\Longrightarrow$ number (molecules)

SOLUTION

$$25.0 \text{ g N}_2O_5 \times \frac{1 \text{ mol}}{108.0 \text{ g N}_2O_5} \times \frac{6.022 \times 10^{23} \text{ molecules}}{\text{mol}} = \underline{1.39 \times 10^{23} \text{ molecules}}$$

ANALYSIS

Building on the ideas in this example, notice that the question asks for number of molecules. Immediately realize the necessity of Avogadro's number. Then the given information notes the presence of 25.0 g of N_2O_5. Grams indicate molar mass, specifically of the N_2O_5. Having both these quantities is now sufficient for the solution to be put together.

SYNTHESIS

The terms *grams per mole* and *molecules per mole* sound very scientific, and the concept of the mole is difficult at first. Always remember that the mole is just a number of particles and as such is treated the same way as the terms *dozen* (12), *score* (20), or *gross* (144). The example above conceptually is no different from asking how many doughnuts are in 6 lb if a dozen doughnuts weighs 2 lb. That's easy. At 2 lb per dozen, 6 lb is 3 doz doughnuts, and 3 doz is 36. If you can make the same conceptual connections in the problems, they'll work much more easily.

The term a *mole of molecules* does sound confusing. It is somewhat unfortunate that the counting unit (mole) and the fundamental particle that is being counted (molecule) read so similarly. Try to remember that the *molecule* is the tiny, fundamental particle of a compound that is being counted. A *mole* is an incredibly large number of molecules (i.e., 6.022×10^{23}).

Another point of caution concerns nonmetal elements, such as chlorine (Cl_2), that exist in nature as molecules rather than solitary atoms. For example, if we report the presence of "one mole of chlorine," it is not always obvious what is meant. Do we have one mole of chlorine atoms (6.022×10^{23} atoms, 35.45 g) or one mole of chlorine molecules (6.022×10^{23} molecules, 70.90 g)? Note that there are *two Cl atoms per molecule* and *two moles of Cl atoms per mole of Cl₂ molecules*. (See Figure 5-4.) In future discussions, one mole of chlorine or any of the diatomic elements will refer to the molecules. Thus it is important to be specific in cases where we are using one mole of atoms. This situation is analogous to the difference between a dozen sneakers and a dozen pairs of sneakers. Both the mass and the actual number of the dozen pairs of sneakers are double those of a dozen sneakers.

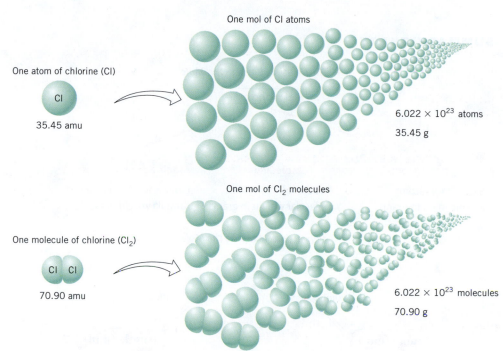

One atom of chlorine (Cl)

One mol of Cl atoms

6.022×10^{23} atoms

35.45 g

One molecule of chlorine (Cl_2)

One mol of Cl_2 molecules

6.022×10^{23} molecules

70.90 g

FIGURE 5-4 Atoms, Molecules, and Moles A molecule is a discrete entity. A mole is a huge number of these entities

▶ **ASSESSING THE OBJECTIVE FOR SECTION 5-3**

EXERCISE 5-3(a) LEVEL 1: Calculate the formula weights of the following:
(a) $MgBr_2$ **(b)** $Ba(OH)_2 \cdot 8H_2O$ **(c)** $CuSO_4 \cdot 5H_2O$

EXERCISE 5-3(b) LEVEL 2: How many grams and how many formula units are in 0.500 mole of CaF_2?

EXERCISE 5-3(c) LEVEL 2: Determine the answer, with the proper unit, of the following.
(a) formula weight of magnesium nitride
(b) molar mass of sulfur hexafluoride

EXERCISE 5-3(d) LEVEL 2: How many moles of Cl_2O and how many individual molecules are present in 438 g of Cl_2O?

EXERCISE 5-3(e) LEVEL 3: How many formula units are in 25.8 mg of sodium phosphate?

EXERCISE 5-3(f) LEVEL 3: Which of the possible iron chlorides [iron(II) chloride or iron(III) chloride] contains 1.00×10^{24} formula units in 269 g?

For additional practice, work chapter problems 5-26, 5-30, and 5-31.

PART A SUMMARY

KEY TERMS

5-2.1   The unit of quantity in chemistry is known as the **mole (mol)** and represents **Avogadro's number** of particles. p. 151

5-2.2   The **molar mass of an element** is the atomic weight in grams. p. 152

5-3.1   The **formula weight** of a compound is the sum of the atomic weights of its component atoms expressed in amu. p. 156

5-3.3   The **molar mass of a compound** is the formula weight expressed in grams. p. 157

SUMMARY CHART

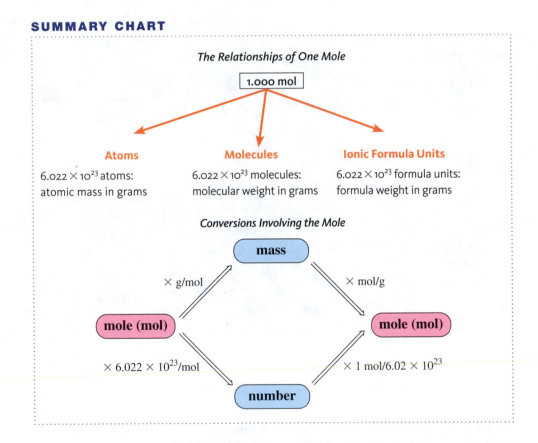

The Relationships of One Mole

1.000 mol

Atoms
6.022×10^{23} atoms: atomic mass in grams

Molecules
6.022×10^{23} molecules: molecular weight in grams

Ionic Formula Units
6.022×10^{23} formula units: formula weight in grams

Conversions Involving the Mole

mass

$\times$ g/mol $\times$ mol/g

mole (mol)   mole (mol)

$\times 6.022 \times 10^{23}$/mol $\times 1$ mol/6.02×10^{23}

number

Part B

The Component Elements of Compounds

SETTING A GOAL

■ You will learn the relationship between the formula of a compound and its elemental composition.

OBJECTIVES

5-4 Given the formula of a compound, determine the mole, mass, and percent composition of its elements.

5-5 Given the percent or mass composition determine the empirical formula and also the molecular formula if the molar mass is provided.

5-4 The Composition of Compounds

▶ OBJECTIVES
FOR SECTION 5-4
Given the formula of a compound,
determine the mole, mass, and percent
composition of its elements.

LOOKING AHEAD! The formula of a compound relays a wealth of information about the component elements. We will now look deeper into how the component parts of a compound relate to its formula. ■

One plus two equals one? At first glance this seems like bad math. However, we do know that *one* frame plus *two* wheels equals *one* bicycle. So in the context of the relationship of parts to the whole, the math works. *Likewise,* one *carbon atom plus two oxygen atoms equals* one *molecule of carbon dioxide.* Now consider the compound sulfuric acid (H_2SO_4). In Table 5-2, we have illustrated the relation of one mole of compound to all its component parts. All of these relationships can be used to construct conversion factors between the compound and its elements. We will use similar conversion factors in sample problems to separate compounds into their mole components, mass components, and, finally, percent composition.

5-4.1 The Mole Composition of a Compound

Our first task is simply to separate a quantity of a compound expressed in moles into the number of moles of each element that is part of the compound. To do this, we will establish a **mole ratio**, a conversion factor that shows *the relationship between the number of moles of a particular element, and a mole of the compound containing that element.* The formula of the compound provides this ratio. This calculation is illustrated in Example 5-10.

TABLE 5-2

The Composition of One Mole of H_2SO_4

NUMBER OF ATOMS		MOLES OF ATOMS		MASS OF ATOMS
1.204×10^{24} H atoms	←	2 mol H	→	2.016 g H
6.022×10^{23} S atoms	←	1 mol S	→	32.07 g S
2.409×10^{24} O atoms	←	4 mol O	→	64.00 g O
Totals: 4.215×10^{24} atoms in		7 mol atoms in		
6.022×10^{23} molecules	←	one mole of molecules	→	98.09g

EXAMPLE 5-10   Converting Moles of a Compound into Moles of Its Component Elements

How many moles of each atom are present in 0.345 mol of $Al_2(CO_3)_3$? What is the total number of moles of atoms present?

PROCEDURE

In this problem, note that there are two Al atoms, three C atoms, and nine O atoms in each formula unit of compound. Therefore, in 1 mol of compound, there are 2 mol of Al, 3 mol of C, and 9 mol of O atoms. These can all be expressed as mole ratios in factor form.

$$\frac{2 \text{ mol Al}}{\text{mol Al}_2(CO_3)_3} \qquad \frac{3 \text{ mol C}}{\text{mol Al}_2(CO_3)_3} \qquad \frac{9 \text{ mol O}}{\text{mol Al}_2(CO_3)_3}$$

SOLUTION

$$\text{Al:} \quad 0.345 \text{ mol Al}_2(CO_3)_3 \times \frac{2 \text{ mol Al}}{\text{mol Al}_2(CO_3)_3} = \underline{0.690 \text{ mol Al}}$$

$$\text{C:} \quad 0.345 \text{ mol Al}_2(CO_3)_3 \times \frac{3 \text{ mol C}}{\text{mol Al}_2(CO_3)_3} = \underline{1.04 \text{ mol C}}$$

$$\text{O:} \quad 0.345 \text{ mol Al}_2(CO_3)_3 \times \frac{9 \text{ mol O}}{\text{mol Al}_2(CO_3)_3} = \underline{3.11 \text{ mol O}}$$

Total moles of atoms present:

$$0.690 \text{ mol Al} + 1.04 \text{ mol C} + 3.11 \text{ mol O} = \underline{4.84 \text{ mol atoms}}$$

ANALYSIS

A problem such as this works both ways, of course. We can hypothetically "build" molecules from the moles of its component elements. How many moles of aluminum carbonate could be built out of 8 moles of aluminum and sufficient amounts of carbon and oxygen? Use the mole ratio between aluminum and aluminum carbonate, but inverted so that you're solving for the moles of aluminum carbonate.

$$8 \text{ mol Al} \quad \times \quad \frac{1 \text{ mol Al}_2(CO_3)_3}{2 \text{ mol Al}} \quad = \quad 4 \text{ mol Al}_2(CO_3)_3$$

SYNTHESIS

Mole ratios like those above will play an important role in *stoichiometry*, a topic explored in the next chapter. Notice that the total moles of elements present is 4.84, which is 14 times the number of moles of compound. This corresponds to 14 atoms of elements per molecule of compound.

5-4.2 The Mass Composition of a Compound

We can now extend this concept into the component masses of the elements. In the following example, we will calculate the mass of each element present in a specific mole quantity of limestone.

EXAMPLE 5-11 Converting Moles of a Compound into Masses of Its Component Elements

What masses of each element are present in 2.36 mole of limestone ($CaCO_3$)?

PROCEDURE

In this problem, convert the moles of each element present in the given quantity of the compound to the mass of that element. This is a two-step conversion as follows.

$$\times \frac{\text{mol element}}{\text{mol compound}} \qquad \times \frac{g}{\text{mol}}$$

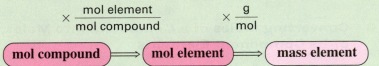

SOLUTION

$$\text{Ca: } 2.36 \text{ mol CaCO}_3 \times \frac{1 \text{ mol Ca}}{\text{mol CaCO}_3} \times \frac{40.08 \text{ g Ca}}{\text{mol Ca}} = \underline{94.6 \text{ g Ca}}$$

$$\text{C: } 2.36 \text{ mol CaCO}_3 \times \frac{1 \text{ mol C}}{\text{mol CaCO}_3} \times \frac{12.01 \text{ g C}}{\text{mol C}} = \underline{28.3 \text{ g C}}$$

$$\text{O: } 2.36 \text{ mol CaCO}_3 \times \frac{3 \text{ mol O}}{\text{mol CaCO}_3} \times \frac{16.00 \text{ g O}}{\text{mol O}} = \underline{113 \text{ g O}}$$

ANALYSIS

The two-step nature of this problem comes from the desire to calculate mass (which requires the molar mass as a conversion factor) and the focus on a single element of a compound (mole ratio between the element and the compound) in each case.

SYNTHESIS

Could you go backward with this problem? Could you calculate the mass of limestone that could be made from 5.50 moles of calcium? What conversion factors are needed to solve this problem? Because you want the mass of limestone, you would need its molar mass. Because the problem asks you to relate limestone and calcium, you would also use the mole ratio between them.

$$5.50 \; \text{mol Ca} \times \frac{1 \; \text{mol CaCO}_3}{1 \; \text{mol Ca}} \times \frac{100.1 \; \text{g CaCO}_3}{1 \; \text{mol CaCO}_3} = 551 \; \text{g CaCO}_3$$

5-4.3 The Percent Composition of a Compound

Perhaps the most common way of expressing the composition of elements in a compound is percent composition. **Percent composition** *expresses the mass of each element per 100 mass units of compound.* For example, if there is an 82-g quantity of nitrogen present in each 100 g of ammonia (NH_3), the percent composition is expressed as 82% nitrogen.

The mass of one mole of carbon dioxide is 44.01 g, and it is composed of one mole of carbon atoms (12.01 g) and two moles of oxygen atoms ($2 \times 16.00 = 32.00$ g). The percent composition is calculated by dividing the total mass of each component element by the total mass (molar mass) of the compound and then multiplying by 100%.

$$\frac{\text{total mass of component element}}{\text{total mass (molar mass)}} \times 100\% = \text{percent composition}$$

In CO_2, the percent composition of C is

$$\frac{12.01 \; \text{g C}}{44.01 \; \text{g CO}_2} \times 100\% = 27.29\% \; \text{C}$$

and the percent composition of O is

$$\frac{32.00 \; \text{g O}}{44.01 \; \text{g CO}_2} \times 100\% = 72.71\%$$

EXAMPLE 5-12   Calculating the Percent Composition of Limestone ($CaCO_3$)

What is the percent composition of all of the elements in limestone ($CaCO_3$)?

PROCEDURE

Find the molar mass and convert the *total* mass of each element to percent of molar mass.

SOLUTION

One mol of $CaCO_3$ contains

1 mol of Ca (40.08 g)

1 mol of C (12.01 g)

$$3 \; \text{mol of O} \left(3 \; \text{mol O} \times \frac{16.00 \; \text{g}}{\text{mol O}} = 48.00 \; \text{g} \right)$$

The molar mass = 40.08 g + 12.01 g + 48.00 g = 100.09 g = 100.1 g.

$$\% \; \text{Ca:} \; \frac{40.08 \; \text{g Ca}}{100.1 \; \text{g CaCO}_3} \times 100\% = \underline{40.04\%}$$

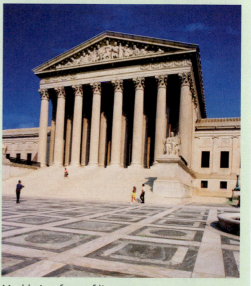

Marble is a form of limestone.

$$\% \text{ C:} \frac{12.01 \text{ g C}}{100.1 \text{ g CaCO}_3} \times 100\% = \underline{12.00\%}$$

$$\% \text{ O:} \frac{48.00 \text{ g O}}{100.1 \text{ g CaCO}_3} \times 100\% = \underline{47.95\%}$$

ANALYSIS

It is easier to find the percent composition of oxygen by subtracting the first two from 100%. That is, $100.00\% - 40.04\% - 12.00\% = 47.96\%$. The difference is due to rounding off. However, it is still best to calculate all three values individually and then check to see that they add to 100%. Notice that the percentages add to 99.99% rather than 100%. This, again, is just a function of rounding with significant figures.

SYNTHESIS

Can you think of some applications for information like this? Why would it be important to know the percentage of an element in a particular compound? How would diverse areas such as mining, agriculture, drug companies, and petroleum producers benefit from knowing the percentage of a given element in a particular compound? Check out Making It Real, "Calcium in Our Diets—How Much Are You Getting?," for one specific example.

EXAMPLE 5-13   Calculating the Percent Composition of Borazine

What is the percent composition of all the elements in borazine ($B_3N_3H_6$)?

PROCEDURE

Find the molar mass and convert the *total* mass of each element to percent of molar mass.

SOLUTION

One mole of $B_3N_3H_6$ contains

$$3 \text{ mol of B} \left(3 \text{ mol B} \times \frac{10.81 \text{ g}}{\text{mol B}} = 32.43 \text{ g} \right)$$

$$3 \text{ mol of N} \left(3 \text{ mol N} \times \frac{14.01 \text{ g}}{\text{mol N}} = 42.03 \text{ g} \right)$$

$$6 \text{ mol of H} \left(6 \text{ mol H} \times \frac{1.008 \text{ g}}{\text{mol H}} = 6.048 \text{ g} \right)$$

The molar mass $= 32.43 \text{ g} + 42.03 \text{ g} + 6.048 \text{ g} = 80.51 \text{ g}$

$$\% \text{ B:} \frac{32.43 \text{ g B}}{80.51 \text{ g B}_3\text{N}_3\text{H}_6} \times 100\% = \underline{40.28\%}$$

$$\% \text{ N:} \frac{42.03 \text{ g N}}{80.51 \text{ g B}_3\text{N}_3\text{H}_6} \times 100\% = \underline{52.20\%}$$

$$\% \text{ H:} \frac{6.048 \text{ g H}}{80.51 \text{ g B}_3\text{N}_3\text{H}_6} \times 100\% = \underline{7.51\%}$$

ANALYSIS

The general procedure presented here works for both molecular and ionic compounds. It works whether there are two elements present or eight elements present. Science is all about learning a few general principles and being able to adjust to a variety of related situations. What if just the name, not the formula, was provided? Use the skills you developed in Chapter 4 to determine the correct formula, and then use the above procedure to get the percentages. Being able to combine multiple skills is also a hallmark of science.

SYNTHESIS

Relative amounts of a substance are most conveniently expressed as percentages. *Percent* literally means "parts per hundred." When the amounts of substances are smaller, other relative measures are used. Parts per thousand, million, or billion can be used for relative amounts much smaller than percentages. Pollutants in water are often measured in ppm (parts per million), as are many trace elements in Earth's crust. When converting a percent to a decimal, you've been conditioned to move the decimal point two places (e.g., 20% = 0.20). To convert ppm, you just move it six places (e.g., 20 ppm = 0.000020). So, if the barium concentration limit in drinking water is 6.5 ppm, this would correspond to only 0.00065%. We will consider ppm and ppb (parts per billion) in more detail in Chapter 12.

MAKING IT REAL

Calcium in Our Diets—How Much Are You Getting?

Calcium is a major component of bones. Our bones are continuously undergoing change. Certain hormones cause bone to be destroyed, while others cause it to be rebuilt. When the process gets out of balance and bone loss exceeds bone formation, a condition called *osteoporosis* may result. Bones can become brittle and subject to fracture or compression, a condition common in elderly people. Women are especially prone to this potentially debilitating disease, which is treated in two ways. One popular drug interferes with the bone-destruction process, resulting in increased bone density. More commonly (and economically), osteoporosis is treated with calcium supplements. Three compounds that are used are calcium carbonate, calcium citrate, and calcium gluconate. These three compounds are all ionic, which means they are solids and can be taken in the form of a tablet. But which actually contains the most calcium in a 500-mg pill?

In Example 5-12, we calculated that calcium carbonate is 40.04% calcium. Similar calculations indicate that calcium citrate is 24.1% calcium and calcium gluconate is 9.31% calcium. The accompanying table summarizes the calculations indicating the amount of calcium in one 500-mg tablet.

CALCIUM SUPPLEMENTS

COMPOUND	FORMULA	%CALCIUM	Mg Ca/ 500-Mg Pill
Calcium carbonate	$CaCO_3$	40.0	205 mg
Calcium citrate	$Ca_3(C_6H_5O_7)_2$	24.1	121 mg
Calcium gluconate	$Ca(C_6H_{11}O_7)_2$	9.31	46.6 mg

It would seem that calcium carbonate is the best. (It is also the cheapest.) However, calcium carbonate is also used as an antacid, so it decreases the acidity in the stomach. Since it needs excess stomach acid to dissolve, nutritionists recommend that it be taken with meals. (Meals cause stomach acid to be produced for digestion.) The total amount of calcium available is not the only important factor in choosing a supplement. How much calcium is actually absorbed during digestion is also a factor. The other two compounds have lower amounts of calcium in one pill, but that calcium is actually better absorbed in the small intestine. Also, they do not have to be taken with meals. The decision of which supplement one takes depends on convenience, side effects, and the advice of a physician.

▶ **ASSESSING THE OBJECTIVE FOR SECTION 5-4**

EXERCISE 5-4(a) LEVEL 1: What are the mole ratios of elements present in K_2CrO_4?

EXERCISE 5-4(b) LEVEL 2: How many moles of H atoms will react with 2.5 moles of N to make NH_3?

EXERCISE 5-4(c) LEVEL 2: Fill in the blanks.

Vitamin C (ascorbic acid) is a compound with the formula $C_6H_8O_6$. The formula weight of the compound is _____ amu. In a 0.650-mol quantity of ascorbic acid, there is _____ mol of carbon, which equals _____ g of carbon. The compound is _____ % carbon.

EXERCISE 5-4(d) LEVEL 2: What is the percent composition by mass of all the elements in $(NH_4)_2Cr_2O_7$?

EXERCISE 5-4(e) LEVEL 3: What mass of aluminum is found in 285 kg of Al_2O_3?

EXERCISE 5-4(f) LEVEL 3: How many moles of chlorine (Cl) are found in 45.0 g of aluminum chloride?

EXERCISE 5-4(g) LEVEL 3: How many grams of nitrogen (N) are found in 3.74×10^{24} formula units of calcium nitrate?

For additional practice, work chapter problems 5-36, 5-38, 5-40, 5-44, and 5-46.

▶ **OBJECTIVE FOR SECTION 5-5**
Given the percent or mass composition determine the empirical formula and also the molecular formula if the mass is provided.

5-5 Empirical And Molecular Formulas

LOOKING AHEAD! From the formula of a compound, we can calculate either the mass composition of the elements or their percent composition. In this section we will do just the opposite—that is, from the mass composition or the percent composition, we will obtain the ratio of the atoms of the elements in a compound. ∎

5-5.1 Empirical Formulas

Two common hydrocarbons (carbon–hydrogen compounds) are known as acetylene (C_2H_2) and benzene (C_6H_6). Actually, except for their classification as hydrocarbons, these two compounds have very few properties in common. Acetylene is a gas used in welding, and benzene is a liquid solvent. The formulas of these two compounds are known specifically as the molecular formula. *The **molecular formula** of a compound is the actual number of atoms present in the compound.*

What these two compounds have in common is their percent composition. This is because they have the same empirical formula. *The **empirical formula** is the simplest whole-number ratio of atoms in a compound.* Most ionic compounds are represented by empirical formulas by virtue of the way that we write their formula units. The formula of a molecular compound, however, represents the actual number of atoms present in one molecule. One benzene molecule is composed of six carbon and six hydrogen atoms, but the empirical formula (the ratio of atoms present) is obtained by dividing the subscripts by 6. The same empirical formula (CH) is obtained for acetylene by dividing the subscripts by 2.

5-5.2 Calculating Empirical Formulas

The key to determining an empirical formula is to realize exactly what a formula tells us. When we picture individual molecules, then our interpretation of a chemical formula is an accounting of the atoms in that molecule. When we expand our concept to a mole of molecules, then we realize that the chemical formula is also the number of moles of each element present in one mole of the compound. So calculating the empirical formula requires us to calculate the relative number of moles of each element. Thus, to calculate the empirical formula of a compound from mass composition, we follow two steps.

1. Convert mass to moles of each element.
2. Find the whole-number ratio of the moles of different elements.

EXAMPLE 5-14   Calculating the Empirical Formula from Mass Composition

Pure magnetite is a mineral composed of an iron–oxygen binary compound. A 3.85-g sample of magnetite is composed of 2.79 g of iron. What is the empirical formula of magnetite?

PROCEDURE

(a) Find the mass of oxygen by subtracting the mass of the iron from the total mass.
(b) Convert the masses of iron and oxygen to moles.
(c) Find the whole-number ratio of moles of iron and oxygen. Divide by the smaller number of moles.

SOLUTION

(a) 3.85 g − 2.79 g (Fe) = 1.06 g (O)

(b) $2.79 \text{ g Fe} \times \dfrac{1 \text{ mol Fe}}{55.85 \text{ g Fe}} = 0.0500 \text{ mole Fe}$

$1.06 \text{ g O} \times \dfrac{1 \text{ mol O}}{16.00 \text{ g O}} = 0.0663 \text{ mol O}$

(c) Fe: $\dfrac{0.0500}{0.0500} = 1.00$ O: $\dfrac{0.0663}{0.0500} = 1.33$

We're not quite finished, since $FeO_{1.33}$ still has a fractional number that must be cleared. (You should keep at least two decimal places in these numbers; do not round off a number like 1.33 to 1.3 or 1.) This fractional number can be cleared by multiplying both subscripts by an integer that produces whole numbers. In this case, 1.33 is equivalent to 4/3, so both subscripts (the 1 implied for Fe and the 1.33 for O) can be multiplied by 3 to clear the fraction.

$$Fe_{(1 \times 3)}O_{(1.33 \times 3)} = \underline{Fe_3O_4}$$

ANALYSIS

A decimal value of 0.50 can be multiplied by 2; values of 0.33 and 0.67 can be multiplied by 3; values of 0.25 and 0.75 can be multiplied by 4; and values of 0.20, 0.40, 0.60, and 0.80 can be multiplied by 5. There are few examples that are more complex than these. Since the last decimal place is estimated, values such as 0.49 should be rounded off to 0.50.

SYNTHESIS

So what would you do if, upon analysis of a compound containing C, H, and O, the moles calculated for each element were C = 1.33 mol, H = 2.50 mol, and O = 1 mol? Since the carbon is multiplied by 3 and the hydrogen is multiplied by 2, the common multiplier is 6. Therefore, everything should be multiplied by 6. The formula is $C_8H_{15}O_6$. Notice that something like this happens whenever the smallest subscript in the empirical formula is greater than 1.

Magnetite is a mineral composed of iron and oxygen.

Empirical formulas are more often calculated from the percent composition of the elements. In this case the percent must be converted to a mass. To do this, it is easiest to assume that we have exactly 100 g of a compound. Thus the percent becomes the mass expressed in grams. We then proceed as in the previous example.

EXAMPLE 5-15   Calculating the Empirical Formula from Percent Composition

What is the empirical formula of laughing gas, which is composed of 63.6% nitrogen and 36.4% oxygen by mass?

PROCEDURE

Remember that *percent* means "parts per 100." Therefore, if we simply assume that we have a 100-g quantity of compound, the percents convert to specific masses as follows.

$$63.6\% \text{ N} = \frac{63.6 \text{ g N}}{100 \text{ g compound}} \qquad 36.4\% \text{ O} = \frac{36.4 \text{ g O}}{100 \text{ g compound}}$$

We now convert the masses to moles of each element.

$$63.6 \text{ g N} \times \frac{1 \text{ mol N}}{14.01 \text{ g N}} = 4.54 \text{ mol N in 100 g compound}$$

$$36.4 \text{ g O} \times \frac{1 \text{ mol O}}{16.00 \text{ g O}} = 2.28 \text{ mol O in 100 g compound}$$

The ratio of N to O atoms will be the same as the ratio of N to O moles. The formula cannot remain fractional, since only whole numbers of atoms are present in a compound. To find the whole-number ratio of moles, the mathematically simplest thing to do is to divide through by the smaller number of moles, which in this case is 2.28 mol of O.

SOLUTION

$$\text{N: } \frac{4.54}{2.28} = 1.99 \approx 2.00 \qquad \text{O: } \frac{2.28}{2.28} = 1.00$$

The empirical formula of the compound is

$$\underline{N_2O}$$

ANALYSIS

Remember that what we are calculating here is the empirical formula. The actual formula could be N_4O_2, N_6O_3, or any other where the ratio is 2:1. More information is needed to determine the exact molecular formula.

SYNTHESIS

This is a very common type of calculation for a chemist. Once a new compound is formed in the lab, an analysis is the next step. This is usually done by a commercial laboratory that specializes in elemental analysis. Knowing what elements to test for is key, but the list is typically evident based on the starting materials used. The analysis is reported in the percentages of each element that was tested for. It is then up to the chemist to establish the empirical formula from this data.

5-5.3 Determining the Molecular Formula of a Compound

The molecular formula may or may not be the same as the empirical formula. The molecular formula of water is H_2O, which is the same as the empirical formula. The molecular formula of hydrogen peroxide, however, is H_2O_2 which is not the same as its empirical formula (HO). Notice that there are two empirical units of HO in one molecule of hydrogen peroxide.

To determine the molecular formula of molecular compounds, we need to know how many empirical units are present in each molecular unit. Thus, it is necessary to know both the molar mass of the compound (g/mol) and the mass of one empirical unit (g/emp. unit). The ratio of these two quantities must be a whole number (represented as a below) and represents the number of empirical units in one mole of compound.

$$a = \frac{\text{molar mass}}{\text{emp. mass}} = \frac{X \text{ g/mol}}{Y \text{ g/emp. unit}} = 1, 2, 3, \text{etc., emp. unit/mol}$$

The molecular formula is obtained by multiplying the subscripts of the empirical formula by a. For example, the empirical formula of borazine (from Example 5-13) is BNH_2. The mass of one empirical unit is $[10.81 + 14.01 + (2 \times 1.008)] = 26.84$ g/emp. unit. Its molar mass is 80.5 g/mol.

$$a = \frac{80.5 \text{ g/mol}}{26.84 \text{ g/emp. unit}} = 3 \text{ emp. unit/mol}$$

The molecular formula is 3 times the subscripts of the empirical formula.

$$B_{(1 \times 3)}N_{(1 \times 3)}H_{(2 \times 3)} = B_3N_3H_6$$

EXAMPLE 5-16   Determining the Molecular Formula from the Empirical Formula and Molar Mass

A phosphorus–oxygen compound has an empirical formula of P_2O_5. Its molar mass is 283.9 g/mol. What is its molecular formula?

PROCEDURE

Find the number of empirical units in each mole of compound a. The empirical mass of the compound is found by adding two moles of phosphorus (30.97 g/mol) and five moles of oxygen (16.00 g/mol). Compare this to the given molar mass to determine the whole-number multiplier.

$$(2 \times 30.97 \text{ g}) + (5 \times 16.00 \text{ g}) = 141.9 \text{ g/emp. Unit}$$

SOLUTION

$$a = \frac{283.9 \text{ g/mol}}{141.9 \text{ g/emp. unit}} = 2 \text{ emp. units/mol}$$

The molecular formula is

$$P_{(2 \times 2)}O_{(2 \times 5)} = \underline{P_4O_{10}}$$

ANALYSIS

When an empirical formula is calculated, particularly in a laboratory setting, masses are sometimes not determined exactly due to experimental errors. Therefore, subscripts that are calculated rarely come out to a whole number. A modest amount of rounding is required. However, once an empirical formula is determined, there are techniques that can determine molar mass with very little error at all. The ratio of empirical unit masses and molar masses will be exactly whole numbers. Anything off by even a small amount of that should be a red flag that there's an error in the calculation or formula.

SYNTHESIS

Now the challenge is to calculate an empirical formula and then to compare that formula with a known molar mass, to determine the actual molecular formula. To determine whether you can do this, check out Exercise 5-5(d) in Assessing the Objectives for Section 5-5.

We will learn in more detail in Chapter 9 of the structural differences between ionic and molecular compounds. At this point, though, it is useful to note that the formula of an ionic compound, identified as a formula unit, is always the smallest whole-number ratio of atoms found in the compound. Thus, by definition, it is an empirical formula. However, the formula of a molecular compound lists the exact number of atoms found in one molecule, and may or may not be the smallest whole-number ratio. It is with molecular compounds that we must be concerned with the distinction between empirical and molecular formulas.

The types of calculations performed in this section are of fundamental importance to the advancement of science. Over 10 million unique compounds have been reported and registered by the American Chemical Society, which serves as the world registry of compounds. To establish that a new discovery is a unique compound, several steps must be followed. *First,* it must be shown that the compound is indeed a pure substance. This is established by a study of its physical properties. For example, pure substances have definite and unchanging melting and boiling points, as was discussed in Chapter 3. *Second,* the formula of the new compound must be determined by elemental analysis. *Third,* the molecular formula of the new compound is calculated from the molar mass and the empirical formula. The molar mass may also be obtained commercially or measured experimentally by methods that will be discussed in Chapter 12. *Finally,* the structural formula, which is the order and arrangement of the atoms in the molecule, is determined. Various instruments available in most laboratories can provide information of this nature. The procedure may take anywhere from a few minutes for simple molecules to a few weeks for very complex molecules such as those associated with life processes.

In many cases, new compounds that are identified by these procedures are tested for anticancer or anti-AIDS activity. Or maybe the new compound has an application as a herbicide, an insecticide, a perfume, a food preservative, or any number of other possibilities. In any case, many of the new compounds that are reported each year have some practical application that improves the quality of our lives.

▶ ASSESSING THE OBJECTIVE FOR SECTION 5-5

EXERCISE 5-5(a) LEVEL 1: Which of these is not an empirical formula?
(a) N_2O_5 **(b)** CCl_4 **(c)** $C_6H_{12}O_6$ **(d)** C_2H_6O **(e)** H_2O_2

EXERCISE 5-5(b) LEVEL 2: A compound contains 5.65 g iron and 4.85 g sulfur. What is its empirical formula?

EXERCISE 5-5(c) LEVEL 3: A compound is composed of 12.7% aluminum, 19.7% nitrogen, and the remainder oxygen. What is the empirical formula? Is this compound likely to have a molecular formula different from its empirical formula?

EXERCISE 5-5(d) LEVEL 2: A compound with the empirical formula C_2H_4S has a measured molar mass of 180.2 g/mol. What is its molecular formula?

EXERCISE 5-5(e) LEVEL 3 A molecular compound is composed of boron and hydrogen. It is 84.4% boron by mass and has a molar mass of 76.96 g/mol. What is its molecular formula?

For additional practice, work chapter problems 5-56, 5-58, 5-60, and 5-68.

PART B SUMMARY

KEY TERMS

5-4.1   The **mole ratio** can be used to determine the moles of an element present in a given number of moles of compound. p. 161

5-4.3   The **percent composition** of a compound is a convenient way to represent the distribution of elements in a compound. p. 163

5-5.1   The **molecular formula** of a compound is the actual formula of a compound. p. 166

5-5.1   The **empirical formula** of a compound can be calculated from the percent composition. p. 166

SUMMARY CHART

Composition of compounds

Mass of compound $\longrightarrow$ Moles of compound $\longrightarrow$ Moles of elements $\longrightarrow$ Mass of elements $\longrightarrow$ % composition

Empirical and Molecular Formulas

% Composition $\longrightarrow$ Mass of elements $\longrightarrow$ Moles of elements $\longrightarrow$ Mole ratio of elements

$\longrightarrow$ Whole-number mole ratio $\longrightarrow$ Empirical formula $\longrightarrow$ Molecular formula

CHAPTER 5 SYNTHESIS PROBLEM

In the next chapter we will illustrate how we represent chemical changes with balanced chemical equations. We will then organize some chemical reactions into several different categories. In this chapter, we can put to use one familiar type of reaction that is used to determine the empirical formulas of compounds that contain carbon and hydrogen. When compounds containing these two elements undergo combustion (burning) in an excess of oxygen, all of the carbon in the original compound is converted to carbon dioxide (CO_2) and all of the hydrogen is converted to water. So, by converting the mass of CO_2 produced to moles of CO_2 and the mass of water to moles of hydrogen, we can establish the ratio of carbon to hydrogen in the original compound. For example, the combustion of one mole of methane (CH_4) would produce one mole of CO_2 and two moles of H_2O.

PROBLEM	SOLUTION
a. The combustion of an unknown hydrocarbon (composed of hydrogen and carbon only) produces 31.5 g of CO_2 and 8.60 g of H_2O. How many moles of carbon and hydrogen are in the original compound?	**a.** The molar mass of CO_2 is 12.01 g + (2 × 16.00 g) = 44.01 g/mol. The molar mass of H_2O is (2 × 1.008 g) + 16.00 g = 18.02 g/mol. $$31.5 \text{ g } CO_2 \times \frac{1 \text{ mol } CO_2}{44.01 \text{ g } CO_2} \times \frac{1 \text{ mol C}}{\text{mol } CO_2} = \underline{0.716 \text{ mol C}}$$ $$8.60 \text{ g } H_2O \times \frac{1 \text{ mol } H_2O}{18.02 \text{ g } H_2O} \times \frac{2 \text{ mol H}}{\text{mol } H_2O} = \underline{0.954 \text{ mol H}}$$
b. What is the empirical formula of the original compound?	**b.** $\dfrac{0.954 \text{ mol H}}{0.716 \text{ mol C}} = \dfrac{1.33 \text{ mol H}}{\text{mol C}} = CH_{1.33}$. Recall that 1.33 in fraction form is 4/3. So multiply through by 3 to get the empirical formula. $$CH_{4/3} = \underline{C_3H_4}$$
c. What is its molecular formula if the molar mass of the compound is 120.2 g/mol?	**c.** The molar mass of the compound is an exact multiple of the empirical mass. Empirical mass is (3 × 12.01) + (4 × 1.008) = 40.06 g/mol. $$\frac{120.2 \text{ g/mol}}{40.06 \text{ g/emp. unit}} = 3 \text{ emp. unit/mol}$$ The molecular formula is $C_{(3\times3)}H_{(4\times3)} = \underline{C_9H_{12}}$.
d. How many individual molecules are in 31.5 g of CO_2?	**d.** $31.5 \text{ g } CO_2 \times \dfrac{1 \text{ mol } CO_2}{44.01 \text{ g } CO_2} \times \dfrac{6.022 \times 10^{23} \text{ molecules}}{\text{mol } CO_2} =$ $\underline{4.31 \times 10^{23} \text{ molecules}}$
e. What is the percent by mass of carbon in the original hydrocarbon?	**e.** The mass of carbon in one mole of the hydrocarbon is (9 × 12.01g) = 180.1 g C The percent carbon is 108.1 g/120.2 g × 100% = $\underline{89.93\% \text{ C}}$

YOUR TURN

An unknown hydrocarbon undergoes combustion in excess oxygen producing 16.9 g CO_2 and 4.32 g of H_2O.

    a. What is the empirical formula of the hydrocarbon?

    b. If the molar mass of the compound is 106.2 g/mol, what is the compound's molecular formula?

    c. How many individual molecules are 4.32 g of H_2O?

    d. What is the percent composition of hydrogen in the hydrocarbon?

Answers are on p 173.

CHAPTER SUMMARY

The atomic masses of the elements can be put to greater use than just comparing the masses of individual atoms. By using atomic masses as ratios, we can measure equivalent numbers of atoms without knowing the actual value of the number. The most useful measure of a number of atoms is obtained by expressing the atomic mass of an element in grams. This is referred to as the **molar mass of the element**, and it is the mass of a specific number of atoms. This number is known as the **mole (mol)**. The mole represents 6.022×10^{23} atoms or other individual particles, a quantity known as **Avogadro's number**.

The concept of the mole is then extended to compounds. The **molar mass of a compound** is the **formula weight** of the compound expressed in grams and is the mass of Avogadro's number of molecules or ionic formula units.

One important purpose of this chapter is for you to become comfortable with the conversions among moles, mass, and numbers of atoms or molecules. The two conversion factors that are used for this are the molar mass and Avogadro's number.

The formula of a compound implies a **mole ratio**. It can be used to determine the number of moles, the masses, and the **percent composition** of each element in a compound.

In the reverse calculation, the **empirical formula** of a compound can be determined from its mass composition or percent composition. One needs to know the molar mass of a compound to determine the **molecular formula** from the empirical formula. The molecular formula is determined from the percent composition and the molar mass.

OBJECTIVES

SECTION	YOU SHOULD BE ABLE TO...	EXAMPLES	EXERCISES	CHAPTER PROBLEMS
5-1	Calculate the masses of equivalent numbers of atoms of different elements.	5-1	1a, 1b, 1c, 1d, 1e	5, 6, 8, 10
5-2	Define the mole and relate this unit to numbers of atoms and to the atomic masses of the elements.	5-2, 5-3, 5-4, 5-5	2a, 2b, 2c	14, 16, 18, 22, 23
5-3	Perform calculations involving masses, moles, and number of molecules or formula units for a compound.	5-6, 5-7, 5-8, 5-9	3a, 3b, 3c, 3d	25, 29, 31, 33
5-4	Given the formula of a compound, determine the mole, mass, and percent composition of its elements.	5-10, 5-11, 5-12, 5-13	4a, 4b, 4c, 4d, 4e	35, 36, 39, 41, 42, 46, 48, 53
5-5	Given the percent or mass composition determine the empirical formula and also the molecular formula if the molar mass is provided.	5-14, 5-15, 5-16	5a, 5b, 5c, 5d	56, 59, 60, 62, 64, 67, 69, 71, 73, 74

▶ ANSWERS TO ASSESSING THE OBJECTIVES

Part A

EXERCISES

5-1(a) (a) 35.453 g (b) 22.99 g (c) 207.2 g

5-1(b) (a) 14:12, or 7:6 (b) 1:19 (c) 32:32, or 1:1

5-1(c) 28.6 g of oxygen

5-1(d) 23.0 g of fluorine

5-1(e) No, you wouldn't use them all. The ratio of S to Fe shows that 8.00 g of S would react with only 13.9 g of Fe. There would be 1.1 g of Fe left over.

5-2(a) Molar mass converts grams to moles. Avogadro's number converts molecules to moles.

5-2(b) (a) 3.0 moles C (b) 20 g Ca (c) 1.2×10^{24} molecules Na (d) 100 moles Sn (e) 0.250 moles Ar

5-2(c) 8.00 moles. (*Note:* This answer is the same, regardless of what element you use.) The mass would be 408 g. (This result is specific to vanadium.)

5-2(d) 7.14×10^{22} atoms of Zn (d) < 8.50 g of C (b) < 0.280 moles of Ca (c) < 4.55×10^{22} atoms of Pb (a)

5-2(e) Silver (Ag)

5-3(a) (a) 184.1 amu (b) 315.5 amu (c) 249.7 amu

5-3(b) 39.0 g and 3.01×10^{23} molecules

5-3(c) (a) 101.0 amu (b) 146.1 g/mol

5-3(d) 5.04 moles and 3.04×10^{24} molecules

5-3(e) 9.48×10^{19} formula units

5-3(f) The number represents 1.66 moles; 269 g represents 1.66 moles of $FeCl_3$ (molar mass = 162 g/mol)

Part B
EXERCISES

5-4(a) 2 mol K/1 mol K_2CrO_4, 1 mol Cr/1 mol K_2CrO_4, 4 mol O/1 mol K_2CrO_4

5-4(b) 7.5 moles of H

5-4(c) Vitamin C (ascorbic acid) is a compound with the formula $C_6H_8O_6$. The formula weight of the compound is 176.1 amu. In a 0.650-mol quantity of ascorbic acid, there is 3.90 mol of carbon, which equals 46.8 g of carbon. The compound is 40.9% carbon.

5-4(d) In order of abundance, 44.4% O, 41.3% Cr, 11.1% N, and 3.2% H.

5-4(e) 151 kg of Al

5-4(f) The molar mass is 133.3 g/mol. The number of moles of $AlCl_3$ is 0.338 mol. The number of moles of Cl is 1.01 mol.

5-4(g) 174 g of N

5-5(a) (c) $C_6H_{12}O_6$ simplifies to CH_2O and (e) H_2O_2 reduces to HO

5-5(b) Fe_2S_3

5-5(c) AlN_3O_9. As this molecule has both metal and nonmetal components, it is an ionic compound (most likely aluminum nitrate, $Al(NO_3)_3$). Its formula is automatically the smallest ratio of atoms.

5-5(d) $C_6H_{12}S_3$

5-5(e) The empirical formula is BH_2. The molecular formula is B_6H_{12}.

ANSWERS TO CHAPTER 5 SYNTHESIS PROBLEM

a. C_4H_5 **b.** C_8H_{10} **c.** 1.44×10^{23} **d.** 9.50%

CHAPTER PROBLEMS

Throughout the text, answers to all exercises in color are given in Appendix E. The more difficult exercises are marked with an asterisk.

Relative Masses of Particles
(SECTION 5-1)

5-1. One penny weighs 2.47 g and one nickel weighs 5.03 g. What mass of pennies has the same number of coins as there are in 12.4 lb of nickels?

5-2. A small glass bead weighs 310 mg and a small marble weighs 8.55 g. A quantity of small glass beads weighs 5.05 kg. What does the same number of marbles weigh?

5-3. A piece of pure gold has a mass of 145 g. What is the mass of the same number of silver atoms?

5-4. A large chunk of pure aluminum has a mass of 212 lb. What is the mass of the same number of carbon atoms?

5-5. A piece of copper wire has a mass of 16.0 g; the same number of atoms of a precious metal has a mass of 49.1 g. What is the metal?

5-6. In the compound CuO, what mass of copper is present for each 18.0 g of oxygen?

5-7. In the compound NaCl, what mass of sodium is present for each 425 g of chlorine?

5-8. In a compound containing one atom of carbon and one atom of another element, it is found that 25.0 g of carbon is combined with 33.3 g of the other element. What is the element? What is the formula of the compound?

*5-9. In the compound $MgBr_2$, what mass of bromine is present for each 46.0 g of magnesium? (Remember, there are two bromines per magnesium.)

*5-10. In the compound SO_3, what mass of sulfur is present for each 60.0 lb of oxygen?

The Magnitude of the Mole (SECTION 5-2)

5-11. If you could count two numbers per second, how many years would it take you to count Avogadro's number? If you were helped by the whole human race of 6.8 billion people, how long would it take?

5-12. A small can of soda contains 355 mL. Planet Earth contains 326 million cubic miles (3.5×10^{20} gal) of water. How many cans of soda would it take to equal all the water on Earth? How many moles of cans is this?

5-13. What would the number in one mole be if the atomic mass were expressed in kilograms rather than grams? In milligrams?

Moles of Atoms of Elements (SECTION 5-2)

5-14. Fill in the blanks. Use dimensional analysis to determine the answers.

Element	Mass in Grams	Moles of Atoms	Number of Atoms
S	8.00	0.250	1.50×10^{23}
(a) P	14.5	_____	_____
(b) Rb	_____	1.75	_____
(c) Al	_____	_____	6.02×10^{23}
(d) _____	363	_____	3.01×10^{24}
(e) Ti	_____	_____	1

5-15. Fill in the blanks. Use dimensional analysis to determine the answers.

Element	Mass in Grams	Moles of Atoms	Number of Atoms
(a) Na	0.390	_____	_____
(b) Ca	_____	_____	6.02×10^{24}
(c) ____	43.2	_____	2.41×10^{24}
(d) K	4.25×10^{-6}	_____	_____
(e) Ne	_____	_____	3.66×10^{21}

5-16. What is the mass in grams of each of the following?

(a) 1.00 mol of Cu   (c) 6.02×10^{23} atoms of Ca

(b) 0.50 mol of S

5-17. What is the mass in grams of each of the following?

(a) 4.55 mol of Be   (c) 3.40×10^5 mol of B

(b) 6.02×10^{24} atoms of Ca

5-18. How many individual atoms are in each of the following?

(a) 32.1 mol of sulfur   (c) 32.0 g of oxygen

(b) 32.1 g of sulfur

5-19. How many moles of atoms are in each of the following?

(a) 281 g of Si   (c) 19.0 atoms of fluorine

(b) 7.34×10^{25} atoms of phosphorus

5-20. Which has more atoms, 50.0 g of Al or 50.0 g of Fe?

5-21. Which contains the most Ni: 20.0 g, 2.85×10^{23} atoms, or 0.450 mol?

5-22. Which contains the most Cr: 0.025 mol, 6.0×10^{21} atoms, or 1.5 g of Cr?

5-23. A 0.251-g sample of a certain element is the mass of 1.40×10^{21} atoms. What is the element?

5-24. A 0.250-mol sample of a certain element has a mass of 49.3 g. What is the element?

Formula Weight (SECTION 5-3)

5-25. What is the formula weight of each of the following? Express your answer to four significant figures.

(a) $KClO_2$ (d) H_2SO_4 (g) $Fe_2(CrO_4)_3$

(b) SO_3 (e) Na_2CO_3

(c) N_2O_5 (f) CH_3COOH

5-26. What is the formula weight of each of the following?

(a) $CuSO_4 \cdot 6H_2O$ (Include H_2O's; they are included in one formula unit.)

(b) Cl_2O_3 (d) $Na_2BH_3CO_2$

(c) $Al_2(C_2O_4)_3$ (e) P_4O_6

5-27. A compound is composed of three sulfate ions and two chromium ions. What is the formula weight of the compound?

5-28. What is the formula weight of strontium perchlorate?

Moles of Compounds (SECTION 5-3)

5-29. Fill in the blanks. Use dimensional analysis to determine the answers.

Molecules	Mass in Grams	Number of Moles	Number of Molecules or Formula Units
N_2O	23.8	0.542	3.26×10^{23}
(a) H_2O	_____	10.5	_____
(b) BF_3	_____	_____	3.01×10^{21}
(c) SO_2	14.0	_____	_____
(d) K_2SO_4	_____	1.20×10^{-4}	_____
(e) SO_3	_____	_____	4.50×10^{24}
(f) $N(CH_3)_3$	0.450	_____	_____

5-30. Fill in the blanks. Use dimensional analysis to determine answers.

Molecules	Mass in grams	Number of moles	Number of Molecules or Formula units
(a) O_3	176	_____	_____
(b) NO_2	_____	3.75×10^{-3}	_____
(c) Cl_2O_3	_____	_____	150
(d) UF_6	_____	_____	8.50×10^{22}

5-31. Tetrahydrocannabinol (THC) is the active ingredient in marijuana. A 0.0684-mol quantity of THC has a mass of 21.5 g. What is the molar mass of THC?

5-32. A 0.158-mol quantity of a compound has a mass of 7.28 g. What is the molar mass of the compound?

5-33. A 287-g quantity of a compound is the mass of 1.07×10^{24} molecules. What is the molar mass of the compound?

5-34. A 0.0812-g quantity of ethylene glycol (antifreeze) represents the mass of 7.88×10^{20} molecules. What is the molar mass of the compound?

The Composition of Compounds (SECTION 5-4)

5-35. How many moles of each type of atom are in 2.55 mol of grain alcohol, C_2H_6O? What is the total number of moles of atoms? What is the mass of each element? What is the total mass?

5-36. How many moles are in 28.0 g of $Ca(ClO_3)_2$? How many moles of each element are present? How many total moles of atoms are present?

5-37. How many moles are in 84.0 g of $K_2Cr_2O_7$? How many moles of each element are present? How many total moles of atoms are present?

5-38. What mass of each element is in 1.50 mol of H_2SO_3?

5-39. What mass of each element is in 2.45 mol of boric acid (H_3BO_3)?

5-40. How many moles of O_2 are in 1.20×10^{22} O_2 molecules? How many moles of oxygen atoms? What is the mass of oxygen molecules? What is the mass of oxygen atoms?

5-41. How many moles of Cl_2 molecules are in 985 g of Cl_2? How many moles of Cl atoms are present? What is the mass of Cl atoms present?

5-42. What is the percent composition of a compound composed of 1.375 g of N and 3.935 g of O?

5-43. A sample of a compound has a mass of 4.86 g and is composed of silicon and oxygen. What is the percent composition if 2.27 g of the mass is silicon?

5-44. The mass of a sample of a compound is 7.44 g. Of that mass, 2.88 g is potassium, 1.03 g is nitrogen, and the remainder is oxygen. What is the percent composition of the elements?

5-45. What is the percent composition of all the elements in the following compounds?

(a) C_2H_6O (c) C_9H_{18} (e) $(NH_4)_2CO_3$
(b) C_3H_6 (d) Na_2SO_4

5-46. What is the percent composition of all the elements in the following compounds?

(a) H_2CO_3 (c) $Al(NO_3)_3$
(b) Cl_2O_7 (d) $NH_4H_2PO_4$

5-47. What is the percent composition of all the elements in borax ($Na_2B_4O_7 \cdot 10H_2O$)?

5-48. What is the percent composition of all the elements in acetaminophen ($C_8H_9O_2N$)? (Acetaminophen is an aspirin substitute.)

5-49. What is the percent composition of all the elements in saccharin ($C_7H_5SNO_3$)?

5-50. What is the percent composition of all the elements in amphetamine ($C_9H_{13}N$)? (Amphetamine is a stimulant.)

*5-51. What mass of carbon is in a 125-g quantity of sodium oxalate?

*5-52. What is the mass of phosphorus in a 25.0-lb quantity of sodium phosphate?

*5-53. What mass of chromium is in a 275-kg quantity of chromium(III) carbonate?

*5-54. Iron is recovered from iron ore, Fe_2O_3. How many pounds of iron can be recovered from each ton of iron ore? (1 ton = 2000 lb)

Empirical Formulas (SECTION 5-5)

*5-55. Which of the following are not empirical formulas?

(a) N_2O_4 (c) $H_2S_2O_3$ (e) Mn_2O_7
(b) Cr_2O_3 (d) $H_2C_2O_4$

*5-56. Convert the following mole ratios of elements to empirical formulas.

(a) 0.25 mol of Fe and 0.25 mol of S
(b) 1.88 mol of Sr and 3.76 mol of I
(c) 0.32 mol of K, 0.32 mol of Cl, and 0.96 mol of O
(d) 1.0 mol of I and 2.5 mol of O
(e) 2.0 mol of Fe and 2.66 mol of O
(f) 4.22 mol of C, 7.03 mol of H, and 4.22 mol of Cl

*5-57. Convert the following mole ratios of elements to empirical formulas.

(a) 1.20 mol of Si and 2.40 mol of O
(b) 0.045 mol of Cs and 0.022 mol of S
(c) 1.00 mol of X and 1.20 mol of Y
(d) 3.11 mol of Fe, 4.66 mol of C, and 14.0 mol of O

5-58. What is the empirical formula of a compound that has the composition 63.2% oxygen and 36.8% nitrogen?

5-59. What is the empirical formula of a compound that has the composition 41.0% K, 33.7% S, and 25.3% O?

5-60. In an experiment it was found that 8.25 g of potassium combines with 6.75 g of O_2. What is the empirical formula of the compound?

5-61. Orlon is composed of very long molecules with a composition of 26.4% N, 5.66% H, and 67.9% C. What is the empirical formula of Orlon?

5-62. A compound is 21.6% Mg, 21.4% C, and 57.0% O. What is the empirical formula of the compound?

5-63. A compound is composed of 9.90 g of carbon, 1.65 g of hydrogen, and 29.3 g of chlorine. What is the empirical formula of the compound?

5-64. A compound is composed of 0.46 g of Na, 0.52 g of Cr, and 0.64 g of O. What is the empirical formula of the compound?

*5-65. A compound is composed of 24.1% nitrogen, 6.90% hydrogen, 27.6% sulfur, and the remainder oxygen. What is the empirical formula of the compound?

*5-66. Methyl salicylate is also known as oil of wintergreen. It is composed of 63.2% carbon, 31.6% oxygen, and 5.26% hydrogen. What is its empirical formula?

*5-67. Nitroglycerin is used as an explosive and a heart medicine. It is composed of 15.9% carbon, 18.5% nitrogen, 63.4% oxygen, and 2.20% hydrogen. What is its empirical formula?

Molecular Formulas (SECTION 5-5)

5-68. A compound has the following composition: 20.0% C, 2.2% H, and 77.8% Cl. The molar mass of the compound is 545 g/mol. What is the molecular formula of the compound?

5-69. A compound is composed of 1.65 g of nitrogen and 3.78 g of sulfur. If its molar mass is 184 g/mol, what is its molecular formula?

5-70. A compound has a composition of 18.7% B, 20.7% C, 5.15% H, and 55.4% O. Its molar mass is about 115 g/mol. What is the molecular formula of the compound?

5-71. A compound reported in 1970 has a composition of 34.9% K, 21.4% C, 12.5% N, 2.68% H, and 28.6% O. It has a molar mass of about 224 g/mol. What is the molecular formula of the compound?

5-72. Fructose is also known as fruit sugar. It has a molar mass of 180 g/mol and is composed of 40.0% carbon, 53.3% oxygen, and 6.7% hydrogen. What is the molecular formula of the compound?

5-73. A compound reported in 1982 has a molar mass of 834 g/mol. A 20.0-g sample of the compound contains 18.3 g of iodine and the remainder carbon. What is the molecular formula of the compound?

5-74. Quinine is a compound discovered in the bark of certain trees. It is an effective drug in the treatment of malaria. Its molar mass is 324 g/mol. It is 74.0% carbon, 7.46% hydrogen, 8.64% nitrogen, and 9.86% oxygen. What is the molecular formula of quinine?

General Problems

5-75. The U.S. national debt was about $4.5 trillion in 1994. How many moles of pennies would it take to pay it off?

5-76. A compound has the formula MN, where M represents a certain unknown metal. A sample of the compound weighs 1.862 g; of that, 0.443 g is nitrogen. What is the identity of the metal, M?

5-77. What would be the number of particles in one mole if the atomic mass were expressed in ounces rather than grams? (28.375 g = 1 oz)

5-78. A certain alloy of copper has a density of 3.75 g/mL and is 65.0% by mass copper. How many copper atoms are in 16.8 cm^3 of this alloy?

5-79. The element phosphorus exists as P_4. How many moles of molecules are in 0.344 g of phosphorus? How many phosphorus atoms are present in that amount of phosphorus?

5-80. Rank the following in order of increasing mass.

(a) 100 hydrogen atoms

(b) 100 moles of hydrogen molecules

(c) 100 grams of hydrogen

(d) 100 hydrogen molecules

5-81. Rank the following in order of increasing number of atoms.

(a) 100 lead atoms   **(c)** 100 grams of lead

(b) 100 moles of helium   **(d)** 100 grams of helium

5-82. Pyrite, a mineral containing iron, has the formula FeS_x. A quantity of 2.84×10^{23} formula units of pyrite has a mass of 56.6 g. What is the value of x?

5-83. A compound has the formula $Na_2S_4O_6$.

(a) What ions are present in the compound? (The anion is all one species.)

(b) How many grams of sulfur are present in the compound for each 10.0 g of Na?

(c) What is the empirical formula of the compound?

(d) What is the formula weight of the compound?

(e) How many moles and formula units are present in 25.0 g of the compound?

(f) What is the percent by mass of oxygen in the compound?

5-84. Glucose (blood sugar) has the formula $C_6H_{12}O_6$. Calculate how many moles of carbon, individual hydrogen atoms, and grams of oxygen are in a 10.0-g sample of glucose.

5-85. A compound has the formula N_2O_x. It is 36.8% nitrogen. What is the name of the compound?

5-86. A compound has the formula SF_x. One sample of the compound contains 0.356 mol of sulfur and 8.57×10^{23} atoms of fluorine. What is the name of the compound?

5-87. Dioxin is a compound that is known to cause cancer in certain laboratory animals. 4.55×10^{22} molecules of this compound have a mass of 24.3 g. Analysis of a sample of dioxin indicated the sample contained 0.456 mol of carbon, 0.152 mol of hydrogen, 0.152 mol of chlorine, and 0.076 mol of oxygen. What is the molecular formula of dioxin?

5-88. A certain compound has a molar mass of 166 g/mol. It is composed of 47.1% potassium, 14.4% carbon, and the remainder oxygen. What is the name of the compound?

5-89. A compound has the general formula $Cr(ClO_x)_y$ and is 14.9% Cr, 30.4% chlorine, and the remainder oxygen. What is the name of the compound?

5-90. Epsom salts has the formula $MgSO_4 \cdot xH_2O$. Calculate the value of x if the compound it 51.1% by mass water.

5-91. Potassium carbonate forms a hydrate, $K_2CO_3 \cdot xH_2O$. What is the value of x if the compound is 20.7% by mass water?

5-92. Nicotine is a compound containing carbon, hydrogen, and nitrogen. Its molar mass is 162 g/mol. A 1.50-g sample of nicotine is found to contain 1.11 g of carbon. Analysis of another sample indicates that nicotine is 8.70% by mass hydrogen. What is the molecular formula of nicotine?

***5-93.** A hydrocarbon (a compound that contains only carbon and hydrogen) was burned, and the products of the combustion reaction were collected and weighed. All the carbon in the original compound is now present in 1.20 g of CO_2. All the hydrogen is present in 0.489 g of H_2O. What is the empirical formula of the compound? (*Hint:* Remember that all the moles of C atoms in CO_2 and H atoms in H_2O came from the original compound.)

***5-94.** A 0.500-g sample of a compound containing C, H, and O was burned, and the products were collected. The combustion reaction produced 0.733 g of CO_2 and 0.302 g of H_2O. The molar mass of the compound is 60.0 g/mol. What is the molecular formula of the compound? (*Hint:* Find the mass of C and H in the original compound; the remainder of the 0.500 g is oxygen.)

STUDENT WORKSHOP

Empirical Formulas

Purpose: To calculate empirical formulas and percent composition for hypothetical compounds. (Work in groups of three or four. Estimated time: 15 min.)

The following activity is a great way for a study group to practice calculations involving percent composition and empirical formulas.

1. Each person in the group makes up the formula of an imaginary chemical compound.

   - Choose three arbitrary elements from the periodic table.

   - Assign subscripts to each. (You should limit yourself to values of 1–5.)

2. Calculate the percent composition of each element in your compound.

3. On a slip of paper, write out the elements present and their percent compositions, and trade this with another person in your group.

4. From the data provided you, calculate the empirical formula of your classmate's compound.

5. Confirm with your classmate the compound's formula based on his or her original work.

Chemical Reactions

I n the rain forest a cycle of life occurs. Green leaves, powered by sunlight, maintain life. The decay of dead trees returns the ingredients for life to the air and Earth.

SETTING THE STAGE

The weather seems to be crazy lately. More violent storms, record floods, melting at the poles, and even severe droughts seem to be happening regularly. Indeed, the past 20 years or so have been the warmest on record. In fact, 2010 tied the record for the highest average global temperature. All of this may be related to what we call global warming. Most scientists agree that the surface of the planet is warming, and at least part of that is due to the increase in carbon dioxide in the atmosphere.

Carbon dioxide is naturally removed from the atmosphere by the oceans and ocean life and also by trees and other vegetation. In a complex series of chemical reactions called *photosynthesis*, green leaves use energy from the sun, carbon dioxide, and water to produce carbohydrates (carbon, hydrogen, and oxygen compounds) along with elemental oxygen. These compounds can then undergo a type of chemical reaction called *combustion*. In this reaction, the carbon compounds from trees or fossil fuels (coal, oil, and natural gas) combine with oxygen in the air to re-form carbon dioxide and water and release the heat energy that originated from the sun. This fascinating cycle of chemical reactions maintains life on this beautiful planet.

The problem is that things have gotten out of balance in the past 160 years since the Industrial Revolution. There is considerably more combustion going on in our world than photosynthesis, so the level of carbon dioxide in the atmosphere is definitely increasing. Carbon dioxide is a gas that traps heat in the atmosphere much as heat is trapped inside a closed automobile on a sunny day. This is referred to as the *greenhouse effect*. The existence of nearly 7 billion people on our globe, all burning fuel and needing space at the expense of forests, assures that the balance will remain in favor of combustion.

Photosynthesis and combustion are only two types of chemical changes. In fact, the atoms of the elements combine, separate, and recombine in millions of ways, all indicating unique chemical changes. The rusting of iron, the decay of a fallen tree in the forest, the formation of muscle in our bodies, the cooking of food—all of these everyday, common occurrences are chemical changes.

Many of the more important chemical reactions on this planet occur in a water (aqueous) medium. Water has a unique ability to hold ionic and polar compounds in solution. Much of the chemistry that occurs in our bodies is a result of the interaction of these compounds circulating in our bloodstream. So it seems appropriate to take a deeper look at the particular chemistry that occurs in aqueous solution.

In Part A, we will introduce the chemical equation, how it is balanced, and three simple types of reactions. Types of reactions are extended in Part B to include the many important reactions that occur in water.

Part A

The Representation of Chemical Changes and Three Types of Changes

SETTING A GOAL

■ You will begin to use the symbolic language of chemistry by writing balanced chemical equations for several identifiable reaction types.

OBJECTIVES

6-1 Write balanced chemical equations for simple reactions from inspection.

6-2 Classify certain chemical reactions as being combustion, combination, or decomposition reactions.

▶ **OBJECTIVE FOR SECTION 6-1**

Write balanced chemical equations for simple reactions from inspection.

6-1 Chemical Equations

LOOKING AHEAD! The main rocket thruster of the space shuttle uses a simple but powerful chemical reaction. Hydrogen combines with oxygen to form water and a lot of heat energy. The way chemical reactions such as this are symbolized is the topic of this section. ■

6-1.1 Constructing an Equation

A chemical change can be illustrated symbolically by a chemical equation. A **chemical equation** is the *representation of a chemical reaction using the symbols of elements and the formulas of compounds*. In the following discussion, we will focus on the matter that undergoes a change in the reaction. In Chapter 7, we will include the heat energy involved.

Let's start with a simple and fundamental reaction. We will build the chemical equation one step at a time that represents the reaction between hydrogen and oxygen to form water. This reaction powers the space shuttle. First, let's represent with symbols the reaction of hydrogen combining with oxygen to form water.

$$H + O \longrightarrow H_2O$$

In a chemical equation, *the original reacting species are shown to the left of the arrow and are called the* **reactants**. *The species formed as a result of the reaction are to the right of the arrow* and are called the **products**. In this format, note that the phrase *combines with* (or *reacts with*) is represented by a plus sign (+). When there is more than one reactant or product, the symbols or formulas on each side of the equation are separated by a +. The word *produces* (or *yields*) may be represented by an arrow (⟶). Note in Table 6-1 that there are other representations for the yield sign, depending on the situation.

The chemical equation shown above tells us only about the elements involved. However, we know more information than just the elements involved. First, if an element exists as molecules under normal conditions, then the formula of the molecule is shown. Recall from Chapter 4 that both hydrogen and oxygen exist as diatomic molecules under normal conditions. Including this information, the equation is

$$\underset{\text{reactants}}{H_2 + O_2} \longrightarrow \underset{\text{products}}{H_2O}$$

An important duty of a chemical equation is to demonstrate faithfully the law of conservation of mass, which states that *mass can be neither created nor destroyed*. In Dalton's atomic theory, this law was explained for chemical reactions. He suggested that reactions are simply rearrangements of the same number of atoms. A close look at the equation above shows that there are two oxygen atoms on the left but only one on the right. To conform to the law of conservation of mass, an equation

TABLE 6-1

Symbols in the Chemical Equation

SYMBOL	USE
+	Between the symbols and/or formulas of reactants or products
$\longrightarrow$	Means "yields" or "produces"; separates reactants from products
=	Same as arrow
$\rightleftharpoons$	Used for reversible reactions in place of a single arrow (see Chapter 14)
(g)	Indicates a gaseous reactant or product
↑	Sometimes used to indicate a gaseous product
(s)	Indicates formation of a solid reactant or product
↓	Sometimes used to indicate formation of a solid product
(l)	Indicates a liquid reactant or product
(aq)	Indicates that the reactant or product is in aqueous solution (dissolved in water)
$\xrightarrow{\Delta}$	Indicates that heat must be supplied to reactants before a reaction occurs
$\xrightarrow{MnO_2}$	An element or compound written above the arrow is a *catalyst; a catalyst speeds up a reaction but is not consumed in the reaction.* It may also indicate the solvent, such as water.

must be balanced. *A **balanced equation** has the same number and type of atoms on both sides of the equation.* An equation is balanced *by introducing* **coefficients**. Coefficients are whole numbers in front of the symbols or formulas. The equation in question is balanced in two steps. If we introduce a 2 in front of the H_2O, we have equal numbers of oxygen atoms, but the number of hydrogen atoms is now unbalanced.

$$H_2 + O_2 \longrightarrow 2H_2O$$

This problem can be solved rather easily. Simply return to the left and place a coefficient of 2 in front of the H_2. The equation is now completely balanced.

$$2H_2 + O_2 \longrightarrow 2H_2O$$

Note that equations cannot be balanced by changing or adjusting the subscripts of the elements or compounds. For example, the original equation could seem to be balanced in one step if the H_2O were changed to H_2O_2. However, H_2O_2 is a compound known as hydrogen peroxide. This is a popular antiseptic but definitely not water.

A powerful chemical reaction blasts the space shuttle into Earth orbit.

Finally, the physical states of the reactants and products under the reaction conditions are sometimes added in parentheses after the formula for each substance. Hydrogen and oxygen are gases, and water is a liquid at room temperature. Using the proper letters shown in Table 6-1, we have the balanced chemical equation in proper form.

$$\mathbf{2H_2}(g) + \mathbf{O_2}(g) \longrightarrow \mathbf{2H_2O}(l)$$

Note that if we describe this reaction in words, we have quite a bit to say. "Two molecules of gaseous hydrogen react with one molecule of gaseous oxygen to produce two molecules of liquid water."

6-1.2 Rules for Balancing Equations

Properly balanced equations are a necessity when we consider the quantitative aspects of reactants and products, as we will do in Chapter 7. Before we consider some guidelines in balancing equations, there are three points to keep in mind concerning balanced equations.

1. The subscripts of a compound are fixed; they cannot be changed to balance an equation.
2. The coefficients used should be the smallest whole numbers possible.
3. The coefficient multiplies all of the number of atoms in the formula. For example, $2K_2SO_3$ indicates the presence of four atoms of K, two atoms of S, and six atoms of O.

In this chapter, equations will be balanced by *inspection*. Certainly, many complex equations are extremely tedious to balance by this method, but such equations will be discussed in Chapter 14, where more systematic methods can be employed. The following rules are helpful in balancing simple equations by inspection.

1. In general, it is easiest to consider balancing elements other than hydrogen or oxygen first. Look to the compound on either side of the equation that contains the greatest number of atoms of an element other than oxygen or hydrogen. Balance the element in question on the other side of the equation.
2. If polyatomic ions appear unchanged on both sides of the equation, consider them as single units.
3. Balance all other elements except hydrogen and oxygen, except those that appear as free elements (not as part of a compound).
4. Balance hydrogen or oxygen next. Choose the one that is present in the fewer number of compounds first. (Usually, that is hydrogen.)
5. Finally, balance any free element.
6. Check to see that the atoms of all elements are balanced. The final balanced equation should have the smallest whole-number ratio of coefficients.

Sometimes fractional coefficients (e.g., 3/2) are used initially, especially with regard to balancing O_2. In this case, multiply the whole equation through by the denominator (usually 2) to clear the fraction and produce only whole numbers. (This is illustrated in Example 6-3.)

EXAMPLE 6-1   **Balancing a Simple Equation**

Ammonia is an important industrial commodity that is used mainly as a fertilizer. It is manufactured from its constituent elements. Write a balanced chemical equation from the following word equation: "Nitrogen gas reacts with hydrogen gas to produce ammonia gas."

PROCEDURE

Refer to Table 6-1 for the proper symbols. Apply the rules above as needed.

SOLUTION

The unbalanced chemical equation using the proper formulas of the elements and compound is

$$N_2(g) + H_2(g) \longrightarrow NH_3(g)$$

First consider the N_2 molecule, since it has the most atoms of an element other than hydrogen or oxygen. Balance the N's on the other side by adding a coefficient of **2** in front of the NH_3.

$$N_2(g) + H_2(g) \longrightarrow 2NH_3(g)$$

Now consider hydrogen atoms. We have "locked in" six hydrogen atoms on the right, so we will need six on the left. By adding a coefficient of 3 in front of the H_2, we have completed the balancing of the equation. (See Figure 6-1.)

$$\underline{N_2(g) + 3H_2(g) \longrightarrow 2NH_3(g)}$$

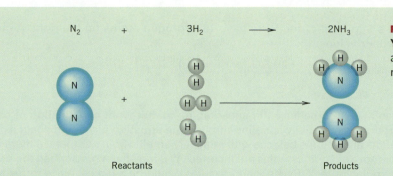

FIGURE 6-1 Nitrogen Plus Hydrogen Yields Ammonia In a chemical reaction, the atoms are simply rearranged into different molecules.

Reactants   Products

ANALYSIS

Many students new to this topic prefer to keep what is called an atom inventory during the course of balancing. In this way, they can see what coefficients are needed and how the addition of one coefficient changes the number of atoms of other elements. In each of the three equations in the solution, write underneath it the number of N atoms and the number of H atoms present on both the reactant side of the equation and the product side.

$N_2(g) + H_2(g) \longrightarrow NH_3(g)$

N 2   N 1
H 2   H 3

$N_2(g) + H_2(g) \longrightarrow 2NH_3(g)$

N 2   N 2
H 2   H 6

$N_2(g) + 3H_2(g) \longrightarrow 2NH_3(g)$

N 2   N 2
H 6   H 6

Notice that in the final solution, the number of atoms on both the reactant and the product side must be identical. This technique can be abandoned once you become proficient in balancing reactions.

SYNTHESIS

Deciding on the proper order in which to balance the elements is the difference between smoothly solving a problem and having to attempt the same problem several times before success is achieved. In what order should the atoms in the following equation be checked and balanced?

$$PbO_2 + HCl \longrightarrow PbCl_2 + Cl_2 + H_2O$$

Balance lead first, then oxygen, hydrogen, and finally chlorine. The chlorine is last, because it is in its free state, Cl_2. The lead is first, since it is not hydrogen or oxygen. Now, why choose the oxygen second and the hydrogen third? Because the oxygen is bonded to the lead in PbO_2. Once we know the correct number of reactant lead atoms, we also know the correct number of reactant oxygen atoms and can predict the coefficient needed on the product side.

EXAMPLE 6-2   Balancing an Equation

Boron hydrogen compounds are being examined as a possible way to produce hydrogen for automobiles that will run on fuel cells. Although B_2H_6 is not the compound that will be used (i.e., it ignites spontaneously in air), the following equation illustrates how boron–hydrogen compounds react with water to produce hydrogen. Balance the following equation.

$$B_2H_6(g) + H_2O(l) \longrightarrow H_3BO_3(aq) + H_2(g)$$

PROCEDURE

Decide in what order the atoms should be balanced. Change coefficients in front of one molecule at a time and notice the effect on the other atoms in that molecule.

SOLUTION

First consider the B_2H_6 molecule, since it has the most atoms of an element other than hydrogen or oxygen. Balance the B by adding a coefficient of **2** in front of the H_3BO_3 on the right.

$$B_2H_6(g) + H_2O(l) \longrightarrow \mathbf{2}H_3BO_3(aq) + H_2(g)$$

Next, we notice that oxygen is in the fewer number of compounds, so we balance it next. Since there are 6 oxygen atoms in **2**H_3BO_3, place a **6** before the H_2O on the left. Finally, balance hydrogen. There are 18 on the left that are "locked in." Thus we need a **6** in front of the H_2 to have 18 hydrogen atoms on the right. A quick check confirms that we have a balanced equation.

$$B_2H_6(g) + 6H_2O(l) \longrightarrow 2H_3BO_3(aq) + 6H_2(g)$$

ANALYSIS

You may run into a case where you can't seem to balance the equation, no matter what combination of coefficients you use. Or perhaps you find yourself continually spiraling through the equation again and again as the numbers get larger and larger. Most chemical reactions balance with relatively small whole numbers. If there is a problem, it very well may be that one or more of your formulas are incorrect. You may have transcribed the problem inaccurately or written a wrong formula yourself. Rechecking formulas would be a logical place to restart.

SYNTHESIS

Technically, the coefficients should be the smallest set of whole numbers possible. Therefore $4\,A + 6\,B \longrightarrow 2\,C + 8\,D$ should reduce to $2\,A + 3\,B \longrightarrow C + 4\,D$. It *is*, however, acceptable to use fractions when balancing, and they can be quite useful for several types of problems whose solutions might not be apparent otherwise. These fractions shouldn't be interpreted as a fraction of a molecule, and if the equation is being used to determine how many molecules of each type react together, then any equation with fractions should be cleared to whole numbers. See Example 6-3 in Section 6-2.1 for this type of problem.

▶ **ASSESSING THE OBJECTIVE FOR SECTION 6-1**

EXERCISE 6-1(a) LEVEL 1: Fill in the blanks.

A chemical reaction is represented with symbols and formulas by means of a chemical _____. The arrow in an equation separates the _____ on the left from the _____ on the right. To conform to the law of conservation of mass, an equation must be _____. This is accomplished by introducing _____ in front of formulas rather than changing subscripts in a formula.

EXERCISE 6-1(b) LEVEL 2: Write unbalanced chemical equations for the following.
(a) Lithium and oxygen react to form lithium oxide.
(b) Aluminum and sulfuric acid yield aluminum sulfate and hydrogen.
(c) Copper reacts with nitric acid to produce copper(II) nitrate, nitrogen monoxide, and water.

EXERCISE 6-1(c) LEVEL 2: In what order should the atoms in the following equations be balanced?
(a) $H_3BCO + H_2O \longrightarrow H_3BO_3 + CO + H_2$
(b) $NH_3 + O_2 \longrightarrow NO + H_2O$
(c) $I_2 + Na_2S_2O_3 \longrightarrow NaI + Na_2S_4O_6$

EXERCISE 6-1(d) LEVEL 2: Balance the following reactions.
(a) $Cr + O_2 \longrightarrow Cr_2O_3$
(b) $Co_2S_3 + H_2 \longrightarrow H_2S + Co$
(c) $C_3H_8 + O_2 \longrightarrow CO_2 + H_2O$

EXERCISE 6-1(e) LEVEL 3: Write a balanced equation including symbols from the following descriptions.
(a) Solid ammonium carbonate decomposes into gaseous ammonia, carbon dioxide, and steam.
(b) Gaseous ammonia reacts with liquid chlorine trifluoride to form nitrogen gas, chlorine gas, and hydrogen fluoride gas.

For additional practice, work chapter problems 6-2, 6-4, 6-10, and 6-11.

6-2 Combustion, Combination, and Decomposition Reactions

► OBJECTIVE
FOR SECTION 6-2
Classify certain chemical reactions as being combustion, combination, or decomposition reactions.

LOOKING AHEAD! All the millions of known chemical changes can be represented by balanced equations. Many of these chemical reactions have aspects in common, so they can be grouped into specific classifications. In the remainder of this chapter, we do this with five basic types of reactions. We will notice that each type has a characteristic chemical equation. These five types are not the only ways that reactions can be grouped. In later chapters we will find other convenient classifications that will suit our purpose at that time. The first three types are the simplest and will be considered in this section. ■

Fire is certainly dramatic evidence of the occurrence of a chemical reaction. What we see as fire is the hot, glowing gases of a combustion reaction. The easiest way to put out a fire is to deprive the burning substance of a reactant (oxygen) by dousing it with water or carbon dioxide from a fire extinguisher. The reaction of elements or compounds with oxygen is the first of three types of reactions that we will discuss in this section.

6-2.1 Combustion Reactions

One of the most important types of reactions that we may refer to in the future is known as a **combustion reaction**. This type of reaction refers specifically to the reaction of an element or compound with elemental oxygen (O_2). Combustion usually liberates considerable heat energy and is accompanied by a flame. It is typically referred to as "burning." When elements undergo combustion, generally only one product (the oxide) is formed. Examples are the combustion of carbon and aluminum shown here.

$$C(s) + O_2(g) \longrightarrow CO_2(g)$$
$$4Al(s) + 3O_2(g) \longrightarrow 2Al_2O_3(s)$$

When compounds undergo combustion, however, two or more combustion products are formed. When carbon–hydrogen or carbon–hydrogen–oxygen compounds undergo combustion in an excess of oxygen, the combustion products are carbon dioxide and water. (Combustion reactions involving carbon–hydrogen or carbon–hydrogen–oxygen compounds are balanced in the order C, H, O).

$$CH_4(g) + 2O_2(g) \longrightarrow CO_2(g) + 2H_2O(l)$$

The metabolism of glucose ($C_6H_{12}O_6$, blood sugar) occurs in our bodies to produce the energy to sustain our life. We will discuss the energy liberated by this reaction in Chapter 7. But we are now concerned with the product compounds. This combustion reaction occurs at a steady, controlled rate.

$$C_6H_{12}O_6(aq) + 6O_2(g) \longrightarrow 6CO_2(g) + 6H_2O(l)$$

When insufficient oxygen is present (as in the combustion of gasoline, C_8H_{18}, in an automobile engine), some carbon monoxide (CO) also forms. This incomplete combustion reaction is shown below.

$$2C_8H_{18}(l) + 17O_2(g) \longrightarrow 16CO(g) + 18H_2O(l)$$

EXAMPLE 6-3   Balancing a Combustion Equation

Most fuels are composed of carbon and hydrogen (known as a hydrocarbon), and some may also contain oxygen. When they react with oxygen gas (burn), they form as products carbon dioxide gas and water. Write a balanced equation showing the burning of liquid rubbing alcohol (C_3H_8O).

PROCEDURE

First, represent the names of the species involved as reactants and products with formulas and indicate their physical states. Remember that oxygen gas is diatomic. Then balance the atoms in the order of C, H, and O. Use fractions where necessary, and scale up.

SOLUTION

The unbalanced equation is written as follows:

$$C_3H_8O(l) + O_2(g) \longrightarrow CO_2(g) + H_2O(l)$$

Now, balance carbon. Place a coefficient of **3** in front of CO_2 to balance the carbons in C_3H_8O. Next, balance the 8 hydrogen atoms in C_3H_8O by adding a coefficient of **4** in front of H_2O.

$$C_3H_8O(l) + O_2(g) \longrightarrow \mathbf{3}CO_2(g) + \mathbf{4}H_2O(l)$$

Notice that there are 10 oxygen atoms on the right. On the left 1 oxygen is in C_3H_8O, so 9 are needed from O_2. To get an odd number of oxygen atoms from O_2, we need to use a fractional coefficient, in this case $\frac{9}{2}$ (i.e., $\frac{9}{2} \times 2 = 9$)

$$C_3H_8O(l) + \frac{9}{2} O_2(g) \longrightarrow \mathbf{3}CO_2(g) + \mathbf{4}H_2O(l)$$

Finally, we need to clear the fraction so that all coefficients are whole numbers. Multiply the whole equation through by 2 and do a quick check.

$$\underline{2C_3H_8O(l) + 9O_2(g) \longrightarrow 6CO_2(g) + 8H_2O(l)}$$

ANALYSIS

Generally, in a combustion reaction, the total number of oxygen atoms on the product side of an equation will be odd or even. If the number is odd, as in the above example, the use of a fraction makes the balancing easier. The fraction will be some half value (3/2, 7/2, 9/2, etc.). Afterwards, simply double each coefficient to arrive at the appropriate balanced equation. If the number of oxygen atoms is even, it is even easier. A whole number is all that is required to balance things from the start.

SYNTHESIS

During any combustion reaction, oxygen combines with the elements present in the fuel. In cleaner-burning fuels, containing only carbon and hydrogen, and possibly oxygen, the only products of *complete* combustion are carbon dioxide and water. When there isn't enough oxygen present (such as in a poorly performing combustion engine), incomplete combustion results, with products containing fewer oxygen atoms than are optimal. This typically means CO (carbon monoxide—a deadly gas) and C (in the form of smoke, soot, or ash). When dirtier fuels are used, ones containing contaminants such as high-sulfur coal, the oxygen combines with these elements, too, to form by-products. SO_2 is the unpleasant result of combustion of sulfur-containing compounds. Combustion at high enough temperatures in the presence of nitrogen in the air can lead to the formation of nitrogen oxides such as NO and NO_2, components of smog, which themselves have harmful health effects.

FIGURE 6-2 Combination or Combustion Reaction When magnesium burns in air, the reaction can be classified as either a combination or a combustion reaction.

6-2.2 Combination Reactions

The chemical properties of an element describe how it does, or in some cases does not, combine with other elements or compounds. One type of reaction, known as a **combination reaction**, concerns *the formation of one compound from two or more elements and/or simpler compounds.* For example, an important chemical property of the metal magnesium is that it reacts with elemental oxygen to form magnesium oxide. (See Figure 6-2.) The synthesis (i.e., production of a substance) of MgO is represented at the end of this paragraph by a balanced equation and an illustration of the magnesium and oxygen atoms in the reaction. Notice in the reaction that an ionic compound is formed from neutral atoms. The Mg^{2+} cation is smaller than the parent atom, while the O^{2-} anion is larger than the parent atom. The reason for this is discussed in Chapter 8. Since one of the reactants is elemental oxygen, this reaction, and others like it, can also be classified as combustion reactions.

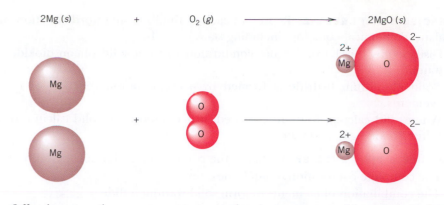

The following equations represent some other important combination reactions.

$$2Na(s) + Cl_2(g) \longrightarrow 2NaCl(s)$$
$$C(s) + O_2(g) \longrightarrow CO_2(g)$$
$$CaO(s) + CO_2(g) \longrightarrow CaCO_3(s)$$

6-2.3 Decomposition Reactions

A chemical property of a compound may be its tendency to decompose into simpler substances. This type of reaction is simply the reverse of combination reactions. That is, one compound is decomposed into two or more elements or simpler compounds. Many of these reactions take place only when heat is supplied, which is indicated by a Δ above the arrow. An example of this type of reaction is the decomposition of carbonic acid (H_2CO_3). This **decomposition reaction** causes the fizz in carbonated beverages. The reaction is illustrated here and is followed by other examples. (See also Figure 6-3.)

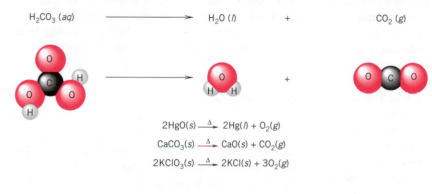

$$2HgO(s) \xrightarrow{\Delta} 2Hg(l) + O_2(g)$$
$$CaCO_3(s) \xrightarrow{\Delta} CaO(s) + CO_2(g)$$
$$2KClO_3(s) \xrightarrow{\Delta} 2KCl(s) + 3O_2(g)$$

▶ **ASSESSING THE OBJECTIVE FOR SECTION 6-2**

EXERCISE 6-2(a) LEVEL 1: Label each statement as applying to a decomposition, combination, and/or combustion reaction.

(a) Oxygen is a reactant.   **(d)** The equation must be balanced.
(b) There are two or more products. **(e)** Reaction usually requires heating to occur.
(c) There are two or more reactants.

EXERCISE 6-2(b) LEVEL 1: Label each of the following reactions as being an example of a combustion, combination, or decomposition reaction.

(a) $Cu(OH)_2(s) \longrightarrow CuO(s) + H_2O(l)$ *Decomp*
(b) $2CH_4O(l) + 3O_2(g) \longrightarrow 2CO_2(g) + 4H_2O(g)$ *combustion*
(c) $Ba(s) + F_2(g) \longrightarrow BaF_2(s)$ *combination*
(d) $(NH_4)_2Cr_2O_7(s) \longrightarrow N_2(g) + 4H_2O(g) + Cr_2O_3(s)$ *decomp*
(e) $Na_2O(s) + H_2O(l) \longrightarrow NaOH(aq)$ *Comb*

FIGURE 6-3 Decomposition Reaction The fizz of carbonated water is the result of a decomposition reaction.

EXERCISE 6-2(c) LEVEL 2: Represent each of the following word equations with a balanced chemical equation, including state symbols.

(a) Disilane gas (Si_2H_6) undergoes combustion to form solid silicon dioxide and water.

(b) Solid aluminum hydride is formed by a combination reaction of its two elements.

(c) When solid calcium bisulfite is heated, it decomposes to solid calcium oxide, sulfur dioxide gas, and water.

EXERCISE 6-2(d) LEVEL 3: Write out the complete reaction for the following:

(a) The complete combustion of liquid hexane.

(b) The combination of elements to form solid sodium iodide.

(c) The decomposition of aqueous sulfurous acid into a common liquid and gas.

For additional practice, work chapter problems 6-16, 6-18, and 6-20

MAKING IT REAL

Life Where the Sun Doesn't Shine—Chemosynthesis

Miles deep in the ocean a phenomenal discovery profoundly changed the way we view the life forms on this planet and perhaps other planets far away. Previously, we understood life as totally dependent on energy from the sun. The chlorophyll in plants produces carbohydrates in a process known as *photosynthesis*, as illustrated below. (Carbohydrates are represented by their empirical formula, CH_2O.)

$$(\text{solar energy}) + CO_2(g) + H_2O(l) \longrightarrow CH_2O(s) + O_2(g)$$

We use these carbohydrates from plants, or animals that eat these plants, to sustain life. So our energy is indirectly solar energy.

In 1977, scientists in the research submarine *Alvin* were exploring the mile-deep ocean bottom near the Galapagos Islands in the Pacific. Suddenly, they came upon an unbelievable diversity of life that existed in scalding water (350°C). The colony of life extended from bacteria to giant clams, mussels, and tubeworms 8 feet long. The *Alvin* had come upon ocean vents discharging huge amounts of hot water from deep in the Earth. The water was rich in minerals and chemicals.

One of those chemicals was hydrogen sulfide (H_2S). This gas is very poisonous to surface life but turned out to be the main source of energy for the bacteria, which serve as the bottom of the food chain. These bacteria are able to produce carbohydrates with the chemical energy from hydrogen sulfide rather than the light energy from the sun. This process is called *chemosynthesis*. The equation illustrating this reaction to produce a carbohydrate is

$$2O_2(g) + 8H_2S(aq) + 2CO_2(aq) \longrightarrow$$
$$2CH_2O(aq) + 2S_8(s) + 6H_2O(l)$$

In the late 1990s, even more startling discoveries were reported. Living bacteria have been found 3000 feet deep in the solid earth. These life-forms seem to use the chemical energy from H_2 to maintain life. They do not even need O_2, which is produced as a by-product of photosynthesis, as was found in the chemosynthesis in the deep-sea vents.

Life is much more resilient than we ever thought. Because of these discoveries, many scientists feel much more confident that life could exist on other planets, especially Mars.

Currently, we are intensely studying the conditions on that planet. Recent discoveries of the presence of large amounts of water at the poles and buried beneath the surface of Mars have heightened expectations. The Mars explorers have more recently found that bodies of water once existed on the surface of this planet. More exciting discoveries about the robustness of life await us.

KEY TERMS

6-1.1 Chemical reactions are represented by **chemical equations**. p. 180

6-1.1 **Reactants** and **products** in an equation are separated by an arrow. p. 180

6-1.1 Equations are **balanced** by use of **coefficients** in front of a compound or element. p. 181

6-2.1 **Combustion reactions** involve oxygen as a reactant. p. 185

6-2.2 **Combination reactions** have one product. p. 186

6-2.3 **Decomposition reactions** have one reactant. p. 187

SUMMARY CHART

Sequence of Balancing Equations

1. Most complex element $\longrightarrow$ 2. Any element (or polyatomic ion) other than H and O $\longrightarrow$

3. Usually H next if present in the most species $\longrightarrow$ 4. Finally balance O $\longrightarrow$ 5. Check

Three Types of Reactions

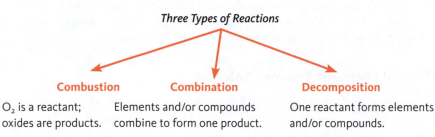

Combustion	**Combination**	**Decomposition**
O_2 is a reactant; oxides are products.	Elements and/or compounds combine to form one product.	One reactant forms elements and/or compounds.

Part B

Ions In Water and How They React

SETTING A GOAL

■ You will learn how ions are formed in aqueous solution and how these ions interact in two important types of chemical reactions.

OBJECTIVES

6-3 Write the ions formed when ionic compounds or acids dissolve in water.

6-4 Given the activity series, complete several single-replacement reactions as balanced molecular, total ionic, and net ionic equations.

6-5 Write balanced molecular, total ionic, and net ionic equations for precipitation reactions.

6-6 Write balanced molecular, total ionic, and net ionic equations for neutralization reactions.

6-3 The Formation of Ions in Water

▶ **OBJECTIVE FOR SECTION 6-3**
Write the ions formed when ionic compounds or acids dissolve in water.

LOOKING AHEAD! Before we discuss some general types of reactions that occur in aqueous solution we will examine how ions are formed in water. ■

6-3.1 Salts in Aqueous Solution

When one adds a sprinkle of table salt (sodium chloride) to water, the salt soon disappears into the aqueous medium. We observe that the table salt is soluble in water forming a homogeneous mixture known as a *solution,* which was first described

FIGURE 6-4 **A Soluble Compound** Table salt dissolves as it is added to hot water.

in Chapter 3. (See Figure 6-4.) However, if we add some powdered chalk (calcium carbonate) to water, it forms a suspension that slowly settles to the bottom of the container without apparent change. *When an appreciable amount of a substance dissolves in a liquid medium, we say that the substance is **soluble**. If very little or none of the substance dissolves, we say that the compound is **insoluble**.* Recall that *a **solution** is composed of a **solvent** (usually a liquid such as water) and a **solute** (a solid, a liquid, or even a gas).*

In Chapters 2 and 4, we described table salt (NaCl) as an ionic compound composed of Na^+ cations and Cl^- anions in the solid state. When an ionic compound dissolves in water, in nearly all cases the compound is separated into individual ions. The solution of NaCl in water can be represented by the equation

$$NaCl(s) \xrightarrow{H_2O} Na^+(aq) + Cl^-(aq)$$

The H_2O shown above the arrow indicates the presence of water as a solvent. The solution of another ionic compound, calcium perchlorate, can be represented by the equation

$$Ca(ClO_4)_2\,(s) \xrightarrow{H_2O} Ca^{2+}(aq) + 2ClO_4^-(aq)$$

Notice that all the atoms in the perchlorate ion remain together in solution as a complete entity. That is, they do not separate into chlorine and oxygen atoms. This is true for all of the polyatomic ions that we have mentioned.

6-3.2 Strong Acids in Aqueous Solution

In Chapter 4, a class of compounds known as acids was introduced and named. These are molecular compounds, but like ionic compounds, they also produce ions when dissolved in water. In this case, the neutral molecule is "ionized" by the water molecules. (We will examine how this happens in a later chapter.) Acids are so named because *they all form the H^+ (aq) ion when dissolved in water.* **Strong acids**, such as hydrochloric acid (HCl), *are completely ionized in water.* We will consider only the action of the six common strong acids (HCl, HBr, HI, HNO_3, $HClO_4$, and H_2SO_4) in water at this time. The solutions of the strong acids, hydrochloric and nitric acid, are illustrated by the equations

$$HCl(aq) \xrightarrow{H_2O} H^+(aq) + Cl^-(aq)$$
$$HNO_3(aq) \xrightarrow{H_2O} H^+(aq) + NO_3^-(aq)$$

EXAMPLE 6-4   **Forming Ions in Aqueous Solution**

Write equations illustrating the solution of the following compounds in water: **(a)** Na_2CO_3, **(b)** $CaCl_2$, **(c)** $(NH_4)_2Cr_2O_7$, **(d)** $HClO_4$, and **(e)** H_2SO_4. (Although it is not exactly correct, consider sulfuric acid as completely separated into two $H^+(aq)$ ions in aqueous solution.)

PROCEDURE

If necessary, refer to Table 4-2 or the periodic table for the charges on the ions.

SOLUTION

(a) Na_2CO_3 is composed of Na^+ and CO_3^{2-} ions.

$$Na_2CO_3 \longrightarrow 2Na^+(aq) + CO_3^{2-}(aq)$$

(b) $CaCl_2$ is composed of Ca^{2+} and Cl^- ions.

$$CaCl_2 \longrightarrow Ca^{2+}(aq) + 2Cl^-(aq)$$

(c) $(NH_4)_2Cr_2O_7$ is composed of NH_4^+ and $Cr_2O_7^{2-}$ ions.

$$(NH_4)_2Cr_2O_7 \longrightarrow 2NH_4^+(aq) + Cr_2O_7^{2-}(aq)$$

(d) $HClO_4$ is a strong acid that produces $H^+(aq)$ and ClO_4^- ions.

$$HClO_4 \longrightarrow H^+(aq) + ClO_4^-(aq)$$

(e) H_2SO_4 is a strong acid that produces $H^+(aq)$ and SO_4^{2-} ions.

$$H_2SO_4 \longrightarrow 2H^+(aq) + SO_4^{2-}(aq)$$

ANALYSIS

The meaning of the subscript again becomes crucial to your ability to perform this task. Distinguish between the 3 in CO_3^{2-} and the 3 in $FeCl_3$. In the carbonate ion, the subscript 3 is a part of the formula of the ion. You could almost consider it part of the "symbol" of carbonate. The symbol for chlorine is just Cl. Therefore, the subscript 3 in $FeCl_3$ tells us how many ions of chloride there are. To show three ions of carbonate, the entire formula would have to be in parenthesis with a subscripted 3 on the outside.

SYNTHESIS

We will learn in Chapter 12 that the number of aqueous ions an ionic compound breaks into can affect some of that solution's properties. Order the following compounds by number of ions they'll form in solution: $Al(ClO_3)_3$, Na_2CO_3, NH_4Cl, $Fe_2(SO_4)_3$. The correct order is NH_4Cl (with two), Na_2CO_3 (three), $Al(ClO_3)_3$ (four), and $Fe_2(SO_4)_3$ (five).

▶ **ASSESSING THE OBJECTIVES FOR SECTION 6-3**

EXERCISE 6-3(a) LEVEL 1: List the cations and the anions present when the following compounds dissolve in water.
(a) $HClO_4$ **(b)** K_3PO_4 **(c)** NH_4NO_3 **(d)** $Zn(ClO_4)_2$

EXERCISE 6-3(b) LEVEL 2: How many ions will the following species form when one formula unit of each dissolves in water?
(a) $MgBr_2$ **(b)** Na_3PO_4 **(c)** $Fe(NO_3)_3$ **(d)** $KClO_3$

EXERCISE 6-3(c) LEVEL 2: Write the equations illustrating the formation of solutions of the following compounds. Indicate state symbols.
(a) Na_2SO_4 **(b)** HNO_3 **(c)** $Al(C_2H_3O_2)_3$ **(d)** $Ca(OH)_2$

For additional practice, work chapter problems 6-24, 6-25, and 6-26.

6-4 Single-Replacement Reactions

LOOKING AHEAD! Solutions containing metal cations may undergo chemical reactions when a solid sample of some other metal comes into contact with this solution. In this section, we will examine this type of reaction. ∎

▶ **OBJECTIVE FOR SECTION 6-4**
Given the activity series, represent several single-replacement reactions with balanced molecular, total ionic, and net ionic equations.

An interesting thing happens when we immerse a strip of zinc metal in a blue aqueous solution of a copper(II) salt such as $CuSO_4(aq)$. When we remove the zinc, it now looks as if it has changed into copper. The blue solution has also lost some of its color. Actually, what happened is that a coating of copper formed on the metal strip. Silver and gold will also form a coating on a strip of zinc immersed in solutions of compounds containing these metal ions. (See Figure 6-5.) These and similar reactions are known as **single-replacement reactions**. In this type of reaction, which most often occurs in aqueous solution, one free element substitutes for another element already combined in a chemical compound. The replacement of zinc ions for the copper ions and the copper metal for the zinc metal is illustrated here; other examples follow.

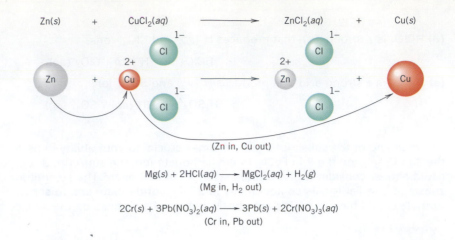

$$Zn(s) + CuCl_2(aq) \longrightarrow ZnCl_2(aq) + Cu(s)$$

(Zn in, Cu out)

$$Mg(s) + 2HCl(aq) \longrightarrow MgCl_2(aq) + H_2(g)$$
(Mg in, H_2 out)

$$2Cr(s) + 3Pb(NO_3)_2(aq) \longrightarrow 3Pb(s) + 2Cr(NO_3)_3(aq)$$
(Cr in, Pb out)

FIGURE 6-5 Single-Replacement Reaction The formation of a layer of copper on a piece of zinc is a single-replacement reaction.

6-4.1 Types of Equations

The equation illustrating the replacement of Zn by Cu is known as the molecular form of the equation. *In a* **molecular equation**, *all reactants and products are shown as neutral compounds.* To represent the nature of soluble ionic compounds and strong acids in water, it is helpful to show the separate ions as they actually exist in aqueous solution. *When the cations and anions of a compound in solution are shown separately, the resulting equation is known as a* **total ionic equation**. It is illustrated as follows.

(molecular equation) $Zn(s) + CuCl_2(aq) \longrightarrow ZnCl_2(aq) + Cu(s)$

(total ionic equation) $Zn(s) + Cu^{2+}(aq) + 2Cl^-(aq) \longrightarrow Zn^{2+}(aq) + 2Cl^-(aq) + Cu(s)$

Notice that in this equation, the Cl^- ions appear on both sides of the equation unchanged. Their role is to provide the anions needed to counteract the positive charge of the cations. The presence of the Cl^- ions is certainly necessary because any compound, whether in the pure state or dispersed in solution, must be electrically neutral. However, they do not actually affect the chemical change that is occurring. *Ions that are in an identical state on both sides of a total ionic equation are called* **spectator ions**. *If spectator ions are subtracted from both sides of the equation, the remaining equation is known as the* **net ionic equation**. This equation focuses only on the species that have undergone a change in the reaction. By subtracting the two Cl^- spectator ions from both sides of the equation we have the net ionic equation. It is the net ionic equation that actually shows us the reaction that is occurring in the beaker.

(net ionic) $Zn(s) + Cu^{2+}(aq) \longrightarrow Zn^{2+}(aq) + Cu(s)$

A reaction is spontaneous in one direction only, so the reverse reaction does not occur. That is, if we were to immerse a copper strip in a $ZnCl_2$ solution, a coating of Zn would not form on the copper. (See Figure 6-6.)

FIGURE 6-6 Copper in a Zinc Sulfate Solution When a strip of copper is immersed in a zinc ion solution, no reaction occurs.

6-4.2 The Activity Series

Series of experiments performed on different metal–solution combinations allow us to compare the ability of one metal to replace the ions of other metals in solution. The **activity series** shown in Table 6-2 *lists some common metals in decreasing order of their ability to replace metal ions in aqueous solution.* (The metal cations present in solution are shown to the right of the metal.) The metal at the top (K) replaces all of the metal ions below it and is therefore the *most* reactive. The second metal replaces all below it, while its ions will be replaced only by K. In fact, K, Na, Mg, and Al are all so reactive that they react with water itself. (Aluminum *appears* unreactive with water because it is coated with Al_2O_3, which protects the metal from contact with water.) They replace other metal ions only when the solids are mixed, not when the ions are in water solution. The following balanced equation illustrates such a reaction.

$$2Al(s) + Fe_2O_3(s) \longrightarrow Al_2O_3(s) + 2Fe(l)$$

The previous reaction is known as the *thermite reaction* because it liberates so much heat that the iron formed is molten. As a result, this reaction has application in welding.

In the activity series H_2 is treated as a metal and $H^+(aq)$ (from one of the six strong acids) as its metal ion. The activity series can be used to predict which reactions are expected to occur. For example, notice in Table 6-2 that nickel is ranked higher than silver. This allows us to predict that elemental nickel metal replaces the Ag^+ ion in solution. Thus, if we immerse a strip of nickel in a solution of aqueous $AgNO_3$, we find that a coating of silver forms on the nickel. The balanced molecular equation and net ionic equation illustrating this reaction are as follows. Note that the nitrate ion (NO_3^-) is a spectator ion.

$$Ni(s) + 2AgNO_3(aq) \longrightarrow 2Ag(s) + Ni(NO_3)_2(aq)$$
$$Ni(s) + 2Ag^+(aq) \longrightarrow 2Ag(s) + Ni^{2+}(aq) \text{ (net ionic)}$$

TABLE 6-2

The Activity Series

	METAL	METAL ION
(most active)	K	K^+
	Na	Na^+
	Ca	Ca^{2+}
	Mg	Mg^{2+}
	Al	Al^{3+}
	Zn	Zn^{2+}
	Cr	Cr^{3+}
	Fe	Fe^{2+}
	Ni	Ni^{2+}
	Sn	Sn^{2+}
	Pb	Pb^{2+}
	H_2	H^+
	Cu	Cu^{2+}
	Ag	Ag^+
(least active)	Au	Au^{3+}

EXAMPLE 6-5 Predicting Spontaneous Single-Replacement Reactions

Consider the following two possible reactions. If a reaction does occur, write the balanced molecular, total ionic, and net ionic equations illustrating the reactions.

(a) A strip of tin metal is placed in an aqueous $AgNO_3$ solution.

(b) A strip of silver metal is placed in an aqueous perchloric acid solution.

PROCEDURE

(a) Notice in the activity series that Sn is higher in the series than Ag. Therefore Sn replaces Ag^+ ions from solution.

(b) In the activity series, H_2 is higher than Ag. Therefore, H_2 replaces Ag^+, but the reverse reaction, the replacement of H^+ by Ag, does not occur.

SOLUTION

(a)
$$Sn(s) + 2AgNO_3(aq) \longrightarrow 2Ag(s) + Sn(NO_3)_2(aq) \text{ (molecular)}$$
$$Sn(s) + 2Ag^+(aq) + 2NO_3^-(aq) \longrightarrow 2Ag(s) + Sn^{2+}(aq) + 2NO_3^-(aq) \text{ (total ionic)}$$
$$Sn(s) + 2Ag^+(aq) \longrightarrow 2Ag(s) + Sn^{2+}(aq) \text{ (net ionic)}$$

(b) No reaction occurs.

ANALYSIS

To say that a metal is reactive means that it has a tendency to exist as an ion, not a free metal. To say that a metal is stable means that it is more often found in its metallic, or free, state. So compare Al and Pb. Which is more reactive? Which is more likely to be found in its free state? Which will replace which in a single-replacement reaction? The aluminum is more reactive. The lead is more stable. Therefore, solid aluminum will switch places with, or replace, aqueous lead ion in a single-replacement reaction. It's worth noting that Table 6-2 also reflects the ease with which metals are recovered in mining. You can proceed up the list from the bottom and have a roughly accurate historical record of the order in which important metals were put into service by various cultures.

SYNTHESIS
....
According to Table 6-2, which metals won't react with acid? Cu, Ag, and Au. How are the uses of these metals related to this chemical property? We use them to make metallic objects that we wouldn't want to corrode or dissolve. Obvious examples are coins, wiring, and jewelry. Why, then, is iron used so extensively in structural building? It is in the middle of the chart, and fairly reactive. It's because it is strong, abundant, and easy to mine and process. We then have to make allowances for the fact that we'll need to protect or replace structures made from iron, and its alloy steel, on a regular basis because of the constant chemical corrosion that occurs. If iron is exposed to the weather, for instance, it will need periodic painting.

▶ **ASSESSING THE OBJECTIVE FOR SECTION 6-4**

EXERCISE 6-4(a) LEVEL 1: Choose the metal that will replace the other one (as an ion) in solution.
(a) Ag or Ni   **(b)** Al or Cu   **(c)** Cr or Mg   **(d)** Sn or acid

EXERCISE 6-4(b) LEVEL 2: Given the following molecular equations, write the total and the net ionic equations for the reactions:
(a) $Cu(NO_3)_2(aq) + Zn(s) \longrightarrow Zn(NO_3)_2(aq) + Cu(s)$
(b) $2Al(s) + 3PbCl_2(aq) \longrightarrow 2AlCl_3(aq) + 3Pb(s)$

EXERCISE 6-4(c) LEVEL 2: Complete balanced molecular equations for the following reactants.
(a) $Mg(s) + AgNO_3(aq) \longrightarrow$
(b) $Sn(s) + KI(aq) \longrightarrow$
(c) $Na(s) + HCl(aq) \longrightarrow$

EXERCISE 6-4(d) LEVEL 3: Write total and net ionic equations for the following reactants.
(a) $CuCl_2(aq) + Fe(s) \longrightarrow$
(b) $Mg(s) + Au(NO_3)_3(aq) \longrightarrow$

For additional practice, work chapter problems 6-28, 6-30, and 6-31.

▶ **OBJECTIVE FOR SECTION 6-5**
Write balanced molecular, total ionic, and net ionic equations for precipitation reactions.

6-5 Double-Replacement Reactions—Precipitation

LOOKING AHEAD! In a single-replacement reaction, only cations are involved. In a double-replacement reaction, both cations and anions are involved. The driving force of these reactions is the formation of a product from the exchange of ions that is insoluble in water, is a molecular compound, or, in a few cases, is both. The first type of double-replacement reaction, formation of a solid, is discussed in this section. ■

6-5.1 Soluble and Insoluble Ionic Compounds

Marble statues have suffered the ravages of weather, for thousands of years in some cases. Marble, limestone, and chalk are essentially the same compound, calcium carbonate. Obviously, this ionic compound is insoluble in water. It formed when calcium ions (Ca^{2+}) and carbonate ions (CO_3^{2-}) present in some ancient sea came together to form a solid deposit. Before we look at this type of reaction, we should bring some order and guidelines to the determination of which ionic compounds are soluble in water and which are not. In Table 6-3, some rules for the solubility of compounds containing common anions are listed. Although this table focuses only on anions, it will help to know that compounds formed from cations of Group IA (i.e., Na^+, K^+) and the NH_4^+ ion are all water soluble regardless of the anion.

TABLE 6-3

Solubility Rules for Some Ionic Compounds

ANION	SOLUBILITY RULE
Mostly Soluble	
Cl^-, Br^-, I^-	All cations form *soluble* compounds except Ag^+, Hg_2^{2+}, and Pb^{2+}. ($PbCl_2$ and $PbBr_2$ are slightly soluble.)
NO_3^-, ClO_4^-, $C_2H_3O_2^-$	All cations form *soluble* compounds. ($KClO_4$ and $AgC_2H_3O_2$ are slightly soluble.)
SO_4^{2-}	All cations form *soluble* compounds except Pb^{2+}, Ba^{2+}, and Sr^{2+}. (Ca^{2+} and Ag^+ form slightly soluble compounds.)
Mostly Insoluble	
CO_3^{2-}, PO_4^{3-}	All cations form *insoluble* compounds except Group IA metals and NH_4^+.
S^{2-}	All cations form *insoluble* compounds except Group IA and IIA metals and NH_4^+.
OH^-	All cations form *insoluble* compounds except Group IA metals, Ba^{2+}, Sr^{2+}, and NH_4^+. [$Ca(OH)_2$ is slightly soluble.]

EXAMPLE 6-6   Predicting Whether an Ionic Compound is Soluble

Use Table 6-3 to predict whether the following compounds are soluble or insoluble.

(a) NaI   **(b)** CdS   **(c)** $Ba(NO_3)_2$ **(d)** $SrSO_4$

PROCEDURE

Apply the solubility rules to the anion. Check to see that the cation is mentioned as an exception.

SOLUTION

(a) According to Table 6-3, all alkali metal (Group IA) compounds of the anions listed are soluble. Therefore, NaI is soluble.
(b) All S^{2-} compounds are insoluble except those formed with Group IA and IIA metals and NH_4^+. Since Cd is in Group IIB, it is not one of the exceptions. CdS is insoluble.
(c) All NO_3^- compounds are soluble. Therefore, $Ba(NO_3)_2$ is soluble.
(d) The Sr^{2+} ion is one of the exceptions to soluble SO_4^{2-} compounds. Therefore, $SrSO_4$ is insoluble.

ANALYSIS

In order for an ionic compound to be insoluble, both the cation and the anion have to form at least some insoluble compounds. If either of them forms only soluble compounds, then the specific compounds containing these cations or anions are soluble (with the noted exceptions). Consider the following compounds: NH_4Cl, Na_3PO_4, $Fe(NO_3)_3$, and $Cu(OH)_2$. In NH_4Cl, Cl^- is a listed soluble anion, and NH_4^+ is one of the universally soluble cations. Clearly the compound is soluble. In the case of Na_3PO_4, the presence of Na^+ (a soluble cation) overrides the presence of the typically insoluble PO_4^{3-}. Just the reverse is true in $Fe(NO_3)_3$, where the presence of the soluble NO_3^- makes the compound soluble. $Cu(OH)_2$, however, is insoluble, since the Cu^{2+} ion is not one of the exceptions.

SYNTHESIS

Occasionally it is useful to remove a particular cation or anion from solution by formation of a solid compound containing that ion. What anion could you add to a solution containing Zn^{2+} to make it insoluble? There are several possibilities. Anything in the lower half of Table 6-3 would work to precipitate the zinc—S^{2-}, for example. This leads us to the discussion of these types of reactions.

6-5.2 Formation of a Precipitate

What happens when we mix solutions containing soluble ionic compounds? It depends. If we mix a solution of $CuCl_2$ (green) and KNO_3 (clear), we simply have a mixture of the four ions in solution, as illustrated in Figure 6-7; no cation is associated with a particular anion.

Now let's consider a case of two solutions, one containing the soluble compound $CuCl_2$ and the other containing the soluble compound $AgNO_3$. When we mix these two solutions, something obviously occurs. The mixture immediately becomes cloudy, and eventually a white solid settles to the bottom of the container. *The solid that is formed by the reaction of the two solutions is called a* **precipitate**. In this case it is the insoluble compound, $AgCl$. In fact, whenever Ag^+ and Cl^- are mixed into the same solution, they come together to form solid $AgCl$. This leaves the Cu^{2+} (which forms a blue color) and the NO_3^- ions in solution, since $Cu(NO_3)_2$ is soluble. (See Figure 6-8.)

The formation of a precipitate by mixing solutions of two soluble compounds is known as a **precipitation reaction**. This is one of three types of **double-replacement reactions** where the two cations involved exchange anions. The reaction is illustrated by the following molecular equation.

$$2AgNO_3(aq) + CuCl_2(aq) \longrightarrow 2AgCl(s) + Cu(NO_3)_2(aq)$$

In the total ionic equation, the soluble ionic compounds are represented as separate ions on both sides of the equation, but the solid precipitate (i.e., $AgCl$) is shown as a neutral compound since the two ions come together to produce the insoluble solid.

$$2Ag^+(aq) + 2NO_3^-(aq) + Cu^{2+}(aq) + 2Cl^-(aq) \longrightarrow$$
$$2AgCl(s) + Cu^{2+}(aq) + 2NO_3^-(aq)$$

FIGURE 6-7 A Mixture of $CuCl_2$ and KNO_3 Solutions No reaction occurs when these solutions are mixed.

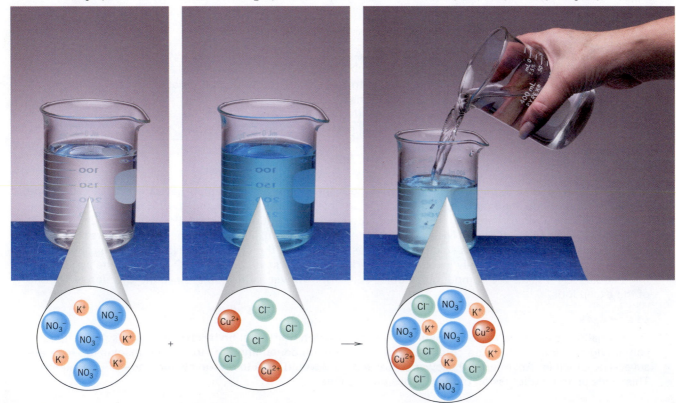

$KNO_3(aq)$ $CuCl_2(aq)$ $K^+(aq)$, $Cu^{2+}(aq)$, $Cl^-(aq)$, $NO_3^-(aq)$

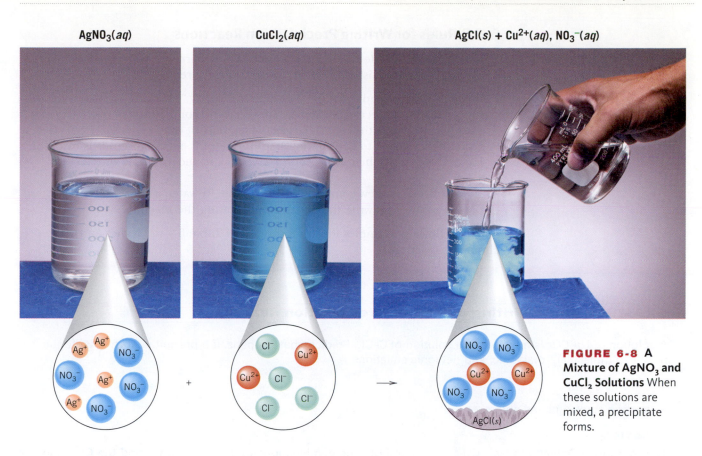

AgNO₃(aq)   **CuCl₂(aq)**   **AgCl(s) + Cu²⁺(aq), NO₃⁻(aq)**

FIGURE 6-8 A **Mixture of AgNO₃ and CuCl₂ Solutions** When these solutions are mixed, a precipitate forms.

Notice that there are two spectator ions, Cu^{2+} and NO_3^-. After spectator ions are subtracted from the equation, we have the net ionic equation for the reaction. As before, the net ionic equation shows us the real reaction occurring in the system.

$$Ag^+\,(aq) + Cl^-\,(aq) \longrightarrow AgCl(s)$$

As in the single-replacement reactions discussed previously, the net ionic equation focuses on the driving force for the reaction, which is the formation of solid AgCl from two soluble compounds.

There are some very practical applications of precipitation reactions in industry as well as in the laboratory. Our example is of particular value. Silver is widely used in the development of film. It is obviously worthwhile to recover this precious metal whenever possible. Silver metal in film can be dissolved in aqueous nitric acid to form the water-soluble compound $AgNO_3$. [This is not a single-replacement reaction (which by itself would not work) but involves the nitrate ion in a more complex reaction.] Although the solution contains many other dissolved substances, addition of a soluble compound containing the Cl^- ion (e.g., NaCl) leads to the formation of solid AgCl, as shown in Figure 6-8. As you notice from Table 6-3, very few other cations form precipitates with Cl^-, so this is a reaction more or less specific to removing Ag^+ from aqueous solution. The AgCl can then be filtered from the solution and silver metal eventually recovered.

In other precipitation reactions, we may want to recover the soluble compound and discard the insoluble compound. In such a case, we would remove the precipitate by filtration and then recover the soluble compound by boiling away the solvent water. See the "Making It Real: Hard Water and Water Treatment" for a discussion on how this concept can be used to improve the quality of a community's water supply.

6-5.3 Rules for Writing Precipitation Reactions

By careful use of Table 6-3, we can predict the occurrence of many precipitation reactions. To accomplish this, we follow this procedure:

1. Write the compounds produced in the reaction by "switching partners," changing subscripts as necessary to make sure that the compounds have the correct formulas based on the ions' charges (which do not change).
2. Examine Table 6-3 to determine whether one of these compounds is insoluble. Label the soluble chemicals (*aq*) for *aqueous* and the insoluble ones (*s*) for *solid*.
3. If one of the two new compounds is insoluble in water, a precipitation reaction occurs and we can write the equation illustrating the reaction.

The following examples illustrate the use of Table 6-3 to predict and write precipitation reactions.

EXAMPLE 6-7　**Writing a Possible Precipitation Reaction**

A solution of Na_2CO_3 is mixed with a solution of $CaCl_2$. Predict what happens. If a precipitate forms, write the balanced molecular, total ionic, and net ionic equations.

PROCEDURE

Follow the steps outlined above.

SOLUTION

The four ions involved are Na^+, CO_3^{2-}, Ca^{2+}, and Cl^-. The combinations of the Na^+ and Cl^- and the Ca^{2+} and CO_3^{2-} produce the compounds NaCl and $CaCO_3$. If both of these compounds are soluble, no reaction occurs. In this case, however, reference to Table 6-3 indicates that $CaCO_3$ is insoluble. Thus, a reaction occurs that we can illustrate with a balanced reaction written in molecular form.

$$Na_2CO_3(aq) + CaCl_2(aq) \longrightarrow CaCO_3(s) + 2NaCl(aq)$$

The equation written in total ionic form is

$$2Na^+(aq) + CO_3^{2-}(aq) + Ca^{2+}(aq) + 2Cl^-(aq) \longrightarrow CaCO_3(s) + 2Na^+(aq) + 2Cl^-(aq)$$

Note that the Na^+ and the Cl^- ions are spectator ions. Elimination of the spectator ions on both sides of the equation leaves the net ionic equation.

$$Ca^{2+}(aq) + CO_3^{2-}(aq) \longrightarrow CaCO_3(s)$$

(This would be an example of the formation of limestone in an ancient sea.)

ANALYSIS

Working as many of these types of problems as possible is the key to seeing the patterns that develop. Typically, the anion of one of the reactants will combine with the cation of the second to form the insoluble product. A quick glance at the anions present will give you an idea of the likelihood that they will form an insoluble compound. Then analyze the cation of the other reactant. Will it combine with the first anion to precipitate? Consider $Ni(NO_3)_2$ and Li_3PO_4. Do we expect an insoluble product to form? Of the two anions, the phosphate is the one likely to form a precipitate. The cation of the first compound, Ni^{2+}, is not in the Group IA column, so it, too, is a good candidate. We'd expect the combination of the two, $Ni_3(PO_4)_2$ (nickel(II) phosphate), to be a precipitate.

SYNTHESIS

Can you go further and find a compound that would react with a given reactant? What compound (and there are several) would form a precipitate with KOH? Here it is the anion that likely will precipitate. Let's pick a cation that will do the job. Cu^{2+} would work. Now we need an anion that will ensure the Cu^{2+} to be initially soluble. Nitrate, chloride, and acetate fit the bill. $Cu(NO_3)_3$, for one, would dissolve in water and then form a precipitate with KOH. Following that same reasoning, any of these would also work: $FeCl_3$, $Al(C_2H_3O_2)_3$, $NiSO_4$, and many others. Consider the following example.

EXAMPLE 6-8   Writing a Possible Precipitation Reaction

A solution of KOH is mixed with a solution of MgI_2. Predict what happens. If a precipitate forms, write the balanced molecular, total ionic, and net ionic equations.

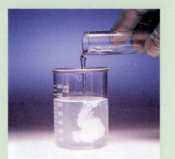

SOLUTION

The four ions involved are K^+, OH^-, Mg^{2+} and I^-. An exchange of ions in the reactants produces the compounds KI and $Mg(OH)_2$. Reference to Table 6-3 indicates that Mg^{2+} forms an insoluble compound with OH^-. Therefore, a precipitation reaction does occur and is illustrated with the following molecular equation.

$$2KOH(aq) + MgI_2(aq) \longrightarrow Mg(OH)_2(s) + 2KI(aq)$$

The total ionic equation is

$$2K^+(aq) + 2OH^-(aq) + Mg^{2+}(aq) + 2I^-(aq) \longrightarrow Mg(OH_2)(s) + 2K^+(aq) + 2I^-(aq)$$

Elimination of spectator ions gives the net ionic equation.

$$Mg^{2+}(aq) + 2OH^-(aq) \longrightarrow Mg(OH)_2(s)$$

ANALYSIS

When writing the total ionic equation, simply go through your equation to all the ionic compounds or acids labeled "aqueous," and break them apart into their ions. Any other chemical, labeled as "solid," "liquid," or "gas," should be written exactly the same way it was in the molecular equation.

SYNTHESIS

It is almost always the case that the combination of the four possible ions from a double-replacement reaction will at most form one insoluble compound. In several cases, *no* insoluble compound forms and there is no reaction (see the next example). By studying Table 6-3, particularly the exceptions, can you come up with a combination of two soluble compounds that will react to form two insoluble products? There are very few choices. Notice that $Ba(OH)_2$ is soluble, whereas most hydroxides are insoluble. Furthermore, $BaSO_4$ is insoluble, whereas most sulfates are soluble. Does that give you an idea? As an example, how about combining solutions of $Ba(OH)_2(aq)$ and $CuSO_4(aq)$? What would that net ionic equation look like?

$$Ba^{2+}(aq) + 2OH^-(aq) + Cu^{2+}(aq) + SO_4^{2-}(aq) \longrightarrow BaSO_4(s) + Cu(OH)_2(s)$$

EXAMPLE 6-9   Writing a Possible Precipitation Reaction

A solution of KBr is mixed with a solution of $Sr(ClO_4)_2$. Predict what happens. If a precipitate forms, write the balanced molecular, total ionic, and net ionic equations.

SOLUTION

The four ions involved are K^+, Br^-, Sr^{2+}, and ClO_4^-. An exchange of ions produces the compounds $KClO_4$ and $SrBr_2$. Both of these compounds are soluble, so no reaction occurs. The solution contains a mixture of these four ions.

ANALYSIS

Visually, what would we expect to see in these examples? In the case of the precipitation reactions, we initially begin with two clear, though not necessarily colorless, solutions. Upon mixing, cloudiness appears due to particles that should eventually settle to the bottom of the reaction vessel. For this last reaction, the two solutions are simply mixed together into one large vessel; they look essentially the same as before mixing. If one solution was colored, the combination might appear to be a lighter hue, but this is just due to diluting. It doesn't indicate a chemical reaction. (See also the example in Figure 6-7.)

SYNTHESIS

The following reaction works, for reasons to be discussed in the following section. What would you predict to be the visual indication of a reaction?

$$Fe(OH)_3 + 3HCl \longrightarrow 3H_2O + FeCl_3$$

According to the solubility rules, $Fe(OH)_3$ is a solid. HCl and $FeCl_3$ are in aqueous solutions. H_2O is, of course, a liquid. So we are starting off with a solid and aqueous solution, and ending with a mixture of a soluble salt in water. Therefore, upon the addition of hydrochloric acid, we should see our solid "disappear," as its soluble product dissolves in water.

MAKING IT REAL

Hard Water and Water Treatment

Hard water is water that contains significant concentrations of Fe^{3+}, Ca^{2+} and Mg^{2+}. These are picked up when rainwater filters down through soil rich in these ions. Most areas of the country, other than the South, East Coast, and Pacific Northwest, suffer from very hard water. Hard water has a strong tendency to form insoluble compounds with the compounds found in soap. This reduces the soap's ability to clean by forming an insoluble residue that deposits on sinks, tubs, dishes, and clothes. This leaves them gray, gritty, and dingy. Insoluble carbonates may also precipitate within water pipes and water heaters, forming what is called *scale*. Scale reduces water flow, insulates the water in heaters from the heat source, and lowers the lifetime of the plumbing.

Hard water can be fixed by using detergents with added softening agents. These compounds remain soluble despite the presence of the hard ions, making lathering and cleaning more effective. The instillation of a water-softening system in a house also solves the problem of hard water. This system exchanges the soft soluble cation Na^+ for the hard ions that form precipitates in the water. (Recall that sodium compounds are all soluble.) This must be done to preserve the total charge, so it takes two sodium ions to remove one calcium ion or one magnesium ion, and three sodium ions to remove one iron ion [e.g., $3Na^+(aq)$ has the same total charge as one Fe^{3+}]. However, the large amount of sodium added to the water can make it unhealthy, especially for those on a low-sodium diet. So water-softening systems are installed only on pipes leading to showers and other cleaning devices. Typically the kitchen and bathroom sinks receive the harder, but healthier, water.

Municipalities in areas that have hard water try to reduce the problem by removing some of the ions at the source before the water goes to commercial and residential areas. They do this by adding CaO and Na_2CO_3 to the water. These compounds cause the hard ions to precipitate at the site as their insoluble hydroxide or carbonate salts. These salts are then allowed to settle out of the water and are periodically removed.

Each treatment option described relies on chemists and engineers understanding the chemical reactions that lead to the problems caused, and then employing other reactions, based on solubility, that allow them to alleviate the problem.

▶ **ASSESSING THE OBJECTIVE FOR SECTION 6-5**

EXERCISE 6-5(a) LEVEL 1: Do the following refer to soluble or insoluble compounds?
(a) It forms a precipitate.
(b) It is labeled "(aq)."
(c) It dissolves in water.
(d) Its formation is the driving force in the reaction.
(e) It breaks apart into its individual cations and anions.

EXERCISE 6-5(b) LEVEL 2: Label the following as being soluble (aq) or insoluble (s).
(a) $(NH_4)_2SO_4$ **(b)** $Sn(OH)_2$ **(c)** $Pb(NO_3)_2$
(d) $Fe_2(SO_4)_3$ **(e)** $MgCO_3$ **(f)** Ag_3PO_4

EXERCISE 6-5(c) LEVEL 2: Determine whether the following reactants will form a precipitate.
(a) K_2CO_3 and Na_2SO_4 **(b)** $AlBr_3$ and $NaOH$ **(c)** $(NH_4)_2S$ and CaI_2

EXERCISE 6-5(d) LEVEL 2: Write the balanced molecular formula, complete with the state symbols (*aq*) and (*s*), for the following reactants.
(a) $CrI_3(aq) + Li_3PO_4(aq) \longrightarrow$ **(b)** $Cs_2CO_3(aq) + FeBr_3(aq) \longrightarrow$

EXERCISE 6-5(e) LEVEL 3: A solution of $AgNO_3$ is mixed with a solution of K_2S. Write the molecular, total ionic, and net ionic equations illustrating the reaction.

EXERCISE 6-5(f) LEVEL 3: What soluble compounds could you mix together to create a precipitate of:
(a) $CaCO_3$ **(b)** $FePO_4$

For additional practice, work chapter problems 6-34, 6-37, 6-39, 6-44, and 6-46.

6-6 Double-Replacement Reactions–Neutralization

▶ **OBJECTIVE**
FOR SECTION 6-6
Write balanced molecular, total ionic, and net ionic equations for neutralization reactions.

LOOKING AHEAD! In a second type of double-replacement reaction, the ions combine to form a molecular compound. Although there are several examples of this, we will focus on the formation of the simple molecular compound water. This is discussed next. ■

6-6.1 Strong Acids and Strong Bases

Strong acids and bases are two compounds that can be difficult, if not dangerous, to handle. They are very corrosive and can cause severe burns. (See Figure 6-9.) When carefully (i.e., slowly) mixed together in the right proportions, however, they become harmless. The corrosive properties of both are "neutralized." In Section 4-6 we defined a class of compounds known as strong acids. They are characterized by their complete ionization in water to form $H^+(aq)$ and an anion. A second class of compounds, **strong bases**, *dissolve in water to form the hydroxide ion [$OH^-(aq)$]*. Unlike acids, these compounds are ionic in the solid state and the ions are simply separated in the aqueous solution as in any soluble ionic compound. The strong bases are hydroxides formed by the Group IA and IIA metal ions (except for Be^{2+}). Examples include sodium hydroxide (lye) and barium hydroxide. Their solution in water is illustrated by the following equations.

$$NaOH(s) \xrightarrow{H_2O} Na^+(aq) + OH^-(aq)$$
$$Ba(OH)_2(s) \xrightarrow{H_2O} Ba^{2+}(aq) + 2OH^-(aq)$$

FIGURE 6-9 Strong Acids and Bases Containers of these compounds usually include a warning about their corrosive properties.

6-6.2 Neutralization Reactions

When solutions of strong acids and bases are mixed, the $H^+(aq)$ from the acid combines with the $OH^-(aq)$ from the base to form the molecular compound water. *The reaction of an acid and a base is known as a* **neutralization reaction**. The neutralization of hydrochloric acid and sodium hydroxide is illustrated below with the molecular, total ionic, and net ionic equations. The ionic compound remaining in solution, NaCl, is known as salt. *A* **salt** *is formed from the cation of the base and the anion of the acid.* If the salt is soluble, its ions are spectator ions and are subtracted from the equation to form the net ionic equation.

$$HCl(aq) + NaOH(aq) \longrightarrow NaCl(aq) + H_2O(l)$$
$$H^+(aq) + Cl^-(aq) + Na^+(aq) + OH^-(aq) \longrightarrow Na^+(aq) + Cl^-(aq) + H_2O(l)$$
$$H^+(aq) + OH^-(aq) \longrightarrow H_2O(l)$$

Unlike precipitation reactions, the net ionic equation for all neutralization reactions between strong acids and strong bases is the same. The driving force for these reactions is the formation of water, a molecular compound, from two ions. The balanced molecular equations for two additional neutralization reactions follow.

$$HBr(aq) + KOH(aq) \longrightarrow KBr(aq) + H_2O(l)$$
$$2HNO_3(aq) + Ca(OH)_2(aq) \longrightarrow Ca(NO_3)_2(aq) + 2H_2O(l)$$

The interactions of other types of acids and bases, known as weak acids and bases, are somewhat more involved and are discussed in more detail in Chapter 13.

EXAMPLE 6-10 **Writing Neutralization Reactions**

Write the molecular, total ionic, and net ionic equations for the neutralization of **(a)** $HClO_4$ and LiOH and **(b)** H_2SO_4 and KOH.

PROCEDURE

Follow the steps outlined in Section 6-5.3. Water is a product in neutralization reactions and should be labeled with an (l) and not ionized.

SOLUTION

To balance the equations, one should make sure that there is one $H^+(aq)$ for each $OH^-(aq)$. Another way is to write the formula of the salt formed and then balance the number of reactant cations (from the base) and reactant anions (from the acid).

(a)
$$HClO_4(aq) + LiOH(aq) \longrightarrow LiClO_4(aq) + H_2O(l)$$
$$H^+(aq) + ClO_4^-(aq) + Li^+(aq) + OH^-(aq) \longrightarrow Li^+(aq) + ClO_4^-(aq) + H_2O(l)$$
$$H^+(aq) + OH^-(aq) \longrightarrow H_2O(l)$$

(b)
$$H_2SO_4(aq) + 2KOH(aq) \longrightarrow K_2SO_4(aq) + 2H_2O(l)$$
$$2H^+(aq) + SO_4^{2-}(aq) + 2K^+(aq) + 2OH^-(aq) \longrightarrow 2K^+(aq) + SO_4^{2-}(aq) + 2H_2O(l)$$
$$H^+(aq) + OH^-(aq) \longrightarrow H_2O(l)$$

ANALYSIS

The water that is produced in the reaction is in addition to the water used as a solvent to dissolve the reactants. So, unlike a precipitation reaction, where the formation of the solid shows us that there is an obvious reaction occurring, neutralization reactions are harder to recognize in the lab. There are many times where no discernible physical change occurs. If the concentrations of reactants are high enough, a significant amount of heat can be produced in the neutralization process and the solution's temperature will rise. Alternatively, you can follow the course of the reaction with chemical indicators that change color as the reactants are neutralized, or the reaction can be followed by electrical meters that track the concentrations of ions in solution. Beyond that, it's just our knowledge of the chemistry of the reaction that tells us that anything is occurring.

SYNTHESIS

It will be useful in upcoming chapters to know the balancing coefficients for chemical reactions. Neutralization reactions provide us with a shortcut to determine the needed coefficients for the reactants. Every acid provides a given number of H^+'s. Every base provides a given number of OH^-'s. Together they'll combine in a 1-to-1 ratio to form water. So how many of the acid molecules and how many of the base molecules are needed to produce the same number of H^+'s and OH^-'s? Consider a reaction between H_3PO_4 and $Ba(OH)_2$. What will be the balancing coefficient in front of each molecule?

H_3PO_4 delivers $3H^+$. $Ba(OH)_2$ delivers $2OH^-$. Two acids and three bases will produce a total of six waters. The balancing coefficients are 2 and 3, respectively.

6-6.3 Gas-Forming Neutralization Reactions

Calcium carbonate is the solid ingredient in several products commonly advertised that effectively neutralizes stomach acid. Although it is not a strong base, as we described earlier, it does do the job. In this case the double-displacement reaction produces a molecular compound other than water. The molecular compound formed in this case is carbonic acid, which then decomposes to carbon dioxide gas and water as described earlier in this chapter. The salt acts as a base-neutralizing stomach acid (HCl) as shown below.

$$2HCl(aq) + CaCO_3(s) \longrightarrow CaCl_2(aq) + H_2CO_3(aq)$$

The carbonic acid formed is a molecular compound that then decomposes to carbon dioxide gas and water.

$$H_2CO_3 \longrightarrow CO_2(g) + H_2O(l)$$

The complete reaction is illustrated below.

$$2HCl(aq) + CaCO_3(s) \longrightarrow CaCl_2(aq) + CO_2(g) + H_2O(l)$$

The net ionic equation for this reaction is

$$CO_3^{2-}(aq) + 2H^+(aq) \longrightarrow CO_2(g) + H_2O(l)$$

Salts containing the following anions neutralize strong acids by forming a gas.

- HCO_3^- and CO_3^{2-} salts produce H_2CO_3, which decomposes to $CO_2(g) + H_2O$.
- HSO_3^- and SO_3^{2-} salts produce H_2SO_3, which decomposes to $SO_2(g) + H_2O$.
- HS^- and S^{2-} salts produce $H_2S(g)$.
- CN^- salts produce $HCN(g)$.

Another example for a **gas-forming reaction** is shown below.

$$K_2S(aq) + 2HNO_3(aq) \longrightarrow 2KNO_3(aq) + H_2S(g)$$

The net ionic equation for this reaction is

$$S^{2-}(aq) + 2H^+(aq) \longrightarrow H_2S(g)$$

The hydrogen sulfide gas formed is poisonous but in very small amounts produces the obnoxious smell of rotten eggs.

▶ **ASSESSING THE OBJECTIVE FOR SECTION 6-6**

EXERCISE 6-6(a) LEVEL 1: Write out how the following compounds exist when placed in water. Use appropriate state symbols.
(a) KOH   (b) $Fe(OH)_3$ (c) HI   (d) $NaHSO_3$

EXERCISE 6-6(b) LEVEL 2: Write out the balanced molecular equation, complete with the state symbols (aq) and (l), for the following reactants.
(a) $H_2SO_4(aq) + LiOH(aq) \longrightarrow$
(b) $Ca(OH)_2(aq) + HClO_4(aq) \longrightarrow$
(c) $HCl(aq) + NaHCO_3(aq) \longrightarrow$
(d) $K_2S(aq) + HBr(aq) \longrightarrow$

EXERCISE 6-6(c) LEVEL 3: Write the balanced molecular, total ionic, and net ionic equations illustrating the neutralization of nitric acid with strontium hydroxide.

EXERCISE 6-6(d) LEVEL 3: The lethal gas used in the gas chamber is hydrogen cyanide (HCN). It is produced by adding a strong acid to a solid salt. Write a balanced molecular equation illustrating such a reaction using sulfuric acid along with its net ionic equation.

For additional practice, work chapter problems 6-55, 6-57, and 6-60.

PART B SUMMARY

KEY TERMS

6-3.1	Ionic compounds may be **soluble** or **insoluble** in water. p. 190
6-3.1	A **solution** is formed by a **solute** dissolving in a **solvent**. p. 190
6-3.2	**Strong acids** are completely ionized in aqueous solution. p. 190
6-4	A **single-replacement reaction** involves the exchange of a free metal for a different metal ion. p. 191
6-4.1	Reactions in aqueous solution can be represented by **molecular** and **total ionic equations**. p. 192
6-4.1	When **spectator ions** are subtracted out, a **net ionic equation** results. p. 192
6-4.2	The **activity series** relates the ability of a metal to replace other metal ions. p. 192
6-4.2	A **precipitate** is a solid formed from solution. p. 196
6-5	A **precipitation reaction** is a type of **double-replacement reaction**. p. 196
6-6.1	**Strong bases** are ionic compounds that produce the hydroxide ion in solution. p. 201
6-6.2	A **neutralization reaction** is a second type of double-replacement reaction that produces a **salt** and water. p. 201
6-6.3	A type of neutralization reaction called a **gas-forming reaction** involves the combination of a strong acid with certain salts to produce a gas. p. 203

SUMMARY CHART

Three More Types of Reactions

Single Replacement

$$M(s) + AX\,(aq) \longrightarrow MX(aq) + A(s)$$

Double Replacement

Precipitation	**Neutralization**
$MY(aq) + AX(aq) \longrightarrow$ $MX(aq) + AY(s)$	$MOH(aq) + HY(aq) \longrightarrow$ $MX(aq) + H_2O$

CHAPTER 6 SYNTHESIS PROBLEM

Copper metal is a stable metal along with other coinage metals such as silver and gold. As mentioned in this chapter, even most acids do not react with these metals. The one exception is nitric acid, which reacts with copper but not by an ordinary single-replacement reaction as described here. The reaction is somewhat more complex and will be discussed further in Chapter 13. In a well-known laboratory experiment we can take metallic copper, dissolve it in nitric acid and through a series of reactions that we have discussed in this chapter, return it to its metallic form. There are five reactions involved in this series.

PROBLEM	SOLUTION
a. Copper dissolves in nitric acid forming copper(II) nitrate and toxic nitrogen dioxide: $Cu(s) + HNO_3(aq) \longrightarrow Cu(NO_3)_2(aq) + NO_2(g) + H_2O(l)$ Balance this equation by inspection using the guidelines described.	**a.** The copper is balanced, so consider the nitrogens. On the right there are three, so balance with three nitric acids on the left. $Cu(s) + 3HNO_3(aq) \longrightarrow Cu(NO_3)_2(aq) + NO_2(g) + H_2O(l)$ Now consider the hydrogens. To balance water we obviously need an even number of nitric acids. If we add one nitric acid on the left, we can now balance the hydrogens on the right. $Cu(s) + 4HNO_3(aq) \longrightarrow Cu(NO_3)_2(aq) + NO_2(g) + 2H_2O(l)$ Notice that we now have one extra nitrogen and two extra oxygens on the left. We can fix this easily by adding one more nitrogen dioxide on the right and we are finished. $Cu(s) + 4HNO_3(aq) \longrightarrow Cu(NO_3)_2(aq) + 2NO_2(g) + 2H_2O(l)$

PROBLEM	SOLUTION
b. Aqueous sodium hydroxide reacts with the copper(II) nitrate solution from above to form copper(II) hydroxide. Write the balanced molecular, total ionic, and net ionic equations for this reaction. Identify the type of reaction.	**b.** $Cu(NO_3)_2(aq) + 2NaOH(aq) \longrightarrow Cu(OH)_2(s) + 2NaNO_3(aq)$ $Cu^{2+}(aq) + 2NO_3^-(aq) + 2Na^+(aq) + 2OH^-(aq) \longrightarrow$ $Cu(OH)_2(s) + 2NO_3^-(aq) + 2Na^+(aq)$ $Cu^{2+}(aq) + 2OH^-(aq) \longrightarrow Cu(OH)_2(s)$ This is a precipitation reaction.
c. The solid copper(II) hydroxide is filtered and then is heated, releasing water. Write the balanced equation for this reaction. Identify the type of reaction.	**c.** $Cu(OH)_2(s) \longrightarrow CuO(s) + H_2O(l)$ By removing a H_2O from $Cu(OH)_2$ we have CuO. This is a decomposition reaction.
d. A solution of sulfuric acid reacts with the solid copper(II) oxide, producing copper(II) sulfate. Write the balanced equation illustrating this reaction. Identify the type of reaction.	**d.** $CuO(s) + H_2SO_4(aq) \longrightarrow CuSO_4(aq) + H_2O(l)$ This is a neutralization reaction where solid CuO can be considered the base. Recall that $CuSO_4$ is soluble.
e. Aluminum metal is added to the copper(II) sulfate solution, re-forming metallic copper. Write the balanced molecular, total ionic, and net ionic equations for this reaction. Identify the type of reaction.	**e.** $2Al(s) + 3CuSO_4(aq) \longrightarrow Al_2(SO_4)_3(aq) + 3Cu(s)$ $2Al(s) + 3Cu^{2+}(aq) + 3SO_4^{2-}(aq) \longrightarrow$ $2Al^{3+}(aq) + 3SO_4^{2-}(aq) + 3Cu(s)$ $2Al(s) + 3Cu^{2+}(aq) \longrightarrow 2Al^{3+}(aq) + 3Cu(s)$ This is a single-replacement reaction.

YOUR TURN

Complete where necessary and balance the following reactions.

a. Magnesium metal reacts with a gold(III) nitrate solution in a single-replacement reaction.

b. Magnesium hydroxide reacts with $HClO_4$. Write the balanced molecular, total ionic, and net ionic equations. Identify the type of reaction.

c. C_2H_6O undergoes combustion.

d. A solution of magnesium bromide is mixed with a solution of potassium hydroxide. Write the balanced molecular, total ionic, and net ionic equations. Identify the type of reaction.

e. Fluorine reacts with aluminum. Identify the type of reaction.

Answers are on p. 207.

CHAPTER SUMMARY

A concise statement of a chemical property is relayed by the **chemical equation**. With symbols, formulas, and other abbreviations, a sizable amount of chemical information can be communicated. This includes the elements or compounds involved as **reactants** and **products**, their physical states, and the number of molecules of each compound involved in the reaction. When the numbers of atoms of each element are made the same on both sides of the equation by use of **coefficients**, the equation is considered **balanced**.

In this chapter, we considered five different types of reactions that can be conveniently represented by equations. Each type has a general equation that characterizes that kind of reaction. The first three types discussed in this chapter are **combustion reactions**, **combination reactions**, and **decomposition reactions**.

In addition to these three types of reactions, two other types usually involve reactions that occur in an aqueous solution. When a substance (a **solute**) is dispersed by a liquid (a **solvent**), it forms a homogeneous mixture known as a **solution**. Substances that dissolve in water are said to be **soluble** and those that do not are **insoluble**. Soluble ionic compounds are separated into their individual ions in aqueous solution.

In **single-replacement reactions**, a metal exchanges places with the cation of a different metal. The ability of metals to replace other metal ions can be compared and ranked in the **activity series**. Hydrogen, although not a metal, is usually included in this series.

Double-replacement reactions involve the exchange of ions between two soluble compounds. In a **precipitation reaction**, the two ions combine to form a solid ionic compound known as a **precipitate**, which separates from the solution. In a **neutralization reaction**, the two ions form a molecular compound.

In neutralization reactions, **strong acids** [molecular compounds that dissolve in water to form $H^+(aq)$ ions] react with **strong bases** [ionic compounds that dissolve in water to form $OH^-(aq)$ ions]. The reaction produces water and a **salt**. **Gas-forming reactions** involve the reaction of a strong acid with certain salts to produce molecular compounds that are either gases or decompose to gases.

Single- and double-replacement reactions can be illustrated by three types of equations. In a **molecular equation**, all species are represented as neutral compounds. In a **total ionic equation**, soluble ionic compounds and strong acids are represented as separate ions. If the **spectator ions** (those ions that are not directly involved in the reaction) are removed, the result is a **net ionic equation**. The three types of equations for a typical precipitation reaction are shown here.

$$Pb(NO_3)_2(aq) \quad + \quad 2KCl(aq) \quad \longrightarrow \quad PbCl_2(s) \quad + \quad 2KNO_3(aq)$$

$$Pb^{2+}(aq) + 2NO_3^-(aq) \quad + \quad 2K^+(aq) + 2Cl^-(aq) \quad \longrightarrow \quad PbCl_2(s) \quad + \quad 2K^+(aq) + 2NO_3^-(aq)$$

$$Pb^{2+}(aq) + 2Cl^-(aq) \quad \longrightarrow \quad PbCl_2(s)$$

OBJECTIVES

SECTION	YOU SHOULD BE ABLE TO...	EXAMPLES	EXERCISES	CHAPTER PROBLEMS
6-1	Write balanced chemical equations for simple reactions from inspection.	6-1, 6-2, 6-3	1a, 1b, 1c, 1d, 1e, 2c	2, 3, 4, 5, 10, 11, 12, 13
6-2	Classify certain chemical reactions as being combustion, combination, or decomposition reactions.	6-3	2a, 2b, 2c	14,15, 16, 18, 19, 20, 21, 22, 23
6-3	Write the ions formed when ionic compounds or acids dissolve in water.	6-4	3a, 3b, 3c	24, 25, 26, 27
6-4	Given the activity series, complete several single-replacement reactions as balanced molecular, total ionic, and net ionic equations.	6-5	4a, 4b, 4c, 4d	28, 29, 30, 33
6-5	Write balanced molecular, total ionic, and net ionic equations for precipitation reactions.	6-6, 6-7, 6-8, 6-9	5a, 5b, 5c, 5d, 5e	34, 35, 40, 42, 44, 45, 46, 47, 48, 49
6-6	Write balanced molecular, total ionic, and net ionic equations for neutralization reactions	6-10	6a, 6b	55, 56, 57, 58

ANSWERS TO ASSESSING THE OBJECTIVES

Part A
EXERCISES

6-1(a) A chemical reaction is represented with symbols and formulas by means of a chemical equation. The arrow in an equation separates the reactants on the left from the products on the right. To conform to the law of conservation of mass, an equation must be balanced. This is accomplished by introducing coefficients in front of formulas rather than changing subscripts in a formula.

6-1(b) (a) $Li(s) + O_2(g) \longrightarrow Li_2O(s)$
(b) $Al(s) + H_2SO_4(l) \longrightarrow Al_2(SO_4)_3(aq) + H_2(g)$
(c) $Cu(s) + HNO_3(aq) \longrightarrow Cu(NO_3)_2(aq) + NO(g) + H_2O(l)$
6-1(c) (a) B, C, O, H **(b)** N, H, N(again), O **(c)** S, O, Na, I
6-1(d) (a) $4Cr + 3O_2 \longrightarrow 2Cr_2O_3$
(b) $Co_2S_3 + 3H_2 \longrightarrow 3H_2S + 2Co$
(c) $C_3H_8 + 5O_2 \longrightarrow 3CO_2 + 4H_2O$

6-1(e) (a) $(NH_4)_2CO_3(s) \longrightarrow 2NH_3(g) + CO_2(g) + H_2O(g)$
(b) $2NH_3(g) + 2ClF_3(l) \longrightarrow N_2(g) + Cl_2(g) + 6HF(g)$

6-2(a) (a) combination, combustion **(b)** decomposition, combustion **(c)** combination, combustion **(d)** all reactions **(e)** decomposition

6-2(b) (a) decomposition **(b)** combustion **(c)** combination **(d)** decomposition **(e)** combination

6-2(c) (a) $2Si_2H_6(g) + 7O_2(g) \longrightarrow 4SiO_2(s) + 6H_2O(l)$
(b) $2Al(s) + 3H_2(g) \longrightarrow 2AlH_3(s)$
(c) $Ca(HSO_3)_2(s) \longrightarrow CaO(s) + H_2O(l) + 2SO_2(g)$

6-2(d) (a) $2C_6H_{14}(l) + 19O_2(g) \longrightarrow 12CO_2(g) + 14H_2O(l)$
(b) $2Na(s) + I_2(s) \longrightarrow 2NaI(s)$
(c) $H_2SO_3(aq) \longrightarrow H_2O(l) + SO_2(g)$

Part B
EXERCISES

6-3(a) H^+ and ClO_4^- **(b)** K^+ and PO_4^{3-} **(c)** NH_4^+ and NO_3^- **(d)** Zn^{2+} and ClO_4^-

6-3(b) **(a)** three **(b)** four **(c)** four **(d)** two

6-3(c) **(a)** $Na_2SO_4(s) \longrightarrow 2Na^+(aq) + SO_4^{2-}(aq)$

(b) $HNO_3(l) \longrightarrow H^+(aq) + NO_3^-(aq)$

(c) $Al(C_2H_3O_2)_3(s) \longrightarrow Al^{3+}(aq) + 3C_2H_3O_2^-(aq)$

(d) $Ca(OH)_2(s) \longrightarrow Ca^{2+}(aq) + 2OH^-(aq)$

6-4(a) **(a)** Ni replaces Ag **(b)** Al replaces Cu **(c)** Mg replaces Cr **(d)** Sn replaces H

6-4(b) **(a)** total: $Cu^{2+}(aq) + 2NO_3^-(aq) + Zn(s) \longrightarrow$
$$Zn^{2+}(aq) + 2NO_3^-(aq) + Cu(s)$$
net: $Cu^{2+}(aq) + Zn(s) \longrightarrow Zn^{2+}(aq) + Cu(s)$

(b) total: $2Al(s) + 3Pb^{2+}(aq) + 6Cl^-(aq) \longrightarrow$
$$2Al^{3+}(aq) + 6Cl^-(aq) + 3Pb(s)$$
net: $2Al(s) + 3Pb^{2+}(aq) \longrightarrow 2Al^{3+}(aq) + 3Pb(s)$

6-4(c) **(a)** $Mg(s) + 2AgNO_3(aq) \longrightarrow Mg(NO_3)_2(aq) + 2Ag(s)$

(b) no reaction (Sn is less reactive than K)

(c) $2Na(s) + 2HCl(aq) \longrightarrow 2NaCl(aq) + H_2(g)$

6-4(d) **(a)** total: $Cu^{2+}(aq) + 2Cl^-(aq) + Fe(s) \longrightarrow$
$$Fe^{2+}(aq) + 2Cl^-(aq) + Cu(s)$$
net: $Cu^{2+}(aq) + Fe(s) \longrightarrow Fe^{2+}(aq) + Cu(s)$

(b) total: $3Mg(s) + 2Au^{3+}(aq) + 6NO_3^-(aq) \longrightarrow$
$$2Au(s) + 3Mg^{2+}(aq) + 6NO_3^-(aq)$$
net: $3Mg(s) + 2Au^{3+}(aq) \longrightarrow 2Au(s) + 3Mg^{2+}(aq)$

6-5(a) **(a)** insoluble **(b)** soluble **(c)** soluble **(d)** insoluble **(e)** soluble

6-5(b) **(a)** soluble **(b)** insoluble **(c)** soluble **(d)** soluble **(e)** insoluble **(f)** insoluble

6-5(c) No. Both the K^+ and the Na^+ will produce soluble products.

(b) Yes. $Al(OH)_3$ is insoluble.

(c) Yes. CaS is insoluble.

6-5(d) $CrI_3(aq) + Li_3PO_4(aq) \longrightarrow CrPO_4(s) + 3LiI(aq)$

(b) $3Cs_2CO_3(aq) + 2FeBr_3(aq) \longrightarrow 6CsBr(aq) + Fe_2(CO_3)_3(s)$

6-5(e) molecular: $2AgNO_3(aq) + K_2S(aq) \longrightarrow$
$$Ag_2S(s) + 2KNO_3(aq)$$
total: $2Ag^+(aq) + 2NO_3^-(aq) + 2K^+(aq) + S^{2-}(aq) \longrightarrow$
$$Ag_2S(s) + 2K^+(aq) + 2NO_3^-(aq)$$
net: $2Ag^+(aq) + S^{2-}(aq) \longrightarrow Ag_2S(s)$

6-5(f) There are several possibilities for each product. The following examples illustrate the general concept of the problem.

(a) $CaCl_2(aq) + Na_2CO_3(aq)$ **(b)** $Fe(NO_3)_3(aq) + K_3PO_4(aq)$

6-6(a) **(a)** $K^+(aq) + OH^-(aq)$

(b) $Fe(OH)_3(s)$ **(c)** $H^+(aq) + I^-(aq)$

(d) $Na^+(aq) + HSO_3^-(aq)$

6-6(b) **(a)** $H_2SO_4(aq) + 2LiOH(aq) \longrightarrow Li_2SO_4(aq) + 2H_2O(l)$

(b) $Ca(OH)_2(aq) + 2HClO_4(aq) \longrightarrow$
$$Ca(ClO_4)_2(aq) + 2H_2O(l)$$

(c) $HCl(aq) + NaHCO_3(aq) \longrightarrow NaCl(aq) + H_2O(l) + CO_2(g)$

(d) $K_2S(aq) + 2HBr(aq) \longrightarrow 2KBr(aq) + H_2S(g)$

6-6(c) molecular: $2HNO_3(aq) + Sr(OH)_2(aq) \longrightarrow$
$$Sr(NO_3)_2(aq) + 2H_2O(l)$$
total: $2H^+(aq) + 2NO_3^-(aq) + Sr^{2+}(aq) + 2OH^-(aq) \longrightarrow$
$$Sr^{2+}(aq) + 2NO_3^-(aq) + 2H_2O(l)$$
net: $H^+(aq) + OH^-(aq) \longrightarrow H_2O(l)$

6-6(d) molecular: $2NaCN(s) + H_2SO_4(aq) \longrightarrow$
$$Na_2SO_4(aq) + 2HCN(g)$$
net: $CN^-(aq) + H^+(aq) \longrightarrow HCN(g)$

ANSWERS TO CHAPTER SYNTHESIS PROBLEM

a. $3Mg(s) + 2Au(NO_3)_3(aq) \longrightarrow$
$$3Mg(NO_3)_2(aq) + 2Au(s)$$

b. $Mg(OH)_2(s) + HClO_4(aq) \longrightarrow$
$$Mg(ClO_4)_2(aq) + 2H_2O(l)$$
$Mg(OH)_2(s) + 2H^+(aq) + 2ClO_4^-(aq) \longrightarrow$
$$Mg^{2+}(aq) + 2ClO_4^-(aq) + 2H_2O(l)$$
$Mg(OH)_2(s) + 2H^+(aq) \longrightarrow$
$$Mg^{2+}(aq) + 2H_2O(l)$$
This is a neutralization reaction.

c. $C_2H_6O + 3O_2 \longrightarrow$
$$2CO_2 + 3H_2O$$

d. $MgBr_2(aq) + 2KOH(aq) \longrightarrow$
$$Mg(OH)_2(s) + 2KBr$$
$Mg^{2+}(aq) + 2NO_3^-(aq) + 2K^+(aq) + 2OH^-(aq)$
$\longrightarrow Mg(OH)_2(s) + 2K^+(aq) + 2Br^-(aq)$
$Mg^{2+}(aq) + 2OH^-(aq) \longrightarrow Mg(OH)_2(s)$
This is a precipitation reaction.

e. $2Al(s) + 3F_2(g) \longrightarrow 2AlF_3(s)$ This is a combination reaction. Notice that AlF_3 is ionic so it is a solid.

CHAPTER PROBLEMS

Throughout the text, answers to all exercises in color are given in Appendix E. The more difficult exercises are marked with an asterisk.

Chemical Equations (SECTION 6-1)

6-1. The physical state of an element is included in a chemical equation. Each of the following compounds is a gas, a solid, or a liquid under normal conditions. Indicate the proper physical state by adding (g), (s), or (l) after the formula.

(a) Cl_2 (d) H_2O (g) Br_2 (j) Na

(b) C   (e) P_4 (h) NaBr   (k) Hg

(c) K_2SO_4 (f) H_2 (i) S_8 (l) CO_2

6-2. Balance the following equations.

(a) $CaCO_3 \xrightarrow{\Delta} CaO + CO_2$

(b) $Na + O_2 \longrightarrow Na_2O$

(c) $H_2SO_4 + NaOH \longrightarrow Na_2SO_4 + H_2O$

(d) $H_2O_2 \longrightarrow H_2O + O_2$

6-3. Balance the following equations.

(a) $NaBr + Cl_2 \longrightarrow NaCl + Br_2$

(b) $KOH + H_3AsO_4 \longrightarrow K_2HAsO_4 + H_2O$

(c) $Ti + Cl_2 \longrightarrow TiCl_4$

(d) $Al + H_2SO_4 \longrightarrow Al_2(SO_4)_3 + H_2$

6-4. Balance the following equations.

(a) $Al + H_3PO_4 \longrightarrow AlPO_4 + H_2$

(b) $Ca(OH)_2 + HCl \longrightarrow CaCl_2 + H_2O$

(c) $Mg + N_2 \longrightarrow Mg_3N_2$

(d) $C_2H_6 + O_2 \longrightarrow CO_2 + H_2O$

6-5. Balance the following equations.

(a) $Ca(CN)_2 + HBr \longrightarrow CaBr_2 + HCN$

(b) $C_3H_6 + O_2 \longrightarrow CO + H_2O$

(c) $P_4 + S_8 \longrightarrow P_4S_3$

(d) $Cr_2O_3 + Si \longrightarrow Cr + SiO_2$

6-6. Balance the following equations.

(a) $Mg_3N_2 + H_2O \longrightarrow Mg(OH)_2 + NH_3$

(b) $H_2S + O_2 \longrightarrow S + H_2O$

(c) $Si_2H_6 + H_2O \longrightarrow Si(OH)_4 + H_2$

(d) $C_2H_6 + Cl_2 \longrightarrow C_2HCl_5 + HCl$

6-7. Balance the following equations.

(a) $Na_2NH + H_2O \longrightarrow NH_3 + NaOH$

(b) $CaC_2 + H_2O \longrightarrow C_2H_2 + Ca(OH)_2$

(c) $XeF_6 + H_2O \longrightarrow XeO_3 + HF$

(d) $PCl_5 + H_2O \longrightarrow H_3PO_4 + HCl$

6-8. Balance the following equations.

(a) $B_4H_{10} + O_2 \longrightarrow B_2O_3 + H_2O$

(b) $SF_6 + SO_3 \longrightarrow O_2SF_2$

(c) $CS_2 + O_2 \longrightarrow CO_2 + SO_2$

(d) $BF_3 + NaH \longrightarrow B_2H_6 + NaF$

6-9. Balance the following equations.

(a) $NH_3 + Cl_2 \longrightarrow NHCl_2 + HCl$

(b) $PBr_3 + H_2O \longrightarrow HBr + H_3PO_3$

(c) $Mg + Fe_3O_4 \longrightarrow MgO + Fe$

(d) $Fe_3O_4 + H_2 \longrightarrow Fe + H_2O$

6-10. Write balanced chemical equations from the following word equations. Include the physical state of each element or compound.

(a) Sodium metal plus water yields hydrogen gas and an aqueous sodium hydroxide solution.

(b) Potassium chlorate when heated yields potassium chloride plus oxygen gas. (Ionic compounds are solids.)

(c) An aqueous sodium chloride solution plus an aqueous silver nitrate solution yields a silver chloride precipitate(solid) and a sodium nitrate solution.

(d) An aqueous phosphoric acid solution plus an aqueous calcium hydroxide solution yields water and solid calcium phosphate.

6-11. Write balanced chemical equations from the following word equations. Include the physical state of each element or compound.

(a) Solid phenol (C_6H_6O) reacts with oxygen to form carbon dioxide gas and liquid water.

(b) An aqueous calcium hydroxide solution reacts with gaseous sulfur trioxide to form a solid of calcium sulfate and water.

(c) Lithium is the only element that combines with nitrogen at room temperature. The reaction forms lithium nitride.

(d) Magnesium dissolves in an aqueous chromium (III) nitrate solution to form chromium and a magnesium nitrate solution.

6-12. Nickel (II) nitrate is prepared by heating nickel metal with liquid dinitrogen tetroxide. In addition to the nitrate, gaseous nitrogen monoxide is formed. Write the balanced equation.

6-13. One of the steps in the production of iron involves the reaction of Fe_3O_4 with carbon monoxide to produce FeO and carbon dioxide. Write the balanced equation.

Combustion, Combination, and Decomposition Reactions (SECTION 6-2)

6-14. Which reactions in problems 6-2 and 6-4 can be classified as a combustion, a combination, or a decomposition reaction?

6-15. Which reactions in problems 6-3 and 6-5 can be classified as either a combustion, a combination, or a decomposition reaction?

6-16. Write balanced combustion reactions when the following compounds react with excess oxygen.

(a) $C_7H_{14}(l)$

(b) $LiCH_3(s)$
(a product is Li_2O)

(c) $C_4H_{10}O(l)$

(d) $C_2H_5SH(g)$
(a product is SO_2)

6-17. Write balanced combustion reactions when the following compounds react with excess oxygen.

(a) $C_2H_6O_2(l)$

(b) $B_6H_{12}(g)$ [a product is $B_2O_3(s)$]

(c) $C_6H_{12}(l)$

(d) $Pb(C_2H_5)_4(s)$ [a product is $PbO_2(s)$]

6-18. Write balanced combination reactions that occur when the metal barium reacts with the following nonmetals.

(a) hydrogen

(b) sulfur

(c) bromine

(d) nitrogen

6-19. Write balanced combination reactions that occur when the metal aluminum reacts with the following nonmetals.

(a) hydrogen

(b) oxygen

(c) iodine

(d) nitrogen

6-20. Write balanced decomposition reactions for the following compounds. Recall that ionic compounds are solids.

(a) $Ca(HCO_3)_2$ into calcium oxide, carbon dioxide, and water

(b) Ag_2O into its elements

(c) N_2O_3 gas into nitrogen dioxide gas and nitrogen monoxide gas

6-21. Write balanced decomposition reactions for the following compounds. Recall that ionic compounds are solids.

(a) liquid SbF_5 into fluorine and solid antimony trifluoride

(b) PtO_2 into its elements

(c) gaseous BrF into bromine and gaseous bromine trifluoride

6-22. Write balanced equations by predicting the products of the following reactions. Include the physical state of each element or compound.

(a) the combination of potassium and chlorine

(b) the combustion of liquid benzene (C_6H_6)

(c) the decomposition of gold(III) oxide into its elements by heating

(d) the combustion of propyl alcohol (C_3H_8O)

(e) the combination of phosphorus (P_4) and fluorine gas to produce solid phosphorus pentafluoride

6-23. Write balanced equations by predicting the products of the following reactions. Include the physical state of each element or compound.

(a) the combustion of liquid butane (C_4H_{10})

(b) the decomposition of aqueous sulfurous acid to produce water and a gas

(c) the combination of sodium and oxygen gas to form sodium peroxide (the peroxide ion is O_2^{2-})

(d) the decomposition of copper(I) oxide into its elements by heating

Ions in Aqueous Solution (SECTION 6-3)

6-24. Write equations illustrating the solution of each of the following compounds in water.

(a) Na_2S

(b) Li_2SO_4

(c) $K_2Cr_2O_7$

(d) CaS

(e) $(NH_4)_2S$

(f) $Ba(OH)_2$

6-25. Write equations illustrating the solution of each of the following compounds in water.

(a) $Ca(ClO_3)_2$

(b) $CsBr$

(c) $AlCl_3$

(d) Cs_2SO_3

6-26. Write equations illustrating the solution of the following compounds in water.

(a) HNO_3

(b) $Sr(OH)_2$

6-27. Write equations illustrating the solution of the following compounds in water.

(a) $LiOH$

(b) HI

Single-Replacement Reactions (SECTION 6-4)

6-28. If any of the following reactions occur spontaneously, write the balanced net ionic equation. If not, write "no reaction." (Refer to Table 6-2.)

(a) $Pb + Zn^{2+} \longrightarrow Pb^{2+} + Zn$

(b) $Fe + H^+ \longrightarrow Fe^{2+} + H_2$

(c) $Cu + Ag^+ \longrightarrow Cu^{2+} + Ag$

(d) $Cr + Zn^{2+} \longrightarrow Cr^{3+} + Zn$

6-29. If any of the following reactions occur spontaneously, write the balanced net ionic equation. If not, write "no reaction." (Refer to Table 6-2.)

(a) $Pb + Sn^{2+} \longrightarrow Pb^{2+} + Sn$

(b) $H_2 + Ni^{2+} \longrightarrow H^+ + Ni$

(c) $Cr + Ni^{2+} \longrightarrow Cr^{3+} + Ni$

(d) $H_2 + Au^{3+} \longrightarrow H^+ + Au$

6-30. In the following situations, a reaction may or may not take place. If it does, write the balanced molecular, total ionic, and net ionic equations illustrating the reaction. Assume all involve aqueous solutions.

(a) Some iron nails are placed in a $CuCl_2$ solution.

(b) Silver coins are dropped in a hydrochloric acid solution.

(c) A copper wire is placed in a $Pb(NO_3)_2$ solution.

(d) Zinc strips are placed in a $Cr(NO_3)_3$ solution.

6-31. In the following situations, a reaction may or may not take place. If it does, write the balanced molecular, total ionic, and net ionic equations illustrating the reaction. Assume all involve aqueous solutions.

(a) A solution of nitric acid is placed in a tin can.

(b) Iron nails are placed in a $ZnBr_2$ solution.

(c) A chromium-plated auto accessory is placed in an $SnCl_2$ solution.

(d) A silver bracelet is placed in a $Cu(ClO_4)_2$ solution.

6-32. When heated, sodium metal reacts with solid Cr_2O_3. Write the balanced molecular and the net ionic equations for this single-replacement reaction.

6-33. When heated, aluminum metal reacts with solid PbO. Write the balanced molecular and net ionic equations for this single-replacement reaction.

Solubility and Precipitation Reactions
(SECTION 6-4)

6-34. Referring to Table 6-3, determine which of the following compounds are insoluble in water.

(a) Na_2S (d) Ag_2S

(b) $PbSO_4$ (e) $(NH_4)_2S$

(c) $MgSO_4$ (f) HgI_2

6-35. Referring to Table 6-3, determine which of the following compounds are insoluble in water.

(a) NiS   (d) Rb_2SO_4

(b) Hg_2Br_2 (e) CaS

(c) $Al(OH)_3$ (f) $BaCO_3$

6-36. Write the formulas of the precipitates formed when Ag^+ combines with the following anions.

(a) Br^- (b) CO_3^{2-} (c) PO_4^{3-}

6-37. Write the formulas of the precipitates formed when Pb^{2+} combines with the following anions.

(a) SO_4^{2-} (b) PO_4^{3-} (c) I^-

6-38. Write the formulas of the precipitates formed when CO_3^{2-} combines with the following cations.

(a) Cu^{2+} (b) Cd^{2+} (c) Cr^{3+}

6-39. Write the formulas of the precipitates formed when OH^- combines with the following cations

(a) Ag^+ (b) Ni^{2+} (c) Co^{3+}

6-40. Which of the following chlorides is insoluble in water? (Refer to Table 6-3.)

(a) NaCl   (c) $AlCl_3$

(b) Hg_2Cl_2 (d) $BaCl_2$

6-41. Which of the following sulfates is insoluble in water? (Refer to Table 6-3.)

(a) K_2SO_4 (c) $SrSO_4$

(b) $ZnSO_4$ (d) $MgSO_4$

6-42. Which of the following phosphates is insoluble in water? (Refer to Table 6-3.)

(a) K_3PO_4 (c) $(NH_4)_3PO_4$

(b) $Ca_3(PO_4)_2$ (d) Li_3PO_4

6-43. Which of the following hydroxides is insoluble in water? (Refer to Table 6-3.)

(a) $Mg(OH)_2$ (c) $Ba(OH)_2$

(b) CsOH   (d) NaOH

6-44. Write the balanced molecular equation for any reaction that occurs when the following solutions are mixed. (Refer to Table 6-3.)

(a) KI and $Pb(C_2H_3O_2)_2$ (d) BaS and $Hg_2(NO_3)_2$

(b) $AgClO_4$ and KNO_3 (e) $FeCl_3$ and KOH

(c) $Sr(ClO_4)_2$ and $Ba(OH)_2$

6-45. Write the balanced molecular equation for any reaction that occurs when the following solutions are mixed. (Refer to Table 6-3.)

(a) $Ba(C_2H_3O_2)_2$ and Na_2SO_4 (c) $Mg(NO_3)_2$ and Na_3PO_4

(b) $NaClO_4$ and $Pb(NO_3)_2$ (d) SrS and NiI_2

6-46. Write the total ionic and net ionic equations for any reactions that occurred in problem 6-44.

6-47. Write the total ionic and net ionic equations for any reactions that occurred in problem 6-45.

6-48. Write the total ionic equation and the net ionic equation for each of the following reactions.

(a) $K_2S(aq) + Pb(NO_3)_2(aq) \longrightarrow PbS(s) + 2KNO_3(aq)$

(b) $(NH_4)_2CO_3(aq) + CaCl_2(aq) \longrightarrow$
$$CaCO_3(s) + 2NH_4Cl(aq)$$

(c) $2AgClO_4(aq) + Na_2CrO_4(aq) \longrightarrow$
$$Ag_2CrO_4(s) + 2NaClO_4(aq)$$

6-49. Write the total ionic equation and the net ionic equations for each of the following reactions.

(a) $Hg_2(ClO_4)_2(aq) + 2HBr(aq) \longrightarrow$
$$Hg_2Br_2(s) + 2HClO_4(aq)$$

(b) $2AgNO_3(aq) + (NH_4)_2SO_4(aq) \longrightarrow$
$$Ag_2SO_4(s) + 2NH_4NO_3(aq)$$

(c) $CuSO_4(aq) + 2KOH(aq) \longrightarrow Cu(OH)_2(s) + K_2SO_4(aq)$

*6-50. Write the balanced molecular equations indicating how the following ionic compounds can be prepared by a precipitation reaction using any other ionic compounds. In some cases, the equation should reflect the fact that the desired compound is soluble and must be recovered by vaporizing the solvent water after removal of a precipitate.

(a) $CuCO_3$ (c) Hg_2I_2 (e) $KC_2H_3O_2$

(b) $PbSO_4$ (d) NH_4NO_3

Neutralization Reactions (SECTION 6-5)

6-51. Which of the following is not a strong acid?

(a) HBr   (c) HNO_3

(b) HF   (d) $HClO_4$

6-52. Which of the following is not a strong acid?

(a) HNO_3 (c) H_2SO_4

(b) HI   (d) HNO_2

6-53. Which of the following is not a strong base?

(a) NaOH   (c) $Al(OH)_3$

(b) $Ba(OH)_2$ (d) CsOH

6-54. Which of the following is not a strong base?

(a) $Be(OH)_2$ **(b)** $Ba(OH)_2$ **(c)** LiOH   **(d)** KOH

6-55. Write balanced molecular equations for the neutralization reactions between the following compounds.

(a) HI and CsOH   **(c)** H_2SO_4 and $Sr(OH)_2$

(b) HNO_3 and $Ca(OH)_2$

6-56. Write balanced molecular equations for the neutralization reactions between the following compounds.

(a) $Ca(OH)_2$ and HI   **(c)** $HClO_4$ and $Ba(OH)_2$

(b) H_2SO_4 and LiOH

6-57. Write the total ionic and net ionic equations for the reactions in problem 6-55.

6-58. Write the total ionic and net ionic equations for the reactions in problem 6-56.

*6-59. Magnesium hydroxide is considered a strong base but has very low solubility in water. It is known as milk of magnesia and is used to neutralize stomach acid (HCl). Write the balanced molecular, total ionic, and net ionic equations illustrating this reaction. (Since magnesium hydroxide is a solid, the total and net ionic equations will be somewhat different.)

*6-60. When calcium hydroxide is neutralized with sulfuric acid, the salt produced is insoluble in water. Write the balanced molecular, total ionic, and net ionic equations illustrating this reaction.

General Problems

6-61. Write the balanced equations representing the combustion of propane gas (C_3H_8), butane liquid (C_4H_{10}), octane (C_8H_{18}) in liquid gasoline, and liquid ethyl alcohol (C_2H_5OH) found in alcoholic beverages.

6-62. In the combination reaction between sodium and chlorine and in the combustion reaction of magnesium, why could these also be considered electron exchange reactions?

6-63. Iron replaces gold ions in solution. Can you think of any practical application of this reaction?

6-64. Write balanced equations by predicting the products of the following reactions. Include the physical state of each element or compound.

(a) the combination of barium and iodine

(b) the neutralization of aqueous rubidium hydroxide with hydrobromic acid

(c) a single-replacement reaction of calcium metal with a nitric acid solution

(d) the combustion of solid naphthalene ($C_{10}H_8$)

(e) a precipitation reaction involving aqueous ammonium chromate and aqueous barium bromide

(f) the decomposition of solid aluminum hydroxide into solid aluminum oxide and gaseous water

6-65. Write the total ionic and net ionic equations for parts (b), (c), and (e) of problem 6-64.

6-66. Write balanced equations by predicting the products of the following reactions. Include the physical state of each element or compound.

(a) the decomposition of solid sodium azide (NaN_3) into solid sodium nitride and nitrogen gas

(b) a precipitation reaction involving aqueous potassium carbonate and aqueous copper(II) sulfate

(c) the combustion of solid benzoic acid ($C_7H_6O_2$)

(d) a single-replacement reaction of iron metal and an aqueous gold(III) nitrate solution

(e) the combination of aluminum and solid sulfur (S_8)

(e) the neutralization of aqueous sulfuric acid with aqueous barium hydroxide

6-67. Write the total ionic and net ionic equations for parts (b), (d), and (f) of problem 6-66.

6-68. Consider a mixture of the following ions in aqueous solution: Na^+, H^+, Ba^{2+}, ClO_4^-, OH^-, SO_4^{2-}. Write the net ionic equations for any reaction or reactions that occur between a cation and an anion.

6-69. Consider a mixture of the following ions in aqueous solution: NH_4^+, Mg^{2+}, Ni^{2+}, Cl^-, S^{2-}, CO_3^{2-}. Write the net ionic equations for any reaction or reactions that occur between a cation and an anion.

6-70. Consider a mixture of the following ions in aqueous solution: K^+, Fe^{3+}, Pb^{2+}, I^-, PO_4^{3-}, and S^{2-}. Write the net ionic equations for any reaction or reactions that occur between a cation and an anion.

STUDENT WORKSHOP

Chemical Reactions

Purpose: To write and balance several different classes of chemical reactions. (Work in groups of three or four. Estimated time: 20 min.)

Use the following ions, elements, and molecular compounds to write as many molecular equations as possible:

$$Ba^{2+}, Na^+, H^+, OH^-, CO_3^{2-}, Cl^-, O^{2-},$$
$$Ba, Na, H_2, O_2, CO_2, H_2O$$

- Assign state symbols [(s), (l), (g), and (aq)] to all reactants and products.
- Balance the chemical reactions.
- Label each reaction by its type (decomposition, combination, combustion, single replacement, precipitation, neutralization).

Each group should be allowed to put one unique equation on the board. How many different reactions was the class able to come up with?

Chapter 7

Quantitative Relationships in Chemical Reactions

The burning of charcoal releases the heat energy that grills our food. But the combustion of charcoal and fossil fuels also releases carbon dioxide, which affects our environment. The amount of carbon dioxide and the heat energy released are quantities discussed in this chapter.

SETTING THE STAGE

In the introduction to Chapter 5, we discussed the greenhouse effect and the result of global warming. The presence of increasing amounts of carbon dioxide, as well as other specific gases, is what seems to be a major cause of the problem. However, carbon dioxide is a reactant in photosynthesis, so an increase in carbon dioxide may cause an increase in the amount and rate of production of vegetation (the product). Early experiments indicate that some types of vegetation are noticeably affected and others less so. Increasing just one reactant may not lead to more growth unless all other reactants are also increased. It's like building an automobile. We need one engine and four tires per car. Having eight tires does not mean two cars unless we also have two engines as well as other necessary parts. In chemical reactions such as photosynthesis, the mass ratios of reacting compounds as well as the masses of the products that are formed are extremely important.

A chemical reaction involves not only product compounds but also energy. For example, when we barbecue over a charcoal fire, we are putting to use the large amount of heat energy that is liberated by the combustion reaction. In this case, the chemical energy in the charcoal (which is mostly carbon) is being converted into the heat energy of the fire. As mentioned in Chapter 2, this is an *exothermic* reaction. A reaction that absorbs heat energy is an *endothermic* reaction.

We will examine the mass relationships in Part A of this chapter. In Part B we will include energy as part of the quantitative relationships in chemical reactions.

Part A

Mass Relationships in Chemical Reactions

SETTING A GOAL

- You will gain command over the aspects of stoichiometric calculations involving masses, moles, and molecules under both ideal and realistic conditions.

OBJECTIVES

7-1 Perform stoichiometry calculations using mole ratios from balanced equations.

7-2 Determine the limiting reactant and the yield in a reaction given the masses of two different reactants.

7-3 Calculate the percent yield of a reaction from the measured actual yield and the calculated theoretical yield.

► **OBJECTIVE**
FOR SECTION 7-1
Perform stoichiometry calculations using mole ratios from balanced equations.

7-1 Stoichiometry

LOOKING AHEAD! The balanced chemical equation was introduced in Chapter 6 in terms of individual molecules and formula units. With the help of the mole, we are ready to scale-up our measurements into the macroscopic world of the laboratory. We will make important calculations in this section concerning the mass relationships between and among reactants and products. ■

7-1.1 Overview of Stoichiometry

Any decent cook knows that a good recipe calls for ingredients to be mixed in precise amounts. Likewise, any profitable chemical company producing a weed killer or a fertilizer will not waste expensive chemicals by using more than the recipe requires. For the chemical company, the balanced chemical equation serves as the recipe. *The quantitative relationships among reactants and products are known as* **stoichiometry.**

Probably, the most important industrial chemical process in the world concerns the production of ammonia (NH_3) from its elements, hydrogen and nitrogen. Its use as a fertilizer is absolutely necessary to feed a hungry world. It is also an essential component of explosives. We will use the balanced equation representing this reaction as our example. (See Table 7-1.) The most basic relationship of the balanced equation refers to the ratio of molecules and is shown in line 1. In fact, any multiple of the basic molecular ratios are valid. In lines 2 and 3, we scaled up the molecular ratios between the two reactants and one product by use of the unit "12 in a dozen." For practical use in a laboratory, however, we need to scale the numbers *way up* to amounts that we can see and deal with. Thus in lines 4 and 5 we have shown the relationships in terms of Avogadro's number and the mole. Finally, as we learned in Chapter 5, the mole relates not only to a number (line 4) but also to a mass (line 6).

TABLE 7-1

Production of Ammonia

$N_2(g)$	+	$3H_2(g)$	→	$2NH_3(g)$
1. 1 molecule	+	3 molecules	→	2 molecules
2. 12 molecules	+	36 molecules	→	24 molecules
3. 1 dozen molecules	+	3 dozen molecules	→	2 dozen molecules
4. 6.022×10^{23} molecules	+	18.1×10^{23} molecules	→	12.0×10^{23} molecules
5. 1 mol	+	3 mol	→	2 mol
6. 28 g	+	6 g	→	34 g

7-1.2 The Mole Ratio

The central relationships in the stoichiometry problems that we will soon encounter are found in line 5 in Table 7-1. From the number of moles involved, we can construct ratios between the two reactants and between the reactants and product. *The mole relation factors generated by the balanced equation are referred to as the* **mole ratios**. There are actually six mole ratios that can be constructed from this one simple equation. First, consider the amount of ammonia produced by nitrogen. Next, we will examine the amount of hydrogen that reacts with nitrogen, and finally, the amount of ammonia produced by hydrogen.

$$1 \text{ mol of } N_2 \text{ produces } 2 \text{ mol of } NH_3$$

This can be expressed in ratio form as (1) $\dfrac{1 \text{ mol } N_2}{2 \text{ mol } NH_3}$ or (2) $\dfrac{2 \text{ mol } NH_3}{1 \text{ mol } N_2}$.

$$1 \text{ mol of } N_2 \text{ reacts with } 3 \text{ mol of } H_2$$

This can be expressed in ratio form as (3) $\dfrac{1 \text{ mol } N_2}{3 \text{ mol } H_2}$ or (4) $\dfrac{3 \text{ mol } H_2}{1 \text{ mol } N_2}$.

$$3 \text{ mol of } H_2 \text{ produces } 2 \text{ mol of } NH_3$$

This can be expressed in ratio form as (5) $\dfrac{3 \text{ mol } H_2}{2 \text{ mol } NH_3}$ or (6) $\dfrac{2 \text{ mol } NH_3}{3 \text{ mol } H_2}$.

Note that the coefficients in the balanced equation must be included in the mole ratios. The six mole ratios generated by the sample balanced equation will be used in the stoichiometry problems that follow. In these examples, the central conversion is from moles of one reactant or product to moles of another reactant or product. In Examples 7-2, 7-3, and 7-4, additional conversions are necessary. The mole ratios from above (1 through 6) will be used in the following examples.

7-1.3 Stoichiometry Problems

The examples below illustrate the following conversions.

mole → mole	(Example 7-1)
mass → mole	(Example 7-2)
mass → mass	(Example 7-3)
mass → number	(Example 7-4)

EXAMPLE 7-1   **Converting Moles to Moles**

How many moles of NH_3 can be produced from 5.00 mol of H_2?

PROCEDURE

Convert moles of what's given (mol of H_2) to moles of what's requested (mol NH_3). This requires mole ratio (6) above, which has *mol H_2* in the denominator and *mol NH_3* in the numerator. The unit map for this problem is illustrated below.

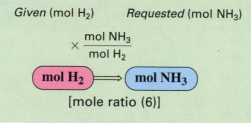

SOLUTION

$$5.00 \; \cancel{\text{mol } H_2} \times \frac{2 \text{ mol } NH_3}{3 \; \cancel{\text{mol } H_2}} = \underline{3.33 \text{ mol } NH_3}$$

EXAMPLE 7-2   Converting Mass to Moles

How many moles of NH_3 can be produced from 33.6 g of N_2?

PROCEDURE

Before we can convert moles of N_2 to moles of NH_3, we must first convert the mass of N_2 to moles of N_2. This means a two-step conversion. In the first step, the mass of N_2 is converted to moles using the molar mass of N_2 as the conversion factor **(a)**. In the second step, moles of N_2 are converted to moles of NH_3 using mole ratio (2) **(b)**. The complete unit map for this problem is illustrated below.

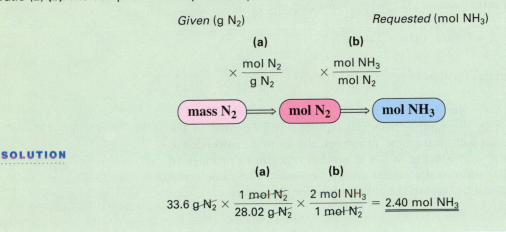

SOLUTION

$$33.6 \; \text{g } \cancel{N_2} \times \frac{1 \; \cancel{\text{mol } N_2}}{28.02 \; \text{g } \cancel{N_2}} \times \frac{2 \text{ mol } NH_3}{1 \; \cancel{\text{mol } N_2}} = \underline{2.40 \text{ mol } NH_3}$$

EXAMPLE 7-3   Converting Mass to Mass

What mass of H_2 is needed to produce 119 g of NH_3?

PROCEDURE

As in the last example, we must first convert the mass of what's given (119 g NH_3) into moles using the molar mass of NH_3 [step **(a)** below]. We then convert moles of NH_3 to moles of H_2 using mole ratio (5) [step **(b)** below]. Finally, since the mass of H_2 is requested, we must convert moles of H_2 to mass using the molar mass of H_2 as a conversion factor [step **(c)** below]. The complete unit map for this three-step problem is illustrated below. Note that this type of conversion is the most common and useful stoichiometry problem.

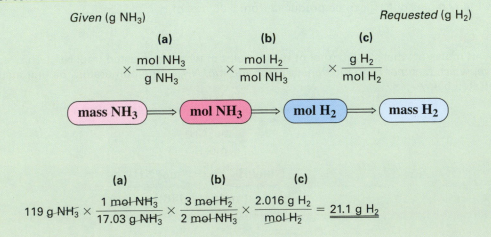

SOLUTION

$$119 \; \text{g } \cancel{NH_3} \times \frac{1 \; \cancel{\text{mol } NH_3}}{17.03 \; \text{g } \cancel{NH_3}} \times \frac{3 \; \cancel{\text{mol } H_2}}{2 \; \cancel{\text{mol } NH_3}} \times \frac{2.016 \text{ g } H_2}{\cancel{\text{mol } H_2}} = \underline{21.1 \text{ g } H_2}$$

EXAMPLE 7-4   Converting Mass to Number

How many molecules of N_2 are needed to react with 17.0 g of H_2?

PROCEDURE

This problem reminds us that the mole relates to a number as well as a mass. In fact, this type of problem is generally not encountered because the actual numbers of molecules involved are not as important as their relative masses. In any case, this problem is much like the previous example except that in the final step moles of N_2 are converted to number of molecules [step **(c)** below]. The complete unit map for this three-step problem is illustrated below.

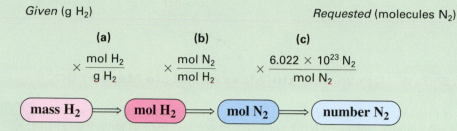

SOLUTION

$$17.0\ \mathrm{g\,H_2} \times \frac{1\ \mathrm{mol\,H_2}}{2.016\ \mathrm{g\,H_2}} \times \frac{1\ \mathrm{mol\,N_2}}{3\ \mathrm{mol\,H_2}} \times \frac{6.022 \times 10^{23}\ \mathrm{molecules\,N_2}}{\mathrm{mol\,N_2}} = 1.69 \times 10^{24}\ \mathrm{molecules\,N_2}$$

ANALYSIS

All of these examples are very similar, but together they illustrate the wide range of possibilities that stoichiometry problems can adopt. The concept of the mole remains central to all quantitative chemistry. We can convert nearly any measured quantity (mass, volume, numbers of molecules) into moles and now, with the use of the mole ratio, into any other compound for which we have a balanced chemical reaction. It is now within our abilities to start with any measure of starting material and calculate any measure of product we wish.

SYNTHESIS

The most important aspect of stoichiometry is that by using the mole ratio, you can interconvert moles of any substance with any other substance within a reaction, given a *balanced* chemical equation. So if you had a reaction with four reactants and six products, you could calculate the amount of each reactant needed for the reaction and the amount of each product produced, while knowing initially the amount of only one of the 10 compounds. How useful would this be to a chemical engineer, for instance? More than likely, an industrial process is required to make a certain amount of a specific compound. It certainly would be useful to know how much of each reactant is required. It would be equally important to know how much other (usually waste) products will be produced. Stoichiometry now provides us the tools for these calculations.

   Before we work through another example of a stoichiometry problem, let's summarize the procedure relating moles, mass, and number of molecules. The general scheme is represented in Figure 7-1 and described in detail here.

1. Write down (A) what is given and (B) what is requested.
2. **a.** If a mass is given, use the molar mass to convert mass to moles of what is given.
   b. If a number of molecules is given, use Avogadro's number to convert to moles of what is given.
3. Using the correct mole ratio from the balanced equation, convert moles of what is given to moles of what is requested.
4. **a.** If a mass is requested, convert moles of what is requested to the equivalent mass.
   b. If a number of molecules is requested, use Avogadro's number to convert to number of molecules of what is requested.

FIGURE 7-1 **General Procedure for Stoichiometry Problems** Any three-step conversion is possible following the proper pathway.

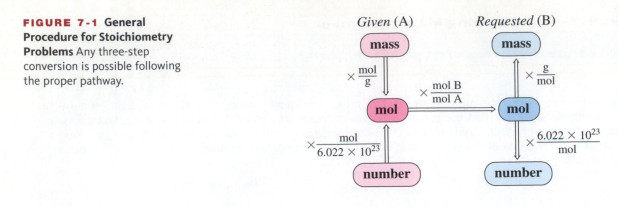

Given (A)   *Requested* (B)

EXAMPLE 7-5   Converting the Mass of FeS$_2$ to Mass of SO$_2$

Some sulfur is present in coal in the form of pyrite (FeS$_2$; also known as "fool's gold"). When it burns, it pollutes the air with the combustion product SO$_2$, as shown by the following balanced equation.

$$4\ FeS_2(s) + 11\ O_2(g) \longrightarrow 2\ Fe_2O_3(s) + 8\ SO_2(s)$$

What mass of SO$_2$ is produced by the combustion of 38.8 g of FeS$_2$?

Pyrite

PROCEDURE

Given: 38.8 g of FeS$_2$. Requested: ? g of SO$_2$. Convert grams of FeS$_2$ to moles of FeS$_2$, moles of FeS$_2$ to moles of SO$_2$, and, finally, moles of SO$_2$ to grams of SO$_2$. The complete unit map for this problem is illustrated below.

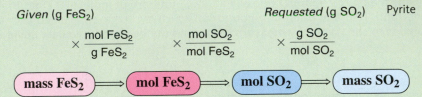

The mole ratio that relates moles of given to moles of requested (in the numerator) is

$$\frac{8\ mol\ SO_2}{4\ mol\ FeS_2}$$

SOLUTION

$$38.8\ g\ FeS_2 \times \frac{1\ mol\ FeS_2}{120.0\ g\ FeS_2} \times \frac{8\ mol\ SO_2}{4\ mol\ FeS_2} \times \frac{64.07\ g\ SO_2}{mol\ SO_2} = \underline{41.4\ g\ SO_2}$$

ANALYSIS

This represents the most common type of stoichiometry problem, called a *mass-to-mass* calculation. Two more quick mass-to-mass calculations can be used to make an important point. What mass of O$_2$ is needed to react completely with the FeS$_2$? Follow the same format as above and confirm that the answer is 28.5 g O$_2$. Finally, what mass of the second product, Fe$_2$O$_3$, should be produced? Again, your calculations should confirm that it is 25.8 g Fe$_2$O$_3$. This then gives us 38.8 g + 28.5 g = 67.3 g of total reactant and 41.4 g + 25.8 g = 67.2 g of total product. The law of conservation of mass is upheld. The small difference is due to rounding off.

SYNTHESIS

Any amount of reactant or product calculated via a stoichiometric setup is strictly theoretical; it represents the maximum amount of product (or the minimum amount of reactant) that can be produced (or is needed). Various real-life situations—such as loss due to transfer or purification, less-than-efficient reaction conditions, and operator error—will lead to actual amounts that are less than what was calculated. Still, these problems can be taken into account, and stoichiometry can give us accurate expectations of what will be produced. In the next two sections, we'll investigate in more detail these variations from our theoretical maximums.

▶ **ASSESSING THE OBJECTIVE FOR SECTION 7-1**

EXERCISE 7-1(a)   LEVEL 1: For the reaction $B_2O_3 + 6NaOH \longrightarrow 2Na_3BO_3 + 3H_2O$, write the corresponding mole ratios for B_2O_3 with the other reactant and with each of the products.

EXERCISE 7-1(b)   LEVEL 2: For the reaction $4NH_3 + 5O_2 \longrightarrow 4NO + 6H_2O$, what mass of NH_3 is needed to produce 75.0 g of H_2O?

EXERCISE 7-1(c)   LEVEL 2: Consider the combustion of methylamine (CH_3NH_2), $4CH_3NH_2(l) + 9O_2(g) \longrightarrow 4CO_2(g) + 2N_2(g) + 10H_2O(l)$.

(a) How many moles of O_2 would react with 6.40 moles of methylamine?

(b) How many moles of N_2 are produced from 322 g of methylamine?

(c) What mass of H_2O is produced along with 6.45 g of CO_2?

(d) How many molecules of O_2 are required to produce 0.568 g of N_2?

EXERCISE 7-1(d)   LEVEL 3: Using the unbalanced reaction $NaN_3 \longrightarrow Na + N_2$, calculate the mass of each product formed from 100.0 g of reactant. Demonstrate that mass is conserved.

EXERCISE 7-1(e)   LEVEL 3: For the neutralization of calcium hydroxide with nitric acid, determine the amount of nitric acid needed to react with 50.0 g of calcium hydroxide. Determine the mass of each product that results. Demonstrate that the total masses of reactants and products are the same.

For additional practice, work chapter problems 7-1, 7-4, 7-8, and 7-10.

7-2 Limiting Reactant

▶ **OBJECTIVE FOR SECTION 7-2**
Determine the limiting reactant and the yield in a reaction given the masses of two different reactants.

LOOKING AHEAD! A common stoichiometry problem involves the conversion of a given mass of reactant to an equivalent mass of product. So far, for a specific amount of one reactant, we have assumed that the needed amounts of all other reactants are present. But how do we determine the amount of product formed if there are specific amounts of each reactant present? This is the subject of this section. ■

7-2.1 The Definition of Limiting Reactant

Assume that you are assembling bicycles from eight wheels and five frames. You can only produce four bicycles, since the number of wheels limits the final number of complete units. You would have one frame in excess. (See Figure 7-2.)

In the case of a chemical reaction, *if specific amounts of each reactant are mixed, the reactant that produces the least amount of product is called the* **limiting reactant.** In other words, the amount of product formed is *limited* by the reactant that is completely consumed. We can illustrate this with the simple example of the production of water from its elements, hydrogen and oxygen.

$$2H_2 + O_2 \longrightarrow 2H_2O$$

The stoichiometry of the reaction tells us that two moles (4.0 g) of hydrogen react with one mole (32.0 g) of oxygen to produce two moles (36.0 g) of water. Thus any time hydrogen and oxygen react in a 4:32 mass ratio, all reactants are consumed

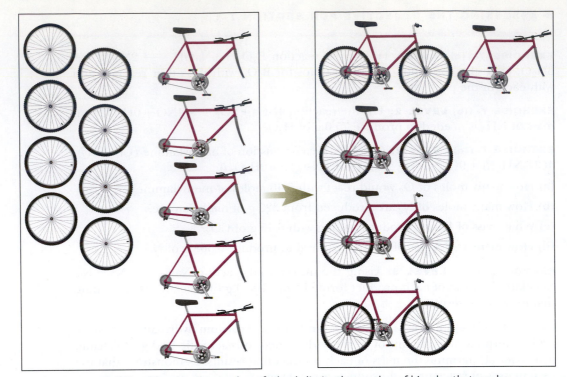

FIGURE 7-2 **Building Bicycles** The number of wheels limits the number of bicycles that can be produced.

and only product appears. *When reactants are mixed in exactly the mass ratio determined from the balanced equation, the mixture is said to be stoichiometric.*

$$4.0 \text{ g } H_2 + 32.0 \text{ g } O_2 \longrightarrow 36.0 \text{ g } H_2O \text{ (stoichiometric)}$$

What if we mix a 6.0-g quantity of H_2 with a 32.0-g quantity of O_2? Do we produce 38.0 g water? No, we still produce only 36.0 g of H_2O using only 4.0 g of the H_2. The extra 2.0 g of H_2 do not react. Thus, H_2 is present in excess, and the amount of product is limited by the amount of O_2 present. In this case, O_2 becomes the limiting reactant.

$$6.0 \text{ g } H_2 + 32.0 \text{ g } O_2 \longrightarrow 36.0 \text{ g } H_2O + 2.0 \text{ g } H_2 \text{ in excess}$$

(H_2 in excess, O_2 the limiting reactant)

If we mix a 4.0-g quantity of H_2 with a 36.0-g quantity of O_2, 36.0 g of H_2O is again produced. In this case, the H_2 is completely consumed and limits the amount of water formed. Thus, H_2 is now the limiting reactant and O_2 is present in excess.

$$4.0 \text{ g } H_2 + 36.0 \text{ g } O_2 \longrightarrow 36.0 \text{ g } H_2O + 4.0 \text{ g } O_2 \text{ in excess}$$

(O_2 in excess, H_2 the limiting reactant)

7-2.2 Solving Limiting Reactant Problems

When quantities of more than one reactant are given, it is necessary to determine which is the limiting reactant. This is accomplished as follows.

1. Convert the amount of *each reactant* to the number of moles (or mass) of product using the general procedure shown in Figure 7-1.
2. The limiting reactant produces the smallest amount of product.

Notice that each *reactant* must be converted into *product* to solve the problem in the manner outlined. So if the reaction contains three different reactants, three individual calculations will need to be performed, and the answers compared. We will illustrate the calculation of limiting reactant with two examples.

EXAMPLE 7-6   Determining the Limiting Reactant

Methanol (CH_3OH) is used as a fuel for racing cars. It burns in the engine according to the equation

$$2CH_3OH(l) + 3O_2(g) \longrightarrow 2CO_2(g) + 4H_2O(g)$$

If 40.0 g of methanol is mixed with 46.0 g of O_2, what is the maximum mass of CO_2 produced?

PROCEDURE

1. Convert mass of CH_3OH to moles of CH_3OH and then to moles of CO_2, and convert mass of O_2 to moles of O_2 and then to moles of CO_2.

2. Convert the smaller number of moles of CO_2 to mass of CO_2.

Race cars at the Indianapolis 500 burn methanol as a fuel.

The complete unit map for this problem is illustrated below. Notice again that it is necessary to convert the masses of both reactants to moles of product.

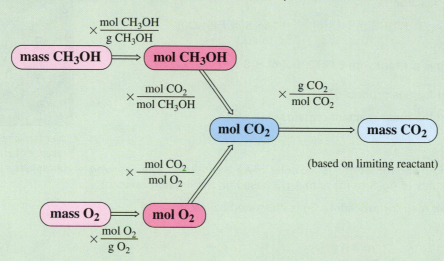

(based on limiting reactant)

SOLUTION

1. CH_3OH: $40.0 \text{ g } CH_3OH \times \dfrac{1 \text{ mol } CH_3OH}{32.04 \text{ g } CH_3OH} \times \dfrac{2 \text{ mol } CO_2}{2 \text{ mol } CH_3OH} = 1.25 \text{ mol } CO_2$

   O_2: $46.0 \text{ g } O_2 \times \dfrac{1 \text{ mol } O_2}{32.00 \text{ g } O_2} \times \dfrac{2 \text{ mol } CO_2}{3 \text{ mol } O_2} = 0.958 \text{ mol } CO_2$

Therefore, O_2 is the limiting reactant.

2. The yield is determined from the amount of product formed *from the limiting reactant*. Thus, we simply convert the 0.958 mol of CO_2 produced by the O_2 to grams of CO_2.

$$0.958 \text{ mol } CO_2 \times \dfrac{44.01 \text{ g } CO_2}{\text{mol } CO_2} = \underline{42.2 \text{ g } CO_2}$$

ANALYSIS

As a rule, whenever the mass of a single reactant is mentioned in the problem, it is assumed that all other reactants are present in excess. However, anytime amounts of two (or more) different compounds are specifically mentioned, it is generally impossible to tell which is in excess without performing the limiting reactant calculation. So anytime there are two given starting masses, expect that the problem requires the calculation of the limiting reactant.

SYNTHESIS

How does this type of problem apply to the everyday world? Again, picture a chemical engineer designing an industrial process. When several reactants are necessary for a multistep process, it makes financial and production sense to design the process so that the most expensive and/or scarcest reactant is the limiting one and all other reactants are present in excess. For instance, it makes no sense to neglect to add enough water to completely react with a more expensive reactant when water is plentiful and cheap. The engineer needs to calculate the amount of product each reactant will produce and ensure that there is plenty of the cheaper reactants to fully complete the process.

EXAMPLE 7-7   Determining the Limiting Reactant

Silver tarnishes (turns black) in homes because of the presence of small amounts of H_2S (a gas that originates from the decay of food and smells like rotten eggs). The reaction is

$$4Ag(s) + 2H_2S(g) + O_2(g) \longrightarrow 2Ag_2S(s) + 2H_2O(l)$$
$$\text{(black)}$$

If 0.145 mol of Ag is present with 0.0872 mol of H_2S and excess O_2,

(a) What mass of Ag_2S is produced?

(b) What mass of the other reactant remains in excess?

Silver tarnishes from the presence of H_2S in the atmosphere.

PROCEDURE (a)

Convert moles of Ag and moles of H_2S to moles of Ag_2S produced. Then convert the number of moles of Ag_2S to mass of Ag_2S based on the limiting reactant.

The complete unit map for procedure (a) is illustrated below.

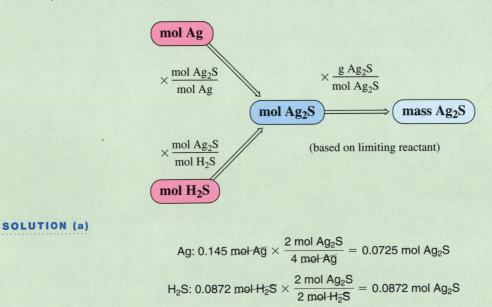

SOLUTION (a)

$$\text{Ag: } 0.145 \ \cancel{\text{mol Ag}} \times \frac{2 \ \text{mol Ag}_2\text{S}}{4 \ \cancel{\text{mol Ag}}} = 0.0725 \ \text{mol Ag}_2\text{S}$$

$$\text{H}_2\text{S: } 0.0872 \ \cancel{\text{mol H}_2\text{S}} \times \frac{2 \ \text{mol Ag}_2\text{S}}{2 \ \cancel{\text{mol H}_2\text{S}}} = 0.0872 \ \text{mol Ag}_2\text{S}$$

Since Ag produces the smaller yield of Ag_2S, *Ag is the limiting reactant.* To find the mass of Ag_2S formed, convert moles of Ag_2S produced by the Ag to mass of Ag_2S.

$$0.0725 \text{ mol Ag}_2\text{S} \times \frac{247.9 \text{ g Ag}_2\text{S}}{\text{mol Ag}_2\text{S}} = \underline{18.0 \text{ g Ag}_2\text{S}}$$

PROCEDURE (b)

To find the mass of H_2S in excess, we first find the mass of H_2S that was consumed in the reaction along with the Ag. Use the mole ratio relating moles of H_2S to moles of Ag. The complete unit map for procedure (b) is illustrated below.

Next, find the moles of H_2S remaining or unreacted by subtracting the moles consumed from the original amount present.

mol H_2S (original) − mol H_2S (consumed) = mol H_2S (in excess)

Finally, convert moles of unreacted H_2S to mass.

$$\boxed{\textbf{mol } H_2S \textbf{ (in excess)}} \Longrightarrow \boxed{\textbf{mass } H_2S}$$

SOLUTION (b)

$$0.145 \text{ mol Ag} \times \frac{2 \text{ mol H}_2\text{S}}{4 \text{ mol Ag}} = 0.0725 \text{ mol H}_2\text{S consumed}$$

$$0.0872 \text{ mol} - 0.0725 \text{ mol} = 0.0147 \text{ mol H}_2\text{S in excess}$$

$$0.0147 \text{ mol H}_2\text{S} \times \frac{34.09 \text{ g H}_2\text{S}}{\text{mol H}_2\text{S}} = \underline{0.501 \text{ g H}_2\text{S in excess}}$$

ANALYSIS

The law of conservation of mass continues to be upheld here as well. All chemical species in the problem converted into masses gives

$$15.6 \text{ g Ag} + 3.0 \text{ g H}_2\text{S} + 1.2 \text{ g O}_2 \text{ (19.8 g total)}$$

$$\text{produces } 18.0 \text{ g Ag}_2\text{S} + 1.3 \text{ g H}_2\text{O} \text{ (19.3 g total)}$$

Why don't the two totals match up this time? The difference is the calculated excess of H_2S, which must be included in the product total. An extra 0.5 g (roughly) mass-balances both sides and keeps conservation of mass alive and well.

SYNTHESIS

These types of problems represent what is potentially the most difficult type of calculation found in an introductory class. In particular, when asked to calculate the amount of excess reactant, it helps to consider another analogy to the limiting reactant. Assume that one is assembling a three-page exam from 85 copies of pages 1 and 2 but only 80 copies of page 3. How many extra sheets of page 1 and page 2 will be left over? We can very quickly answer: 5 pages each. But think more analytically about how you arrived at that conclusion. You first converted the amount of limiting reactant into each of page 1 and page 2, realizing that it would require 80 of each page to complete the test. You then *subtracted* the amount you used from the amount you started with to arrive at the 5 extra pages apiece. This problem does the same types of calculations with the added complication of mole ratios. Can you make the analogy fit step by step?

▶ **ASSESSING THE OBJECTIVE FOR SECTION 7-2**

EXERCISE 7-2(a) LEVEL 1: The recipe for a batch of sugar cookies is

2 cups flour   3 eggs   1 cup sugar   1 stick butter   2 tsp vanilla

In the cupboard are 12 cups of flour, a dozen eggs, 10 cups of sugar, 5 sticks of butter, and 40 tsp of vanilla. What is the limiting reactant, and how many batches of cookies will you make?

EXERCISE 7-2(b) LEVEL 1: For the following reaction, $6Mg + P_4 \longrightarrow 2Mg_3P_2$:

If 12 moles of Mg reacted with 5 moles of P_4,

(a) What is the limiting reactant?

(b) How many moles of magnesium phosphide would be produced?

(c) How many moles of the excess reactant will be left after the reaction?

EXERCISE 7-2(c) LEVEL 2: $PF_3(g) + F_2(g) \longrightarrow PF_5(g)$

If 100.0 g of PF_3 reacts with 50.0 g of F_2, what mass of PF_5 will be produced?

EXERCISE 7-2(d) LEVEL 2: Unbalanced equation:

$Al_2S_3(s) + H_2O(l) \longrightarrow Al(OH)_3(s) + H_2S(g)$

If 48.0 g Al_2S_3 reacts with 32.0 g H_2O,

(a) What is the limiting reactant?

(b) How much $Al(OH)_3$ will be produced?

(c) How much of the excess reactant remains after the reaction is complete?

EXERCISE 7-2(e) LEVEL 3: An engineer is overseeing the production of C_2H_5Cl by mixing equal masses of C_2H_4 and HCl. If she could add only one chemical to increase production, which one should she choose?

For additional practice, work chapter problems 7-21, 7-28, and 7-29.

▶ **OBJECTIVE FOR SECTION 7-3**
Calculate the percent yield of a reaction from the measured actual yield and the calculated theoretical yield.

7-3 Percent Yield

LOOKING AHEAD! There is something else that we have been taking for granted in the problems worked so far. We have assumed that at least one reactant has been completely converted into products and that the reactants form only the products shown. This is not always the case, however. In this section, we will look into incomplete reactions. ∎

An efficient automobile engine burns gasoline (mainly a hydrocarbon, C_8H_{18}) to form carbon dioxide and water. Untuned engines, however, do not burn gasoline efficiently. They produce carbon monoxide and may even exhaust unburned fuel. The two relevant combustion reactions are shown below.

Complete combustion: $2C_8H_{18}(g) + 25O_2(g) \longrightarrow 16CO_2(g) + 18H_2O(l)$

Incomplete combustion: $2C_8H_{18}(g) + 17O_2(g) \longrightarrow 16CO(g) + 18H_2O(l)$

Note that if we were asked to calculate the mass of CO_2 produced from a given amount of C_8H_{18}, our answer would not be correct if all the hydrocarbon were not converted to CO_2. *The measured amount of product obtained in any reaction is known as the* **actual yield.** *The* **theoretical yield** *is the calculated amount of product that would be obtained if all the reactant were converted to a given product. The* **percent yield** *is the ratio of the two, with the actual yield in grams or moles divided by the theoretical yield in grams or moles times 100%.*

$$\frac{\text{actual yield}}{\text{theoretical yield}} \times 100\% = \text{percent yield}$$

There are other types of reactions where reactants are not completely converted into products. These are known as *reversible* reactions where a reverse reaction occurs when products re-form reactants. These reactions reach a *point of equilibrium* where

MAKING IT REAL

Forensic Chemistry: The Limiting Reactant in Breathalyzers

Drunk driving takes as many as 20,000 lives each year in the United States alone. To combat this problem, communities have adopted strict standards by which a driver may be labeled as intoxicated. By 1964, 39 states had adopted a limit of 0.10% blood alcohol concentration as the legal limit. Since then, many states have lowered this to 0.08%. Other countries are stricter. The limit in Ireland and Japan is 0.05%, and in Sweden it is 0.02%. That could be as little as one drink. How do police measure the blood alcohol?

The best method is to take a blood sample. The blood is analyzed by a device that measures the amount of volatile compounds (those that vaporize) in the blood. The device, called a *gas chromatograph*, is highly accurate. However, it could take an hour or more to get blood drawn by a trained professional and have the concentration of alcohol measured. A lot of alcohol can be metabolized in one hour. So it is important to get a reading when the driver is stopped if such a test is suggested. Usually, a suspect is asked to perform some physical tasks such as walking a straight line or standing on one foot. Also, the officer may observe the movement of the eyes. In the latter test, if the driver is under the influence, the eye tends to jerk involuntarily when following a penlight. If the officer suspects intoxication, he or she may subject the suspect to a portable *breathalyzer* test.

When one has been drinking, water and alcohol vapor are expelled from the lungs with each breath. By measuring a specific volume of breath, the amount of alcohol in the breath can be determined. This is directly proportional to the amount of alcohol in the blood. There are several modern devices, including fuel cells, that measure alcohol in the breath, but the original test involved a chemical reaction. The breath is forced through an aqueous

solution where alcohol (C_2H_5OH) reacts according to the following equation.

$$2K_2Cr_2O_7(aq) + 3C_2H_5OH(aq) + 8H_2SO_4(aq) \xrightarrow{AgNO_3}$$
$$2Cr_2(SO_4)_3(aq) + 2K_2SO_4(aq) + 3CH_3COOH(aq) +$$
$$11H_2O(l)$$

In words, ethyl alcohol reacts with a potassium dichromate solution containing excess sulfuric acid. Also present is silver nitrate ($AgNO_3$, shown above the reaction arrow), which acts as a catalyst to speed up the reaction. Acetic acid (CH_3COOH) is formed from the alcohol. Potassium dichromate solutions have a pronounced yellow color. As will be discussed in more detail in the next chapter, the solution reflects yellow light because its complement (blue light) is absorbed. (See Making It Real, p. 231.) So the absorbance of blue light is a measure of the intensity of the yellow reflected light. The absorbance is measured with a device called a *spectrometer*. The limiting reactant in this reaction is the alcohol. Potassium dichromate and sulfuric acid are present in excess. However, the more alcohol present in a sample of breath, the more potassium dichromate that reacts and the less the amount of the light beam absorbed. If all the potassium dichromate is used up (i.e., it becomes the limiting reactant), all the yellow light disappears in the solution and no blue light is absorbed. Additional alcohol will not change anything at this point. By measuring the change in the yellow light in the sample, the amount of alcohol can be determined.

Still, the best evidence of intoxication is from a sample of blood. This is usually done if the suspect fails the breathalyzer test. The best way to beat the rap for drunk driving is to let somebody else drive if you've been drinking.

reactants and products coexist in specific amounts. An example is the reaction illustrated previously in which nitrogen and hydrogen react to produce ammonia. Hydrogen and nitrogen are not completely converted to ammonia because some of the ammonia decomposes back to its two elements. Reversible reactions will be discussed in more detail in Chapters 13 and 15. Since the reactions are not complete, percent yield is also relevant to reversible reactions.

To determine the percent yield, it is necessary to determine the theoretical yield (which is what we've been doing all along) and compare this with the actual yield.

EXAMPLE 7-8   Determining the Percent Yield

In a given experiment, a 4.70-g quantity of H_2 is allowed to react with N_2; a 12.5-g quantity of NH_3 is formed. What is the percent yield based on the H_2? The equation for the reaction is

$$N_2(g) + 3H_2(g) \longrightarrow 2NH_3(g)$$

PROCEDURE

Find the mass of NH_3 that would form if the entire 4.70 g of H_2 was converted to NH_3 (the theoretical yield). Using the actual yield (12.5 g), find the percent yield. The complete unit map for this problem is illustrated below.

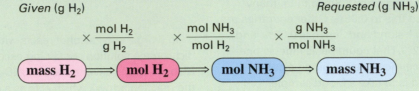

The mole ratio needed is

$$\frac{2 \text{ mol } NH_3}{3 \text{ mol } H_2}$$

SOLUTION

$$4.70 \text{ g } H_2 \times \frac{1 \text{ mol } H_2}{2.016 \text{ g } H_2} \times \frac{2 \text{ mol } NH_3}{3 \text{ mol } H_2} \times \frac{17.03 \text{ g } NH_3}{\text{mol } NH_3} = 26.5 \text{ g } NH_3$$

theoretical yield

$$\frac{12.5 \text{ g}}{26.5 \text{ g}} \times 100\% = \underline{47.2\% \text{ yield}}$$

ANALYSIS

One of the keys to success in working percent yield problems is to be able to identify an amount of compound as being an actual or a theoretical amount. Essentially the terms *actual* and *experimental* are interchangeable, as are the terms *theoretical* and *calculated*. Any value determined by a stoichiometry calculation is theoretical in nature. Once you've positively identified the actual and theoretical amounts, these problems become just a regular mass-to-mass stoichiometry calculation with one extra calculation (percent yield) thrown in.

SYNTHESIS

Typically, chemists would like to try to increase the percent yield of a reaction as much as possible. What types of things might they do to this end, other than use improved experimental technique? Changing reaction conditions is always a good place to start. Varying things like temperature, pressure, solvent, and reaction time can alter percent yield, for better or worse. Experimentation is called for. If the reaction is of the reversible type, increasing the *excess* reactant can actually increase the experimental yield. This, of course, doesn't affect the theoretical yield, which is determined from the limiting reactant. An increase in actual yield while theoretical yield stays the same will increase the percent yield. How reactions are affected by conditions is a subject of Chapter 15.

EXAMPLE 7-9   Determining the Actual Yield from Percent Yield

Zinc and silver undergo a single-replacement reaction according to the equation

$$Zn(s) + 2AgNO_3(aq) \longrightarrow Zn(NO_3)_2(aq) + 2Ag(s)$$

When 25.0 g of zinc is added to the silver nitrate solution, the percent yield of silver is 72.3%. What mass of silver is formed?

PROCEDURE

Find the theoretical yield of silver by converting the mass of zinc to the equivalent mass of silver. Use the percent yield to calculate the actual yield. This can be accomplished algebraically by rearranging the equation to solve for actual yield as follows:

$$\text{actual yield} = \frac{\text{percent yield}}{100\%} \times \text{theoretical yield}$$

The complete unit map for this example is shown below.

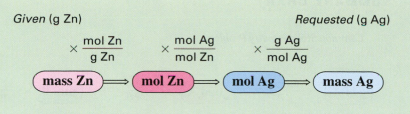

SOLUTION

$$25.0 \text{ g Zn} \times \frac{1 \text{ mol Zn}}{65.39 \text{ g Zn}} \times \frac{2 \text{ mol Ag}}{1 \text{ mol Zn}} \times \frac{107.9 \text{ g Ag}}{\text{mol Ag}} = 82.5 \text{ g Ag}$$
theoretical yield

$$\frac{72.3\%}{100\%} \times 82.5 \text{ g Ag} = \underline{59.6 \text{ g Ag (actual yield)}}$$

ANALYSIS

When we designate something the theoretical yield, we are stating that it is the theoretical *maximum*. It makes sense, then, that the actual yield would be less than that.

SYNTHESIS

A third type of problem involving percent yield is to inquire how much reactant is necessary to produce a given amount of product, assuming a percent yield less than 100%. The trick here is to perform the percent yield calculation first, to calculate theoretical yield, and then use that amount in the mass-to-mass stoichiometry calculation to determine the amount of starting material. For example, if a reaction had a 50% yield, this means that you'd have to use *double* the amount of starting material you'd have calculated for a straightforward stoichiometry problem with 100% yield. For an example, see Exercise 7-3d.

▶ **ASSESSING THE OBJECTIVE FOR SECTION 7-3**

EXERCISE 7-3(a) LEVEL 1: If a chemical reaction theoretically should produce 32.0 g of product, but the actual amount is 24.0 g, what is the percent yield for the reaction?

EXERCISE 7-3(b) LEVEL 1: A chemical reaction has a percent yield of 75%. If you desired an actual yield of 1.8 kg, what mass of product should you *attempt* to make?

EXERCISE 7-3(c) LEVEL 2: For the reaction

$$Zn(s) + 2AgNO_3(aq) \longrightarrow Zn(NO_3)_2(aq) + 2Ag(s)$$

If 10.0 g of Zn produces 25.0 g of Ag, what's the percent yield of the reaction?

EXERCISE 7-3(d) LEVEL 3: For the reaction

$$4KNO_3(s) \longrightarrow 2K_2O(s) + 2N_2(g) + 5O_2(g)$$

What mass of KNO_3 is needed to produce 20.0 g K_2O, assuming the reaction produces an 84.0% yield?

EXERCISE 7-3(e) LEVEL 3: Anyone with experience in a teaching lab knows that students frequently report percent yields in excess of 100%. Could this be accurate? What are some potential explanations for yields this high?

For additional practice, work chapter problems 7-32, 7-35, and 7-37.

PART A SUMMARY

KEY TERMS

7-1.1   **Stoichiometry** calculations involve conversions among amounts of reactants and products. p. 214

7-1.2   **Mole ratios** from balanced equations are the keys to solving stoichiometry problems. p. 215

7-2.1   The **limiting reactant** is completely consumed in a reaction. p. 219

7-3   The **percent yield** is calculated from the **theoretical yield** and **actual yield**. p. 224

SUMMARY CHART

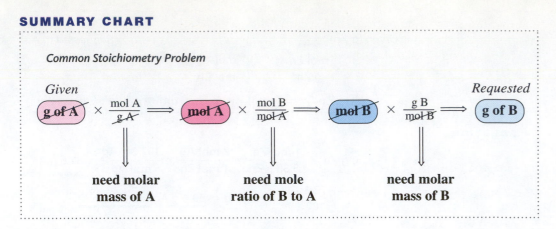

Common Stoichiometry Problem

Part B

Energy Relationships in Chemical Reactions

SETTING A GOAL

- You will extend your knowledge of stoichiometry to include amounts of heat involved in chemical reactions.

OBJECTIVE

7-4 Given the change in enthalpy for a reaction, calculate the amount of heat gained or released by a given mass of reactant.

> ► **OBJECTIVE FOR SECTION 7-4**
> Given the change in enthalpy for a reaction, calculate the amount of heat gained or released by a given mass of reactant.

7-4 Heat Energy in Chemical Reactions

LOOKING AHEAD! Matter is not all that is involved in chemical reactions. The other component of the universe, energy, is also an intimate part of chemical reactions. A definite and measurable amount of energy is involved in any chemical change. ■

The amount of heat energy (measured in kilocalories or kilojoules) involved in a reaction is a constant amount that depends on the amount of reactants consumed. For example, if one mole of hydrogen undergoes combustion to form liquid water, 286 kJ of heat is evolved.

7-4.1 Thermochemical Equations

A balanced equation that includes heat energy is referred to as a **thermochemical equation**. A thermochemical equation can be represented in either of two ways. In the first, the heat is shown separately from the balanced equation using the symbol **ΔH**. This is referred to as heat of the reaction. The symbol is known technically as the *change in* **enthalpy**. By convention, *a negative sign for ΔH corresponds to an exothermic reaction, and a positive sign corresponds to an endothermic reaction.* From the perspective of the reaction, this convention makes sense. Loss of heat energy by the reaction is exothermic and thus negative. Gain of heat energy is endothermic and thus positive. Written in this manner, the thermochemical equation for the combustion of two moles of hydrogen is

$$2H_2(g) + O_2(g) \longrightarrow 2H_2O(l) \quad \Delta H = -572 \text{ kJ}$$

Notice in the preceding equation that the heat evolved is per *two* moles of H_2 consumed or per *two* moles of H_2O formed. This result is general for any reaction. The enthalpy reported for a reaction is for the amount in moles of compounds represented by the balancing coefficients. The second way a thermochemical equation can be represented shows the heat energy as if it were a reactant or product. In exothermic reactions heat energy is a product, and in endothermic reactions heat energy is a reactant. A positive sign is used in either case.

$$2H_2\ (g) + O_2(g) \longrightarrow 2H_2O\ (l) + 572\ kJ \qquad \text{(exothermic reaction)}$$

$$N_2\ (g) + O_2(g) + 181\ kJ \longrightarrow 2NO(g) \qquad \text{(endothermic reaction)}$$

7-4.2 Calculation of Heat Energy in a Reaction

The heat energy involved in a chemical reaction may be treated quantitatively in a manner similar to the amount of a reactant or product. The enthalpy of the reaction, with units of kJ/mol, is analogous to molar mass and its units of g/mol. Where molar mass allows us to convert between grams and moles, reaction enthalpies allow us to convert between energy and moles. The following examples illustrate the calculations implied by a thermochemical equation.

EXAMPLE 7-10   Converting Heat to Mass

Acetylene, which is used in welding torches, undergoes combustion according to the following thermochemical equation:

$$2C_2H_2(g) + 5O_2(g) \longrightarrow 4CO_2(g) + 2H_2O(l) \quad \Delta H = -2602\ kJ$$

If 550 kJ of heat is evolved in the combustion of a sample of C_2H_2, what is the mass of CO_2 formed?

PROCEDURE

Two steps are required, as illustrated below.

$$\times \frac{mol\ CO_2}{kJ} \qquad \times \frac{g\ CO_2}{mol\ CO_2}$$

$$\boxed{kJ} \Longrightarrow \boxed{mol\ CO_2} \Longrightarrow \boxed{mass\ CO_2}$$

The combustion of acetylene produces a flame hot enough to melt iron.

SOLUTION

$$550\ kJ \times \frac{4\ mol\ CO_2}{2602\ kJ} \times \frac{44.01\ g\ CO_2}{mol\ CO_2} = \underline{37.2\ g\ CO_2}$$

ANALYSIS

Notice that the enthalpy of the reaction is −2602 kJ/4 mol CO_2, or −650.5 kJ/mol CO_2. What is the enthalpy expressed relative to the three other species in the reaction? −1301 kJ/mol C_2H_2, −520.4 kJ/mol O_2, and −1301 kJ/mol H_2O. You could then use these values to calculate the amount of water produced or the amounts of C_2H_2 or oxygen needed for the specified amount of heat.

SYNTHESIS

The combustion of all hydrocarbons produces heat, but which burns best with the greatest amount of heat? The answer differs depending on whether you want your answer in units of kJ/mol or kJ/g. When evaluating heats of reaction on a per-mole basis, the bigger, the better. One mole of a large molecule will have many more carbons and therefore can produce much more heat. A better comparison is on a per-gram basis. And when we do it that way, we find that the smallest hydrocarbon, methane, has the largest heat of combustion at −55.6 kJ/g. The heats of combustion per gram are also important for the health conscious. One gram of fat produces about nine Calories (about 36 kJ), whereas one gram of protein or carbohydrate produces about four Calories (about 16 kJ). As you can see, fat contains more than twice as many Calories on a mass basis than proteins or carbohydrates.

EXAMPLE 7-11   Converting Mass to Heat

Carbon dioxide forms from its elements, as illustrated by the following thermochemical equation:

$$C(s) + O_2(g) \longrightarrow CO_2(g) \quad \Delta H = -393.5 \text{ kJ}$$

How much heat will be generated from 50.0 g of C?

PROCEDURE

$$\times \frac{\text{mol C}}{\text{g C}} \qquad \times \frac{\text{kJ}}{\text{mol C}}$$

$$\boxed{\text{mass C}} \Longrightarrow \boxed{\text{mol C}} \Longrightarrow \boxed{\text{kJ}}$$

SOLUTION

$$50.0 \text{ g C} \times \frac{1 \text{ mol C}}{12.01 \text{ g C}} \times \frac{-393.5 \text{ kJ}}{\text{mol C}} = \underline{\underline{-1640 \text{ kJ}}}$$

The negative sign indicates that heat is being released by this reaction.

ANALYSIS

Don't confuse the −393.5 kJ reported in conjunction with the reaction with the 1640 kJ reported as the heat released by the reaction. Officially, the −393.5 kJ is the enthalpy of the reaction and has units of kJ/mol, though those are often not reported, as it can be per mole of any of the reactants. The values are analogous to molar mass and mass, respectively.

SYNTHESIS

Let's continue to examine where calculations like these might be useful. How might someone in an industrial setting benefit from this sort of information? It's clearly important when designing equipment for processes to know whether the reaction being run is endothermic or exothermic. Will you need to supply heat? Or will heat be produced and need to be drawn away with cooling equipment? Exactly how much heat will there be? Is the equipment sturdy enough to withstand the generated heat? Is the cooling system efficient enough to avoid damage, meltdowns, or explosions? These are some fairly important considerations, and they are determined by examining heats of reactions.

▶ **ASSESSING THE OBJECTIVE FOR SECTION 7-4**

EXERCISE 7-4(a) LEVEL 1: The combustion of C_2H_4 releases 556 kJ/mol CO_2. Write two thermodynamic equations for that reaction that express that heat in two different ways.

EXERCISE 7-4(b) LEVEL 2: For the reaction in the previous exercise, what mass of C_2H_4 is required to produce 250 kJ of heat, assuming 100% yield?

EXERCISE 7-4(c) LEVEL 2 The decomposition of iron(III) oxide by the reaction

$2Fe_2O_3 \longrightarrow 4Fe + 3O_2$ has a $\Delta H = +1625$ kJ as written. How much energy is required to promote the decomposition of 125 g of Fe_2O_3?

EXERCISE 7-4(d) LEVEL 3: Consider the reaction $2H_2 + O_2 \longrightarrow 2H_2O$. This is the combination reaction to form water; it is exothermic, with an enthalpy of −572 kJ. Does it follow that the decomposition of water, $2H_2O \longrightarrow 2H_2 + O_2$, must be endothermic, with an enthalpy of +572 kJ?

For additional practice, work chapter problems 7-40, 7-43, and 7-45.

MAKING IT REAL

Hydrogen—The Perfect Fuel

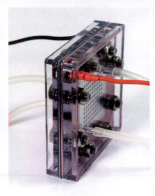

Fuel cells make efficient use of hydrogen.

The 6.8 billion citizens of this planet all need something to burn. We need fuel to cook our food, to keep our homes warm, and to harvest our crops. Of course, many of us drive cars and other vehicles, which takes even more fuel. Throughout most of history, the main fuel was wood. Now, we mainly use fossil fuels such as coal, natural gas, and oil. Many of the planet's forests have already been lost, and there is a limit to the amount of fossil fuels stored in the earth. Nuclear energy can help, but more is needed. Enter the simplest and lightest of the elements—hydrogen.

Why is hydrogen such a good fuel? There are many reasons. First, when it burns it forms only water, which does not accumulate in the atmosphere and cause global warming. Also, it is very light, producing 143 kJ/g compared to 55.6 kJ/g for natural gas and 48.1 kJ/g for gasoline. Hydrogen is already the workhorse fuel of the space program. It is used in the rockets and in the fuel cells that produce electricity in the space shuttles. A fuel cell uses the reaction of hydrogen with oxygen to form water to produce electrical energy rather than heat energy (see Section 13-6). Fuel cells are thought to be the wave of the future, for both utilities and automobiles.

So why not use hydrogen more? First, pure hydrogen is expensive. The cheapest way to produce hydrogen is to extract it from fossil fuels such as methane (CH_4). That is still comparatively expensive, and it produces carbon dioxide, a greenhouse gas. The best hope is in the electrolysis of water. Electrolysis uses energy (i.e., 286 kJ/mol) to break water down into its elements (hydrogen and oxygen). Since it takes so much energy, this process is also expensive. However, there is intensive research to find an inexpensive way to do this. There is hope that solar energy can be harnessed for this task, but it needs to be made more efficient and less expensive.

Another problem is the storage of hydrogen. Its boiling point is −253°C, which means that the liquid state cannot exist at room temperature. Thus, it has to be stored as a gas under high pressure, which is inherently dangerous. One possibility that is being explored is to store the hydrogen in a solid compound. For example, if water is added to sodium borohydride ($NaBH_4$), the reaction generates hydrogen gas, as illustrated by the following equation.

$$NaBH_4(s) + 2H_2O(l) \longrightarrow NaBO_2(s) + 4H_2(g)$$

The storage of hydrogen is a problem that will be solved, however.

PART B SUMMARY

KEY TERMS

7-4.1   **Thermochemical equations** include the changes in **enthalpy (ΔH)** as well as the compounds involved. p. 228

SUMMARY CHART

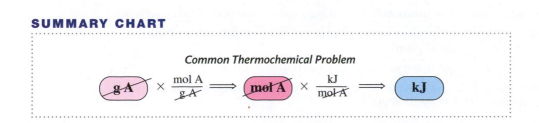

Common Thermochemical Problem

CHAPTER 7 SYNTHESIS PROBLEM

In this and the previous chapter, we have finally arrived at the "meat" of chemistry. That concerns chemical reactions and the equations that represent these reactions. In the first five chapters, we presented most of the basic tools needed to make sense of chemical reactions. Among other concepts, these included handling numbers, matter and energy, formulas, names, and the counting unit of chemistry (the mole). In the previous chapter we tried to bring some order and even predictability to the vast number of chemical reactions. In this chapter, we consider the quantitative relationships inherent in the equations representing these reactions. Future chapters will add important details and expansion of these two chapters. The following problems apply what we have learned in this chapter.

PROBLEM	SOLUTION
a. An aqueous solution of $FeCl_3$ and NaOH is mixed. A precipitate forms. Write the balanced equation illustrating this reaction indicating the precipitate accordingly.	**a.** $FeCl_3(aq) + 3NaOH(aq) \longrightarrow 3NaCl(aq) + Fe(OH)_3 (s)$ From the solubility rules discussed in Chapter 6, notice that only the iron(III) hydroxide is insoluble.
b. Calculate the theoretical yield (in grams) of precipitate formed from the complete reaction of 50.0 g of NaOH.	**b.** The conversion sequence is $g\ NaOH \longrightarrow mol\ NaOH \longrightarrow mol\ Fe(OH)_3 \longrightarrow g\ Fe(OH)_3$ $50.0\ g\ NaOH \times \dfrac{1\ mol\ NaOH}{40.00\ g\ NaOH} \times \dfrac{1\ mol\ Fe(OH)_3}{3\ mol\ NaOH} \times \dfrac{106.9\ g\ Fe(OH)_3}{mol\ Fe(OH)_3}$ $= 44.5\ g\ Fe(OH)_3$
c. A solution of 50.0 g of NaOH is now mixed with a solution of 60.0 g of $FeCl_3$ solution. Which is the limiting reactant and what is the theoretical yield of precipitate from this reaction?	**c.** We must now calculate the theoretical yield from the complete reaction of $FeCl_3$ and compare it to that from **b** above. The one producing the smaller yield is the limiting reactant and produces the theoretical yield from the mixture. The conversion sequence is $g\ FeCl_3 \longrightarrow mol\ FeCl_3 \longrightarrow mol\ Fe(OH)_3 \longrightarrow g\ Fe(OH)_3$ $60.0\ g\ FeCl_3 \times \dfrac{1\ mol\ FeCl_3}{162.2\ g\ FeCl_3} \times \dfrac{1\ mol\ Fe(OH)_3}{mol\ FeCl_3} \times \dfrac{106.9\ g\ Fe(OH)_3}{mol\ Fe(OH)_3}$ $= 39.5\ g\ Fe$ The $FeCl_3$ is the limiting reactant. The theoretical yield from this reaction is 39.5 g $Fe(OH)_3$.
d. In the reaction described in **c** above, what is the percent yield if the actual yield of precipitate is 31.7 g?	**d.** $\dfrac{actual\ yield}{theoretical\ yield} \times 100\% = percent\ yield$ $\dfrac{31.7\ g}{39.5\ g} \times 100\% = \underline{80.3\%\ yield}$
e. If the enthalpy change (ΔH) is -12.8 kJ (per mole of precipitate), what is the amount of heat actually produced by the reaction considering the percent yield?	**e.** Use the actual yield from the reaction. The conversion sequence is $g\ Fe(OH)_3 \longrightarrow mol\ Fe(OH)_3 \longrightarrow kJ$ $31.7\ g\ Fe(OH)_3 \times \dfrac{1\ mol\ Fe(OH)_3}{106.9\ g\ Fe(OH)_3} \times \dfrac{-12.8\ kJ}{mol\ Fe(OH)_3}$ $= \underline{-3.80\ kJ}\ (heat\ evolved)$

YOUR TURN

 a. An aqueous solution of K_2S is mixed with an aqueous solution of $Al(NO_3)_3$. A precipitate forms. Write the balanced equation illustrating this reaction indicating the precipitate accordingly.

 b. What is the theoretical yield of the precipitate from the complete reaction of 145 g $Al(NO_3)_3$?

 c. If a solution of 145 g of $Al(NO_3)_3$ is mixed with a solution of 125 g of K_2S, what is the limiting reactant and the theoretical yield?

 d. If the percent yield is 88.8%, what is the actual yield from the reaction in **c**?

 e. If the enthalpy change (ΔH) for the reaction is -21.5 kJ, what is the amount of heat energy produced by the reaction? (Use the actual yield.)

Answers are on p 234.

CHAPTER SUMMARY

In this chapter, we examined how chemical equations tell us about the relationships between masses of reactants and masses of products. These are calculations of **stoichiometry**. In these problems, the balanced equation provides the **mole ratios** for conversion from moles of one compound in the reaction to moles of another. All of these conversions are summarized in Figure 7-1.

When reactants are mixed in other than stoichiometric amounts, it is necessary to determine the **limiting reactant**. The limiting reactant is completely consumed; thus, it determines the amount of product formed.

Some reactions lead to more than one set of products, and others reach a state of equilibrium before all reactants are converted to products. In either case, the **actual yield** of a product may be less than the **theoretical yield**. The actual yield can be expressed as a percent of the theoretical yield. This is called the **percent yield**.

Finally, we studied the relation of heat energy to chemical reactions. In a **thermochemical equation**, the heat evolved or absorbed is represented as either a reactant or product, or as the heat of the reaction, which is referred to as the change in **enthalpy** and is symbolized as ΔH.

OBJECTIVES

SECTION	YOU SHOULD BE ABLE TO...	EXAMPLES	EXERCISES	CHAPTER PROBLEMS
7-1	Perform stoichiometry calculations using mole ratios from balanced equations.	7-1, 7-2, 7-3, 7-4, 7-5	1a, 1b, 1c, 1d	3, 5, 7, 10, 13, 15, 19
7-2	Determine the limiting reactant and the yield in a reaction given the masses of two different reactants.	7-6, 7-7	2a, 2b, 2c, 2d	23, 24, 27, 30, 31
7-3	Calculate the percent yield of a reaction from the measured actual yield and the calculated theoretical yield.	7-8, 7-9	3a, 3b, 3c, 3d	32, 33, 34, 37
7-4	Given the change in enthalpy for a reaction, calculate the amount of heat gained or released by a given mass of reactant.	7-10, 7-11	4a, 4b, 4c	43, 44, 45, 47

ANSWERS TO ASSESSING THE OBJECTIVES

PART A

EXERCISES

7-1(a) $\dfrac{1 \text{ mol } B_2O_3}{6 \text{ mol NaOH}}, \dfrac{1 \text{ mol } B_2O_3}{2 \text{ mol } Na_3BO_3}, \dfrac{1 \text{ mol } B_2O_3}{3 \text{ mol } H_2O}$

7-1(b) 47.3 g NH_3

7-1(c) (a) 14.4 mol O_2 (b) 5.18 mol N_2 (c) 6.60 g H_2O (d) 5.49×10^{22} molecules of O_2

7-1(d) 35.4 g Na + 64.6 g N_2 = 100 g total product

7-1(e) The balanced equation for the reaction is:

$$Ca(OH)_2 + 2\,HNO_3 \longrightarrow Ca(NO_3)_2 + 2\,H_2O$$

50.0 g of $Ca(OH)_2$ reacts with 85.1 g of HNO_3 to produce 111 g of $Ca(NO_3)_2$ and 24.3 g of H_2O. Total reactant mass and total product mass are both 135 g.

7-2(a) Eggs are the limiting reactant. You can make four batches of cookies.

7-2(b) (a) Mg (b) 4 mol Mg_3P_2 (c) 3 mol P_4 remains

7-2(c) 143 g PF_5

7-2(d) (a) H_2O (b) 46.2 g Al $(OH)_3$ (c) 3.6 g Al_2S_3

7-2(e) The compounds add in a 1:1 mole ratio. Therefore, the lighter chemical has the most moles. The compound with the greater molar mass, the HCl, is the limiting reactant. Add HCl.

7-3(a) 24.0 g/32.0 g $\times$ 100% = 75.0% yield

7-3(b) 2.4 kg of product

7-3(c) The theoretical yield of Ag is 33.0 g. The percent yield is 75.8%.

7-3(d) 51.1 g KNO_3

7-3(e) No, it couldn't be. The sample may be contaminated (unevaporated water is very common). Alternatively, the student may have made an experimental or mathematical error in calculating the yield.

PART B

EXERCISES

7-4(a) (1) $C_2H_4 + 3O_2 \longrightarrow 2CO_2 + 2H_2O + 1112$ kJ; (2) $C_2H_4 + 3O_2 \longrightarrow 2CO_2 + 2H_2O \quad \Delta H = -1112$ kJ

7-4(b) 6.31 g C_2H_4

7-4(c) 636 kJ are required.

7-4(d) Yes! It does. Reverse processes differ only by the sign of their enthalpies. It's analogous to saying that if you had to climb 100 ft to get to the top of a hill, then you must descend 100 ft (-100 ft) in order to get back down.

ANSWERS TO CHAPTER SYNTHESIS PROBLEM

a. $3K_2S(aq) + 2Al(NO_3)_3(aq) \longrightarrow Al_2S_3(s) + 6KNO_3(aq)$ **b.** 51.1 g Al_2S_3 **c.** Since 125 g K_2S produces 56.7 g Al_2S_3, the $Al(NO_3)_3$ is the limiting reactant. The theoretical yield is 51.1 g Al_2S_3.

d. 45.4 g Al_2S_3 **e.** −6.50 kJ (assuming 88.8% yield)

CHAPTER PROBLEMS

Throughout the text, answers to all exercises in color are given in Appendix E. The more difficult exercises are marked with an asterisk.

Stoichiometry (SECTION 7-1)

7-1. Given the balanced equation

$$Mg + 2HCl \longrightarrow MgCl_2 + H_2$$

provide the proper mole ratio for each of the following mole conversions.

(a) Mg to H_2

(b) Mg to HCl

(c) HCl to H_2

(d) $MgCl_2$ to HCl

7-2. Given the balanced equation

$$2C_4H_{10} + 13O_2 \longrightarrow 8CO_2 + 10H_2O$$

provide the proper mole ratio for each of the following mole conversions.

(a) CO_2 to C_4H_{10}

(b) O_2 to C_4H_{10}

(c) CO_2 to O_2

(d) O_2 to H_2O

7-3. Given the balanced equation

$$Cu + 4HNO_3 \longrightarrow Cu(NO_3)_2 + 2NO_2 + 2H_2O$$

provide the proper mole ratio for each of the following mole conversions.

(a) Cu to NO_2

(b) HNO_3 to Cu

(c) H_2O to NO_2

(d) $Cu(NO_3)_2$ to HNO_3

(e) NO_2 to Cu

(f) Cu to H_2O

7-4. The reaction that takes place in the reusable solid rocket booster for the space shuttle is shown by the following equation.

$3Al(s) + 3NH_4ClO_4(s) \longrightarrow$
$\qquad Al_2O_3(s) + AlCl_3(s) + 3NO(g) + 6H_2O(g)$

(a) How many moles of each product are formed from 10.0 moles of Al?

(b) How many moles of each product are formed from 3.00 moles of NH_4ClO_4?

7-5. Phosphine (PH_3) is a poisonous gas once used as a fumigant for stored grain. It is prepared according to the following equation.

$$Ca_3P_2(s) + 6H_2O(l) \longrightarrow 3Ca(OH)_2(s) + 2PH_3(g)$$

(a) How many moles of phosphine are prepared from 5.00 moles of Ca_3P_2?

(b) How many moles of phosphine are prepared from 5.00 moles of H_2O?

7-6. Hydrogen cyanide is an important industrial chemical used to make a plastic, acrylonitrile. HCN is prepared according to the following equation.

$$2NH_3(g) + 3O_2(g) + 2CH_4(g) \longrightarrow 2HCN(g) + 6H_2O(l)$$

(a) How many moles of O_2 and CH_4 react with 10.0 moles of NH_3?

(b) How many moles of HCN and H_2O are produced from 10.0 moles of O_2?

7-7. Iron rusts according to the equation

$$4Fe(s) + 3O_2(g) \longrightarrow 2Fe_2O_3(s)$$

(a) What mass of rust (Fe_2O_3) is formed from 0.275 mole of Fe?

(b) What mass of rust is formed from 0.275 mole of O_2?

(c) What mass of O_2 reacts with 0.275 mole of Fe?

7-8. Glass (SiO_2) is etched with hydrofluoric acid according to the equation

$$SiO_2(s) + 4HF(aq) \longrightarrow SiF_4(g) + 2H_2O(l)$$

If 4.86 moles of HF reacts with SiO_2,

(a) What mass of SiF_4 forms?   **(c)** What mass of SiO_2 reacts?

(b) What mass of H_2O forms?

7-9. Consider the reaction

$$2H_2 + O_2 \longrightarrow 2H_2O$$

(a) How many moles of H_2 are needed to produce 0.400 mole of H_2O?

(b) What mass of H_2O will be produced from 0.640 g of O_2?

(c) What mass of H_2 is needed to react with 0.032 g of O_2?

(d) What mass of H_2O would be produced from 0.400 g of H_2?

7-10. Propane burns according to the equation

$$C_3H_8 + 5O_2 \longrightarrow 3CO_2 + 4H_2O$$

(a) How many moles of CO_2 are produced from the combustion of 0.450 mole of C_3H_8? How many moles of H_2O? How many moles of O_2 are needed?

(b) What mass of H_2O is produced if 8.80 g of CO_2 is also produced?

(c) What mass of C_3H_8 is required to produce 1.80 g of H_2O?

(d) What mass of C_3H_8 is required to react with 160 g of O_2?

(e) What mass of CO_2 is produced by the reaction of 6.38 g of O_2?

(f) How many moles of H_2O are produced if 4.50×10^{22} molecules of CO_2 are produced?

7-11. The alcohol component of gasohol burns according to the equation

$$C_2H_5OH(l) + 3O_2(g) \longrightarrow 2CO_2(g) + 3H_2O(g)$$

(a) What mass of alcohol is needed to produce 98.2 g of H_2O?   *83.7 g*

(b) How many moles of CO_2 are produced along with 155 g of H_2O?   *5.73 g*

(c) What mass of CO_2 is produced from 146 g of C_2H_5OH?   *278.9 g*

(d) What mass of C_2H_5OH reacts with 0.898 g of O_2?

(e) What mass of H_2O is produced from 5.85×10^{24} molecules of O_2?

7-12. In the atmosphere, N_2 and O_2 do not react with each other. In the high temperatures of an automobile engine, however, the following reaction occurs.

$$N_2(g) + O_2(g) \longrightarrow 2NO(g)$$

When the NO reaches the atmosphere through the engine exhaust, a second reaction takes place.

$$2NO(g) + O_2(g) \longrightarrow 2NO_2(g)$$

The NO_2 is a brownish gas that contributes to the haze of smog and is irritating to the nasal passages and lungs. What mass of N_2 is required to produce 155 g of NO_2?

7-13. Elemental iron is produced in what is called the *thermite reaction* because it produces enough heat that the iron is initially in the molten state. The liquid iron can then be used to weld iron bars together.

$$2Al(s) + Fe_2O_3(s) \longrightarrow Al_2O_3(s) + 2Fe(l)$$

What mass of Al is needed to produce 750 g of Fe? How many formula units of Fe_2O_3 are used in the process?

7-14. Antacids, which contain calcium carbonate, react with stomach acid according to the equation

$$CaCO_3(s) + 2HCl(aq) \longrightarrow CaCl_2(aq) + CO_2(g) + H_2O(l)$$

What mass of stomach acid reacts with 1.00 g of $CaCO_3$?

7-15. Elemental copper can be recovered from the mineral chalcocite, Cu_2S. From the following equation, determine what mass of Cu is formed from 4.16 g of O_2.

$$Cu_2S(s) + O_2(g) \longrightarrow 2Cu(s) + SO_2(g)$$

7-16. Fool's gold (pyrite) is so named because it looks much like gold. When it is placed in aqueous HCl, however, it dissolves and gives off the pungent gas H_2S. Gold itself does not react with aqueous HCl. From the following equation, determine how many individual molecules of H_2S are formed from 62.4 g of pyrite (FeS_2).

$$FeS_2(s) + 2HCl(aq) \longrightarrow FeCl_2(aq) + H_2S(g) + S(s)$$

7-17. Nitrogen dioxide may form so-called acid rain by reaction with water in the air according to the equation

$$3NO_2(g) + H_2O(l) \longrightarrow 2HNO_3(aq) + NO(g)$$

What mass of nitric acid is produced from 18.5 kg of NO_2?

7-18. Elemental chlorine can be generated in the laboratory according to the equation

$$MnO_2(s) + 4HCl(aq) \longrightarrow MnCl_2(aq) + 2H_2O(l) + Cl_2(g)$$

What mass of Cl_2 is produced from the reaction of 665 g of HCl?

7-19. The fermentation of sugar to produce ethyl alcohol is represented by the equation

$$C_6H_{12}O_6(s) \longrightarrow 2C_2H_5OH(l) + 2CO_2(g)$$

What mass of alcohol is produced from 25.0 mol of sugar?

7-20. Methane gas can be made from carbon monoxide gas according to the equation

$$2CO(g) + 2H_2(g) \longrightarrow CH_4(g) + CO_2(g)$$

What mass of CO is required to produce 8.75×10^{25} molecules of CH_4?

Limiting Reactant (SECTION 7-2)

7-21. Nitrogen gas can be prepared by passing ammonia over hot copper(II) oxide according to the equation

$$3CuO(s) + 2NH_3(g) \longrightarrow N_2(g) + 3Cu(s) + 3H_2O(g)$$

How many moles of N_2 are prepared from the following mixtures?

(a) 3.00 mol of CuO and 3.00 mol of NH_3

(b) 3.00 mol of CuO and 2.00 mol of NH_3

(c) 3.00 mol of CuO and 1.00 mol of NH_3

(d) 0.628 mol of CuO and 0.430 mol of NH_3

(e) 5.44 mol of CuO and 3.50 mol of NH_3

7-22. How many moles remain of the reactant in excess in problem 7-21(a) and (c)?

7-23. Consider the equation illustrating the combustion of arsenic.

$$4As(s) + 5O_2(g) \longrightarrow 2As_2O_5(s)$$

How many moles of As_2O_5 are prepared from the following mixtures?

(a) 4.00 mol of As and 4.00 mol of O_2

(b) 3.00 mol of As and 4.00 mol of O_2

(c) 5.62 mol of As and 7.50 mol of O_2

(d) 3.86 mol of As and 4.75 mol of O_2

7-24. How many moles remain of the reactant in excess in problem 7-23(a) and (b)?

7-25. Consider the equation

$$2Al + 3H_2SO_4 \longrightarrow Al_2(SO_4)_3 + 3H_2$$

If 0.800 mol of Al is mixed with 1.00 mol of H_2SO_4, how many moles of H_2 are produced? How many moles of one of the reactants remain?

7-26. Consider the equation

$$2C_5H_6 + 13O_2 \longrightarrow 10CO_2 + 6H_2O$$

If 3.44 mol of C_5H_6 is mixed with 20.6 mol of O_2, what mass of CO_2 is formed?

7-27. Elemental fluorine is very difficult to prepare by ordinary chemical reactions. In 1986, however, a chemical reaction was reported that produces fluorine.

$$2K_2MnF_6 + 4SbF_5 \longrightarrow 4KSbF_6 + 2MnF_3 + F_2$$

If a 525-g quantity of K_2MnF_6 is mixed with a 900-g quantity of SbF_5, what mass of F_2 is produced?

7-28. Consider the equation

$$4NH_3(g) + 3O_2(g) \longrightarrow 2N_2(g) + 6H_2O(l)$$

If a 40.0-g sample of O_2 is mixed with 1.50 mol of NH_3, which is the limiting reactant? How many moles of N_2 form?

7-29. Consider the equation

$$2AgNO_3(aq) + CaCl_2(aq) \longrightarrow 2AgCl(s) + Ca(NO_3)_2(aq)$$

If a solution containing 20.0 g of $AgNO_3$ is mixed with a solution containing 10.0 g of $CaCl_2$, which compound is the limiting reactant? What mass of $AgCl$ forms? What mass of one of the reactants remains?

7-30. Limestone ($CaCO_3$) dissolves in hydrochloric acid as shown by the equation

$$CaCO_3(s) + 2HCl(aq) \longrightarrow CaCl_2(aq) + CO_2(g) + H_2O(l)$$

If 20.0 g of $CaCO_3$ and 25.0 g of HCl are mixed, what mass of CO_2 is produced? What mass of one of the reactants remains?

7-31. Consider the balanced equation

$$2HNO_3(aq) + 3H_2S(aq) \longrightarrow 2NO(g) + 4H_2O(l) + 3S(s)$$

If a 10.0-g quantity of HNO_3 is mixed with 5.00 g of H_2S, what are the masses of each product and of the reactant present in excess after reaction occurs?

Theoretical and Percent Yield (SECTION 7-3)

7-32. Sulfur trioxide is prepared from SO_2 according to the equation

$$2SO_2(g) + O_2(g) \longrightarrow 2SO_3(g)$$

This is a reversible reaction in which not all SO_2 is converted to SO_3 even with excess O_2 present. In a given experiment, 21.2 g of SO_3 was produced from 24.0 g of SO_2. What is the theoretical yield of SO_3? What is the percent yield?

7-33. The following is a reversible decomposition reaction.

$$2N_2O_5 \longrightarrow 4NO_2 + O_2$$

When 25.0 g of N_2O_5 decomposes, it is found that 10.0 g of NO_2 forms. What is the percent yield?

7-34. The following equation represents a reversible combination reaction.

$$P_4O_{10} + 6PCl_5 \longrightarrow 10POCl_3$$

If 25.0 g of PCl_5 reacts, there is a 45.0% yield of $POCl_3$. What is the actual yield in grams?

7-35. Octane in gasoline burns in an automobile engine according to the equation

$$2C_8H_{18}(l) + 25O_2(g) \longrightarrow 16CO_2(g) + 18H_2O(g)$$

If a 57.0-g sample of octane is burned, 152 g of CO_2 is formed. What is the percent yield of CO_2?

*7-36. In problem 7-35, the C_8H_{18} that is *not* converted to CO_2 forms CO. What is the mass of CO formed? (CO is a poisonous pollutant that is converted to CO_2 in a car's catalytic converter.)

*7-37. Given the reversible reaction

$$2NO_2 + 4H_2 \longrightarrow N_2 + 4H_2O$$

what mass of hydrogen is required to produce 250 g of N_2 if the yield is 70.0%?

*7-38. When benzene reacts with bromine, the principal reaction is

$$C_6H_6 + Br_2 \longrightarrow C_6H_5Br + HBr$$

If the yield of bromobenzene (C_6H_5Br) is 65.2%, what mass of bromobenzene is produced from 12.5 g of C_6H_6?

*7-39. A second reaction between C_6H_6 and Br_2 (see problem 7-38) produces dibromobenzene ($C_6H_4Br_2$).

(a) Write the balanced equation illustrating this reaction.

(b) If the remainder of the benzene from problem 7-38 reacts to form dibromobenzene, what is the mass of $C_6H_4Br_2$ produced?

Heat in Chemical Reactions (SECTION 7-4)

7-40. When one mole of magnesium undergoes combustion to form magnesium oxide, 602 kJ of heat energy is evolved. Write the thermochemical equation in both forms.

7-41. In problem 7-12, the reaction between N_2 and O_2 was discussed. A 90.5-kJ quantity of heat energy must be supplied per mole of NO formed. Write the balanced thermochemical equation in both forms. Is the reaction exothermic or endothermic?

7-42. To decompose one mole of $CaCO_3$ to CaO and CO_2, 176 kJ must be supplied. Write the balanced thermochemical equation in both forms.

7-43. The complete combustion of one mole of octane (C_8H_{18}) in gasoline evolves 5480 kJ of heat. The complete combustion of one mole of methane in natural gas (CH_4) evolves 890 kJ. How much heat is evolved per 1.00 g by each of these fuels?

7-44. Methyl alcohol (CH_3OH) is used as a fuel in racing cars. It burns according to the equation

$$2CH_3OH(l) + 3O_2(g) \longrightarrow 2CO_2(g) + 4H_2O(l) + 1750 \text{ kJ}$$

What amount of heat is evolved per 1.00 g of alcohol? How does this compare with the amount of heat per gram of octane in gasoline? (See problem 7-43.)

7-45. The thermite reaction was discussed in problem 7-13. For the balanced equation, $\Delta H = -850$ kJ. What mass of aluminum is needed to produce 35.8 kJ of heat energy?

7-46. Photosynthesis is an endothermic reaction that forms glucose ($C_6H_{12}O_6$) from carbon dioxide, water, and energy from the sun. The balanced equation is

$$6CO_2(g) + 6H_2O(l) \longrightarrow C_6H_{12}O_6(aq) + 6O_2(g)$$
$$\Delta H = +2519 \text{ kJ}$$

What mass of glucose is formed from 975 kJ of energy?

7-47. When butane (C_4H_{10}) in a cigarette lighter undergoes combustion, it evolves 2880 kJ per mole of butane. What is the mass of water formed if 1250 kJ of heat evolves?

General Problems

*7-48. Liquid iron is made from iron ore (Fe_2O_3) in a three-step process in a blast furnace as follows.

1. $3Fe_2O_3(s) + CO(g) \longrightarrow 2Fe_3O_4(s) + CO_2(g)$
2. $Fe_3O_4(s) + CO(g) \longrightarrow 3FeO(s) + CO_2(g)$
3. $FeO(s) + CO(g) \longrightarrow Fe(l) + CO_2(g)$

What mass of iron would eventually be produced from 125 g of Fe_2O_3?

7-49. A 50.0-g sample of *impure* $KClO_3$ is decomposed to KCl and O_2. If a 12.0-g quantity of O_2 is produced, what percent of the sample is $KClO_3$? (Assume that all of the $KClO_3$ present decomposes.)

7-50. In Example 7-5, SO_2 was formed from the burning of pyrite (FeS_2) in coal. If a 312-g quantity of SO_2 was collected from the burning of 6.50 kg of coal, what percent of the original sample was pyrite?

*7-51. Copper metal can be recovered from an ore, $CuCO_3$, by the decomposition reaction

$$2CuCO_3(s) \longrightarrow 2Cu(s) + 2CO_2(g) + O_2(g)$$

What is the mass of a sample of *impure* ore if it is 47.5% $CuCO_3$ and produces 350 g of Cu? (Assume complete decomposition of $CuCO_3$.)

*7-52. Consider the equation

$$4NH_3(g) + 5O_2(g) \longrightarrow 4NO(g) + 6H_2O(l)$$

When an 80.0-g quantity of NH_3 is mixed with 200 g of O_2, a 40.0-g quantity of NO is formed. Calculate the percent yield based on the limiting reactant.

*7-53. Consider the equation

$$3K_2MnO_4 + 4CO_2 + 2H_2O \longrightarrow 2KMnO_4 + 4KHCO_3 + MnO_2$$

How many moles of MnO_2 are produced if 9.50 mol of K_2MnO_4, 6.02×10^{24} molecules of CO_2, and 90.0 g of H_2O are mixed?

*7-54. Calcium chloride hydrate ($CaCl_2 \cdot 6H_2O$) is a solid used to melt ice at low temperatures. It is prepared according to the equation

$$CaCO_3(s) + 2HCl(g) + 5H_2O(l) \longrightarrow$$
$$CaCl_2 \cdot 6H_2O(s) + CO_2(g)$$

What mass of the hydrate is prepared from a mixture of 0.250 mole of H_2O, 9.50×10^{22} molecules of HCl, and 15.0 g of $CaCO_3$?

*7-55. A 2.85-g quantity of gaseous methane is mixed with 15.0 g of chlorine to produce a liquid, compound X, and gaseous hydrogen chloride. Compound X is 14.1% C, 2.35% H, and 83.5% Cl. Its molecular formula is the same as its empirical formula.

(a) From the formula of X, write a balanced equation. (*Hint:* Balance hydrogens before chlorines.)

(b) What is the theoretical yield of compound X?

*7-56. A 10.00-g sample of gaseous ethane (C_2H_6) reacts with chlorine to form gaseous hydrogen chloride and a liquid compound (Y) that has a molar mass of 168 g/mol. Compound Y is 14.3% carbon, 84.5% chlorine, and the remainder hydrogen. The reaction produces 12.0 g of compound Y, which is 57.0% yield.

(a) Write the balanced equation illustrating the reaction.

(b) What mass of ethane was required?

*7-57. The remainder of the ethane from problem 7-56 reacts to form a liquid compound (Z) that is 18.0% carbon, 79.8% chlorine, and 2.25% hydrogen. The empirical and molecular formulas of this compound are the same.

(a) Write the balanced equation illustrating the reaction.

(b) What mass of compound Z is formed?

STUDENT WORKSHOP

Limiting Reactants

Purpose: To manipulate a model to evaluate a reaction containing a limiting agent. (Work in groups of three or four. Estimated time: 30 min.)

Equipment: 12 Lego blocks of each of these colors: white, red, blue, and yellow. These will be used to make "molecules" of different combinations of "atoms."

1.   Construct the following molecules: one blue (B) and one white (W) to make a BW; two red (R) and one yellow (Y) to make an R_2Y.

2.   Create as many of these molecules as you can, and discard the excess blocks.
     *   What is the largest number of BW molecules you can make?
     *   What is the largest number of R_2Y's?

3.   Rearrange the given molecules into two new molecules, R_4W and BY_2.
     *   Write out a balanced chemical reaction:

     $$? \text{ BW} + ? \text{ } R_2Y \longrightarrow ? \text{ } R_4W + ? \text{ } BY_2$$

4.   Given the initial amounts of the two reactants:
     *   Which was the limiting reactant?
     *   How many molecules of product were you able to make?
     *   How many of the excess reactants were unreacted?
     *   How many of the original BW and R_2Y molecules reacted?

5. Repeat this procedure for the following reaction:

    $$BR_3 + YW \longrightarrow R_2Y + B_2W_3$$

6. Now, make 12 B's, 6 Y_2's, 12 W's, and 4 R_3's. From the equation:

    $$12 \text{ B} + 3Y_2 + 6 \text{ W} + 4R_3 \longrightarrow 6 \text{ } B_2YR_2W$$

    *   Predict how many B_2YR_2W's you can make from each of the given colors.
    *   Predict which one(s) is (are) the limiting reactant(s).
    *   Predict how many of the excess reactants are going to be left over.
    *   Perform the experiment and see whether you were correct.

Modern Atomic Theory

Fireworks produce spectacular colors. Each color is caused by the presence of a specific element. The information that flows from the emission of color by the atoms of hot, gaseous elements is a topic of this chapter.

SETTING THE STAGE

Dazzling explosions with bursts of bright silver and streaks of green, yellow, red, and blue from spectacular displays of fireworks light up the sky. Many holidays around the world would not be complete without these exciting explosions of colors. In fact, it has been known for hundreds of years that when certain minerals are added to fireworks, specific colors are produced in the hot explosions. We now know that it is the presence of specific elements in these minerals that is responsible for the colors. For example, red is produced by the element strontium, green by barium, blue by copper, and yellow by sodium. In fact, the specific colors emitted by hot, gaseous atoms can be used to identify elements much like fingerprints are used to identify individuals. We even use this method to determine the elemental composition of stars millions of light years away. Of more concern to us is that the study of the colors of these *emission spectra* is what opened the door to a deeper understanding of the chemistry of the elements.

Often in science, when a theory is developed to explain one phenomenon, other mysteries are explained as well. This happened with the theory that explained the emission spectra of the elements—most specifically, hydrogen. The theory, which emphasized the electrons in the atom, led to an understanding of the theoretical basis of the periodic table. As we explained in Chapter 4, this marvelous table displays elements that are chemically related in vertical columns called *groups*. Although the existence of chemically related elements has been known for over 200 years, only since the 1930s have scientists had a feeling for *why* elements are chemically related.

A theory of the atom that emphasized the electrons was first advanced in 1913 by a student of Lord Rutherford (see Section 2-2.3) named Niels Bohr. This had the immediate effect of explaining the emission spectrum of hydrogen, the simplest of the elements. More significant to our purposes, however, is that Bohr's theory eventually led to a new, improved theory that explained and predicted chemical similarities among certain elements.

In Part A in this chapter, we will see how the emission spectra provided the clue for an understanding of the basis of the periodic table. In Part B, we will apply this information to properties of the elements that relate to the periodic table.

Part A

The Energy of the Electron in the Atom

SETTING A GOAL

- You will understand historically how elemental emission spectra were used to determine the structure electrons adopt within an atom.

OBJECTIVES

8-1 Describe how Bohr's model of the atom accounts for observed wavelengths and energies of emission spectra.

8-2 Describe the electronic structure of an atom, including shells, subshells, and various types of orbitals.

▶ **OBJECTIVE**
FOR SECTION 8-1
Describe how Bohr's model of the atom accounts for observed wavelengths and energies of emission spectra.

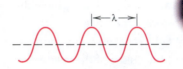

8-1 The Emission Spectra of the Elements and Bohr's Model

LOOKING AHEAD! Hot, gaseous atoms in fireworks give off spectacular light, but any substance glows if it is heated to a high enough temperature. For example, the tungsten filament in an incandescent light bulb emits a bright light as a result of heat generated by the flow of electricity. Understanding the nature of emitted light was the clue that led to an understanding of the periodic table. ■

8-1.1 Electromagnetic Radiation and Wavelength

Light, which is also known as **electromagnetic radiation**, *is a form of energy like heat or electricity.* Light travels through space in tiny packets of energy called *photons* in waves much like the waves moving across a pond or a lake. All light waves travel at the same velocity in a vacuum at the phenomenal rate of 3.0×10^8 m/s (186,000 mi/s). Light waves have properties called *wavelength* and *amplitude.* The amplitude of a wave refers to its height, which in turn relates to the intensity of light. The amplitude, or intensity, of light is not important to our discussion, so it will not be mentioned further. *The* **wavelength** *of light refers to the distance between two adjacent peaks in the wave.* Wavelength is very important for our purposes because it relates to the energy of light. Wavelength is represented by the Greek letter λ (lambda).

8-1.2 The Continuous Spectrum

Consider the white light emitted by a tungsten filament. We can analyze this light by passing it through a glass prism. Specific colors have different velocities when traveling through a medium such as glass. (See Figure 8-1.) As a result of the geometry of the prism, the white light is separated into its component colors, which appear as a continuous range of colors that we associate with a rainbow. *Since one color blends gradually into another, this is known as a* **continuous spectrum**. A rainbow after a rainstorm is a continuous spectrum caused by the separation of sunlight, which contains the complete spectrum of colors, into its component colors by raindrops in the air.

Each color of light in a rainbow has a specific wavelength. Violet light, on one end of the spectrum, has the shortest wavelength [400 nm (1 nm = 10^{-9} m)]. Red, on the other end of the spectrum, has the longest wavelength (750 nm). Important to our consideration of light, however, is that the energy of light is inversely proportional to its wavelength. That is,

$$E \propto \frac{1}{\lambda} \qquad E = \frac{hc}{\lambda}$$

FIGURE 8-1 The Spectrum of Incandescent Light When a narrow beam of light from an ordinary light-bulb is passed through a glass prism, the white light is found to contain all the colors of the visible spectrum.

where λ (lambda) is the symbol of wavelength, *h* is a constant of proportionality known as Planck's constant, and *c* is the velocity of light in a vacuum, which is also a constant.

Red light, with the longest wavelength, has the lowest energy in the visible spectrum. Violet, with the shortest wavelength, has the highest energy. The spectrum of electromagnetic radiation extends in both directions well beyond the wavelengths of visible light. Light that extends somewhat beyond violet is termed *ultraviolet light* (less than 400 nm). It is not visible but is highly energetic. Indeed, it is powerful enough to damage living organisms. For example, UVA (320–400 nm) and UVB (280–320 nm) light from the sun can cause various types of skin cancer. Light with extremely small wavelengths and thus extremely high energies includes X-rays and gamma rays. Only small amounts of this radiation can be tolerated by living systems. (See Figure 8-2.)

Infrared light has wavelengths somewhat longer than 750 nm and is also outside the visible range of the electromagnetic spectrum. Beyond infrared light are the longer wavelengths of radio and TV waves. Light with wavelengths longer than 750 nm is not known to cause damage to living tissue.

MAKING IT REAL

Roses Are Red; Violets Are Blue—But Why?

The three primary colors.

From the beauty of flowers and sky to the brightness of rubies and emeralds, we marvel at the colors of light that lie in the visible part of the spectrum (wavelengths 400–750 nm).

It was mentioned earlier in this chapter that the sun emits the complete spectrum of colors of the rainbow. Blended together, sunlight appears nearly white. Fireworks emit discrete colors depending on the element that is heated in the explosion. But most of what our eyes receive is a result of reflected light rather than emitted light.

When white light is directed at a surface, the amount of reflection determines its shade. If all of the light is reflected, the surface appears white. If all of the light is absorbed, the surface appears black. (That is why light-colored clothing is preferred in the summer, since it reflects the light energy.) If part of the light is absorbed and part is reflected, the surface appears as a shade of gray.

Most of the wonderful colors of our world result from absorption of specific regions of the light spectrum. There are three primary colors of light, as shown in the photograph. Red light comprises the highest third of the wavelengths of visible light (the lowest energies). Green is the middle third of the wavelengths, and blue is the lower third. A blend of all three appears white.* It is only when one region of the spectrum is absorbed that we perceive color. A rose appears red because it absorbs the blue–green part of the spectrum and reflects the complement to these colors, which is red. If only the blue part is absorbed and both red and green are reflected, we see a yellow rose. A violet absorbs the red–green part of the spectrum and reflects blue. Leaves appear green because they absorb the red–blue part of the spectrum. Some of that absorbed light energy is converted into chemical energy in the production of sugars by *photosynthesis*.

Color television works on the same principle. Three different signals are shot through different primary-color filters and then transmitted to the TV set. The three signals are then automatically blended back together to provide the proper color. Older sets had red, green, and blue drives that had to be manually adjusted.

We are lucky. Not all animals see color. Life for us certainly would be dull without the "wonderful world of color."

*The three primary colors of pigment or paint are different. They are red, yellow, and blue. A blend of these three pigments produces white.

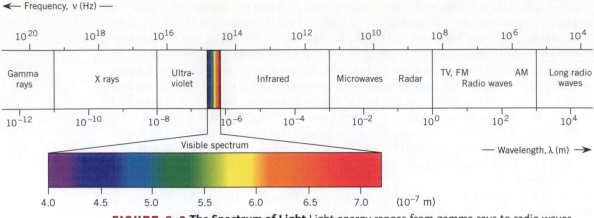

FIGURE 8-2 **The Spectrum of Light** Light energy ranges from gamma rays to radio waves.

8-1.3 The Discrete Spectrum

When an element is heated enough so that the hot, gaseous atoms emit light, the results are different from the continuous spectrum emitted by the sun or a light bulb. Only specific, or discrete, colors are produced. Since the colors are discrete, the energies emitted from the atoms are discrete. The spectrum of each element (called its atomic emission spectrum) is unique to that element and is the reason specific elements are used in fireworks. (See Figure 8-3.) *The atomic emission spectra of elements are referred to as* **discrete**, *or* **line**, **spectra**. In Figure 8-4, the spectrum of hydrogen in the visible range is shown along with the wavelengths of light of the four lines that appear. Energy is supplied to the hydrogen gas by an electrical discharge similar to that in the familiar neon light.

8-1.4 Bohr's Model

By 1913, the nuclear theory of the atom proposed by Lord Rutherford was generally accepted by scientists. In that year, Neils Bohr set out to develop this theory further in order to explain the origin of the discrete spectrum of hydrogen. His main goal was to explain how the electron in the hydrogen atom was responsible for the energy emitted in the form of light. Bohr did not realize that his theory would lay

FIGURE 8-3 The Continuous Spectrum and the Emission Spectra of Sodium and Hydrogen in the Visible Range The light from an incandescent light (top) is continuous, whereas the atomic spectra of sodium (middle) and hydrogen (bottom) are composed of discrete colors.

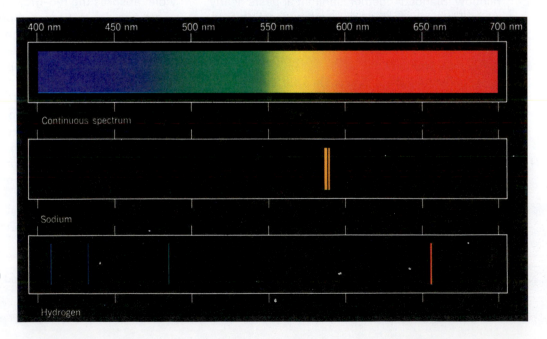

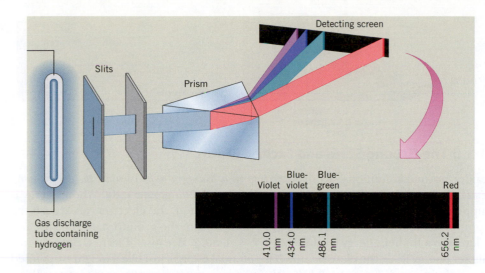

FIGURE 8-4 Production of the Line Spectrum of Hydrogen The light from excited hydrogen atoms is passed through a glass prism. The prism divides the light into four discrete colors in the visible range.

the foundation for the explanation of other phenomena such as the periodicity of elements and the way atoms bond to each other.

First, Bohr proposed that the electron in the hydrogen atom revolves around the nucleus in a stable, circular orbit. The electrostatic attraction of the negative electron for the positive proton is counterbalanced by the centrifugal force of the orbiting electron. These two forces offset each other and keep the electron in a stable orbit. His theory of the hydrogen atom is analogous to the orbiting of the planets around the sun and thus is often referred to as the planetary model. *A model is a description or analogy used to help visualize a phenomenon.* **Bohr's model** does not seem so revolutionary at first glance, but classical physics prohibits such a model. Classical, or Newtonian, physics, which was all that was known at the time, stated that a charged particle (the electron) orbiting around another charged particle (the proton) would lose energy and spin into the nucleus. Bohr sidestepped this problem by postulating (suggesting without proof as a necessary condition) that classical physics does not apply in the small dimensions of the atom and that other laws are appropriate. Eventually, he would be proven correct.

8-1.5 Quantized Energy Levels

Second, Bohr suggested that there are several orbits that the electron may occupy. Furthermore, these orbits that are available to the electron are **quantized**, *which means that they are at definite, or discrete, distances from the nucleus.* Since the energy of a particular orbit depends on its distance from the nucleus, the energy of an electron in such an orbit is also quantized. *The discrete orbits available to an electron in a hydrogen atom are referred to as* **energy levels.** (See Figure 8-5.) *Each energy level in the hydrogen atom is designated by an integral number known as the* **principal quantum number** and given the symbol n. The first energy level is designated $n = 1$; the second level, $n = 2$; and so on.

The quantized energy levels in the hydrogen atom can be compared with a stairway in your home. In Figure 8-6, you can see that between floor A and floor B there are five steps that are analogous to energy levels in an atom. Since you cannot stand (with both feet together) between steps, we can say that each step represents a discrete, or *quantized,* amount of energy. However, a ramp between the two floors represents a *continuous* change in energy. In this case, all energy levels would be possible between floors.

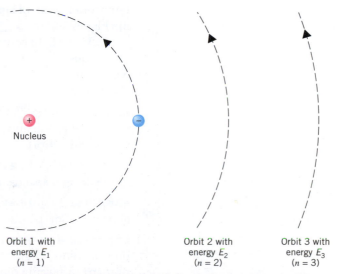

Orbit 1 with
energy E_1
($n = 1$)

Orbit 2 with
energy E_2
($n = 2$)

Orbit 3 with
energy E_3
($n = 3$)

FIGURE 8-5 Bohr's Model of the Hydrogen Atom In this model, the electron can exist only in definite, or discrete, energy states.

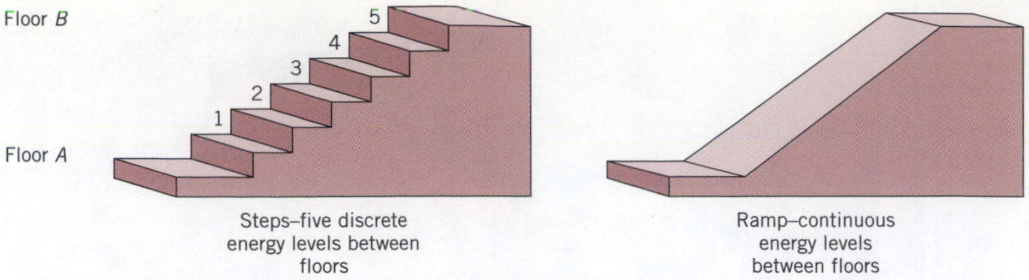

FIGURE 8-6 Discrete versus Continuous Energy Levels The steps represent discrete energy levels; the ramp, continuous energy levels.

Floor B

Floor A

Steps–five discrete energy levels between floors

Ramp–continuous energy levels between floors

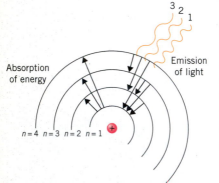

Absorption of energy

Emission of light

$n = 4$ $n = 3$ $n = 2$ $n = 1$

FIGURE 8-7 Light Emitted from a Hydrogen Atom An electron in an excited state emits energy in the form of light when it drops to a lower energy level.

8-1.6 The Ground State and Excited States

Under normal conditions, *the single electron in a hydrogen atom occupies the lowest energy level, which is the orbit closest to the nucleus. This is called the* **ground state**. When energy is supplied to a hydrogen atom, such as when it is heated, the electron can absorb the appropriate amount of energy needed to "jump" from the ground state (i.e., the $n = 1$ level) to a higher energy level (i.e., $n = 2, 3, 4$, etc.). Because of its new position in a higher energy level, the electron now has potential energy, just like a weight suspended above the ground. *Energy levels higher than the ground state are called* **excited states**. What is most significant, according to Bohr, is what happens when the electron "falls" back down to lower excited states or all the way to the ground state. He suggested that the electron gives up this extra energy in the form of light when it falls back. Since energy levels are quantized, the difference in energy between any two levels is also quantized. When an electron falls back to a lower energy level, it must emit a discrete amount of energy. Since this energy is emitted as light, the light would have a discrete energy, a discrete wavelength, and a discrete color (if the light is in the visible region of the spectrum), thus explaining the line spectrum of hydrogen. (See Figure 8-7.)

This is the qualitative explanation of the discrete spectrum. The real significance of Bohr's model is that he was able to calculate the expected wavelength of light in the hydrogen spectrum from the mathematical relationships. The experimental values listed in Figure 8-4 correspond well with those computed by Bohr. Bohr's model worked specifically for the hydrogen atom. Calculations of the wavelengths of the lines in the discrete spectra of other elements had to await more sophisticated models.

▶ **ASSESSING THE OBJECTIVE FOR SECTION 8-1**

EXERCISE 8-1(a) LEVEL 1: Fill in the blanks.

Long wavelengths of light have _____ (high/low) energies. The color of visible light with the highest energy is _____, and the color of visible light with the longest wavelength is _____. White light contains all colors of light, called a _____ spectrum. Light from heated gaseous element sources gives off only discrete energies to form a _____ spectrum. Niels Bohr suggested that the orbits of an electron were _____ or set to specific values. The lowest energy state is the _____ state, and the higher states are the _____ states.

EXERCISE 8-1(b) LEVEL 1: List the parts of the electromagnetic spectrum in order. Label the appropriate ends of the list as high energy/low energy and long wavelength/short wavelength.

EXERCISE 8-1(c) LEVEL 2: In a sentence or two, describe what causes the lines in the hydrogen spectrum.

EXERCISE 8-1(d) LEVEL 3: Bohr's model applied only to the results of the hydrogen spectrum. Is this of any significance when considering other elements?

EXERCISE 8-1(e) LEVEL 3: Neon lights excite electrons in atoms of neon to produce a colored light. If the energy produced is quantized, why can we make lights of many different colors?

For additional practice, work chapter problems 8-1, 8-2, and 8-3.

MAKING IT REAL

Forensic Chemistry: Solving Crimes with Light

Analysis of light can provide forensic scientists with an amazing amount of information about evidence found at a crime scene. Perhaps one of the more important objects found at a scene are glass particles. Maybe a window was broken or someone was killed by a hit-and-run driver. Did the tiny particles of glass found in the clothes of a suspect originate from the location of the crime or did they come from somewhere else? In Chapter 3, we discussed how density is an intensive property that helps connect two fragments as originating from the same pane of glass. Two other measurements can also be important evidence. The elements and their relative abundances present in the glass as well as the refractive index of the glass are also significant clues. First, consider the elements present in a fragment of glass or a paint sample.

As we mentioned, every element has its characteristic line spectrum that is as specific to the element as a fingerprint is to an individual. A sample of glass or paint can be analyzed in a device called an *emission spectrograph*. The instrument contains a source of intense heat, on the order of 7000 to 10,000°C, into which the sample is introduced and vaporized. The emitted light is focused by a lens through a glass prism, which breaks the light into its component lines. (See Figure 8-4.) These are then recorded on a photographic plate. Since many elements are present, a large number of lines result. The process is used more to compare samples rather than to determine exactly what elements are present. If two glass or paint samples match, it is strong evidence that they originated from the same source.

An intensive physical property known as the *refractive index* is also used to compare glass samples. It is the ratio of the velocity of a specific wavelength of light in a vacuum divided by the velocity in the medium. For example, a refractive index of 1.25 means that the light travels 25% faster in a vacuum than through the medium in question. Perhaps you have noticed this effect when you viewed something at the bottom of a container of water. When you reach in for the object, it is not where your eyes tell you it is. This is because the light waves travel slower and have been bent as they pass through the water.

Forensic scientists also use refractive index to compare glass fragments from a crime scene. The actual refractive index of a piece of glass can be hard to measure directly, so it is done by placing the glass in a medium such as a silicone oil. When a transparent solid such as glass is immersed in a liquid, the boundaries of the glass are apparent because of the difference in refractive index of the glass and the liquid. If the oil is heated, its refractive index changes by a known amount, but the refractive index of the solid glass does not change. When the oil and the glass have the same refractive index, the boundaries of the glass are not visible and the glass seems to disappear. At this point, the refractive index of the glass can be determined by reference to a table listing the refractive indexes of the liquid as a function of temperature. This all used to be done visually but now it is carried out automatically with the help of a computer.

When the density and refractive index of two glass samples are very close together, it serves as admissible evidence in court. However, a conviction usually depends on additional evidence, such as hair, fibers, and fingerprints, and, of course, DNA, that can help connect a suspect to a crime scene.

8-2 Modern Atomic Theory: A Closer Look at Energy Levels

▶ **OBJECTIVE FOR SECTION 8-2**
Describe the electronic structure of an atom, including shells, subshells, and various types of orbitals.

LOOKING AHEAD! Bohr's model is still used in the popular press as a convenient, simple way to represent the atom, with electrons as particles circling a nucleus. In the modern scientific view of the atom, however, the characteristics of the electron are much more complex. Therefore, a more sophisticated view of the electron has evolved. In this section, we will examine the results of the modern view of the electron in the hydrogen atom. ■

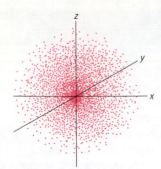

FIGURE 8-8 Electron Density The electron density of the ground state of hydrogen is spherical.

8-2.1 The Wave Mechanical Model

Soon after Bohr's model was presented, experiments confirmed that the electron has properties both of a particle (mass) *and* of light (wave nature). Because of its dual nature, the electron could not be viewed as a simple particle circling the nucleus at a definite distance. Also, if the electron moved with a high velocity, as Bohr had claimed, one could not know its location with much certainty. The faster a particle moves, the less we would know about its location. This is analogous to a fastball thrown by a professional baseball pitcher. The faster the ball is thrown, the less the batter knows about its location and the less likely that the ball will be hit. *The wave nature of the electron and the uncertainty of its location led to a complex mathematical approach to the electron in the hydrogen atom, known as* **wave mechanics**. Using the equations of the wave mechanical model, only the probability of finding the electron in a given region of space at a certain instant could be determined. Bohr calculated the exact radius of the ground-state energy level for the electron in hydrogen. In the wave mechanical model, this radius represents the highest *probability* of finding the electron. We no longer view the electron as orbiting in a specific path around the nucleus. In fact, the motion of an electron in an atom in this model is not understood.

Locating the electron in the vicinity of the nucleus is like trying to locate a fast-moving dancer on a stage with a strobe light. With each flash of the strobe, we find the dancer in a different location. However, if we marked the dancer's position at each flash, eventually we would probably see a pattern and notice that she has a higher probability of being in certain locations on the floor than others. For example, the exact center of the stage would have the highest probability and the wings would have a low probability. A similar situation would occur if we were to try to measure the position of the electron in a hydrogen atom. That is, each time we would locate the electron, it would most likely be in a different location than before. However, if we took enough measurements, a pattern would eventually develop. For example, in Figure 8-8 a typical pattern of electron probability is shown. The figure includes x, y, and z coordinates to emphasize the three-dimensional nature of the points. Notice that this pattern suggests that the probability of finding the electron, known as the *electron density*, is in a spherical region of space around the nucleus for the ground state of hydrogen.

8-2.2 The First Shell and s Orbitals

The region of space where there is the significant probability of finding a particular electron is known as an **orbital**. (Note that this is a different concept from Bohr's orbits.) An orbital is sometimes viewed as an "electron cloud," as if the electron were spread out in a volume of space like a cloud. In some places the cloud is thick and in others it is thin. It is difficult to represent an orbital either as a huge number of dots or as a cloud, so, for convenience, we illustrate the shape of this orbital as a solid sphere, as shown in Figure 8-9a. This sphere is understood to represent a volume of space containing about 90% of the electron density. (We do not envision the electron as moving around on the surface of this sphere.) *A spherical volume of probability (the electron cloud) is known as an* **s orbital**. The energy levels, or orbits, that Bohr described do have meaning in our modern approach but are now known as **shells**. We will refer to the energy levels as shells, which may contain one or more orbitals. The first shell in hydrogen ($n = 1$), which is known as the ground state of hydrogen because it has the lowest energy, contains only the one spherical s orbital. Normally, that's the state occupied by the hydrogen's single electron.

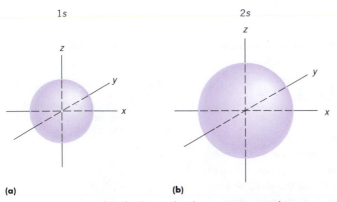

FIGURE 8-9 s Orbitals The s orbitals are represented as spheres. The 2s orbital (b) extends farther from the nucleus than the 1s orbital (a).

8-2.3 The Second Shell and *p* Orbitals

Now let's look at the first excited state of hydrogen, which is the second shell (i.e., $n = 2$). The second shell is normally unoccupied, just like vacant rooms on the second floor of a home. We find in this shell that there are two kinds of orbitals. First, there is a spherical orbital (an *s* orbital) much like the one in the first level except that its maximum electron density (probability) is farther from the nucleus than the *s* orbital in the first shell. In other words, it is a bigger shell. (See Figure 8-9b.)

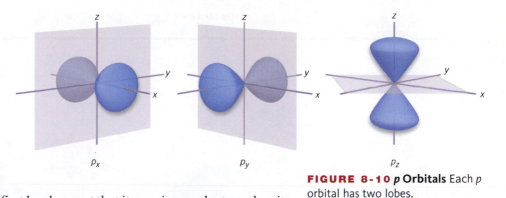

FIGURE 8-10 *p* Orbitals Each *p* orbital has two lobes.

The second kind of orbital present in the second shell is known as a ***p* orbital** *and has two regions of high probability, called lobes, on either side of the nucleus.* The electron distribution in a *p* orbital is not spherical but is shaped much like a weird baseball bat with two fat ends. The difference between an *s* and a *p* orbital is analogous to dorm rooms that have different dimensions or shapes. The first-floor rooms may be perfectly square, but the second-floor rooms may be rectangular. In fact, there are three distinct *p* orbitals lying at 90° angles to each other. In a three-coordinate graph, if we define *x* as the axis along which the two lobes of a *p* orbital are directed, we can designate that *p* orbital as the p_x, orbital. The other two *p* orbitals are each at 90° angles to the p_x orbital and each other. They are referred to as the p_y and the p_z orbitals. The surfaces that represent about 90% of the electron density of the *p* orbitals are shown in Figure 8-10. Again, we do not visualize the electron as a particle moving about the surface of this orbital representation. Instead, for this purpose, it is more useful to picture electrons as clouds occupying a region of space. The second shell has four regions of space that may contain electrons—one *s* orbital and three *p* orbitals. *The orbitals of the same type in each shell make up what is referred to as a* **subshell**. A subshell is labeled with a number corresponding to the shell (1, 2, 3, etc.) and the type of orbital that makes up that subshell (*s, p,* etc.). Thus, the first shell has only a single *s* subshell, which is called the 1*s*. The second shell contains an *s* subshell (the 2*s*) and a *p* subshell (the 2*p*), which is made up of three individual *p* orbitals.

8-2.4 Outer Shells with *d* and *f* Orbitals

Now let's continue outward from the nucleus in the hydrogen atom and consider the third shell (i.e., $n = 3$). Each successive shell is larger and has one additional type of orbital. Thus, the third shell has three types of orbitals, or three subshells. Like the first two, it has a spherical 3*s* orbital and like the second level it has three different 3*p* orbitals. The third type of orbital found in the third shell is known as a ***d* orbital.** There are five *d* orbitals, and their shape is more complex. Most of them have four lobes, as illustrated in Figure 8-11. Three of these *d* orbitals have significant electron density between the axes (d_{xy}, d_{xz}, d_{yz}) and two have significant density along the axes $d_{x^2-y^2}$, d_{z^2}. The third shell with its three subshells and nine orbitals is summarized in Figure 8-12.

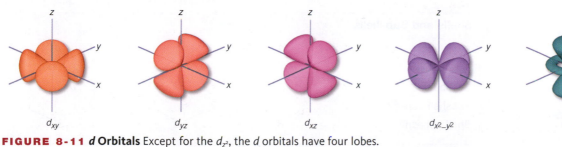

FIGURE 8-11 *d* Orbitals Except for the d_{z^2}, the *d* orbitals have four lobes.

The third shell
(n = 3)

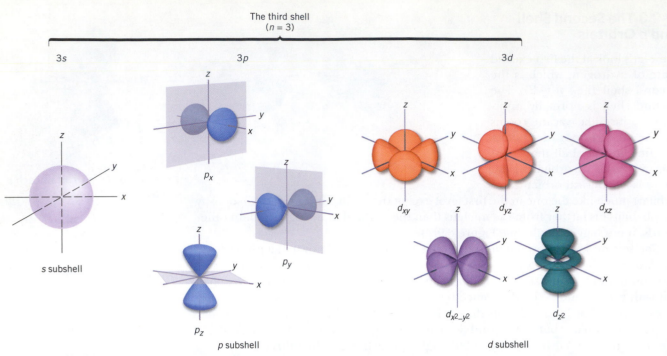

3s

3p

3d

s subshell

p_x

p_y

p_z

d_{xy}

d_{yz}

d_{xz}

d_{x2_y2}

d_{z2}

p subshell

d subshell

FIGURE 8-12 The Third Shell There are three subshells containing a total of nine orbitals in the third shell.

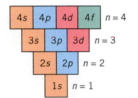

FIGURE 8-13 Subshells Each successive shell has one additional subshell.

The fourth shell ($n = 4$) has all of the same types of orbitals as the third shell (i.e., one 4s, three 4p, and five 4d orbitals) and one additional type known as an **f orbital**. The shapes of the seven 4f orbitals are even more complex than the d orbitals. They are of little consequence to us, since electrons occupying these orbitals are rarely involved in bonding, as we will see in the next chapter. The subshells present in the first four shells are shown in Figure 8-13. Just as successive shells have higher energy, successive subshells also have higher energy. (All the orbitals in a specific subshell have the same energy.) In atoms with more than one electron, the energy of the subshells *within a shell* is not the same and increases in the order

$$s < p < d < f$$
increasing energy

We can now adjust our representation of the shells and subshells to account for the difference in energy of the subshells within a shell. In Figure 8-14 we show the first four shells with their subshells in order of increasing energy. Notice that the 4s energy level appears to be lower in energy than the 3d. This is not a mistake, and we will refer back to this situation in the next section.

The capacity of electrons in a subshell is determined by the number of orbitals in that particular subshell. *Each orbital can hold two electrons.* Thus, the capacity of an s subshell is 2, a p subshell is 6 (two electrons in each of three orbitals), a d subshell is 10, and an f subshell is 14. This information is summarized in Table 8-1. The total

TABLE 8-1

Shells and Subshells

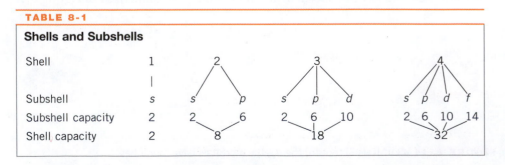

Shell	1	2		3			4			
Subshell	s	s	p	s	p	d	s	p	d	f
Subshell capacity	2	2	6	2	6	10	2	6	10	14
Shell capacity	2		8		18				32	

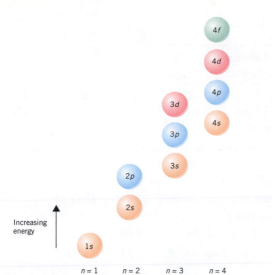

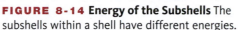

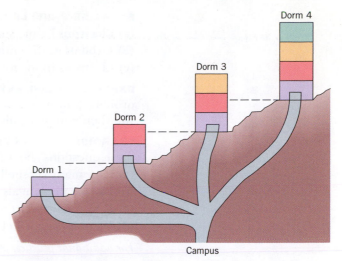

FIGURE 8-14 Energy of the Subshells The subshells within a shell have different energies.

FIGURE 8-15 Floors within a Dorm The floors represent different subdivisions of energy within each dorm, analogous to the subshells within each shell.

electron capacity of a shell, including all the orbitals in all the subshells, works out to be conveniently equal to $2n^2$. Thus, the first shell ($n = 1$) holds 2 electrons, the second shell ($n = 2$) holds 8, the third shell holds 18, and so on.

Consider an analogy to shells and subshells. An orbital is analogous to a dorm room, which is a region of space with the highest probability of finding a particular student. That is, the student spends more time there than in any other single place. Assume that a dormitory is analogous to a shell and the floors within that dorm as analogous to subshells. We will assume that there are four dorms, with each dorm located farther and farther up a hill. The dorm lowest on the hill is the lowest in energy, since it is the easiest to reach. In Figure 8-15 the dorms are labeled 1 through 4. As we go to higher-energy dorms up the hill, they become larger, with each having one additional floor. (Recall that each successive shell has one additional subshell.) Just as the dorms are at different energy levels, so are the floors within a dorm. (In real-world dorms, students are quite aware of the difference in energy of the floors, since the elevator is often out of order or wedged open somewhere.) Obviously, the lower the floor in a specific dorm, the lower the energy. In summary, individual dorms are analogous to specific shells that represent different energy levels and distances from the nucleus in an atom. Floors within a dorm are analogous to subshells that represent different energy levels *within* the shell. Rooms on a floor are analogous to orbitals. All rooms on the same floor have the same energy, as do all orbitals in the same subshell in an atom. We'll continue with this analogy later when we include electrons in our model.

▶ **ASSESSING THE OBJECTIVE FOR SECTION 8-2**

EXERCISE 8-2(a) LEVEL 1: Fill in the blanks.

Wave mechanics tell us that the electron exists in a region of space known as an _____. The four types of these regions are labeled _____, _____, _____, and _____. The lowest in energy of these four has a _____ shape. A principal energy level, called a _____, can be subdivided into _____. Such a subdivision contains all of the same type of _____ in that particular energy level.

EXERCISE 8-2(b) LEVEL 2: Evaluate the $n = 3$ shell.
(a) What is its capacity (how many electrons can be held)?
(b) What subshells are present?
(c) What are the capacities of each subshell?
(d) How many orbitals are present in each subshell?

EXERCISE 8-2(c) LEVEL 2: How many . . .

(a) Electrons in an *f* subshell?

(b) Orbitals in the entire third shell?

(c) Electrons in the fourth shell?

EXERCISE 8-2(d) LEVEL 3: Based on the descriptions of the other types of orbitals, how many lobes would you predict an *f*-type orbital to have? How might you then symmetrically arrange that?

EXERCISE 8-2(e) LEVEL 3: Use the patterns established in this section to answer these questions about the fifth shell of an atom.

(a) How many subshells should it have?

(b) How many orbitals are there in the *g* subshell (the one after *f*)?

(c) How many electrons should the *g* subshell hold?

(d) What is the total electron capacity of the fifth shell?

For additional practice, work chapter problems 8-6, 8-12, 8-13, and 8-15.

PART A SUMMARY

KEY TERMS

8-1.1	The energy of **electromagnetic radiation** is proportional to its **wavelength**. p. 240
8-1.2	The **continuous spectrum** in the visible region contains all colors of the rainbow. p. 240
8-1.3	Hot, gaseous elements give off a **discrete** or **line spectrum**. p. 242
8-1.4	The **Bohr model** of the hydrogen atom describes the location of the electron in a discrete energy level. p. 243
8-1.5	In the Bohr model, the **energy level** of the electron is **quantized** and indicated by a **principal quantum number**. p. 243
8-1.6	The electron in the hydrogen atom is either in the **ground state** or higher **excited states**. p. 244
8-2.1	In **wave mechanics** the energy of the electron is described in terms of probability. p. 246
8-2.2	The electron in an atom exists in a region of space called an **orbital**. The first **shell** contains only a single **s orbital**. p. 246
8-2.3	The second shell also contains three **p orbitals**, which constitute a **subshell**. p. 247
8-2.4	The third shell also contains a **d orbital** and the fourth shell also contains an **f orbital**. pp. 247–248

SUMMARY CHART

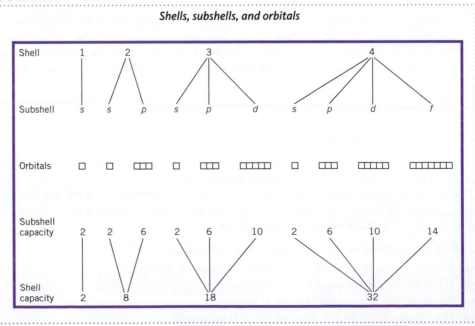

Shells, subshells, and orbitals

Part B

The Periodic Table and Electron Configuration

SETTING A GOAL

■ You will have a detailed understanding of the structure of electrons in atoms and how this affects that element's properties.

OBJECTIVES

8-3 Using the periodic table, write the outer electron configuration of the atoms of a specific element.

8-4 Using orbital diagrams, determine the distribution of electrons in an atom.

8-5 Using the periodic table, predict trends in atomic and ionic radii, ionization energy, and electron affinity.

8-3 Electron Configurations of the Elements

LOOKING AHEAD! Fortunately, all the elements have the same types of orbitals as the hydrogen atom. The assignment of electrons to specific shells and subshells makes the basis of the periodic table apparent. Once this is accepted, the periodic table is all that is needed to establish the electron configuration of any element. ■

▶ **OBJECTIVE FOR SECTION 8-3**
Using the periodic table, write the outer electron configuration of the atoms of a specific element.

8-3.1 The Aufbau Principle and Electron Configuration

In our analogy to the dorm, an orbital is analogous to a room on a floor. A floor may have only one room analogous to an *s* subshell or be composed of several rooms (analogous to *p*, *d*, or *f* subshells). Assigning electrons to orbitals in an atom is much like assigning students to rooms in a dormitory. Students will occupy the available rooms on the floors that require the least energy to get to. *In the case of atoms, electrons occupy the available orbitals in the subshells of lowest energy.* This is known as the **Aufbau principle** (from the German for "building up"). *The assignment of all the electrons in an atom into specific shells and subshells is known as the element's* **electron configuration**. In this section, we will consider only the electron's shell and subshell designations. In the next section, we will also consider the orbital assignment. In the dorm analogy, it is like considering only the dorm and floor assignment but not the specific room on a floor.

8-3.2 Assignment of Electrons into Shells and Subshells

A shell is designated by the principal quantum number (*n*) and the subshell by the appropriate letter. The number of electrons in that subshell can be displayed with an appropriate superscript number. For example, the existence of three electrons in the *p* subshell of the fourth shell (*n* = 4) is shown as follows:

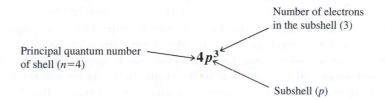

Applying the Aufbau principle, we begin with the ground state of the one electron in the simplest and first element, hydrogen. The electron is in the *n* = 1 shell, which has an *s* subshell. The electron configuration for H is

$$1s^1$$

The next element, helium, has two electrons. The $1s$ subshell has a capacity of two electrons, so He has the configuration

$$1s^2$$

Next comes Li, with three electrons. The first two electrons fill the first shell, but the third must be in the second shell ($n = 2$). The second shell has two subshells, the s and the p, but the s is lower in energy, so it fills first. The next element is Be. Its four electrons must also be in the $1s$ and $2s$ subshells. The electron configurations of Li and Be are

$$\text{Li:} \quad 1s^2 2s^1 \qquad \text{Be:} \quad 1s^2 2s^2$$

The next element is B. Four of its five electrons have the same configuration as Be, but the fifth electron begins the filling of the next subshell, the $2p$. This is analogous to having to proceed to a second-floor room because the first-floor room is full. The next five elements, C through Ne, complete the filling of the $2p$ subshell.

$$\text{B:} \quad 1s^2 2s^2 2p^1 \qquad \text{N:} \quad 1s^2 2s^2 2p^3 \qquad \text{F:} \quad 1s^2 2s^2 2p^5$$
$$\text{C:} \quad 1s^2 2s^2 2p^2 \qquad \text{O:} \quad 1s^2 2s^2 2p^4 \qquad \text{Ne:} \quad 1s^2 2s^2 2p^6$$

With the element neon, all the orbitals in the second shell are full. We now continue with the element sodium, which has 11 electrons. The first 10 electrons fill the first and second shells as before, so the 11th electron is assigned to the third shell. The third shell has three subshells, the $3s$, $3p$, and $3d$. The lowest in energy is the $3s$ subshell, so the 11th electron is in the $3s$ subshell. At this point, it becomes somewhat tedious to write all of the filled subshells. A shorthand method of writing configurations is to represent all the filled subshells of a noble gas by the symbol of that noble gas in brackets (e.g., [Ne]). Thus, in the following electron configurations, we will assume

$$[\text{Ne}] = 1s^2 2s^2 2p^6$$

We chose a noble gas to use in our shorthand notation for a good reason. As we will see in the next chapter, electrons in noble gas cores are particularly stable and not involved in bonding, so we can treat them as a group. It is like a closed club of 10 people where no one drops out and no new members are accepted. Using the [Ne] core symbolism, the electron configurations of the next eight elements after neon are shown as follows.

$$\text{Na:} \quad [\text{Ne}]3s^1 \qquad \text{Si:} \quad [\text{Ne}]3s^2 3p^2 \qquad \text{Cl:} \quad [\text{Ne}]3s^2 3p^5$$

$$\text{Mg:} \quad [\text{Ne}]3s^2 \qquad \text{P:} \quad [\text{Ne}]3s^2 3p^3 \qquad \text{Ar:} \quad [\text{Ne}]3s^2 3p^6$$

$$\text{Al:} \quad [\text{Ne}]3s^2 3p^1 \qquad \text{S:} \quad [\text{Ne}]3s^2 3p^4$$

At this point, it helps to pay close attention to the location of elements in the periodic table. Notice that the electron configuration of Li is similar to that of Na (i.e., $[\text{He}]2s^1$ and $[\text{Ne}]3s^1$), Be to Mg, B to Al, and so forth. Notice also that these pairs of elements are in the same groups in the periodic table. This is very significant. In fact, we can make a general statement for this observation. *The electron configuration of the elements is a periodic property.* Elements in the same group have the same outer subshell electron configurations but in different shells. As we will see in the next chapter, these similar electron configurations are a major reason that the elements in a group have similar chemical properties. The existence of subshells composed of orbitals is a result of our understanding of wave mechanics. By placing electrons in these subshells, we have actually developed the theoretical basis for the periodic table. Now, however, we will continue with the electron configurations of elements beyond argon. Again, we will use the noble gas core shorthand with [Ar] representing the 18 electrons in the orbitals of argon.

The next element after Ar presents a question. The third subshell in the third shell ($3d$) is still empty and available, so at first we may be inclined to assign the 19th electron in K to the $3d$ subshell. However, now refer to the periodic table on the inside back cover. Notice that both Li and Na have an s^1 configuration and K is right under Na. The location of K (under Na) indicates that it should have its outermost electron in the $4s$ subshell. This is indeed the case. Using noble gas shorthand notation, the configuration of K is

$$[\text{Ar}]4s^1$$

To understand how the $4s$ fills before the $3d$, let's return to the analogy of the dorms, as shown in Figure 8-15. Although dorm 4 as a whole is at a higher level than dorm 3, note that not all floors in dorm 3 are lower than those in dorm 4. In fact, it is easier to proceed farther up the hill to occupy the lowest floor in dorm 4 than to go all the way up to the third floor in dorm 3. In the case of the assignment of the electrons in atoms, a similar phenomenon is true. As shown in Figure 8-14, the $4s$ subshell is lower in energy than the $3d$ subshell; thus the $4s$ fills first. After two electrons are assigned to the $4s$, the $3d$ begins filling with the element Sc. The next nine elements after Sc also involve the filling of the $3d$ subshell. After the $3d$ is filled at Zn, the next higher energy subshell is the $4p$. This subshell is completely filled at the next noble gas, Kr.

K:	$[\text{Ar}]4s^1$		Ga:	$[\text{Ar}]4s^23d^{10}4p^1$
Ca:	$[\text{Ar}]4s^2$			$\vdots$
Sc:	$[\text{Ar}]4s^23d^1$		Kr:	$[\text{Ar}]4s^23d^{10}4p^6$
$\vdots$				
Zn:	$[\text{Ar}]4s^13d^{10}$			

8-3.3 Exceptions to the Normal Order of Filling

The electron configurations of many elements are not always exactly as we would predict by following the usual order of filling. For example, there are two notable exceptions among the transition elements when the d subshells are filling. The actual electron configurations of chromium (#24) and copper (#29) are

$$\text{Cr: } [\text{Ar}]4s^13d^5 \qquad \text{Cu: } [\text{Ar}]4s^13d^{10}$$

If the normal order had been followed, we would predict the configuration of Cr to be $[\text{Ar}]4s^23d^4$ and Cu to be $[\text{Ar}]4s^23d^9$. These exceptions are explained on the basis of a particular stability for a half-filled and a completely filled subshell (the $3d$).

There are many other exceptions, especially where d and f subshells are concerned. *We should be aware that such exceptions exist, but it is not important or necessary to know them all.*

8-3.4 Using the Periodic Table to Determine Electron Configuration

In Chapter 4, we first presented the periodic table as a valuable tool in locating periods and groups, gases and solids, nonmetals and metals. We have now developed the theoretical basis of the periodic table with the concept of electron configuration. It is appropriate, then, that we use the periodic table itself as a guide to the electron configuration of an element. If we superimpose electron configurations onto the periodic table, the periodic relationship becomes apparent.

We can make some generalizations that may help. The period number is the same as the shell number for the outermost s and p electrons in that atom. The transition metals are characterized by the filling of d subshells. They appear one number down from their period number (i.e., the $4d$ subshell is in the fifth period). The lanthanides and actinides (inner transition elements) correspond to the f orbitals. The f subshells appear two numbers down from their period number (i.e., the $4f$ subshell is in the 6th period).

In Figure 8-16, we show the periodic table with the subshell configuration common to a group at the top of each vertical column. The specific subshell representation is shown within each period. The electron configurations of the sub-shell that is in the process of filling are shown in more detail for the fourth period from K to Ar. With some practice using a periodic table along with Figure 8-16, we can predict the expected outer electron configuration of any element.

8-3.5 Electron Configurations of Specific Groups

As we will see in the next chapter, the electrons beyond the previous noble gas core are the ones commonly involved in the formation of chemical bonds. It will be these electrons that we will be mostly concerned with. To focus on these outer electrons, we use the noble gas shorthand in expressing electron configurations. In the following discussion, we designate a general noble gas configuration as [NG] and a general shell as n. For example, $[NG]ns^1$ means a noble gas configuration followed by one electron in the n shell and the s subshell.

Representative or Main Group Elements: $[NG]ns^x np^y$

These are the elements shown in blue in Figure 8-16. Four of the groups within this category have family names. The general electron configurations of each group are as follows.

IA (1): $[NG]ns^1$ Except for hydrogen, these are the *alkali metals*. All have one electron beyond a noble gas configuration. *Notice that the numbering for the s subshell begins at 1 for hydrogen and is consecutive down the table.*

IIA (2): $[NG]ns^2$ These are the *alkaline earth metals*. All have two electrons beyond a noble gas. The numbering begins at 2 in this column with Be. He ($1s^2$) is located to the far right with the noble gases.

IIIA (13): $[NG]ns^2 np^1$ Notice that boron (B) is the first element with an electron in a p subshell. *The numbering of the p subshells begins with 2 p at B and continues consecutively down the table.* The elements Ga (#31) and In (#49), as well as all of those to the right of these elements, also have filled d subshells beyond the previous noble gas. Tl (#81) and the elements directly to the right

IA s^1	IIA											IIIA p^1	IVA p^2	VA p^3	VIA p^4	VIIA p^5	VIIIA p^6
1s	s^2																$1s^2$
2s	2s											2p	2p	2p	2p	2p	2p
3s	3s	IIIB d^1	IVB d^2	VB d^3	VIB d^{5*}	VIIB d^5	d^6	d^7	d^8	IB d^{10*}	IIB d^{10}	3p	3p	3p	3p	3p	3p
19 K $4s^1$	20 Ca $4s^2$	21 Sc $3d^1$	22 Ti $3d^2$	23 V $3d^3$	24 Cr $4s^1 3d^5$	25 Mn $3d^5$	26 Fe $3d^6$	27 Co $3d^7$	28 Ni $3d^8$	29 Cu $4s^1 3d^{10}$	30 Zn $3d^{10}$	31 Ga $4p^1$	32 Ge $4p^2$	33 As $4p^3$	34 Se $4p^4$	35 Br $4p^5$	36 Kr $4p^6$
5s	5s	4d	4d	4d	4d	4d	4d	4d	4d	4d	4d	5p	5p	5p	5p	5p	5p
6s	6s	5d	5d	5d	5d	5d	5d	5d	5d	5d	5d	6p	6p	6p	6p	6p	6p
7s	7s	6d	6d	6d	6d	6d	6d	6d	6d	6d	6d	7p	7z	7	7	7	7 id

	f^1	f^2	f^3	f^4	f^5	f^6	f^7	f^8	f^9	f^{10}	f^{11}	f^{12}	f^{13}	f^{14}
	4f	4f	4f	4f	4f	4f	4f	4f	4f	4f	4f	4f	4f	4f
	5f	5f	5f	5f	5f	5f	5f	5f	5f	5f	5f	5f	5f	5f

*These groups have s^1 configuration rather than s^2.

FIGURE 8-16 Electron Configuration and the Periodic Table The electron configuration of an element can be determined from its position in the periodic table. The value of n shown in each box is the shell of the outermost subshell.

of Tl also have filled $4f$ subshells as well as a filled $6d$ subshell beyond the noble gas. The electron configurations of In (#49) and Tl (#81) are shown below. So as to emphasize the outermost shell, the *filled* inner subshells are sometimes listed first in order of n, the shell number.

$$\text{In: } [\text{Kr}]4d^{10}5s^25p^1 \qquad \text{Tl: } [\text{Xe}]4f^{14}5d^{10}6s^26p^1$$

IVA (14): [NG]ns^2np^2

VA (15): [NG]ns^2np^3

VIA (16): [NG]ns^2np^4   These elements are sometimes referred to as the *chalcogens*, but this title is rarely used anymore. They are all nonmetals except for Po.

VIIA (17): [NG]ns^2np^5   These elements are known as the *halogens*. The halogens are all nonmetals and one electron short of having a noble gas configuration.

VIIIA (18): Noble Gases: [NG]ns^2np^6

This category of elements is shown in green in Figure 8-16. These elements are so named because they rarely form chemical bonds. They are characterized by filled outer s and p subshells. Helium has only $1s^2$.

Transition Metals: [NG]$ns^2(n-1)d^x$

These 40 elements are shown in yellow in Figure 8-16. The element scandium (#21) is the first element with an electron in a d subshell, which is in the third shell. *Therefore, the numbering for the d subshell begins at 3.* Thus Sc has a $4s^23d^1$ configuration, and the next element in Group IIIB has a $5s^24d^1$ configuration, and so forth down the group. As mentioned previously, there are many exceptions to the expected order of filling in the transition metals, but two important examples involve the VIB (6) and IB (11) elements. For example, Ag (#47) has the configuration $[\text{Kr}] 5s^14d^{10}$. The $6d$ series of elements are not naturally occurring but have been synthesized in various laboratories. In most cases, only a few atoms of each element have been produced. Still, there is enough evidence to indicate that these elements belong in their appropriate positions in the periodic table.

Inner Transition Metals: [NG]$ns^2(n-1)d^1(n-2)f^x$

These elements are shown in orange in Figure 8-16. The orange bar in the transition metals indicates where they would fit in if the periodic table were extended fully to include these elements. *These elements involve the filling of the* f *subshell, which begins with shell 4.* There are two series of these elements: the *lanthanides*, which have the general configuration $[\text{Xe}]6s^25d^14f^x$, and the *actinides*, which have the configuration $[\text{Rn}]7s^26d^15f^x$.

The filling of the f orbitals is somewhat unusual and needs elaboration. Notice in Figure 8-3 that the 57th electron of La and the 89th electron of Ac go into the $5d$ and $6d$ subshells, respectively, rather than the $4f$ and $5f$. Their correct configurations are

$$\text{La: } [\text{Xe}]6s^25d^1 \qquad \text{Ac: } [\text{Rn}]7s^26d^1$$

Thus, these two elements are shown in the periodic table in Group IIIB under Sc and Y, which also have one electron in a d subshell. Notice that the next two elements after La and Ac are shown as inner transition elements, however, which are characterized by the filling of the $4f$ and $5f$ subshells, respectively. These two electron configurations can be written as

$$\text{Ce: } [\text{Xe}]6s^25d^14f^1 \qquad \text{Th: } [\text{Rn}]7s^26d^15f^1$$

In fact, the configurations of these elements may not be exactly as shown here because the outer d and f subshells are very close in energies. This leads to various deviations from an orderly filling of the f subshells. In any case, the elements Lu and Lr complete the filling of the $4f$ and $5f$ subshells. The next elements after these, Hf and Rf, return to the $5d$ and $6d$ subshells, respectively. If we use the periodic table to determine electron configuration and follow the atomic numbers faithfully, we don't have to remember the exceptions relating to d versus f orbitals.

EXAMPLE 8-1   Writing Complete Electron Configuration

Write the complete electron configuration of iron.

PROCEDURE

Use the periodic table as a reference. Note the atomic number of iron and build up to there. Start at the beginning of the table and write the subshells in the order of filling plus a running summation of the total number of electrons involved with respect to iron.

SOLUTION

From the table inside the front cover, we find that iron has an atomic number of 26, which means that the neutral atom has 26 electrons.

SUBSHELL	NUMBER OF ELECTRONS IN SUBSHELL	TOTAL NUMBER OF ELECTRONS
$1s$	2	2
$2s$	2	4
$2p$	6	10
$3s$	2	12
$3p$	6	18 [Ar]
$4s$	2	20
$3d$	6	26

Iron has 6 electrons past the filled $4s$ subshell, but they do not completely fill the $3d$ subshell, which could accommodate 10 electrons. The complete electron configuration of iron is

$$1s^2 2s^2 2p^6 3s^2 3p^6 4s^2 3d^6$$

ANALYSIS

Full electron configurations are tedious and repetitive; they need to be done only by students beginning their study of electron configurations. Once you understand the general principles and patterns behind configurations, noble gas shorthand notations are all that's required to get a complete analysis of the electronic structure of elements. That's because they distinguish between *core* electrons, which are not involved in chemical bonding and are incorporated into the noble gas symbol, and the electrons that are the ones listed beyond the noble gas and have a profound effect on the chemistry of that particular element.

SYNTHESIS

Based on this configuration, can we now understand why iron favors formation of +2 and +3 ions? If you had to remove two electrons to make a +2 cation, which two would you remove? Everything up to the $3p$ is a core electron and so shouldn't be affected. That leaves 2 electrons in the $4s$ and 6 in the $3d$. Removing the two electrons from the $4s$ gives us a configuration that looks closer to a noble gas than taking only two of the six from the $3d$, so that's where these two electrons come from. For the +3 ion, an additional electron needs to be removed, this time from the $3d$, leaving 5. This, too, is advantageous, as 5 is half-filled and has some additional stability. Many charges of nonrepresentative elements that were previously a mystery become more understandable when viewed in reference to electron configurations.

EXAMPLE 8-2   Writing Electron Configurations Using Noble Gas Shorthand

Write the electron configuration for (a) rhodium and (b) bismuth using the noble gas shorthand.

PROCEDURE

Using the periodic table, locate the noble gas directly preceding the elements in question. Proceed with the configurations as before.

SOLUTION

(a) Rhodium
   Locate rhodium (#45) in the periodic table on the inside back cover. Now locate Rh in the periodic table shown in Figure 8-16. Rh is the next element below Co, so the highest energy subshell for this element is

the 4d. Counting over from the first 4d element (Y), the configuration is 4d^7. Starting then with the previous noble gas, notice that the 5s fills before the 4d. The configuration becomes

$$[Kr]5s^2 4d^7$$

(b) Bismuth

The electron configuration of bismuth using the noble gas shorthand that begins with the noble gas xenon (with 54 electrons) is

$$[Xe]6s^2 4f^{14} 5d^{10} 6p^3$$

ANALYSIS

The completeness of the examples above is provided as an instructional tool. Most students, after a little practice, can turn out accurate electron configurations with only the periodic table and a pencil. List the noble gas preceding the element in question, and begin with the next s subshell down the column. What you are then listing are the outermost electrons in an atom.

SYNTHESIS

What would the electron configurations of the elements directly below rhodium or directly above bismuth look like? The noble gas core would change, of course. And so would the numbers on the shell designations. They'd advance or lower by 1. But the numbers of electrons in each subshell would be identical for each set, as they would for all other elements in that column. Note the presence of the 4f subshell below rhodium and its absence above bismuth.

Rh: $[Kr]5s^2 4d^7$ 　　　Sb: $[Kr]4d^{10} 5s^2 5p^3$

Ir: $[Xe]4f^{14} 6s^2 5d^7$ 　　　Bi: $[Xe]4f^{14} 5d^{10} 6s^2 6p^3$

Mendeleev originally organized the periodic table so that elements in the same column had similar chemical properties. We are now beginning to see the structural basis for this consistency in reactivity.

EXAMPLE 8-3　　Identifying an Element on the Table from its Electron Configuration

What element has the electron configuration $[Kr]4d^{10} 5s^2 5p^5$?

PROCEDURE

Begin with the noble gas and work across the next period of the table until you reach the fifth element of the p subshell (column VIIA).

SOLUTION

The element after Kr in the p^5 column in Figure 8-16 has atomic number 53, which is iodine.

ANALYSIS

This is a good time to draw specific parallels between the periodic table and the electronic structure of atoms. Electrons in atoms are organized into shells, subshells, and orbitals. On the periodic table, the shells translate, roughly, into rows or periods. There are seven rows on the periodic table, and there are seven shells in the ground state of the largest atoms. The regions of the periodic table, then, represent different subshells. The IA and IIA columns represent the s subshell. The remaining six columns of representative elements fill the p subshell. The ten columns of transition metals represent the ten electrons it takes to fill up a d subshell, and the 14 lanthanides and actinides have their outermost electrons in f subshells. The s and p subshells appear in the row for which they are numbered: the s beginning in the first row and the p in the second row. Due to the increases in energy, the d subshell falls one period and the f subshell falls two. Specifically, elements filling the 3d subshell are found in the fourth period and those filling the 4f subshell are in the sixth period.

SYNTHESIS

We've already worked our way through the 4p (elements 31 through 36). From there the order continues as we've laid it out above: 5s, 4d, 5p, 6s, 4f, 5d, 6p, 7s, 5f, 6d, 7p. It would be a worthwhile exercise to open your text to the periodic table inside the back cover and confirm that you can duplicate that series from the beginning of the table (1s) right through to the known end (7p).

▶ **ASSESSING THE OBJECTIVE FOR SECTION 8-3**

EXERCISE 8-3(a) LEVEL 1: Fill in the blanks.

The electron configuration $4d^5$ indicates that there are _____ electrons in the _____ subshell in the _____ shell.

EXERCISE 8-3(b) LEVEL 2: Using the periodic table only, give the electron configurations for the elements Zn (#30) and W (#74).

EXERCISE 8-3(c) LEVEL 2: Using the periodic table only, give the symbol of the element with the electron configuration $[Ar]4s^2 3d^{10} 4p^3$

EXERCISE 8-3(d) LEVEL 3: Give the general electron configurations for the elements in Groups VIA and VIB.

For additional practice, work chapter problems 8-23, 8-25, 8-30, 8-32, 8-33, 8-34, and 8-35.

▶ **OBJECTIVE FOR SECTION 8-4**
Using orbital diagrams, determine the distribution of electrons in an atom.

8-4 Orbital Diagrams of the Elements

LOOKING AHEAD! The number of bonds an atom can be involved in sometimes relates to how many orbitals contain a single electron. In this section, we will take a closer look at how electrons exist within a specific subshell. ■

Recall that subshells are composed of one or more orbitals, which are regions of space in which electrons reside. Each orbital can hold a maximum of two electrons. The orbitals in a subshell are analogous to the rooms on a certain floor. The dorm and the floor within the dorm are analogous to the shell and subshell. (See Figure 8-17.) In the following scheme, we will represent an individual orbital by a box. Therefore, an *s* subshell will have one box, and the three orbitals in a *p* subshell will be represented by three boxes. Likewise, a *d* subshell will be represented by five boxes and an *f* subshell by seven. Individual electrons will be represented by arrows.

8-4.1 The Pauli Exclusion Principle and Orbital Diagrams

As mentioned earlier, an electron has a dual nature. In some respects it has properties of a wave, and in other respects it has properties of a particle. One particle property is that the electron behaves like a charged particle, spinning in either a clockwise or a counterclockwise direction. A spinning charged particle is like a tiny magnet. What if there are two electrons in the same orbital? *The* **Pauli exclusion principle** *states that no two electrons in the same orbital can have the same spin.*

We will represent the electrons in orbitals by means of orbital diagrams. *The* **orbital diagram** *of an element represents the orbitals in a subshell as boxes and its electrons as arrows. The spin of an electron is indicated by the direction of the arrow, pointing either up or down. Two electrons with opposite spins in the same orbital are said to be* paired. Thus, a doubly occupied orbital is represented as follows:

FIGURE 8-17 Dorm Rooms and Orbitals The rooms represent different regions of space on each floor of a dorm, analogous to the orbitals of each subshell.

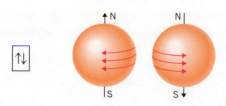

We will now expand on the electron configuration of the first five elements by including their orbital diagrams.

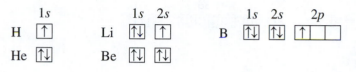

8-4.2 Hund's Rule

Before going on, we need to consider the placement of the sixth electron in carbon. There are three possibilities. Does it pair with the first $2p$ electron in the same orbital or have an opposite spin in a different orbital, or does it go into a different orbital with the same spin? We have one more rule to guide us. **Hund's rule** *states that electrons occupy separate orbitals in the same subshell with parallel (i.e., identical) spins.* At least part of this rule is understandable. Since electrons have the same charge, they will repel each other to different regions of space. Electrons "want their space," so they prefer separate orbitals rather than pairing in the same orbital. This is analogous to students wanting their own room rather than having to tolerate a roommate. Pairing occurs when separate empty orbitals in the same subshell are not available. With Hund's rule in mind, we can now write the orbital diagrams of the next five elements.

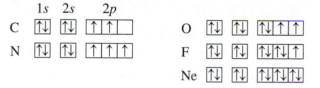

A similar phenomenon occurs with elements that have electrons occupying d or f orbitals. For example, Mn (#25) has the following electron configuration and orbital diagram.

$$\text{Mn: } [\text{Ar}]4s^2 3d^5 \qquad [\text{Ar}]\ \boxed{\uparrow\downarrow}\ \boxed{\uparrow\ \uparrow\ \uparrow\ \uparrow\ \uparrow}$$

Although we will not pursue the topic in this text, orbital diagrams give us important information about the magnetic properties of elements. Orbital diagrams are also relevant to the types of bonds that a particular element forms.

EXAMPLE 8-4 **Writing Orbital Diagrams**

Determine the electron configuration and write the orbital diagram of the electrons beyond the noble gas core for the elements S, Ni, and As.

PROCEDURE

Use Figure 8-16 to determine the electron configuration of the element as in Example 8-2. Identify the type of subshells that exist beyond the noble gas core for each element. Draw the number of boxes indicated for each subshell (1 for *s*, 3 for *p*, 5 for *d*, 7 for *f*). Fill up all boxes with paired arrows until the outermost subshell. Apply Hund's rule for the remaining electrons in the outermost shell.

SOLUTION

S: $[Ne]3s^23p^4$ [Ne] 3s $\uparrow\downarrow$ 3p $\uparrow\downarrow$$\uparrow$$\uparrow$

Ni: $[Ar]4s^23d^8$ [Ar] 4s $\uparrow\downarrow$ 3d $\uparrow\downarrow$$\uparrow\downarrow$$\uparrow\downarrow$$\uparrow$$\uparrow$

As: $[Ar]4s^23d^{10}4p^3$ [Ar] 4s $\uparrow\downarrow$ 3d $\uparrow\downarrow$$\uparrow\downarrow$$\uparrow\downarrow$$\uparrow\downarrow$$\uparrow\downarrow$ 4p $\uparrow$$\uparrow$$\uparrow$

ANALYSIS

These orbital diagrams give us an intuitive feel for why filled and half-filled orbitals have an energetic stability associated with them. It's as if we wouldn't want to disturb the nice symmetrical arrangement that nature has provided. Chemical reactivity involves adjustments in electron configurations. More than likely, those electrons will come from subshells that have yet to achieve a stable configuration.

SYNTHESIS

We will find out later how bonds are able to form by using electrons in half-filled orbitals. How many bonds would we predict oxygen might form? The orbital diagram for oxygen:

2s $\uparrow\downarrow$ 2p $\uparrow\downarrow$$\uparrow$$\uparrow$

indicates that there is the opportunity for two bonds to form, one to each unpaired electron. How many would you predict for hydrogen? Answer: just one possible bond. We will deal with bond formation in detail in the next chapter.

▶ **ASSESSING THE OBJECTIVE FOR SECTION 8-4**

EXERCISE 8-4(a) LEVEL 1: What is the name of each of the following rules/principles?
(a) The spin of the electrons in the same orbital must be paired.
(b) Electrons go into separate orbitals with parallel spins.
(c) Fill orbitals starting with the lowest energy orbital first.

EXERCISE 8-4(b) LEVEL 2: Show the electron configuration and orbital diagram of all of the electrons beyond the previous noble gas for Nb (#41).

EXERCISE 8-4(c) LEVEL 2: Using orbital diagrams, determine the number of unpaired electrons in:
(a) Te   **(b)** Sr   **(c)** Ir

EXERCISE 8-4(d) LEVEL 3: Carbon is known to form four bonds. Bonds form when half-filled orbitals overlap. Draw the orbital diagram for carbon. How will it be possible for carbon to make four bonds?

For additional practice, work chapter problems 8-60, 8-61, and 8-64.

▶ **OBJECTIVE FOR SECTION 8-5**

Using the periodic table, predict trends in atomic and ionic radii, ionization energy, and electron affinity.

8-5 Periodic Trends

LOOKING AHEAD! Besides electron configuration, there are other characteristics of the atoms of an element that follow trends in the periodic table. These properties relate directly to the chemical properties of elements that we will pursue in the next chapter. The first characteristic that we will consider in this section is the size of neutral atoms. ■

We have seen that electron configurations correspond to the position of an element in the periodic table. But there is much more. The number and types of chemical bonds that an element forms are perhaps the most fundamental properties that the periodic table can help us predict and understand. The basis for these trends, however, lies first in the size, or radius, of the atoms of an element and then in other considerations that relate directly to the atomic size.

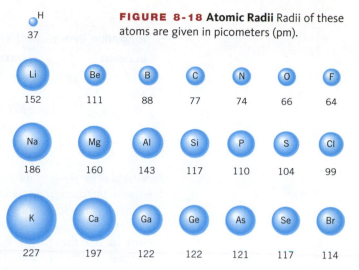

FIGURE 8-18 Atomic Radii Radii of these atoms are given in picometers (pm).

8-5.1 Atomic Radius

*The **radius** of an atom is the distance from the nucleus to the outermost electrons.* It is not an easy task to measure the radius of an atom. There are both experimental and theoretical problems (e.g., electrons are not at a fixed distance from the nucleus). Despite all these difficulties, consistent values for the radii of neutral atoms have been compiled. Some generally accepted values are shown in Figure 8-18 for the first three rows of representative elements. Two units of measurement are most often used for such small distances—the *nanometer* (1 nm = 10^{-9} m) and the *picometer* (1 pm = 10^{-12} m). We will use picometers because the numbers are easier to express and compare (e.g., 37 pm rather than 0.037 nm).

Notice that there is a general decrease in the radii from left to right across representative element groups and an increase in radii down a group. Not surprisingly, the radii of atoms increase down a group simply because the outermost electron is in a shell farther from the nucleus. On the other hand, it may seem surprising that the radii *decrease* from Li to F and from Na to Cl, even though the atoms are heavier, since they have more electrons and nucleons as we move to the right. Since electrons are being added to the same outer subshell as we move from Li to F (i.e., the 2*p*), one may at first predict that the size of the atoms would not change. However, an additional proton (positive charge) is also being added to each successive nucleus. This increased positive charge increases the attraction between the nucleus and all the negatively charged electrons. As a result, all the outer electrons are drawn in tighter around the nucleus and we observe a decrease in radii of the atoms of successive elements as a subshell is filling.

These same trends are also demonstrated in the transition metals. The radius of Sc (#21), with one 3*d* electron, is 144 pm, and the radius of Zn (#30), with ten 3*d* electrons (and nine more protons), is 125 pm.

8-5.2 Ionization Energy

The energy required to form a cation from an atom is an important periodic property. **Ionization energy (I.E.)** *is the energy required to remove an electron from a gaseous atom to form a gaseous ion.* Since the outermost electron is generally the least firmly attached, it will be the first to go.

$$M(g) \longrightarrow M^+(g) + e^-$$

$M(g)$, a gaseous atom, forms a gaseous cation and an electron.

Because the electrons are held in the atom by electrostatic attractions for the nucleus, it requires energy (an *endothermic* process) to remove an electron. The ionization energy generally increases across a period but decreases down a group. The ionization energies for the second and third periods are shown in Table 8-2. The energy unit is abbreviated kJ/mol, which stands for kilojoules per mole, the energy for a defined quantity of atoms. Notice that the trends in ionization energy follow from the discussion of atomic radius. That is, the smaller the atom, the harder it is to remove an electron. This makes sense according to Coulomb's law. **Coulomb's law**

TABLE 8-2

Ionization Energy of Some Elements

ELEMENT	I.E. (KJ/MOL)
Li	520
Be	900
B	801
C	1086
N	1402
O	1314
F	1681
Na	496
Mg	738
Al	578
Si	786
P	1102
S	1000
Cl	1251

TABLE 8-3

Ionization Energies (kJ/mol)

ELEMENT	FIRST I.E.	SECOND I.E.	THIRD I.E.	FOURTH I.E.
Na	496	4565	6912	9,540
Mg	738	1450	7732	10,550
Al	577	1816	2744	11,580

states that the forces of attraction increase as the charges increase but decrease as the distance between the changes increases. In smaller atoms, the positive protons and negative electrons are closer together and are thus attracted to each other more strongly.

Notice that ionization energy generally decreases to the left and down. It is no coincidence that this is the same direction as increasing metallic properties. *In fact, the most significant chemical property of metals is that they lose electrons relatively easily to form cations in compounds.* However, some metals have considerably higher ionization energies than others. The chemical reactivity of a metal relates to this energy. For example, the ionization energy of sodium is 496 kJ/mol and of gold is 890 kJ/mol. As a result, sodium is a very reactive metal (i.e., it reacts dramatically with water), yet gold is called the *eternal metal* because of its relative unreactivity. Nonmetals also have very high ionization energies and so do not form positive ions in compounds. Instead they form negative ions, as we will discuss next.

The second ionization energy for an ion involves removal of an electron from a +1 ion to form a +2 ion.

$$M^+ (g) \longrightarrow M^{2+}(g) + e^-$$

In a similar manner, the third ionization energy forms a +3 ion, and so forth. In all cases it becomes increasingly difficult to remove each succeeding electron. The trends in consecutive ionization energies for Na through Al are shown in Table 8-3.

Note that the second I.E. for Na, the third I.E. for Mg, and the fourth I.E. for Al are all very large compared to the preceding number. For example, it takes 2188 kJ (1450 + 738) to remove the first two electrons from Mg (to form Mg^{2+}) but about three times as much energy to remove the third electron (7732 kJ to form Mg^{3+}). Why is there such a big jump? The answer lies in the electron configuration of magnesium, which is $[Ne]3s^2$. The first two electrons are removed from the outer 3s subshell. To form Mg^{3+}, however, the third electron must be removed from the filled inner shell of the neon configuration (the 2p). Because the second shell is closer to the nucleus than the third shell, this is very difficult and requires a large amount of energy. The same reasoning holds for the second electron from Na and the fourth from Al. We will refer to this observation in the next chapter. For now, it is important to note that a filled shell represents a very stable arrangement. Thus, the magnitude of the positive charge that a representative metal can form is limited by the number of electrons beyond a noble gas configuration.

8-5.3 Electron Affinity

Atoms can also gain electrons as well as lose them. *The tendency of a gaseous atom to gain an electron is termed* **electron affinity (E.A.)** (*affinity* means "an attraction for something"). This process is represented symbolically as

$$X(g) + e^- \longrightarrow X^-(g)$$

Whereas ionization energy is always an endothermic process, electron affinity can also be exothermic, meaning a favorable process. In this case, energy is released as an electron is added to an orbital of an atom. The added electron will enter into an empty orbital according to Hund's rule in the same manner as the other electrons in the subshell.

The larger negative values (the *higher* electron affinities) are found to the upper right on the periodic table. This is the direction of nonmetallic behavior. This confirms what we already knew—that the representative element nonmetals have a tendency to form anions. They do so because there are one to three vacancies in their outermost p orbitals. Table 8-4 lists the electron affinities for the second period of the table.

Beryllium and neon show positive electron affinities because they each have a filled subshell ($2s$ and $2p$, respectively), and so they actually have no tendency to add an electron. Nitrogen also has a positive affinity because its $2p$ subshell is half-filled and somewhat stable. An additional electron would disrupt this. Otherwise, there is a general trend to increase electron affinity from left to right, corresponding to the greater number of protons pulling on the electron and the smaller size of the atom. Electron affinity also decreases going down in the periodic table.

TABLE 8-4

Electron Affinities (kJ/mol)

ELEMENT	E.A. (KJ/MOL)
Li	−60
Be	>0
B	−27
C	−122
N	>0
O	−141
F	−328
Ne	>0

8-5.4 Electron Configurations of Ions

In previous chapters, we learned that the charge acquired by ions formed by the representative elements is a periodic property. Metals or nonmetals in the same group formed ions with the same charge. In the last two subsections we found out why. Metals can lose electrons because of low ionization energies while nonmetals (except the noble gases) can gain electrons because of favorable electron affinities. The number of electrons lost or gained depends on the number of electrons in the outer s and p subshells of the representative element. As we will expand upon in the next chapter, these elements tend to form ions that are isoelectronic with a noble gas. **Isoelectronic** *means having the same number of electrons and having the same electron configuration.* Consider the three nonmetals and the three metals before and after the noble gas neon in the periodic table. First consider the three metals, Na, Mg, and Al. These three metals have one, two, and three electrons respectively beyond the noble gas neon. By forming the Na^+, Mg^{2+} and Al^{3+} ions they lose all of their outer electrons and are now isoelectronic with neon. Now consider the nonmetals, N, O, and F. Notice that these three nonmetals are three, two, and one electron respectively short of the neon noble gas configuration. By forming the N^{3-}, O^{2-} and F^- ions the added electrons also make these ions isoelectronic with neon. All six ions have the same electron configuration of neon, which is

$$[\text{He}]\ 2s^2 2p^6$$

This is also a periodic property in that the metals in the same groups and the nonmetals in the same groups also form ions that are isoelectronic with a noble gas. Note that IIIA metals below Al form +3 cations that also have a filled d subshell in addition to the noble gas core. In this case, the filled d subshell can be considered as core electrons. In the next chapter, we will see how this tendency to lose or gain electrons so as to be isoelectronic with a noble gas explains the formulas of many ionic compounds.

Although we will be emphasizing the representative element ions in the next chapter we will mention transition metal cations here. The tendency to be isoelctronic with a noble gas does not generally extend to transition metal cations. An exception is for the IIIB metals, which form +3 ions by losing two outer s electrons and their one d electron to have a noble gas configuration. Otherwise, there is very little periodicity among the transition metal groups. However, all transition metals can lose their outer s electrons to form +1 or +2 ions. They may also lose one or more d electrons to form ions with higher charges. For example the iron atom and its two ions have the following electron configurations.

$$\text{Fe: } [\text{Ar}]4s^2\,3d^6 \qquad \text{Fe}^{2+}\text{: } [\text{Ar}]3d^6 \qquad \text{Fe}^{3+}\text{: } [\text{Ar}]3d^5$$

FIGURE 8-19 Ionic Radii A positive ion is always smaller than the original neutral atom. A negative ion is always larger than the original atom.

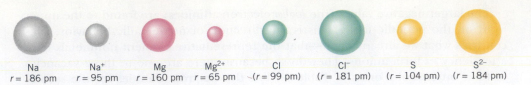

Na	Na$^+$	Mg	Mg^{2+}	Cl	Cl$^-$	S	S^{2-}
r = 186 pm	r = 95 pm	r = 160 pm	r = 65 pm	(r = 99 pm)	(r = 181 pm)	(r = 104 pm)	(r = 184 pm)

8-5.5 Ionic Radii

When an electron is removed from an atom, the resulting cation is smaller than the parent atom. The loss of an electron from an outer subshell results in an increased attraction between the nucleus and the remaining electrons. As a result, the loss of electrons results in a contraction. Also, if the electrons in the outer shell are removed, the remaining inner shells reside closer to the nucleus. If all the electrons in the outermost shell are removed, the resulting ion will be considerably smaller, as is the case of the two cations shown in Figure 8-19. Just as a cation is smaller than its parent atom, an anion is larger than its neutral parent atom. When an electron is added to an atom, the added repulsions between the electrons cause the radius of the ion to expand. Just as a metal cation becomes smaller as the positive charge increases, a nonmetal anion becomes larger as the negative charge increases.

▶ **ASSESSING THE OBJECTIVE FOR SECTION 8-5**

EXERCISE 8-5(a) LEVEL 1: Where on the periodic table (upper/lower, right/left) would you find atoms with the following?
(a) largest atomic radius **(b)** smallest ionization energy **(c)** largest electron affinity

EXERCISE 8-5(b) LEVEL 2: Given the following four elements: Ga, Ge, In, and Sn:
(a) Which has the smallest radius?
(b) Which has the lowest first ionization energy?
(c) Which is most likely to form a positive ion?

EXERCISE 8-5(c) LEVEL 2: Cs$^+$, Ba^{2+}, I$^-$, Te^{2-}, and Xe represent an isoelectronic series (they all have 54 electrons). Rank them from largest to smallest.

EXERCISE 8-5(d) LEVEL 3: Explain why aluminum's first ionization energy is lower. than silicon's. Explain why aluminum's first ionization energy is lower than boron's. Explain why aluminum's first ionization energy is lower than magnesium's.

EXERCISE 8-5(e) LEVEL 3: Only the first E.A. (adding the first electron to form the −1 anion) of most atoms is exothermic. The second, third, and higher are all endothermic. Why?

EXERCISE 8-5(f) LEVEL 3: Write the electron configuration of the following ions: Se^{2-}, Rb$^+$, and Zn^{2+}.

EXERCISE 8-5(g) LEVEL 3: Based on electron configurations, what charges would you assign to the monatomic ions formed from the following, and why?
(a) sulfur **(b)** copper **(c)** gallium

For additional practice, work chapter problems 8-67, 8-70, 8-73, 8-76, 8-77, and 8-81.

PART B SUMMARY

KEY TERMS

8-3.1	**Electron configuration** is determined as a result of the **Aufbau principle**. p. 251
8-4.1	**Orbital diagrams** are written with direction from the **Pauli exclusion principle**. p. 258
8-4.2	**Hund's rule** determines the spin orientation of electrons in the same subshell. p. 259
8-5.1	The **atomic radius** of an element is the distance to the outermost electrons. p. 261

8-5.2   **Ionization energy (I.E.)** relates to the relative ease of formation of positive ions. Coulomb's law states that the forces of attraction increase as the charges increase but decrease as the distance between the changes increases. p. 261

8-5.3   **Electron affinity (E.A.)** relates to the formation of negative ions. p. 262

8-5.4   **Isoelectric** ions or atoms have the same number of electrons and electron configurations. p. 263

SUMMARY CHART

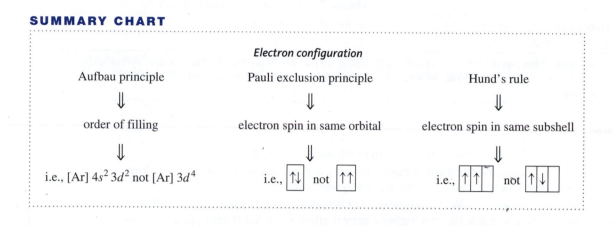

Electron configuration

Aufbau principle	Pauli exclusion principle	Hund's rule
⇓	⇓	⇓
order of filling	electron spin in same orbital	electron spin in same subshell
⇓	⇓	⇓
i.e., [Ar] $4s^2 3d^2$ not [Ar] $3d^4$	i.e., [↑↓] not [↑↑]	i.e., [↑|↑] not [↑|↓]

CHAPTER 8 SYNTHESIS PROBLEM

Cesium and gold are both metals in the sixth period that form +1 ions. That is about where the similarities end. Gold is very unreactive and is known as the _eternal_ metal. We commonly find it in the earth as a free metal. Cesium, however, is very reactive and is found in nature only as a part of ionic compounds. Why is that? The concepts in this chapter provide us with answers. This and the next chapter take us to the theoretical side of chemistry and some answers to _why_ questions. Scientists get excited about discovering answers to why things are as they are.

PROBLEM	SOLUTION
a. Write the electron configurations of cesium and gold using noble gas shorthand.	**a.** cesium: [Xe] $6s^1$ gold: [Xe] $6s^1 5d^{10}$
b. A cesium atom has one electron in its outermost shell. Suppose this electron resides in a _p_ sub-shell. Describe the state of the cesium atom. Also describe any process that will occur in the cesium atom.	**b.** If the electron is in a _p_ subshell, it is in an excited state. When the electron drops back to the $6s$ subshell (the ground state) it emits energy in the form of a discrete wavelength and color of light.
c. Write the orbital diagrams of the electrons beyond the noble gas for cesium and gold.	**c.** Cs: [↑] Au: [↑] [↑↓|↑↓|↑↓|↑↓|↑↓] $6s$ $6s$ $5d$
	d. Cesium has the larger atoms and the lower first ionization energy. Gold has smaller atoms because the filling of the $5d$ subshell causes a gradual shrinkage of the atoms across the transition series. The smaller the atom, the stronger is the hold on its outer electrons and the higher the ionization energy. Gold is unreactive for several reasons but the high ionization energy is very important. In order for metals to become part of compounds at least one electron must be removed. Cesium would have the higher second ionization energy because the second electron would have to come out of an inner subshell.

PROBLEM	SOLUTION
e. Silver is in the same group (IB) as gold but in the preceding period. According to the trends discussed, gold atoms should be larger than silver atoms. In fact, they are almost the same size. Considering what subshells are filled between silver and gold, explain the similarity in size.	**e.** What is significant between silver and gold is the presence of the 4f subshell. As the 14 electrons are added across this inner transition series the atoms in general are getting smaller. By the time we get back to filling the 5d subshell at hafnium, the higher energy shell has been counteracted by the shrinkage from the 4f subshell. This process actually has a name and is referred to as the "lanthanide contraction."
f. Write the electron configurations of the +1 ions formed by cesium and gold.	**f.** Cs^+: $5s^2 4d^{10} 5p^6 = $ [Xe] Au^+: [Xe]$5d^{10}$
g. Gold forms an unusual ion that is isoelectronic with zinc. What is the charge on this ion? Write the electron configuration of this ion.	**g.** This is the Au^- ion. (It is actually known as the "auride ion." The electron configuration for this ion is [Xe] $6s^2 5d^{10}$.

YOUR TURN

a. Write the electron configuration for selenium (#34) and chlorine (#17).

b. Write the orbital diagrams for the electrons beyond the previous noble gas for selenium and chlorine. How many unpaired electrons are in each atom?

c. Which of the two elements has the larger atoms? Which has the higher first ionization energy? Which do you think has the higher electron affinity (i.e, the most negative value)?

d. Write the electron configurations for the ions of each element that are isoelectronic with a noble gas.

e. List the following four species in order of increasing size: Se and Cl atoms and the two ions formed by these elements.

Answers are on p. 268.

CHAPTER SUMMARY

This chapter starts us on a journey that ultimately leads to an explanation of why and how atoms of elements bond to each other. The periodic basis of chemical bonding lies in the nature of the electrons and the various regions of space within atoms where they exist. Our journey started with a theoretical explanation of the **discrete**, or **line**, **spectrum** of **electromagnetic radiation** (light) emitted by hot, gaseous hydrogen atoms. This is quite unlike the **continuous spectrum** seen in a rainbow. The **wavelength** of light is inversely related to its energy. In 1913, Niels Bohr proposed a theory, known as **Bohr's model**, in which the hydrogen electron orbits the nucleus in **quantized energy levels**. He assigned a **principal quantum number** (n) to each energy level, in which the first energy level ($n = 1$) is lowest in energy and known as the **ground state**. In hydrogen, the energy levels higher than $n = 1$ are the **excited states**. Light is emitted from an atom when an electron falls from an excited state to a lower state. The difference in energy between the states becomes the energy of the light wave. If the energy of the light is within the visible part of the spectrum, we see a specific color.

Eventually, Bohr's model had to be adjusted and then mostly discarded as newer, more inclusive theories were advanced. Modern atomic theory, which is known as **wave mechanics**, takes into account the wave nature of the electron. Thus, the electron is viewed as having a certain probability of existing in a region of space known as an **orbital**. There are four different types of orbitals, known as *s*, *p*, *d*, and *f* **orbitals**. All four have distinctive shapes for the regions of highest probability. The simplest are *s* orbitals, which have a spherical shape. If the principal energy levels are designated as **shells**, then the orbitals of one type within a particular shell make up what is known as a **subshell**. Just as the shells increase in energy from $n = 1$ on, the subshells within a shell increase in energy in the order *s*, *p*, *d*, and *f*.

Each shell holds $2n^2$ electrons and has n different subshells. The electrons in any atom can be assigned to a given shell and subshell. This is known as the element's **electron configuration**. Electrons fill subshells according to the **Aufbau principle**, which simply means that the lowest energy subshells fill first. As we proceed through the electron configurations of the elements, one fact makes itself apparent. Atoms of elements in vertical columns or groups have the same outermost subshell configuration. The next higher shell is indicated as one goes down the table. Since the periodic table and electron configuration are interrelated, we now put the periodic table to work in writing electron configurations. We see that various groups can be identified by their specific configurations as well as their properties.

By using **orbital diagrams**, we can expand the representation of electron configuration to include assignment of electrons to orbitals. Two other rules are required to do this successfully. The **Pauli exclusion principle** relates to the spin of electrons in the same orbital, and **Hund's rule** relates to the electron distribution in separate orbitals of the same subshell.

The periodic table tells us even more. There are general trends in the **radii** of atoms. In general, atomic radii decrease up and to the right on the table. The radius of an atom is related to the shell of the outermost electrons and to the number of electrons in the outermost subshell. The higher the energy of the shell and the fewer electrons in the outermost subshell, the larger the atom. Thus, atoms to the lower left are the largest atoms.

The opposite trend occurs if we compare the energy required to remove the outermost electron, which is known as the element's **ionization energy (I.E.)**. The same factors that affect the size of an atom affect the ionization energy. That is, in general, larger atoms have lower ionization energies than smaller atoms. Metal atoms tend to have larger atoms than nonmetal atoms, so they have lower ionization energies. How many electrons can be easily removed from a metal is determined by its outer electron configuration.

Opposite to ionization energy is **electron affinity (E.A.)**, the energy released when an electron is added to an atom. The general trend is that smaller atoms will release more energy than larger atoms as an electron is added, meaning that the smaller nonmetals form anions more easily.

Finally, the radius of an ion depends on the position of the element in the periodic table, the sign of the charge, and the magnitude of the charge. Generally, the more positive the ion, the smaller its radius. If we compare species in the same period, a monatomic anion is larger than a neutral atom, which in turn is larger than a monatomic positive ion.

OBJECTIVES

SECTION	YOU SHOULD BE ABLE TO...	EXAMPLES	EXERCISES	CHAPTER PROBLEMS
8-1	Describe how Bohr's model of the atom account for observed wavelengths and energies of emission spectra.		1a, 1b, 1c, 1d, 1e	1, 3, 4
8-2	Describe the electronic structure of an atom, including shells, subshells and various types of orbitals.		2a, 2b, 2c, 2d	6, 8, 9, 10, 11, 12, 13, 15, 16, 17
8-3	Using the periodic table, write the outer electron configuration of the atoms of a specific element.	8-1, 8-2, 8-3	3a, 3b, 3c, 3d, 3e	21, 23, 26, 28, 32, 33, 36, 39, 40, 44, 45, 55
8-4	Using orbital diagrams, determine the distribution of electrons in an atom.	8-4	4a, 4b, 4c	60, 61, 64, 65, 66
8-5	Using the periodic table, predict trends in atomic and ionic radii, ionization energy, and electron affinity.		5a, 5b, 5c, 5d, 5e	68, 70, 71, 73, 75, 77, 81, 82

▶ ANSWERS TO ASSESSING THE OBJECTIVES

Part A

EXERCISES

8-1(a) Long wavelengths of light have <u>low</u> (high/low) energies. The color of visible light with the highest energy is <u>violet</u>, and the color of visible light with the longest wavelength is <u>red</u>. White light contains all colors of light, called a <u>continuous</u> spectrum. Light from heated gaseous element sources gives off only discrete energies to form a <u>line</u> spectrum. Niels Bohr suggested that the orbits of an electron were <u>quantized</u>, or set to specific values. The lowest energy state is the <u>ground</u> state, and the higher states are the <u>excited</u> states.

8-1(b) (high energy/short wavelength) gamma rays—X-rays—ultraviolet—visible—infrared—microwaves—radio waves (low energy/long wavelength)

8-1(c) Electrons absorb energy and jump a quantized distance to an energy level in an excited state. When the electrons return to lower energy states, this excess energy is released at a specific value that shows up as a single wavelength in a line spectrum. Different lines represent transitions between different energy states.

8-1(d) It is reasonable to assume that all atoms are fundamentally designed similarly. They differ only in numbers of nucleons and electrons. So the structure found in the hydrogen atom is going to be similar to the structure of larger atoms. Understanding hydrogen allows us to understand these larger atoms as well.

8-1(e) We use different elements to make different colors. For example, argon produces a red color.

8-2(a) Wave mechanics tell us that the electron exists in a region of space known as an <u>orbital</u>. The four types of these regions are labeled <u>s</u>, <u>p</u>, <u>d</u>, and <u>f</u>. The lowest in energy of these four has a <u>spherical</u> shape. A principal energy level, called a <u>shell</u>, can be subdivided into <u>subshells</u>. Such a subdivision contains all of the same type of <u>orbital</u> in that particular energy level.

8-2(b) (a) 18 (b) s, p, and d (c) 2, 6, and 10, respectively (d) 1, 3 and 5, respectively

8-2(c) (a) 7 orbitals × 2 electrons each = 14 electrons
(b) 1 s + 3 p's + 5 d's = 9 total orbitals (c) 1 s + 3 p's + 5 d's + 7 f's = 16 orbitals × 2 electrons each = 32 electrons

8-2(d) The pattern of one lobe for s, two for p, and four for d lets us assume that the f orbital has 8 lobes (doubling each time). Eight lobes would be arranged most symmetrically by having them point toward the corners of a cube. The actual f orbitals are a small modification of this.

8-2(e) (a) 5 (b) 9 (c) 18 (d) 50

Part B
EXERCISES

8-3(a) The electron configuration $4d^5$ indicates that there are <u>five</u> electrons in the <u>d</u> subshell in the <u>fourth</u> shell.

8-3(b) Zn: [Ar] $4s^2 3d^{10}$ W: [Xe] $4f^{14} 6s^2 5d^4$

8-3(c) As (arsenic)

8-3(d) Group VIA: [NG] $ns^2 (n-1) d^{10} np^4$; Groups VIB: [NG] $ns^2 (n-1) d^5$

8-4(a) (a) Pauli exclusion principle (b) Hund's rule (c) Aufbau principle

8-4(b) Electron configuration: [Kr]$5s^2 4d^3$ orbital diagram:

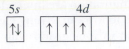

8-4(c) (a) 2   (b) 0   (c) 3

8-4(d) Carbon:

If one electron is moved from the $2s$ to the $2p$, that will create four half-filled orbitals.

8-5(a) (a) lower left (b) lower left (c) right

8-5(b) (a) Ge (b) In (c) In

8-5(c) $Te^{2-} > I^- > Xe > Cs^+ > Ba^{2+}$

8-5(d) Aluminum's first ionization energy is lower than silicon's because aluminum has one fewer proton in the nucleus attracting its outer electron, making that electron easier to remove. Aluminum's first I.E. is lower than boron's because aluminum has an extra shell that places its outer electrons farther away from its nucleus, reducing the attraction and making it easier to remove. Aluminum's first I.E. is lower than magnesium's because its outer electron is in a $3p$ subshell, rather than in the more stable $3s$ like magnesium. It's harder to remove the electron from the more stable subshell.

8-5(e) Once an electron is added, the resulting species has a negative charge. This charge repels additional electrons, making them progressively harder to add.

8-5(f) Se^{2-}: [Ar] $4s^2 3d^{10} 4p^6$ or [Kr]
Rb^+: [Kr]
Zn^{2+}: [Ar] $3d^{10}$

8-5(g) (a) −2. It takes two more electrons to fill up the $3p$ subshell. (b) +1. Lose one electron from the $4s$ subshell. (It also forms a +2 ion that can't be explained by subshell structure.) (c) +3. Remove the electrons from the $4p$ and the $4s$ subshell, leaving a filled $3d$.

ANSWERS TO CHAPTER SYNTHESIS PROBLEM

a. Se: [Ar]$4s^2 3d^{10} 4p^4$ Cl: [Ne]$3s^2 3p^5$

b. Se: ⬆⬇ ⬆⬇⬆⬇⬆⬇⬆⬆ ⬆⬇⬆⬆
      $4s$ $3d$ $4p$
Cl: ⬆⬇ ⬆⬇⬆⬇⬆
      $3s$ $3p$
Se has two unpaired electrons and chlorine has one unpaired electron.

c. Selenium has the larger atoms so chlorine has the higher ionization energy. Chlorine has the higher electron affinity (i.e., forms a negative ion more easily.)

d. Selenium forms a −2 ion that is isoelectronic with krypton. That is: [Ar]$4s^2 3d^{10} 4p^6$. Chlorine forms a −1 ion that is isoelectronic with argon: That is: [Ne]$3s^2 3p^6$.

e. $Cl < Se < Cl^- < Se^{2-}$

CHAPTER PROBLEMS

Throughout the text, answers to all exercises in color are given in Appendix E. The more difficult exercises are marked with an asterisk.

Atomic Theory and Orbitals (SECTION 8-1)

8-1. Ultraviolet light causes sunburns, but visible light does not. Explain.

8-2. The $n = 8$ and $n = 9$ energy levels are very close in energy. Using the Bohr model, describe how the wave lengths of light compare as an electron falls from these two energy levels to the $n = 1$ energy level.

8-3. The $n = 3$ energy level is of considerably greater energy than the $n = 2$ energy level. As an electron falls from these two energy level, use the Bohr model to describe:

(a) how the wavelengths of light compare

(b) how the energies of light compare

8-4. An electron in the lithium atom is in the third energy level. Is the atom in the ground or excited state? Can the atom emit light? If so, how?

8-5. Which of the following types of orbitals do not exist?

      $6s$ $3p$ $1p$ $4d$ $3f$

8-6. Which of the following types of orbitals do not exist?

      $5f$ $2d$ $7p$ $6d$ $1s$

8-7. What is meant by the "shape" of an orbital?

8-8. Describe the shape of the $4p$ orbitals.

8-9. Describe the shape of the 3s orbital. How does it differ from that of a 2s orbital?

8-10. Describe the shape of the $4d$ orbital. How does it differ from a $4p$ orbital?

8-11. How many total orbitals are in the third energy level?

8-12. How many total orbitals are in the fourth energy level?

Shells and Subshells (SECTION 8-2)

8-13. How many orbitals are in the following subshells?

(a) $3p$ **(b)** $4d$ **(c)** $6s$

8-14. How many orbitals are in the following subshells?

(a) $5f$ **(b)** $6p$ **(c)** $2d$

8-15. What are the total electron capacities of the $n = 3$ and the $n = 2$ shells?

8-16. What is the total electron capacity of the fourth shell?

8-17. What is the total electron capacity of each subshell in the fourth shell?

8-18. What is the total electron capacity of the fifth shell?

8-19. The fifth shell (theoretically) contains a fifth subshell designated the $5g$ subshell. What is the total electron capacity of the g subshell?

8-20. How many orbitals are in the $5g$ subshell? (See problem 8-19.)

Electron Configuration (SECTION 8-3)

8-21. Which subshell always fills first?

8-22. Which subshell fills second and which subshell fills third?

8-23. Explain how a subshell in a particular shell can be lower in energy than a subshell in a lower shell.

8-24. Which of the following subshells fills first? (Refer to Figure 8-16.)

(a) $6s$ or $6p$ **(b)** $5d$ or $5p$ **(c)** $4p$ or $4f$ **(d)** $4f$ or $4d$

8-25. Which of the following subshells is lower in energy? (Refer to Figure 8-16.)

(a) $6s$ or $5p$ **(c)** $5s$ or $4d$ **(e)** $4f$ or $6d$

(b) $6s$ or $4f$ **(d)** $4d$ or $5p$ **(f)** $3d$ or $4p$

8-26. Write the following subshells in order of increasing energy. (Refer to Figure 8-16.)

      $4s$, $5p$, $4p$, $5s$, $4f$, $4d$, $6s$

8-27. Write the total electron configuration for each of the following elements. (Refer to Figure 8-16.)

(a) Mg   **(b)** Ge   **(c)** Cd   **(d)** Si

8-28. Write the total electron configuration for each of the following elements. (Refer to Figure 8-16.)

(a) B   **(b)** Ag   **(c)** Se   **(d)** Sr

8-29. Write the electron configuration that is implied by [Ar]. (Refer to Figure 8-16.)

8-30. Write the electron configuration that is implied by [Xe]. (Refer to Figure 8-16.)

8-31. Using the noble gas shorthand, write the electron configurations for the following elements. (Refer to Figure 8-16.)

(a) S   **(b)** Zn   **(c)** Pu   **(d)** I

8-32. Using the noble gas shorthand, write the electron configurations for the following elements. (Refer to Figure 8-16.)

(a) Sn   **(b)** Ni   **(c)** Cl   **(d)** Au

8-33. Write the symbol of the first element that has the following.

(a) a $5p$ electron   **(c)** a $4f$ electron

(b) a $4d$ electron   **(d)** a filled $3p$ subshell

8-34. Write the symbol of the first element that has the following.

(a) a filled $4d$ subshell   **(c)** a $6s$ electron

(b) a half-filled $4p$ subshell   **(d)** a filled fourth shell

8-35. Which subshell begins to fill for each of the following elements?

(a) Al   (b) In   (c) La   (d) Rb

8-36. Which subshell begins to fill for each of the following elements?

(a) Y   (b) Th   (c) Cs   (d) Ga

8-37. What do Groups IIIA and IIIB have in common? How are they different?

8-38. What do the elements Al and Ga have in common? How are they different?

8-39. Which elements have the following electron configurations? (Use only the periodic table.)

(a) $1s^2 2s^2 2p^5$ (d) [Xe] $6s^2 5d^1 4f^7$

(b) [Ar] $4s^2 3d^{10} 4p^1$ (e) [Ar]$4s^1 3d^{10}$ (exception to rules)

(c) [Xe]$6s^2$

8-40. Which elements have the following electron configurations?

(a) [He] $2s^2 2p^3$ (d) [Rn] $7s^2 6d^1$

(b) [Kr]$5s^2 4d^2$ (e) [Xe]$6s^2 5d^{10} 4f^{14} 6p^1$

(c) [Ar]$4s^2 3d^{10} 4p^6$ (f) [Ar]$4s^2 3d^{10}$

8-41. If any of the elements in problem 8-39 belong to a numerical group (e.g., Group IA) in the periodic table, indicate the group.

8-42. Write the number designation of the two groups that may have five electrons beyond a noble gas configuration.

8-43. Write the number designation of a group that has 2 electrons beyond a noble gas configuration. Write the number designation of a group with 12 electrons beyond a noble gas configuration.

8-44. Which two configurations belong to the same group in the periodic table?

(a) [Kr]$5s^2$ (d) [Ar]$4s^2 3d^2$

(b) [Kr]$5s^2 4d^{10} 5p^2$ (e) [Ne]$3s^2 3p^2$

(c) [Xe]$6s^2 5d^1 4f^2$

8-45. Which two configurations belong to the same group in the periodic table?

(a) [Kr]$5s^2 4d^{10}$ (d) [Xe]$6s^2$

(b) [Ar]$4s^1 3d^{10}$ (e) [Xe]$6s^2 4f^{14} 5d^{10}$

(c) [Ne]$3s^1$

8-46. Which group in the periodic table has the following general electron configuration? (n is the principal quantum number.)

(a) $ns^2 np^2$ (c) $ns^1 (n-1) d^{10}$ (e.g., $4s^1 3d^{10}$)

(d) $ns^2 np^6$ (d) $ns^2 (n-1)d^1 (n-2)f^2$ (two elements)

8-47. Write the general electron configurations for the following groups.

(a) IIA   (b) IIB   (c) VIA   (d) IVB

8-48. How does He differ from the other elements in Group VIIIA?

8-49. Which of the following elements fits the general electron configuration $ns^2 (n-1)d^{10} np^4$

(a) Cr   (b) Te   (c) S   (d) O   (e) Si

8-50. What is the electron configuration of the noble gas at the end of the third period?

8-51. What is the electron configuration of the noble gas at the end of the fourth period?

*8-52.** What would be the atomic number of an element with one electron in the $5g$ subshell?

*8-53.** Element number 114 forms a comparatively long-lived isotope. What would be its electron configuration and in what group would it be?

*8-54.** How many elements are in the seventh period?

*8-55.** How many elements would be in a complete eighth period? (*Hint:* Consider the $5g$ subshell.)

*8-56.** Classify the following electron configurations into one of the four main categories of elements.

(a) [Ar]$4s^2 3d^2$ (c) [Ar]$4s^2 3d^{10} 4p^6$

(b) [Kr]$5s^2 4d^{10} 5p^5$ (d) [Rn] $7s^2 6d^1 5f^3$

*8-57.** Classify the following electron configurations into one of the four main categories of elements.

(a) [Ne]$3s^2 3p^6$ (c) [Xe]$6s^2 4f^{14} 5d^2$

(b) [Ne]$3s^2 3p^5$ (d) [Xe]$6s^2 5d^1 4f^7$

*8-58.** What would be the atomic number and group of the next nonmetal after element number 112 (Cn)?

*8-59.** What is the atomic number of the heaviest metal that would appear in the periodic table after element number 112 (Cn)?

Orbital Diagrams (SECTION 8-4)

8-60. Which of the following orbital diagrams is excluded by the Aufbau principle? Which by the Pauli exclusion principle? Which by Hund's rule? Which is correct? Explain how a principle or rule is violated in the others.

8-61. Write the orbital diagrams for electrons beyond the previous noble gas core for the following elements. (Refer to Figure 8-16.)

(a) S   (b) V   (c) Br   (d) Pm

8-62. Write the orbital diagrams for electrons beyond the previous noble gas core for the following elements. (Refer to Figure 8-16.)

(a) As   (b) Ar   (c) Tc   (d) Tl

8-63. How many unpaired electrons are in the atoms of elements in Groups IIB, VB, VIA, and VIIA? An atom with electron configuration given by $[ns^1 (n-1)d^5]$; and Pm (atomic number 61)?

8-64. The atoms of the elements in three groups in the periodic table have all their electrons paired. What are the groups?

8-65. What group or groups have two unpaired p electrons?

8-66. What group or groups have five unpaired d electrons?

Periodic Trends (SECTION 8-5)

8-67. Which of the following elements has the larger radius?

(a) As or Se   (b) Ru or Rh   (c) Sr or Ba   (d) F or I

8-68. Which of the following elements has the larger radius?

(a) Tl or Pb   (b) Sc or Y   (c) Al or Si   (d) As or P

8-69. Four elements have the following radii (in pm): 117, 122, 129, and 134. The elements, in random order, are V, Cr, Nb, and Mo. Which element has a radius of 117 pm? Which has a radius of 134 pm?

8-70. Four elements have the following radii: 171 pm, 162 pm, 154 pm, and 140 pm. The elements, in random order, are In, Sn, Tl, and Pb. Which element has a radius of 140 pm? Which has a radius of 171 pm?

8-71. Which of the following elements has the higher ionization energy?

(a) Ti or V   (c) Mg or Sr   (e) B or Br

(b) P or Cl   (d) Fe or Os

8-72. Four elements have the following first ionization energies (in kJ/mol): 762, 709, 579, and 558. The elements, in random order, are Ga, Ge, In, and Sn. Which element has an ionization energy of 558 kJ/mol? Which element has an ionization energy of 762 kJ/mol?

8-73. Four elements have the following first ionization energies (in kJ/mol): 869, 941, 1010, and 1140. The elements, in random order, are Se, Br, Te, and I. Which element has an ionization energy of 869 kJ/mol? Which element has an ionization energy of 1140 kJ/mol?

8-74. The first four ionization energies for Ga are 578.8, 1979, 2963, and 6200 kJ/mol. How much energy is required to form each of the following ions: Ga^+, Ga^{2+}, Ga^{3+}, and Ga^{4+}? Why does the formation of Ga^{4+} require a comparatively large amount of energy?

8-75. Which of the following monatomic cations is the easiest to form, and which is the hardest to form?

(a) Cs^+ (b) Rb^{2+} (c) Ne^+ (d) Sc^{3+}

8-76. Which of the following atoms would most easily form a cation?

(a) B   (b) Al   (c) Si   (d) C

8-77. Which of the following atoms would most likely form an anion?

(a) Be   (b) Al   (c) Ga   (d) I

8-78. Noble gases form neither anions nor cations. Why?

8-79. Which of the following would have the highest second ionization energy?

(a) Rb   (b) Pb   (c) Ba   (d) Al   (e) Be

8-80. Which of the following would have the lowest third ionization energy?

(a) Na   (b) B   (c) Ga   (d) Mg   (e) N

8-81. Arrange the following ions and atoms in order of increasing radii.

(a) Mg   (c) S   (e) K^+

(b) S^{2-} (d) Mg^{2+} (f) Se^{2-}

8-82. Arrange the following ions and atoms in order of increasing radii.

(a) Br   (c) K   (e) Br^-

(b) K^+ (d) I^- (f) Ca^{2+}

*8-83. Zirconium and hafnium are in the same group and have almost the same radius despite the general trend down a group. As a result, the two elements have almost identical chemical and physical properties. The fact that these elements have almost the same radius is due to what is referred to as the *lanthanide contraction*. With the knowledge that atoms get progressively smaller as a subshell is being filled, can you explain this phenomenon? (*Hint:* Follow all the expected trends between the two elements.)

8-84. The first five ionization energies for carbon are 1086, 2353, 4620, 6223, and 37,830 kJ/mol. How much energy is required to form the following ions: C^+, C^{2+}, C^{3+}, C^{4+}, and C^{5+}? In fact, even C^+ does not form in compounds. Compare the energy required to form this ion with that needed to form some metal ions. Explain.

General Problems

8-85. Write the symbol of the element that corresponds to the following.

(a) the first element with a p electron

(b) the first element with a filled $4p$ subshell

(c) three elements with only one electron in the $4s$ subshell

(d) the first element with one p electron that also has a filled d subshell

(e) the element after Xe that has two electrons in a d subshell

8-86. Write the symbol of the element that corresponds to the following.

(a) the first element with a half-filled p subshell

(b) the element with only three electrons in a $4d$ subshell

(c) the first two elements with a filled $3d$ subshell

(d) the element with only three electrons in a $5p$ subshell

8-87. Write the symbol of the element that corresponds to the following.

(a) a nonmetal with only one electron in a p subshell

(b) an element that is a liquid at room temperature that has five p electrons

(c) a metalloid that has three p electrons

(d) a metalloid with two p electrons and no d electrons

8-88. Write the symbol of the element that corresponds to the following.

(a) a transition metal in the fifth period with three unpaired electrons

(b) a representative true metal with two unpaired electrons

(c) an element that is a liquid at room temperature with no unpaired electrons

(d) a metalloid with two unpaired electrons and a filled d subshell

8-89. Identify the first element that has the following total number of electrons and indicate whether the element is a metal or nonmetal, its category, and its group number.

(a) 10 s electrons   (b) 28 d electrons   (c) 15 p electrons

8-90. Identify the first element that has the following total number of electrons and indicate whether the element is a metal or nonmetal, its category, and its group number.

(a) 15 f electrons   (b) 36 d electrons   (c) 24 p electrons

8-91. A certain isotope has a mass number of 30. The element has three unpaired electrons. What is the element?

8-92. A certain isotope has a mass number of 196. It has two unpaired electrons. What is the element?

8-93. Given two elements, X and Z, identify them from the following information.

(a) They are both metals, but one is a representative element.

(b) The first ionization energy of Z is greater than that of X.

(c) They are both in the same period that does not have f electrons.

(d) They both have two unpaired electrons and neither is used in jewelry.

(e) A nonmetal in the same period is a diatomic solid.

8-94. Given two elements, Q and R, identify them from the following information.

(a) One is a gas and one is a solid.

(b) One forms a +1 ion and the other does not.

(c) Q is larger than R, but both elements are the first elements that have full shells.

(d) One is used in coins and the other in fluorescent lights.

8-95. Which of the following monatomic cations would require a particularly large amount of energy to form? If a certain ion requires a large amount of energy to form, give the reason,

(a) In^{3+} (b) I^+ (c) Ca^{2+} (d) K^+ (e) B^{3+}

8-96. Which of the following atoms would not be likely to form a +2 cation? Explain.

(a) Sr   (b) Li   (c) B   (d) Ba

8-97. Chemical reactivity relates to the size of the atom or ion. In the following pairs, which has the larger radius?

(a) Be or Ca   (d) S^{2-} or Cl^-

(b) Br or Br^- (e) Na^+ or Mg^{2+}

(c) Cl or S

8-98. In the following pairs, which has the larger radius?

(a) Na^+ or K^+ (d) Si or Ge

(b) O^{2-} or Se^{2-} (e) K^+ or K

(c) Ga or Ge

*__8-99.__ On the planet Zerk, the periodic table of elements is slightly different from ours. On Zerk, there are only two p orbitals, so a p subshell holds only four electrons. There are only four d orbitals, so a d subshell holds only eight electrons. Everything else is the same as on Earth, such as the order of filling ($1s$, $2s$, etc.) and the characteristics of noble gases, metals, and nonmetals. Construct a Zerkian periodic table using numbers for elements up to element number 50. Then answer these questions.

(a) How many elements are in the second period? In the fourth period?

(b) What are the atomic numbers of the noble gases at the ends of the third and fourth periods?

(c) What is the atomic number of the first inner transition element?

(d) Which element is more likely to be a metal: element number 5 or element number 11? element number 17 or element number 27?

(e) Which element has the larger radius: element number 12 or element number 13? element number 6 or element number 12?

(f) Which element has a higher ionization energy: element number 7 or element number 13? element number 7 or element number 5? element number 7 or element number 9?

(g) Which ions are reasonable?

(1) 16^{2+} (3) 7^+ (5) 17^{4+} (7) 1^-

(2) 9^{2+} (4) 13^- (6) 15^+

STUDENT WORKSHOP

Plotting Periodic Trends

Purpose: To graphically represent several periodic trends and predict expected behavior based on those trends. (Work in groups of three or four. Estimated time: 45 min.)

Equipment: Several sheets of graph paper; several different colors of pens or markers.

Procedure: We will make line graphs even though the data more properly belongs on bar graphs. This way we can put two sets of data on the same graph.

Graph 1: First Ionization Energies of the Second- and Third-Period Elements

1. Refer to the data in Table 8-2. The *x*-axis will be seven separate points labeled for the seven columns on the periodic table, IA through VIIA. The *y*-axis needs to range from 0 kJ/mol to 1700 kJ/mol. Scale accordingly.

2. In one color ink, plot the points for the elements Li through F from Table 8-2. Connect the points in that color.

3. In a second color ink, plot the points for the elements Na through Cl, and connect those points as well.

4. Answer the following questions:
    *   Which series of elements has the higher ionization energies? Why?
    *   What is the general trend as you move from left to right? Why?
    *   There are two dips in the graph. Can you explain why they are there?

Graph 2: First to Fourth Ionization Energies for Third-Period Elements

1. Refer to the data in Table 8-3. The *x*-axis will be four separate points labeled I, II, III, and IV. The *y*-axis will range from 0 kJ/mol to 12,000 kJ/mol. Scale as needed.

2. In one color ink, plot the points for the first through fourth ionization energies of Na.

3. In separate color inks, do the same thing for Mg and Al.

4. Answer the following questions:
    *   Is it clear from this graph what the most favorable number of electrons to remove from each atom is?
    *   How does that relate to the electronic structure of the atom?
    *   How does it relate to the atom's position on the periodic table?

Graph 3: Atomic Radii for Second- and Third-Period Elements

1. Refer to Figure 8-18 for the data. The *x*-axis will be the same as in graph 1. The *y*-axis can range from 50 pm to 200 pm (graphs do not necessarily have to start at 0).

2. Plot the data using the same color ink as in graph 1. Make one line for the data of Li through Ne and a second line for Na through Ar.

3. Answer the following questions:
    *   How does graph 3 compare to graph 1?
    *   Is this a coincidence, or are the two trends related? If so, why?

9

The Chemical Bond

This is our home as seen from far-out space. Its surface and atmosphere are composed of some free elements as well as ionic and molecular compounds. We look deeper into the nature of compounds in this chapter.

SETTING THE STAGE

Earth is a complex world of chemicals. Consider the air. It is composed of molecular elements—nitrogen and oxygen—as well as smaller quantities of the solitary atoms of noble gases—argon, helium, and neon. In addition there are trace amounts of several molecular compounds such as carbon dioxide and water. The surface of Earth is largely composed of water containing dissolved compounds. The solid surface of Earth contains compounds of living things, rocks, and fossils.

We will focus on two compounds that occur in nature and are necessary in life processes—salt and water. These two compounds are examples of the two basic types of compounds: ionic and molecular. Salt (specifically, sodium chloride) is typical of ionic compounds. It is a hard, brittle solid with a high melting point. Water, however, is typical of many molecular compounds in that it is a liquid at or near room temperature. What is there about these two compounds that make them so distinct? Even though both are composed of just two elements, there actually is an important difference. Sodium chloride is a binary compound formed from a metal, sodium, and a nonmetal, chlorine. Water is also a binary compound, but it is formed from two nonmetals, hydrogen and oxygen. In this chapter we will discover why the combination of a metal and a nonmetal results in an ionic compound, whereas the combination of two nonmetals results in a molecular compound.

In Part A in this chapter, we will examine the ionic compounds formed from the combination of metals and nonmetals. Nonmetals combine to form molecular compounds. The structure of these compounds is the subject of Part B. Finally, in Part C, we will examine some of the many implications that follow from the structure of molecular compounds.

Part A

Chemical Bonds and the Nature of Ionic Compounds

SETTING A GOAL

■ You will learn how the octet rule is used to determine the charge on ions and the formulas of ionic compounds.

OBJECTIVES

9-1 Write the Lewis dot structure of the atoms and monatomic ions by following the octet rule.

9-2 Using the Lewis dot structure and the octet rule, predict the charges on the ions of representative elements and the formulas of binary ionic compounds.

▶ **OBJECTIVE
FOR SECTION 9-1**
Write the Lewis dot symbols of atoms and monatomic ions by following the octet rule.

9-1 Bond Formation and Representative Elements

LOOKING AHEAD! All the elements in the periodic table interact with other elements to form compounds except for some of the noble gases. Why these elements do not form bonds with other elements provides a clue to understanding the bonding of the representative (main group) elements. ■

9-1.1 The Octet Rule

Argon, neon, and other noble gases exist in nature as solitary atoms, since they have little or no tendency to form chemical bonds. What is unique about the noble gases that makes this so? The answer relates to their electron configurations. It has been known for some time that the formation of chemical bonds involves changes in the electron configurations of the atoms involved. Apparently, the noble gases have stable electron configurations with little or no tendency to change. All the noble gases except for helium have filled outer s and p subshells (e.g., ns^2np^6). Since this is a total of eight electrons, it is referred to as an *octet* of electrons. An octet of outer electrons forms a particularly stable configuration, as we mentioned in the previous chapter. In fact, the bonding we will discuss in this chapter is dictated by the **octet rule**, *which states that elements of the representative elements form bonds so as to have access to eight outer electrons.* Note that helium is a noble gas but has only a filled $1s$ subshell containing two electrons. Thus, helium has a stable electron configuration with only two electrons compared to the other noble gases with an octet of electrons.

9-1.2 Valence Electrons and Dot Symbols

The outer s and p electrons in the atoms of a representative element are referred to as the **valence electrons**. Since only the valence electrons are involved in bonding, we can focus on these electrons exclusively. **Lewis dot symbols** *represent valence electrons as dots around the symbol of the element.* Table 9-1 shows the Lewis dot symbols for the first four periods of representative elements and noble gases. The first four valence electrons are shown placed around the symbol, one dot at a time, on each of the four sides. The next four valence electron dots then pair up with each of the originals. Notice that the elements in each group have the same number of valence electrons and thus the same number of dots.

The Roman numeral of the group number also represents the number of valence electrons (dots) for a neutral atom. In some periodic tables, the groups are labeled 1 through 18. In this case the last digit of the group number represents the number of valence electrons. For example, Group 14 has four valence electrons.

If an octet of electrons is a particularly stable configuration, how can the atoms of elements other than the noble gases alter their electron configuration so as to obtain an octet? There are three ways.

TABLE 9-1

Lewis Dot Symbols[a]

IA	IIA	IIIA	IVA	VA	VIA	VIIA	VIIIA
Ḣ							He̤
Li	Be·	Ḃ·	·Ċ·	·N̈·	·Ö:	:F̈:	:N̈e:
Na	Mg·	Al·	·Si·	·P̈·	·S̈:	:C̈l:	:Är:
K̇	Ca·	Ga·	·Ge·	·Äs·	·S̈e:	:Br̈:	:K̈r:

[a]Named after the American Chemist G. N. Lewis (1875–1946), who developed this theory of bonding.

1. A metal may *lose* one to three electrons to form a cation with the electron configuration of the previous noble gas (i.e., the one with the next-lowest atomic number). The loss of electrons results in a positive ion (i.e., a cation).

2. A nonmetal may *gain* one to three electrons to form an anion with the electron configuration of the next noble gas (i.e., the one with the next-highest atomic number). The gain of electrons results in a negative ion (i.e., an anion).

3. Atoms (usually two nonmetals) may *share* electrons with other atoms to obtain access to the number of electrons in the next noble gas.

Notice that the octet rule must be modified somewhat for the elements H, Li, and Be. They can obtain the stable noble gas configuration of helium with only two electrons. For these elements, it would be more appropriate to call it a "duet" rule.

Processes 1 and 2 complement each other in the formation of ionic compounds. Process 3 produces molecular compounds. We will consider the formation of ions first and molecular compounds in Part B of this text.

▶ **ASSESSING THE OBJECTIVE FOR SECTION 9-1**

EXERCISE 9-1(a) LEVEL 1: How many valence electrons are found in the following atoms:
(a) Ga   (b) S   (c) Kr

EXERCISE 9-1(b) LEVEL 1: What is the Lewis dot symbol for the following?
(a) Mg   (b) N   (c) I

EXERCISE 9-1(c) LEVEL 2: Where does the reference to eight in the "octet rule" come from?

EXERCISE 9-1(d) LEVEL 3: In advanced chemistry courses, students study complicated molecules that follow the "18-electron rule." Which subshells need to be filled in these compounds?

For additional practice, work chapter problems 9-1, 9-2, and 9-4.

9-2 Formation of Ions and Ionic Compounds

LOOKING AHEAD! In Chapter 4, we made use of the periodic table to assign charges to representative metals and nonmetals. In this section we will describe how the octet rule leads us to the expected charge on representative element ions.

▶ **OBJECTIVE FOR SECTION 9-2**
Using the Lewis dot structure and the octet rule, predict the charges on the ions of representative elements and the formulas of binary ionic compounds.

9-2.1 Formation of Binary Ionic Compounds

FIGURE 9-1 Reaction of Sodium with Chlorine Sodium reacts with chlorine to form sodium chloride.

When a small piece of sodium metal is placed in a bottle containing chlorine gas, a chemical change is obvious. (See Figure 9-1.) The sodium ignites in the chlorine, and a white coating of sodium chloride forms on the sides of the bottle. The changes that the elements undergo in this and other reactions are subjects of this section.

In earlier chapters, we were able to establish the formulas of many ionic compounds simply by balancing charges. In those chapters, however, we presented the charges that specific atoms achieve as facts, without explanations (e.g., halogens form −1 ions). By expanding on the material in the previous chapter, we are now ready to reexamine the formation of ionic compounds with the theoretical explanations that make these formulas logical.

First, we will examine how metal cations are formed. In Chapter 8, we indicated that representative metals have low ionization energies and can lose outer electrons relatively easily. The octet rule tells us how many electrons will be lost and from this the magnitude of the positive charge. *If a representative metal loses all its valence electrons, it acquires the octet of the previous noble gas.* We also found in the previous chapter that the ionization energies of core electrons are very high, so these electrons are not lost in compound formation. Thus, only valence electrons are lost in the formation of positive ions. We can illustrate the octet rule and cation formation using the Lewis dot symbols of the first three metals in the third period—Na, Mg, and Al.

$$\overset{\cdot}{Na} \longrightarrow Na^+ + e^-$$
$$[Ne]3s^1 \qquad [Ne]$$

$$\overset{\cdot}{Mg}\cdot \longrightarrow Mg^{2+} + 2e^-$$
$$[Ne]3s^2 \qquad [Ne]$$

$$\cdot\overset{\cdot}{Al}\cdot \longrightarrow Al^{3+} + 3e^-$$
$$[Ne]3s^2 3p^1 \qquad [Ne]$$

The Lewis representations of the three cations above do not show any electrons (dots) because the octets of electrons of the ions occupy filled inner subshells (the $2s$ and $2p$). All the metals in Group IA have analogous dot symbols to Na, so each can lose one electron to form a +1 ion with the electron configuration of the preceding noble gas. The ionizations of IIA metals are all analogous to Mg to form +2 ions and the ionizations of IIIA metals to form a +3 ion are analogous to Al.

Now we turn our attention to the formation of negative ions, which must be present in ionic compounds to balance the positive charge. In Chapter 2, we found that the atoms of some elements gain electrons to form negative ions. In the previous chapter, we found that these elements are all nonmetals in specific groups.

First, we will focus on ions formed by the VIIA nonmetals such as chlorine. The halogens are one electron short of a noble gas configuration. Thus, an octet of electrons can be achieved by adding one electron. The result is an anion with a −1 charge and the electron configuration of the next noble gas.

$$e^- + \quad :\overset{\cdot\cdot}{\underset{\cdot\cdot}{Cl}}\cdot \longrightarrow :\overset{\cdot\cdot}{\underset{\cdot\cdot}{Cl}}:^- \qquad \text{(chloride ion)}$$
$$[Ne]3s^2 3p^5 \qquad [Ne]3s^2 3p^6 = [Ar]$$

Next, consider the VIA nonmetals such as oxygen. An oxygen atom needs two electrons to achieve an octet and form an anion with a −2 charge.

$$2e^- + \quad :\overset{\cdot}{\underset{\cdot\cdot}{O}}\cdot \longrightarrow :\overset{\cdot\cdot}{\underset{\cdot\cdot}{O}}:^{2-} \qquad \text{(oxide ion)}$$
$$[He]2s^2 2p^4 \qquad [He]2s^2 2p^6 = [Ne]$$

Finally, we move to the VA nonmetals. Actually, only N and P are known to form −3 ions such as shown here.

$$3e^- + \cdot \ddot{N} \cdot \longrightarrow \; :\!\ddot{N}\!:^{3-} \qquad \text{(nitride ion)}$$

$$[He]2s^2 2p^3 \qquad [He]2s^2 2p^6 = [Ne]$$

Now we bring the electron losers (metals) together with the electron gainers (nonmetals) to form ionic compounds as we did with sodium and chlorine gas. We will see how the formulas of these compounds follow from this electron exchange. When we bring sodium in contact with chlorine, an exchange of the single valence electron occurs. The exchange of the electron means that both of the ions formed have octets of electrons. As mentioned, the sodium now has a +1 charge balanced by the −1 charge on the chlorine.

$$Na + \; :\!\ddot{Cl}\!: \longrightarrow Na^+ \; :\!\ddot{Cl}\!:^- \qquad \text{(sodium chloride)}$$

Formula = NaCl

Next, we will look at some other reactions such as when lithium metal comes into contact with the oxygen in the air. Since a lithium atom can lose only one electron, two lithium atoms are needed to supply the two electrons needed by one oxygen atom. Recall that the two +1 ions balance the charge of the one −2 ion [i.e., 2(+1) + (−2) = 0]. The chemical formula of the compound lithium oxide is therefore Li_2O.

$$Li\cdot$$
$$+ \; \cdot \ddot{O}\!: \longrightarrow 2(Li^+) \; :\!\ddot{O}\!:^{2-} \qquad \text{(lithium oxide)}$$
$$Li\cdot \qquad \text{Formula} = Li_2O$$

The opposite situation occurs for the formation of $CaBr_2$ where one calcium atom loses two electrons to two bromine atoms. Finally, we will include a slightly more complex case involving the reaction of aluminum with oxygen to form aluminum oxide. Group IIIA metals (Al and lower) can lose three electrons in order to form an octet of electrons.* In this reaction, two Al atoms give up six electrons, which are then accepted by three O atoms. The formula is thus Al_2O_3, and the charges cancel [2(+3) + 3(−2) = 0].

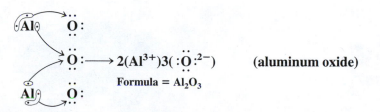

$$\longrightarrow 2(Al^{3+})3(\;:\!\ddot{O}\!:^{2-}) \qquad \text{(aluminum oxide)}$$

Formula = Al_2O_3

9-2.2 Other Representative Element Ions

Group IVA metals such as tin and lead have four electrons in their outer subshells. Loss of all four of these electrons to produce a +4 ion requires a rather large amount of energy. Instead, these metals can lose two of their four outer electrons to form a +2 ion that does not follow the octet rule. They do form compounds

---

*Ions such as Tl^{3+} and Ga^{3+} have a filled *d* subshell in addition to a noble gas configuration. This is sometimes referred to as a *pseudo–noble gas* configuration. The filled *d* subshell does not seem to affect the stability of these ions. In this text, we do not distinguish between noble gas and pseudo–noble gas electron configurations. Transition metals also form positive ions, but for the most part these ions do not relate to a noble gas configuration. Some of these ions were discussed in Chapter 4.

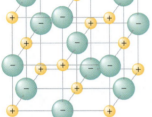

where all four of their outer electrons are involved, but the bonding in these compounds is best described by electron sharing rather than ion formation. For the most part, Group IVA nonmetals also bond by electron sharing rather than forming monatomic ions. Although there is some evidence for a C^{4-} ion with an octet of electrons, formation of such highly charged ions is an energetically unfavorable process.

In Group VA, bismuth forms a +3 ion that does not follow the octet rule.

9-2.3 The Physical State of Ionic Compounds

Much of the solid, hard surface of our Earth is composed of ionic compounds that tend to have high melting points and are usually hard and brittle. If we look into the basic structure of a crystal of table salt, we can see why. Ionic compounds do not exist as discrete molecular units with one Na^+ attached to one Cl^-. As shown in Figure 9-2, each Na^+ is actually surrounded by six Cl^- ions, and each Cl^- ion is surrounded by six Na^+ ions in a *three-dimensional array of ions called a* **lattice**. Recall from Chapter 8 that cations are smaller than their parent atoms, whereas anions are larger. Thus, in most cases we can assume that the anion is larger than the cation. The lattice is held together strongly and rigidly by electrostatic interactions. *These electrostatic attractions are known as* **ionic bonds**. There are several other arrays of ions (lattices) used to accommodate both the size differences and varying ratios of cations to anions found in all of the possible ionic compounds. For example, in CsCl both the Cs^+ and the Cl^- are surrounded by eight oppositely charged ions.

Besides the monatomic ions, polyatomic ions exist where two or more atoms are bound together by electron sharing, but the total entity carries a net charge [e.g., the carbonate ion $(CO_3{}^{2-})$]. These species exist as ions because they have an imbalance of electrons compared to the total number of protons in their nuclei. We will discuss the bonding within a polyatomic ion later in this chapter, but for now we acknowledge their presence in ionic compounds.

FIGURE 9-2 An Ionic Solid Table salt (photo) is composed of cations and anions in a specific geometric lattice.

Rubies and sapphires are hard substances found in nature. They are ionic compounds.

EXAMPLE 9-1   Writing Formulas of Binary Ionic Compounds

Using the Lewis dot symbols, determine the formulas of the ionic compounds formed between **(a)** aluminum and fluorine and **(b)** barium and sulfur.

PROCEDURE

Use the octet rule and the Lewis dot symbol to determine the number of electrons to be lost by the metal and to be gained by the nonmetal. Determine the correct ratio of cation to anion so that the total number of electrons transferred from cations to anions is the same.

SOLUTION

(a) Aluminum is in Group IIIA and fluorine is in Group VIIA. Their dot symbols are

$$\text{Al}\cdot \qquad \cdot\ddot{\text{F}}:$$

To have a noble gas configuration (an octet), the Al, a metal, must lose all three outer electrons to form a +3 ion. Fluorine can add only one electron, which forms a −1 ion and gives the F⁻ ion an octet. Three fluorine atoms are needed to balance the charge. The compound formed is

$$\text{Al}^{3+}\ 3(:\ddot{\text{F}}:^{-}) = \underline{\underline{\text{AlF}_3}} \qquad \text{(aluminum fluoride)}$$

(b) Barium is in Group IIA and sulfur is in Group VIA, and they have the dot symbols

$$\text{Ba}\cdot \qquad \ddot{\text{S}}:$$

One Ba atom gives up two electrons, and one S atom takes up two electrons, forming the compound

$$\text{Ba}^{2+}:\ddot{\text{S}}:^{2-} = \underline{\underline{\text{BaS}}} \qquad \text{(barium sulfide)}$$

ANALYSIS

This example is less about teaching you something new than it is about providing stronger evidence for something we've already learned. In Chapters 2 and 4 we illustrated how to write the formulas of ionic compounds by assuming that certain elements would have specific charges. Now, based on the octet rule, valence electrons, and Lewis dot formulas, we provide the underlying rationale for these charges that previously we asked you to accept without discussion.

SYNTHESIS

Tying this in with other previously learned material, we realize that the strength of the ionic bonds affects whether an ionic solid is soluble or insoluble in water. When the ions are held together more tightly, it becomes harder to break them apart, and so harder to get them to dissolve. Certainly other factors are present, but the strength of the ionic bond is a major component to the calculation. Would you expect bigger or smaller ions to be held together more tightly? +1's with −1's or +2's with −2's? In fact, the higher charges lead to a stronger attraction between the ions in the lattice (known as the *lattice energy*). Less intuitively, smaller ions generate higher attractions than larger ones because the ions can get closer to each other. According to Coulomb's law, the closer charges (ions) are to each other, the greater the attraction and the harder to break apart.

▶ **ASSESSING THE OBJECTIVE FOR SECTION 9-2**

EXERCISE 9-2(a) LEVEL 1: What are the charges on the ions formed by Be and As?

EXERCISE 9-2(b) LEVEL 2: What is the formula of the compound formed when Be combines with As?

EXERCISE 9-2(c) LEVEL 2: What are the formulas of the *ionic* compounds formed between chlorine and each of lead, tin, and bismuth?

EXERCISE 9-2(d) LEVEL 3: What is the general formula ($M_?X_?$) when a IIIA metal **M** bonds with a VIA nonmetal **X**?

EXERCISE 9-2(e) LEVEL 3: When we write a formula for an ionic compound such as potassium nitride, what do we mean when we write K_3N?

For additional practice, work chapter problems 9-8, 9-9, 9-14, 9-19, 9-22, and 9-25.

PART A SUMMARY

KEY TERMS

9-1.1 The formulas of binary ionic compounds can be understood by reference to the **octet rule**. p. 276

9-1.2 The **valence electrons** of the atoms of representative elements can be displayed as **Lewis dot symbols**. p. 276

9-2.3 The **lattice** of an ionic compound is the arrangement of ions in a crystal. p. 280

9-2.3 The electrostatic interactions of the oppositely charged ions in a crystal are known as **ionic bonds**. p. 280

The summary of ionic compounds is included in the Summary Chart after Part B.

Part B

Chemical Bonds and the Nature of Molecular Compounds

SETTING A GOAL

■ You will learn how to apply the octet rule to draw structures that form the basis of our understanding of bonding in most molecular compounds.

OBJECTIVES

9-3 Apply the octet rule to determine the number of bonds formed in simple compounds.

9-4 Draw Lewis structure of a number of molecular compounds and polyatomic ionic compounds.

9-5 Write multiple Lewis structures for molecules capable of resonance.

9-6 Determine the validity of a Lewis structure based on formal charge considerations.

▶ **OBJECTIVE FOR SECTION 9-3**
Apply the octet rule to determine the number of bonds formed in simple compounds.

9-3 The Covalent Bond

LOOKING AHEAD! We are now ready to turn our attention to compounds that may exist as gases and liquids as well as solids. Since metals are not involved in these compounds, an exchange of electrons does not occur. In these cases, the octet rule applies to the sharing of electrons. ■

In binary ionic compounds, octets are achieved when metals lose electrons and nonmetals gain electrons. But what if both elements are nonmetals, so neither has a tendency to give up electrons? In that case, octets are obtained by electron sharing. The shared electrons are counted toward each element's octet of electrons. For example, hydrogen and oxygen (both nonmetals) combine in a dramatic chemical reaction (i.e., an explosion) to form water if initiated by a spark. In the formation of water, electrons are not completely exchanged but instead are shared between each of the two hydrogen atoms and the oxygen atom, resulting in a neutral molecule. Compounds composed of neutral molecules such as water and carbon dioxide have properties very different from compounds composed of ions such as sodium chloride.

9-3.1 Lewis Structures

The fluorine atom (Group VIIA) has seven valence electrons. Fluorine can achieve an octet in two ways. With a metal, it can gain one electron to form the fluoride ion as described previously, or with another nonmetal it can share one electron with another atom to have access to eight electrons. For example, two fluorine atoms can achieve an octet of electrons by sharing two electrons, one from each fluorine. *The sharing of two electrons between two atoms is known as a* **covalent bond**. The bonding in the F_2 molecule, and all of the other diatomic halogens, is illustrated as follows.

$$: \ddot{F} \cdot \longrightarrow : \ddot{F} \! : \! \ddot{F} :$$

Shared pair of electrons
(one from each F)

It is easy to appreciate how a complete exchange of electrons can satisfy the octet rule, but the concept of electron sharing and the octet rule is more subtle. A simple analogy may help. Assume we have a young man (Henry) who has $7. Henry wishes to have access to $8 and, for some strange reason, no more than $8. Now let Henry happen on to an even weirder person who also wishes to have no more than $8 but instead has $9. An exchange of $1 to Henry leaves them both happy. This is analogous to a metal and a nonmetal forming an ionic bond, with the metal giving its extra electron to the nonmetal. In a second situation, assume that two people have only $7 each and, again, both wish access to $8. There is a solution to the dilemma. If they keep $6 in their own pockets and contribute $1 each (for a total of $2) to a joint checking account, then both can claim access to $8 (but no more than $8). That is, each has $6 plus access to the $2 in the joint account.

We can now extend the concept of Lewis dot representations to molecular compounds containing covalent bonds. *A* **Lewis structure** *for a molecule shows the order and arrangement of atoms in a molecule (the structural formula) as well as all of the valence electrons for the atoms involved.* There are several variations of how Lewis structures represent molecules. A pair of electrons is sometimes shown as a pair of dots (:) or as a dash (—). In this text, we use a pair of dots to represent unshared pairs (also called *lone pairs*) of electrons on an atom and a dash to represent a pair of electrons that are shared between atoms. In this way, shared and unshared electrons can be distinguished.

The Lewis structure of $F_2{}^*$ is illustrated as follows:

Total of 14 outer electrons (7 from each F)

$$: \ddot{F} \! - \! \ddot{F} \! : \quad \text{Three lone pairs on each F}$$

Two shared electrons in a covalent bond

Similarly, other halogens exist as diatomic molecules like F_2 and have the same Lewis structures. Hydrogen, which forms the simplest of all molecules, also exists as a diatomic gas with one covalent bond between atoms:

$$H—H$$

Recall that hydrogen follows a duet rule in order to attain the noble gas configuration of He.

It is important to note that two atoms of hydrogen are more stable bonded together than as two separate atoms. Thank goodness! If the chemical bond were not stable and strong, the Earth and even all of us would not exist. The universe would just be composed of individual atoms. What holds the two hydrogen atoms together? The reason goes back to modern atomic theory discussed in the previous chapter. A simple way of viewing a covalent bond is as arising from the overlapping of half-filled atomic orbitals, one from each atom. To be more descriptive, two hydrogen

---

[*]An F atom has one unpaired electron in a $2p$ orbital. Formation of a covalent bond pairs the electrons in the two F atoms so that the F_2 molecule has no unpaired electrons. Although most atoms of the representative elements have unpaired electrons, most molecules or ions formed from these elements do not have unpaired electrons.

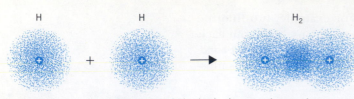

FIGURE 9-3 A Covalent Bond The high electron density between the two nuclei holds the two atoms together.

atoms alone have one electron in a $1s$ orbital, which is spherically diffuse. The electron has a probability of existing in any direction from the nucleus in the separate atoms. When two hydrogen atoms come together, however, the two electrons become more localized between the two nuclei in a region where the two $1s$ orbitals overlap. Each positive hydrogen nucleus is attracted to two negative electrons between the atoms, rather than just one. Although there are also forces of repulsion between the two electrons, the mutual attraction of two nuclei for two electrons predominates and holds them together. (See Figure 9-3.) All other covalent bonds can also be viewed as arising from the overlap of atomic orbitals.

9-3.2 Hydrogen Compounds with other Nonmetals

Just as we were able to justify the formulas of simple binary ionic compounds by the octet rule, we can do the same with simple binary molecular compounds. In fact, this works so well that we can predict the formulas of compounds based on the octet rule.

First, we will consider the compounds formed by hydrogen with the halogens in Group VIIA. For example, consider the compound formed from hydrogen and fluorine, which has the formula HF. A shared pair of electrons (one from each atom) gives both atoms access to the same number of electrons as a noble gas.

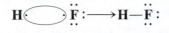

Now consider the compounds formed between hydrogen and the Group VIA elements. Our primary example, of course, is water. Since one oxygen atom needs access to two more electrons to have an octet, two hydrogen atoms are required to form two covalent bonds to one oxygen atom.

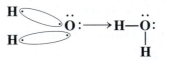

As we move across the periodic table to consider the hydrogen compounds formed between Group VA and Group IVA nonmetals, we see that the octet rule serves us well. Three hydrogen atoms are needed by N (Group VA) and four by C (Group IVA). Recall that hydrogen is written second in binary compounds with Group IVA and VA elements but is written first with Group VIA and VIIA elements. (See also some hydrogen compounds of third-period elements in Figure 9-4.)

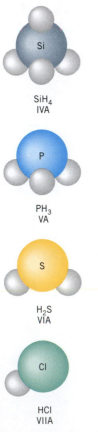

SiH₄
IVA

PH₃
VA

H₂S
VIA

HCl
VIIA

FIGURE 9-4 Formulas of Hydrogen Compounds The formulas of some simple hydrogen compounds can be predicted from the octet rule.

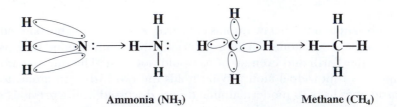

Ammonia (NH₃)   Methane (CH₄)

Let's try to predict the formula of the simplest compound formed between phosphorus (Group VA) and fluorine (Group VIIA). Since the dot symbol of P indicates that it needs access to three more electrons and F needs access to one more, the solution points to one P sharing a pair of electrons with each of three different F atoms.

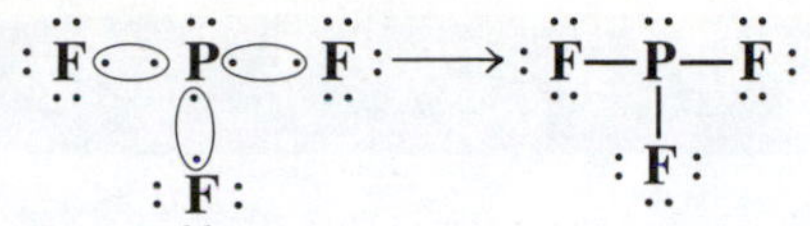

$$: \ddot{F} \ominus \ddot{P} \ominus \ddot{F} : \longrightarrow : \ddot{F} - P - \ddot{F} :$$

$$\underset{\ddot{\ddot{F}}}{\overset{\circ}{\underset{\cdot\cdot}{|}}} \qquad \underset{\ddot{\ddot{F}}}{\overset{|}{\underset{\cdot\cdot}{|}}}$$

Simplest formula = PF$_3$

9-3.3 The Electrons in Polyatomic Ions

In addition to molecular compounds, atoms within polyatomic ions share electrons in covalent bonds. For example, consider the hypochlorite ion (ClO$^-$). The -1 charge on the ion tells us that there is one more electron present in this species than the valence electrons provided by one Cl atom and one O atom. The total number of electrons is calculated as follows.

$$
\begin{aligned}
\text{from a neutral Cl} &= 7 \\
\text{from a neutral O} &= 6 \\
\text{additional electron indicated by charge} &= \underline{1} \\
\text{total number of electrons} &= 14
\end{aligned}
$$

Two atoms bonded together with 14 electrons have a Lewis structure like F$_2$, which also has 14 electrons.

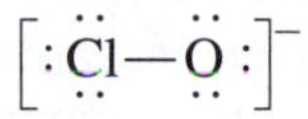

$$\left[: \ddot{Cl} - \ddot{O} : \right]^-$$

The brackets indicate that the total ion has a -1 charge. The extra electron has not been specifically identified because all electrons are identical and belong to the ion as a whole.

9-3.4 Double and Triple Bonds

In the examples illustrated so far, two atoms share one pair of electrons. There are also examples, especially among the second-period nonmetals (B through O), where two or even three pairs of electrons are shared between two atoms. *The sharing of two pairs of electrons between the same two atoms is known as a* **double bond**. *The sharing of three pairs is known as a* **triple bond**. A double bond is illustrated as ═, and a triple bond is illustrated as ≡. The use of a double bond to satisfy the octet rule is analogous to two people who wish to have access to $8 but only have a total of $12 between them. In this case, each person could have $4 in a private account while sharing $4 in a joint account. A triple bond would be analogous to the two people having only $10 between them. Each could have only $2 in a private account while sharing $6 in the joint account.

The math involved and the octet rule for multiple bonds is illustrated here.

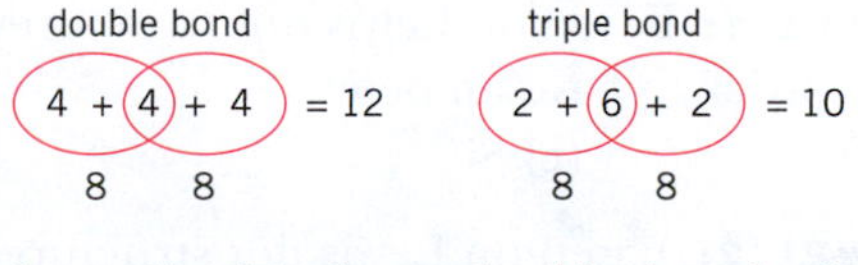

<table>
<tr><td>double bond</td><td>triple bond</td></tr>
<tr><td>4 + ④ + 4 = 12</td><td>2 + ⑥ + 2 = 10</td></tr>
<tr><td>8   8</td><td>8   8</td></tr>
</table>

The molecules of carbon dioxide have double bonds. In the Lewis structure shown below, notice that the octets of both carbon and oxygen are satisfied by the sharing of two pairs of electrons in each carbon–oxygen bond. Elemental nitrogen, N$_2$, is an example of a molecule that satisfies the octet rule with a triple bond.

$$\ddot{O} = C = \ddot{O} \qquad\qquad : N \equiv N :$$

MAKING IT REAL

Nitrogen: From the Air to Proteins

Field of soybeans.

Our bodies are literally held together by proteins. In the form of enzymes, proteins are involved in virtually every chemical reaction that occurs inside us. Proteins, which are composed of various amino acids such as glycine (H_2NCH_2COOH), contain nitrogen. Therefore, nitrogen is essential to life. Fortunately, we literally live in a sea of nitrogen, since about 80% of our atmosphere is in the form of N_2. Unfortunately, the change from the elemental form of nitrogen into the chemical bonds in proteins is not an easy path, since N_2 is very hard to break down.

Elemental nitrogen (N_2) contains a triple bond. Triple bonds between the same two elements are stronger than double bonds and much stronger than single bonds. The triple bond must be broken, however, before nitrogen can be taken from the air and incorporated into compounds. Nature breaks down nitrogen so it can form compounds in two ways, both of which are known generally as *nitrogen fixation*.

One way that nitrogen can be fixed is with the energy from a powerful bolt of lightning strong enough to break its triple bond. When the N_2 bond is broken, the free nitrogen atoms combine with oxygen in the air to form nitric oxide (NO), which eventually falls to Earth dissolved in the rain. However, this path does not provide nearly enough fixed nitrogen for plants and animals. The second,

more important, way is through the action of certain plants, such as beans and peas (legumes), that have bacteria attached to their roots. These bacteria contain an enzyme known as *nitrogenase*, which has the unusual ability to break the N_2 triple bond and produce nitrogen compounds that can be used by their host plants. This is how nature fertilizes crops. With nearly 7 billion people on Earth, however, nature isn't nearly enough, so farmers must now supplement their crops with huge amounts of ammonia (NH_3). This compound is manufactured from its elements in a difficult and expensive process that is discussed in Chapter 15.

Both academic and industrial chemists are searching for an inexpensive and continuous way to fix nitrogen. The goal is to duplicate the action of nitrogenase in plants, but the process is very complex and only partially understood. Nevertheless, there is active research at various laboratories where scientists are hoping to find an agent (known as a catalyst) that will absorb atmospheric nitrogen, form bonds between nitrogen and other elements (i.e., hydrogen and oxygen), release the new compound, and then regenerate the original agent. The right process has been elusive, but the benefits of future success are enormous.

Imagine the formation of a solid compound that could be spread on the surface of a field or garden that would use solar energy to continually generate nitrogen fertilizers directly from the air. Crop yields would increase, and overused and misused farmland could be made fertile again. The payoff for such a substance would be immeasurable.

▶ **ASSESSING THE OBJECTIVE FOR SECTION 9-3**

EXERCISE 9-3(a) LEVEL 1: Based on Lewis dot structures, how many bonds must the following make in order to have an octet?
(a) P   **(b)** Br   **(c)** Si

EXERCISE 9-3(b) LEVEL 2: Based on Lewis dot structures, what is the formula of the simplest molecule that can be formed from the following?
(a) As and H   **(b)** C and Cl   **(c)** Cl and F   **(d)** S and H

EXERCISE 9-3(c) LEVEL 3: We've seen examples of single, double, and triple bonds. Could a quadruple bond exist?

For additional practice, work chapter problems 9-26, 9-27, and 9-28.

9-4 Writing Lewis Structures

LOOKING AHEAD! The Lewis structures of simple binary compounds with single covalent bonds can be written without too much difficulty by following the octet rule. The writing of other structures can be more complex, so a set of rules or guidelines is most helpful. In this section, we will see how a sequential approach will make this important task quite manageable. ■

The octet rule is the key that allows us to write the correct Lewis structures for many compounds and polyatomic ions. From this we not only justify the formulas of compounds (e.g., H_2O, not H_3O) but can predict other formulas as well. Other features of a compound follow from the correct Lewis structure. For example, later in this chapter we will use the Lewis structure to predict the geometry of some simple molecules. Many physical and chemical properties of a compound are directly related to its geometry. As we progress in the study of chemistry, we will continually refer to the Lewis structures of many compounds. Therefore, writing Lewis structures correctly is considered a fundamental skill to be acquired early in the study of chemistry.

9-4.1 Rules for Writing Lewis Structures

Writing Lewis structures according to the octet rule is quite straightforward when certain guidelines or rules are systematically applied. These rules, of course, require considerable practice in their application. Starting with the formula of a compound (either ionic or molecular), the rules are as follows.

1. Check to see whether any ions are involved in the compounds. Write any ions present.

   a. Metal–nonmetal binary compounds are mostly ionic.

   b. If Group IA or Group IIA or Al metals (except Be) are part of the formula, ions are present. For example, KClO is K^+ ClO^- because *K is a Group IA element and forms only a +1 ion*. If K is +1, the ClO must be −1 to have a neutral compound. Likewise, $Ba(NO_3)_2$ contains ions because *Ba is a Group IIA element and forms only a +2 ion*. The formula also indicates the presence of two nitrate ions. The ions are represented as Ba^{2+} and $2(NO_3^-)$.

   c. Compounds composed of nonmetals contain only covalent bonds.

2. For a molecule, add all of the outer (valence) electrons of the neutral atoms. Recall that the number of valence electrons is the same as the column number (or last digit when columns are labeled 1–18). For an ion, add (if negative) or subtract (if positive) the number of electrons indicated by the charge.

3. Write the symbols of the atoms of the molecule or ion in a skeletal arrangement.

   a. A hydrogen atom can form only one covalent bond and therefore bonds to only one atom at a time. Hydrogen atoms are situated on the periphery of the molecule.

   b. The atoms in molecules and polyatomic ions tend to be arranged symmetrically around a central atom. The central atom is generally a nonmetal other than oxygen or hydrogen. Oxygen atoms usually do not bond to each other. Thus SO_3 has an S surrounded by three O's.

$$\begin{array}{c} O \\ S \\ O \quad O \end{array}$$

rather than such structures as

$$S\ O\ O\ O \qquad O\ S\ O\ O \qquad \begin{array}{cc} S & O \\ O & O \end{array}$$

In most cases, the first atom in a formula is the central atom, and the other atoms are bound to it. Usually the central atom is the atom found farther to the left or lower down on the periodic table (e.g., CO_2, PCl_3, OF_2).

4. Place a dash representing a shared pair of electrons between adjacent atoms that have covalent bonds (not between ions). Subtract the electrons used for this (two for each bond) from the total calculated in step 2.

5. Distribute the remaining electrons among the atoms so that no atom has more than eight electrons, starting with the atoms on the outside of the structure first.

6. Check all atoms for an octet (except H). If an atom has access to fewer than eight electrons (usually the central atom), put an electron pair from an adjacent atom into a double bond. Each double bond increases by two the number of electrons available to the atom needing electrons without taking them away from the other atom. Remember that you cannot satisfy an octet for an atom by adding any electrons at this point.

EXAMPLE 9-2   Writing the Lewis Structure of a Molecular Compound

Write the Lewis structure for NCl_3.

PROCEDURE

Follow the procedure as outlined above.

SOLUTION

1. This is a binary compound between two nonmetals. Therefore, it is not ionic.

2. The total number of electrons available for bonding is

$$N\ 1 \times 5 = 5$$
$$Cl\ 3 \times 7 = \underline{21}$$
$$Total = 26$$

3. The skeletal arrangement is

$$Cl \quad N \quad Cl$$
$$Cl$$

4. Use 6 electrons to form bonds.

$$Cl—N—Cl$$
$$|$$
$$Cl$$

5. Distribute the remaining 20 electrons (26 − 6 = 20).

$$:\overset{..}{Cl}—N—\overset{..}{Cl}:$$
$$|$$
$$:\overset{..}{Cl}:$$

6. Check to make sure that all atoms satisfy the octet rule.

EXAMPLE 9-3   Writing the Lewis Structure of an Ion

Write the Lewis structure of the cyanide ion (CN⁻).

SOLUTION

1. This is an ion.

2. The total number of electrons available is

$$
\begin{aligned}
\text{N } 1 \times 5 &= 5 \\
\text{C } 1 \times 4 &= 4 \\
\text{From charge} &= \underline{1} \\
\text{Total} &= 10
\end{aligned}
$$

3. The skeletal arrangement is

$$\text{C \quad N}$$

4. Use two electrons to form a bond.

$$\text{C—N}$$

5. Distribute the remaining eight electrons ($10 - 2 = 8$).

$$\left[:\overset{..}{\text{C}}-\overset{..}{\text{N}}: \right]^{-}$$

6. Notice that both carbon and nitrogen have access to only six electrons each. Use two electrons from the carbon and two electrons from nitrogen to make a triple bond. Now the octet rule is satisfied.

$$\left[:\text{C}\equiv\text{N}: \right]^{-}$$

EXAMPLE 9-4   Writing the Lewis Structure of an Ionic Compound

Write the Lewis structure for $CaCO_3$.

SOLUTION

1. This is an ionic compound composed of Ca^{2+} and CO_3^{2-} ions. (Since you know that Ca is in Group IIA, it must have a +2 charge; therefore, the polyatomic anion must be –2.) A Lewis structure can be written for CO_3^{2-}.

2. For the CO_3^{2-} ion, the total number of outer electrons available is

$$
\begin{aligned}
\text{C } 1 \times 4 &= 4 \\
\text{O } 3 \times 6 &= 18 \\
\text{From charge} &= \underline{2} \\
\text{Total} &= 24
\end{aligned}
$$

3, 4. The skeletal structure with bonds is

$$
\begin{array}{ccc}
\text{O} & & \text{O} \\
 & \text{C} & \\
 & \text{O} &
\end{array}
$$

5. Add the remaining 18 electrons ($24 - 6 = 18$).

6. The C needs two more electrons, so one double bond is added using one lone pair from one oxygen.

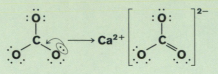

EXAMPLE 9-5   Writing the Lewis Structure of an Oxyacid

Write the Lewis structure for H_2SO_4.

SOLUTION

1. All three atoms are nonmetals, which means that all bonds are covalent.

2. The total number of outer electrons available is

$$
\begin{aligned}
&\text{H } 2 \times 1 = 2 \\
&\text{S } 1 \times 6 = 6 \\
&\text{O } 4 \times 6 = \underline{24} \\
&\phantom{\text{O } 4 \times 6} \text{Total} = 32
\end{aligned}
$$

3. In most molecules containing H and O, the H is bound to an O and the O to some other atom, which in this case is S. The skeletal structure is

$$
\begin{array}{ccccc}
 & & \text{O} & & \\
\text{H} & \text{O} & \text{S} & \text{O} & \text{H} \\
 & & \text{O} & &
\end{array}
$$

4. Use 12 electrons for the six bonds.

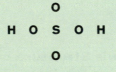

5. Add the remaining 20 electrons ($32 - 12 = 20$).

$$
\text{H}-\ddot{\text{O}}-\overset{\displaystyle :\ddot{\text{O}}:}{\underset{\displaystyle :\ddot{\text{O}}:}{\text{S}}}-\ddot{\text{O}}-\text{H}
$$

6. All octets are satisfied.

ANALYSIS

The Lewis structure of a molecule clearly shows how atoms are attached to each other and with what types of bonds. It also tells us how many unshared electrons are found in the valence shell of each atom in the structure. Later in this chapter, we'll examine the other information contained in a Lewis structure that begins to build on itself and serves as a basis for the next several chapters of the text. Your success in studying chemistry from this point onward is going to depend on your ability to accurately draw Lewis structures. You're encouraged to work as many examples as are necessary to become proficient at the procedure.

9-4.2 Exceptions to the Octet Rule

The octet rule is very useful in describing the bonding in most compounds of the representative elements. A significant number of compounds, however, do not follow the octet rule. For example, a few molecules have an odd number of valence electrons, such as nitric oxide (NO—11 valence electrons) and nitrogen dioxide (NO_2—17 valence electrons). In both of these cases the nitrogen has access to fewer than 8 electrons. *When there are an odd number of electrons in a molecule it is known as a* **free radical**. It turns out that free radicals such as these are very reactive because they are missing the stability associated with a filled octet and paired electrons.

In other molecules, a Lewis structure may be written that follows the octet rule, but other evidence suggests that the situation is more complex. For example, the ordinary O_2 molecule with 12 valence electrons could be written with a double bond as shown below. This representation implies that all electrons are in pairs. However, we know from experiments that O_2 is also a free radical with two unpaired electrons. We just need to remember that the Lewis structure is simply the representation of a theory; in the case of O_2, the theory doesn't work perfectly. There are other theories of bonding that explain the case of O_2 quite well, although these theories are more complicated. Another example of the conflict between theory and experiments is illustrated by the molecule BF_3. Experiments indicate that the B—F bond has little or no double-bond character. Thus, the correct structure shows the boron with access to only 6 electrons.

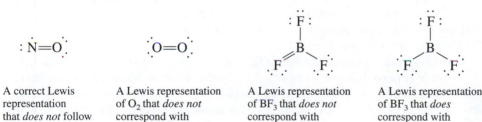

| A correct Lewis representation that *does not* follow octet rule | A Lewis representation of O_2 that *does not* correspond with experiments | A Lewis representation of BF_3 that *does not* correspond with experiments | A Lewis representation of BF_3 that *does* correspond with experiments |

There are a significant number of compounds involving representative elements where the central atoms are in the third and higher periods. In some of these compounds the central atom has access to more than eight electrons (e.g., SF_4 and ClF_5). The bonding in these compounds will not be discussed in this text.

▶ **ASSESSING THE OBJECTIVE FOR SECTION 9-4**

EXERCISE 9-4(a) LEVEL 1: How many valence electrons are in the following compounds or ions?

(a) CS_2 **(b)** NO_3^- **(c)** $SOCl_2$

EXERCISE 9-4(b) **LEVEL 1:** Draw Lewis structures for the following.

(a) SF_2 **(b)** NO_2Cl (with N as the central atom)
(c) $Mg(ClO_2)_2$

EXERCISE 9-4(c) **LEVEL 2:** Draw Lewis structures for the following.

(a) HCN   **(b)** C_2H_4 **(c)** PO_3^{3-}

EXERCISE 9-4(d) **LEVEL 3:** Two compounds of oxygen are named oxygen difluoride and dioxygen difluoride. Write the Lewis structures of these two compounds. (The latter compound contains an O—O bond.) Oxygen is usually written and named second in binary compounds. Why not here?

For additional practice, work chapter problems 9-33, 9-34, and 9-35.

▶ **OBJECTIVE**
   FOR SECTION 9-5
Write multiple Lewis structures for
molecules capable of resonance.

9-5 Resonance Structures

LOOKING AHEAD! Glance again at the Lewis structure of the carbonate ion shown in Example 9-4. The structure that is displayed implies that the three carbon–oxygen bonds are not all identical in that one bond is double and the other two are single. Is that true? Actually, the answer is no, but we need to explore this phenomenon in more detail. ∎

If we compare a double bond to a single bond between the same two elements, we find there are significant differences. The sharing of four electrons holds two atoms together more strongly, and thus more closely, than the sharing of two electrons. Likewise, a triple bond is even stronger and shorter than a double bond. The one Lewis structure of the CO_3^{2-} ion, shown in Example 9-4, implies that one carbon–oxygen bond is shorter and stronger than the other two. We know from experiment, however, that the ion is perfectly symmetrical, meaning that all three bonds are identical and just as likely to contain the extra electrons. Experiments also tell us that the lengths of the three identical bonds are somewhere between those expected for a single and a double bond.

One Lewis representation of the CO_3^{2-} ion does not convey this information, but three representations (connected by double-headed arrows) illustrate that all three bonds are just as likely to contain the extra pair of electrons. *The three structures as shown below are known as* **resonance structures**. The actual structure of the molecule can be viewed as a **resonance hybrid**, or average, of the three structures. Our understanding, then, is that the extra pair of electrons is spread evenly over the three carbon-oxygen bonds in the molecule, making each equivalent to 1 1/3 bonds.

$$\left[\begin{array}{c} :\ddot{O}: \\ | \\ :\ddot{O}\diagup{C}\diagdown\ddot{O}: \end{array} \right]^{2-} \longleftrightarrow \left[\begin{array}{c} :\ddot{O}: \\ \| \\ :\ddot{O}\diagdown{C}\diagup\ddot{O}: \end{array} \right]^{2-} \longleftrightarrow \left[\begin{array}{c} :\ddot{O}: \\ | \\ :\ddot{O}\diagdown{C}\diagup\ddot{O}: \end{array} \right]^{2-}$$

Resonance structures exist for molecules where equally correct Lewis structures can be written without changing the basic skeletal geometry or the position of any atoms. An example of a molecule with resonance structures is a form of elemental oxygen known as ozone (O_3). (Different forms of the same element are known as *allotropes*.) Ozone has very different properties than the

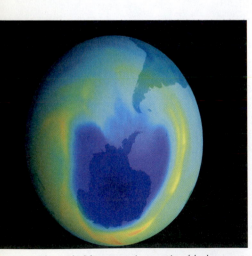

The pale blue over Antarctica (dark blue) in October 1999 shows the area of ozone depletion.

other allotrope of oxygen, O_2, which maintains life. Ozone is a critical component in the stratosphere because it absorbs ultraviolet light from the sun. The resonance structures of ozone are illustrated in the following example.

EXAMPLE 9-6   Writing Resonance Structures

Write a Lewis structure and any equivalent resonance structures for ozone (O_3), where one oxygen serves as the central atom.

PROCEDURE

Follow the procedure for writing Lewis structures. When the time comes to form double or triple bonds, draw all possibilities separately, and connect the structures with double-headed arrows.

SOLUTION

1. No ions are involved.

2. There are 18 (3×6) outer electrons.

3. The skeletal structure is

O

O   O

(Later in the chapter, we will discuss why we write the basic structure as bent rather than linear.)

4. Use 4 electrons for the two bonds.

5. Add the remaining 14 electrons.

:O̤—O̤—O̤:

6. Notice that the central oxygen does not have an octet. A double bond must be made from one of the outer oxygen atoms. Notice that it doesn't matter which oxygen is chosen; the two structures are equivalent. Draw both possibilities, which are the two resonance structures. The resonance structures indicate that each oxygen–oxygen bond is a hybrid between a single and a double bond.

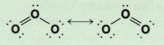

ANALYSIS

Molecules with double or triple bonds somewhere in the structure have the possibility of resonance. Once the skeletal structure is in place, see if the double or triple bonds can be placed in the structure in more than one way to make a valid Lewis structure. If so, the molecule has resonance structures.

SYNTHESIS

Resonance is an important concept for chemists because it is used to explain many observations in the lab. For example, it is found that molecules and ions that have resonance structures are inherently more stable than molecules that do not. Where normal bonds hold only two atoms together (the two that share the electrons), the electrons used in resonating bonds can hold three, four, or more atoms together at the same time. A chain is only as strong as its weakest link, and a molecule is only as stable as its weakest bond. A bond is stronger if it is halfway between a single and a double bond rather than just a single bond. Chemists often invoke resonance to explain why some molecules' formation is more favorable than others. When the choice is between two, the one that has resonance is highly favored.

The word *resonance* is often associated with vibration or a constantly changing situation. That concept can be misleading in this case. The two oxygen–oxygen bonds in ozone are not changing rapidly back and forth between a single and a double bond. In fact, both bonds exist *at all times* as intermediate between a single and a double bond. It is much like a large, sweet hybrid tomato that you grow in the garden or buy in the grocery store. This tomato is a hybrid of a large tomato and a small but sweet tomato. It isn't changing rapidly back and forth between these two forms but has properties intermediate between the two original species of tomatoes. When we view resonance structures, we try to visualize the molecule as a combination of the two or more structures. This is analogous to trying to taste the hybrid tomato by combining a piece of the large tomato with a piece of the small, sweet tomato in your mouth at the same time.

▶ **ASSESSING THE OBJECTIVE FOR SECTION 9-5**

EXERCISE 9-5(a) LEVEL 2: Draw four resonance structures for N_2O_3. Notice that not all of these structures are equivalent. The skeletal structure is

$$O-N-N\begin{matrix}O\\\\O\end{matrix}$$

EXERCISE 9-5(b) LEVEL 2: Which of the following molecules have resonance structures?
(a) HCN   **(b)** SO_2 **(c)** PCl_3 **(d)** OCN^-

EXERCISE 9-5(c) LEVEL 3: A possible Lewis structure for CO_2 involves a triple bond and a single bond between C and the O's. Write the two resonance structures involving the triple bond. What is implied about the nature of the C—O bond by these two structures? How does this relate to the common Lewis structure for CO_2 involving two double bonds?

For additional practice, work chapter problems 9-42, 9-43, and 9-46.

▶ **OBJECTIVE**
FOR SECTION 9-6
Determine the validity of a Lewis structure based on formal charge considerations.

9-6 Formal Charge

LOOKING AHEAD! Sometimes a properly written Lewis structure may not actually represent the true structure of the molecule. How would we know this? It turns out that there is a way we can determine whether a structure is indeed correct. The concept of formal charges in the molecule can give us information in these cases. ■

9-6.1 Resonance Structures that Violate the Octet Rule

The representation of the sulfuric acid molecule (H_2SO_4), as shown in Example 9-5, follows the octet rule quite nicely. However, if one proceeds to the study of organic chemistry and biochemistry, the molecule is usually represented with two double bonds in violation of the octet rule, as shown here.

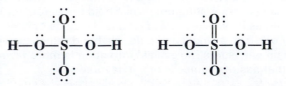

Octet rule   Alternate representation

Why represent the molecule in this manner? The answer is that the latter structure represents a more favorable formal charge distribution. **Formal charge** *is the charge that each atom in a molecule would have if the electrons in the bonds were divided equally between the two atoms.* In other words, all bonds are treated as if they were nonpolar. Formal charge is a method of electron bookkeeping. It is not meant to imply that it is the actual charge on the atom in question. Formal charge is calculated by subtracting the number of lone-pair electrons on the atom in question and half of the shared electrons from the group number. (The group number represents the number of valence electrons in a neutral atom.) For example, consider the structure above, which follows the octet rule.

- The formal charge on the S is [6 (group number) − 1/2 × 8 (four bonded pairs) = +2].

- The two O atoms bonded to both S and H have formal charges of zero [i.e., 6−4 (two lone pairs) − 1/2 × 4 (two bonded pairs) = 0].

- The two O atoms bonded only to S each have a formal charge of −1 [i.e., 6−6 (three lone pairs) − 1/2 × 2 (one bonded pair) = −1].

The formal charge is represented as the charge in a circle, as follows. Zero formal charge is not shown. *All the formal charges add to zero for a molecule or to the charge on an ion.*

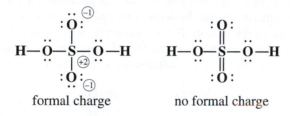

formal charge   no formal charge

Now consider the alternate structure. None of the atoms have any formal charge; that is, for the S, $6 − 1/2 × 12 = 0$, and for the O atoms with double bonds, $6 − 4$ (two lone pairs) $−1/2 × 4$ (two bonded pairs) $= 0$. The other two oxygen atoms also have no formal charge. Many chemists consider the structure with the least amount of formal charge to be the more favorable structure. The point remains controversial, however.

Other sulfur and phosphorus compounds are often represented by favorable formal charge structures at the expense of the octet rule. For example, SO_2 and the PO_4^{3-} ion can be represented with a structure that follows the octet rule or has a favorable formal charge distribution that violates the octet rule. Note that ozone, O_3 (Example 9-6), which has the same geometry as SO_2, cannot have the more favorable formal charge representation. Oxygen, which is in the second period, cannot have access to more than eight electrons. As mentioned earlier, only elements in the third period and higher can form compounds where the central atom has access to more than eight electrons.

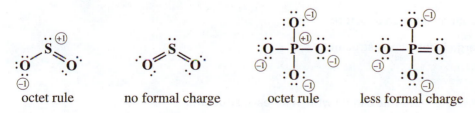

octet rule   no formal charge   octet rule   less formal charge

9-6.2 Other Applications of Formal Charge

Formal charge has one other very useful application. It can be very helpful in answering questions regarding the ordering of bonds. In other cases, it can help us decide whether a specific Lewis structure is legitimate. Just because we can write a structure that follows the octet rule does not necessarily mean that the structure actually represents the bonding in the molecule. For example, consider the compound nitrosyl chloride, which has the formula NOCl. (Despite the way it is written, N is the central atom, which also can be verified by formal charge.) One can write two resonance structures for this compound.

$$\ddot{O}\!=\!\ddot{N}\!-\!\ddot{Cl}\!: \qquad :\!\ddot{O}\!-\!\ddot{N}\!=\!\ddot{Cl}\!:$$
$$\overset{-1}{}\overset{+1}{}$$

no formal charge   has formal charge

If both were important, we would expect both bonds to have double-bond character (i.e., length). In the structure on the left, there are no formal charges on the atoms. In the structure on the right, there is a formal charge of −1 on the oxygen and +1 on the chlorine. Recall that structures with the least amount of formal charge are most important in determining the actual bonding. Thus, the structure with the formal charge does not represent the actual bonding in the molecule. Experiments tell us that only the structure on the left, which has an N=O double bond and an N—Cl single bond, represents the bonding in NOCl. The resonance structure on the right does not contribute to the true resonance hybrid of the molecule.

For another example of the utility of formal charge, consider the free radical NO. NO has an important role in the biological chemistry of the body. It bonds at the location of the unpaired electron. Is the unpaired electron on the nitrogen or on the oxygen atom? The structure with the least formal charge points to the answer. The Lewis structures of the two possibilities are shown here.

$$:\!\dot{N}\!=\!\ddot{O}\!: \qquad :\!\ddot{N}\!=\!\dot{O}\!:$$
$$\overset{-1}{}\overset{+1}{}$$

no formal   has formal
charge   charge

The structure on the left has no formal charge, indicating that the unpaired electron is on the nitrogen. Experiments indicate that this is the case.

EXAMPLE 9-7   Using Formal Charge

Cyanogen $(CN)_2$ is known as a *pseudohalogen* because it has some properties similar to halogens. It is composed of two CN's joined together. Do the two CN's join through the carbon or the nitrogen (i.e., C—N—N—C or N—C—C—N)?

PROCEDURE

1. Draw the two Lewis structures following the rules given in Section 9-4.

2. Assign formal charges to the atoms in each structure.

3. The one with less formal charge is the actual structure.

SOLUTION

1. Only one structure that follows the octet rule can be written for each structure.

$$:\!C\!\equiv\!N\!-\!N\!\equiv\!C\!: \qquad :\!N\!\equiv\!C\!-\!C\!\equiv\!N\!:$$
$$\overset{-1}{}\overset{+1}{}\overset{+1}{}\overset{-1}{}$$

high formal   no formal
charge   charge

2. In the structure on the left, the formal charge on the C is $2 - 1/2(6) = -1$. The formal charge on the N is $5 - 1/2(8) = +1$. In the structure on the right, the formal charge on the C is $4 - 1/2(8) = 0$. The N also has a formal charge of zero.

3. In cyanogen, the two CN's are bonded through the carbons.

ANALYSIS

With formal charge, we are determining the best distribution of electrons within a molecule. Some molecules will show formal charges regardless of how we draw them. Our task is to minimize the amount of formal charge present. The formal charges on all the atoms of a molecule will add to the charge on that molecule. So an ion such as OCN^- will show at least a single negative formal charge somewhere.

SYNTHESIS

Notice that whether we choose to view bonds as single or double to satisfy either the octet rule or the formal charge, it does not change our determination of molecular geometries or bond angles. How would perchlorate (ClO_4^-) look when satisfying the octet rule? As we will see in Part C, both structures have the same geometry.

▶ **ASSESSING THE OBJECTIVE FOR SECTION 9-6**

EXERCISE 9-6(a) LEVEL 1: For the molecule nitrosyl chloride (NOCl), which was discussed previously, show why the oxygen cannot be the central atom using the Lewis structure shown here. What are the formal charges on each atom?

$$\ddot{N} = \ddot{O} - \ddot{\underset{\cdot\cdot}{Cl}}:$$

EXERCISE 9-6(b) LEVEL 2: In HClO, use formal charge to predict which is the central atom Cl or O.

EXERCISE 9-6(c) LEVEL 2: In $COCl_2$, use formal charge to determine whether the necessary double bond is between the C and O or the C and a Cl. (C is the central atom.)

EXERCISE 9-6(d) LEVEL 3: Draw a Lewis structure of the molecule SO_2Cl_2 (S is the central atom) that follows the octet rule and assign formal charges to the atoms. Write a second structure that does not follow the octet rule but has the best distribution of formal charge.

EXERCISE 9-6(e) LEVEL 3: Carbon monoxide is known to form a strong bond to a number of compounds such as the iron in hemoglobin. Use formal charge to decide whether the new bond forms to the carbon or the oxygen.

For additional practice, work chapter problems 9-45 and 9-51.

PART B SUMMARY

KEY TERMS

9-3.1   A **covalent bond** between two atoms consists of a shared pair of electrons and is represented as a dash in a **Lewis structure**. p. 283

9-3.4   Two pairs of electrons are shared in a **double bond** and three pairs are shared in a **triple bond**. p. 285

9-4.2   A **free radical** is a molecule with unpaired electrons. p. 291

9-5   **Resonance structures** indicate the presence of a **resonance hybrid** structure. p. 292

9-6.1   Assignment of **formal charge** is useful in determining important resonance structures. p. 295

SUMMARY CHART

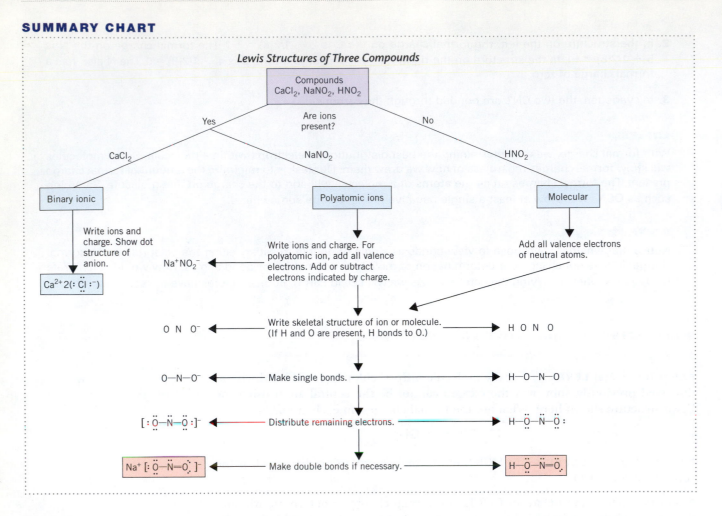

Lewis Structures of Three Compounds

Part C

The Distribution of Charge in Chemical Bonds

SETTING A GOAL

■ You will learn how the correct Lewis structure of a compound allows chemists a thorough understanding of its properties.

OBJECTIVES

9-7 Classify a bond as being nonpolar, polar, or ionic.

9-8 Determine the bond angles and geometries present in simple molecules or ions from the Lewis structure.

9-9 Classify a molecule as polar or nonpolar based on geometry and electronegativity.

▶ **OBJECTIVE FOR SECTION 9-7**
Classify a bond as being nonpolar, polar, or ionic.

9-7 Electronegativity and Polarity of Bonds

LOOKING AHEAD! In the formation of a covalent bond, electrons are shared. However, atoms of different elements rarely share electrons equally in a covalent bond. This unequal sharing leads to a partial charge, as we will see in this section. ■

9-7.1 Electronegativity

You are probably aware that *sharing* a carton of popcorn at a movie rarely means *equal sharing*. The hungrier, faster popcorn eater usually gets the lion's share. Likewise, in a chemical bond between the atoms of two different elements, the pair of electrons is not shared equally, and one atom gets a larger share of the electrons. *The ability of an atom of an element to attract electrons to itself in a covalent bond is known as the element's* **electronegativity**. The value assigned for the electronegativity of many elements is shown in Figure 9-5. The most electronegative element is fluorine, which is assigned an electronegativity value of 4.0. Notice that nonmetals tend to have higher values of electronegativity than metals. Electronegativity is considered a periodic property in that, generally, the closer the element is to fluorine in the periodic table, the more electronegative it is. The values shown in Figure 9-5 were first calculated by Linus Pauling (winner of two Nobel Prizes). Although more refined values are now available, the actual numbers are not as important as how the electronegativity of one element compares with that of another to which it is bonded.

Increases →						
1 H 2.1						
3 Li 1.0	4 Be 1.5	5 B 2.0	6 C 2.5	7 N 3.0	8 O 3.5	9 F 4.0
11 Na 0.9	12 Mg 1.2	13 Al 1.5	14 Si 1.8	15 P 2.1	16 S 2.5	17 Cl 3.0
19 K 0.8	20 Ca 1.0	31 Ga 1.6	32 Ge 1.8	33 As 2.0	34 Se 2.4	35 Br 2.8
37 Rb 0.8	38 Sr 1.0	49 In 1.7	50 Sn 1.8	51 Sb 1.9	52 Te 2.1	53 I 2.5
55 Cs 0.7	56 Ba 0.9	81 Tl 1.8	82 Pb 1.8	83 Bi 1.9	84 Po 2.0	85 At 2.2
87 Fr 0.7	88 Ra 0.9					

(Decreases ↓)

FIGURE 9-5 Electronegativity

9-7.2 Representing Polar Bonds

Electrons carry a negative charge. When there is a complete exchange of an electron between atoms, as in the formation of an ionic bond, one atom acquires a full negative charge. In a covalent bond between two atoms of different electronegativity, the more electronegative atom attracts the electrons in the bond partially away from the other atom and thus acquires a *partial* negative charge (symbolized by δ^-). This leaves the less electronegative atom with a partial positive charge (symbolized by δ^+).

A covalent bond that has a partial separation of charge due to the unequal sharing of electrons is known as a **polar covalent bond** (or simply, a **polar bond**). A polar bond has a negative end and a positive end and is said to contain a **dipole** (two poles). A polar bond is something like Earth itself, which contains a magnetic dipole with a north and south magnetic pole. (The poles in a bond dipole are electrostatic rather than magnetic.) The dipole of a bond is represented by an arrow pointing from the positive to the negative end ($\longmapsto$).

The polarity of bonds has a significant effect on the chemical properties of compounds. For example, the polarity of the H—O bonds in water accounts for many of its familiar properties that we take for granted. We will discuss the chemistry of water in more detail in Chapter 11.

9-7.3 Predicting the Polarity of Bonds

When electrons are shared between atoms of the same element, they are obviously shared equally. *If electrons are shared equally, the bond is known as a* **nonpolar bond**. The greater the difference in electronegativity between two elements, the more polar is the bond between them. In fact, if the difference in electronegativity is 1.8 or greater on the Pauling scale, it indicates that one atom has gained complete control of the pair of electrons. In other words, the bond is most likely ionic.

In summary, when two atoms compete for a pair of electrons in a bond, there are three possibilities for the pair of electrons.

1. Both atoms share the electrons equally, forming a nonpolar bond (an electronegativity difference between the two atoms of zero or near zero).
2. The two atoms share electrons unequally, forming a polar bond. This is intermediate between purely ionic and equal sharing (an electronegativity difference of less than 1.8).

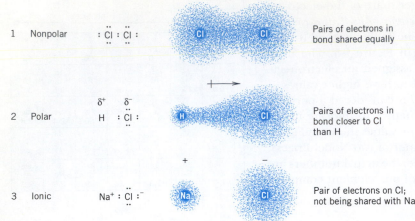

1	Nonpolar	$:\!\ddot{C}l\!:\!\ddot{C}l\!:$	Pairs of electrons in bond shared equally
2	Polar	$\overset{\delta^+}{H}\ \overset{\delta^-}{:\!\ddot{C}l\!:}$	Pairs of electrons in bond closer to Cl than H
3	Ionic	$Na^+\ :\!\ddot{C}l\!:^-$	Pair of electrons on Cl; not being shared with Na

FIGURE 9-6 Nonpolar, Polar, and Ionic Bonds In case 1, the bond is nonpolar covalent; in case 2, the bond is polar covalent; in case 3, the bond is ionic.

3. The electron pair is not shared, since one atom acquires the electrons. This is an ionic bond, in which each atom acquires a complete charge (an electronegativity difference greater than about 1.8).

These three cases are illustrated in Figure 9-6. The bond in Cl_2 is nonpolar (case 1), since both atoms are identical. To determine the charge on each Cl, we will assign electrons to the two Cl atoms. In this case, each Cl is assigned the six electrons from its three lone pairs. Since the two atoms share the pair evenly, we can assign exactly one-half of the shared electrons to each Cl for a total of seven each [6 + (1/2 × 2) = 7]. Since Cl is in Group VIIA, seven valence electrons leave each Cl exactly neutral. Thus there is not a positive and a negative end.

The molecule HCl illustrates case 2. There is a significant difference in electronegativity between H and Cl (0.9), indicating a polar bond, but not so much as to indicate an ionic bond. Since the Cl is more electronegative than H, it has a partial negative charge. In this case, the Cl is still assigned six electrons from its three lone pairs but more than one-half of the pair of electrons in the bond. Since it is assigned more than seven electrons but fewer than eight, it acquires a *partial negative charge*. The hydrogen, with less than one-half of the electron pair, has less than one electron and thus has an equal but opposite *partial positive charge*. However, the difference in electronegativity between carbon (2.5) and hydrogen (2.1) is relatively small (0.4), so the bond is considered essentially nonpolar. The fact that the carbon–hydrogen bond is nearly nonpolar is very important in the chemistry of living systems.

The bond in NaCl illustrates case 3. As mentioned earlier in this chapter, there is a complete exchange of the valence electron from Na to Cl, producing charged ions. Eight electrons on Cl form an anion with a −1 charge. Notice that the difference in electronegativity between Na and Cl is large (2.1), predicting an ionic nature.

EXAMPLE 9-8   Comparing the Polarity of Bonds

Referring to Figure 9-5, rank the following bonds in order of increasing polarity. The positive end of the dipole is written first. On the basis of electronegativity differences, indicate whether any of the bonds are predicted to be ionic.

$$Ba-Br \quad C-N \quad Be-F \quad B-H \quad Be-Cl$$

PROCEDURE

Determine the difference in electronegativity between the elements. The smaller the difference, the less polar the bond. Higher differences make the bonds more ionic.

SOLUTION

$$Ba-Br \quad 2.8 - 0.9 = 1.9$$
$$C-N \quad 3.0 - 2.5 = 0.5$$
$$Be-F \quad 4.0 - 1.5 = 2.5$$

$$B-H \quad 2.1 - 2.0 = 0.1$$
$$Be-Cl \quad 3.0 - 1.5 = 1.5$$
$$B-H < C-N < Be-Cl < Ba-Br < Be-F$$

ANALYSIS

The differences in electronegativity suggest that Ba—Br and Be—F have ionic bonds. Notice that the difference in electronegativity suggests that the Be—Cl bond is polar covalent rather than ionic. This is confirmed by a profound difference in the physical properties of Be compounds containing these two bonds. Despite being a binary compound between a metal and a nonmetal, the $BeCl_2$ species has a molecular nature rather than an ionic one.

SYNTHESIS

It is important to realize that the distinctions *nonpolar*, *polar*, and *ionic* do not represent three cut-and-dried ways of sharing electrons, but rather a continuum of increasingly unequal sharing from the even 50/50 sharing found in purely nonpolar molecules all the way to the 100/0 sharing (or lack of it) found in the most ionic substances—and everything in between. While we give approximate cutoffs for electronegativity differences between polar and ionic bonds, these are guidelines only. Two elements whose difference is 1.5, for example, can be expressed as having a certain amount of ionic character and a certain amount of covalent character. Experimental evidence is the final arbiter.

▶ **ASSESSING THE OBJECTIVE FOR SECTION 9-7**

EXERCISE 9-7(a) LEVEL 1: Fill in the blanks.

The most electronegative element is: _____. The second most electronegative element is: _____. In a covalent bond between unlike atoms, the more electronegative element has a partial _____ charge. This type of bond is said to be a _____ covalent bond. A large difference in electronegativity between two atoms leads to an _____ bond.

EXERCISE 9-7(b) LEVEL 2: Indicate with a dipole arrow from the partially positive atom to the partially negative atom if any of these bonds is polar.
(a) Al—Se   (b) As—S   (c) S—S   (d) F—Br   (e) C—H

EXERCISE 9-7(c) LEVEL 2: Based on electronegativity differences, label each bond as being nonpolar, polar, or ionic.
(a) N—Cl   (b) Ti—O   (c) Si—S   (d) C—I

EXERCISE 9-7(d) LEVEL 3: Explain how differences in electronegativity play a roll in whether a bond is ionic, nonpolar covalent, or polar covalent.

For additional practice, work chapter problems 9-55, 9-58, and 9-59.

9-8 Geometry of Simple Molecules

▶ **OBJECTIVE FOR SECTION 9-8**
Determine the bond angles and geometries present in simple molecules or ions from the Lewis structure.

LOOKING AHEAD! Because the covalent bonds in most molecules are polar, does that mean the molecule itself is polar? Surprisingly, the answer is not necessarily. The polarity of a molecule depends on its geometry as well as the polarity of its bonds. In this section, we will use Lewis structures to tell us about the geometry of some simple molecules and then use this information to discuss molecular polarity in the next section. ■

9-8.1 VSEPR Theory

In order to metabolize ordinary sugar, we need a molecule known as an *enzyme* that breaks the sugar molecule into two components. The enzyme and the sugar molecules require specific molecular geometries so that the two molecules can connect. The geometrical connection is sometimes likened to the way a specific key fits into a specific lock. In another example of the importance of molecular geometry, consider ordinary water. If water were a linear molecule rather than bent, it would drastically affect its properties. For example, water would be a gas at room temperature rather than a liquid. Life as we know it could not exist under those conditions. When properly interpreted, however, the Lewis structure of water indicates its bent nature. The approximate geometry of the atoms around a central atom can be predicted by the **valence shell electron-pair repulsion theory (VSEPR)**. *This theory tells us that electron groups on an atom, either unshared pairs or electrons localized in a bond, repel each other to the maximum extent.* In other words, the negatively charged electron groups get as far away from each other as possible (without breaking the bonds) because of their electrostatic repulsion. *The geometry assumed by the unshared electron pairs and the electrons in the bond is known as the* **electronic geometry**.

Consider the CO_2 molecule. Its Lewis structure indicates two double bonds to each of the two oxygen atoms. There are two groups of electrons surrounding the carbon, with each group containing four electrons. To place the two groups as far apart as possible, the electrons in the bonds must lie on opposite sides of the carbon atom at an angle of 180°. This creates an electronic geometry of the molecule that is said to be *linear*. Another linear molecule is HCN, shown below. Notice that *double and triple bonds are treated the same as single bonds when determining the geometry of a molecule.* That is, all three types of bonds are counted as one group of electrons (even though double and triple bonds constitute larger groups).

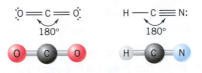

The BF_3 molecule has three groups of electrons. In this case, the three groups get as far away from each other as possible by assuming the geometry of an equilateral triangle with an F—B—F angle of 120°. The electronic geometry of this molecule is said to be *trigonal planar*.

Finally, consider the molecule CCl_4. This molecule consists of four bonds around the central atom, the most for a molecule following the octet rule. The farthest apart four groups of electrons can move is a three-dimensional structure called a *tetrahedron* with an internal angle of 109.5°.

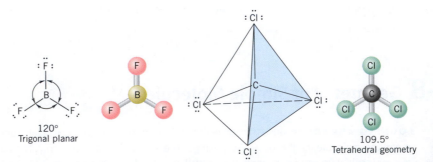

The electronic geometry of a molecule, then, is directly related to the number of groups surrounding the central atom of a molecule. The three electronic geometries discussed are listed in Table 9-2.

9-8.2 Molecular Geometry

Now consider molecules where there are one or two unshared electron pairs in place of covalent bonds. The SO_2 molecule, which has three electron groups, has the same *electronic* geometry as BF_3, described eariler. The central sulfur atom is bonded to two oxygen atoms, and it has one unshared pair of electrons. (Remember that the two S—O bonds are actually identical, since two resonance structures can be written.) The O—S—O angle is approximately that of an equilateral triangle. *The* **molecular geometry** *of a molecule is the geometry described by the bonded atoms and does not include the unshared pairs of electrons.* It would be as if the unshared pair or pairs are invisible when we view the actual atom although they do affect the geometry. Thus we describe the molecular geometry of the three atoms in SO_2 as *V-shaped* or *bent.*

TABLE 9-2

Electronic Geometries

NUMBER OF ELECTRON GROUPS	ELECTRONIC GEOMETRIES	BOND ANGLE
2	linear	180°
3	trigonal planar	120°
4	tetrahedral	109.5°

117°
V-shaped

Ammonia (NH_3) has four groups of electrons, so it has the same electronic geometry as CCl_4. In the case of NH_3, however, one group is an unshared pair of electrons. The H—N—H angle is found to be 107°, which is in good agreement with the angle predicted by this theory. (The angle is somewhat less than 109° because lone pairs of electrons take up more space than bonded pairs.) The molecular geometry of the NH_3 molecule is described as *trigonal pyramidal.* Finally, the familiar H_2O molecule also has four groups but with two unshared pairs of electrons. The H—O—H angle is known to be 105°, which also agrees with this theory. The molecular geometry of H_2O is described as *V-shaped* or *bent.* (In the V-shaped structure of SO_2, the angle of 117° is near the expected trigonal angle of 120°; in the V-shaped structure of H_2O, the angle of 105° is near the tetrahedral angle of 109°.)

Notice that when a molecule has no unshared electrons, the electronic and molecular geometry are the same. *When discussing the geometry of a molecule in the future, however, we will be referring to the molecular geometry.*

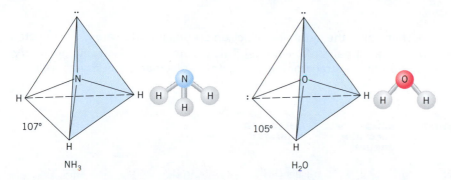

107°

NH₃

105°

H₂O

The molecular geometries for all of these molecules are summarized in Table 9-3.

TABLE 9-3

Molecular Geometry

NUMBER OF ELECTRON GROUPS ON CENTRAL ATOM (BONDED ATOMS + UNSHARED PAIRS)	NUMBER OF ATOMS BONDED TO CENTRAL ATOM	NUMBER OF UNSHARED PAIRS OF ELECTRONS ON CENTRAL ATOM	MOLECULAR GEOMETRY	MODEL
2	2	0	linear	
3	3	0	trigonal planer	
3	2	1	V-shaped (near 120)	
4	4	0	tetrahedral	
4	3	1	trigonal pyramidal	
4	2	2	V-shaped (near 109)	

EXAMPLE 9-9   Predicting the Geometry of Molecules

What is the electronic and molecular geometry of each of the following ions and molecules?

(a) OCN^- **(b)** H_2CO **(c)** $HClO$ **(d)** $SiCl_4$

In **(a)** and **(b)**, the middle atom in the formula is the central atom. In **(c)**, oxygen is the central atom.

PROCEDURE

First write the correct Lewis structure. Then count the number of electron groups [bonded atoms (connected to the central atom) + the number of unshared pairs of electrons]. This provides the electronic geometry. For atoms with unshared electron pairs on the central atom, evaluate the molecular geometry as well.

SOLUTION

	LEWIS STRUCTURE	NUMBER OF ELECTRON GROUPS	MOLECULAR GEOMETRY	
(a) OCN^-	$\ddot{O}=C=\ddot{N}\!:^-$	2	linear (180°)	
(b) H_2CO	$\overset{H}{\underset{H}{>}}C=\ddot{O}\!:$	3	trigonal planer	

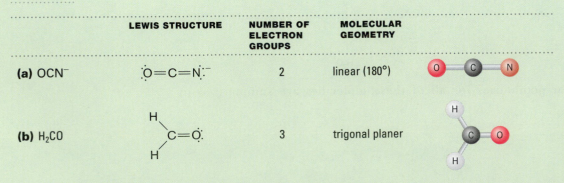

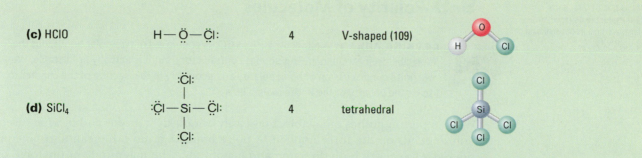

| **(c)** HClO | H—Ö—Cl: | 4 | V-shaped (109) | |
| **(d)** SiCl₄ | :Cl:
\|
:Cl—Si—Cl:
\|
:Cl: | 4 | tetrahedral | |

ANALYSIS

The molecular geometries where there are unshared pairs of electrons are best reasoned out rather than memorized. Learn and picture the three basic electronic geometries, and then ask yourself what shape is left as one or two bonds are "removed" from the geometry (and replaced with an unshared pair). The resulting molecular geometries readily become apparent.

   Also, consider the answer in (d) above. Based on electron repulsions, what's the farthest apart that four groups of electrons should be able to get? Many students immediate response is 90°, which is 360° divided by 4. This would be a square plane with a central atom. What we fail to realize is that all three dimensions are in play, and the bonds can actually spread out even farther than that. In three-dimensional space, the maximum bond angle that can be achieved is 109.5°, and the electronic geometry created is described as tetrahedral, a result directly from Euclidean geometry.

SYNTHESIS

While it is beyond the scope of this book to provide a detailed description of geometries resulting from molecules with more than an octet of electrons, a quick look into one possible geometry is instructive as regards the application of VSEPR theory. What should the bond angles be for a molecule with six pairs of electrons on a central atom? What's the farthest apart you can place six groups on a sphere? The answer is along each of the three mathematical axes, the *x*, *y*, and *z*. These axes are 90° apart from each other, and so we'd predict 90° bond angles in the molecule. What would the *molecular* geometries look like for one or two sets of unshared electrons in these types of molecules? Picture a three-dimensional axis, and remove one bond. That geometry is described as being *square pyramidal*. Removal of the second bond creates a *square planar* arrangement. Can you picture these structures?

▶ **ASSESSING THE OBJECTIVE FOR SECTION 9-8**

EXERCISE 9-8(a) LEVEL 1: How many groups of electrons are found in the following, for the purpose of determining geometries? [*Note:* At first glance (a) may be considered as an ionic compound. However, after comparing the electronegativities of Be and Cl, you can understand that it is actually most likely a molecular compound.]
(a) Cl—Be—Cl   **(b)** H—S—H   **(c)** O—N—O⁻

EXERCISE 9-8(b) LEVEL 2: From the Lewis structure, what is the bond angle in PBr₃?

EXERCISE 9-8(c) LEVEL 2: What is the electronic geometry of COCl₂ (C is the central atom)?

EXERCISE 9-8(d) LEVEL 2: What is the molecular geometry of SCl₂?

EXERCISE 9-8(e) LEVEL 3: Give the bond angle and the electronic and molecular geometries for the following.
(a) BeCl₂ (a molecular compound)   **(b)** H₂S   **(c)** NO₂⁻

For additional practice, work chapter problems 9-64, 9-65, and 9-67.

► OBJECTIVE
   FOR SECTION 9-9
Classify a molecule as polar or
nonpolar based on geometry and
electronegativity.

9-9 Polarity of Molecules

LOOKING AHEAD! A molecule such as CO_2 is composed of two polar bonds, but the molecule itself is nonpolar. This sounds contradictory, but it is actually predictable. We will unite our discussion of bond polarity and geometry in this section of the chapter to discuss the polarity of the entire molecule. ■

When a person pushes on a table, how easily or quickly it moves and in what direction depends on the force applied. A polar bond in a molecule can also be considered a force with both direction and magnitude. The magnitude of the force of the polar bond depends on the degree of polarity, which relates to the difference in electronegativity of the two atoms in the bond. The greater the difference in electronegativities, the greater the partial charges on the two atoms in the bond and the larger the magnitude of the dipole.

9-9.1 Nonpolar Molecules

Force has direction as well as magnitude. Consider what would happen if someone of exactly equal strength was pushing on the table described above in exactly the opposite direction (i.e., an angle of 180°). The two forces would exactly cancel and there would be no movement of the table. In such a shoving match, it would be a standoff. In a molecule, *the combined effects of the bond dipoles (the net dipole) are known collectively as the* **molecular dipole**. If the geometry of the molecule is such that equal dipoles cancel, *there is no molecular dipole, which means that the molecule is nonpolar.* For example, in CO_2 the bond dipoles are equal and are orientated in exactly opposite directions. CO_2 is thus a nonpolar molecule. The same is true for trigonal planar (e.g., BF_3) and tetrahedral molecular geometries (e.g., CCl_4), where all terminal atoms are the same. (See Figure 9-7.)

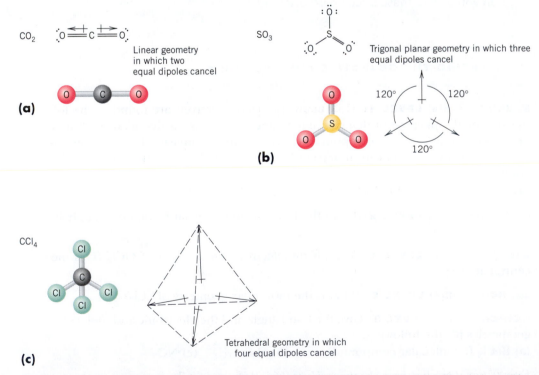

A molecule is also nonpolar if it is made entirely of nonpolar bonds, regardless of its geometry. Since we view the C—H bond as being essentially nonpolar, then any molecule composed entirely of carbon and hydrogen (a *hydrocarbon*) needs to be considered nonpolar as well. This fact is very important in the chemistry of living systems.

9-9.2 Polar Molecules

Polar molecules occur when the dipole forces do not cancel. There are two reasons that the dipole forces do not cancel.

(1) The molecule has a symmetrical geometry (i.e., linear, trigonal pyramid, or tetrahedral), but the terminal atoms are not all the same. This would be the case where two people of different strengths were in a tug of war. The stronger predominates, and the weaker is forced to move in the direction opposite to the force he or she is applying. For example, HCN is linear and $CHCl_3$ (chloroform) is tetrahedral, but both molecules are polar because the unequal dipoles in the bonds do not cancel.

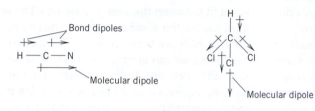

(2) The geometry of the molecule is V-shaped or trigonal pyramid. The geometry of the molecule leads to forces that do not cancel. Such a thing occurs in V-shaped or trigonal pyramidal molecules regardless of the identity of the terminal atoms.

In the analogy of the forces applied by two people on a table, consider what would happen if the forces were at an angle such as from adjacent corners of the table. In this case, the table will move. The direction of movement and the amount of movement depend on the strength of the two people pushing and the angle between them.

The most important example of a molecule that is polar because it is V-shaped is water. The two O—H bonds are at an angle of about 105°, so the equal bond dipoles do not cancel. As we will see again in later chapters, the fact that water is polar (has a molecular dipole) is of critical importance to its role as a room-temperature liquid that can dissolve many ionic compounds.

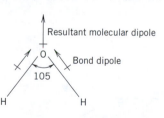

▶ **ASSESSING THE OBJECTIVE FOR SECTION 9-9**

EXERCISE 9-9(a) LEVEL 2: Indicate whether the following molecules are polar or nonpolar.
(a) CH_2Cl_2 **(b)** NI_3 **(c)** $BeCl_2$ **(d)** SiH_4

EXERCISE 9-9(b) LEVEL 2: Rank the following molecules by increasing polarity.
(a) CH_2O **(b)** HF   **(c)** CS_2 **(d)** SO_2

EXERCISE 9-9(c) LEVEL 3: One piece of evidence we have for the tetrahedral geometry for four electron pairs over a square arrangement is the polarity of molecules. Consider the molecule CH_2F_2. What is its polarity? If the square arrangement existed, when would it be polar?

For additional practice, work chapter problems 9-69, 9-70, 9-72, and 9-75.

MAKING IT REAL

Enzymes—The Keys of Life

Enzymes work like a lock and key.

You can't blink your eye or digest a slice of pizza without the help of enzymes. Enzymes demonstrate quite dramatically the importance of geometry in how they function while doing so many jobs. Enzymes are rather complex molecules made from proteins that serve as catalysts. A catalyst is an agent that causes or speeds up a chemical reaction but is not itself consumed in the reaction. Catalysts will be discussed in more detail in later chapters. In effect, an enzyme is much like a specific key that unlocks a specific door. If the shape of the key is not exact, it does not work. The action of an enzyme in doing a particular job has to do with the geometry and polarity of a part of the molecule known as its *active site*.

Glucose, blood sugar, is the only sugar that we can use in our body in metabolism. It is used in the combustion reaction that provides the energy to keep us alive. Carbohydrates and other sugars, including sucrose (table sugar), must be broken down, or converted, into glucose before they can be used. Sucrose is actually composed of two simpler sugars, fructose and glucose, chemically bonded

together. An enzyme known as *sucrase* is responsible for breaking sucrose into its components. The sucrase molecule has a specific molecular geometry that is complementary to the sucrose molecule. The polarity of the bonds in the enzyme is also such that they attract the sucrose molecule so as to form an exact fit between the two geometries. The enzyme then causes the sucrose to break into its two components, which are then released. The sucrase enzyme can then seek out another sucrose molecule and repeat the process. This is illustrated in the accompanying figure. Other enzymes cause the fructose to be converted into glucose so that it, too, is used in metabolism.

Dairy products contain milk sugar (lactose), which can be broken down into its two component sugars, glucose and galactose. The enzyme that causes this reaction is known as *lactase*. However, some people are lactose intolerant, which means that they lack the lactase enzyme. To avoid discomfort, they must avoid dairy products or take a lactase tablet (i.e., Lactaid) before they can enjoy a dish of ice cream.

The list of enzymes seems endless. There is a specific enzyme in our body for every biochemical function. In fact, we are just now learning about many of the enzymes that make life possible.

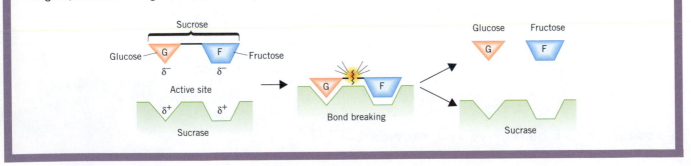

PART C   SUMMARY

9-7.1 **Electronegativity** is a measure of the ability of an atom of an element in a covalent bond to attract electrons to itself. p. 299

9-7.2 **Polar covalent bonds** (or simply **polar bonds**) have a **dipole** that indicates a separation of charge. p. 299

9-7.3   A **nonpolar bond** has no charge separation. p. 299

9-8.1   **VSEPR theory** describes the geometry around a central atom in a molecule. p. 302

9-8.1   The **electronic geometry** of a molecule includes both unshared pairs of electrons and electrons in bonds. p. 302

9-8.2   The **molecular geometry** of a specific molecule includes only the atoms involved. p. 303

9-9.1   A **molecular dipole** is the resultant dipole of all of the bond dipoles. p. 306

SUMMARY CHART

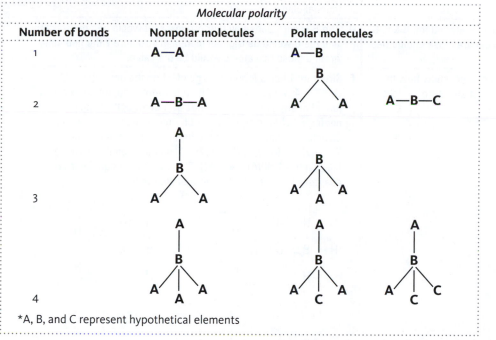

*A, B, and C represent hypothetical elements

CHAPTER 9 SYNTHESIS PROBLEM

The ability to write correct Lewis structures is one of the most important tools in chemistry. Some very important concepts are implied from Lewis structures, such as geometry, bond polarity, and molecular polarity. From these we can determine how molecules interact with each other, which is a topic that we will pursue in the next chapter. Intermolecular interactions are essential information in the study of biochemistry, which relates to life processes. In the following problems we present two ions, each of which has more than one possible Lewis structure. We then use the methods presented in this chapter to determine bond polarity and geometry around central atoms. Just because we can write two resonance structures that follow the octet rule does not mean that both structures truly represent the bonding in molecules or ions. So we will use experimental and theoretical ways, presented in this chapter, to determine which of the two resonance structures is the actual structure.

PROBLEM	SOLUTION
A compound was discovered to have the empirical formula KCH_5BNO. **a.** Write the formula of this compound as ions. How many valence electrons are in the anion?	**a.** When potassium is part of the compound, ions are certainly involved. Since K has only a +1 charge, the anion must have a −1 charge. The ionic form is $K^+CH_5BNO^-$. The anion has $4(C) + (5 \times 1) H + 3(B) + 5(N) + 6(O) + 1$ (negative charge) $= 24$ valence electrons.
b. O The basic structure of the anion is H_3BCNH_2. Write two possible Lewis structures for this species (i.e., two resonance structures). Consider only the arrangement of atoms shown in the basic structure.	**b.** (1) (2)

PROBLEM	SOLUTION
c. What is the direction of polarity of the B—H and the B—C bond?	**c.** H and C are both more electronegative than B, so the dipole arrow would point toward the H and C from the B.
d. The geometry around the N is actually trigonal pyramidal. How does this information indicate which of the two structures is the correct one?	**d.** The geometry around each atom is determined only by the bonds and lone pairs on that particular atom. Structure 1 would have to be the correct structure since it predicts the N to be trigonal pyramidal. Structure 2 predicts the N to be trigonal planar.
e. What is the molecular geometry around the B and C in the correct structure? What are those geometries in the incorrect structure?	**e.** The B is surrounded by four groups, so it would be tetrahedral, and the C would be trigonal planar. In structure 2 the geometry around these two atoms would be the same.
f. Using the concept of formal charge, show how this information points to the correct structure.	**f.** Structure 1 has a formal charge of -1 on the boron [i.e., $(1/2(8) = 4; 3(IIIA) - 4 = -1$] The other atoms do not carry a formal charge. Since the ion has a -1 charge, there must be a net formal charge equal to the charge on the ion as a whole. Structure 2 shows a -1 on the boron, a -1 on the oxygen [i.e., $1/2(2) + 6 = 7; 6(VIA) - 7 = -1$], and a $+1$ on the nitrogen [i.e., $1/2(8) = 4; 5(VA) - 4 = +1$]. This structure would be less desirable since it has more formal charge than 1.

YOUR TURN

A compound has the empirical formula $Ca(CH_2NO_2)_2$.

   a. Write this compound as ions. How many valence electrons are in each anion?

   b. Write three possible Lewis structures for the anion. Use only the order of bonds shown.

   O O
   ‖ ‖
   The basic structure is H C N H.

   c. What is the polarity of the C—N and the N—H bonds?

   d. The geometry around the N is trigonal pyramidal. What does this information tell us about the true structures?

   e. What is the geometry around the carbon atom in each of the three structures?

   f. Using the concept of formal charge, show how this information points to the correct structure.

Answers are on p. 313.

CHAPTER SUMMARY

A simple observation followed by a hypothesis leads us to a solid understanding of why representative elements interact with each other as they do. The observation is that noble gas elements exist as solitary atoms, and the hypothesis is that they are that way because their electron configuration (an octet of electrons) is stable. From this we can rationalize why the representative elements adjust their electron configurations to follow the **octet rule**. For example, a metal and a nonmetal combine by an exchange of electrons, thereby achieving an octet for both.

The use of **Lewis dot symbols** of the elements aids us in focusing on the **valence electrons**. The ions formed by the electron exchange are arranged in a **lattice** held together by **ionic bonds**. Two nonmetals, on the other hand, follow the octet rule by electron sharing, forming a **covalent bond**. In certain cases, two atoms form **double bonds** or **triple bonds**. Covalent bonds also exist between the atoms comprising polyatomic ions.

Elements	Type of Bond	Comments
Metal–nonmetal	Ionic	Nonmetal can form −1, −2, or −3 ion. Metal can form +1, +2, or +3 ion.
Nonmetal–nonmetal	Covalent	Single, double, or triple bonds are used to form an octet.

By following certain rules, we can become proficient at writing **Lewis structures** of compounds. Not all molecules follow these rules exactly, such as **free radicals**. Two or more equally correct Lewis structures can be written for a molecule such as ozone (O_3), an *allotrope* of oxygen. These are known as **resonance structures**, and the actual structure is a **resonance hybrid** of all the Lewis structures.

The use of **formal charge** was introduced as an alternative way of representing the bonding involving sulfur and phosphorus. It also is an aid in answering questions regarding bonding that may not be obvious in writing Lewis structures.

Atoms of two different nonmetals do not share electrons equally. **Electronegativity** is a measure of the periodic property of the atoms of an element to attract the electrons in the bond. When electrons are not shared equally, the bond is **polar covalent**, or simply **polar**, and contains a **dipole**. Atoms of the same element share electrons equally, so the bond is **nonpolar**.

Whether a molecule is polar depends not only on the polarity of the bonds in the molecule but also on its **molecular geometry**. The **electronic geometry** and molecular geometry of simple molecules can be determined from their Lewis structures and the **VSEPR theory**. If equal polar bonds are in a geometric arrangement where their dipoles cancel each other, the compound has no **molecular dipole** and is nonpolar. If the bond dipoles are not the same in each direction or they do not cancel, the compound is polar.

OBJECTIVES

SECTION	YOU SHOULD BE ABLE TO...	EXAMPLES	EXERCISES	CHAPTER PROBLEMS
9-1	Write the Lewis dot symbols of atomic and monatomic ions by following the octet rule.		1a, 1b, 1c, 1d	1, 3, 4, 10, 11, 12
9-2	Using a Lewis structure and the octet rule, predict the charges on the ions of representative elements and the formulas of binary ionic compounds.	9-1	2a, 2b, 2c	7, 9, 15, 16, 20, 22
9-3	Apply the octet rule to determine the number of bonds formed in simple compounds.		3a, 3b, 3c	26, 27, 29
9-4	Draw Lewis structures of a number of molecular compounds and polyatomic ions.	9-2, 9-3, 9-4, 9-5	4a, 4b, 4c	32, 33, 34, 35, 36, 37, 38, 39
9-5	Write a multiple Lewis structures for molecules capable of resonance.	9-6	5a, 5b, 5c, 5e	40, 41, 43, 44
9-6	Determine the validity of a Lewis structure based on formal charge considerations.	9-7	6a, 6b, 6c	45, 46, 47, 48, 51
9-7	Classify a bond as being nonpolar, polar, or ionic.	9-8	7a, 7b, 7c, 7d	53, 54, 55, 56
9-8	Determine the bond angles and geometries present in simple molecules or ions from the Lewis structure.	9-9	8a, 8b, 8c, 8d, 8e	60, 61, 62, 63, 64, 65
9-9	Classify a molecule as polar or nonpolar based on geometry and electronegativity.		9a, 9b, 9c	66, 67, 68, 71, 73, 75

▶ANSWERS TO ASSESSING THE OBJECTIVES

Part A

EXERCISES

9-1(a) (a) 3 (b) 6 (c) 8

9-1(b) (a) $Mg\cdot$ (b) $\cdot\ddot{N}\cdot$ (c) $\ddot{\underset{\cdot\cdot}{\ddot{I}}}$

9-1(c) The eight electrons are the ones necessary to fill up the $s(2)$ and the $p(6)$ subshells in the valence shell of an atom.

9-1(d) The $s(2)$, the $p(6)$, and $d(10)$ subshells hold a total 18 electrons.

9-2(a) Be, +2; As, −3.

9-2(b) Be_3As_2

9-2(c) $PbCl_2$; $SnCl_2$; $BiCl_3$

9-2(d) M_2X_3

9-2(e) In the lattice structure of the compound, there is a 1:3 ratio of potassium cations to nitride anions.

Part B

EXERCISES

9-3(a) three (b) one (c) four

9-3(b) (a) AsH_3 (b) CCl_4 (c) ClF (d) H_2S

9-3(c) A quadruple bond actually exists in transition metals but not among the representative elements. There are various complex reasons, but, basically, the sharing of eight electrons would require that all four bonds point in the same direction. This is not possible using just s- and p-type orbitals.

9-4(a) (a) 16 electrons   (b) 24 electrons   (c) 26 electrons

9-4(b) (a) $:\ddot{F}-\ddot{S}-\ddot{F}:$ (b)

$$:\ddot{O}-N=\ddot{O}:$$
$$|$$
$$:\ddot{Cl}:$$

(c) $Mg^{2+}\ 2[:\ddot{O}-\ddot{Cl}-\ddot{O}:]^-$

9-4(c) (a) $H-C\equiv N:$ (b)

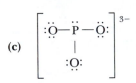

(c)

$$\left[\begin{array}{c} :\ddot{O}-\ddot{P}-\ddot{O}: \\ | \\ :\ddot{O}: \end{array}\right]^{3-}$$

9-4(d) Oxygen difluoride: $:\ddot{F}-\ddot{O}-\ddot{F}:$ Dioxygen difluoride: $:\ddot{F}-\ddot{O}-\ddot{O}-\ddot{F}:$ Typically, the central atom (the least electronegative) is written first, and the outer atoms second. Oxygen is second in electronegativity to fluorine, so only when it is bonded to fluorine is it the central atom.

9-5(a)

$$:O=\ddot{N}-N\overset{\diagup\ddot{O}:}{\diagdown\ddot{O}:} \leftrightarrow :O=\ddot{N}-N\overset{\diagup\ddot{O}:}{\diagdown\ddot{O}:}$$

$$\leftrightarrow :\underset{\cdot}{O}=N=N\overset{\diagup\ddot{O}:}{\diagdown\ddot{O}:} \leftrightarrow :O\equiv N-\ddot{N}\overset{\diagup\ddot{O}:}{\diagdown\ddot{O}:}$$

9-5(b) The molecule (b) SO_2 and the ion (d) OCN^-

9-5(c) $:\ddot{O}-C\equiv O:$ and $:O\equiv C-\ddot{O}:$. These resonance structures imply either that the bonds are unequal (which they are not) or that they average out to two double bonds (which they do). Resonance structures are not necessary in this case

since they lead to the same conclusion as the one structure with two double bonds.

9-6(a) Formal charges on the structure shown: N is −1; O is +1, Cl is 0. The structure shown with N as the central atom has no formal charge.

9-6(b) The O is the central atom. H—O—Cl has no formal charge. The other possibility is H—Cl—O, which has formal charges.

9-6(c) The double bond is between the C and the O.

Lewis structure for octet rule:

$$:\overset{\cdot\cdot}{O}:^{\ominus}$$
$$|$$
$$:\ddot{Cl}-\overset{+2}{S}-\ddot{O}:^{\ominus}$$
$$|$$
$$:\ddot{Cl}:$$

Lewis structure for formal charge:

$$:\overset{\cdot\cdot}{O}:$$
$$||$$
$$:\ddot{Cl}-S=\ddot{O}:$$
$$|$$
$$:\ddot{Cl}:$$

9-6(d)

9-6(e) In carbon monoxide, the carbon has a −1 formal charge and the oxygen has a +1 formal charge. If it forms a bond through the carbon, the formal charge on the carbon is increased to zero. If it forms a bond through the oxygen, the formal charge on the oxygen increases to +2. Bonding through the carbon would be preferred based on formal charges.

Part C

EXERCISES

9-7(a) The most electronegative element is <u>fluorine</u>. The second most electronegative element is <u>oxygen</u>. In a covalent bond between unlike atoms, the more electronegative element has a partial <u>negative charge</u>. This type of bond is said to be a <u>polar</u> covalent bond. A large difference in electronegativity between two atoms leads to an <u>ionic bond</u>.

9-7(b) (a) $Al\overset{\longmapsto}{}Se$ (b) $As\overset{\longmapsto}{}S$ (c) $S—S$ (d) $F\overset{\longleftarrow}{}Br$ (e) $C\overset{\longleftarrow}{}H$

9-7(c) (a) nonpolar   (b) polar (nearly ionic)   (c) polar (d) nonpolar

9-7(d) The larger the difference in electronegativity, the more likely the bond between two atoms is to be an ionic bond. Smaller difference leads to covalent bonds. Electronegativity is a periodic property, so the farther apart two atoms are on the periodic table, the more likely they are to form ionic bonds.

9-8(a) (a) two (b) four (c) three

9-8(b) a little less than 109.5°

9-8(c) trigonal planar

9-8(d) bent or V-shaped

9-8(e) (a) 180°, linear, linear (b) 109.5°, tetrahedral, *V*-shaped (c) 120°, trigonal planar, *V*-shaped

9-9(a) (a) polar (b) polar (c) nonpolar (d) nonpolar

9-9(b) $CS_2 < SO_2 < CH_2O < HF$

9-9(c) CH_2F_2 is a polar molecule. (Making a model helps to visualize this.) If the molecule were square instead, there would be two possibilities. It would be nonpolar if the fluorines were opposite to each other, but polar if they were adjacent to each another.

ANSWERS TO CHAPTER SYNTHESIS PROBLEM

a. Ca^{2+} 2 $CH_2NO_2^-$ There are 24 valence electrons in each anion.

b.

(1)

$$\overset{\displaystyle :\ddot{O}: \quad :\ddot{O}:}{\underset{H-C=N-H}{|\qquad\quad|}}{}^{-}$$

(2)

$$\overset{\displaystyle :\ddot{O}: \quad \cdot\ddot{O}\cdot}{\underset{H-\underset{\displaystyle \cdot\cdot}{C}-N-H}{\|\qquad\quad|}}{}^{-}$$

(3)

$$\overset{\displaystyle \cdot\ddot{O}\cdot \quad :\ddot{O}:}{\underset{H-C-N-H}{|\qquad\quad|}}{}^{-}$$

c. The N has the partial negative charge in the C—N bond. The N also has the partial negative charge in the N—H bond. N is more electronegative than both C and H.

d. Structure 3 is the only one with a trigonal pyramidal geometry around the N. It would be the correct structure. It is trigonal planar in the other two structures.

e. The geometry around the carbon is trigonal planar in the correct structure and also in structure 1. It is trigonal pyramidal in structure 2.

e. Structure 3 has only a −1 formal charge on the oxygen bonded to the nitrogen. All other atoms have zero formal charge. Structure 1 has formal charges of −1 on both O's and a +1 on the N. Structure 2 has a formal charge of −1 on the C and the O bonded to it and a +1 on the N. This all points to structure 3 as the correct one.

CHAPTER PROBLEMS

Throughout the text, answers to all exercises in color are given in Appendix E. The more difficult exercises are marked with an asterisk.

Dot Symbols of Elements (SECTION 9-1)

9-1. Write Lewis dot symbols for

(a) Ca   (d) I   (g) all Group VIA elements

(b) Sb   (e) Ne

(c) Sn   (f) Bi

9-2. Identify the representative element groups from the dot symbols.

(a) $\cdot\dot{M}\cdot$ (b) $\cdot\ddot{X}\cdot$ (c) $\dot{A}\cdot$

9-3. Identify the representative element groups from the dot symbols.

(a) $R\cdot$ (b) $:\ddot{Q}\cdot$ (c) $\cdot\dot{D}\cdot$

9-4. Why are only outer electrons represented in do symbols?

9-5. Which of the following dot symbols are incorrect?

(a) $P\dot{b}\cdot$ (c) $:\ddot{H}e:$ (e) $\cdot\ddot{T}e\cdot$

(b) $\cdot\ddot{B}i\cdot$ (d) $Cs\cdot$ (f) $\cdot\ddot{T}l\cdot$

Binary Ionic Compounds (SECTION 9-2)

9-6. From the periodic table, predict which pairs of elements can combine to form ionic bonds.

(a) H and Cl   (c) H and K   (e) B and Cl

(b) S and Sr   (d) Al and F   (f) Xe and F

9-7. From the periodic table, predict which pairs of elements can combine to form ionic bonds.

(a) Ba and I   (c) C and O   (e) P and S

(b) Cs and Se   (d) H and Se   (f) Cs and P

9-8. Which ions would not have a noble gas electron configuration?

(a) Sr^{2+} (c) Cr^{2+} (e) In^+ (g) Ba^{2+}

(b) S^- (d) Te^{2-} (f) Pb^{2+} (h) Tl^{3+}

9-9. Which ions would not have a noble gas electron configuration?

(a) Rb^+ (c) Sc^{2+} (e) As^{2-}

(b) Si^{4-} (d) I^{2-} (f) Se^{2-}

9-10. The ions Li^+, Be^{2+}, and H^- do not follow the octet rule. Why?

9-11. Write Lewis dot symbols for the following ions.

(a) K^+ (c) I^- (e) Ba^+ (g) Sc^{3+}

(b) O^- (d) P^{3-} (f) Xe^+

9-12. What is the origin of the octet rule? How does the octet rule relate to s and p subshells and to noble gases?

9-13. Which of the ions listed in problem 9-11 do not follow the octet rule?

9-14. For the following atoms, write the charge that would give the element a noble gas configuration.

(a) Mg   (b) Ga   (c) Br   (d) S   (e) P

9-15. For the following atoms, write the charge that would give the element a noble gas configuration.

(a) Rb   (b) Ba   (c) Te   (d) N

9-16. Write six ions that have the same electron configuration as Ne.

9-17. Write five ions that have the same electron configuration as Kr.

9-18. The Tl^{3+} ion does not have the same electron configuration as Xe, even though it lost its three outermost electrons. Explain.

9-19. Complete the following table with formulas of the ionic compounds that form between the anions and cations shown.

Cation/Anion	Br^-	S^{2-}	N^{3-}
Cs^+	CsBr		
Ba^{2+}			
In^{3+}			

9-20. Complete the following table with formulas of the ionic compounds that form between the anions and cations shown.

Cation/Anion	F^-	Te^{2-}	P^{3-}
Rb^+			
Mg^{2+}			
Cr^{3+}			

9-21. Write the formulas of the compounds formed between the following nonmetals and the metal calcium.

(a) I   (b) O   (c) N   (d) Te   (e) F

9-22. Write the formulas of the compounds formed between the following metals and the nonmetal sulfur.

(a) Be   (b) Cs   (c) Ga   (d) Sr

9-23. Most transition metal ions cannot be predicted by reference to the octet rule. Determine the charge on the metal cation from the charge on the anion for each of the following compounds.

(a) Cr_2O_3 (c) MnS   (e) $NiBr_2$
(b) FeF_3 (d) CoO   (f) VN

9-24. Determine the charge on the metal cation from the charge on the anion for each of the following compounds.

(a) IrO_4 (c) PtF_6 (e) Tc_2O_3
(b) ScN   (d) $CoCl_3$ (f) Ag_2Se

9-25. Why isn't a formula unit of $BaCl_2$ referred to as a molecule?

Lewis Structures of Compounds

(SECTIONS 9-3 AND 9-4)

9-26. From their Lewis dot symbols, predict the formula of the simplest compound formed by the combination of each pair of elements.

(a) H and Se   (c) Cl and F   (e) N and Cl
(b) H and Ge   (d) Cl and O   (f) C and Br

9-27. From their Lewis dot symbols, predict the formula of the simplest compound formed by the combination of each pair of elements.

(a) H and I   (c) Si and Br
(b) Se and Br   (d) H and As

9-28. From a consideration of the octet rule, which of the following compounds are impossible?

(a) PH_3 (c) SCl_2 (e) H_3O
(b) Cl_3 (d) NBr_4

9-29. Which of the following binary compounds does not follow the octet rule?

(a) NI_3 (c) CH_3 (e) FCl
(b) F_3O (d) HSe   (f) SCl_2

9-30. Determine the charge on each polyatomic anion from the charge on the cation.

(a) K_2SO_4 (c) $Al_2(SeO_4)_3$ (e) BaS_2O_3
(b) $Ca(IO_3)_2$ (d) $Ca(H_2PO_4)_2$

9-31. Determine the charge on each polyatomic anion from the charge on the cation.

(a) $NaBrO_2$ (c) $AlAsO_4$
(b) $SrSeO_3$ (d) $Mg(H_2PO_4)_2$

9-32. Write Lewis structures for the following.

(a) C_2H_6 (c) NF_3
(b) H_2O_2 (d) SCl_2
(e) C_2H_6O (There are two correct answers; both have the lone pairs on the oxygen.)

9-33. Write Lewis structures for the following.

(a) N_2H_4 (c) C_3H_8
(b) AsH_3 (d) CH_4O (All unshared electrons are on the O.)

9-34. Write Lewis structures for the following.

(a) CO   (c) KCN
(b) SO_3 (d) H_2SO_3 (H's are on different O's.)

9-35. In the following four ions, the C is the central atom. Write the Lewis structures.

(a) CN_2^{2-} (c) HCO_3^- (H on the O)
(b) HCO_2^- (H is on the C.)   (d) $ClCO^-$

9-36. Write Lewis structures for the following.

(a) N_2O (N is the central atom.)   (d) H_2S
(b) $Ca(NO_2)_2$ (e) CH_2Cl_2
(c) $AsCl_3$ (f) NH_4^+

9-37. Write Lewis structures for the following

(a) Cl_2O (c) C_2H_4 (e) BF_3
(b) SO_3^{2-} (d) H_2CO (f) NO^+

9-38. Write Lewis structures for the following.

(a) CO_2 (d) NO_3^- (g) C_2H_2
(b) H_2NOH (e) HOCN   (h) O_3
(c) $BaCl_2$ (f) $SiCl_4$

9-39. Write Lewis structures for the following.

(a) Cs_2Se

(b) $CH_3CO_2^-$ (All H's are on one C, and both O's are bonded to the other C.)

(c) $LiClO_3$ (d) N_2O_3 (e) PBr_3

Resonance Structures (SECTION 9-5)

9-40. Write all equivalent resonance structures (if any) for the following.

(a) SO_3 (b) NO_2^- (c) SO_3^{2-}

9-41. Write all equivalent resonance structures (if any) for the following.

(a) NO_3^-

(b) N_2O_4 (skeletal geometry)

```
        O            O
            N    N
        O            O
```

9-42. Write the equivalent resonance structures for the $H_3BCO_2^{2-}$ anion. The skeletal structure for the ion is

```
          H            O
      H   B    C
          H            O
```

9-43. What is meant by a resonance hybrid? What is implied about the nature of the C—O bonds from the resonance structures in problem 9-42?

9-44. Write any resonance structures possible for the ions in problem 9-35. What are the implications for any bonds where resonance structures exist?

Formal Charge (SECTION 9-6)

9-45. Nitrogen dioxide is also a free radical. By means of formal charge, assign the unpaired electron as being on the N or O atom

9-46. In hydrogen cyanide, the hydrogen is bonded to the carbon rather than the nitrogen. Write Lewis structures for both possibilities (i.e., HCN or HNC), assign formal charges, and rationalize the known sequence of atoms.

9-47. Assign formal charges to all the atoms in problem 9-34.

9-48. Assign formal charges to all the atoms in problem 9-35.

9-49. Assign formal charges that follow the octet rule to the atoms in the ClO_3^- ion. Write a structure that does not follow the octet rule but has the best formal charge distribution.

9-50. Write the Lewis structure of SO_3 that has the best distribution of formal charge.

9-51. The correct structure of nitrous oxide (N_2O) has an NNO sequence of atoms as opposed to the more symmetrical NON. Write the Lewis structures of both possibilities and rationalize the actual structure with formal charge.

Electronegativity and polarity (SECTION 9-7)

9-52. Rank the following elements in order to increasing electronegativity: B, Ba, Be, C, Cl, Cs, F, O

9-53. For bonds between the following elements, indicate the positive end of the dipole by a δ^+ and the negative end by a δ^-. Also indicate with a dipole arrow the direction of the dipole.

(a) N—H   (d) F—O   (g) C—B

(b) B—H   (e) O—Cl   (h) Cs—N

(c) Li—H   (f) S—Se   (i) C—S

9-54. Rank the bonds in problem 9-46 in order of increasing polarity.

9-55. Which of the following bonds is ionic, which is polar covalent, and which is nonpolar?

(a) C—F   (b) Al—F   (c) F—F

9-56. On the basis of difference in electronegativity, predict whether the following pairs of elements will form an ionic or a polar covalent bond.

(a) Ga and Br   (c) B and Br

(b) H and B   (d) Ca and Cl

9-57. On the basis of difference in electro negativity, predict whether the following pairs of elements will form an ionic or a polar covalent bond.

(a) Al and Cl   (c) K and O

(b) Ca and I   (d) Sn and Te

9-58. Which of the following bonds is nonpolar or nearly nonpolar?

(a) I—F   (c) C—H   (e) B—N

(b) I—I   (d) N—Br

9-59. On the following pairs of bonds, which is the more polar?

(a) N—H or C—H   (c) S—Cl or S—F

(b) Si—O or Si—N   (d) P—Cl or P—I

Geometry of Simple Molecules (SECTION 9-8)

9-60. From the Lewis structure, determine the molecular geometry of the following molecules.

(a) SF_2 (b) CS_2

(c) CCl_2F_2 (C is the central atom.)

(d) NOCl (N is the central atom.)

(e) Cl_2O

9-61. From the Lewis structure, determine the molecular geometry of the following molecules or ions.

(a) BF_2Cl (B is the central atom.)

(b) ClO_3^-

(c) N_2O (N is the central atom.)

(d) $COCl_2$ (C is the central atom.)

(e) SO_3

9-62. What are the approximate carbon-oxygen bond angles in the following?

(a) CO   (b) CO_2 (c) CO_3^{2-}

9-63. Which of the following have bond angles of approximately 109°?

(a) H_2O (c) NH_4^+

(b) HCO_2^- (d) HCN

*9-64. When a molecule or ion has more than one central atom, the geometry is determined at each central atom site. What are the geometries around the two central atoms in the following?

(a) H_2NOH (N and O)

(b) HOCN (O and C)

(c) H_3CCN (two C's)

(d) $H_3BCO_2^{2-}$ (B and C; see problem 9-42.)

*9-65. When a molecule or ion has more than one central atom, the geometry is determined at each central atom site. What are the geometries around the two central atoms in the following?

(a) H_2CCH_2 (2 C's)   (c) $HClO_3$ (Cl and O attached to H)

(b) Cl_2NNO (2 N's)   (d) H_3COCl (C and O)

Molecular Polarity (SECTION 9-9)

9-66. How can a molecule be nonpolar if it contains polar bonds?

9-67. Discuss the molecular polarities of the molecules in problem 9-60.

9-68. Discuss the molecular polarities of the molecules in problem 9-61.

9-69. Write the Lewis structure of SO_2Cl_2 (S is the central atom). Is the molecule polar?

9-70. Compare the expected molecular polarities of H_2O and H_2S. Assume that both molecules have the same angle.

9-71. Compare the expected molecular polarities of CH_4 and CH_2F_2.

9-72. Compare the expected molecular polarities of $CHCl_3$ and CHF_3.

9-73. CO_2 is a nonpolar molecule, but CO is polar. Explain.

9-74. SO_3 is a nonpolar molecule, but SO_2 is polar. Explain.

9-75. Propene has the formula C_3H_6 (H_3CCHCH_2). Write the Lewis structure and determine whether the molecule is polar.

General Problems

9-76. Write the formulas of three binary ionic compounds that contain one cation and one anion. One compound should contain Rb, one Sr, and one N.

9-77. There is a noble gas compound formed by xenon, XeO_3, that follows the octet rule. Write the Lewis structure of this compound. What is the geometry of the molecule? Is the XeO_3 molecule polar?

9-78. Which of the following compounds contains both ionic and covalent bonds?

(a) H_2SO_3 (c) K_2S (e) C_2H_6

(b) K_2SO_4 (d) H_2S

9-79. Write the Lewis structure of H_3BCO (all H's on the B). Can any resonance structures be written? What is the geometry around the B? What is the geometry around the C?

9-80. Carbon suboxide has the formula C_3O_2. All the atoms in the molecule are arranged linearly with an oxygen at each end. Write the Lewis structure.

9-81. A molecule may exist that has the formula N_2O_2. The order of the bonds is O—N—N—O. Write a Lewis structure for the compound that has two N—O double bonds. Write any resonance structures.

9-82. Cyanogen has the formula C_2N_2. The order of the bonds is N—C—C—N. Write a Lewis structure for cyanogen involving a C—C single bond. What is the geometry around a C atom? Draw dipole arrows for the bonds. Is the molecule polar?

9-83. A certain species is composed of one As and four Cl's. It contains 32 valence electrons. Write the Lewis structure of the species. What is the geometry of the species and the approximate Cl—As—Cl bond angle?

9-84. A certain species is composed of P and three O's. It contains 26 valence electrons. Write the Lewis structure of the species. What is the geometry of the species and the approximate O—P—O bond angle?

9-85. A certain species contains one chlorine and two iodines. It has a total of 20 valence electrons. Write the Lewis structure of the species. What is the geometry of the species and the approximate I—Cl—I bond angle?

9-86. A certain species contains one B, three C's, and 9 H's. It has a total of 24 valence electrons. Write the Lewis structure if the B is bonded only to the three C's. What is the geometry around the B and the approximate C—B—C bond angle? What is the geometry around the C's and the approximate H—C—H bond angle? Is the species polar?

*9-87. There are two compounds composed of potassium and nitrogen. Write the formula of the compound expected between potassium and nitrogen. Another compound (KN_3) is named potassium azide. Write the Lewis structure for the azide ion plus any resonance structures. What is the geometry about the central N atom?

*9-88. Write two resonance structures for hydrogen azide HN_3 (HNNN). What is the predicted HNN bond angle in each structure? The actual angle is about 110°. What does this indicate?

9-89. Write the Lewis structure for phosphoric acid (hydrogens bonded to oxygens). What is the geometry around the P? What is the geometry around the oxygen in the H—O—P bond? What is the negative end of the dipole in the P—O bond? Is the molecule polar?

9-90. Cyanic acid has the formula HOCN (atoms bonded in that order). The H—O—C bond angle is about 105°. Write the Lewis structure that demonstrates this fact. What is the geometry around the C?

9-91. Thiocyanic acid has the formula HSCN (atoms bonded in that order). The H—S—C bond angle is about 116°. Write the Lewis structure that demonstrates this fact. What is the geometry around the C?

9-92. A compound has the formula N_2F_2. Write a Lewis structure that contains a N=N bond.

9-93. A second compound of nitrogen and fluorine contains one nitrogen and the expected number of fluorines. A third has the formula N_2F_4. Write the Lewis structures of these two compounds. What is the geometry around the N in each compound? Are either or both compounds polar?

*9-94. Two compounds of oxygen are named oxygen difluoride and dioxygen difluoride. Write the Lewis structures of these two compounds. The latter compound contains an O—O bond. Oxygen is usually written and named second in binary compounds. Why not here? Are either or both of these compounds polar?

*9-95. A compound has the formula $Na_2C_2O_4$. Write a Lewis structure for the compound that contains a C—C bond. Write any resonance structures present.

*9-96. If you have already covered Chapter 6, consider the following problem. A compound is 27.4% Na, 14.3%

C, 57.1% O, and 1.19% H. What is compound? Write the Lewis struct bonded to O). What is the geomet is the approximate O—C—O bond geometry around the H—O—C bon mate angle?

*9-97. If you have already covered C the following problem. A compound ... Li, 75.0% C, and 3.16% H. What is the formula of the compound? Write the Lewis structure of the anion. What is the geometry around the central C and the approximate H—C—C bond angle?

*9-98. If you have already covered Chapter 6, consider the following problem. A compound is 19.8% Ca, 1.00% H, 31.7% S, and 47.5% O. What is the formula of the compound? (*Hint:* The formula of the anion should be reduced to its empirical formula.) What is the name of the compound? (Refer to Table 4-2.) What is the geometry around the S in the anion? What is the approximate H—O—S bond angle?

*9-99. Refer to problem 8-99 in the previous chapter. Use the periodic table from the planet Zerk to answer the following.

1. What are the simplest formulas of compounds formed between the following (Example: between 7 and 7 is 7_2.)

(a) 1 and 7	**(d)** 7 and 9	**(g)** 6 and 7
(b) 1 and 3	**(e)** 7 and 13	**(h)** 3 and 6
(c) 1 and 5	**(f)** 10 and 13	

2. Write Lewis structures for all of the above. Indicate which are ionic. (Remember that on Zerk there will be something different from an octet rule.)

STUDENT WORKSHOP

VSEPR Theory

Purpose: To make models of several small molecules and ions, and to evaluate them using VSEPR theory. (Work in groups of three or four. Estimated time: 20 min.)

Equipment: molecular model kit

*   As a group, draw out the Lewis structures of the following chemicals.

SCN⁻	SO_2	CH_2Cl_2
CF_2O	CS_2	CO_3^{2-}

*   Make the model using the kits provided.

*   Determine bond angle, electronic geometry, and molecular geometry from VSEPR theory. Verify these with the model.

*   Using this model, identify whether the molecule is polar or nonpolar.

*   Indicate whether the chemical has resonance and/or formal charge.

10 The Gaseous State

Our atmosphere serves as a protective blanket of gases.
The behavior of these gases can be predicted by gas laws,
which are discussed in this chapter.

SETTING THE STAGE

Our atmosphere of gases is like the combination of a warm blanket and a protective shield. It nurtures life on this planet in numerous ways. Oxygen supports the metabolism that gives energy to living things. Nitrogen becomes part of proteins and also dilutes the oxygen so that combustion does not occur explosively. Carbon dioxide and water vapor lock in the daytime heat so that the temperature does not become bitterly cold at night. Ozone high in the stratosphere absorbs the harmful portion of the sun's radiation. The thickness of the air burns up most incoming meteorites and other debris from space. Life as we know it requires the constant support and protection of this sea of gases.

Just over 200 years ago, there was little understanding of the nature of the gases that support our existence. It wasn't until the late eighteenth century that the experiments of Antoine Lavoisier in France and Joseph Priestley in England gave us some understanding about the gaseous mixture of the atmosphere. These scientists proved that air was not just one substance, as had been previously thought, but was mainly a mixture of two elements: nitrogen and oxygen. Their work also laid the foundation for the development of the law of conservation of mass. This was the beginning of modern quantitative chemistry.

With a few exceptions, gases are invisible. This makes the study of gases somewhat difficult. Except on a windy day, it is easy to forget that this form of matter surrounds us at all.

There is good news when we study the nature of gases, however. In many ways, they all behave similarly, which allows convenient generalizations, including the gas laws. Solids and liquids are less abstract than gases simply because they are visible. The downside of working with solids and liquids, however, is that few generalizations apply and each substance must be studied individually.

In Part A of this chapter, we will examine how conditions such as pressure and temperature affect the volume of a gas. In Part B, we will apply these relationships to how the volume of a gas relates to the amount present under various conditions. We can then expand the concept of stoichiometry to include gas volumes.

Part A

The Nature of the Gaseous State and the Effects of Conditions

SETTING A GOAL

- You will learn how the molecular nature of gases leads to predictable behavior, known as the gas laws.

OBJECTIVES

10-1 List the general properties of gases based on the postulates of the Kinetic Molecular Theory.

10-2 Calculate the effects changing pressure has on the volume of a gas (Boyle's Law).

10-3 Using the gas laws, perform calculations involving relationships between volume, pressure, and temperature.

▶ OBJECTIVE
FOR SECTION 10-1
List the general properties of gases based on the postulates of the Kinetic Molecular Theory.

10-1 The Nature of Gases and the Kinetic Molecular Theory

LOOKING AHEAD! Living in a sea of gases, we tend to take the properties of the gaseous state for granted. In this section, we will not only examine their common properties but also provide the accepted theory that explains them. ■

10-1.1 The Properties of Gases

Four of the common properties of gases are described below.

1. *Gases are compressible.*
   When a glass of water is full, we mean what we say—no more can be added. But when is an automobile tire full of air? Actually, never. We can always add more air (at least until the tire bursts). The nature of the gaseous state allows us to press a volume of gas into a smaller volume. Liquids and solids are not like that. They are essentially incompressible.

2. *Gases have low densities.*
   The density of a typical liquid or solid is around 2 g/mL. The density of a typical gas is around 2 g/L. This means that the density of a solid or liquid is roughly 1000 times greater than that of a gas. This is a big difference.

3. *Gases mix thoroughly.*
   Nothing compares with a good home-cooked meal. Perhaps the best part is the pleasurable fragrances that we detect as the meal is being prepared. It doesn't take long for those familiar odors to drift from the kitchen throughout the entire home. The vapors from cooking mix thoroughly and rapidly with the surrounding air. This is a property of gases that is unique to their state. In contrast, some liquids do not mix at all (e.g., oil and water). If they do mix (e.g., water and alcohol), the mixing process occurs quite slowly without stirring.

4. *A gas fills a container, uniformly exerting equal pressure in all directions.*
   When we blow air into a round balloon, it remains spherical and uniform. The air in the balloon obviously pushes out equally in all directions. If we were to fill the balloon with water instead (for scientific reasons only, of course), we would notice that the balloon sags. The water accumulates in the bottom of the balloon.

The balloon is spherical because the gas contained inside exerts equal pressure in all directions.

In a room full of air, it would seem that gases defy gravity. The pressure of the atmosphere is the same on the ceiling, walls, and floor.* Liquid water is quite different, however. Scuba divers are well aware of the increase in pressure as they dive deeper and deeper.

10-1.2 The Kinetic Molecular Theory

Obviously, the gaseous state of matter is quite different from the other two physical states. Gases are composed of either individual atoms (in the case of noble gases) or molecules (for all gaseous diatomic elements and other gaseous compounds). This fact and others can be summarized in a set of assumptions collectively known as the **kinetic molecular theory**, or simply the **kinetic theory**. The kinetic theory was advanced in the late nineteenth century to explain the common properties of gases. It allows us to visualize the behavior of gases at the molecular level. The major points of this theory as applied to gases are as follows.

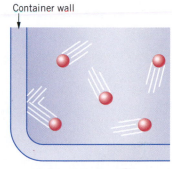

1. A gas is composed of very small particles called molecules, which are widely spaced and occupy negligible volume compared to the volume of the container. A container of gas is comprised mostly of the empty space between molecules.
2. The molecules of a gas are in rapid, random motion, colliding with each other and the sides of the container. Pressure is a result of these collisions with the sides of the container. (See Figure 10-1.)
3. All collisions involving gas molecules are elastic. This means that the total energy of two colliding molecules is conserved. A ball bouncing off the pavement undergoes inelastic collisions; it does not bounce as high each time.
4. Gas molecules have negligible attractive (or repulsive) forces between them.
5. The temperature of a gas is related to the average kinetic energy of the gas molecules. Also, at the same temperature, different gases have the same average kinetic energy (K.E.).

FIGURE 10-1 Gases Gas molecules are in rapid, random motion.

The kinetic theory explains the properties described above as well as the gas laws that follow. For example, since gas molecules themselves have essentially no volume and a gas is mostly empty space, it has a low density and is easily compressed. The rapid motion explains why gases mix rapidly and thoroughly. Because gas molecules are not attracted to each other, they do not "clump together" as do the molecules in liquids and solids. Thus, the molecules in gases spread out and fill a container uniformly, exerting pressure equally in all directions.

10-1.3 Graham's Law

Another consequence of the kinetic theory concerns the relative velocities of gas molecules at the same temperature. The kinetic energy of a gas is given by the equation

$$K.E. = \tfrac{1}{2}mv^2 \qquad \text{where } m = \text{mass} \quad \text{and} \quad v = \text{velocity}$$

The average velocity of a gas molecule relates directly to two other aspects of molecular motion. These are the rates of **diffusion** (*mixing*) and **effusion** (*moving through an opening or hole*). The relationship between diffusion and molar mass was first proposed in 1832 by Thomas Graham. This was actually some time before the acceptance of kinetic theory. **Graham's law** *states that the rates of diffusion of gases at the same temperature are inversely proportional to the square roots of their molar masses.* The average velocities and the rates of effusion can also be substituted for rates of diffusion

---

*In fact, gases do not really defy gravity, since our atmosphere gets thinner as the altitude increases. However, the thinning of the atmosphere occurs on a scale of several kilometers. It would not be noticed on the scale of a few meters in a room.

FIGURE 10-2 Relative Velocities At the same temperature, a helium atom travels almost three times faster than a hydrogen chloride molecule.

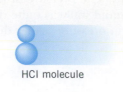

HCl molecule

Helium atom

in the law. Recall, that an inverse relationship exists when one variable goes up as another variable goes down, and vice versa.

Although we will not do specific calculations involving Graham's law in this text, consider the inverse relationship between the molecular velocities of two particles. An HCl molecule (about 36 amu) has about nine times the mass of a helium atom (about 4 amu). However, the helium atom has an average velocity of three times that of the HCl molecule. (See Figure 10-2.) The lighter the molecule, the higher is its average velocity.

Graham's law and gaseous effusion have important applications, such as the enrichment of the isotope of uranium (^{235}U) used for nuclear power or as a nuclear weapon. This procedure is used to enrich the uranium from the naturally occurring 0.7% ^{235}U to 2–3% for nuclear reactors and to around 90% for nuclear weapons. We take advantage of the very slightly increased rate of effusion of gaseous $^{235}UF_6$ compared to $^{238}UF_6$. Due to the small difference in masses, however, it is a long and tedious process to enrich uranium by this method.

You may have noticed Graham's law in action with a helium balloon. Helium balloons shrink much faster than air-filled balloons. (See Figure 10-3.) Since the average helium atoms have over twice the velocities of the nitrogen and oxygen molecules of air, they effuse out of the tiny pores of a balloon faster.

▶ **ASSESSING THE OBJECTIVE FOR SECTION 10-1**

EXERCISE 10-1(a) LEVEL 1: Fill in the blanks.

Unlike solids and liquids, a gas has a _____ density and fills a container _____. The unique behavior of gases can be explained by the _____ theory. In this theory, gas molecules are in _____, _____ motion, exerting pressure by _____ with the walls of the container. Another aspect of this theory is that the temperature relates to the average _____ energy of the gas. At the same temperature, the gas with the larger molar mass will have the _____ velocity. The velocity of a gas relates to the rates of _____ and _____ as expressed by Graham's law.

EXERCISE 10-1(b) LEVEL 1: Indicate whether the following are true or false:
(a) Different gases have different energies at the same temperature.
(b) Different gases have different speeds at the same temperature.
(c) Gases are easier to compress than liquids or solids.
(d) Gases have strong forces between molecules.
(e) Gases are mostly empty space.

EXERCISE 10-1(c) LEVEL 2: How does the average velocity of an SF_6 molecule compare to an N_2 molecule at the same temperature?

EXERCISE 10-1(d) LEVEL 3: It is dangerous to leave aerosol containers in the trunk of a car in the summer. Use the postulates of kinetic molecular theory to explain why.

EXERCISE 10-1(e) LEVEL 3: Argon is used in thermopane windows because it is a better insulator than air between these two glass panes. Based on Graham's law, can you suggest why?

For additional practice, work chapter problems 10-5, 10-7, and 10-9.

FIGURE 10-3 Graham's Law and Helium Balloons The balloon on the left is filled with air, and the one on the right is filled with helium. One day later, the helium balloon has shrunk because the small atoms of helium have effused out of the balloon.

MAKING IT REAL

Ozone—Friend and Foe

Ozone is associated with air pollution.

Oxygen is the element that made animal life possible. Billions of years ago, primitive forms of life on this planet began to use solar energy, carbon dioxide, and water to generate energy-rich compounds with oxygen as a by-product. As the oxygen became more concentrated in the atmosphere, new life-forms appeared that used this oxygen for metabolism. But this new life could not have survived on the surface of Earth if it were not for another role that oxygen plays in a region around 50 km above Earth's surface known as the stratosphere. There, solar radiation is powerful enough to break the bond in O_2 to form oxygen atoms. (Chemical reactions that are initiated by light energy are known as *photochemical reactions*.) The atoms of oxygen combine with molecules of oxygen to form ozone (O_3). (O_2 and O_3 are known as allotropes of oxygen.) These reactions are illustrated below. The solar energy is represented as $h\nu$ in the following equation.

$$O_2(g) + h\nu \longrightarrow O(g) + O(g)$$

$$O(g) + O_2(g) \longrightarrow O_3(g) + \text{heat energy}$$

The solar radiation that breaks the O_2 bond lies in the higher-energy ultraviolet region of the spectrum (around 240 nm). However, there are lower-energy wavelengths of ultraviolet light (UVA and UVB from 240 nm to 310 nm) that would still come through to the surface if it were not for ozone. Ozone in the stratosphere effectively absorbs these damaging wavelengths of ultraviolet radiation. The equation is represented as

$$O_3(g) + h\nu \rightarrow O_2(g) + O(g)$$

Without ozone in the stratosphere, most life on the surface of Earth would be destroyed by ultraviolet radiation. Destruction of ozone by synthetic chemicals is a crucial area of concern and is being addressed. (See Section 15-2.)

In the lower atmosphere, where we live, however, ozone is a problem. Ozone is associated with what is called *photochemical smog*. Automobiles and other heat sources form NO, which then combines with oxygen molecules in the atmosphere to form NO_2. Visible light from the sun then breaks the NO_2 back into NO and oxygen *atoms,* which then lead to ozone. This is illustrated by the following equations.

$$NO_2(g) + h\nu \rightarrow NO(g) + O(g) \quad O(g) + O_2(g) \rightarrow O_3(g)$$

Ozone is toxic and chemically reactive. It causes trouble for anyone with lung disorders or those exercising outdoors. It is considered a serious pollutant. So ozone is a friend when it is high up in the stratosphere but a serious problem near the surface. "Everything in its place" is a saying that seems appropriate.

10-2 The Pressure of a Gas

LOOKING AHEAD! Since kinetic theory makes the nature of gases seem so simple and reasonable, it is sometimes difficult to appreciate the advances made by those who were not aware of the existence of atoms, molecules, or their motions. Before we look at how other conditions affect gases, we will take a closer look at the property known as pressure and how it affects the volume of a gas. ■

▶ **OBJECTIVE FOR SECTION 10-2**
Calculate the effect changing pressure has on the volume of a gas (Boyle's law).

10-2.1 The Barometer and Pressure

A newscast on TV would not be complete without the weather forecast. The weather data probably include the atmospheric pressure as read from a **barometer**. Rising pressure usually means improving weather, and a dropping barometric pressure often means that a storm is approaching.

FIGURE 10-4 The Barometer
When a long tube is filled with mercury and inverted in a bowl of mercury, the atmosphere supports the column to a height of 76.0 cm.

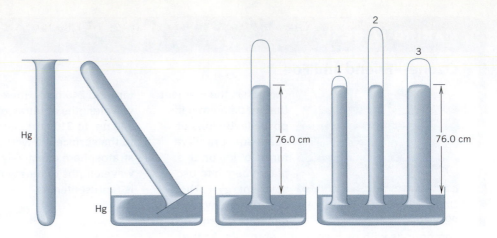

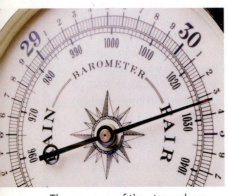

The pressure of the atmosphere is measured by a barometer. The barometric pressure can be used to forecast weather.

The barometer was invented by an Italian scientist, Evangelista Torricelli, in 1643. Torricelli filled a long glass tube with mercury, a dense liquid metal, and then inverted the tube into a bowl of mercury so that no air would enter the tube. Torricelli found that the mercury in the tube seemed to defy gravity by staying suspended to a height of about 76 cm no matter how long or wide the tube. (See Figure 10-4.)

At the time, many scientists thought that since "nature abhors a vacuum," the vacuum created at the top of the tube suspended the mercury. Torricelli suggested instead that it is the weight of the air on the outside that pushes the mercury up. Otherwise, the greater vacuum present in tubes 2 and 3 in Figure 10-4 would support a higher level of mercury. Torricelli also suggested that the thinner atmosphere at higher levels in the mountains would support less mercury. He was correct. The higher the elevation, the lower the level of mercury in the barometer.

The weight of a quantity of matter pressing on a surface exerts a *force*. **Pressure** *is defined as the force applied per unit area.* This can be expressed mathematically as

$$P \text{ (pressure)} = \frac{F \text{ (force)}}{A \text{ (area)}}$$

In a barometer, the pressure exerted by the atmosphere on the outside is balanced by the pressure exerted by the weight of the column of mercury on the inside.

If, at times, it seems like you are under a lot of "pressure," it is not necessarily from the atmosphere. However, you are under a lot of "force" from the atmosphere—over 20,000 lb of it. Fortunately, this force is spread out over your entire body surface of about 1500 in.^{2} (with much individual variation). When we divide the actual force on any one person by that person's body area, the resulting pressure comes out the same for everyone—a reasonable 15 lb/in.^{2}

One atmosphere *(1 atm) is defined as the average pressure of the atmosphere at sea level and is the standard of pressure.* As we have seen, this is equivalent to the pressure exerted by a column of mercury 76.0 cm (760 mm) high.

1.00 atm = 76.0 cm Hg = 760 mm Hg = 760 torr

The unit of mm of Hg is also known as the *torr* in honor of Torricelli. In addition to the torr, there are several other units of pressure (e.g., lb/in.^{2}) that have special uses. The relationships of the units to 1 atm and their applications are listed in Table 10-1. An example of a conversion between units is shown in Example 10-1.

TABLE 10-1

One Atmosphere

UNIT OF PRESSURE	SPECIAL USE
760 mm Hg or 760 torr	Most chemistry laboratory measurements for pressures in the neighborhood of one atmosphere
14.7 lb/in.^{2} (psi)	U.S. pressure gauges
29.9 in. Hg	U.S. weather reports
101.325 kPa (kilopascals)	The metric unit of pressure [1 N (newton)/m^2]
1.013 bars	Used in physics and astronomy mainly for very low pressures (millibars) or very high pressures (kilobars)

EXAMPLE 10-1 **Converting Between Units of Pressure**

What is 485 torr expressed in atmospheres?

PROCEDURE

The conversion factors between atm and torr are

$$\frac{760 \text{ torr}}{\text{atm}} \quad \text{and} \quad \frac{1 \text{ atm}}{760 \text{ torr}}$$

SOLUTION

$$485 \text{ torr} \times \frac{1 \text{ atm}}{760 \text{ torr}} = \underline{\underline{0.638 \text{ atm}}}$$

ANALYSIS

This is a good simple example to review the use of dimensional analysis to solve mathematical problems. While pressure readings come from a number of different sources, and therefore in a number of different units, you'll find it most convenient to convert all pressure units into atmospheres for purposes of comparison and canceling.

SYNTHESIS

Generally we fill our tires up to around 35 psi (lbs/in.^{2}), according to our tire pressure gauge. This means that at 14.7 psi/atm, the gauge is reading greater than 2 atm. Does that mean that there is a little more than 2 atm of pressure in the tire? What if your gauge reads zero? Does that mean that there is *no* pressure inside the tire? No. It means that there is *equal* pressure inside and out. The gauge reads the *difference* in pressure between the inside of a tire and the outside. *Gauge pressure* is always 1 atm (or whatever the external pressure) less than the actual pressure inside a container. Therefore, our tires have slightly more than 3 atm of pressure inside of them, not two. When you get a chance, check your tire pressure gauge for units. Chances are that along with psi, there's also a listing for kPa and perhaps bars as well.

10-2.2 Boyle's Law

Every breath we take illustrates the interaction between the volume and the pressure of a gas. When the diaphragm under our rib cage relaxes, it moves up, squeezing our lungs and decreasing their volume. The decreased volume increases the air pressure inside the lungs relative to the outside atmosphere, and we expel air. When the diaphragm contracts, it moves down, increasing the volume of the lungs and, as a result, decreasing the pressure. Air rushes in from the atmosphere until the pressures are equal.

The relationship between the volume and pressure of a sample of gas under constant-temperature conditions was first described in 1660 by the English scientist Robert Boyle. In an apparatus similar to that shown in Figure 10-5, Boyle found that the decrease in volume can be predicted by the amount of pressure increase. This observation is now expressed as **Boyle's law**, *which states that there is an inverse relationship between the pressure exerted on a quantity of gas and its volume if the temperature is held constant.*

The inverse relationship suggested by Boyle is represented as

$$V \propto \frac{1}{P}$$

or as an equality with k the constant of proportionality:

$$V = \frac{k}{P} \quad \text{or} \quad PV = k$$

Boyle's law can be illustrated with the apparatus shown in Figure 10-5. In experiment 1, a certain quantity of gas ($V_1 = 10.0$ mL) is trapped in a U-shaped tube by some mercury. Since the level of mercury is the same in both sides of the tube and

FIGURE 10-5 Boyle's Law **Apparatus** Addition of mercury in the apparatus causes an increase in pressure on the trapped gas. This leads to a reduction in the volume.

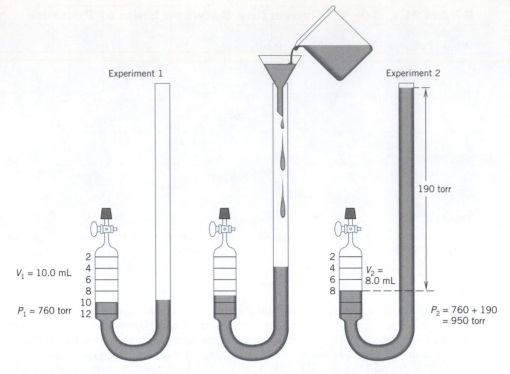

Experiment 1   Experiment 2

190 torr

$V_1 = 10.0$ mL

$P_1 = 760$ torr

$V_2 = 8.0$ mL

$P_2 = 760 + 190 = 950$ torr

the right side is open to the atmosphere, the pressure on the trapped gas is the same as the atmospheric pressure ($P_1 = 760$ torr). In this experiment,

$$P_1 V_1 = 760 \text{ torr} \times 10.0 \text{ mL} = 7600 \text{ torr} \cdot \text{mL} = k$$

When mercury is added to the tube, the pressure on the trapped gas is increased to 950 torr (760 torr originally plus 190 torr from the added mercury). Note in experiment 2 that the increase in pressure has caused a decrease in volume to 8.0 mL. Applying Boyle's law to this experiment, we have

$$P_2 V_2 = 950 \text{ torr} \times 8.0 \text{ mL} = 7600 \text{ torr} \cdot \text{mL} = k$$

As predicted by Boyle's law, PV equals the same value in both experiments. Therefore, for a quantity of gas under two sets of conditions at the same temperature,

$$P_1 V_1 = P_2 V_2 = k$$

We can use this equation to calculate how a volume of gas changes when the pressure changes. For example, if V_2 is the new volume that we are to find at a given new pressure, P_2, the equation becomes

$$V_2 = V_1 \times \frac{P_1}{P_2}$$

If a series of measurements are made of volume and pressure, the results can be graphed. A graph constructed of a few points allows us to determine a large number of volumes from specific pressures. Boyle's law is another example of an inverse relationship. That is, one variable goes up as the other goes down. Such a relationship produces a nonlinear graph as illustrated in Figure 10-6.

At higher altitudes, or in an airplane where the external pressure is lower, the bag of potato chips expands.

EXPERIMENT	PRESSURE (TORR)	VOLUME (L)
1	500	15.0
2	600	12.5
3	760	9.8
4	900	8.3
5	1200	6.2
6	1600	4.7

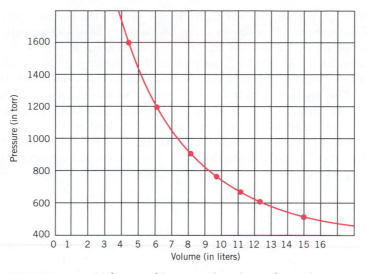

FIGURE 10-6 Volume and Pressure The volume of a gas decreases as the pressure increases. This an inverse relationship.

EXAMPLE 10-2 Applying Boyle's Law

Inside a certain automobile engine, the volume of a cylinder is 475 mL when the pressure is 1.05 atm. When the gas is compressed, the pressure increases to 5.65 atm at the same temperature. What is the volume of the compressed gas?

PROCEDURE

First, identify and list the initial and final conditions. Make sure the units of pressure are the same in the initial and final conditions.

SOLUTION

Initial Conditions	Final Conditions
$V_1 = 475$ mL	$V_2 = ?$
$P_1 = 1.05$ atm	$P_2 = 5.65$ atm

$$V_2 = V_1 \times \frac{P_1}{P_2}$$

$$V_2 = 475 \text{ mL} \times \frac{1.05 \text{ atm}}{5.65 \text{ atm}} = \underline{88.3 \text{ mL}}$$

ANALYSIS

Problems of this sort go by the descriptive name among students of "plug and chug." The equation for the relationship has four variables, and the problem identifies three of them. Simply rearrange the equation for the target variable, plug in the given values in the problem, and mathematically chug through the calculation for the answer. Gas law problems can also be worked by logic, since we know qualitatively how a gas should react to a change of conditions. In this example, we should predict that an increase in pressure ($P_2 > P_1$) would lead to a decrease in volume ($V_2 < V_1$). We therefore choose a pressure ratio that will do the job. This requires the use of a fraction less than 1 (known as a proper fraction) with the lower pressure in the numerator.

SYNTHESIS

As noted, Boyle's Law demonstrates an inverse relationship among variables. That is why the product, $P \times V$, always gives the same constant value. Other simple mathematically inverse relationships exist between speed and time. The higher the speed, the shorter the time to travel a set distance. Mass and acceleration are also inversely related. For a given force, the larger mass will be accelerated more slowly. There are also some less mathematical inverse relationships, such as weight and exercise, or consumed alcohol and driving ability. Can you think of other examples?

Although it wasn't known in the 1660s, Boyle's law is a natural consequence of the kinetic theory of gases. In order to focus on how a change in volume affects the pressure, we will assume that there is an average molecule of a gas at a given temperature. In Figure 10-7, the path of this one average molecule is traced. The pressure exerted by this molecule is a result of collisions with the sides of the container. The distance traveled in a given unit of time is represented by the length of the path. Since this is an average molecule, it travels the same distance per unit time in both the high-volume situation (on the left) and the low-volume situation (on the right). Note that, on the right, the lower volume leads to an increased number of collisions with the sides of the container. More frequent collisions mean higher pressure.

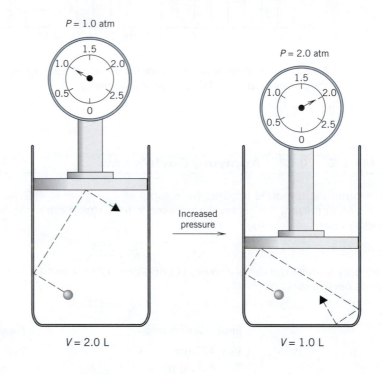

FIGURE 10-7 Boyle's Law and Kinetic Molecular Theory When the volume decreases, the pressure increases because of more frequent collisions with the walls of the container.

▶ **ASSESSING THE OBJECTIVE FOR SECTION 10-2**

EXERCISE 10-2(a) LEVEL 1: Fill in the blanks.

Atmospheric pressure is read from a device called a _____. Pressure is defined as _____ per unit _____. One atmosphere of pressure is equivalent to _____ torr. An increase in pressure on a volume of gas causes the volume to _____,

EXERCISE 10-2(b) LEVEL 2: What is 0.650 atm expressed in **(a)** torr, **(b)** kPa, and **(c)** lb/in.^{2}?

EXERCISE 10-2(c) LEVEL 2: The volume of 0.0388 mol of gas is 870 mL at a pressure of 1.00 atm and a temperature of 0°C. What is the volume at 1.32 atm and 0°C?

EXERCISE 10-2(d) LEVEL 2: A piston with a volume of 330.0 mL at 625 torr is compressed until the volume is 50.0 mL. What is the new pressure, measured in kPa?

EXERCISE 10-2(e) LEVEL 3: How does the kinetic molecular theory explain Boyle's law?

For additional practice, work chapter problems 10-10, 10-12, 10-16, and 10-19.

10-3 Charles's, Gay-Lussac's, and Avogadro's Laws

LOOKING AHEAD! We probably sense that all matter expands when it is heated. Only gases expand by the same amount, however, regardless of the identity of the gas. One hundred years after Boyle's relationship was advanced, the quantitative relationship between volume and temperature was studied. Other relationships discussed in this section involve temperature, pressure, and the amount of gas present. ■

▶ OBJECTIVE
FOR SECTION 10-3
Using the gas laws, perform calculations involving the relationships volume, pressure, and temperature.

10-3.1 Charles's Law

Very few sights appear more tranquil than brightly colored hot-air balloons drifting across a blue summer sky. Except for the occasional "swoosh" of gas burners, they cruise by in majestic silence. What keeps them afloat? It must have something to do with the gas burners. When air is heated, it expands and becomes less dense than the surrounding cooler air and thus rises. By confining the trapped hot air, the balloon and attached gondola become "lighter than air" and lift into the sky. The quantitative effect of temperature on the volume of a sample of gas at constant pressure was first advanced by a French scientist, Jacques Charles, in 1787. Charles showed that any gas expands by a definite fraction as the temperature rises. He found that the volume increases by a fraction of 1/273 for each 1°C rise in the temperature. (See Figure 10-8.)

Let's assume that we have made four measurements of the volume of a gas at four temperatures at constant pressure. The results of these experiments are listed in the table on the left in Figure 10-9. The four points are also plotted in a graph on the right. When the four points in the graph are connected, a straight line results. The direct relationship between volume and temperature is known as a *linear relationship*. In the graph, the straight line has been extended to what the volume would be at temperatures lower than those in the experiment. (The procedure of extending data beyond experimental results is known as *extrapolation*.) If extended all the way to where the gas would theoretically have zero volume, the line would intersect the temperature axis at −273.15°C. Certainly, matter could never have zero volume, and it is impossible to cool gases indefinitely. At some point, all gases condense to become liquids or solids, so gas laws no longer apply. However, this temperature does have significance, because it is the lowest temperature possible. As noted in kinetic molecular theory, the average kinetic energy or velocity of molecules is related to the temperature. *The lowest possible temperature, known as* **absolute zero**, *is the temperature at which translational motion (motion from point to point) ceases. The* **Kelvin scale** *starts at absolute zero.*

Thus, there are no negative values on the Kelvin scale, just as there are no negative values on any pressure scale and no such thing as a negative volume. Since the magnitudes of Celsius and Kelvin degrees are the same, we have the following simple relationship between scales, where $T(K)$ is the temperature in kelvins and $T(°C)$ represents the Celsius temperature. In our calculations, we will use the value of the lowest temperature expressed to three significant figures (i.e., −273°C).

$$T(K) = T(°C) + 273$$

Hot-air balloons are practical applications of Charles's law.

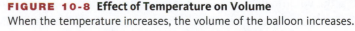

$V = 10.0$ L
$T = 0°C$

Increased temperature

$V = 13.7$ L
$T = 100°C$

FIGURE 10-8 Effect of Temperature on Volume
When the temperature increases, the volume of the balloon increases.

FIGURE 10-9 Volume and Temperature The volume of a gas increases by a definite amount as the temperature increases. This is a direct relationship.

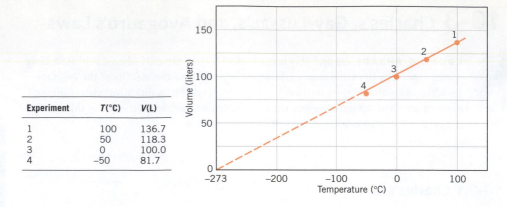

Experiment	$T(°C)$	$V(L)$
1	100	136.7
2	50	118.3
3	0	100.0
4	−50	81.7

The relationship between volume and temperature noted by Charles can now be restated in terms of the Kelvin scale. **Charles's law** *states that the volume of a gas is directly proportional to the Kelvin temperature* (T) *at constant pressure.* (From this point on, *T* will represent the Kelvin temperature exclusively.) This can be expressed mathematically as a proportion or as an equality, with *k* again serving as the symbol for the constant of proportionality.

$$V \propto T \qquad V = kT \qquad \text{or} \qquad \frac{V}{T} = k$$

For a quantity of gas at two temperatures at the same pressure,

$$\frac{V_1}{T_1} = \frac{V_2}{T_2}$$

This equation can be used to calculate how the volume of gas changes when the temperature changes. For example, if V_2 is a new volume that we are to find at a given new temperature T_2, the equation becomes

$$V_2 = V_1 \times \frac{T_2}{T_1}$$

EXAMPLE 10-3 Applying Charles's Law

A given quantity of gas in a balloon has a volume of 185 mL at a temperature of 52°C. What is the volume of the balloon if the temperature is lowered to −17°C? Assume that the pressure remains constant.

PROCEDURE

As before, identify the initial and final conditions. Remember, we must use the Kelvin temperature scale in all gas law calculations.

SOLUTION

Initial Conditions
$V_1 = 185$ mL
$T_1 = (52 + 273) = 325$ K

Final Conditions
$V_2 = ?$
$T_2 = (-17 + 273) = 256$ K

$$V_2 = V_1 \times \frac{T_2}{T_1}$$

$$V_2 = 185 \text{ mL} \times \frac{256 \text{ K}}{325 \text{ K}} = \underline{146 \text{ mL}}$$

ANALYSIS

Rather than just substituting numbers, we can also set this problem up with logic. Note that since the temperature decreases, the volume decreases. The temperature factor must therefore be less than 1 (again a proper fraction).

Qualitatively, is our answer reasonable? The volume has decreased about 40 mL, or close to 1/5 of its volume. This must be due to an accompanying 1/5 decrease in the temperature. Looking at the Celsius temperatures, 52°C down to −17°C, it's hard to say if that's our 1/5 decrease. But looking at the Kelvin temperatures, 325 down to 256, we see a drop of about 70°, or close to 1/5 of the total initial temperature. The decreases in volume and temperature are proportional, and the answer is reasonable. This is why temperature must be expressed in the Kelvin scale. Also, a Celsius reading with two significant figures (e.g., −17°C) becomes a Kelvin reading with three significant figures (e.g., 256 K).

SYNTHESIS

The existence of a lowest possible temperature (absolute zero) was first conceived as a result of the relationships found in Charles's law. While there are half a dozen ways of calculating the value of absolute zero, historically it was the concept of extrapolation down to zero volume and notation of the corresponding temperature that gave us the earliest reported value. Prior to observations based on Charles's law, there was no real concept that matter couldn't just keep getting colder but was instead limited. The same result was achieved years later with more accuracy by employing the relationship expressed in Gay-Lussac's law, as discussed in the next section.

10-3.2 Gay-Lussac's Law

We have now expressed as laws how the volume of a gas is affected by a change in pressure and a change in temperature. Now imagine that we have a confined or constant volume of gas that is subjected to a change in temperature. If the volume cannot change, what happens to the pressure? The answer is implied on any pressurized can such as hairspray or deodorant: DO NOT INCINERATE OR STORE NEAR HEAT. This practical warning hints at what will happen. When a confined quantity of gas is heated, the pressure increases. Eventually, the seals on the can break and an explosion follows. The relationship between temperature and pressure was proposed in 1802 by Joseph Louis Gay-Lussac. **Gay-Lussac's law** *states that the pressure is directly proportional to the Kelvin temperature at constant volume.* The law can be expressed mathematically as follows.

$$P \propto T \qquad P = kT$$

For a sample of gas under two sets of conditions at the same volume,

$$\frac{P_1}{T_1} = \frac{P_2}{T_2} \qquad P_2 = P_1 \times \frac{T_2}{T_1}$$

Heat increases the pressure of gas in a confined volume. If heated enough, the can will explode.

EXAMPLE 10-4   Applying Gay-Lussac's Law

A quantity of gas in a steel container has a pressure of 760 torr at 25°C. What is the pressure in the container if the temperature is increased to 50°C?

PROCEDURE

Identify the initial and final conditions and substitute. Again, the temperature must be expressed in kelvins in the calculation.

SOLUTION

Initial Conditions
$P_1 = 760$ torr
$T_1 = (25 + 273) = 298$ K

Final Conditions
$P_2 = ?$
$T_2 = (50 + 273) = 323$ K

$$P_2 = P_1 \times \frac{T_2}{T_1}$$

$$P_2 = 760 \text{ torr} \times \frac{323 \text{ K}}{298 \text{ K}} = \underline{824 \text{ torr}}$$

ANALYSIS

We can again apply logic to this problem. Since the temperature increases, the pressure increases. The temperature correction factor must therefore be greater than 1. This is known as an improper fraction.

Again, we can do a quick estimation regarding the size of our answer. The initial analysis leads us to believe that the answer should be greater than the starting pressure, and it is. Is the *size* of the increase appropriate? We're increasing from 25°C to 50°C. But we also know that it's the Kelvin temperatures that matter, so the appropriate increase is from 298 K to 323 K, less than a 10% rise. The pressure increases from 760 torr to 824 torr, also less than a 10% rise. Our answer appears to be correct in both the direction and magnitude of the change.

SYNTHESIS

When one variable affects another, it can do so in a number of ways. We saw the inverse relationship in Boyle's law. A second common relationship is a *directly proportional* one. With variables that are related directly, *when the value of one increases, the other also increases proportionately*. Temperature and pressure are directly related as is the diameter and circumference of a circle or the voltage and current in a circuit. Less mathematically related would be study time and grades, or caloric intake and weight. Can you find other examples?

Both Charles's law and Gay-Lussac's law follow naturally from kinetic theory. In Figure 10-10, we again focus on one average molecule of gas moving in a container. In the center, we have the situation before any changes are made. To the left, we have raised the temperature at constant pressure. The molecule travels faster on average at the higher temperature, as shown by its longer path. It also collides with more force on the walls of the container. In order for the pressure to remain constant, the volume must expand correspondingly, which confirms Charles's law. To the right, we have again raised the temperature but this time at constant volume. The more frequent and more forceful collisions mean that the pressure is now higher, which explains Gay-Lussac's law.

10-3.3 The Combined Gas Law

What if we have a quantity of gas where both temperature and pressure change? Of course, we could calculate the new volume of the gas by first correcting for the pressure change (Boyle's law) and then correcting for the temperature change (Charles's law). Or we could construct a new relationship that includes both laws and even Gay-Lussac's law as well. This is known, logically, as the **combined gas law**

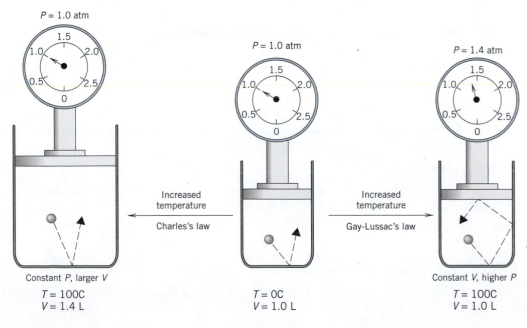

FIGURE 10-10 Charles's Law and Gay-Lussac's Law When temperature increases, molecules travel faster. If the pressure is constant, the volume increases in order to permit the frequency of collisions to remain the same. If the volume is constant, the pressure increases as a result of the more frequent and more forceful collisions.

P = 1.0 atm

Constant *P*, larger *V*
T = 100C
V = 1.4 L

Increased temperature
Charles's law

P = 1.0 atm

T = 0C
V = 1.0 L

Increased temperature
Gay-Lussac's law

P = 1.4 atm

Constant *V*, higher *P*
T = 100C
V = 1.0 L

and is expressed as follows, first as a proportionality and second for a quantity of gas under two sets of conditions. This is not actually a new gas law but a combination of Boyle's, Charles's, and Gay-Lussac's laws.

$$\frac{PV}{T} = k \qquad \frac{P_1 V_1}{T_1} = \frac{P_2 V_2}{T_2}$$

We can now work problems involving simultaneous changes of two conditions. *Notice in the following problems that the calculations require that the units of temperature be in kelvins. The units of pressure and volume can vary but must be the same on both sides in the calculations (i.e., mL or L for volume, torr or atm for pressure).*

Obviously, a major property of a gas is that its volume is very dependent on temperature and pressure. Yet gas is sold by the local gas utility in volume units of cubic feet.

Does that mean we get more or less gas in the warm summer than in the cold winter? Actually, we get the same amount because the volume of the gas is measured under certain universally accepted conditions known as **standard temperature and pressure (STP)**.

> *Standard temperature:* 0°C or 273 K
> *Standard pressure:* 760 torr or 1 atm

Example 10-6 illustrates the use of STP in the combined gas law.

EXAMPLE 10-5   Applying the Combined Gas Law

A 25.8-L quantity of gas has a pressure of 690 torr and a temperature of 17°C. What is the volume if the pressure is changed to 1.85 atm and the temperature to 345 K?

PROCEDURE

Identify the two sets of conditions as initial or final. Make sure the units of pressure are the same and temperature is expressed in kelvins.

Initial Conditions	Final Conditions
$V_1 = 25.8$ L	$V_2 = ?$
$P_1 = 690 \text{ torr} \times \dfrac{1 \text{ atm}}{760 \text{ torr}}$	$P_2 = 1.85$ atm
$= 0.908$ atm	
$T_1 = (17 + 273)$ K $= 290$ K	$T_2 = 345$ K

SOLUTION

$$\frac{P_1 V_1}{T_1} = \frac{P_2 V_2}{T_2}$$

$$V_2 = V_1 \times \frac{P_1}{P_2} \times \frac{T_2}{T_1}$$

$$V_2 = 25.8 \text{ L} \times \frac{0.908 \text{ atm}}{1.85 \text{ atm}} \times \frac{345 \text{ K}}{290 \text{ K}} = \underline{15.1 \text{ L}}$$

ANALYSIS

$$\text{final volume} = \text{initial volume} \times \begin{array}{c} \text{pressure} \\ \text{correction} \\ \text{factor} \end{array} \times \begin{array}{c} \text{temperature} \\ \text{correction} \\ \text{factor} \end{array}$$

Notice that pressure increases, so its correction factor should be less than 1, but the temperature inc̶ so its correction factor should be greater than 1. In this case, whether the final answer is larger or than the original depends on the relative changes of each. Notice that the pressure nearly doubled. The temperature had a modest increase of less than 20%. The pressure effect should be greater, and it is. The volume decreases.

SYNTHESIS

What would have to happen to the pressure and temperature of a gas to produce an increase in volume? What would have to happen to have a decrease in volume? For the gas to expand, the pressure would have to fall and/or the temperature rise. For the gas to contract, just the opposite would have to occur. If both the temperature and pressure rose or fell, then we have the situation as in this example. Which effect is more pronounced? That will be the one having the greater effect on gas volume. It is hard to predict without actually doing the calculation.

EXAMPLE 10-6   Applying the Combined Gas Law and STP

A 5850-ft^3 quantity of natural gas measured at STP was purchased from the gas company. Only 5625 ft^3 was received at the house. Assuming that all the gas was delivered, what was the temperature at the house if the delivery pressure was 1.10 atm?

PROCEDURE

Notice that the initial conditions are STP and the final conditions are different. List the conditions and substitute into the combined gas law.

Initial Conditions	Final Conditions
$V_1 = 5850$ ft^3	$V_2 = 5625$ ft^3
$P_1 = 1.00$ atm	$P_2 = 1.10$ atm
$T_1 = 273$ K	$T_2 = ?$

SOLUTION

$$\frac{P_1 V_1}{T_1} = \frac{P_2 V_2}{T_2} \qquad T_2 = T_1 \times \frac{P_2}{P_1} \times \frac{V_2}{V_1}$$

$$T_2 = 273 \text{ K} \times \frac{1.10 \text{ atm}}{1.00 \text{ atm}} \times \frac{5625 \text{ ft}^3}{5850 \text{ ft}^3} = 289 \text{ K}$$

$$T(^\circ\text{C}) = 289 \text{ K} - 273 = \underline{16^\circ\text{C}}$$

ANALYSIS

In this case, the final temperature is corrected by a pressure factor and a volume factor. Since the final pressure is higher, the pressure factor must be greater than 1 to increase the temperature. The final volume is lower, so the volume factor must be less than 1 to decrease the temperature.

SYNTHESIS

In the course of studying chemistry, it should be a goal of the student to attempt to memorize as little as possible but instead learn applicable concepts that work in the widest number of cases. If you are going to memorize a gas law, make it the combined gas law.

$$\frac{PV}{T} = k \qquad \frac{P_1 V_1}{T_1} = \frac{P_2 V_2}{T_2}$$

Notice, then, that if temperature is not part of the problem (or is being held constant), it's no longer part of the equation, and the result is Boyle's law. Similarly, if pressure is no longer a variable, this reduces to Charles's law; if volume is constant, it disappears, and Gay-Lussac's law is the result. You can know four gas laws by knowing just one!

10-3.4 Avogadro's Law

The more air we blow into a balloon, the larger it gets. This very obvious statement is actually the basis of still another gas law. **Avogadro's law** *states that equal volumes of gases at the same pressure and temperature contain equal numbers of molecules.* Another way of stating this law is that *the volume of a gas is proportional to the number of molecules (moles) of gas present at constant pressure and temperature.* Mathematically, this can be stated in three ways, where n represents the number of moles of gas.

$$V \propto n \quad V = kn \quad \frac{V_1}{n_1} = \frac{V_2}{n_2}$$

In Figure 10-11, the expansion of a balloon by adding solid carbon dioxide (dry ice), which sublimes to a gas, is illustrated. The following example shows the use of Avogadro's law in a sample calculation.

FIGURE 10-11 Avogadro's Law The sublimation of the solid carbon dioxide (dry ice) in the balloon increases the number of gas molecules, which increases the volume.

EXAMPLE 10-7 Applying Avogadro's Law

A uninflated balloon that is full of air has a volume of 275 mL and contains 0.0120 mol of air. As shown in Figure 10-11, a piece of dry ice (solid CO_2) weighing 1.00 g is placed in the balloon and the neck is tied. (Assume that the initial amount of air doesn't change.) What is the volume of the balloon after the dry ice has vaporized? (Assume constant T and P.)

PROCEDURE

Identify the initial and final conditions. The mass of carbon dioxide must be converted to moles and added to the original number of moles present to find the final number of moles of gas.

Initial Conditions
$V_1 = 275$ mL
$n_1 = 0.0120$ mol

Final Conditions
$V_2 = ?$
$n_2 = $ mol air + mol CO_2

$$= 0.0120 + \left(1.00 \text{ g } CO_2 \times \frac{1 \text{ mol}}{44.01 \text{ g } CO_2}\right)$$

$$= 0.0120 + 0.0227 = 0.0347 \text{ mol}$$

SOLUTION

$$\frac{V_1}{n_1} = \frac{V_2}{n_2}$$

$$V_2 = V_1 \times \frac{n_2}{n_1} = 275 \text{ mL} \times \frac{0.0347 \text{ mol}}{0.0120 \text{ mol}}$$

$$V_2 = \underline{795 \text{ mL}}$$

ANALYSIS

Notice that the final condition of moles is the original amount plus the added amount. Since more gas has been added, the volume should increase and the mole factor is greater than 1.

SYNTHESIS

An interesting consequence of Avogadro's law is that it tells us that the volume of a gas at a given temperature and pressure is dependent only on the total number of moles and not on their identity. It doesn't matter if the gas particles themselves are big or small. One mole of very tiny H_2 molecules and one mole of gigantic (relatively speaking) CO_2 molecules occupy exactly the same amount of space. This is counterintuitive at first, until we remember from kinetic molecular theory that gas particles are widely spaced apart and that gas volumes are mostly empty space. So when 1 L of oxygen gas completely reacts with 2 L of hydrogen gas, it is this nature of gases that allows us to say that the resulting water must have a formula of H_2O.

▶ **ASSESSING THE OBJECTIVE FOR SECTION 10-3**

EXERCISE 10-3(a) LEVEL 1: Fill in the blanks.

An increase in the pressure of a gas causes the volume to _____, whereas an increase in temperature causes the same volume to _____. Gay-Lussac's law tells us that if the temperature of a confined volume of gas increases, the pressure _____. To calculate the effect of two changes in conditions, the _____ is used. Increasing the number of moles of gas present will _____ the volume if other conditions are constant.

EXERCISE 10-3(b) LEVEL 2: The volume of 0.0227 mol of gas is 550 mL at a pressure of 1.25 atm and a temperature of 22°C.

(a) What is the volume at 1.25 atm and 44°C?

(b) What is the volume if 0.0115 mol of gas is added at the same T and P?

(c) What is the pressure at 122°C if the volume remains 550 mL?

(d) What is the pressure at 22°C if the volume expands to 825 mL?

(e) What is the volume when you convert to STP conditions?

(f) At what temperature (in °C) would the pressure change to 85.0 kPa?

EXERCISE 10-3(c) LEVEL 1: Name the law used in each of the calculations in Exercise 10-3(b).

EXERCISE 10-3(d) LEVEL 3: Decide which law predicts the behavior of gas in each of the following conditions.

(a) a balloon in the winter being moved from indoors to outdoors

(b) a car tire pressure fluctuating over the course of a year

(c) dry ice being used in a fog machine to cover a stage with mist

(d) piston movement in an internal combustion engine

For additional practice, work chapter problems 10-24, 10-25, 10-31, 10-33, 10-40, 10-43, 10-51, and 10-54.

PART A   SUMMARY

KEY TERMS

10-1.2	The properties of gases are explained by the assumptions of the **kinetic molecular theory**, or simply the **kinetic theory**. p. 321
10-1.3	**Graham's law** follows from kinetic theory and relates the rates of **diffusion** and **effusion** of a gas to its molar mass. p. 321
10-2.1	The **barometer** was the first instrument to measure atmospheric pressure. p. 323
10-2.1	**Pressure** is defined as force per unit area. p. 324
10-2.1	The pressure of **one atmosphere** equals 760 mm Hg or 760 torr. p. 324
10-2.2	**Boyle's law** relates to the effect of pressure on the volume of a gas. p. 325
10-3.1	**Charles's law** relates the volume of a gas to the **Kelvin** scale, which begins at **absolute zero**. pp. 329–330
10-3.2	**Gay-Lussac's law** relates the pressure of a gas to the temperature. p. 331
10-3.3	The **combined gas law** combines Boyle's, Charles's, and Gay-Lussac's laws. p. 332
10-3.3	**Standard temperature and pressure (STP)** defines specific conditions for a gas (i.e., 0°C, 1 atm). p. 332
10-3.4	**Avogadro's law** relates the moles of gas present to the volume. p. 335

SUMMARY CHART

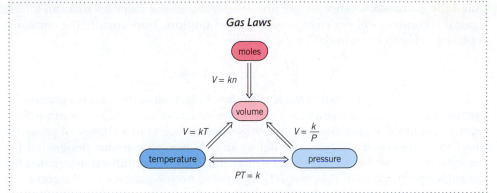

Gas Laws

moles

$V = kn$

volume

$V = kT$ $V = \dfrac{k}{P}$

temperature ⇄ pressure

$PT = k$

Part B

Relationships among Quantities of Gases, Conditions, and Chemical Reactions

SETTING A GOAL

■ You will learn to apply a generalized gas law to a variety of quantitative gas problems.

OBJECTIVES

10-4 Using the ideal gas law, calculate one condition of a gas given the other stated conditions.

10-5 Using Dalton's law, calculate the partial pressure of a gas.

10-6 Convert between volume, moles and density of a gas measured at STP.

10-7 Perform stoichiometric calculations involving gas volumes.

10-4 The Ideal Gas Law

LOOKING AHEAD! In this section, we will find that we can arrive at the ultimate general gas law by putting the information in the last two sections together in the form of one equation. This very important equation leads us to other general properties of gases. ■

▶ **OBJECTIVE FOR SECTION 10-4**
Using the ideal gas law, calculate one condition of a gas given the other stated conditions.

10-4.1 Derivation of the Ideal Gas Law

In Part A, we discussed three ways in which we could enlarge a balloon. First, we could take it up a mountain to a higher elevation. The lower atmospheric pressure would allow it to expand (Boyle's law). Second, we could raise the temperature. If we placed the balloon in boiling water, it would certainly expand (Charles's law). And finally, we could just do the obvious and blow more air into it (Avogadro's law). We can summarize these three relationships as follows.

$$V \propto \frac{1}{P} \qquad V \propto T \qquad V \propto n$$

We can put all three individual relationships into one general relationship as follows.

$$V \propto \frac{nT}{P}$$

This relationship is an extended version of the combined gas law that now includes Avogadro's law. As before, we can change a proportionality to an equality by introducing a constant. The constant used in this case is designated **R** and is called the **gas constant**. Since we are now examining *all* the variables that can affect a gas, this constant will not change under any condition. Traditionally, the constant is placed between the n and T in the equation.

$$V = \frac{nRT}{P} \qquad \text{or} \qquad PV = nRT$$

This relationship is known as the **ideal gas law**. In fact, all of the gas laws previously discussed (except for Graham's law) can be derived from this one relationship. For example, consider a confined quantity of gas that is subject to a change of pressure when the temperature is constant. Before the change (conditions designated by the subscript 1), we have $P_1V_1 = n_1RT_1$. After the change (conditions designated by the subscript 2), we have $P_2V_2 = n_2RT_2$. By solving both equations for the constant R and setting them equal to each other we have

$$R = \frac{P_1V_1}{n_1T_1} = \frac{P_2V_2}{n_2T_2}$$

Since $n_1 = n_2$ and $T_1 = T_2$, these two variables can be ignored (i.e., they cancel out). The equation above can thus be simplified to Boyle's law.

$$P_1V_1 = P_2V_2$$

Similar exercises lead to Charles's, Gay-Lussac's, and the combined gas laws. However, the ideal gas law has a more important application than the exercises we have covered so far, where the effect of changing conditions is calculated. With this law we can calculate one condition (e.g., the volume) of a gas if the other three are known (i.e., n, P, and T). For these calculations, however, we need the value of the gas constant, which is

$$0.082057 \; \frac{\text{L} \cdot \text{atm}}{\text{K} \cdot \text{mol}}$$

For the calculations used in this text, we will use the value of R expressed to three significant figures (e.g., 0.0821). Note that the units of R are a result of specific units of P, V, and T. Thus, in all calculations that follow, the pressure must be expressed in *atmospheres*, the temperature in *kelvins*, and the volume in *liters*. The following three examples illustrate the use of the ideal gas law with a missing variable under one set of conditions.

10-4.2 The Meaning of an Ideal Gas

The ideal gas law follows from the assumptions of the kinetic theory, which describes an "ideal" gas. In an ideal gas, the molecules are assumed to have negligible volume and no attraction or repulsion between molecules. The molecules of a "real" gas obviously do have a volume, and there is some interaction between molecules. This is especially true at high pressures (the molecules are pressed close together and the volume of the particles is no longer negligible) and low temperatures (the molecules move more slowly, allowing for attractive forces between them).

Fortunately, at normal temperatures and pressures found on the surface of Earth, most gases have close to ideal behavior. That is, the particles are far enough apart so that the gas molecules themselves essentially take up no space and they move past one another so fast that interactions between molecules can be ignored. The latter situation is like a young man and young woman rapidly passing each other in jet airliners, unaware of each other's presence. In contrast, assume they pass while walking in the park, where they may very likely slow down as they exchange glances indicating attraction (or, conversely, speed up if they are repelled).

The use of the ideal gas law is justified for any problems that we may generally encounter. However, regions of the atmosphere of the planet Jupiter are composed of cold gases under very high pressures. The ideal gas law would not provide acceptable values under these conditions (i.e., *PV* does not exactly equal *nRT*). Under such nonideal, conditions other, more complex equations would have to be used that take into account the volume of the molecules and the interactions between molecules. These will not be discussed here.

EXAMPLE 10-8 Calculating Pressure Using the Ideal Gas Law

What is the pressure of a 1.45-mol sample of a gas if the volume is 20.0 L and the temperature is 25°C?

PROCEDURE
Write down conditions, making sure the units correspond to those in *R*. Solve the ideal gas law for *P* by dividing both sides of the equation by *V*.

$$P = ? \quad V = 20.0 \text{ L} \quad n = 1.45 \text{ mol} \quad T = (25 + 273) \text{ K} = 298 \text{ K}$$

$$PV = nRT \quad P = \frac{nRT}{V}$$

SOLUTION

$$P = \frac{nRT}{V} = \frac{1.45 \text{ mol} \times 0.0821 \frac{\text{L} \cdot \text{atm}}{\text{K} \cdot \text{mol}} \times 298 \text{ K}}{20.0 \text{ L}}$$

$$= 1.77 \text{ atm}$$

EXAMPLE 10-9 Calculating Volume Using the Ideal Gas Law

What is the volume of 1.00 mol of gas at STP?

PROCEDURE
List the variables given and solve for *V* in the ideal gas law.

$$P = 1.00 \text{ atm} \quad V = ? \quad T = 273 \text{ K} \quad n = 1.00 \text{ mol}$$

SOLUTION

$$PV = nRT$$

$$V = \frac{nRT}{P}$$

$$V = \frac{1.00 \text{ mol} \times 0.0821 \frac{\text{L} \cdot \text{atm}}{\text{K} \cdot \text{mol}} \times 273 \text{ K}}{1.00 \text{ atm}} = 22.4 \text{ L}$$

EXAMPLE 10-10 Calculating Temperature Using the Ideal Gas Law

What is the Celsius temperature of a 1.10-g quantity of oxygen in a 4210-mL container at a pressure of 170 torr?

PROCEDURE
Convert mass of oxygen to moles, mL to L, and torr to atm to be consistent with the units expressed for *R*. Solve for *T* by dividing both sides of the ideal gas equation by *nR*. Substitute variables in the equation and solve for *T* (i.e., *T* = *PV/RT*). Finally, convert *T* to *T*(°C).

SOLUTION

$$n = 1.10 \text{ g O}_2 \times \frac{1 \text{ mol O}_2}{32.00 \text{ g O}_2} = 0.0344 \text{ mol O}_2$$

$$P = \frac{170 \text{ torr}}{760 \text{ torr/atm}} = 0.224 \text{ atm} \qquad V = 4210 \text{ mL} \times \frac{1 \text{ L}}{10^3 \text{ mL}} = 4.21 \text{ L}$$

$$T = \frac{PV}{nR} = \frac{0.224 \text{ atm} \times 4.21 \text{ L}}{(0.0344 \text{ mol}) \times \left(0.0821 \dfrac{\text{L} \cdot \text{atm}}{\text{K} \cdot \text{mol}}\right)} = 334 \text{ K}$$

$$^\circ\text{C} = 334 - 273 = \underline{61^\circ\text{C}}$$

ANALYSIS

In all the previous cases of applications of gas laws, we were merely converting gas conditions from one set of variables to another. The reason is that we were always missing two of the four variables (volume, pressure, temperature, and moles). But in these last three problems, three of the four variables are given. With the knowledge of the proportionality constant, R, we are in a position to calculate specifics of a particular example rather than just comparing how a sample might change.

SYNTHESIS

There might be initial confusion when faced with a typical gas law problem. Which of the several possible laws is the one to use? A careful accounting of the specific variables mentioned in the problem shows the right path. A problem providing two volumes and a pressure while asking for a second pressure requires which law to solve? Boyle's law. A problem giving a volume, pressure, and number of moles (or some value, like grams, easily converted to moles) while asking for a temperature? The ideal gas law.

▶ **ASSESSING THE OBJECTIVE FOR SECTION 10-4**

EXERCISE 10-4(a) LEVEL 2: What is the Celsius temperature of 2.46×10^{-4} g of N_2O if it is present in a 2.35-mL container at a pressure of 0.0468 atm?

EXERCISE 10-4(b) LEVEL 2: How many particles of gaseous air molecules are found in a 500.0-mL flask at room temperature (25°C) and 745 torr?

EXERCISE 10-4(c) LEVEL 2: What is the volume of one mole of any gas at STP conditions?

EXERCISE 10-4(d) LEVEL 2: What is the pressure, in psi, of 0.045 g of gaseous hydrogen in a 250-mL flask at 38°C?

EXERCISE 10-4(e) LEVEL 2: What is the pressure, in psi, of 0.045 g of gaseous hydrogen in a 250-mL flask at 38°C?

For additional practice, work chapter problems 10-55, 10-57, and 10-59.

▶ **OBJECTIVE FOR SECTION 10-5**

Using Dalton's law, calculate the partial pressure of a gas.

10-5 Dalton's Law of Partial Pressures

LOOKING AHEAD! Kinetic theory tells us that gas molecules behave independently. This property has several consequences. Avogadro's law illustrates that equal volumes of gases contain equal numbers of molecules regardless of their masses or nature. In this section, we will see that pressures are also independent of the identity of the gases involved. ■

About 78% of the volume of the atmosphere is composed of nitrogen, 21% is oxygen, and a little less than 1% is argon. From the ideal gas law, we can also conclude that 21% of the molecules as well as 21% of the pressure we feel from the atmosphere are attributed to oxygen. In terms of the mass of the atmosphere oxygen does not represent 21%, because of the different molar masses of the gaseous components. John Dalton, author of modern atomic theory, was the first to suggest that pressure, like volume, does not depend on the identity of the gas. The resulting principle, known as **Dalton's law**, *states that the total pressure of a gas in a system is the sum of the partial pressures of each component gas.* The partial pressure of a specific gaseous element or compound is defined as the pressure due to that substance alone. Mathematically, this can be stated as $P_{\text{tot}} = P_1 + P_2 + P_3 + \dots$ (P_1 is the pressure due to gas 1, etc.).

EXAMPLE 10-11   Calculating the Total Pressure of a Mixture of Gases

Three gases, Ar, N_2, and H_2, are mixed in a 5.00-L container. Ar has a pressure of 255 torr, N_2 has a pressure of 228 torr, and H_2 has a pressure of 752 torr. What is the total pressure in the container?

PROCEDURE

According to Dalton's law, the total pressure is the sum of the partial pressures of each gas. Add the three partial pressures together.

SOLUTION

$$P_{\text{Ar}} = 255 \text{ torr} \qquad P_{N_2} = 228 \text{ torr} \qquad P_{H_2} = 752 \text{ torr}$$

$$P_{\text{tot}} = P_{\text{Ar}} + P_{N_2} + P_{H_2} = 255 \text{ torr} + 228 \text{ torr} + 752 \text{ torr}$$

$$= \underline{1235 \text{ torr}}$$

ANALYSIS

While the pressures of each of the gases are different, the volume of each is 5.00 L. Each gas must also necessarily be at the same temperature, 25°C. By using the ideal gas law, we can calculate the number of moles of each. Doing so, we realize that in a sample of gas that is a mixture, the number of moles is directly proportional to each gas's partial pressure. So, for our sample above, there are close to the same number of moles of argon and nitrogen, and three times as many moles of hydrogen as either of the others.

SYNTHESIS

One of the important applications of Dalton's law occurs when we collect gas samples. How can we fill a container full of gas? The process is called "collecting over water." Fill the container with water first, and then allow the gas to bubble into the inverted container, placed in a sink. The gas displaces the water. This process, though, results in two gases actually being in the container—the gas we're collecting and water vapor. By Dalton's law, we have to subtract the vapor pressure of water from the total pressure if we are to know the pressure of the gas we were hoping to isolate. (See the photo below)

We can now apply Dalton's law to the composition of our atmosphere. Since 21% of the molecules of the atmosphere are oxygen, 21% of the pressure of the atmosphere is due to oxygen. The partial pressure of oxygen can be calculated by multiplying the percent in decimal form by the total pressure.

$$P_{O_2} = 0.21 \times 760 \text{ torr} = 160 \text{ torr}$$

Most of us function best breathing this partial pressure of oxygen. When we live at higher elevations, the partial pressure of oxygen (as well as the number of molecules) is lower and our bodies adjust by producing more red blood cells, which transport oxygen. On top of the highest mountain, Mt. Everest, the total atmospheric pressure is 270 torr, so the partial pressure of oxygen is only 57 torr, or about one-third of normal. A human cannot survive for long at such a low pressure of oxygen. At that altitude, even conditioned climbers must use an oxygen tank and mask, which give an increased partial pressure of oxygen to the lungs.

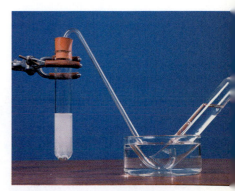

Gases collected over water are actually a mixture of the gas and water vapor.

EXAMPLE 10-12   Calculating the Partial Pressure of a Gas

On a humid summer day, the partial pressure of water in the atmosphere is 18 torr. If the barometric pressure on this day is a high 766 torr, what are the partial pressures of nitrogen and of oxygen if 78.0% of the dry atmosphere is composed of nitrogen molecules and 21.0% is oxygen? (1% of the dry atmosphere is composed of all other gases.)

PROCEDURE

1. Find the pressure of the dry atmosphere by subtracting the pressure of water from the total. This is an application of Dalton's law (i.e., $P_{tot} = P_{N_2} + P_{O_2} + P_{H_2O} + P_{other}$).

2. Find the partial pressures of N_2 and of O_2 from the total pressure of the dry atmosphere and the percents.

SOLUTION

1. P (dry atmosphere) $= P_{tot} - P_{H_2O} = 766 - 18 = 748$ torr

2. $P_{N_2} = 0.780 \times 748$ torr $= \underline{583\ torr}$

   $P_{O_2} = 0.210 \times 748$ torr $= \underline{157\ torr}$

ANALYSIS

The percentages we are supplying for the gases in the atmosphere are *mole* percentages as opposed to *mass* percentages. The percent composition of a mixture could be presented in either fashion, depending on which amount (mass or number of particles) is more important. For applications involving gas theory and specifically Dalton's law, mole percentages (or *mole fractions*) are the necessary values. They are calculated like any other percentage; that is, the number of moles of the component is divided by the total number of moles present.

SYNTHESIS

Water vapor is a variable component of the atmosphere and can range from almost 0% in dry climates to as much as 4% in humid areas. We don't typically include it in our discussion of atmospheric gases, which are reported under dry conditions. The amount of water vapor in the air is also limited by the temperature. The warmer it is, the more water vapor the atmosphere can hold. What time of year do your lips chap the worst? Why? Typically it's in the winter, which, despite snow and ice, has very low amounts of water vapor in the air; therefore, your lips dry out. We will discuss this in more detail in the next chapter.

The total pressure of a mixture of gases can be extended to the ideal gas law. For example, for three gases in one container,

$$P_{tot} = P_1 + P_2 + P_3 = \frac{n_1 RT}{V} + \frac{n_2 RT}{V} + \frac{n_3 RT}{V} = \frac{n_{tot} RT}{V}$$

We see that, at a given temperature, the pressures of gases depend only on the total number of moles present. This is illustrated in Figure 10-12 and Example 10-13.

FIGURE 10-12 Pressures of a Pure Gas and a Mixture of Gases Pressure depends only on the number of molecules at a certain temperature, not on their identity.

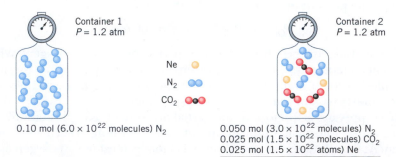

Container 1
$P = 1.2$ atm

Ne
N_2
CO_2

Container 2
$P = 1.2$ atm

0.10 mol (6.0×10^{22} molecules) N_2

0.050 mol (3.0×10^{22} molecules) N_2
0.025 mol (1.5×10^{22} molecules) CO_2
0.025 mol (1.5×10^{22} atoms) Ne

0.100 mol (6.0×10^{22} particles) total

EXAMPLE 10-13   Calculating the Total Pressure of a Mixture of Gases

What is the pressure (in atmospheres) exerted by a mixture of 12.0 g of N_2 and 12.0 g of O_2 in a 2.50-L container at 25°C?

PROCEDURE

1. Find the number of moles of N_2 and O_2 present.

2. Use the total number of moles, temperature, and volume in the ideal gas law.

SOLUTION

1. $n_{O_2} = 12.0 \text{ g } O_2 \times \dfrac{1 \text{ mol } O_2}{32.00 \text{ g } O_2} = 0.375 \text{ mol } O_2$

   $n_{N_2} = 12.0 \text{ g } N_2 \times \dfrac{1 \text{ mol } N_2}{28.02 \text{ g } N_2} = 0.428 \text{ mol } N_2$

   $n_{tot} = 0.375 \text{ mol} + 0.428 \text{ mol} = 0.803 \text{ mol of gas}$

2. $V = 2.50 \text{ L} \qquad T = (25 + 273) = 298 \text{ K} \qquad R = 0.0821 \dfrac{\text{L} \cdot \text{atm}}{\text{K} \cdot \text{mol}}$

   $PV = nRT$

   $P = \dfrac{nRT}{V}$

   $P = \dfrac{0.803 \text{ mol} \times 0.0821 \dfrac{\text{L} \cdot \text{atm}}{\text{K} \cdot \text{mol}} \times 298 \text{ K}}{2.50 \text{ L}} = \underline{7.86 \text{ atm}}$

ANALYSIS

Notice that the volume of each gas is taken to be the volume of the container (2.50 L), reinforcing the notion that each gas acts independently when mixed, expanding to fill the container completely. Furthermore, all gases act the same way, so we can add the moles of two different gases together and treat them as one.

SYNTHESIS

This example has applications during pressure-changing events like scuba diving, ballooning, or mountain climbing. It also explains why airlines pressurize cabins before flights. The percent of each component of a gas will remain constant, even as the pressure changes. This leads to a lower partial pressure of oxygen when ascending in the atmosphere or a higher partial pressure when descending (while scuba diving, for instance). If we are to get the necessary amount of oxygen into our system during these various events, sometimes extra precautions are needed, such as tanks of pure O_2 to supplement the lack of O_2 in the air or, on the other end of the spectrum, diving tables that instruct us how to limit the amount of pressurized N_2 that we're exposed to during submersion.

▶ ASSESSING THE OBJECTIVE FOR SECTION 10-5

EXERCISE 10-5(a) LEVEL 1: For a gas collected over water at a total pressure of 750 torr, if the vapor pressure of the water is 35 torr, what's the pressure of the gas?

EXERCISE 10-5(b) LEVEL 2: What is the partial pressure of O_2 if 0.0450 mol of O_2 is mixed with 0.0328 mol of N_2 and the total pressure in the container is 596 torr?

EXERCISE 10-5(c) LEVEL 2: A mixture of 1.00 g of H_2 and 10.0 g of CO_2 are mixed in a 3.00-L container at 25.0°C. What is the total pressure of the container?

EXERCISE 10-5(d) LEVEL 3: If you had equal masses of helium, methane (CH_4), and carbon dioxide, which would exert the highest pressure?

For additional practice, work chapter problems 10-64, 10-66, and 10-68.

▶ **OBJECTIVE FOR SECTION 10-6**
Convert between volume, moles and density of a gas measured STP.

10-6 The Molar Volume and Density of a Gas

LOOKING AHEAD! In previous chapters, we found that one mole of a compound represents a set number of molecules and a specific mass. If the compound also happens to be a gas, we find that one mole also represents a set volume under specified conditions. We explore this next. ■

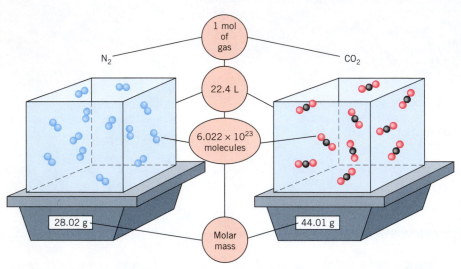

FIGURE 10-13 Moles of Two Different Gases At STP, one mole of gases such as N_2 and CO_2 have the same volume, the same number of molecules, but different masses.

10-6.1 The Molar Volume

In Example 10-9, we calculated the volume of one mole of a gas at standard temperature and pressure. This result represents the volume of one mole of *any* gas, or even one mole of a mixture of gases, at STP. This phenomenon is certainly not true of equal volumes of liquids and solids. *The volume of one mole of gas at STP, 22.4 L, is known as the* **molar volume**. This corresponds to about 6 gallons. We can now expand on the significance of the mole that was described in Chapter 5. One mole of a gas refers to three quantities: (1) the molar volume (22.4 L/mol, or 22.414 L/mol to be more precise) and (2) Avogadro's number (6.022×10^{23} molecules/mol), both of which are independent of the identity of the gas, and (3) the molar mass (g/mol), which does depend on the identity of the gaseous compound or element. (See Figure 10-13.) By using the molar volume relationship, scientists were able to calculate the relative masses of various gaseous compounds. This occurred well before they even knew about the existence of the protons and neutrons that provide the actual masses.

Moles of a gas relate directly to volume at STP using the molar volume relationship. At any other conditions, moles relate to the volume using the ideal gas law.

EXAMPLE 10-14   Calculating Mass from Volume at STP

What is the mass of 4.55 L of O_2 measured at STP?

PROCEDURE

1. Convert volume to moles using the molar volume relationship. The two possible conversion factors for converting volume to moles are

$$\frac{1\ mol}{22.4\ L} \quad and \quad \frac{22.4\ L}{mol}$$

2. Convert moles to mass of O_2 using the molar mass of O_2. The complete unit map is

$$\boxed{\text{volume (STP)}} \Longrightarrow \boxed{\text{moles}} \Longrightarrow \boxed{\text{mass}}$$

SOLUTION

$$4.55\ \cancel{L} \times \frac{1\ \cancel{mol}}{22.4\ \cancel{L}} \times \frac{32.00\ g}{\cancel{mol}} = \underline{6.50\ g}$$

ANALYSIS

The value of 22.4 L/mol is the STP condition. More realistically, gases would be evaluated closer to room temperature conditions (eg., at 25°C (298 K) and 0.987 atm). The molar volume at any set of conditions can be calculated using the ideal gas law.

$$V = \frac{nRT}{P} \qquad V = \frac{1.00 \text{ mol} \times 0.0821 \frac{L \cdot atm}{K \cdot mol} \times 298 \text{ K}}{0.987 \text{ atm}}$$

$$= 24.8 \text{ L}$$

This value can now be used for calculations at these conditions.

SYNTHESIS

This problem illustrates how much gas can be compressed into a container as a liquid. Whether in a butane lighter, a carbon dioxide fire extinguisher, or a propane tank for a gas grill, one mole of the liquid will expand to a volume of 22.4 L at STP. Since we usually use these gases at temperatures higher than 0°C, the volume of one mole is even higher—closer to 25 L per mole. That is a dramatic increase in volume and why caution should be exercised every time you use gases under compressed conditions. The possibility of a rapid expansion (or explosion) is very real if the regulators on these tanks fail.

EXAMPLE 10-15   Calculating Molar Mass from Volume at STP

A sample of gas has a mass of 3.20 g and occupies 2.00 L at 17°C and 380 torr. What is the molar mass of the gas?

PROCEDURE

1. Convert the given volume to the corresponding volume at STP using the combined gas law.

Initial Conditions	Final Conditions
$V_1 = 2.00$ L	$V_2 = ?$
$P_1 = 380$ torr	$P_2 = 760$ torr
$T_1 = (17 + 273)$ K $= 290$ K	$T_2 = 273$ K

2. Convert volume at STP to moles of gas using the molar volume relationship.
3. Convert moles (n) and mass to molar mass (represented in equations as MM) as follows.

$$n = \frac{mass}{MM} \qquad MM = \frac{mass}{n}$$

SOLUTION

1. $\dfrac{P_1 V_1}{T_1} = \dfrac{P_2 V_2}{T_2}$

$$V_2 = V_1 \times \frac{P_1}{P_2} \times \frac{T_2}{T_1}$$

$$V_2 = 2.00 \text{ L} \times \frac{380 \text{ torr}}{760 \text{ torr}} \times \frac{273 \text{ K}}{290 \text{ K}} = \underline{0.941 \text{ L (STP)}}$$

2. $0.941 \text{ L} \times \dfrac{1 \text{ mol}}{22.4 \text{ L}} = 0.0420 \text{ mol of gas}$

3. molar mass $= \dfrac{mass}{n} = \dfrac{3.20 \text{ g}}{0.0420 \text{ mol}} = \underline{76.2 \text{ g/mol}}$

ANALYSIS

There is actually a more direct way of solving this problem using the ideal gas law. First, use the ideal gas law to find the number of moles of gas at the given conditions. Then use the relationship for molar mass (i.e., $n =$ mass/MM) to find the molar mass of the gas.

$$PV = nRT \qquad n = \frac{PV}{RT} \qquad T = 17 + 273 = 290 \text{ K}$$

$$P = \frac{380 \text{ torr}}{760 \text{ torr/atm}} = 0.500 \text{ atm}$$

$$n = \frac{0.500 \text{ atm} \times 2.00 \text{ L}}{0.0821 \dfrac{\text{L} \cdot \text{atm}}{\text{K} \cdot \text{mol}} \times 290 \text{ K}} = 0.0420 \text{ mol}$$

$$n = \frac{\text{mass}}{\text{MM}} \qquad \text{MM} = \frac{\text{mass}}{n} = \frac{3.20 \text{ g}}{0.0420 \text{ mol}} = 76.2 \text{ g/mol}$$

SYNTHESIS

The calculations involved in this example are really quite remarkable. They provide a way of determining the molar mass of a compound without any knowledge of its chemical composition. The only criteria are that the compound's mass is known and it can be converted into a gas. The French chemist Jean-Baptiste Dumas pioneered this technique in the early 1800s. He was able to determine relative molecular weights even before the concept of the mole had been developed.

MAKING IT REAL

Defying Gravity—Hot-Air Balloons

Hot air is less dense than cool air.

In July 2002, Steve Fossett became the first person to fly solo around the globe in a balloon using hot air. However, hot-air balloons have a long history. Ancient peoples probably realized that hot air rose by observing the ascending embers of a campfire. The first creatures to go aloft in a hot-air balloon were a duck, a chicken, and a sheep on September 19, 1783. The inventors of this device were the Montgolfier brothers in France. Convinced that the apparatus worked, two men made the first manned ascent into the sky later that year. Since the 1960s, hot-air ballooning has become increasingly popular with the use of newer, lightweight materials. Now we have the magnificent displays of balloons of all shapes and sizes floating with the wind over the countryside.

How do they work? They are actually an application of Charles's law, which states that the hotter a gas becomes, the more volume it requires at the same pressure. The expansion of a gas lowers its density, so it becomes buoyant in the surrounding air. Thus, it rises, just like a cork floats in water, because it is less dense. The normal density of air at 25°C is about 1.18 g/L. (It is 1.29 g/L at STP.) If the air is heated to 125°C, the expansion causes the density to decrease to 0.88 g/L. The difference is the lifting power of the balloon. Thus a 1-L balloon containing the hot gas could lift 0.30 g. In order to lift the balloon, a gondola, a burner, and a couple of people, the balloon must lift about 1000 lb. That would take a volume of over 1.5×10^6 L (about 53,000 ft^3). A larger balloon or hotter air could lift more weight or go higher.

The other way to have a gas less dense than the surroundings is to use a gas with a low molar mass such as helium (4 g/mol) or hydrogen (2 g/mol). For 1 L of gas at 25°C, this means that He can lift 1.0 g, and hydrogen, 1.1 g. Helium is preferred since it is noncombustible. Hydrogen was used until the 1930s disaster at Lakehurst, New Jersey, when the German airship *Hindenburg* caught fire. (Recent evidence, though, suggests that it was, in fact, the skeletal structure of the airship, not the hydrogen that initially ignited.)

In fact, Fossett's lighter-than-air ship used a combination of hot air and helium. It ascended as high as 35,000 feet to the altitude of jet airliners. It was a marvelous feat that he accomplished on the sixth try. It took only 219 years after the first two men left the bounds of Earth.

10-6.2 Density of a Gas at STP

In Chapter 3, the densities of solids and liquids were given in units of g/mL. Since gases are much less dense, units in this case are usually given in g/L (STP). The density of a gas at STP can be calculated by dividing the molar mass by the molar volume.

$$CO_2: 44.01 \text{ g/mol} = 22.4 \text{ L/mol}$$

$$\frac{44.01 \text{ g/mol}}{22.4 \text{ L/mol}} = 1.96 \text{ g/L(STP)}$$

The densities of several gases are given in Table 10-2. According to kinetic molecular theory, two different gases under identical conditions with the same volume contain the same number of particles. But if one gas has a higher molar mass, then it has a higher mass per unit volume, or density. The density of air (a mixture) is 1.29 g/L. Gases such as He and H_2 have low molar masses and are thus less dense than air. Gases less dense than air rise in the air, just as solids or liquids less dense than water float on water. Helium is used as the gas in blimps to make the whole craft "lighter than air." (See chapter problem 10-62.)

TABLE 10-2	
Densities of Some Gases	
GAS	**DENSITY [g/L (STP)]**
H_2	0.090
He	0.179
N_2	1.25
Air (average)	1.29
O_2	1.43
CO_2	1.96
CF_2Cl_2	5.40
SF_6	6.52

▶ **ASSESSING THE OBJECTIVE FOR SECTION 10-6**

EXERCISE 10-6(a) LEVEL 1: Fill in the blanks.

One mole of any gas occupies _____ L at STP; this is known as the _____. The density of a gas at STP is obtained by dividing the _____ by the molar volume.

EXERCISE 10-6(b) LEVEL 2: What is the volume occupied by 142 g of SF_4 at STP?

EXERCISE 10-6(c) LEVEL 2: What is the density of SF_4 at STP?

EXERCISE 10-6(d) LEVEL 2: What is the molar mass of a gas where 11.9 g occupies 8.33 L at STP? What diatomic element could it be?

EXERCISE 10-6(e) LEVEL 3: The density of a hydrocarbon is 1.80 g/L at 1 atm and 25.0°C. What is the identity of the hydrocarbon?

EXERCISE 10-6(f) LEVEL 3: It is well known that breathing helium makes your voice higher. This is due to the lighter, less dense atoms of helium causing your vocal cords to vibrate faster and at a higher pitch. Which other gases on the periodic table would have the same effect? What gases in Table 10-2 would lower the pitch of your voice? (This is certainly not an experiment that one would actually try.)

For additional practice, work chapter problems 10-75, 10-78, 10-80, and 10-83.

10-7 Stoichiometry Involving Gases

▶ **OBJECTIVE FOR SECTION 10-7**
Perform stoichiometric calculations involving gas volumes.

LOOKING AHEAD! Gases are formed or consumed in many chemical reactions. Since the volumes of gases relate directly to the number of moles, we can use volume as we did mass or number of molecules in previous stoichiometric calculations discussed in Chapter 7. ∎

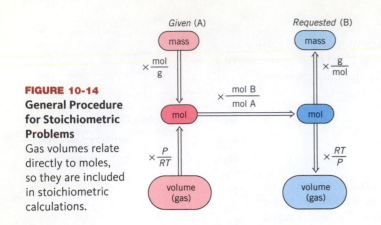

FIGURE 10-14
General Procedure for Stoichiometric Problems
Gas volumes relate directly to moles, so they are included in stoichiometric calculations.

Recall that gas volumes relate to moles using the ideal gas law. At standard conditions, it is more convenient to use the molar volume relationship [i.e., $n = V(STP)/22.4$ L/mol]. We can now include gas volumes with mass and number of molecules (see Figure 7-1) in the general procedure for solving stoichiometry problems. (See Figure 10-14.) We have excluded the number of molecules from this scheme since problems involving actual numbers of molecules are rarely encountered.

The following two examples illustrate the relationship of the gas laws to stoichiometry.

EXAMPLE 10-16   Calculating Mass of a Reactant from the Volume of a Product

Automobiles in the future may use a fuel cell running on hydrogen gas. Storing the hydrogen and dispensing it conveniently, however, remain problems. One possible method involves the addition of water to a solid compound such as sodium borohydride, which then liberates hydrogen. The balanced equation illustrating the generation of hydrogen by this reaction is

$$NaBH_4(s) + 2H_2O(l) \longrightarrow NaBO_2(aq) + 4H_2(g)$$

What mass of $NaBH_4$ is needed to produce 50.0 L of H_2 measured at STP?

PROCEDURE

The general procedure is shown below. Since the conditions of this case are at STP, it is easier to use the molar volume as a conversion factor between volume and moles. The complete unit map for this problem is illustrated below.

Given [L(g) H_2]   *Requested* (g $NaBH_4$)

$$\times \frac{mol\ H_2}{22.4\ L} \qquad \times \frac{mol\ NaBH_4}{mol\ H_2} \qquad \times \frac{g\ NaBH_4}{mol\ NaBH_4}$$

volume H_2 $\Longrightarrow$ mol H_2 $\Longrightarrow$ mol $NaBH_4$ $\Longrightarrow$ mass $NaBH_4$

SOLUTION

$$50.0\ L\ (STP) \times \frac{1\ mol\ H_2}{22.4\ L\ (STP)} \times \frac{1\ mol\ NaBH_4}{4\ mol\ H_2} \times \frac{37.83\ g\ NaBH_4}{mol\ NaBH_4} = \underline{21.1\ g\ NaBH_4}$$

EXAMPLE 10-17   Calculating Volume of a Product from Mass of a Reactant

Given the balanced equation

$$4NH_3(g) + 5O_2(g) \longrightarrow 4NO(g) + 6H_2O(l)$$

what volume of NO gas measured at 0.724 atm and 25°C will be produced from using 19.5 g of O_2?

PROCEDURE

Convert mass of O_2 to moles of O_2, then moles of O_2 to moles of NO, and finally to volume of NO at the stated conditions. The complete unit map for this problem is illustrated below.

Given (g) O_2)   *Requested* [L(g) NO]

$$\times \frac{mol\ O_2}{g\ O_2} \qquad \times \frac{mol\ NO}{mol\ O_2} \qquad \times \frac{RT}{P}$$

mass O_2 $\Longrightarrow$ mol O_2 $\Longrightarrow$ mol NO $\Longrightarrow$ volume NO

SOLUTION

$$19.5 \text{ g } O_2 \times \frac{1 \text{ mol } O_2}{32.00 \text{ g } O_2} \times \frac{4 \text{ mol NO}}{5 \text{ mol } O_2} = 0.488 \text{ mol NO}$$

Using the ideal gas law

$$V = \frac{nRT}{P} = \frac{0.488 \text{ mol} \times 0.0821 \dfrac{\text{L} \cdot \text{atm}}{\text{K} \cdot \text{mol}} \times 298 \text{ K}}{0.724 \text{ atm}} = \underline{16.5 \text{ L}}$$

ANALYSIS

An alternate path is to (1) convert the moles of NO to the volume at STP then (2) convert from STP to the stated conditions using the combined gas law.

1. $0.488 \text{ mol} \times 22.4 \text{ L/ mol} = 10.9 \text{ L (STP)}$

2. $V_2 = V_1 \times \dfrac{P_1}{P_2} \times \dfrac{T_2}{T_1}$

Initial Conditions	Final Conditions
$V_1 = 10.9$ L	$V_2 = ?$
$P_1 = 760$ torr	$P_2 = 550$ torr
$T_1 = 273$ K	$T_2 = (25 + 273)$ K $= 298$ K

$$V_2 = 10.9 \text{ L} \times \frac{760 \text{ torr}}{550 \text{ torr}} \times \frac{298 \text{ K}}{273 \text{ K}} = \underline{16.4 \text{ L}}$$

(The slight difference in the two values is due to rounding off.)

SYNTHESIS

Gas stoichiometry has real-world significance if you have been unfortunate enough to have been in an automobile when the airbag deployed. The bag fills with nitrogen gas (N_2) produced from the decomposition of solid sodium azide (NaN_3). The decomposition of sodium azide is illustrated by the equation

$$2NaN_3(s) \longrightarrow 2Na(s) + 3N_2(s)$$

Can you estimate how much NaN_3 it would take to prime an airbag? Let's assume an airbag of 50 L. Under conditions close to STP, that's about 2 mol of gas. It takes 2 mol of NaN_3 to produce 3 mol of N_2, or 1 mol of NaN_3 to make 1 1/2 mol of N_2. The molar mass of NaN_3 is 65 grams per mole, so it will take between 80 and 90 g of NaN_3 to inflate the bag.

▶ **ASSESSING THE OBJECTIVE FOR SECTION 10-7**

EXERCISE 10-7(a) LEVEL 1: Fill in the blanks.

In stoichiometry calculations, the volume of a gas can be converted to number of moles by use of the _____ _____ if the volume is measured at STP. If other conditions are present, the volume is related to the number of moles by the _____ _____ _____.

EXERCISE 10-7(b) LEVEL 2: Humans obtain energy from the combustion of blood sugar (glucose) according to the equation

$$C_6H_{12}O_6(aq) + 6O_2(g) \longrightarrow 6CO_2(g) + 6H_2O(l)$$

(a) What volume of O_2 measured at STP is required to react with 0.122 mol of glucose?

(b) What volume of CO_2 measured at 25°C and 1.10 atm pressure is released from the combustion of 228 g of glucose?

(c) What mass of glucose is required to react with 2.48 L of oxygen measured at 22°C and 655 torr?

EXERCISE 10-7(c) LEVEL 3: What volume of H_2 reacts with 15.0 L of O_2 measured at STP to form water? How many grams of water will be produced?

EXERCISE 10-7(d) LEVEL 3: NO reacts with O_2 to form NO_2. All three are gases. What happens to the volume if you begin with 2 L of NO and add 1 L of O_2 at the same temperature and pressure?

For additional practice, work chapter problems 10-75, 10-78, 10-80, and 10-83.

PART B SUMMARY

KEY TERMS

10-4.1	The **ideal gas law** includes all the variables that affect a volume of gas. p. 338	
10-4.1	The **gas constant**, **R**, uses specific units for temperature, volume, and pressure. p. 338	
10-5	**Dalton's** law tells us that gas partial pressures are additive. p. 341	
10-6.1	The **molar volume** of a gas is 22.4 L under STP conditions. p. 344	

SUMMARY CHART

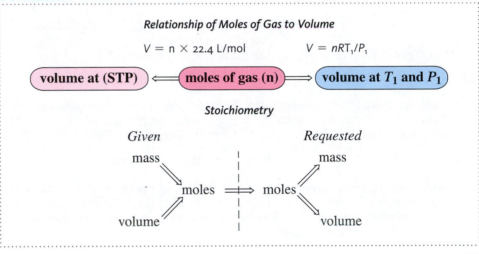

Relationship of Moles of Gas to Volume

$V = n \times 22.4$ L/mol $V = nRT_1/P_1$

(**volume at (STP)**) $\Longleftarrow$ (**moles of gas (n)**) $\Longrightarrow$ (**volume at T_1 and P_1**)

Stoichiometry

Given		*Requested*
mass		mass
	moles $\Longrightarrow$ moles	
volume		volume

CHAPTER 10 SYNTHESIS PROBLEM

Sulfur dioxide is a noxious pollutant, which is dumped into our atmosphere from the combustion of coal containing pyrite (FeS_2) as an impurity. In complex reactions sulfur dioxide in the atmosphere reacts with oxygen to form sulfur trioxide. The sulfur trioxide ends up as acid rain (sulfuric acid), which is very harmful to the environment. The combustion of pyrite is represented by the following equation.

$$4FeS_2(s) + 11O_2(g) \longrightarrow 2Fe_2O_3(s) + 8SO_2(g)$$

The following questions relate to the formation of 288 L of SO_2 measured at STP.

PROBLEM	SOLUTION
a. What is the volume of this gas at 956 torr?	**a.** (Boyle's Law) $V_2 = V_1 \times \dfrac{P_1}{P_2} = 288$ L $\times \dfrac{760 \text{ torr}}{956 \text{ torr}} = \underline{229 \text{ L}}$
b. What is the volume of the gas at 225°C?	**b.** (Charles's Law) $V_2 = V_1 \times \dfrac{T_2}{T_1} = 288$ L $\times \dfrac{(225 + 273) \text{ K}}{273 \text{ K}} = \underline{525 \text{ L}}$

PROBLEM	SOLUTION
c. What is the pressure of the gas if the temperature is increased to 375 K?	**c.** (Gay-Lussac's law) $$P_2 = P_1 \times \frac{T_2}{T_1} = 1.00 \text{ atm} \times \frac{375 \text{ K}}{273 \text{ K}} = \underline{1.38 \text{ atm}}$$
d. How many moles of SO_2 are present?	**d.** (molar volume) $288 \text{ L (STP)} \times \dfrac{1 \text{ mol}}{22.4 \text{ L (STP)}} = \underline{12.9 \text{ mol } SO_2}$
e. The pressure is increased to 1.55 atm by adding N_2 to the same volume of SO_2. What mass of N_2 was added?	**e.** (Dalton's law) $P_{N_2} = P_{total} - P_{SO_2} = 1.55 - 1.00 = \underline{0.55 \text{ atm}}$ due to N_2 The number of moles of each gas present is proportional to their respective partial pressures since they are additive. $$n_{N_2} = \frac{P_{N_2}}{P_{O_2}} \times n_{O_2} = \frac{0.55 \text{ atm}}{1.00 \text{ atm}} \times 12.9 \text{ mol} = 7.1 \text{ mol } N_2$$ $$7.1 \text{ mol } N_2 \times \frac{28.02 \text{ g } N_2}{\text{mol } N_2} = \underline{199 \text{ g } N_2}$$
f. What volume of O_2 measured at 37°C and 927 torr pressure is required to produce the original amount of SO_2?	**f.** (Ideal Gas Law) $12.9 \text{ mol } SO_2 \times \dfrac{11 \text{ mol } O_2}{8 \text{ mol } SO_2} = 17.7 \text{ mol } O_2$ $T = 37 + 273 = 310 \text{ K} \quad P = \dfrac{927 \text{ torr}}{760 \text{ torr/atm}} = 1.22 \text{ atm}$ $$V = \frac{nRT}{P} = \frac{17.7 \text{ mol} \times 0.0821 \frac{\text{L} \cdot \text{atm}}{\text{K} \cdot \text{mol}} \times 310 \text{ K}}{1.22 \text{ atm}} = 369 \text{ L}$$

YOUR TURN

We will use the same reaction as above, which represents the combustion of pyrite.

$$4FeS_2(s) + 11O_2(g) \longrightarrow 2Fe_2O_3(s) + 8SO_2(g)$$

The following questions relate to the combustion of 288 g of pyrite.

- **a.** What volume of SO_2 is formed measured at STP?
- **b.** What volume of SO_2 is formed measured at 102°C and 1 atm pressure?
- **c.** What volume of SO_2 is formed measured at 0°C and 540 torr?
- **d.** What is the temperature of this volume of gas (in °C) if the pressure is increased under STP conditions to 2.00 atm?
- **e.** How many moles of SO_2 are formed?
- **f.** If 220 g of N_2 is added to the SO_2, what is the new pressure?
- **g.** What volume of O_2 measured at 127°C and 627 torr is required to react with that amount of pyrite?

Answers on p. 354

CHAPTER SUMMARY

Our modern theories about matter depended on some basic understanding about the most abstract state of matter—gases. Beginning in the 1600s, the quantitative laws that we know as the gas laws began to be advanced. These laws and other observations of the nature of gases led to the accepted model of behavior known as the **kinetic molecular theory**. One important result of **kinetic theory** is the relationship between temperature and the average kinetic energy of molecules. The velocity of the molecules, rate of **effusion**, and rate of **diffusion** all relate to the molar mass of the gas and are described by **Graham's law**.

An understanding of the gaseous state began in the mid-1600s with Torricelli. His studies of the **barometer** led to a description of **pressure** as a force per unit area exerted by gases of the atmosphere. The principal unit of gas pressure is the **atmosphere** (atm), which is usually expressed in torr. The gas laws now seem quite reasonable and predictable from kinetic theory. Except for average velocity and density, which depend on the mass of the molecules, the other gas laws are independent of the identity of the gas. Avogadro advanced the observation that equal volumes of gases contain equal numbers of molecules under the same conditions, and Dalton observed that pressures depended on the total amount of gas present.

The gas laws require the use of the **Kelvin scale**, which begins at **absolute zero**. The volumes of gases are often described under **standard temperature and pressure (STP)** conditions.

Gas Law	Relationship	Meaning	Constant Conditions	Application
Graham's	$v \propto \dfrac{1}{\sqrt{MM}}$	$v\uparrow MM\downarrow$	T	Relates MM and v of two different gases at a specific T
Boyle's	$V \propto \dfrac{1}{P}$	$V\uparrow P\downarrow$	T, n	Relates V and P of a gas under two different sets of conditions
Charles's	$V \propto T$	$V\uparrow T\uparrow$	P, n	Relates V and T of a gas under two different sets of conditions
Gay-Lussac's	$P \propto T$	$P\uparrow T\uparrow$	V, n	Relates P and T of a gas under two different sets of conditions
Combined	$PV \propto T$	$PV\uparrow T\uparrow$	n	Relates P, V, and T of a gas under two different sets of conditions
Avogadro's	$V \propto n_{tot}$	$V\uparrow n_{tot}\uparrow$	P, T	Relates V and n of gas under two different sets of conditions
Ideal	$PV \propto n_{tot}T$	$PV\uparrow n_{tot}T\uparrow$	—	Relates P, V, T, or n to the other three variables
Dalton's	$P_{tot} = P_1 + P_2$, etc.	$P_{tot}\uparrow n_{tot}\uparrow$	T, V	Relates P_{tot} to partial pressures of component gases

V = volume	P_{tot} = total pressure	v = average velocity
T = Kelvin temperature	n = moles	$\uparrow$ = quantity increases
P = pressure	n_{tot} = total moles	$\downarrow$ = quantity decreases
	MM = molar mass	

At STP, the volume of one mole of a gas is 22.4 L and is known as the **molar volume**. The molar volume can be used to calculate the density of a gas at STP, or it can be used directly as a conversion factor between moles and volume at STP. To convert between moles and volume under other conditions, the use of the ideal gas law is convenient. In the **ideal gas law** ($PV = nRT$), R is known as the **gas constant**.

In the final section, we related the volume of a gas to stoichiometry as summarized in Figure 10-13.

OBJECTIVES

SECTION	YOU SHOULD BE ABLE TO...	EXAMPLES	EXERCISES	CHAPTER PROBLEMS
10-1	List the general properties of gases based on the postulates of the Kinetic Molecular Theory.		1a, 1b, 1c, 1d, 1e, 2e	1, 3, 7, 8, 9, 20, 27, 36
10-2	Calculate the effect changing pressure has on the volume of a gas (Boyle's Law).	10-1, 10-2	2a, 2b, 2c, 2d	10, 12, 15, 17, 18, 19, 21
10-3	Using the gas laws, perform calculations involving the relationships between volume, pressure, and temperature.	10-3, 10-4, 10-5, 10-6, 10-7	3a, 3b, 3c, 3d	24, 25, 28, 31, 33, 35, 39, 41, 43, 44, 46, 50, 52, 54
10-4	Using the ideal gas law, calculate one condition of a gas given the other stated conditions.	10-8, 10-9, 10-10	4a, 4b, 4c, 4d	55, 56, 58, 59, 61, 63
10-5	Using Dalton's law, calculate the partial pressure of a gas.	10-11, 10-12, 10-13	5a, 5b, 5c, 5d	65, 66, 67, 68, 69, 71, 74
10-6	Convert between volume, moles and density of a gas measured at STP.	10-14, 10-15	6a, 6b, 6c, 6d, 6e, 6f	75, 76, 78, 81, 83, 86
10-7	Perform stoichiometric calculations involving gas volumes.	10-16, 10-17	7a, 7b, 7c, 7d	87, 89, 90, 91, 92

▶ANSWERS TO ASSESSING THE OBJECTIVES

Part A
EXERCISES

10-1(a) Unlike solids and liquids, a gas has a <u>low</u> density and fills a container <u>completely</u>. The unique behavior of gases can be explained by the <u>kinetic molecular</u> theory. In this theory, gas molecules are in <u>constant</u>, <u>random</u> motion, exerting pressure by <u>collisions</u> with the walls of the container. Another aspect of this theory is that the temperature relates to the average <u>kinetic</u> energy of the gas. At the same temperature, the gas with the larger molar mass will have the <u>lower</u> velocity. The velocity of a gas relates to the rates of <u>effusion</u> and <u>diffusion</u> as expressed by Graham's law.

10-1(b) (a) False (b) True (c) True (d) False (e) True

10-1(c) Since the molar mass of the SF_6 (146 g/mol) is larger, it must have a lower velocity than the N_2 (28.0 g/mol).

10-1(d) The trunk of a car can heat up rapidly. At the higher temperatures, the gas particles are moving faster, colliding with the walls of the cylinder more frequently and with more energy, thereby increasing the pressure. Ultimately the can will explode.

10-1(e) The molar mass of argon is 40 g/mol while the average molar mass of air is 29 g/mol. Argon atoms move slower at the same temperature than air molecules. It takes longer for the argon atoms to move (and transmit heat or cold) from the outside pane to the inside pane.

10-2(a) Atmospheric pressure is read from a device called a <u>barometer</u>. Pressure is defined as <u>force</u> per unit <u>area</u>. One atmosphere of pressure is equivalent to <u>760</u> torr. An increase in pressure on a volume of gas causes the volume to <u>decrease</u>.

10-2(b) (a) 494 torr (b) 65.9 kPa (c) 9.56 lb/in^2

10-2(c) 659 mL

10-2(d) 550 kPa

10-2(e) By decreasing the volume, you decrease the average distance between the particles of gas and the walls of the container. Since the temperature doesn't change, the particles are moving at the same speed and will therefore strike the walls of the container more frequently, leading to an increase in pressure.

10-3(a) An increase in the pressure of a gas causes the volume to <u>decrease</u>, whereas an increase in temperature causes the same volume to <u>increase</u>. Gay-Lussac's law tells us that if the temperature of a confined volume of gas increases, the pressure <u>increases</u>. To calculate the effect of two changes in conditions, the <u>combined gas law</u> is used. Increasing the number of moles of gas present will <u>increase</u> the volume if other conditions are constant.

10-3(b) (a) 591 mL (b) 829 mL (c) 1.67 atm (d) 0.833 atm (e) 636 mL (f) −75°C

10-3(c) (a) Charles's law (b) Avogadro's law (c) Gay-Lussac's law (d) Boyle's law (e) combined gas law (f) Gay-Lussac's Law

10-3(d) (a) Volume is affected by temperature: Charles's law. (b) Tire pressure is affected by temperature: Gay-Lussac's law. (c) Volume is determined by the number of moles at constant T and P. Avogadro's law. (d) The piston moves from an increase in temperature and internal pressure: combined gas law.

Part B
EXERCISES

10-4(a) −33°C

10-4(b) 1.20×10^{24} particles

10-4(c) 22.4 L

10-4(d) 34 psi

10-4(e) $P/n = k$ or $P_1/n_1 = P_2/n_2$

10-5(a) 715 torr

10-5(b) 345 torr

10-5(c) 5.90 atm

10-5(d) Helium has the highest number of particles (moles) per mass unit. So in order of decreasing pressure, $He > CH_4 > CO_2$.

10-6(a) One mole of any gas occupies <u>22.4</u> L at STP; this is known as the <u>molar volume</u>. The density of a gas at STP is obtained by dividing the <u>molar mass</u> by the molar volume.

10-6(b) 29.4 L

10-6(c) 4.83 g/L

10-6(d) molar mass = 32.0 g/mol. The diatomic element is O_2.

10-6(e) molar mass = 44.0 g/mol. The hydrocarbon is C_3H_8, propane.

10-6(f) raising the pitch: H_2, He, N_2(barely), Ne; lowering the pitch: O_2 (barely), F_2 (barely, and includes death), Cl_2 (more death), Ar, Kr, Xe, and Rn (radiation poisoning).

10-7(a) In stoichiometry calculations, the volume of a gas can be converted to number of moles by use of the <u>molar volume</u> if the volume is measured at STP. If other conditions are present, the volume is related to the number of moles by the <u>ideal gas law</u>.

10-7(b) (a) 16.4 L (b) 169 L (c) 2.65 g

10-7(c) 30.0 L of H_2, 24.1 g of H_2O

10-7(d) The total volume decreases. In gas stoichiometry, for gases under the same conditions, volumes react according to their mole ratios. In the balanced equation, $2NO(g) + O_2(g) \longrightarrow 2NO_2(g)$. Three total moles of gas reactant become two total moles of gas product, or 3 total liters of reactants become 2 total liters of product. So if you start with 2 liters of NO and add 1 liter of O_2, you'll end up with 2 liters of NO_2.

CHAPTER PROBLEMS

Throughout the text, answers to all exercises in color are given in Appendix E. The more difficult exercises are marked with an asterisk.

The Kinetic Theory of Gases (SECTION 10-1)

10-1. It is harder to move your arms in water than in air. Explain on the basis of kinetic molecular theory.

10-2. A balloon filled with water is pear-shaped, but a balloon filled with air is spherical. Explain.

10-3. When a gasoline tank is filled, no more gasoline can be added. When a tire is "filled," however, more air can be added. Explain.

10-4. The pressure inside an auto tire is the same regardless of the location of the nozzle (i.e., up, down, or to the side). Explain.

10-5. A sunbeam forms when light is reflected from dust suspended in the air. Even if the air is still, the dust particles can be seen to bounce around randomly. Explain.

***10-6.** A bowling ball weighs 6.00 kg and a bullet weighs 1.50 g. If the bowling ball is rolled down an alley at 20.0 mi/hr, what is the velocity of a bullet having the same kinetic energy?

10-7. Arrange the following gases in order of increasing rate of effusion at the same temperature.

(a) CO_2 (c) N_2 (e) N_2O

(b) SO_2 (d) SF_6 (f) H_2

10-8. Arrange the following gases in order of increasing rate of effusion at the same temperature

(a) argon   (d) methane

(b) carbon monoxide   (e) chlorine dioxide

(c) chlorine   (f) neon

***10-9.** The kinetic theory assumes that the volume of molecules and their interactions are negligible for gases. Explain why these assumptions may not be true when the pressure is very high and the temperature is very low.

Units of Pressure and Boyle's Law (SECTION 10-2)

10-10. Make the following conversions.

(a) 1650 torr to atm   (d) 5.65 kPa to atm

(b) 3.50×10^{-5} atm to torr   (e) 190 torr to lb/in.^{2}

(c) 185 lb/in.^{2} to torr   (f) 85 torr to kPa

10-11. Make the following conversions.

(a) 30.2 in. of Hg to torr   (c) 57.9 kPa to lb/in.^{2}

(b) 25.7 kilobars to atm   (d) 0.025 atm to torr

10-12. Complete the following table.

torr	lb/in.^{2}	in. Hg	kPa	atm
455	___	___	___	___
___	2.45	___	___	___
___	___	117	___	___
___	___	___	783	___
___	___	___	___	0.0768

10-13. The atmospheric pressure on the planet Mars is 10.3 millibars. What is this pressure in Earth atmospheres?

10-14. The atmospheric pressure on the planet Venus is 0.0920 kilobars. What is this pressure in Earth atmospheres?

10-15. The density of water is 1.00 g/mL. If water is substituted for mercury in the barometer, how high (in feet) would a column of water be supported by 1 atm? A water well is 40 ft deep. Can suction be used to raise the water to ground level?

10-16. A gas has a volume of 6.85 L at a pressure of 0.650 atm. What is the volume of the gas if the pressure is decreased to 0.435 atm?

10-17. If a gas has a volume of 1560 mL at a pressure of 81.2 kPa, what is its volume if the pressure is increased to 2.50 atm?

10-18. At sea level, a balloon has a volume of 785 mL. What is its volume if it is taken to a place in Colorado where the atmospheric pressure is 610 torr?

10-19. A gas has a volume of 125 mL at a pressure of 62.5 torr. What is the pressure if the volume is decreased to 115 mL?

10-20. How does the kinetic molecular theory explain Boyle's law?

10-21. A few miles above the surface of Earth, the pressure drops to 1.00×10^{-5} atm. What would be the volume of a 1.00-L sample of gas at sea-level pressure (1.00 atm) if it were taken to that altitude? (Assume constant temperature.)

*10-22. A gas in a piston engine is compressed by a ratio of 15:1. If the pressure before compression is 0.950 atm, what pressure is required to compress the gas? (Assume constant temperature.)

*10-23. The volume of a gas is measured as the pressure is varied. The four measurements are reported as follows.

Experiment	Volume (mL)	Pressure (torr)
1	125	450
2	145	385
3	175	323
4	220	253

Make a graph using volume on the x-axis and pressure on the y-axis. Is the graph linear?

Charles's Law (SECTION 10-3)

10-24. A balloon has a volume of 1.55 L at 25°C. What would be the volume if the balloon is heated to 100°C? (Assume constant P.)

10-25. A sample of gas has a volume of 677 mL at 63°C. What is the volume of the gas if the temperature is decreased to 46°C?

10-26. A balloon has a volume of 325 mL at 17°C. What is the temperature (in °C) if the volume increases to 392 mL?

10-27. How does the kinetic molecular theory explain Charles's law?

10-28. A quantity of gas has a volume of 3.66×10^4 L. What will be the volume if the temperature is changed from 455 K to 50°C?

*10-29. The temperature of a sample of gas is 0°C. When the temperature is increased, the volume increases by a factor of 1.25 (i.e., $V_2 = 1.25 V_1$.) What is the final temperature in degrees Celsius?

*10-30. The volume of a gas is measured as the temperature is varied. The four measurements are reported as follows.

Experiment	Volume (L)	Temperature (°C)
1	1.54	20
2	1.65	40
3	1.95	100
4	2.07	120

Make a graph of the volume on the x-axis and the Kelvin temperature on the y-axis. Is the graph linear?

Gay-Lussac's Law (SECTION 10-3)

10-31. A confined quantity of gas is at a pressure of 2.50 atm and a temperature of −22°C. What is the pressure if the temperature increases to 22°C?

10-32. A quantity of gas has a volume of 3560 mL at a temperature of 55°C and a pressure of 850 torr. What is the temperature (in °C) if the volume remains unchanged but the pressure is decreased to 0.652 atm?

10-33. A metal cylinder contains a quantity of gas at a pressure of 558 torr at 25°C. At what temperature (in °C) does the pressure inside the cylinder equal 1 atm pressure?

10-34. An aerosol spray can has gas under pressure of 1.25 atm at 25°C. The can ruptures when the pressure reaches 2.50 atm. At what temperature will this happen? (Do not throw these cans into a fire!)

10-35. The pressure in an automobile tire is 28.0 lb/in.^{2} on a chilly morning of 17°C. After it is driven a while, the temperature of the tire rises to 40°C. What is the pressure in the tire if the volume remains constant?

10-36. How does the kinetic molecular theory explain Gay-Lussac's law?

*10-37. The pressure of a confined volume of gas is measured as the temperature is raised. The four measurements are reported as follows:

Experiment	Pressure (torr)	Temperature (K)
1	550	295
2	685	372
3	745	400
4	822	445

Make a graph of the pressure on the x-axis and the temperature on the y-axis. Is the graph linear?

The Combined Gas Law (SECTION 10-3)

10-38. Which of the following are legitimate expressions of the combined gas law?

(a) $PV = kT$

(b) $PT \propto V$

(c) $\dfrac{P_1 T_1}{V_1} = \dfrac{P_2 T_2}{V_2}$

(d) $\dfrac{P}{T} \propto \dfrac{1}{V}$

(e) $VT \propto P$

10-39. Which of the following are not STP conditions?

(a) T = 273 K

(b) P = 760 atm

(c) T = 0 K

(d) P = 1 atm

(e) T°(C) = 273°C

(f) P = 760 torr

(g) T°(C) = 0°C

10-40. In the following table, indicate whether the pressure, volume, or temperature increases or decreases.

Experiment	P	V	T
1	increases	constant	_____
2	constant	_____	decreases
3	_____	decreases	constant
4	increases	increases	_____

10-41. In the following table, indicate whether the pressure, volume, or temperature increases or decreases.

Experiment	P	V	T
1	decreases	_____	constant
2	constant	_____	T(initial) = 350 K
			T(final) = 40°C
3	P(initial) = 1.75 atm	constant	_____
	P(final) = 2200 torr		
4	_____	increases	decreases

10-42. A 5.50-L volume of gas has a pressure of 0.950 atm at 0°C. What is the pressure if the volume decreases to 4.75 L and the temperature increases to 35°C?

10-43. A quantity of gas has a volume of 17.5 L at a pressure of 6.00 atm and temperature of 100°C. What is its volume at STP?

10-44. A quantity of gas has a volume of 88.7 mL at STP. What is its volume at 0.845 atm and 35°C?

10-45. A quantity of gas has a volume of 4.78×10^{-4} mL at a temperature of -50°C and a pressure of 78.0 torr. If the volume changes to 9.55×10^{-5} mL and the pressure to 155 torr, what is the temperature?

10-46. A gas has a volume of 64.2 L at STP. What is the temperature if the volume decreases to 58.5 L and the pressure increases to 834 torr?

10-47. A quantity of gas has a volume of 6.55×10^{-5} L at 7°C and 0.882 atm. What is the pressure if the volume changes to 4.90×10^{-3} L and the temperature to 273 K?

10-48. A balloon has a volume of 1.55 L at 25°C and 1.05 atm pressure. If it is cooled in the freezer, the volume shrinks to 1.38 L and the pressure drops to 1.02 atm. What is the temperature in the freezer?

10-49. A bubble from a deep-sea diver in the ocean starts with a volume of 35.0 mL at a temperature of 17°C and a pressure of 11.5 atm. What is the volume of the bubble when it reaches the surface? Assume that the pressure at the surface is 1 atm and the temperature is 22°C.

Avogadro's Law (SECTION 10-3)

10-50. A 0.112-mol quantity of gas has a volume of 2.54 L at a certain temperature and pressure. What is the volume of 0.0750 mol of gas under the same conditions?

10-51. A balloon has a volume of 188 L and contains 8.40 mol of gas. How many moles of gas must be added to expand the balloon to 275 L? Assume the same temperature and pressure in the balloon.

10-52. A balloon has a volume of 275 mL and contains 0.0212 mol of CO_2. What mass of N_2 must be added to expand the balloon to 400 mL?

10-53. A balloon has a volume of 75.0 mL and contains 2.50×10^{-3} mol of gas. What mass of N_2 must be added to the balloon for the volume to increase to 164 mL at the same temperature and pressure?

10-54. A 48.0-g quantity of O_2 in a balloon has a volume of 30.0 L. What is the volume if 48.0 g of SO_2 is substituted for O_2 in the same balloon?

The Ideal Gas Law (SECTION 10-4)

10-55. What is the temperature (in degrees Celsius) of 4.50 L of a 0.332-mol quantity of gas under a pressure of 2.25 atm?

10-56. A quantity of gas has a volume of 16.5 L at 32°C and a pressure of 850 torr. How many moles of gas are present?

10-57. What mass of NH_3 gas has a volume of 16,400 mL, a pressure of 0.955 atm, and a temperature of -23°C?

10-58. What is the pressure (in torr) exerted by 0.250 g of O_2 in a 250-mL container at 29°C?

10-59. A container of Cl_2 gas has a volume of 750 mL and is at a temperature of 19°C. If there is 7.88 g of Cl_2 in the container, what is the pressure in atmospheres?

10-60. What mass of Ne is contained in a large neon light if the volume is 3.50 L, the pressure 1.15 atm, and the temperature 23°C?

10-61. A sample of H_2 is collected in a bottle over water. The volume of the sample is 185 mL at a temperature of 25°C. The pressure of H_2 in the bottle is 736 torr. What is the mass of H_2 in the bottle?

*10-62. A blimp has a volume of about 2.5×10^7 L. What is the mass of He (in lb) in the blimp at 27°C and 780 torr? The average molar mass of air is 29.0 g/mol. What mass of air (in lb) would the blimp contain? The difference between these two values is the lifting power of the blimp. What mass could the blimp lift? If H_2 is substituted for He, what is the lifting power? Why isn't H_2 used?

*10-63. A good vacuum pump on Earth can produce a vacuum with a pressure as low as 1.00×10^{-8} torr. How many molecules are present in each milliliter at a temperature of 27.0°C?

Dalton's Law (SECTION 10-5)

10-64. Three gases are mixed in a 1.00-L container. The partial pressure of CO_2 is 250 torr, that of N_2 is 375 torr, and that of He 137 is torr. What is the pressure of the mixture of gases?

10-65. The total pressure in a cylinder containing a mixture of two gases is 1.46 atm. If the partial pressure of one gas is 750 torr, what is the partial pressure of the other gas?

10-66. Air is about 0.90% Ar. If the barometric pressure is 756 torr, what is the partial pressure of Ar?

10-67. A sample of oxygen is collected in a bottle over water. The pressure inside the bottle is made equal to the barometric pressure, which is 752 torr. When collected over water, the gas is a mixture of oxygen and water vapor. The partial pressure of water (known as the vapor pressure) at that temperature is 24 torr. What is the pressure of the pure oxygen?

10-68. A container holds two gases, A and B. Gas A has a partial pressure of 325 torr and gas B has a partial pressure of 488 torr. What percent of the molecules in the mixture is gas A?

10-69. A volume of gas is composed of N_2, O_2, and SO_2. If the total pressure is 1050 torr, what is the partial pressure of each gas if the gas is 72.0% N_2 and 8.00% O_2?

*10-70. A mixture of two gases is composed of CO_2 and O_2. The partial pressure of O_2 is 256 torr, and it represents 35.0% of the molecules of the mixture. What is the total pressure of the mixture?

*10-71. A volume of gas has a total pressure of 2.75 atm. If the gas is composed of 0.250 mol of N_2 and 0.427 mol of CO_2, what is the partial pressure of each gas?

*10-72. The following gases are all combined into a 2.00-L container: a 2.00-L volume of N_2 at 300 torr, a 4.00-L volume of O_2 at 85 torr, and a 1.00-L volume of CO_2 at 450 torr. What is the total pressure?

*10-73. The total pressure of a mixture of two gases is 0.850 atm in a 4.00-L container. Before mixing, gas A was in a 2.50-L container and had a pressure of 0.880 atm. What is the partial pressure of gas B in the 4.00-L container?

*10-74. What is the pressure (in atm) in a 825-mL container at 33°C if it contains 6.25 g of N_2 and 12.6 g of CO_2?

Molar Volume and Density (SECTION 10-6)

10-75. What is the volume of 15.0 g of CO_2 measured at STP?

10-76. What is the mass (in kilograms) of 850 L of CO measured at STP?

10-77. What is the volume of 3.01×10^{24} molecules of N_2 measured at STP?

10-78. A 6.50-L quantity of a gas measured at STP has a mass of 39.8 g. What is the molar mass of the compound?

10-79. What is the mass of 6.78×10^{-4} L of NO_2 measured at STP?

10-80. What is the density in g/L (STP) of B_2H_6?

10-81. What is the density in g/L (STP) of BF_3?

10-82. A gas has a density of 1.52 g/L (STP). What is the molar mass of the gas?

10-83. A gas has a density of 6.14 g/L (STP). What is the molar mass of the gas?

*10-84. A gas has a density of 3.60 g/L at a temperature of 25°C and a pressure of 1.20 atm. What is its density at STP?

*10-85. What is the density (in g/L) of N_2 measured at 500 torr and 22°C?

*10-86. What is the density (in g/L) of SF_6 measured at 0.370 atm and 37°C?

Stoichiometry Involving Gases (SECTION 10-6)

10-87. Limestone is dissolved by CO_2 and water according to the equation

$$CaCO_3(s) + H_2O(l) + CO_2(g) \longrightarrow Ca(HCO_3)_2(aq)$$

What volume of CO_2 measured at STP would dissolve 115 g of $CaCO_3$?

10-88. Magnesium once used in flashbulbs, burns according to the equation

$$2Mg(s) + O_2(g) \longrightarrow 2MgO(s)$$

What mass of Mg combines with 5.80 L of O_2 measured at STP?

10-89. Oxygen gas can be prepared in the laboratory by decomposition of potassium nitrate according to the equation

$$2KNO_3(s) \longrightarrow 2KNO_2(s) + O_2(g)$$

What mass of KNO_2 forms along with 14.5 L of O_2 measured at 1 atm and 25°C?

10-90. Acetylene (C_2H_2) is produced from calcium carbide as shown by the reaction

$$CaC_2(s) + 2H_2O(l) \longrightarrow Ca(OH)_2(s) + C_2H_2(g)$$

What volume of acetylene measured at 25°C and 745 torr would be produced from 5.00 g of H_2O?

10-91. Nitrogen dioxide is an air pollutant. It is produced from NO (from car exhaust) as follows.

$$2NO(g) + O_2(g) \longrightarrow 2NO_2(g)$$

What volume of NO measured at STP is required to react with 5.00 L of O_2 measured at 1.25 atm and 17°C?

10-92. Butane (C_4H_{10}) burns according to the equation

$$2C_4H_{10}(g) + 13O_2(g) \longrightarrow 8CO_2(g) + 10H_2O(l)$$

(a) What volume of CO_2 measured at STP would be produced by 85.0 g of C_4H_{10}?

(b) What volume of O_2 measured at 3.25 atm and 127°C would be required to react with 85.0 g of C_4H_{10}?

(c) What volume of CO_2 measured at STP would be produced from 45.0 L of C_4H_{10} measured at 25°C and 0.750 atm?

10-93. In March 1979, a nuclear reactor overheated, producing a dangerous hydrogen bubble at the top of the reactor core. The following reaction occurring at the high temperature (about 1500°C) accounted for the hydrogen. (Zr alloys hold the uranium pellets in long rods.)

$$Zr(s) + 2H_2O(g) \longrightarrow ZrO_2(s) + 2H_2(g)$$

If the bubble had a volume of about 28,000 L at 250°C and 70.0 atm, what mass (in kg and tons) of Zr had reacted?

10-94. Nitric acid is produced according to the equation

$$3NO_2(g) + H_2O(l) \longrightarrow 2HNO_3(aq) + NO(g)$$

What volume of NO_2 measured at −73°C and 1.56×10^{-2} atm would be needed to produce 4.55×10^{-3} mol of HNO_3?

*10-95. Natural gas (CH_4) burns according to the equation

$$CH_4(g) + 2O_2(g) \longrightarrow CO_2(g) + 2H_2O(l)$$

What volume of CO_2 measured at 27°C and 1.50 atm is produced from 27.5 L of O_2 measured at −23°C and 825 torr?

General Problems

10-96. A column of mercury (density 13.6 g/mL) is 15.0 cm high. A cross-section of the column has an area of 12.0 cm^2. What is the force (weight) of the mercury at the bottom of the tube? What is the pressure in grams per square centimeter and in atmospheres?

10-97. A tube containing an alcohol (density 0.890 g/mL) is 1.00 m high and has a cross-section of 15.0 cm^2. What is the total force at the bottom of the tube? What is the pressure? Assuming the same cross-section, how high would an equivalent amount of mercury be?

10-98. A 1.00-L volume of a gas weighs 8.37 g. The gas volume is measured at 1.45 atm pressure and 35°C. What is the molar mass of the gas?

10-99. A gaseous compound is 85.7% C and 14.3% H. A 6.58-g quantity of this gas occupies 4500 mL at 77.0°C and a pressure of 1.00 atm. What is the molar mass of the compound? What is its molecular formula?

10-100. What is the volume at STP of a mixture of 10.0 g each of Ar, Cl_2, and N_2? What is the partial pressure of each gas?

10-101. What is the volume occupied by a mixture of 0.265 mol of O_2, 9.88 g of N_2, and 9.65×10^{22} molecules of CO_2 at a temperature of 37°C and a pressure of 2.86 atm? What is the partial pressure of each gas?

10-102. Molecular clouds in space contain 30,000 molecules/mL at a temperature of 10 K. What is the pressure in atmospheres?

*10-103. What is the molar volume of an ideal gas (i.e., volume of one mole) at 25°C and 1.25 atm? What is the density of CO_2 under these conditions?

*10-104. A hot-air balloon rises because the heated air trapped in the balloon is less dense than the surrounding air. What is the density of air (assume an average molar mass of 29.0 g/mol) at 400°C and 1 atm pressure? *(Hint:* Calculate the volume of one mole of gas under these conditions.) Compare this to the density of air at STP.

10-105. A compound is 80.0% carbon and 20.0% hydrogen. Its density at STP is 1.34 g/L. What is its molecular formula?

10-106. Given the following *unbalanced* equation

$$H_3BCO(g) + H_2O(l) \longrightarrow B(OH)_3(aq) + CO(g) + H_2(g)$$

A 425-mL quantity of H_3BCO measured at 565 torr and 100°C was allowed to react with excess H_2O. What volume of gas was produced measured at 25°C and 0.900 atm?

10-107. Given the following *unbalanced* equation

$$C_3H_8O(g) + O_2(g) \longrightarrow CO_2(g) + H_2O(l)$$

What mass of water forms if 6.50 L of C_3H_8O measured at STP is allowed to react with 42.0 L of O_2 measured at 27°C and 1.68 atm pressure? Assume that this is the only reaction that occurs.

10-108. Given the following *unbalanced* equation

$$Al(s) + F_2(g) \longrightarrow AlF_3(s)$$

An 8.23-L quantity of F_2 measured at 35°C and 725 torr was allowed to react with some Al. At the end of the reaction, 3.50 g of F_2 remained. What mass of AlF_3 formed?

10-109. Sulfuric acid is made from SO_3, which is obtained from the combustion of sulfur according to the following *unbalanced* equations.

$$S(s) + O_2(g) \longrightarrow SO_2(g)$$
$$SO_2(g) + O_2(g) \longrightarrow SO_3(g)$$

What volume of SO_3 measured at 2.75 atm and 400°C is prepared from 50.0 kg of sulfur?

10-110. Liquid N_2O_3 decomposes according to the equation

$$N_2O_3(l) \longrightarrow NO_2(g) + NO(g)$$

What is the total volume of gas measured at 1.58 atm and 35°C produced by the decomposition of 2.54×10^{24} molecules of N_2O_3?

*10-111. Calcium bicarbonate is formed according to the following equation.

$$CaO(s) + H_2O(l) + 2CO_2(g) \longrightarrow Ca(HCO_3)_2(s)$$

If 80.0 g of CaO is mixed with 7.85×10^{23} molecules of H_2O and 30.0 L of CO_2 measured at 25°C and 820 torr, what mass of calcium bicarbonate is formed?

10-112. What volume of water vapor (gas) measured at 22.0 torr and 25°C contains the same number of molecules as 15.0 mL of ice if the density of ice is 0.917 g/mL?

10-113. The following equation represents what happens in swimming pools when ammonia (from people) reacts with sodium hypochlorite (used as a disinfectant). The N_2H_4 formed has a seriously bad odor.

$$2NH_3(aq) + NaOCl(aq) \longrightarrow N_2H_4(g) + NaCl(aq) + H_2O$$

What volume of N_2H_4 (in mL) measured at STP is produced if 20.0 mL of NH_3 gas measured at 1.20 atm and 25°C is dissolved in water?

*10-114. A 1.000-g sample of a gaseous compound composed of only nitrogen and fluorine contains 0.269 g of nitrogen. The gas has a density of 4.25 g/L measured at room temperature (25°C) and standard pressure. What is the formula of the gas?

*10-115. A liquid compound composed of only nitrogen and oxygen is 69.6% oxygen. Decomposition of 0.0220 mole of this compound produces 2.03 g of a single gas that has a volume of 1.05 L measured at standard temperature and 715 torr. What is the formula of the original compound?

10-116. Steering on space vehicles is provided by a propulsion system that produces gaseous products when two liquids are mixed. It can produce short bursts of gases. The reaction is

$$H_2NN(CH_3)_2(l) + 2N_2O_4(l) \longrightarrow$$
$$3N_2(g) + 4H_2O(g) + 2CO_2(g)$$

What volume of gas measured at 1.75 atm and 120°C is produced if 125 g of each of the two reactants are mixed?

10-117. Magnesium and lithium both react with elemental nitrogen to form their respective nitrides. Write the balanced equations illustrating these reactions and determine what mass of each metal would react with 256 L of nitrogen measured at 985 torr and 373 K.

STUDENT WORKSHOP

Gas Properties

Purpose: To apply the laws governing the behavior of gases to some simple demonstrations. (Work in groups of three or four. Estimated time: Predictions—10 min.; Activities—30 min.)

Read the following activities. Based on the concepts of this chapter, predict what should happen. If time and space permit, perform the activities in your group. Were your predictions accurate?

1. Candle, matches, beaker, $NaHCO_3$, dilute HCl:
   - Light the candle.
   - In the beaker, mix a spatula full of solid $NaHCO_3$ with 10 mL of HCl.
   - Hold the beaker at an angle over the flame.

2. Balloon, string, ruler:
   - Take a deep breath and blow once into a balloon.
   - Pinch the end and wrap the string around the balloon.
   - Measure the length of the string with a ruler.
   - Without releasing any air, repeat four more times.
   - Plot the number of breaths against the length of string squared.

3. Soda bottle, balloon, beaker, ice:
   - Fill the bottle with hot water.
   - Fill the beaker with ice water.
   - After one minute, empty the bottle and stretch the balloon over the mouth.
   - Place the bottle into the ice water.

Chapter 11

The Solid and Liquid States

Life can flourish on this planet because all three physical states of water can exist. Ice and snow form from the condensation of the vapor or the freezing of the liquid. The solid and liquid states of water are featured in this chapter.

SETTING THE STAGE

Ice cubes floating in a glass of water—what could be more familiar? What we may not appreciate in this common sight is that it represents very unusual behavior for a compound. Usually the solid state of a substance is denser than the liquid state, so the solid sinks to the bottom rather than floating on top. We are very fortunate that water is unique. Life on Earth could not exist as we know it if water and ice behaved as do most other liquids and their solid forms. If the ice were to sink as it forms, lakes would freeze solid in winter. The hot summer sun would thaw only the top layer of a lake, so very little life could survive. Heat could not be distributed from warm to cold climates by ocean currents. Nothing would be the same on this planet. In addition to this property, the presence of water in the liquid state under conditions found on this planet is another key to our very existence. Scientists speculate that liquid water was the necessary medium allowing the complex chemical reactions connected with life processes to occur. In the next chapter, we will emphasize the properties of water as a solvent in which chemical reactions take place. Life on this planet is also dependent on the energy released or absorbed by water as it changes between its three physical states.

Water is certainly our most familiar liquid. Water and other liquids and their solid forms are, in a way, easier to study than the gaseous state. These forms of matter are more concrete—we can see them, feel them, and conveniently isolate and measure them. However, there is a disadvantage in our work with solids and liquids compared to gases. By their very nature, the condensed phases of matter do not lend themselves to such simplifying generalizations as the gas laws.

In Part A in this chapter, we will examine the general properties of solids and liquids, with emphasis on the solid state. In Part B, we will consider properties of the liquid state and the energies involved in phase transitions.

Part A

The Properties of Condensed States and the Forces Involved

SETTING A GOAL

■ You will become familiar with the various forces between molecules and ions and how these forces affect the condensed states of matter.

OBJECTIVES

11-1 Use kinetic molecular theory to distinguish the physical properties of solids and liquids from gases.

11-2 Describe the types and relative strengths of intermolecular forces that can occur between two molecules.

11-3 Classify a solid as ionic, molecular, network, or metallic based on its physical properties.

▶ **OBJECTIVE
FOR SECTION 11-1**
Use kinetic molecular theory to distinguish the physical properties of solids and liquids from gases.

11-1 Properties of the Solid and Liquid States

LOOKING AHEAD! Our first question concerns the difference between the liquid or solid states and the gaseous state at the molecular level. In this section, we will see where the kinetic theory, introduced in the previous chapter, does and does not apply to liquids and solids. ■

11-1.1 Properties of the Condensed States

Most of the properties of the solid and liquid states are obvious to us. Still, it is worthwhile to note these common properties in order to visualize the actions and interactions of the ions or molecules that comprise these states.

1. *They have high density.*
   Solids and liquids are about 1000 times denser than a typical gas.

2. *They are generally considered incompressible.*
   A tall building can be supported by bricks and other solids because they can't be compressed to any extent as would a gas. Likewise, a hydraulic jack uses a liquid to support weights such as that of a huge truck. Unlike the behavior of a gas, an increase in pressure on a solid or liquid does not result in a significant decrease in volume.

3. *They undergo little thermal expansion.*
   When a bridge is constructed, a small space must be left between sections for expansion on a hot day. Still, this space amounts to only a few centimeters for a bridge span many meters long. In addition, the degree of expansion varies for different solids and liquids. Gases, however, expand significantly as the temperature rises, and all gases expand by the same factor.

4. *They have a fixed volume.*
   The volume of a gas is that of the container. Also, the volume is the same for the same number of particles under the same conditions. There is no such convenient relationship for solids and liquids. The same volumes of different solids or liquids have no relationship to the number of molecules present.

In addition to these common characteristics, solids and liquids differ with regard to shape. Solids are rigid and thus have a definite shape. Liquids flow and thus do not have a definite shape that is determined by the shape of the container.

11-1.2 The Condensed States and Kinetic Theory

The characteristics of gases are adequately explained by the kinetic molecular theory. Two of the basic assumptions of the kinetic theory are also applicable to the other states. That is, solids and liquids are composed of basic particles that have kinetic energy. The average kinetic energy of the particles is related to the temperature. However, to explain the characteristics of the other two states, there are obviously some assumptions related to gases that no longer apply. In the solid and liquid states the following circumstances exist.

1. The basic particles (molecules or ions) have significant attractions for each other and so are held close together.

2. Since the basic particles are close together, the particles occupy a significant portion of the volume occupied by the substance.

3. The basic particles are not in random motion; their motion is restricted by interactions with other, neighboring particles.

Many properties of the solid and liquid states can be understood on the basis of these assumptions. Since the basic particles are in close contact, they cannot be pressed together easily, so the substances are incompressible and have high densities. In fact, both solids and liquids are referred to as *condensed states*. The attraction of the particles for each other holds them together, which counteracts the tendency of kinetic energy to move them apart. Thus, solids and liquids undergo little thermal expansion.

In Figure 11-1 we illustrate the fundamental differences in the behavior of molecules in the three states of matter. We use the water molecule as an example. This simple but amazing compound exists in all three physical states on Earth: vapor (the gaseous state of a substance normally in the liquid or solid state is sometimes referred to as vapor), liquid, and solid (ice). In fact, in a Thermos containing ice water, all three states exist at once, although the presence of some H_2O molecules in the gaseous state above the ice may not be obvious.

First, let's consider ice, the solid state of water. In this case, the forces of attraction between molecules hold them in fixed positions relatively close together. In any physical state, the molecules have kinetic energy, meaning that

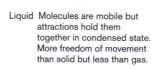

FIGURE 11-1 Physical States of Water Interactions between water molecules are different in the three physical states.

Solid Molecules in fixed positions, motion within a confined volume highly ordered.

Liquid Molecules are mobile but attractions hold them together in condensed state. More freedom of movement than solid but less than gas.

Gas Random motion, very weak interactions, disorder.

they have motion. But in the solid state the motion is restricted to various types of vibrations about a fixed point. This is much like a view of a crowded dance floor, with all the dancers shaking and vibrating but not moving around the dance floor. In the liquid state, the water molecules are also held close together by forces of attraction, but the molecules are not confined to fixed positions and thus have more freedom of motion than in the solid state. That is, individual molecules or groups of molecules have translational motion as well as vibrational motion. This is like viewing the same crowded dance floor and noticing the dancers are still close together but are now moving around the floor as well as shaking and vibrating. Since the molecules can move past one another, liquids can flow and take the shape of the bottom of the container. Finally, we are already familiar with the behavior of water molecules in the gaseous state, as discussed in the preceding chapter. In this case, the molecules have so much translational motion that they move freely throughout the whole container and are unaffected by attractions to other molecules. Their motion and collisions scatter them as far apart as possible. In the dance floor analogy, it's like the dance floor has greatly expanded. The dancers can now move around freely, eventually taking up all the added space. And, they are totally oblivious of any other dancers.

▶ **ASSESSING THE OBJECTIVE FOR SECTION 11-1**

EXERCISE 11-1(a) LEVEL 1: Are the following properties of gases, or condensed phases, or both?
(a) are highly compressible
(b) are in constant random motion
(c) made from small, individual particles
(d) have low densities
(e) composed of mostly empty space
(f) have strong intermolecular forces
(g) volume expands proportionately with heating

EXERCISE 11-1(b) LEVEL 3: Review the densities of the solids listed in Table 3-1 in Section 3-2.1. In general, which have higher densities, solids or liquids? Why is this so? What does this say about the forces between individual molecules of solids versus liquids?

For additional practice, work chapter problems 11-1, 11-2, and 11.3.

▶ **OBJECTIVE FOR SECTION 11-2**
Describe the types and relative strengths of intermolecular forces that can occur between two molecules.

11-2 Intermolecular Forces and Physical State

LOOKING AHEAD! The ions and molecules in the condensed states obviously "stick together." In this section, we will explore the attractive forces at work in molecular compounds and free elements that cause the molecules to stick together. The magnitude of these forces determines whether a compound is a gas, liquid, or solid at a particular temperature. ∎

At room temperature methane is a gas, water is a liquid, and table sugar is a solid. Yet all three are molecular compounds. The physical state of molecular compounds at a specific temperature depends on how strongly the molecules are attracted to each other. These interactions are known as **intermolecular forces**.

Molecular compounds are held together by attractions called London forces. Some molecules may have additional attractions, called dipole–dipole or hydrogen-bonding forces, which may add to the basic London force. First, consider London forces.

11-2.1 London Forces

Since atoms and molecules are surrounded by negatively charged electrons, it may seem reasonable that one molecule would *repel* another because of repulsions of like charges. In fact, it is just the opposite. There are electrostatic forces of attraction between molecules. These forces are known as *London dispersion forces* or simply **London forces**. London forces also have the rather imposing name of *instantaneous dipole–induced dipole forces*. Let's see if we can make some sense of all this. In a molecule, positively charged nuclei do exist within negatively charged electron clouds. However, according to modern atomic theory, the electrons in these clouds have probabilities of being in various locations. Because of this, a molecule may have an imbalance of electron charge on one side of the atom or molecule at a given instant. For that instant, the molecule becomes polar (i.e., forms a dipole with a negative side and an opposite positive side). In other words, the molecule achieves an instantaneous dipole. A neighboring molecule is influenced by this *instantaneous dipole*, and that molecule also becomes *polarized*. That is, a dipole is *induced* in the second molecule. Now a force of attraction forms between the negative side of the instantaneous dipole on one molecule and the positive side of the induced dipole on another. Recall from Section 9-8 that δ^+ and δ^- represent partial positive and negative charges, respectively. (See Figure 11-2).

In larger molecules with more electrons, instantaneous dipoles become more common and stronger, so London forces become more significant. Also, larger molecules are generally more *polarizable* than smaller ones because they are surrounded by larger, more diffuse electron clouds. One molecule can be larger than another by having more atoms, or by being made from atoms further down the periodic table. *Since molar mass usually indicates a larger molecule, we can state that the heavier the molecule, the greater the London forces and the more likely we are to find it in the liquid or solid state at a given temperature.* For example, at room temperature, natural gas [CH_4 (molar mass = 16 g/mol)] is, of course, a gas. A major component of gasoline, called octane [C_8H_{18} (molar mass = 114 g/mol)], is a liquid, and paraffin [$C_{24}H_{50}$ (molar mass = 338 g/mol)] is a solid. The trend of gas → liquid → solid corresponds to the magnitude of the London forces for the three molecules (i.e., $CH_4 < C_8H_{18} < C_{24}H_{50}$). All three of these molecules are nonpolar, and for nonpolar molecules, London forces are the only type of force present.

It should be mentioned that this generalization (i.e., the larger the nonpolar molecule, the more likely it is to be a solid) does not always hold. Some compounds are gases at higher temperatures than we may predict. This is especially true when fluorine atoms which are highly electronegative, are the peripheral atoms in the molecule.

δ^+ δ^- δ^+ δ^-

Unpolarized atom   Unpolarized atom

Instantaneous dipole   Induced dipole

FIGURE 11-2 London Forces
An instantaneous dipole in one atom or molecule creates an induced dipole in a neighbor.

11-2.2 Dipole–Dipole Attractions

Covalent bonds between any unlike atoms in a molecule are polar, at least to some extent, but the molecule itself may not be. This apparent contradiction was discussed in Chapter 9. With the help of the Lewis structures and VSEPR theory, we were able to predict that certain molecules may be linear (e.g., CO_2), trigonal planar (e.g., BF_3), or tetrahedral (e.g., CH_4). In these highly symmetrical structures, the equal bond dipoles cancel, and the molecules as a whole are nonpolar. Consider the case of CO_2. (O=C=O). Since the two polar carbon–oxygen bonds lie at a 180° angle, the individual bond dipoles cancel and the molecule is nonpolar. In contrast, carbonyl sulfide (O=C=S) is also linear, like CO_2, but since the bond dipoles are unequal and do not cancel the molecule is polar. The SO_2 molecule is also polar. Drawing the Lewis structure illustrates a lone pair of electrons on the central sulfur, so the geometry is V-shaped rather than linear. Thus, the two equal bond dipoles are at an angle and do not cancel. The OCS and SO_2 molecules both have a permanent dipole, meaning that they have sites with permanent partially negative and

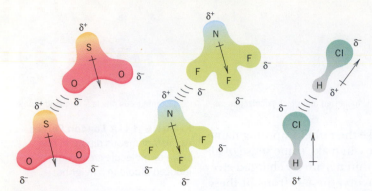

FIGURE 11-3 Dipole–Dipole Attractions SO_2, NF_3, and HCl are polar molecules and have dipole–dipole attractions. The negative end (δ^-) of one molecule is attracted to the positive end (δ^+) of another.

positive charges in contrast to the temporary dipoles that exist in nonpolar molecules. (See Section 9-8.2.) *Molecules with a permanent dipole can align themselves so that the negative end of one molecule is attracted to the positive end of another.* (See Figure 11-3.) *These intermolecular interactions are known as* **dipole–dipole attractions** *and add to the effect of the London forces. Thus, if we have two compounds of similar molar mass (meaning similar London forces) but one is composed of polar molecules and one of nonpolar molecules, the compound with polar molecules is more likely to be in a condensed state at a given temperature.* Having additional dipole–dipole forces of attraction for molecules is like getting 5 extra-credit points on a 100-point test. It may not change anything, but it may be just enough to get a higher grade. In the case of compounds, the added attractions may be just enough to hold the molecules together in a condensed state. For example, at room temperature, CO_2 (44 g/mol), a compound with nonpolar molecules, is a gas, whereas CH_3CN (41 g/mol), a compound with polar molecules, is a liquid. It is difficult to compare heavy nonpolar molecules with lighter polar molecules. Generally, the mass of the molecules and the London forces are more important than the additional dipole–dipole forces.

11-2.3 Hydrogen Bonding

In Section 8-5, we discussed trends in the size of atoms. Atoms of the elements to the upper right in the periodic table are the smallest. In Chapter 9, we also mentioned that these same atoms (fluorine, oxygen, and nitrogen in particular) are the most electronegative elements. This means that these small, highly electronegative atoms tend to attract a significant amount of negative charge to themselves when chemically bonded to other atoms. As an example, consider molecules of water. There is a large difference in electronegativity between oxygen and hydrogen ($3.5 - 2.1 = 1.4$). This difference is not enough to indicate an ionic bond, but it does point to a highly polar covalent bond. In Chapter 9, we discussed the geometry of water molecules. According to VSEPR theory, the mutually repulsive effect of the two electron pairs on the oxygen atom and the two bonded pairs causes the electron pairs and the hydrogen atoms to be located at the corners of a rough tetrahedron. (See Figure 11-4a.) Since this structure gives H_2O a V-shaped molecular geometry, the bond dipoles do not cancel and the water molecule is significantly polar. (See Figure 11-4b.) The partial positive charge is centered on the hydrogen atoms at two of the corners of the tetrahedron, and the partial negative charge is centered on the electron pairs at the other two corners. As shown in Figure 11-5 in solid ice, hydrogen atoms on two different water molecules interact with the two electron pairs on one water molecule. In liquid water, the structure is less orderly because the interactions between molecules are more random. Individual and groups of water molecules can slide past one another in the liquid state.

In the solid state, the H_2O molecules have a relatively open structure compared to the liquid state. The more compact structure in the liquid state accounts for its higher density and thus why ice cubes float in water. (See Figure 11-5.)

The interaction between a partially positive hydrogen atom on one molecule and the electron pair of the oxygen on another molecule is an example of what is known as a hydrogen bond. *A* **hydrogen bond** *is generally restricted to molecules that have an N—H, O—H, or F—H bond, where the hydrogen in these bonds interacts with an unshared electron pair on an F, O, or N atom of another molecule.* Hydrogen bonding usually involves only

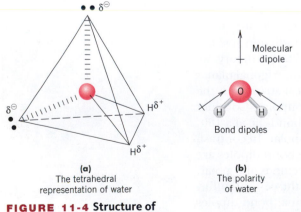

(a)
The tetrahedral representation of water

(b)
The polarity of water

Molecular dipole

Bond dipoles

FIGURE 11-4 Structure of Water Water molecules are polar because they are V-shaped.

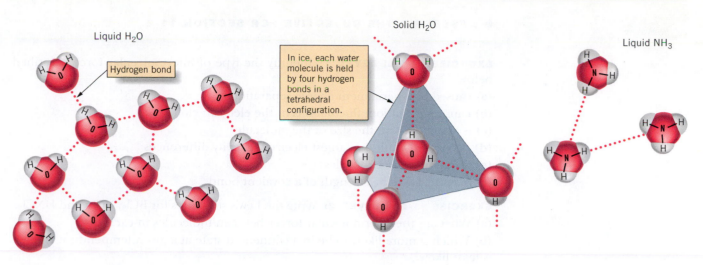

FIGURE 11-5 Hydrogen Bonding H_2O molecules in both liquid and solid states, as well as NH_3 molecules, are attracted by hydrogen bonds.

these three atoms because of their high electronegativities and small size. This interaction at first may seem like a case of an extreme dipole–dipole attraction. In fact, it is more complex than that. A hydrogen bond is not nearly as strong as a regular covalent bond (about 1/10 or less as strong), but it is considerably stronger than typical dipole–dipole attractions. In fact, *hydrogen bonding has a significant effect on the properties of a compound.* Dipole–dipole attractions usually do not have a large effect on the physical properties of the compound. (In case of hydrogen bonding, it is like getting 15–20 extra-credit points on a 100-point test. It would most likely make a difference.) Consider the water molecule. It has a very small molar mass of 18 g/mol. If only London forces were present, it would be a gas at temperatures as low as −75°C. Even if regular dipole–dipole attractions were present, it would still boil at a very low temperature. The presence of hydrogen bonding between water molecules provides the "glue" that holds the molecules together so that the substance exists as a liquid and even a solid under conditions found on this planet. Other compounds whose properties are altered considerably by the presence of hydrogen bonding are ammonia (NH_3) and hydrogen fluoride (HF).

Hydrogen bonding is also important in the huge, complex molecules on which life is based. Consider, for example, a molecule of DNA, the "messenger of life." DNA is an extremely large molecule composed of two strands of covalently bonded atoms offset from each other in a helical arrangement (somewhat like the railings on each side of a spiral staircase). One of the wonders of life is that these two strands can separate from each other and replicate themselves from smaller molecules. They can do this because the strands are held together by hydrogen bonds, much like Velcro holds two pieces of clothing together. The hydrogen bonds are not as strong as the covalent bonds within the strands (see Figure 11-6), so while the two strands are generally held together, they can break apart under the right circumstances. When DNA unravels, the hydrogen bonds are broken but not the covalent bonds.

As we proceed in this chapter, we will notice that hydrogen bonding plays an important role in many physical properties such as melting point and boiling point.

Hydrogen bonds also play an important role in the structure of proteins. Proteins are made of amino acids, which contain N—H bonds, and also have oxygen atoms with unshared electron pairs for the hydrogen bonds to form. A protein is made up of thousands of individual amino acids, so there are potentially thousands of sites along a protein where hydrogen bonds can occur. The fact that proteins can form into sheets to build muscle, or twist into specific shapes to make enzymes, is due in large part to their ability to join together into hydrogen bonds.

A typical hydrogen-bonding interaction between strands

Two strands of DNA. The horizontal lines represent hydrogen bonds.

FIGURE 11-6 Hydrogen Bonding in DNA The double-helix structure of DNA consists of two strands twisted about each other. The strands are connected by hydrogen bonds.

▶ **ASSESSING THE OBJECTIVE FOR SECTION 11-2**

EXERCISE 11-2(a) LEVEL 1: Identify the type of intermolecular force described below.

(a) caused by a permanent charge separation
(b) caused by random fluctuations in the electron cloud
(c) is dependent on the size of the molecule
(d) occurs only with the largest electronegativity differences
(e) found in polar molecules
(f) is about 1/10 the strength of a covalent bond

EXERCISE 11-2(b) LEVEL 2: Write the Lewis structures for BCl_3, NCl_3, and $HNCl_2$.
(a) What are the intermolecular forces between molecules in each case?
(b) Which is more likely to be in a condensed state at a given temperature? Which is least likely?

EXERCISE 11-2(c) LEVEL 2: Rank the following molecules on the basis of increasing intermolecular forces: CH_3F, CH_2F_2, C_2H_6, CH_3NH_2.

EXERCISE 11-2(d) LEVEL 3: Of CF_4, SiF_4, and PF_3, which should be the correct order of increasing strength of forces between molecules? Why?

EXERCISE 11-2(e) LEVEL 3: Of Cl_2, Br_2, and I_2, at room temperature one is a solid, one is a liquid, and one is a gas. Which is which? On what basis did you decide?

For additional practice, work chapter problems, 11-6, 11-9, 11-14, and 11-17.

▶ **OBJECTIVE FOR SECTION 11-3**
Classify a solid as ionic, molecular, network, or metallic based on its physical properties.

11-3 The Solid State: Melting Point

LOOKING AHEAD! Heat is a measure of kinetic energy, the energy of motion. Therefore, heating a solid substance increases the motion of the molecules or ions. Eventually, the particles' motion will overcome the forces holding the molecules or ions in place and the lattice begins to come apart or melts. We will discuss different types of solids in this section and how intermolecular forces affect the temperature at which they melt. ■

On a hot summer afternoon, ice cream melts too fast and water evaporates quickly. In fact, we may all feel as if we are "melting." The process of melting is understandable from kinetic theory. As the temperature increases, the average kinetic energy of molecules increases, which means that they move faster and faster. The increased motion of the molecules eventually overcomes whatever forces are holding the solid or liquid molecules together, and a phase change occurs. It's like a feeling that most of us have experienced. Sometimes we just get too "fidgety" to stay seated and we have to move around.

Why do some solids melt at a certain temperature while others require a higher temperature to melt? The situation is analogous to what happens in an earthquake that begins slowly and intensifies. At first, only the flimsiest buildings collapse. As the earthquake intensifies, stronger and sturdier buildings are affected and collapse. The buildings with the strongest structure may collapse only in the strongest quake. Heating solids is much like subjecting them to an earthquake. A rising temperature causes the molecules or ions of a solid to vibrate more and more vigorously. Solids whose basic particles are not strongly attracted to each other are the first to come apart, so they melt or vaporize at the lower temperatures. At higher temperatures, solids whose particles have stronger attractive forces change to the liquid state. The stronger the forces holding the particles together, the higher the temperature needed to cause melting.

Aquamarine   Fluorite   Dolomite

FIGURE 11-7 Crystalline Solids The minerals shown here are crystalline solids that are ionic compounds. Their symmetrical shapes reflect the ordered geometry of the ions within.

11-3.1 Amorphous and Crystalline Solids

There are basically two general categories of solids: amorphous and crystalline. **Amorphous solids** *are so named because they have no defined shape.* The basic particles in amorphous solids are not located in any particular positions. Examples of amorphous solids are glass, rubber, and many plastics. (Although a pane of glass has a defined shape when it is shattered, no pieces are the same.) *In* **crystalline solids**, *the molecules or ions are arranged in a regular, symmetrical structure called a* **crystal lattice**. A salt crystal, a piece of quartz, and many minerals found on Earth naturally form solids with defined geometric patterns. These symmetrical shapes reflect the ordered arrangements of the molecules or ions that lie within the basic structure. (See Figure 11-7.)

In this section, we are interested in how heat causes a change from the solid to the liquid state. *The temperature at which a pure crystalline solid melts (the melting point) is a definite and constant physical property.* When pure crystalline solids melt, the added heat causes a phase change and the temperature remains constant. When amorphous solids change to the liquid state, a gradual softening occurs over a temperature range. Think of how glass or most plastics become more pliable as they are heated. There is, however, no one point where they transition between solid and liquid.

Different types of solids have different melting characteristics. There are four types of crystalline solids, and we will examine each individually.

11-3.2 Ionic Solids

Ionic solids *are crystalline solids in which ions are the basic particles making up the crystal lattice.* (See Figure 11-8a.) Forces between oppositely charged ions are quite strong, especially when compared to forces between individual molecules. The strong ion–ion forces result in solid compounds that have relatively high melting points. They can be as high as 3000°C (e.g., ZrN). As mentioned in Chapters 2 and 9, ionic

(a) An ionic solid

(b) A molecular solid

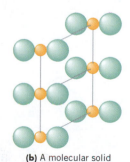

FIGURE 11-8 Ionic and Molecular Solids (a) In ionic compounds, ions occupy lattice points. (b) In molecular solids, molecules occupy lattice points.

compounds comprise much of the solid surface of Earth. Ionic compounds are also very hard and brittle (they shatter into pieces when struck).

11-3.3 Molecular Solids

In **molecular solids**, *the basic particles of the crystal lattice are individual molecules, which are held together by London forces and, in some cases, dipole–dipole attractions or hydrogen bonding.* (See Figure 11-8b.) These attractions are not nearly as strong as the ion–ion attractions found in ionic compounds. As a result, molecular solids have low melting points compared to ionic solids. These melting points range from very low for small, nonpolar molecules such as N_2 and CH_4 to well above room temperature for large molecules such as table sugar, where large molecules are attracted to each other by hydrogen bonds.

11-3.4 Network Solids

There are a few examples of solids where the atoms are covalently bonded throughout the entire sample of the solid. These are known as **network solids** and generally have very high melting points. Examples of such solids are the two familiar allotropes of carbon—*diamond* and *graphite*. (See Figure 11-9.) In diamond, each carbon is in the center of a tetrahedron and is bonded to four other carbons at the corners of the tetrahedron. These carbons are bonded to four other carbons and so forth throughout the entire crystal. The compact arrangement of atoms makes diamond the hardest material known. To melt diamond, a large number of covalent bonds must be broken, which requires a high temperature to supply the large amount of energy needed. The melting point of diamond is so high (over 4100°C) as to be difficult to establish.

FIGURE 11-9 The Allotropes of Carbon There are three allotropes of carbon—diamond, graphite, and buckminsterfullerene.

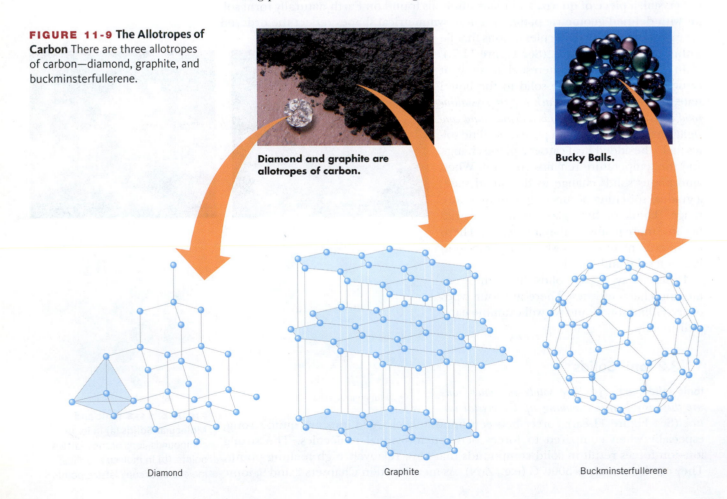

Diamond and graphite are allotropes of carbon.

Bucky Balls.

Diamond

Graphite

Buckminsterfullerene

It is hard to believe that an ordinary chunk of charcoal is the same element as a precious diamond. The charcoal is mostly composed of graphite, which is another allotrope of carbon. (Diamond is formed deep in Earth's mantle from graphite at extremely high temperatures and pressures.) In graphite, the carbons are arranged in parallel planar sheets. Since these sheets can slip past one another, graphite is used as a lubricant and as the "lead" in pencils. A third allotrope of carbon, buckminsterfullerene, was not identified until 1985 and is composed of carbon atoms bonded in spherical shapes somewhat like a soccer ball. It was named after the architect Buckminster Fuller, the designer of the geodesic dome that also reminds one of this allotrope. These "bucky balls," as they are sometimes called, come in a variety of sizes. The two most familiar have the formulas C_{60} and C_{70}. Although an important application of this discovery has not yet been forthcoming, we can expect to hear more about this form of carbon in the future. Graphite, like diamond, is a network solid and has a very high melting point, but buckminsterfullerene is a molecular solid.

MAKING IT REAL

The Melting Point of Iron and the World Trade Center

Iron has been the most important metal for thousands of years. Of all of the metals, how did iron gain this exalted position? First, iron ore is plentiful and widely available. Second, the reduction of the ore can be accomplished with charcoal, limestone, and a hot fire. But most importantly, it is a strong metal that can support many times its weight in other materials. All large structures like skyscrapers and dams are supported by iron (actually an alloy, known as steel) or concrete reinforced with iron bars.

But as was tragically illustrated in the destruction of the World Trade Center towers on September 11, 2001, it loses strength at high temperatures. The melting point of iron is 1535°C, which is quite high. However, changes in the internal structure of the metal begin around 900°C, and it begins to lose its strength. When the airplanes struck the towers, the strong steel beams, reinforced with concrete, did their job and withstood these powerful collisions. But large amounts of jet fuel ignited inside the buildings, raising the temperature to an estimated 1000°C. Eventually, the heat caused the metal to soften and collapse from the weight of the floors.

How can iron be protected from high temperatures? In the past, it was coated with asbestos.* Asbestos is a fibrous form of silica that occurs naturally in the earth. It has a high melting point and is a poor conductor of heat. Iron coated with asbestos or similar materials is effectively insulated from heat at least long enough for a typical fire to burn out. In the case of the World Trade Center, the iron was indeed covered by a protective coating, but the force of the collisions blew it away, exposing the metal to the extreme heat.

The space shuttle must also be protected from heat during its reentry into the atmosphere. As the shuttle encounters air at 17,000 mi/hr, the friction heats the bottom surface to 1650°C. This temperature would melt most metals, including iron. *Refractory ceramic* tiles are used in this case. A ceramic is a mineral or a mixture of minerals (e.g., quartz and clays) that have been mixed and heated to about 900°C. Dishes and vases are two of many ceramics used in our daily lives. Refractory ceramics have particularly high melting points (above 2000°C) and incorporate metal oxides such as aluminum oxide. Bricks of these ceramics protect the underside of the spacecraft, which is exposed to the heat.

Iron will continue to support our infrastructure. Research continues, however, on various ways to make it stronger and protect it from the effects of extreme heat.

*Asbestos coatings are no longer applied because they have been found to cause lung disease, including lung cancer, when the small fibers are inhaled over a period of time.

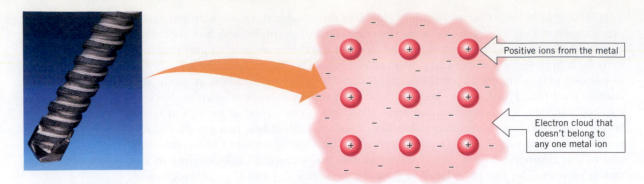

FIGURE 11-10 Metallic Solids In metals, positive ions occupy the lattice points.

Another example of a network solid is quartz. It is the major component of ordinary sand and also forms the backbone structure of several gemstones, such as amethyst. The formula of quartz, SiO_2, represents an empirical formula, since the oxygen atoms in this compound are each attached to two silicon atoms, forming a network throughout the crystal. Quartz also has a high melting point (1610°C).

11-3.5 Metallic Solids

Metals are also crystalline solids. As mentioned in Section 8-5, the outer electrons of metals are loosely held. *In **metallic solids**, the positive metal ions occupy regular positions in the crystal lattice, with the valence electrons moving freely among these positive ions.* (See Figure 11-10.) This is the reason metals are good conductors of electricity. All metals except mercury (melting point −39°C) are solids at room temperature. Some, like the alkali metals, are soft and melt at comparatively low temperatures. Others, such as iron, are hard and have high melting points. Tungsten has one of the highest melting points of any substance known (3380°C). Metals are generally ductile (can be drawn into wires) and malleable (can be pounded into sheets).

▶ **ASSESSING THE OBJECTIVE FOR SECTION 11-3**

EXERCISE 11-3(a) LEVEL 1: Fill in the blanks.
The two types of solids are _____ and _____. _____ solids have a definite melting point. Of the four types of crystalline solids, _____ solids are composed of ions and are always solids at room temperature. _____ solids have covalent bonds through-out the crystal. An allotrope of carbon that has a very high melting point is _____. Molecular solids are usually soft and melt at _____ temperatures. _____ solids have positive ions in a sea of _____.

EXERCISE 11-3(b) LEVEL 1: Identify the type(s) of solids that fit the following descriptions.
(a) easily malleable   (d) wide range of melting points
(b) generally low melting   (e) brittle
(c) high melting   (f) ductile

EXERCISE 11-3(c) LEVEL 2: Choose the one in the pair with the higher melting point.
(a) W or Fe   (c) SiO_2 or CO_2
(b) CaF_2 or SeF_2 (d) Sn or Hg

EXERCISE 11-3(d) LEVEL 3: Which type of molecular solid is each of the following chemicals? Estimate the order of increasing melting points among them: $H_2O(s)$, $C(s)$, $CS_2(s)$, KF, Sn.

For additional practice, work chapter problems 11-19, 11-20, 11-24, and 11-25.

KEY TERMS

11-2   Molecules in the condensed states are held together by **intermolecular forces**. p. 364

11-2.1 All molecules are attracted to each other by **London forces**, but in nonpolar compounds these act exclusively. p. 365

11-2.2 Polar compounds have **dipole–dipole attractions** in addition to London forces. p. 366

11-2.3 An interaction involving an electropositive hydrogen and highly electronegative atoms is known as **hydrogen bonding**. p. 366

11-3.1 The molecules in **amorphous solids** have no order, but the particles in **crystalline solids** exist in a **crystal lattice**. p. 369

11-3.2 The lattice positions in an **ionic solid** are occupied by ions. p. 369

11-3.3 The lattice positions in **molecular solids** are occupied by molecules. p. 370

11-3.4 The bonding in **network solids** extends throughout the entire crystal. p. 370

11-3.5 A **metallic solid** can be thought of as metal cations in a sea of electrons. p. 372

SUMMARY CHART

Solids and Forces

Type of Solid	Forces	Compounds	Examples
	London	All molecules	CCl_4, CO_2
Molecular	Dipole–dipole	Polar molecules	SO_2, CH_3Cl
	Hydrogen bonding	Mainly O—H and N—H	NH_2Cl, HOCl
Ionic	Ion–ion	Ionic compounds	NaCl, K_2SO_4
Network	Covalent bonds		SiO_2, diamond
Metallic	Cation–electron	Solid metals	Fe, Ni, Co

Part B

The Liquid State and Changes in State

SETTING A GOAL

■ You will learn about properties of the liquid state and examine how energy is associated with phase changes.

OBJECTIVES

11-4 Describe the relationship between intermolecular forces and the physical properties of the liquid state.

11-5 Given the appropriate heats of fusion and vaporization, calculate the energy required to melt and vaporize a given compound.

11-6 Calculate the energy absorbed or released as a substance progresses along a heating curve.

11-4 The Liquid State: Surface Tension, Viscosity, and Boiling Point

▶ **OBJECTIVE FOR SECTION 11-4**

Describe the relationship between intermolecular forces and the physical properties of the liquid state.

LOOKING AHEAD! All the free elements and compounds, even those that make up our atmosphere, can be found in the solid state if the temperature is low enough. As the temperature rises, these solids, one by one, melt to form the liquid state. We now shift our focus to some properties of the liquid state. ■

Intermediate between the complete disorder of molecules in the gaseous state and the highly ordered crystal lattice of the solid state lies the liquid state. Here, the basic particles are still held closely together, so liquids remain condensed as in the solid state. Unlike the solid state, however, the basic particles can move past one another. In this respect, the liquid state is like the gaseous state. At a lower temperature, liquids freeze to the solid state, and at some higher temperature, liquids vaporize to the gaseous state. Freezing point and boiling point are two properties of liquids that are dependent on their intermolecular forces. Other physical properties, too, are also related to these forces, such as surface tension and viscosity. We will discuss surface tension first.

11-4.1 Surface Tension

Have you ever noticed that certain insects can walk on water? Also, if one carefully sets a needle or a small metal grate on water, it remains on the surface despite the fact that the metal is much denser than water. (See Figure 11-11.) This tells us that there is some tendency for the surface of the water to stay together. **Surface tension** *is the force that causes the surface area of a liquid to contract.* Because of surface tension, drops of liquid are spherical. We can see why this happens. A molecule within the body of the liquid is equally attracted in all directions by the intermolecular forces. Molecules on the surface are attracted to the side and downward, but not upward. The unequal attraction results in surface molecules contracting, leading to a minimum amount of surface area. A liquid placed on a flat surface draws itself into a "bead," or into a sphere if it is suspended in space. Although raindrops are not completely spherical because of gravity, water released in the space shuttle forms perfectly spherical drops. The force that is required to break through a surface relates to its surface tension. Water obviously has a high enough surface tension to support small insects. Other liquids, where the intermolecular forces are greater, have higher surface tensions.

Dissolved soaps reduce the surface tension of water, thus allowing the surface to expand and "wet" clothes or skin rather than form beads. An insect would do well to avoid trying to walk on soapy water. Because of the reduced surface tension, the bug would sink. In this context, we think of "wetness" as the ability of a liquid to cover a surface. While we generally think of water or any liquid as being wet, some are wetter than others. If you've ever examined liquid mercury (hopefully not for very long), you would notice just how wet it is *not*. The strong metallic bonds, while not strong enough to make it a solid at room temperature like other metals, do give it a high surface tension.

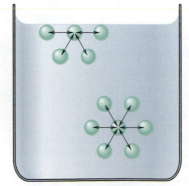

(a) Molecules on the surface are pulled down and to the sides.

(b) An insect can walk on water.

(c) The metal grid floats on the surface.

FIGURE 11-11 Surface Tension The unequal attraction for molecules on the surface accounts for surface tension.

11-4.2 Viscosity

We all know it seems to take forever to pour ketchup on french fries when we are hungry. Ketchup is more viscous than water. *The **viscosity** of a liquid is a measure of its resistance to flow.* Water and gasoline flow freely because they have low viscosity. Motor oil and syrup flow slowly because they have high viscosity. The viscosity of a liquid depends to some extent on the intermolecular forces between molecules. Strong intermolecular attractions usually mean a more viscous liquid. Water is an exception—even though its molecules interact relatively strongly, it has a comparatively low viscosity. Compounds with long, complex molecules tend to form viscous liquids because the molecules tangle together much like strands of spaghetti in a bowl.

A breakfast of pancakes with syrup on a cold morning is hard to beat. Unfortunately, the syrup barely moves when it is cold. A little warming of the syrup solves the problem. All liquids become less viscous as the temperature increases. The higher kinetic energy of the molecules counteracts the intermolecular forces, allowing the molecules to move past one another more easily. (See Figure 11-12.)

FIGURE 11-12 Viscosity Syrup is a viscous liquid.

11-4.3 Vapor Pressure

We all know that wet things eventually become dry. *Liquid changes to the gaseous state in a process known as **vaporization**. When vaporization occurs at temperatures lower than the boiling point, it is known as **evaporation**.* In order for a molecule of a liquid to escape to the vapor state, however, it must overcome the intermolecular forces attracting it to its neighbors in the liquid. Two conditions allow a molecule in a liquid to escape the liquid state to the gaseous state. First, it must be at or near the surface of the liquid. Second, it must have at least the minimum amount of kinetic energy needed to overcome the intermolecular forces. So why don't all molecules on the surface escape? To answer this, we must recall that at a given temperature, the molecules have a wide range of kinetic energies. The temperature relates to the *average* kinetic energy. The distribution of kinetic energies of the molecules at two temperatures (T_1 and T_2) can be represented graphically as in Figure 11-13. The unbroken vertical lines represent the average kinetic energy at each of these two temperatures. T_2 represents a higher temperature than T_1 because it has a higher average. The broken vertical line represents the minimum kinetic energy that a molecule must have in order to escape from the surface. Notice that a small fraction of molecules has the minimum energy at T_1, but a larger fraction of molecules has the minimum energy at the higher temperature, T_2. As a result, a liquid evaporates faster at a higher temperature.

If only the molecules with the highest kinetic energy escape to the vapor state, what effect does that have on the liquid left behind? The effect is similar to what would happen if the top 10% of the grades were left out when the average of the last chemistry test was computed. The remaining average would be lower. When the molecules with the highest kinetic energy escape, the average kinetic energy of the molecules remaining in the liquid state is lowered. This means that the liquid water will be cooled and the gas above the water will be correspondingly heated.

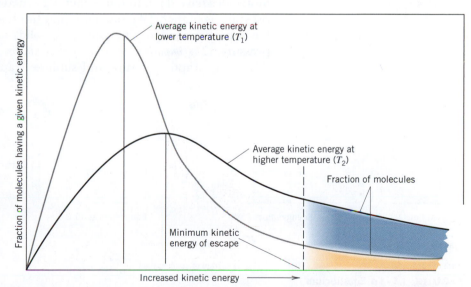

FIGURE 11-13 Distribution of Kinetic Energies at Two Temperatures The average kinetic energy is higher at T_2 than at T_1.

The cooling effect of evaporating water is important to health maintenance in warm climates. Perspiration covers our bodies with a layer of water when it is warm. The evaporation of this liquid cools the water left on our bodies and us along with it. The cool feeling after a hot shower is not just a feeling but a reality. Our perspiration cools us but can make life more miserable for the other people in a crowded room. Evaporation cools the liquid but heats the air. If water is allowed to evaporate under a vacuum, the evaporation process occurs faster. If the water cools enough, it freezes. Certain food products, such as some instant coffees, advertise that they are "freeze dried." These are made from a solution of coffee and then removing the water by evaporation, leaving coffee crystals. Boiling the coffee solution to remove the water presumably affects the taste. Thus, removing the water at lower temperatures (freeze drying) should preserve the flavor.

Now, instead of letting the water vapor escape to the surroundings, we can measure the buildup of pressure by placing a beaker of water in a closed glass container so that the vapor molecules are trapped. Attached to the apparatus is an open-ended mercury manometer for measuring the increase in pressure within the container. We will assume that initially the air is completely dry, meaning that any water vapor will come from our beaker of water. Before any water evaporates, the manometer indicates the same pressure inside and outside the container. As the water begins to evaporate, the additional molecules in the gas above the beaker cause the total pressure to increase (recall Dalton's law). The pressure increases rather rapidly at first but then increases more and more slowly until it remains steady. As the number of molecules in the gaseous state increases, some molecules collide with the surface of the liquid and return to that state. *The change of state from the gaseous to the liquid state is known as* **condensation**. As more liquid molecules enter the gaseous state, more gaseous molecules enter the liquid state. Eventually, a point is reached where the rate of evaporation equals the rate of condensation, and the pressure remains constant. That is, for every molecule that goes from the liquid to the gaseous state, a molecule goes from the gaseous to the liquid state. The system is said to have reached a point of *equilibrium*. (See Figure 11-14).

The height difference between the levels of the manometer at the dry state and when the pressure has reached equilibrium is the pressure exerted by the water vapor. *The pressure exerted by the vapor above its liquid at a given temperature is called its* **equilibrium vapor pressure** *or simply* **vapor pressure**. At a higher temperature, more molecules have the minimum energy needed to escape and, as a result, the vapor pressure is higher. (We say the liquid has become more *volatile*, meaning that it has a higher tendency to vaporize.) Solids may also have an equilibrium vapor pressure. *The vaporization of a solid is known as* **sublimation**. Dry ice (solid CO_2) has a high vapor pressure and sublimes rapidly. Ordinary ice also has a small but significant vapor pressure. A thin layer of snow will vaporize (i.e., sublime) away even if the temperature remains below the melting point of ice.

In the experiment described above, the vapor pressure at first increases rapidly but then slowly reaches the equilibrium vapor pressure. We notice this same effect on humid summer days. The closer the actual vapor pressure of water is to the equilibrium vapor pressure, the slower the rate of evaporation. The relative humidity measures how near the air is to saturation with vapor at a particular temperature. For example, if the relative humidity is 60% and the equilibrium vapor pressure of water at

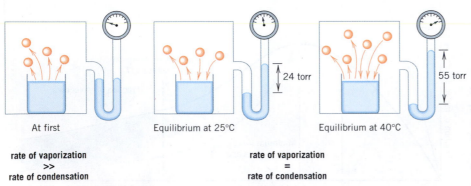

At first

rate of vaporization
>>
rate of condensation

Equilibrium at 25°C

24 torr

rate of vaporization
=
rate of condensation

Equilibrium at 40°C

55 torr

FIGURE 11-14 Equilibrium Vapor Pressure A given fraction of molecules escape to the vapor state above a liquid.

this temperature is 30 torr, the actual vapor pressure of water in the atmosphere is about 0.60×30 torr $= 18$ torr. The higher the humidity, the slower the evaporation and the less efficient is our personal air-conditioning system provided by perspiration. "It's not the heat, it's the humidity." In fact, both factors affect our comfort level. On the weather report, the combined effect of heat and humidity is called the *heat index*. This tells us how much hotter it actually feels because of the humidity.

How does the vapor pressure for various liquids at a given temperature relate to their intermolecular forces? In fact, molecules that are held more strongly to the surface will require greater energies to vaporize. Thus, at a given temperature a smaller percentage of surface molecules will have the required energy to escape, which in turn means a lower vapor pressure and a slower rate of evaporation. In summary, liquids with weak intermolecular forces have higher vapor pressures and liquids with stronger intermolecular forces have lower vapor pressures (i.e., are less volatile) at the same temperature. It is no wonder that the odor of gasoline can be found in a warm garage after a recent fill-up. The smell of acetone fills the bathroom if it is being used to remove fingernail polish. Both of these liquids have weak forces and relatively high vapor pressures at room temperature.

Solid naphthalene (moth balls) sublimes. The vapor kills moths.

11-4.4 The Boiling Point

The vapor pressure of liquids varies regularly as a function of temperature. This is represented graphically in Figure 11-15. The vapor pressure curves of water, diethyl ether, and ethyl alcohol are included. Notice that the vapor pressure of water at 100°C is equal to the pressure of the atmosphere (760 torr). *When the vapor pressure of a liquid equals the restraining pressure, bubbles of vapor form in the liquid and rise to the surface.* This is the **boiling point** of the liquid. The boiling point depends on the restraining pressure. *The* **normal boiling point** *of a liquid is the temperature at which the vapor pressure is equal to 760 torr (1 atm).* Diethyl ether and ethyl alcohol are more volatile than water, which means that their vapor pressures reach 1 atm at temperatures below that of water. They boil at around 34°C and 78°C, respectively, at 1 atm pressure. At other atmospheric pressures, the boiling points of liquids are different. For example, on Pikes Peak in Colorado, water boils at 86°C because the atmospheric pressure is only 450 torr. The highest point on Earth is the top of Mt. Everest, where water boils at 76°C. Cooking takes much longer at high elevations because of the lower boiling point and may not even kill all harmful bacteria. Appliances called pressure cookers retain some of the water vapor, which increases the pressure inside the cooker and increases the boiling point.

The temperature at which a liquid begins to boil also depends on the attractive forces between the basic particles. Ionic compounds naturally have very high boiling points and molecular compounds have much lower ones. Of the molecular compounds, those that are composed of large molecules and those that have hydrogen bonding have the highest boiling points. The effect of hydrogen bonding has a rather dramatic effect on boiling points. For example, consider the boiling points of the Group VIA hydrogen compounds: H_2O, H_2S, H_2Se, and H_2Te. The molar masses of these compounds increase in the order listed. Thus London forces should increase in the same order and, as a result, so should the intermolecular attractions. From this result alone, we may predict steadily higher boiling points for these compounds. Note in Table 11-1 that this is indeed the case for H_2S, H_2Se, and H_2Te. However, the boiling point of H_2O, by far the highest, is way out of line. The explanation for the unusually high boiling point of water is that hydrogen bonding provides significant intermolecular interactions. The other three compounds are polar, but the normal

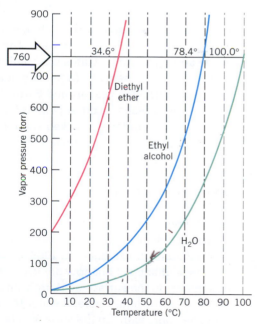

FIGURE 11-15 Vapor Pressure and Temperature The boiling points of these liquids are the temperatures at which the vapor pressures equal 760 torr.

TABLE 11-1

Boiling Points (°C) of Some Binary Hydrogen Compounds

GROUP	SECOND PERIOD		THIRD PERIOD		FOURTH PERIOD		FIFTH PERIOD	
VIIA	HF	17	HCl	−84	HBr	−70	HI	−37
VIA	H_2O	100	H_2S	−61	H_2Se	−42	H_2Te	−2.0
VA	NH_3	−33	PH_3	−88	AsH_3	−62	SbH_3	−18

dipole–dipole attractions that result do not seem to have a major effect on the boiling points compared with the trend in London forces. In Table 11-1, the boiling points of Group VA and Group VIIA hydrogen compounds are also listed. Hydrogen bonding explains the unusually high boiling points of NH_3 and HF compared with other hydrogen compounds in their group.

MAKING IT REAL

The Oceans of Mars

Water once flowed on Mars and maybe still does.*

This magnificent planet where we live exists at a perfect distance from the sun. We receive just the right amount of energy so that water can exist in the liquid state. Of course, it is solid in the colder regions, and the discomfort of a humid day tells us it is also present in the gaseous state. The exchange of water among the three states is a continual process powered by the energy from the sun. It also moderates and distributes heat around Earth.

Are there oceans on other planets? In the 1970s the scientific world was shocked when spacecraft sent back photos of what appeared to be dry riverbeds on the surface of our neighboring planet, Mars. Scientists have been intrigued and excited ever since about the possibility of oceans of water on Mars. The existence of liquid water implies the possibility that life may have existed there (and maybe still does). More recent visits, including that of the Phoenix polar lander in 2008, confirmed the existence of ancient but now dry lakebeds. Current conditions on the surface of Mars would not be suitable for the liquid state of water to exist. The atmospheric pressure on the surface is between 6 and 8 torr, which is about 1% of the pressure on Earth. At this low pressure, water would boil at about 5°C,

so the liquid range would be only from 0°C to 5°C. The temperature actually varies from about −75°C at the poles to around 10°C at the equator. Even if the liquid state existed at 3°C or 4°C, it would quickly evaporate in the thin atmosphere and freeze from the evaporative process.

Apparently, 1 to 2 billion years ago, the temperature and pressure could have been much like Earth's, meaning that liquid water could have existed in many places. What happened to all the water on Mars? For a long time, we thought it all must have escaped into space. Data reported in 2007 indicated that much of it could still be there. It is either frozen at the surface at its poles or lies in layers beneath the surface. Very recent evidence seems to suggest that some sort of liquid water, perhaps very acidic or briny, has flowed on the surface within our lifetimes. These chemicals would lower the melting point of the water so that, perhaps for brief periods, it could exist on the surface. We will know much more in the next few years as more space probes orbit the planet and some make soft landings.

Farther out in space lies a moon of Saturn known as Titan. As mentioned in the Setting the Stage in Chapter 3, the *Cassini* spacecraft showed a rugged terrain that appeared to be eroded by flowing liquids. It has a thick, hazy atmosphere where the cold temperature (−180°C) and pressure of 1.6 atm are just right for oceans of methane (CH_4). It is speculated that there may be weather patterns there similar to those on Earth, except that the rain in this plain is mainly methane.

* In 2011, photos from a Mars orbiter indicated that salty water flows down a slope in the Martian summer.

▶ **ASSESSING THE OBJECTIVE FOR SECTION 11-4**

EXERCISE 11-4(a) LEVEL 1: Fill in the blanks.
In the liquid state, _____ causes water to form spherical drops. _____ is a property relating to the rate of flow. When evaporation of a liquid occurs, the remaining liquid becomes _____. The equilibrium _____ ____ is the pressure exerted by the gas at a given _____. The vaporization of a solid is known as _____. The normal boiling point is the temperature at which the _____ _____ is equal to 760 torr.

EXERCISE 11-4(b) LEVEL 1: Do the following properties result from strong or weak intermolecular forces?
(**a**) a high boiling point   (**c**) a high surface tension   (**e**) a wetter liquid
(**b**) a low viscosity   (**d**) a high vapor pressure

EXERCISE 11-4(c) LEVEL 2: Refer to Figure 11-15 to answer the following.
(**a**) What is the boiling point of water at an altitude where the pressure is 550 torr?
(**b**) At what pressure will diethyl ether boil at 20°C?

EXERCISE 11-4(d) LEVEL 3: At a given temperature, substance A has a vapor pressure of 850 torr and substance B has a vapor pressure of 620 torr. What can you say about the physical states of the two substances at this temperature? What can you reasonably conclude about their respective viscosities and surface tensions?

EXERCISE 11-4(e) LEVEL 3: In Table 11-1, the boiling points of NH_3, H_2O, and HF are listed. All three substances have hydrogen bonding. Explain the order. Why is HF higher than NH_3? Why is H_2O dramatically larger than either?

For additional practice, work chapter problems 11-29, 11-32, 11-37, 11-38, and 11-41.

11-5 Energy and Changes in State

LOOKING AHEAD! The energy required to cause a phase change is specific for a particular substance. The amount of that energy is the subject of this section. We will see that the energy required to melt a solid or vaporize a liquid depends on the nature of the solid or liquid. ■

▶ **OBJECTIVE FOR SECTION 11-5**
Given the appropriate heats of fusion and vaporization, calculate the energy required to melt and vaporize a given compound.

Large lakes and oceans moderate the climate in a number of ways. When one of the American Great Lakes freezes in winter, heat is released to the surroundings by this process. This heat has the same effect as a giant natural furnace and helps keep the air temperature from falling as much as it otherwise would. Since the Great Lakes do not usually completely freeze over, heat is released all through the winter from the freezing process. On the downside, spring is delayed in this region because the melting ice absorbs heat and keeps the temperature from rising as much as it otherwise would. Regions of Siberia, in Russia, lie about as far north as Minnesota but have few large lakes. As a result, it is much colder there in the winter but also much hotter in the summer.

An ice cube at its melting point of 0°C remains in the solid state indefinitely unless energy is supplied. The addition of sufficient energy causes the solid to change to the liquid state yet remain at 0°C. The same is true for a pure liquid at its boiling point. That is, a specific amount of energy must be supplied to change the liquid to vapor at its boiling point. The opposite processes (i.e., fusion and condensation) *release* the same amount of energy as was absorbed in the reverse processes. (See Figure 11-16.)

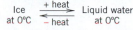
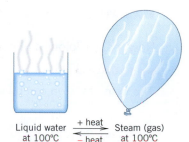

Ice at 0°C $\xrightarrow{\text{+ heat}}$ $\xleftarrow{\text{- heat}}$ Liquid water at 0°C

Liquid water at 100°C $\xrightarrow{\text{+ heat}}$ $\xleftarrow{\text{- heat}}$ Steam (gas) at 100°C

FIGURE 11-16 Heat and Changes in State Application or removal of heat causes a change in state at the melting point or boiling point of a substance.

TABLE 11-2

Heats of Fusion and Melting Points

COMPOUND	TYPE OF COMPOUND	HEAT OF FUSION (cal/g)	(J/g)	MELTING POINT (°C)
NaCl	ionic	124	519	801
H_2O	polar covalent (hydrogen bonding)	79.8	334	0
ethyl alcohol	polar covalent (hydrogen bonding)	24.9	104	−114
ethyl ether	polar covalent	22.2	92.5	−116
benzene	nonpolar covalent	30.4	127	5.5
carbon tetrachloride	nonpolar covalent	4.2	17.6	−24

We understand that the temperature at which a substance melts and boils depends on the forces between the basic particles. The amount of energy it takes to cause changes in physical state also depends on the magnitude of these forces. We will first consider the energy involved in transitions between the solid and the liquid states (melting and freezing) and then transitions between the liquid and the gaseous states (boiling and condensation).

11-5.1 Melting and Freezing

We are all aware that ice melts in the hot sun and that water changes to ice in the freezer. Melting [i.e., $H_2O(s) \longrightarrow H_2O(l)$] is an endothermic process—so heat must be supplied as in the hot sun. Freezing [i.e., $H_2O(l) \longrightarrow H_2O(s)$] is an exothermic process—so heat must be removed as in a freezer. The same amount of heat energy is released when a given amount of liquid freezes as would be required if the same amount of solid were to melt at the melting point. Each compound requires a specific amount of heat energy to melt a specific mass of sample. *The* **heat of fusion** *of a substance is the amount of heat in calories or joules required to melt one gram of the substance.* Table 11-2 lists the heats of fusion and the melting points of several substances. Note that sodium chloride, which is ionic, has the strongest attractions between particles of those listed and thus has the highest melting point and the highest heat of fusion.

Nonpolar compounds of low molar mass have comparatively small heats of fusion. Water, because of hydrogen bonds, has a rather high heat of fusion for such a light molecule. In fact, it is this relatively high heat of fusion that helps make water so effective at moderating climate, such as described for the Great Lakes in the winter.

EXAMPLE 11-1   Calculating the Heat Released by Freezing

How many kilojoules of heat are released when 185 g of water freezes?

PROCEDURE

The heat of fusion can be used as a conversion factor relating mass in grams to joules.

$$\text{mass} \times \text{heat of fusion} = \text{heat}$$

SOLUTION

$$185 \text{ g} \times \frac{334 \text{ J}}{\text{g}} \times \frac{1 \text{ kJ}}{10^3 \text{ J}} = \underline{61.8 \text{ kJ}}$$

11-5.2 Boiling and Condensation

If we want to boil water, we would naturally place the container of water on the stove or in the microwave. One would not be thinking very clearly if it was placed in the freezer to boil. The vaporization of a liquid [e.g., $H_2O(l) \longrightarrow H_2O(g)$] requires energy and so is also an endothermic process. The opposite process, condensation, is exothermic. As with melting and freezing, the same amount of energy is released when a given amount of vapor condenses [e.g., $H_2O(g) \longrightarrow H_2O(l)$] as would be required if the same amount of liquid were to vaporize at the boiling point. Each compound also requires a specific amount of heat energy to vaporize a specific mass of the sample. *The **heat of vaporization** of a substance is the amount of heat in calories or joules required to vaporize one gram of the substance.* The heats of vaporization and the boiling points of several substances are given in Table 11-3.

Note again that water has an unusually high heat of vaporization compared with other molecular compounds. Once again, the strength and number of hydrogen bonds between H_2O molecules are responsible.

The heats of fusion and vaporization are a result of how heat can cause phase changes at a constant temperature. When only one phase is present, the application of heat results in an increased temperature of the phase. All pure solids, liquids, and gases have a physical property that we previously defined as *specific heat.* (See Table 3-2 in Section 3-5.1.) *Specific heat refers to the amount of heat required to raise one gram of a substance one degree Celsius.* We observed that water also has an unusually high specific heat compared with other compounds or elements. The specific heats for the solid and gaseous states of water are not the same as for the liquid state, however.

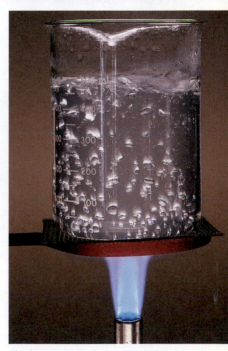

The heat required to vaporize water is comparatively large.

TABLE 11-3

Heats of Vaporization and Boiling Points

COMPOUND	TYPE OF COMPOUND	HEAT OF VAPORIZATION (cal/g)	(J/g)	NORMAL BOILING POINT (°C)
NaCl	ionic	3130	13,100	1465
H_2O	polar covalent (hydrogen bonding)	540	2260	100
ethyl alcohol	polar covalent (hydrogen bonding)	204	854	78.5
ethyl ether	polar covalent	89.6	375	34.6
benzene	nonpolar covalent	94	393	80
carbon tetrachloride	nonpolar covalent	46	192	76

EXAMPLE 11-2　Calculating the Heat Released by Condensation and Cooling

Steam causes more severe burns than an equal mass of water at the same temperature. Compare the heat released in calories when 3.00 g of water at 100°C cools to 60°C with the heat released when 3.00 g of steam at 100°C condenses and then cools to 60°C.

PROCEDURE

There are two processes to consider. The first is the heat released when the liquid is cooled from 100°C to 60°C, or a change of 40°C. The second is the heat released when the gas condenses at 100°C.

Use the specific heat of water as a conversion factor to convert mass and temperature change to calories.

$$\text{mass} \times \text{specific heat} \times \Delta T = \text{heat}$$

Use the heat of vaporization as a conversion factor to convert mass to calories.

$$\text{mass} \times \text{heat of vaporization} = \text{heat}$$

SOLUTION

Cooling of water:

$$3.00 \text{ g} \times (100 - 60)\,°C \times \frac{1.00 \text{ cal}}{g \cdot °C} = \underline{120 \text{ cal}}$$

Condensation of steam:

$$3.00 \text{ g} \times \frac{540 \text{ cal}}{g} = 1620 \text{ cal}$$

Total heat released from condensation and cooling $= 1620 + 120 = \underline{1740 \text{ cal}}$

ANALYSIS

Note that almost 15 times more heat energy is released by the steam at 100°C than by the water at the same temperature.

SYNTHESIS

There are many applications of the fact that steam at 100°C carries with it so much energy. Steam radiators allow steam to pass by its own power through buildings and carry large quantities of heat with it compared to hot water. Steamed vegetables cook quickly and thoroughly. Steam cleaning dislodges dirt that vacuuming could not remove. Initially, car manufacturers built steam-powered automobiles, which were sold until the internal combustion engine became popular. Can you think of other uses?

▶ ASSESSING THE OBJECTIVE FOR SECTION 11-5

EXERCISE 11-5(a) LEVEL 1: What are the units of heat of vaporization, heat of fusion, and specific heat?

EXERCISE 11-5(b) LEVEL 2: What mass of ethyl ether can be melted with 1.00 kJ of heat energy? (Refer to Table 11-2.)

EXERCISE 11-5(c) LEVEL 2: How much energy is required to vaporize 35.0 g of benzene? (Refer to Table 11-3.)

EXERCISE 11-5(d) LEVEL 3: Compare the values of the heats of fusion in Table 11-2 with the heats of vaporization in Table 11-3. Notice that the heats of vaporization are anywhere from 3 to 20 times higher than the heats of fusion. Use kinetic molecular theory to explain why this is so.

EXERCISE 11-5(e) LEVEL 3: Refrigerators cool because a liquid extracts heat when it is vaporized. Before the synthesis of Freon (CF_2Cl_2), ammonia (NH_3) was used. The heat of vaporization of NH_3 is 1.36 kJ/g and that of Freon is 161 J/g. Which is the better refrigerant on a mass basis? Which is the better refrigerant on a mole basis? What might be a reason we don't use ammonia today?

For additional practice, work chapter problems 11-47, 11-54, 11-55, and 11-56

MAKING IT REAL

Solar Energy and the Heat of Fusion

Wind and solar energy are the two most available sources of nonpolluting and renewable energy. Wind energy in the form of windmills is producing more and more energy but still supplies only a fraction of the power needs in the United States. Of course, the wind is temperamental and can't be counted on for steady generation. It can merely supplement conventional power sources. In the U.S. Southwest, steady sunlight is more dependable but obviously not available at night. So, how can we use these energy resources when the wind doesn't blow or the sun doesn't shine?

How many of us as amateur scientists burned a hole in a piece of paper with a magnifying glass and a beam of sunlight? As we may have discovered, focusing the sun's rays can produce an intense heat. Hopefully, no harm was done. This same principle is now being used as a renewable and environmentally friendly way to generate electricity.

A unit of a Spanish company, Abengoa Solar Inc., may have some answers. This company is planning to begin construction in 2011 of a solar energy facility at a site near Phoenix, Arizona, at a cost of about $2 billion. It will not use ordinary solar cells, which directly convert light energy into electrical energy. (They remain comparatively expensive.) As seen in the photo, huge numbers of solar panels simply reflect and concentrate light energy onto a central tower. The intense heat generated is more than enough to convert water into steam, which then runs a conventional turbine to generate electricity.

But how do they plan to generate electricity well into the nighttime hours? Enter ordinary table salt

(sodium chloride) and its very high heat of fusion. (See Table 11-2.) Excess heat generated during the day will be used to melt large quantities of salt at its melting point (801°C) and store the molten salt in an insulated building. After the sun sets, the molten salt is allowed to crystallize, liberating 519 J/g to be used to change water to steam and generate electricity as before. (Notice from Table 11-3 that it would take the crystallization of about four grams of salt to vaporize one gram of water.) The process is proven and very efficient. Given enough salt storage capacity it should be able to generate power for eight to ten hours after sundown. The ability to generate electricity during peak nighttime hours makes these expensive projects more practical. Expect more in the future.

The initial projects are very expensive and far away from the main energy users on the East and West Coasts. However, the use of molten salts to store energy for later use is one way to make alternative energy sources more practical.

11-6 Heating Curve of Water

▶ OBJECTIVE
FOR SECTION 11-6
Calculate the energy absorbed or released as a substance progresses along a heating curve.

LOOKING AHEAD! In the final section of this chapter, we will summarize what happens when heat is applied to a chunk of ice from the freezer so that it eventually changes from a cold solid to a hot vapor. The added heat will cause three temperature changes and two phase changes. As each of these changes occurs, we will try to visualize how the molecules of H_2O are affected. ∎

Assume that we take an ice cube cooled to −10°C from the freezer compartment of the refrigerator. Imagine that the ice crystal could be magnified so that the motions and positions of the individual molecules could be seen. We would notice

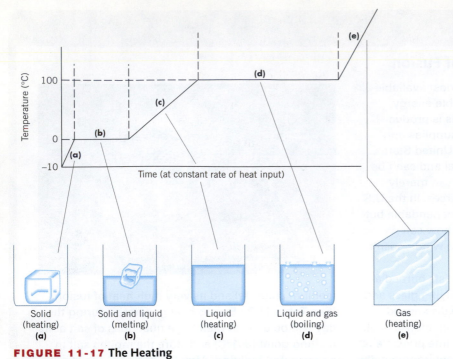

FIGURE 11-17 The Heating Curve of Water The heating curve illustrates the temperature changes by heating the solid state through the liquid state to the gaseous state.

that all the molecules are in fixed positions and that ice is a crystalline molecular solid. The molecules are certainly in motion, but the motion consists mainly of vibrations about their fixed locations. We are now going to supply heat at a constant rate to the ice cube and observe the changes that occur. *The graphical representation of the temperature as a solid is heated through the two phase changes plotted as a function of the time of heating is called the* **heating curve**. (See Figure 11-17. Note that it is really a series of straight lines despite being called a curve.) There are five regions of interest in the curve.

11-6.1 Heating Ice

The amount of heat energy to change the temperature of a pure substance such as ice is determined by its *specific heat*. Recall from Chapter 3 that the unit of specific heat is read as joules per gram per degree Celsius (or Kelvin) [i.e., $J/(g \cdot {}^{\circ}C)$]. Starting from $-10{}^{\circ}C$, the added heat in our sample of ice causes an increase in kinetic energy in the crystal in the form of increased vibrations and rotations of the molecules about the fixed positions in the crystal lattice. Since only one phase is present, the temperature of the ice rises corresponding to the increase in kinetic energy. Figure 11-17a represents the temperature change as the ice is heated.

11-6.2 Melting Ice

At $0{}^{\circ}C$ (and 1 atm pressure), a phase change begins. The vibrations of the molecules become so great that some hydrogen bonds break and the molecules begin to move out of their fixed positions in the crystal lattice. What we notice is that the solid ice begins to melt but the temperature of the ice–water mixture remains constant. Not all the hydrogen bonds break, however, so groups of molecules still stick together in the liquid state. When particles that attract each other move apart, there is an increase in the potential energy (analogous to suspending a weight above the ground or pulling magnets apart). Changes in potential energy do not relate to temperature, so the temperature remains constant while the ice is melting. A constant temperature as a solid melts is a property of pure matter. (See Figure 11-17b.) How long it takes a given solid to melt depends on the *heat of fusion*.

11-6.3 Heating Water

Between $0{}^{\circ}C$ and $100{}^{\circ}C$, only the liquid phase is present, so the added heat causes the temperature of the liquid to rise. Unlike in the solid state, the motion of water molecules in the liquid state includes translational motion (movement from point to point). The rate at which the temperature of a liquid rises depends on the *specific heat* of the liquid. Notice from the slope of the lines that the temperature does not appear to rise as fast in the liquid as in the solid. This is because the specific heat of water [$4.184 \, J/(g \cdot {}^{\circ}C)$] is about twice that of ice [$2.05 \, J/(g \cdot {}^{\circ}C)$]. The higher the specific heat, the slower is the rise in temperature. (See Figure 11-17c.)

11-6.4 Vaporizing Water

At 100°C (and 1 atm pressure), a second phase change begins. As the temperature of the liquid increases from 0°C to 100°C, the vapor pressure of the liquid has also been rising. At this temperature, the vapor pressure of water is equal to the atmospheric pressure, and the water begins to boil. The kinetic energy of the water molecules has now become great enough to break the remaining hydrogen bonds holding them in the liquid state. Just as in the melting process, the added heat is now causing molecules to move apart rather than increasing the kinetic energy of the molecules. As with melting, the added heat causes an increase in potential energy, and the temperature remains constant. How long it takes to vaporize a given liquid depends on the *heat of vaporization* of that liquid. Notice that it takes much longer to vaporize the water than to melt it. This is because the heat of vaporization of water ($2260 \, J/g \cdot °C$) is almost seven times that of the heat of fusion ($334 \, J/g \cdot °C$). This makes sense. The amount of energy needed to allow molecules to slide past one another (melting) should be just a fraction of the energy needed for those molecules to completely pull apart from one another (boiling). (See Figure 11-17d.)

11-6.5 Heating Vapor

Above 100°C, we again have only one phase, so the added heat causes the temperature of the gas to increase. In the gaseous state, the kinetic energy of the gas molecules is great enough to overcome all the interactions between molecules. As a result, water vapor acts as any other gas and is subject to the gas laws discussed in the previous chapter. (See Figure 11-17e.) The vapor absorbs heat at its own unique rate, different from that of the liquid or solid. In the case of steam, the *specific heat* is again about half of water's [$2.16 \, J/(g \cdot °C)$]

EXAMPLE 11-3   Calculating the Heat Involved in the Heating Curve

How many kilojoules of heat are required to convert 250 g of ice at −15°C to steam at 100°C?

PROCEDURE
There are four processes to consider. First (1) is the heating of the ice from −15°C to 0°C, or a total of 15°C. Second (2) is the melting of the ice at 0°C. Third (3) is the heating of the liquid from 0°C to 100°C, or a total of 100°C. And finally, fourth (4) is the vaporization of the liquid water to steam at 100°C. In summary, our processes are

heat ice (1), melt ice (2) heat liquid (3), vaporize liquid (4)

$$(1) \qquad (2) \qquad (3) \qquad (4)$$

$$-15°C \longrightarrow 0°C \longrightarrow 0°C \longrightarrow 100°C \longrightarrow 100°C$$

The heats of fusion and vaporization as well as the specific heats of ice and water (Table 3-2) are needed.

1. To calculate the heat required to heat the ice to 0°C, use the specific heat of ice to convert mass and temperature change to heat.

$$\text{mass} \times \Delta T \times \text{sp. heat (ice)} = \text{heat}$$

2. To calculate the heat required to melt the ice, use the heat of fusion to convert mass to heat.

$$\text{mass} \times \text{heat of fusion} = \text{heat}$$

3. To calculate the heat required to heat the water, use the specific heat of water to convert mass and temperature change to heat.

$$\text{mass} \times \Delta T \times \text{sp. heat (water)} = \text{heat}$$

4. To calculate the heat required to vaporize the water, use the heat of vaporization to convert mass to heat.

$$\text{mass} \times \text{heat of vaporization} = \text{heat}$$

SOLUTION

To heat ice from −15°C to its melting point (0°C), or a total of 15°C:

$$250 \text{ g} \times 15°C \times 2.06 \frac{J}{g \cdot °C} \times \frac{1 \text{ kJ}}{10^3 \text{ J}} = 7.7 \text{ kJ}$$

To melt the ice at 0°C:

$$250 \text{ g} \times 334 \frac{J}{g} \times \frac{1 \text{ kJ}}{10^3 \text{ J}} = 83.5 \text{ kJ}$$

To heat the water from 0°C to its boiling point (100°C), or a total of 100°C:

$$250 \text{ g} \times 100°C \times 4.18 \frac{J}{g \cdot °C} \times \frac{1 \text{ kJ}}{10^3 \text{ J}} = 105 \text{ kJ}$$

To vaporize the water at 100°C:

$$250 \text{ g} \times 2260 \frac{J}{g} \times \frac{1 \text{ kJ}}{10^3 \text{ J}} = 565 \text{ kJ}$$

$$\text{total} = 7.7 + 83.5 + 105 + 565 = \underline{761 \text{ kJ}}$$

ANALYSIS

The trouble that many students have with a problem of this type is the difficulty in viewing it as five separate calculations, although all have been solved previously. There's a tendency to look for one overriding equation where all variables can be plugged in simultaneously. But because the curve is not really a single curve, but five discrete lines, we *must* see it as five individual calculations, each one dictated by a different conversion factor. The heat changes of the single phases (solid, liquid, and gas) are determined by the specific heats of each phase. The phase changes themselves are determined by the heats of fusion and vaporization. Note also the units of each factor. Heats of fusion and vaporization are J/g, while specific heats are J/(g · °C). There is a temperature factor here because temperature changes. There is no temperature factor for fusion or vaporization because these processes occur at constant temperature.

SYNTHESIS

What would the heating curve of ethyl alcohol look like compared to that of water? The intermolecular forces in the alcohol are slightly weaker. Is the slope of the liquid line greater or smaller than in water? Are the lengths of the phase changes longer or shorter than in water? Would the phase changes occur at higher or lower temperatures?

▶ ASSESSING THE OBJECTIVE FOR SECTION 11-6

EXERCISE 11-6(a) LEVEL 1: Are the following processes endothermic or exothermic?
(a) melting   (b) boiling   (c) freezing   (d) evaporating   (e) condensing

EXERCISE 11-6(b) LEVEL 1: In which of the following does the potential energy of the molecules increase? In which does the kinetic energy increase?
(a) melting   (c) freezing   (e) condensing
(b) boiling   (d) heating a liquid   (f) cooling a solid

EXERCISE 11-6(c) LEVEL 2: Calculate the energy needed to melt 40.0 g of ice at 0°C and then raise its temperature to 50°C.

EXERCISE 11-6(d) LEVEL 3: How much water at 25°C can be frozen if 32.0 kJ is removed?

EXERCISE 11-6(e) LEVEL 3: If 75.0 g of water at 50°C is poured onto 50.0 g of ice at 0°C, will all the ice melt?

For additional practice, work chapter problems 11-70 and 11-76.

PART B SUMMARY

KEY TERMS

11-4.1	The measure of how difficult it is to break the surface of a liquid is its **surface tension**. p. 374
11-4.2	The measure of how rapidly a liquid flows is its **viscosity**. p. 375
11-4.3	When a liquid has an **equilibrium vapor pressure** it undergoes **vaporization** or **evaporation**. pp. 375–376
11-4.3	Gases undergo **condensation** to the liquid state. The solid state may undergo **sublimation** to the gaseous state. p. 376
11-4.4	The temperature at which the vapor pressure of a liquid equals the restraining pressure is the **boiling point**. The temperature at which the vapor pressure of the liquid is equal to one atmosphere is the **normal boiling point**. p. 377
11-5.1	A measure of the amount of energy needed to cause melting is the **heat of fusion**. p. 380
11-5.2	A measure of the amount of energy needed to cause vaporization is the **heat of vaporization**. p. 381
11-6	The **heating curve** of a compound traces the changes in phases and the changes in temperature of the phases. p. 384

SUMMARY CHART

What Happens When Heat is Applied to a Phase or Phases

Phase	What Happens	Energy Change	Temperature Change	Relevant Thermo-dynamic Equation
Solid	Particles move faster about fixed positions.	K. E.	*T* increases	Sp. heat of solid
Solid, liquid	Particles break attractions and move past each other.	P. E.	*T* constant	Heat of fusion
Liquid	Particles move faster, which includes translational motion.	K. E.	*T* increases	Sp. heat of liquid
Liquid, vapor	Particles break final attractions. as they escape to vapor	P. E.	*T* constant	Heat of vaporization
Vapor	Particles move faster in gaseous state.	K. E.	*T* increase	Sp. heat of gas

---

CHAPTER 11 SYNTHESIS PROBLEM

In this chapter, we have used as an example of properties and phase changes the most important liquid in our lives. That, of course, is water. In the problem that follows, we will use two compounds that are liquids at room temperature—acetone and isopropyl alcohol—as examples. Acetone (C_3H_6O) is a popular solvent used in the home to remove nail polish. Isopropyl alcohol (C_3H_8O) also finds use around the house as "rubbing alcohol" and as a solvent. We will use some important tools presented in Chapter 9 and this chapter to explain and predict some properties.

PROBLEM	SOLUTION
a. Write the Lewis structures of these two compounds. The basic skeletal structures of the two molecules are shown below. acetone: $$\text{H}_3\text{CCCH}_3$$ (with O double bonded above central C) isopropyl alcohol: $$\text{CH}_3$$ $$\text{CH}_3\text{CHOH}$$	**a.**
b. What is the geometry around the central C in acetone and the O in the alcohol?	**b.** In acetone the geometry around the central C is trigonal planar. In isopropyl alcohol the geometry around the O is V-shaped.
c. What intermolecular forces are present in the two compounds? How do these forces compare between the two compounds?	**c.** In acetone, there are London forces and dipole-dipole forces since the molecule is polar. In isopropyl alcohol there are London forces and hydrogen bonding. The hydrogen bonding occurs between the oxygen lone pair on one alcohol and a hydrogen attached to an oxygen on another molecule. In both cases the hydrogens bonded to carbons are not involved in hydrogen bonding because the C—H bond is essentially nonpolar. Since the molar masses of both compounds are nearly the same (i.e., 58 g/mol for acetone and 60 g/mol for the alcohol) the London forces are nearly the same. The hydrogen bonding in the alcohol, however, is considerably stronger than the dipole–dipole forces in acetone.
d. In their solid form, what type of solid would they be?	**d.** As they are each made of discrete molecules, they would form low-melting molecular solids.
e. The boiling points of the two liquids are about 56°C and 82°C. Assign the boiling points to the appropriate liquid. Which of the two compounds is the more volatile (i.e., evaporates more easily)?	**e.** Isopropyl alcohol has the higher boiling point of 82°C because of hydrogen bonding. Acetone is more volatile.
f. The heat of vaporization of acetone is 551 J/g and of isopropyl alcohol is 702 J/g. The specific heat of liquid acetone is 2.17 J/g · °C and of isopropyl alcohol is 2.58 J/g · °C. How many kJ of heat is required to change 40.0 g of each liquid originally at 25°C to form a vapor at its boiling point?	**f.** acetone: $56 - 25 = 31°C$ heating to the boiling point. $$40.0 \text{ g} \times 31°C \times \frac{2.17 \text{ J}}{\text{g} \cdot °C} \times \frac{1 \text{ kJ}}{10^3 \text{ J}} = 2.7 \text{ kJ (to heat the liquid)}$$ $$40.0 \text{ g} \times \frac{551 \text{ J}}{\text{g}} \times \frac{1 \text{ kJ}}{10^3 \text{ J}} = 22.0 \text{ kJ (to vaporize)}$$ Total heat $= 2.7 + 22.0 = \underline{24.7 \text{ kJ}}$ Isopropyl alcohol: $82 - 25 = 57°C$ heating $$40.0 \text{ g} \times 57°C \times \frac{2.58 \text{ J}}{\text{g} \cdot °C} \times \frac{1 \text{ kJ}}{10^3 \text{ J}} = 5.9 \text{ kJ (to heat the liquid)}$$ $$40.0 \text{ g} \times \frac{702 \text{ J}}{\text{g}} \times \frac{1 \text{ kJ}}{10^3 \text{ J}} = 28.1 \text{ kJ}$$ Total heat $= 5.9 + 28.1 = \underline{34.0 \text{ kJ}}$

YOUR TURN

In this problem, we will use as examples two other simple molecular compounds. The compounds are dimethyl ether and dimethyl amine.

    a. Write the Lewis structures of these two compounds using the basic skeletal geometry shown below.

    dimethyl ether: H_3COCH_3 dimethyl amine: H_3CNCH_3
                                                                                                          H

    b. What is the geometry around the O in the ether and the N in the amine?

    c. What intermolecular forces are present in each molecule and how do they compare?

d. The two boiling points are −25°C and 6.6°C. Which compound has the higher boiling point and why? What are the physical states of these two compounds at room temperature?

e. The heats of vaporization of the two compounds are 581 J/g and 492 J/g . Which compound has the higher heat of vaporization? How many kJ is required to vaporize 60.0 g of each compound at their boiling points?

Answers are on p. 391.

CHAPTER SUMMARY

This chapter gives the liquid and solid state their proper consideration. The most significant difference between these "condensed" states and the gaseous state is that cohesive forces hold the basic particles together and the basic particles occupy a significant part of the volume. Motions are confined to vibrations about fixed positions in the solid state or vibrational and translational motion in the liquid state. For this reason, solids and liquids are essentially incompressible, undergo little thermal expansion, and have definite volumes and high densities.

The temperature at which a molecular substance undergoes a phase transition (melting or boiling) depends on the **intermolecular forces** between molecules. **London forces** occur between all molecules but act exclusively for nonpolar molecules. These forces depend to a large extent on the size of the molecules. Compounds composed of heavy molecules generally have higher melting points than lighter ones. If a molecule has a permanent dipole, it has **dipole–dipole attractions** between molecules in addition to London forces. Compounds composed of polar molecules usually have higher melting and boiling points than nonpolar compounds of similar molar mass. **Hydrogen bonding** is a particular type of electrostatic interaction that has a significant effect on physical properties. It involves hydrogen bonded to an F, O, or N interacting with an electron pair on an F, O, or N on another molecule. Hydrogen bonding is responsible for many of the unusual properties of water.

We first examined the solid state of matter. Solids may be either **amorphous** or **crystalline**. Most solids are crystalline, which means they have atoms, ions, or molecules occupying fixed positions in a **crystal lattice**. They display sharp and definite melting points. There are four types of crystalline solids. **Ionic solids** are composed of ions as the basic particles in the crystal lattice. The ion–ion attractions are strong, so ionic compounds are solids at room temperature. **Molecular solids** are composed of molecules in the crystal lattice. The melting points of such solids depend on the intermolecular forces between molecules. Generally, they are soft solids melting at lower temperatures than ionic solids. **Network solids** are somewhat rare but generally melt at

high temperatures because covalent bonds must be broken to cause melting. Of the three allotropes of carbon, diamond and graphite melt at high temperatures. The third allotrope of carbon, buckminsterfullerene, is a molecular solid. **Metallic solids** have metal cations at the crystal lattice positions surrounded by a sea of electrons. These solids have melting points that range from low to very high temperatures.

The liquid state is intermediate between the solid and gaseous states. The cohesive forces between the basic particles in the liquid state are evident because of the liquid's **surface tension** and **viscosity**. Some molecules in the liquid state may overcome these cohesive forces and escape to the gaseous state. When the process of **vaporization** is exactly counteracted by the process of **condensation**, an **equilibrium vapor pressure** above the liquid is established. The **boiling point** of the liquid relates to its vapor pressure. The **normal boiling point** is the temperature at which the vapor pressure equals 1 atm. **Evaporation** occurs when the liquid vaporizes below the boiling point. A solid may also vaporize in a process known as **sublimation**.

Melting and boiling each take place at a specific temperature for most substances. It also takes a specific amount of energy to change one gram of solid to liquid, which is known as the **heat of fusion**, and a specific amount of energy to change one gram of liquid to vapor, which is known as the **heat of vaporization**. The magnitudes of these quantities depend on the forces between the basic particles of the solid or liquid. A substance with a high melting point usually has a high heat of fusion and a high heat of vaporization. Much of the information on phase transitions can be shown in a **heating curve**. The rate of heating of a single phase is dependent on the specific heat of that phase. The times required for a solid to melt and a liquid to boil at a given pressure depend on the magnitudes of the heat of fusion and vaporization, respectively. The heating curve of a pure substance shows that, when two phases are present during a phase change, the temperature remains constant. When melting or boiling is occurring, the applied heat energy is increasing the potential energy of the system rather than the kinetic energy.

OBJECTIVES

SECTION	YOU SHOULD BE ABLE TO...	EXAMPLES	EXERCISES	CHAPTER PROBLEMS
11-1	Use kinetic molecular theory to distinguish the physical properties of solids and liquids from gases.		1a, 1b	1, 2, 4
11-2	Describe the types and relative strengths of intermolecular forces that can occur between two molecules.		2a, 2b, 2c, 2d, 2e	6, 7, 8, 9, 12, 13, 14, 17
11-3	Classify a solid as ionic, molecular, network, or metallic based on its physical properties.		3a, 3b, 3c, 3d	19, 20, 21, 22, 24, 25
11-4	Describe the relationship between intermolecular forces and physical properties of liquid state.		4a, 4b, 4c, 4d, 4e	27, 29, 31, 32, 35, 37, 41, 42, 45, 46
11-5	Given the appropriate heats of fusion and vaporization, calculate the energy required to melt and vaporize a given compound.	11-1, 11-2	5a, 5b, 5c, 5d, 5e	52, 54, 56, 58, 62, 64
11-6	Calculate the energy absorbed or released as a substance progresses along a heating curve.	11-3	6a, 6b, 6c, 6d, 6e	66, 68, 70, 71, 72, 75, 77

►ANSWERS TO ASSESSING THE OBJECTIVES

Part A

EXERCISES

11-1(a) (a) gases (b) gases (c) both
(d) gases (e) gases (f) condensed phases (g) gases

11-1(b) Generally, solids have higher densities. The particles in solids are held together more tightly than in liquids. This implies that the forces between individual molecules in a solid are greater than those in a liquid.

11-2(a) (a) dipole–dipole force (b) London force (c) London force (d) hydrogen bonding (e) dipole–dipole force hydrogen bonding

11-2(b)

(a) BCl_3—London force only; NCl_3—Dipole–dipole force (and London force); $HNCl_2$—Hydrogen bonds (and London force) (b) $HNCl_2$ has the strongest force and so is most likely to be in a condensed phase. BCl_3 has the weakest force and so is the most likely to be gaseous.

11-2(c) C_2H_6 $(-88°C) < CH_3F$ $(-78°C) < CH_2F_2$ $(-52°C) < CH_3NH_2$ $(-6°C)$

11-2(d) The molar masses and polarities of each are as follows:

CF_4, 88 g/mol, nonpolar $< PF_3$, 88 g/mol, polar $< SiF_4$, 104 g/mol, nonpolar.

The largest molecule has the strongest force. Of the two similarly sized molecules, the polar one has the stronger force. This can be confirmed by their relative boiling points, $-128°C$, $-102°C$, and $-86°C$, respectively.

11-2(e) All three are nonpolar molecules. Therefore, the only intermolecular forces present are London forces. Based on molecular size, we'd predict that I_2 would have the strongest forces and Cl_2 would have the weakest. The stronger the force, the more likely a molecule will be found in a condensed phase, so I_2 is the solid, Br_2 is the liquid, and Cl_2 is the gas.

11-3(a) The two types of solids are <u>crystalline</u> and <u>amorphous</u>. Crystalline solids have a definite melting point. Of the four types of crystalline solids, <u>ionic</u> solids are composed of ions and are always solids at room temperature. <u>Network</u> solids have covalent bonds throughout the crystal. An allotrope of carbon that has a very high melting point is <u>diamond</u>. Molecular solids are usually soft and melt at <u>low</u> temperatures. <u>Metallic</u> solids have positive ions in a sea of <u>electrons</u>.

11-3(b) (a) metallic (b) molecular (c) ionic, network (d) metallic (e) ionic, network, molecular (f) metallic

11-3(c) (a) W (b) CaF_2 (c) SiO_2 (d) Sn

11-3(d) $H_2O(s)$ is a molecular solid. $C(s)$ is a network solid. $CS_2(s)$ is a molecular solid. KF is an ionic solid. Sn is a metallic solid.

The network and ionic solids should have the highest melting points, and the molecular solids should have the lowest. $H_2O(s)$ has hydrogen bonding, a stronger force than the London forces in $CS_2(s)$. Therefore, the order of increasing melting points should be $CS_2 < H_2O < Sn < KF < C$.

Part B

EXERCISES

11-4(a) In the liquid state, <u>surface tension</u> causes water to form spherical drops. <u>Viscosity</u> is a property relating to the rate of flow. When evaporation of a liquid occurs, the remaining liquid becomes <u>cooler</u>. The equilibrium <u>vapor pressure</u> is the pressure exerted by the gas at a given <u>temperature</u>. The vaporization of a solid is known as <u>sublimation.</u> The normal boiling point is the temperature at which the <u>equilibrium vapor pressure</u> is equal to 760 torr.

11-4(b) (a) strong (b) weak (c) strong (d) weak (e) weak

11-4(c) (a) about 92°C (b) about 450 torr

11-4(d) Substance A is a gas at that temperature. Substance B is a liquid. Because substance B has the stronger intermolecular forces, it has both a higher viscosity and a greater surface tension.

11-4(e) All three have similar molecular weights. HF is higher than NH_3 due to the stronger dipole. Water is the highest because of the larger number of hydrogen bonds it can make per molecule—two. HF is limited to one because of a single H, and NH_3 is limited to one because of only one lone pair of electrons.

11-5(a) Heat of vaporization and heat of fusion are in J/g. They can also be expressed in kJ/mol. Specific heat has units of $J/(g \cdot °C)$.

11-5(b) 10.8 g

11-5(c) 13.8 kJ

11-5(d) When a substance boils, all of the intermolecular forces between the particles must be fully broken, so that they may move individually and randomly. This requires a great deal of energy, particularly when the forces are strong. In order to melt the solid, just the rigid structure must be overcome, but the most of the intermolecular forces remain in place. This requires much less energy.

11-5(e) NH_3 has a higher heat of vaporization by mass and can therefore extract more heat than Freon. Converted into kJ/mol, it's closer (23.1 kJ/mol for NH_3 and 19.5 kJ/mol for CF_2Cl_2), but NH_3 is still the better refrigerant. However, NH_3 is noxious-smelling and toxic. A leak would be unpleasant and perhaps dangerous.

11-6(a) (a) endothermic (b) endothermic (c) exothermic (d) endothermic (e) exothermic

11-6(b) Melting (a) and boiling (b) lead to an increase in potential energy for the molecules. Heating a liquid (d) leads to an increase in kinetic energy of the molecules.

11-6(c) 13.4 kJ to melt and 8.4 kJ to heat up, for a total of 21.8 kJ

11-6(d) 72.9 g of water

11-6(e) It takes 16.7 kJ of energy to melt all the ice. The warm water releases 15.7 kJ of energy as it cools to 0°C. It would require 1.0 kJ more to melt all the ice. 3.0 g out of the original 50.0 g remains as ice.

ANSWERS TO CHAPTER SYNTHESIS PROBLEM

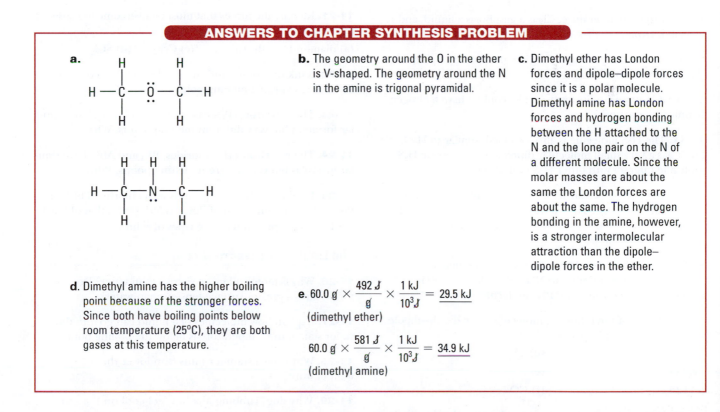

a.

b. The geometry around the O in the ether is V-shaped. The geometry around the N in the amine is trigonal pyramidal.

c. Dimethyl ether has London forces and dipole–dipole forces since it is a polar molecule. Dimethyl amine has London forces and hydrogen bonding between the H attached to the N and the lone pair on the N of a different molecule. Since the molar masses are about the same the London forces are about the same. The hydrogen bonding in the amine, however, is a stronger intermolecular attraction than the dipole–dipole forces in the ether.

d. Dimethyl amine has the higher boiling point because of the stronger forces. Since both have boiling points below room temperature (25°C), they are both gases at this temperature.

e. $60.0 \, \cancel{g} \times \dfrac{492 \, \cancel{J}}{\cancel{g}} \times \dfrac{1 \, kJ}{10^3 \, \cancel{J}} = \underline{29.5 \, kJ}$
(dimethyl ether)

$60.0 \, \cancel{g} \times \dfrac{581 \, \cancel{J}}{\cancel{g}} \times \dfrac{1 \, kJ}{10^3 \, \cancel{J}} = \underline{34.9 \, kJ}$
(dimethyl amine)

CHAPTER PROBLEMS

Throughout the text, answers to all exercises in color are given in Appendix E. The more difficult exercises are marked with an asterisk.

Nature of the Solid and Liquid States (SECTION 11-1)

11-1. Liquids mix more slowly than gases. Why?

11-2. Why is a gas compressible, whereas a liquid is not compressible?

11-3. Describe why a drop of food coloring in a glass of water slowly becomes evenly distributed without the need for stirring.

11-4. What properties do liquids have in common with solids? With gases?

11-5. Review the densities of the solids listed in Table 2-1. In general, which have higher densities—solids or liquids? Why is this so?

Intermolecular Forces (SECTION 11-2)

11-6. Arrange the following in order of increasing magnitude of the London forces between their molecules.

(a) CH_4 **(b)** CCl_4 **(c)** $GeCl_4$

11-7. Arrange the following in order of increasing magnitude of the London forces between their molecules or atoms.

(a) Ne   **(b)** Xe   **(c)** N_2 **(d)** SF_6

11-8. At room temperature, Cl_2 is a gas, Br_2 is a liquid, and I_2 is a solid. Explain the trend.

11-9. At room temperature, CO_2 is a gas and CS_2 is a liquid. Why is this reasonable?

11-10. If H_2O were a linear molecule, could it have hydrogen-bonding interactions?

11-11. The H_2S molecule is also V-shaped, similar to H_2O, but it has a very small molecular dipole. How does one H_2S molecule interact with other H_2S molecules?

11-12. Write the Lewis structure of NH_3, with the electron pair and the hydrogens in a tetrahedral arrangement. (See Figure 11-4a.) How does one NH_3 molecule interact with other NH_3 molecules?

11-13. The PH_3 molecule can be represented in a tetrahedral arrangement similar to NH_3. How does one PH_3 molecule interact with other PH_3 molecules?

11-14. Which of the following molecules have dipole–dipole attractions in a condensed state?

(a) HBr   **(d)** BF_3

(b) SO_2 **(e)** N_2

(c) CO_2 **(f)** CO

11-15. Which of the following molecules have dipole–dipole interactions in a condensed state?

(a) SCl_2 **(d)** O_3

(b) PH_3 **(e)** FCl

(c) CCl_4

11-16. What is the difference between a covalent bond and hydrogen bonding?

11-17. Which of the following molecules can have hydrogen bonding in the liquid state?

(a) HF   **(f)**

(b) NCl_3

(c) H_2NCl

(d) H_2O **(g)** CH_3Cl (tetrahedral)

(e) CH_4 (tetrahedral)

11-18. Which should have stronger hydrogen bonding, NH_3 or H_2O?

The Solid State (SECTION 11-2)

11-19. Identify the forces that must be overcome to cause melting in the following solids.

(a) KF   **(b)** HF   **(c)** HCl   **(d)** F_2

11-20. Rank the compounds in problem 11-19 in order of increasing expected melting points.

11-21. Identify the forces that must be overcome to cause melting in the following solids.

(a) diamond   **(b)** CF_4 **(c)** CrF_2 **(d)** SCl_2

11-22. Rank the compounds in problem 11-21 in order of increasing expected melting points.

11-23. The two Group IVA oxides, CO_2 and SiO_2, have similar formulas but very different melting points. Why?

11-24. The two Group IIA fluorides, BF_3 and AlF_3, have similar formulas but very different melting points. Why?

11-25. Lead forms two compounds with chlorine, $PbCl_2$ and $PbCl_4$. The melting point of $PbCl_2$ is 501°C and that of $PbCl_4$ is −15°C. Interpret in terms of types of solids.

The Liquid State (SECTION 11-4)

11-26. Why is motor oil more viscous than water? Does motor oil have a greater surface tension than water?

11-27. Explain how molecules of a liquid can go into the vapor state if the temperature is below the boiling point.

11-28. Why does a summer rainstorm lower the temperature?

11-29. Why does rubbing alcohol feel cool on the skin even if the alcohol is at room temperature when first applied?

11-30. Ethyl chloride boils at 12°C. When it is sprayed on the skin, it freezes a small part of the skin and thus serves as a local anesthetic. Explain how it cools the skin.

11-31. You have a sample of water at 90°C and a sample of water at 30°C. In which liquid does the temperature change at a faster rate when both are allowed to evaporate?

11-32. A beaker of ethyl ether with a vapor pressure of 550 torr at 25°C is set alongside a beaker of water and both are allowed to evaporate. In which liquid does the temperature change at a faster rate? Why?

11-33. What is implied by the word *equilibrium* in *equilibrium vapor pressure*?

11-34. What is the difference between boiling point and normal boiling point?

11-35. A liquid has a vapor pressure of 850 torr at 75°C. Is the substance a gas or a liquid at 75°C and 1 atm pressure?

11-36. If the atmospheric pressure is 500 torr, what are the approximate boiling points of water, ethyl alcohol, and ethyl ether? (Refer to Figure 11-15.)

11-37. How can the boiling point of a pure liquid be raised?

11-38. On top of Mt. Everest, the atmospheric pressure is about 260 torr. What is the boiling point of ethyl alcohol at that pressure? If the temperature is 10°C, is ethyl ether a gas or a liquid under conditions on Mt. Everest? (Refer to Figure 11-15.)

11-39. The boiling point of water in Death Valley, California, is about 100.2°C. Why is the actual boiling point higher than the normal boiling point?

11-40. Propane is used as a fuel to heat rural homes where natural gas pipelines are not available. It is stored as a liquid under normal temperature conditions, although its normal boiling point is −42°C. How can propane remain a liquid in a tank when the temperature is well above its normal boiling point?

11-41. The normal boiling point of neon is −246°C and that of argon is −186°C. What accounts for this order of boiling points?

11-42. The boiling point of HCl is −84°C and that of HBr is −70°C. Why is this order reasonable on one account but opposite from what one would expect from a consideration of polarity? Which trend is more important in this case?

11-43. On Mars the temperature can reach as high as a comfortable 50°F (10°C) at the equator. The atmospheric pressure is about 8 torr on Mars, however. Can liquid water exist on Mars under these conditions? What would the atmospheric pressure have to be before liquid water could exist at this temperature? What would happen to a glass of water set out on the surface of Mars under these conditions? (Refer to Figure 11-15.)

Energy and Changes of State (SECTION 11-6)

11-44. At 10°C, liquid hexane has a vapor pressure of 75 torr and water has a vapor pressure of 9 torr. Which liquid probably has the lower boiling point? Which probably has the lower heat of vaporization?

11-45. Ethyl ether $(C_2H_5OC_2H_5)$ and ethyl alcohol (C_2H_5OH) are both polar covalent molecules. What accounts for the considerably higher heat of vaporization for alcohol?

11-46. The heats of vaporization of liquid O_2, liquid Ne, and liquid CH_3OH are in the order Ne $<$ O_2 $<$ CH_3OH. Why?

11-47. A given compound has a heat of fusion of about 600 cal/g. Is it likely to have a comparatively high or low melting point?

11-48. A given compound has a boiling point of −75°C. Is it likely to have a comparatively high or low heat of vaporization?

11-49. A given compound has a boiling point of 845°C. Is it likely to have a comparatively high or low melting point?

11-50. Graph the data in Table 11-1 for the Group VIA hydrogen compounds. Plot the boiling points on the y-axis and the molar masses of the compounds on the x-axis. What would be the expected boiling point of H_2O if only London forces were important, as is the case with the other compounds in this series? (Determine from the graph.)

11-51. Graph the data in Table 11-1 for the Group VA hydrogen compounds. Plot the boiling points on the y-axis and the molar masses of the compounds on the x-axis. What would be the expected boiling point of NH_3 if only London forces were important, as is the case with the other compounds in this series?

11-52. How many kilojoules are required to vaporize 3.50 kg of H_2O at its boiling point? (Refer to Table 11-3.)

11-53. How many joules of heat are released when an 18.0-g sample of benzene condenses at its boiling point? (Refer to Table 11-3.)

11-54. Molten ionic compounds are used as a method to store heat. How many kilojoules of heat are released when 8.37 kg of NaCl solidifies at its melting point? (Refer to Table 11-2.)

11-55. If 850 J of heat is added to solid H_2O, NaCl, and benzene at their respective melting points, what mass of each is changed to a liquid? (Refer to Table 11-2.)

11-56. What mass of carbon tetrachloride can be vaporized by addition of 1.00 kJ of heat energy to the liquid at its boiling point? What mass of water would be vaporized by the same amount of heat? (Refer to Table 11-3.)

11-57. When a 25.0-g quantity of ethyl ether freezes, how many calories are liberated? When 25.0 g of water freezes, how many calories are liberated? In a large lake, which liquid would be more effective in modifying climate? How many joules of heat are required to melt 125 g of ethyl alcohol at its melting point? (Refer to Table 11-2.)

11-58. Air conditioners and refrigerators cool by vaporization of Freon. How many kcal of heat are removed by the vaporization of 1.00 kg of Freon? Will this be enough to freeze one can (325 mL) of soda? Assume the soda is essentially pure water at 0°C. (The heat of vaporization of Freon is 38.5 cal/g.)

11-59. Refrigerators cool because a liquid extracts heat when it is vaporized. Before the synthesis of Freon (CF_2Cl_2), ammonia was used. Freon is nontoxic, and NH_3 is a pungent and toxic gas. The heat of vaporization of NH_3 is 1.36 kJ/g and that of Freon is 161 J/g. How many joules can be extracted by the vaporization of 450 g of each of these compounds? Which is the better refrigerant on a mass basis?

11-60. The heat of vaporization of BCl_3 is 4.5 kcal/mol and that of PCl_3 is 7.2 kcal/mol. There are two reasons that account for this order. What are they?

11-61. How many joules of heat are released when 275 g of steam at 100.0°C is condensed and cooled to room temperature (25.0°C)? (Refer to Table 11-3.)

11-62. How many kilocalories of heat are required to melt 0.135 kg of ice and then heat the liquid water to 75.0°C? (Refer to Table 11-2.)

11-63. How many calories of heat are needed to heat 120 g of ethyl alcohol from 25.5°C to its boiling point and then vaporize the alcohol? [The specific heat of alcohol is 0.590 cal/(g · °C). Refer to Table 11-3.]

11-64. Rubbing alcohol (isopropyl alcohol) helps reduce fevers by cooling the skin when it is rubbed on. How much heat is removed by the vaporization of 25.0 mL of alcohol? (The density of the alcohol is 0.786 g/mL and the heat of vaporization is 705 cal/g.)

11-65. How many calories of heat are required to change 132 g of ice at −20.0°C to steam at 100.0°C? [The specific heat of ice is 0.492 cal/(g · °C). Refer to Tables 11-2 and 11-3.]

11-66. How many kilojoules of heat are released when 2.66 kg of steam at 100.0°C is condensed, cooled, frozen, and then cooled to −25.0°C? [The specific heat of ice is 2.06 J/(g · °C). Refer to Tables 11-2 and 11-3.]

*11-67. A sample of steam is condensed at 100.0°C and then cooled to 75.0°C. If 28.4 kJ of heat is released, what is the mass of the sample? (Refer to Table 11-3.)

*11-68. What mass of ice at 0°C can be changed into steam at 100°C by 2.00 kJ of heat? (Refer to Tables 11-2 and 11-3.)

*11-69. A 10.0-g sample of benzene is condensed from the vapor at its boiling point, and the liquid is allowed to cool. If 5000 J is released, what is the final temperature of the liquid benzene? [The specific heat of benzene is 1.72 J/(g · °C). Refer to Table 11-3.]

The Heating Curve (SECTION 11-6)

11-70. Which of the following processes are endothermic?

(a) freezing   **(c)** boiling

(b) melting   **(d)** condensation

11-71. Which has the higher kinetic energy: H_2O molecules in the form of ice at 0°C or in the form of liquid water at 0°C?

11-72. Which has the higher potential energy: H_2O molecules in the form of ice at 0°C or in the form of liquid water at 0°C?

11-73. Which has the higher potential energy: H_2O molecules in the form of steam at 100°C or in the form of liquid water at 100°C?

11-74. If water is boiling and the flame supplying the heat is turned up, does the water become hotter? What happens?

11-75. How many kilocalories are released when 18.0 g of steam at 100°C is condensed and cooled to ice at 0°C?

*11-76. Construct a heating curve for carbon tetrachloride. Refer to Tables 11-2 and 11-3. The specific heat of the liquid is 1.51 J/g °C. Assume that the specific heats of both the gas and solid are about half that of the liquid. How should the time of constant temperature for melting compare with that for boiling?

*11-77. Construct a heating curve for ethyl ether. Refer to Tables 11-2 and 11-3. How should the time of constant temperature for melting compare with that for boiling? (Assume that the specific heats of the three phases of water are about four times those of ethyl ether.)

General Problems

11-78. The following three compounds have similar molar masses: $C_2H_5NH_2$, CH_3OCH_3, and CO_2. The temperatures at which these compounds boil are −78°C, −25°C, and 17°C. Match the boiling point with the compound and give the respective intermolecular forces that account for this order.

11-79. At room temperature, SF_6 is a gas and SnO is a solid. Both have similar formula weights. What accounts for the difference in physical states of the two compounds?

11-80. At room temperature, CH_3OH is a liquid and H_2CO is a gas. Both are polar and have similar molar masses. What accounts for the difference in physical states of the two compounds?

11-81. CH_3F and CH_3OH have almost the same molar mass, and both are polar compounds. Yet CH_3OH boils at 65°C and CH_3F at −78°C. What accounts for the large difference?

11-82. The heat of fusion of gold is 15.3 cal/g and that of silver is 25.0 cal/g, yet it takes more heat to melt the same volume of gold as silver. Calculate the heat required to melt 10.0 mL each of silver and gold. (The density of silver is 10.5 g/mL and that of gold is 19.3 g/mL.)

11-83. SiH_4, PH_3, and H_2S melt at −185°C, −133°C, and −85°C, respectively. Since all have about the same molar mass, what accounts for the order?

11-84. The boiling point of F_2 is −188°C and that of Cl_2 is −34°C, yet the boiling point of HF is much higher than that of HCl. Explain.

11-85. Nitrogen gas and carbon monoxide have the same molar masses. Carbon monoxide boils at a slightly higher temperature, however (−191°C versus −196°C for N₂). Account for the difference.

11-86. Liquid sodium metal is used as a coolant in a certain type of experimental nuclear reactor. How many kilojoules does it take to heat 1.00 mol of sodium from a solid at its melting point of 98°C to vapor at its boiling point of 883°C? [For sodium, the heat of fusion is 113 J/g, the heat of vaporization is 3.90 kJ/g, and the specific heat of liquid sodium is 1.18 J/(g · °C).]

11-87. The heat of fusion of iron is 266 J/g. Iron is formed in industrial processes in the molten state and is solidified with water. The water vaporizes when it comes in contact with the liquid iron. What mass of water is needed to solidify or freeze 1.00 ton (2000 lb) of iron? (Assume that the water is originally at 25.0°C and that the steam remains at 100.0°C.)

*11-88. On a hot, humid day the relative humidity is 70% of saturation. If the temperature is 34°C, the vapor pressure of water at 100% of saturation is 39.0 torr. What mass of water is in each 100 L of air under these conditions?

11-89. Chromium metal, which is used to make stainless steel, is obtained by reaction of chromium (III) oxide with elemental aluminum. How much heat energy is required to melt the chromium produced from 1.25 kg of chromium (III) oxide? The heat of fusion of chromium is 21.0 kJ/mol.

11-90. The heat of fusion of chromium is 21.0 kJ/mol, molybdenum is 28.0 kJ/mol, and tungsten is 35.0 kJ/mol. What mass (in kg) of each of these elements can be melted by the same amount of heat required to change 1.00 kg of liquid water at 25°C to steam at 100°C?

11-91. A certain element melts at −39°C and boils at 357°C. The specific heat of the liquid is 0.139 J/g · °C. The element forms a +2 ion. What is the element? What mass of the element can be melted at −39°C, heated to its boiling point, and then vaporized at its boiling point by 15.00 kJ of heat? The heat of fusion of the element is 11.5 J/g, and the heat of vaporization is 29.5 J/g.

11-92. A common element has a melting point of 114°C. The element forms a −1 ion. What is the element? It has a vapor pressure of 90.5 torr at its melting point. If 50.0 g of the solid element were placed in a 1.00-L container at 114°C and 1.00 atm pressure, what percent of the element would eventually be in the vapor state?

11-93. Ice has a vapor pressure of 2.50 torr at −5°C. Will 50.0 g of ice completely vaporize if placed in a room at −5°C and 1.00 atm pressure with a volume 20.0 kL? Assume that the air in the room was originally dry.

11-94. Liquid air is composed mainly of N₂ and O₂. At −196°C, the vapor pressure of N₂ is 760 torr. Would you expect the vapor pressure of O₂ to be larger or smaller than 760 torr at this temperature? Would liquid O₂ boil at a higher or lower temperature than liquid N₂? Explain.

STUDENT WORKSHOP

Constructing a Heating Curve for Ethanol

Purpose: To plot the change of temperature due to the absorption of energy for a chemical as it changes phase from solid to liquid to gas. (Work in groups of three or four. Estimated time: 25 min.)

We will create a plot of temperature versus added energy in joules for a 1.0-g sample of ethanol over a temperature range of −180°C to 180°C. The relevant values for your calculations are:

sp. heat (solid) = 0.54 J/g · °C

sp. heat (liquid) = 2.44 J/g · °C

sp. heat (gas) = 0.81 J/g · °C

melting point = −114°C

boiling point = 78°C

heat of fusion = 104 J/g

heat of vaporization = 854 J/g

1. Calculate the following amounts of energy:

    (a) joules to heat 1.0 g of solid ethanol from −180°C to −114°C

    (b) joules to melt 1.0 g of solid ethanol

    (c) joules to heat 1.0 g of liquid ethanol from −114°C to 78°C

    (d) joules to boil 1.0 g of liquid ethanol

    (e) joules to heat 1.0 g of gaseous ethanol from 78°C to 180°C

2. On graph paper (turned sideways), scale the *x*-axis in joules from 0 to 1600 (this should be the approximate total energy required). Scale the *y*-axis from −180 to 180 (about 10°C per square should do).

3. Make your plot with five distinct straight lines:

    - Line 1: Joules increase by the amount in 1(a); temperature increases from −180°C to −114°C.
    - Line 2: Joules increase by the amount in 1(b); no temperature increase (a straight line).
    - Line 3: Joules increase by the amount in 1(c); temperature increases from −114°C to 78°C.
    - Line 4: Joules increase by the amount in 1(d); no temperature increase.
    - Line 5: Joules increase by the amount in 1(e); temperature increases from 78°C to 180°C.

4. Label the lines a, b, c, d, and e, respectively.

5. Answer the following questions:

    - Why is line d significantly longer than line b?
    - Why does line c slope more than lines a or e?
    - How does your graph compare to the graph in Figure 11-6? What are the similarities? What are the differences?
    - Over what range on your graph would you represent the energy needed to completely boil 1.0 g of ethanol that was originally at 25°C?

Chapter 12

Aqueous Solutions

The woman floats like a cork in the Dead Sea in Israel. The water in this lake contains high concentrations of solutes, so it is denser than pure water. The properties of aqueous solutions are discussed in this chapter.

SETTING THE STAGE

The sound and sensation that we feel from the rhythmic pounding of waves against a silvery ocean beach are among nature's most tranquil gifts. This endless body of liquid seems to extend forever past the horizon. Actually, the ocean covers two-thirds of the surface of Earth and in places it is nearly 6 miles deep. But we can't drink this water, nor can we use it to irrigate crops. This water is far from pure; it contains a high concentration of dissolved compounds as well as suspended matter. The blood coursing in our veins is an example of how our life also depends on this same ability of water to dissolve substances. This red waterway dissolves oxygen and nutrients for life processes and then carries away dissolved waste products for removal by the lungs and kidneys. In fact, water is indispensable for any life-form as we know it.

The ability of water to act as a solvent is the emphasis of this chapter. In the last chapter, we examined the unusual properties of this important compound. Most of the unique properties of water can be attributed to hydrogen bonding between molecules.

In Part A in this chapter, we will examine how water acts as such a versatile solvent for many ionic and polar covalent compounds. The importance of measuring the amount of solute present in a given amount of solvent and how this relates to the stoichiometry of reactions that take place in aqueous solution will be detailed. In Part B we will discuss how the presence of a solute changes the physical properties of a solvent.

Part A

Solutions and the Quantities Involved

SETTING A GOAL

- You will learn how water acts to dissolve specific types of compounds and how concentrations of solutes are expressed and used.

OBJECTIVES

12-1 Describe the forces that interact during the formation of aqueous solutions of ionic compounds, strong acids, and polar molecules.

12-2 Describe the effects of temperature and pressure on the solubility of solids and gases in a liquid.

12-3 Perform calculations of concentration involving percent by mass, ppm, and ppb.

12-4 Perform calculations of concentration involving molarity and the dilution of solution.

12-5 Perform calculations involving titrations and other solution stoichiometry.

▶ **OBJECTIVE FOR SECTION 12-1**
Describe the forces that interact during the formation of aqueous solutions of ionic compounds, strong acids, and polar molecules.

12-1 The Nature of Aqueous Solutions

LOOKING AHEAD! The composition of a solution is familiar territory. In Chapter 3, we defined a solution as a homogeneous mixture of a solute (that which dissolves) in a solvent (the medium that dissolves the solute—usually a liquid). In this section we will look into the process of how and why a solute is dispersed into a solvent. ■

12-1.1 Mixtures of Two Liquids

Consider two possibilities when two liquids are mixed. *If homogeneous mixing occurs when the solvent and solute are both liquids, we say that the two liquids are* **miscible**. *If two liquids do not mix to form a solution, we say that they are* **immiscible** and remain as a heterogeneous mixture with two liquid phases. Oil and water, for example, are immiscible and thus form a heterogeneous mixture with two visible phases. Alcohol and water are miscible and form a solution with one phase. This was previously discussed in Section 3-3.2 and illustrated in Figure 3-5. In Chapter 10, we discussed mixtures of gases, which are also homogeneous. Mixtures of gases are subject to the same gas laws as pure gases. Alloys are solids that are mainly homogeneous mixtures of metals.

12-1.2 The Formation of Aqueous Solutions of Ionic Compounds

Water is a fantastic solvent. It has a unique ability to dissolve many solids and liquids. (See Figure 12-1.) First, we will consider the aqueous solution of an ionic solid such as table salt, NaCl. It is a typical example of how ionic compounds dissolve in water. Recall that solid NaCl is composed of alternating Na^+ ions and Cl^- ions in a three-dimensional crystal lattice. It is held together by the rather strong electrostatic attractions known as ion-ion forces. Water is a polar molecular compound with positive dipoles located on the hydrogen atoms and negative dipoles located on the unshared pairs of electrons on the oxygen. In pure water, the molecules are moderately attracted to each other by hydrogen bonds. When solid NaCl is placed in water, competing forces develop between the ions and the dipoles of water. (See Figure 12-2.) *The interactions between the ions and dipoles of solvent molecules are referred to as* **ion–dipole forces**. An individual ion–dipole force is not as strong as one ion–ion force, but there are many ion–dipole forces at work. As a result, a tug-of-war

FIGURE 12-1 A Solution
The red-tinted solution is a homogeneous mixture of solute (the solid, red chromium compound) and solvent (water). The solution is in the same physical state as the solvent.

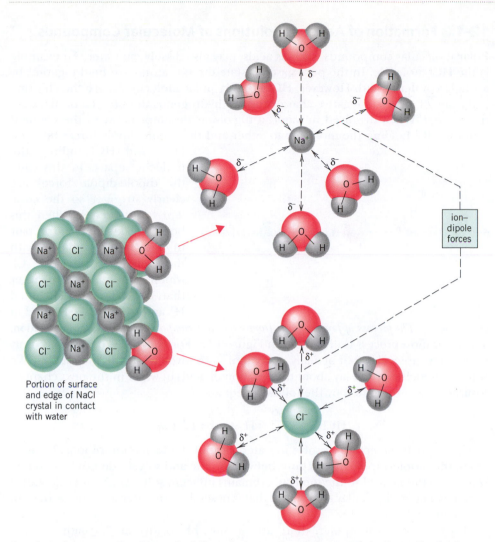

FIGURE 12-2 Interaction of Water and Ionic Compounds
There is an electrostatic interaction between the polar water molecules and the ions. This is an ion–dipole force.

ion–dipole forces

Na⁺ Cl⁻ Na⁺
Cl⁻ Na⁺ Cl⁻
Na⁺ Cl⁻ Na⁺
Cl⁻ Na⁺ Cl⁻

Portion of surface and edge of NaCl crystal in contact with water

develops between the forces holding the crystal together and the ion–dipole forces that are pulling the ions into solution. In the case of NaCl, the ion-dipole forces are stronger and the ions on the surface of the solid are separated from the rest of the crystal lattice and pulled into the aqueous medium. Since the ions in solution are now surrounded by water molecules, we say that the ions are *hydrated*. In solution, the positive and negative ions are no longer associated with one another. In fact, the charges on the ions are diminished or insulated from each other by the "escort" of hydrating water molecules around each ion. As before, we represent the solution of the ionic compound as going from the ions in the crystal lattice on the left to the hydrated ions on the right. Two examples of the solution of ionic compounds are as follows.

$$NaCl(s) \xrightarrow{H_2O} Na^+(aq) + Cl^-(aq)$$

$$K_2SO_4(s) \xrightarrow{H_2O} 2K^+(aq) + SO_4^{2-}(aq)$$

The H_2O (above the reaction arrow) represents a large, undetermined number of water molecules needed to break up the crystal and hydrate the ions. The (*aq*) after the symbol of each ion indicates that it is hydrated in the solution. Of course, not all ionic compounds dissolve to an appreciable extent in water. In these cases, the ion–ion forces are considerably stronger than the ion–dipole attractions, so the solid crystal remains largely intact when placed in water. For example, ancient marble (i.e., $CaCO_3$) sculptures have suffered the indignities of weather for centuries with little deterioration.

12-1.3 Formation of Aqueous Solutions of Molecular Compounds

Polar molecular compounds, such as acids, may also dissolve in water. An example is the HCl molecule. In the pure gaseous state the two atoms are held together by a single covalent bond. However, HCl is a very polar molecule where the chlorine is the site of a partial negative charge and the hydrogen is the site of a partial positive charge. When dissolved in water, a tug-of-war develops between the covalent bond in HCl holding the molecule together and the dipole-dipole forces between

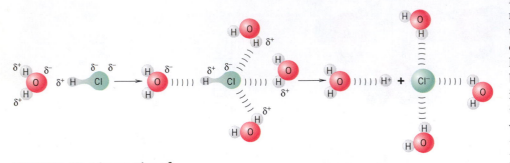

H_2O and HCl pulling the molecule apart. In this case, the dipole-dipole forces are clearly stronger so the covalent bond breaks. When this bond breaks the electron pair in the bond remains with the more electronegative Cl, which produces ions rather than neutral atoms. Hydrated H^+ and Cl^- ions are formed in

FIGURE 12-3 Interaction of HCl and H₂O The dipole–dipole interaction between H_2O and HCl leads to breaking of the HCl bond.

the process. *The process of forming ions from molecular compounds is known as* **ionization**. The ionization process is illustrated in Figure 12-3. For clarity, we have shown only four water molecules, but actually many more are involved in the interaction. We will have much more to say about the reaction of acids in water in the next chapter. Ionization is represented by the following equation.

$$HCl(g) \xrightarrow{\text{H}_2\text{O}} H^+ (aq) + Cl^- (aq)$$

Other polar molecules dissolve in water without the formation of ions. In these cases, the dipole–dipole interactions between solute and solvent do not lead to ionization, so the neutral solute molecule remains intact in solution. Methyl alcohol, a liquid, is a polar molecular compound that is miscible with water without ionization. (See Figure 12-4.)

If a molecule is nonpolar, there are no sites of positive and negative charge to interact with the dipoles of water. Since the water molecules are moderately attracted to one another by hydrogen bonding, there is no tendency for the water molecules to interact with a solute molecule. *As a result, most nonpolar compounds do not dissolve in polar solvents such as water.* Nonpolar compounds do tend to dissolve in nonpolar solvents such as benzene or carbon tetrachloride. This occurs because neither solvent nor solute molecules are held together by particularly strong forces, so they mix freely. For example, if we wish to dissolve a grease stain (i.e., grease is composed of nonpolar molecules), we know that plain water does not work but a nonpolar solvent such as gasoline does. Within chemistry circles, a well-known phrase sums up the tendency of one chemical to dissolve another: "Like dissolves like." This simply means that, in most cases, *polar solvents dissolve ionic or polar molecular compounds, and nonpolar solvents dissolve nonpolar compounds.*

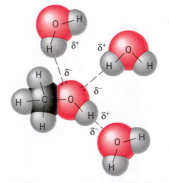

FIGURE 12-4 Methyl Alcohol in Water For some polar covalent molecules in water, there are only dipole–dipole attractions between solute and water molecules.

Why should we be concerned with the dissolving of substances? A great deal of time and effort are spent trying to find the right solvent for all sorts of chemical compounds in industry. Chemical reactions occur more readily when reactants are in solution. The close packing of molecules in a solid hinders the reacting molecules or ions from coming together. The proper solvent is needed so the solid can enter into the liquid phase and its molecules or ions mix thoroughly.

▶ **ASSESSING THE OBJECTIVE FOR SECTION 12-1**

EXERCISE 12-1(a) LEVEL 1: Are the following compounds soluble in water?
(a) NaI   **(b)** H_2SO_4 **(c)** CCl_4 **(d)** NH_2CH_3 **(e)** $FeCl_3$

EXERCISE 12-1(b) LEVEL 2: Write symbols for the aqueous form of each compound, including state symbols [for example, $Na^+ (aq) + Cl^- (aq)$]. All dissolve in water.

(a) $(NH_4)_2Cr_2O_7$ (b) HNO_3 (c) $CH_3-\overset{\overset{\displaystyle O}{\|}}{C}-CH_3$ (d) $ZnBr_2$

EXERCISE 12-1(c) LEVEL 2: What kinds of forces form between the following?
(a) water and $Zn(NO_3)_2$ (b) water and CH_3CH_2OH (c) CCl_4 and C_6H_{14}

EXERCISE 12-1(d) LEVEL 3: When a compound like $KF(s)$ dissolves in water, we say that it dissociates. When a compound like $HNO_3(l)$ dissolves in water, we say that it ionizes. What do these terms mean, and why is there a difference?

For additional practice, work chapter problems 12-1, 12-3, 12-4, and 12-6.

12-2 The Effects of Temperature and Pressure on Solubility

▶ **OBJECTIVE
FOR SECTION 12-2**
Describe the effects of temperature and pressure on solubility of solids and gases in a liquid.

LOOKING AHEAD! The terms used to define the relative amount of a substance that dissolves almost constitute a separate vocabulary. Besides the terms that we use, in this section we are also interested in how temperature affects the solubility of compounds. ∎

12-2.1 The Solubility of Compounds

The **solubility** *of a solute is defined as the maximum amount that dissolves in a given amount of solvent at a specified temperature.* The terms *soluble* and *insoluble* were introduced in Section 5-5.1. In fact, these terms are somewhat arbitrary since every compound has a maximum solubility and many compounds that we designate as insoluble actually do have a limited solubility. So there is no clear boundary between soluble and insoluble compounds.

Now consider the amount of solute that dissolves in a solvent. *When a defined amount of solvent contains the maximum amount of dissolved solute, the solution is considered to be* **saturated**. *If less than the maximum amount is present, the solution is* **unsaturated**. In certain unusual situations, *an unstable condition may exist in which there is actually more solute present in solution than its solubility would indicate. Such a solution is said to be* **supersaturated**. They are formed when saturated solutions are slowly cooled. Supersaturated solutions often shed the excess solute if a tiny "seed" crystal of solute is added or if the solution is shaken. The excess solute rapidly solidifies and collects at the bottom of the container as a precipitate.

Given that certain compounds are soluble in water, how do we express the **concentration**, *which is the amount of solute present in a given amount of solvent or solution?* Several units are available, and the one we choose depends on the application involved. Some units emphasize the mass of solute and some the number of particles (moles) of solute. Some relate the solute to the mass of solvent, the mass of the solution, or the volume of the solution. We will introduce these units later as they are applied.

The first concentration unit that we will discuss is used to compare solubilities of different compounds. This unit expresses the amount of solute present in a saturated solution as mass of solute present in 100 g of solvent (g solute/100 g H_2O in the case of aqueous solutions). The solubilities of several compounds in water are listed in Table 12-1. For example, note that 205 g of sugar dissolved in 100 g (100 mL) of water produces a saturated solution at 20°C. Also notice that $PbSO_4$, $Mg(OH)_2$, and AgCl have very low

TABLE 12-1

Solubilities of Compounds (at 20°C)

COMPOUND	SOLUBILITY (g SOLUTE/100 g H_2O)
sucrose (table sugar)	205
HCl	63
NaCl	36
KNO_3	28
$PbSO_4$	0.04
$Mg(OH)_2$	0.01
AgCl	1.9×10^{-4}

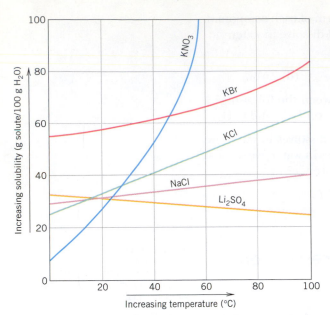

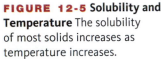

FIGURE 12-5 Solubility and Temperature The solubility of most solids increases as temperature increases.

solubilities and thus are usually considered as insoluble (or, more accurately, *sparingly soluble*). In Chapter 6, we indicated that whenever both ions involved in one of these insoluble compounds (e.g., Mg^{2+} and OH^-) are mixed from separate solutions, a precipitate of that compound forms [e.g., $Mg(OH)_2$]. The other compounds listed are considered *soluble*, although to various extents. Table 5-3 can be used as an aid in determining whether compounds of some common anions are considered soluble or generally insoluble.

12-2.2 The Effect of Temperature on the Solubility of Solids in Water

It is apparent that the nature of a solute affects its solubility in water. Temperature is also an important factor. From practical experience, most of us know that more sugar or salt dissolves in hot water than in cold. This is generally true. *Most solids and liquids are more soluble in water at higher temperatures.* In Figure 12-5, the solubilities of several ionic compounds are graphed as a function of temperature. Note that all except Li_2SO_4 are more soluble as the temperature increases. The information shown in Figure 12-5 has important laboratory implications. An impure solid can be purified by a process called **recrystallization**. In this procedure, a solution is saturated with the solute to be purified at a high temperature, such as the boiling point of the solution. Insoluble impurities are then filtered from the hot solution. As the solution is allowed to cool, the solvent can hold less and less solute. The excess solute precipitates from the solution as it cools (if it does not become supersaturated). This solid, now more pure, can then be filtered from the cold solution. The soluble impurities, as well as some of the original compound still in solution, pass through the filter. For example, if 100 g of water is saturated with KBr at its boiling point near 100°C, it contains 85 g of dissolved KBr. If the solution is cooled to 0°C in an ice bath, the water can now contain only 55 g of dissolved salt. The difference (i.e., 85 g − 55 g = 30 g) forms a precipitate. (See Figure 12-6.)

12-2.3 The Effect of Temperature on the Solubility of Gases in Water

Despite their nonpolar nature, many gases also dissolve in water to a small extent (e.g., O_2, N_2, and CO_2). Indeed, the presence of dissolved oxygen in water provides the means of life for fish and other aquatic animals. *Unlike solids, gases become less*

FIGURE 12-6 Recrystallization A hot, saturated solution of KBr is being prepared on the left. A precipitate has formed in the cold solution on the right because of its lower solubility.

soluble as the temperature increases. This can be witnessed by observing water being heated. As the temperature increases, the water tends to fizz somewhat as the dissolved gases are expelled. High temperatures in lakes can be a danger to aquatic animals and may cause fish kills. The lower solubility of oxygen at the higher temperatures can lead to an oxygen-depleted lake.

Fish maintain life by extracting dissolved oxygen from water.

12-2.4 The Effect of Pressure on Solubility

The solubility of solids and liquids in a liquid is not affected to any extent by the external pressure. However, this is not true of the solubility of gases in liquids. The solubility of a gas in a liquid relates to the partial pressure of that gas above the liquid. This is known as **Henry's law**, *which states that the solubility of a gas is proportional to the partial pressure of that gas above the solution.*

$$\text{solubility} = k\,P_{\text{gas}}$$

If you have ever opened a carbonated beverage, you have experienced Henry's law. A sealed can of soda contains carbon dioxide gas above the liquid under modest pressure. When the can is opened, the pressure of carbon dioxide above the liquid decreases as it escapes. The lowering of the partial pressure of the carbon dioxide results in the decreased solubility of carbon dioxide, and bubbles of CO_2

MAKING IT REAL

Hyperbaric Therapy

Hyperbaric cylinder.

A scuba diver needs to be very much aware of Henry's law. As a diver descends in the ocean, the pressure of the water increases. As a result, the diver must inhale air under high pressure from a tank. However, the high pressure of air that is inhaled increases the concentration of dissolved N_2 in the blood. If the diver comes to the surface too quickly (high pressure to low pressure), the lower pressure in the lungs causes N_2 to come out of solution and form bubbles of N_2 in the blood and cells. This is a condition known as *the bends* and can be fatal if the lungs are not quickly repressurized. A careful diver knows to return to the surface very slowly, pausing at specific intervals until the excess nitrogen is expelled from the blood.

Treatment of the bends requires that the patient be placed in a hyperbaric chamber as soon as possible. The person is completely enclosed in the chamber, where pressures of 2–3 atm of pure oxygen are applied. The higher pressure forces the nitrogen back into solution in the blood, and the pure oxygen speeds the purging of the system of nitrogen. The pressure in the chamber is slowly brought back to normal.

Hyperbaric chambers are also used in the treatment of carbon monoxide and cyanide poisoning. Both of these substances attach to the hemoglobin in the blood, thus blocking it from attaching to molecular oxygen in the lungs. In effect, the victim suffocates. The partial pressure of oxygen in the chamber is 10 to 15 times the normal partial pressure in the atmosphere (about 0.20 atm). Under these conditions, the concentration of dissolved oxygen in the blood is from 10 to 15 times normal. Oxygen is transported to the tissues in solution, thus bypassing the normal hemoglobin transport. The high concentration of oxygen in the blood also helps purge the system of carbon monoxide. Eventually, the patient is revived through blood transfusions or the regeneration of new red blood cells.

Hyperbaric treatment has a large variety of other uses, especially in treating hard-to-heal wounds, such as those found in diabetics, or wounds in remote areas, such as the ankles and feet. The high concentration of oxygen destroys bacteria, which thrive in a low-oxygen environment. This treatment is becoming more popular to treat more and more conditions that seem to respond to a high concentration of oxygen.

Bottled carbonated beverages fizz when the bottle is opened because the sudden drop in pressure causes a sudden drop in gas solubility. (Andy Washnik)

gas evolve from the solution. As time goes on, the carbon dioxide will continue to escape and the soda eventually goes flat. (Keeping an open can in a refrigerator will extend its carbonated life—see Section 12-2.3.) An understanding of Henry's law is also very important for scuba divers, as discussed in Making it Real, "Hyperbaric Therapy."

▶ **ASSESSING THE OBJECTIVE FOR SECTION 12-2**

EXERCISE 12-2(a) LEVEL 1: Fill in the blanks.
If a soluble compound dissolves to the limit of its solubility at a given temperature, the solution is _____; if the solute is not dissolved to the limit, the solution is _____. If more solute is present than indicated by its solubility, the solution is_____ and the excess solute may eventually form a _____. Solids generally become _____ soluble at higher temperatures, whereas gases become _____ soluble. An increase in partial pressure of a gas above a liquid _____ its solubility.

EXERCISE 12-2(b) LEVEL 1: Under what conditions of temperature and pressure will the following be most soluble?
(a) Cl_2 **(b)** Na_2SO_4

EXERCISE 12-2(c) LEVEL 2: Refer to Table 12-1. Describe the following as saturated, unsaturated, or supersaturated:
(a) 25.0 g of sugar in 10.0 g of H_2O
(b) 9.0 g of NaCl in 25.0 g of H_2O
(c) 12.0 g of KNO_3 in 50.0 g of H_2O

EXERCISE 12-2(d) LEVEL 3 Referring to Figure 12-5, describe each of the following solutions as saturated, unsaturated, or supersaturated. (All are in 100 g of H_2O.)
(a) 60 g of KBr at 20°C   **(c)** 50 g of KCl at 60°C
(b) 40 g of KNO_3 at 40°C   **(d)** 20 g of NaCl at 40°C

EXERCISE 12-2(e) LEVEL 3: While working in lab, you want to dissolve a large crystal (about 3 g in size) of $CaCl_2$. What types of things could you do to speed the dissolving process?

For additional practice, work chapter problems 12-9, 12-11, and 12-12.

▶ **OBJECTIVE FOR SECTION 12-3**
Perform calculations of concentration involving percent by mass, ppm and ppb.

12-3 Concentration: Percent by Mass

LOOKING AHEAD! A *concentrated* solution means that a large amount of a given solute is present in a given amount of solvent. A *dilute* solution means that comparatively little of the same solute is present. Obviously, we need more quantitative methods of expressing concentrations for laboratory situations. ■

In the previous discussion, the concentration unit (g solute/100 g solvent) related mass of solute to *solvent*. A second type of unit relates mass of solute to mass of *solution*, which contains both solute and solvent. **Percent by mass** *expresses the mass of solute per 100 grams of solution.* Consider 100 g of a solution that is 25% by mass HCl. There are 25 g of HCl and 75 g of H_2O present. The unit allows us to convert between mass of solute, mass of solvent, and mass of solution. We actually introduced mass percent in Section 3-3.3, which included Examples 3-5 and 3-6. However, we will also review the unit in this section and include an additional worked-out example as well as chapter-end practice problems. The formula for percent by mass is

$$\text{\% by mass (solute)} = \frac{\text{mass of solute}}{\text{mass of solute} + \text{mass of solvent}} \times 100\%$$

$$= \frac{\text{mass of solute}}{\text{mass of solution}} \times 100\%$$

EXAMPLE 12-1   Calculating the Mass of Solution from Mass Percent

A solution is 14.0% by mass H_2SO_4. There are 0.221 moles of H_2SO_4 in the solution. What is the mass of the solution?

PROCEDURE

1. Convert moles of H_2SO_4 to mass.

2. Convert mass of H_2SO_4 to mass of the solution.

SOLUTION

1. The molar mass is

$$2.016 \text{ g (H)} + 32.07 \text{ g (S)} + 64.00 \text{ g (O)} = 98.09 \text{ g/mol}$$

$$0.221 \text{ mol } H_2SO_4 \times \frac{98.09 \text{ g } H_2SO_4}{\text{mol } H_2SO_4} = 21.7 \text{ g } H_2SO_4$$

2. $21.7 \text{ g } H_2SO_4 \times \dfrac{100 \text{ g solution}}{14.0 \text{ g } H_2SO_4} = \underline{155 \text{ g solution}}$

ANALYSIS

Notice we are using the conversion factor

$$\frac{100 \text{ g solution}}{14.0 \text{ g solute}}$$

as we were given the mass of solute and were looking for the total mass of solution. What conversion factor would you use if the question asked for the mass of solvent instead? It would be

$$\frac{86.0 \text{ g solvent}}{14.0 \text{ g solute}}$$

SYNTHESIS

Percent solutions can also be described by the term "parts per hundred," or pph. In this example, 14 mass units out of every 100 mass units in the solution are due to H_2SO_4. The other 86 mass units are due to water molecules. As solutions become more and more dilute, and the mass percent of solute becomes smaller and smaller, other concentration units become necessary. Read on.

12-3.1 Parts per Million and Parts per Billion

Percent by mass is equivalent to *parts per hundred*. When concentrations are very low, however, two closely related units become more convenient. These units are **parts per million (ppm)** and **parts per billion (ppb)**. For example, the concentration of carbon dioxide in the atmosphere is currently 0.0380%. It is more conveniently represented as 380 ppm. (It is unfortunately increasing by about 3 ppm each year.) Dioxin, a synthetic chemical linked to cancer, is measured in the soil in units of ppb.

Parts per million is obtained by multiplying the ratio of the mass of solute to mass of solution by 10^6 ppm rather than 100%. Parts per billion is obtained by multiplying the same ratio by 10^9 ppb. For example, if a solution has a mass of 1.00 kg and contains only 3.0 mg of a solute, it has the following concentration in percent by mass, ppm, and ppb.

$$\frac{3.0 \times 10^{-3} \text{ g (solute)}}{1.0 \times 10^3 \text{ g (solution)}} \times 100\% = \underline{3.0 \times 10^{-4}\%}$$

$$\frac{3.0 \times 10^{-3} \text{ g}}{1.0 \times 10^3 \text{ g}} \times 10^6 \text{ ppm} = \underline{3.0 \text{ ppm}}$$

$$\frac{3.0 \times 10^{-3} \text{ g}}{1.0 \times 10^3 \text{ g}} \times 10^9 \text{ ppb} = \underline{3.0 \times 10^3 \text{ ppb}}$$

In this case, the most convenient expression of concentration is in units of ppm. In the examples cited, the units, ppm and ppb, refer to the mass of the solution. In the case of the concentration of CO_2 in the atmosphere, the ppm refers to 1 million particles (molecules of compounds or atoms of noble gases) in the atmosphere. Thus, there are currently 380 CO_2 molecules per million molecules.

EXAMPLE 12-2 **Calculating ppm and ppb of a Solution**

What is the concentration in ppm and ppb of a solution if it is $8.8 \times 10^{-5}\%$ by mass solute?

PROCEDURE

Find the mass of solute per gram of solution and multiply by 10^6 for ppm or 10^9 for ppb.

SOLUTION

$$\frac{8.8 \times 10^{-5}\%}{100\%} = \frac{8.8 \times 10^{-7} \text{ g solute}}{\text{g solution}}$$

$$\frac{8.8 \times 10^{-7} \text{ g solute}}{\text{g solution}} \times 10^6 \text{ ppm} = \underline{0.88 \text{ ppm}}$$

$$\frac{8.8 \times 10^{-7} \text{ g solute}}{\text{g solution}} \times 10^9 \text{ ppb} = \underline{880 \text{ ppb}}$$

ANALYSIS

The units of ppm and ppb can look quite imposing, but if you're comfortable using percents, then conversions become natural. The procedures are virtually identical. If you scored 80% on a test, and the test had 60 questions, you'd multiply $60 \times \frac{80\%}{100\%}$, to see how many you got right (48 questions). If you had 25 g of water that had an 880-ppb concentration of lead in it, you'd multiply $25g \times \frac{880 \text{ ppb}}{10^9 \text{ ppb}}$ to see the mass of toxin (2.2×10^{-5} g of Pb).

SYNTHESIS

Choosing the appropriate concentration unit (%, ppm, ppb, or even ppt!) is a choice left to the scientist to discern the most reasonable unit. People are most comfortable dealing with numbers that fall between 1 and 1000. Beyond that range, they begin to look too large and unwieldy. Once they become decimals, their significance diminishes with size. But in the 1 to 1000 range, we are mentally prepared to compare and react to them in comparison to other data. As a rule of thumb, then, most scientists, agencies, health professionals, and others presenting data will choose units so that the data fall comfortably in that range. With our choices above, concentrations of most things that we measure can be manipulated into one of those units.

▶ **ASSESSING THE OBJECTIVE FOR SECTION 12-3**

EXERCISE 12-3(a) LEVEL 1: What are the three ratios that can be written involving the masses of solute, solvent, and solution for an 8.0% aqueous solution of KBr?

EXERCISE 12-3(b) LEVEL 2: What is the percent by mass if 15.0 g of NH_4Cl is dissolved in 45.0 g of water?

EXERCISE 12-3(c) LEVEL 2: What mass of solvent is needed to make a 14.0% by mass solution using 25.0 g of solute?

EXERCISE 12-3(d) LEVEL 2: What mass of *solute* is required to make 300.0 g of a 6.0% solution?

EXERCISE 12-3(e) LEVEL 3: A sample of river water contained 2.50 mg of Ba^{2+} contaminant in a 10.0-kg sample. What is the amount of Ba^{2+} expressed in the following measurements?
(a) percent   **(b)** ppm   **(c)** ppb
Which of these is the most appropriate way to report the concentration?

EXERCISE 12-3(f) LEVEL 3: The EPA allows 15 ppb lead in drinking water. How many mg of lead are allowed in 50.0 kg of water?

For additional practice, work chapter problems 12-13, 12-16, 12-17, 12-18, and 12-19.

12-4 Concentration: Molarity

▶ OBJECTIVE
FOR SECTION 12-4
Perform calculations involving molarity and the dilution of solutions.

LOOKING AHEAD! In the laboratory, it is more convenient to measure the volume of a solution rather than its mass. In stoichiometry, recall that we are interested primarily in the mole relationships, so it is also important to have a unit that expresses the solute in moles rather than mass. Such a unit is discussed in this section. ∎

12-4.1 Calculations Involving Molarity

When a substance is in an aqueous solution, the most convenient way to measure the amount is by measuring the volume of the solution. Graduated cylinders, burets, or other readily available laboratory apparatus allow this to be a convenient measurement. **Molarity (M)** *is defined as the number of moles of solute* (n) *per volume in liters* (V) *of solution.* Although molarity may be shown without units, it is understood to have units of *mol/L.* This is expressed as follows in equation form.

$$M = \frac{n \text{ (moles of solute)}}{V \text{ (liters of solution)}}$$

The following examples illustrate the calculation of molarity and its use in determining the amount of solute in a specific solution.

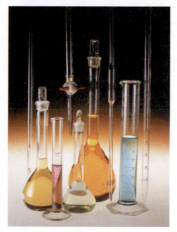

Solution volume is measured by burets, pipettes, and volumetric flasks.

EXAMPLE 12-3   Calculating Molarity

What is the molarity of H_2SO_4 in a solution if 49.0 g of H_2SO_4 is present in 250 mL of solution?

PROCEDURE

Write down the formula for molarity and what you have been given, and then solve for what's requested. Recall that the volume is expressed in liters in this calculation.

SOLUTION

$$M = \frac{n}{V}$$

$$n = 49.0 \text{ g } H_2SO_4 \times \frac{1 \text{ mol}}{98.09 \text{ g } H_2SO_4} = 0.500 \text{ mol}$$

$$V = 250 \text{ mL} \times \frac{10^{-3} \text{ L}}{\text{mL}} = 0.250 \text{ L}$$

$$\frac{n}{V} = \frac{0.500 \text{ mol}}{0.250 \text{ L}} = \underline{2.00 \text{ M}}$$

EXAMPLE 12-4 Calculating the Amount of Solute from Molarity

What mass of HCl is present in 155 mL of a 0.540-M solution?

PROCEDURE

Volume must be expressed in liters. Molarity has units of mol/L. A quantity in grams requires the molar mass of HCl to convert to moles.

SOLUTION

$$M = \frac{n}{V} \quad n = M \times V$$

$$M = 0.540 \text{ mol/L}$$

$$V = 155 \text{ mL} = 0.155 \text{ L}$$

$$n = 0.540 \text{ mol/L} \times 0.155 \text{ L} = 0.0837 \text{ mol HCl}$$

$$0.0837 \text{ mol HCl} \times \frac{36.46 \text{ g}}{\text{mol HCl}} = \underline{3.05 \text{ g HCl}}$$

EXAMPLE 12-5 Calculating Molarity from Percent by Mass

Concentrated laboratory acid is 35.0% by mass HCl and has a density of 1.18 g/mL. What is its molarity?

PROCEDURE

Since a volume was not given, you can start with any volume you want since this is an intensive property. The molarity will be the *same* for 1 mL as for 25 L. To make the problem as simple as possible, assume that you have exactly 1 L of solution ($V = 1.00$ L) and go from there. We just need to find the number of moles of HCl in that 1 L. This is obtained as follows.

1. Find the mass of 1 L from the density.
2. Find the mass of HCl in 1 L using the percent by mass and the mass of 1 L.
3. Convert the mass of HCl to moles of HCl.

SOLUTION

Assume that $V = 1.00$ L

1. The mass of 1.00 L (10^3 mL) is

$$10^3 \text{ mL} \times 1.18 \text{ g/mL} = 1180 \text{ g solution}$$

2. The mass of HCl in 1.00 L is

$$1180 \text{ g solution} \times \frac{35.0 \text{ g HCl}}{100 \text{ g solution}} = 413 \text{ g HCl}$$

3. The number of moles of HCl in 1.00 L is

$$431 \text{ g} \times \frac{1 \text{ mol}}{36.46 \text{ g}} = 11.3 \text{ mol HCl}$$

$$\frac{n}{V} = \frac{11.3 \text{ mol}}{1.00 \text{ L}} = \underline{\underline{11.3 \text{ M}}}$$

A student adds solvent with a pipette to a volumetric flask.

ANALYSIS

All the examples above are solved by dimensional analysis. If you've mastered that problem-solving technique, then by this time the calculations should be straightforward. Remember to write down all the given information and the units. Also, remember to express molarity as mol/L. Most importantly, remember that the liters in molarity are liters of solution, not solvent. This has implications for the formation of solutions, as we will see in future discussions.

SYNTHESIS

What if you were sent into the laboratory to make a specific molarity of solution? How would you do it? First, of course, you'd calculate the amount of solute you'd need for the required volume. Generally, the volumes we make are very specific, such as 100 mL, 250 mL, 500 mL, or 1000 mL. That is because we use volumetric flasks to ensure that our volumes are as accurate as possible. Then just mix the amount of solute in the required amount of solvent and stir? No! The solid itself takes up some volume. Until it has been made into a solution, we can't be sure how its volume will be accounted for. So you should transfer the solid into the flask as quantitatively as possible (use a funnel, rinse the weighing boat), and then fill up the flask halfway with solvent. This allows you to swirl until *all* the solute is dissolved. Then and only then can you fill the remainder of the flask with solvent. Toward the top, it is crucial that you do not exceed the line marked on the volumetric flask. You can always add more pure solvent, but you won't be able to remove it. Most scientists would use an eyedropper or disposable pipette to get the solvent exactly to the volume line. At this point, simply cap the flask and invert it several times to ensure thorough mixing, and your (very accurate) solution is made.

12-4.2 Dilution of Concentrated Solutions

From soft drinks to medicine, many products are transported as concentrated solutions. Before being sold, these products are carefully diluted to the desired concentration. Dilution is a straightforward procedure requiring a simple calculation. For example, in our laboratory situations, assume we are asked to prepare 200 mL of a 2.0-M solution from a large supply of a 6.0 M-HCl solution. We simply need to know what volume of the more concentrated solution is needed. (See Figure 12-7.)

In the following relationships, the variables of the dilute solution are designated M_d, V_d, and n_d. The variables of the concentrated solution are designated M_c, V_c, and n_c. It is important to understand that in a dilution process the moles of solute present in the dilute solution will be equal to those transferred from the concentrated solution. Thus we have the relationship

$$\text{moles solute} = n_d = n_c$$

FIGURE 12-7 Dilution of Concentrated HCl (*Note:* Never add water directly to concentrated acid, because it may splatter and cause severe burns.)

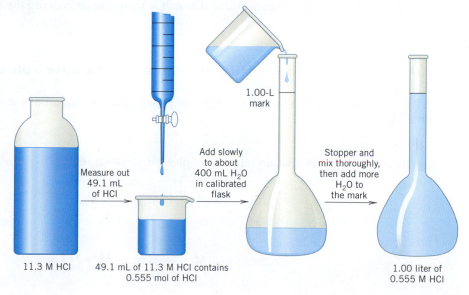

11.3 M HCl

Measure out 49.1 mL of HCl

49.1 mL of 11.3 M HCl contains 0.555 mol of HCl

Add slowly to about 400 mL H₂O in calibrated flask

1.00-L mark

Stopper and mix thoroughly, then add more H₂O to the mark

1.00 liter of 0.555 M HCl

Recall that $M_d \times V_d = n_d$ and $M_c \times V_c = n_c$ Since n_c and n_d are the same quantity, we can construct a simple equation relating the volume and molarity of the dilute solution to the volume and molarity of the concentrated solution:

$$M_c \times V_c = M_d \times V_d$$

EXAMPLE 12-6   Calculating the Volume of a Concentrated Solution

What volume of 11.3 M HCl must be mixed with water to make 1.00 L of 0.555 M HCl? (See also Figure 12-7.)

PROCEDURE

Recall that the same number of moles of HCl will be measured from the concentrated solution as needed in the dilute solution. Thus we can simply rearrange the dilution equation to solve for the requested variable.

$$M_c \times V_c = M_d \times V_d \qquad V_c = \frac{M_d \times V_d}{M_c}$$

SOLUTION

$$M_c = 11.3 \text{ M} \qquad M_d = 0.555 \text{ M}$$

$$V_c = ? \qquad V_d = 1.00 \text{ L}$$

$$V_c = \frac{0.555 \text{ mol/L} \times 1.00 \text{ L}}{11.3 \text{ mol/L}} = 0.0491 \text{ L} = \underline{49.1 \text{ mL}}$$

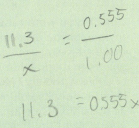

ANALYSIS

We have the same issue here that we had when we were making solutions from solutes directly. The total volume calculated is the volume of the solution. So, in this problem, 49.1 mL of HCl would be measured out (probably with a buret) and then added to a 1.0-L volumetric flask. The volume of water added would then be the amount necessary to bring the total volume up to 1.0 L. This will, of course, be very close to 950.9 mL, but because volumes are not necessarily additive, the exact amount may be a little off from this value. Filling up to the 1.0-L line on the volumetric flask assures that the proper amount is added.

SYNTHESIS

This problem illustrates a concept that you learned in lab: When diluting an acid (or other concentrated compound), do you pour the concentrated solution into the water or pour the water into the concentrated solution? The catchphrase is "add acid." Diluting acids is often very exothermic, and splattering may occur. The proper procedure of pouring the concentrated solution into the solvent ensures that any splatter is from a more dilute solution, rather than a more concentrated one; it is therefore considered the safer route.

EXAMPLE 12-7   Calculating the Molarity of Dilute Solution

What is the molarity of a solution of KCl that is prepared by dilution of 855 mL of a 0.475 M solution to a volume of 1.25 L?

PROCEDURE

Again, we can use the dilution equation. Solve the algebraic equation for M_d.

$$M_c \times V_c = M_d \times V_d \qquad M_d = \frac{M_c \times V_c}{V_d}$$

SOLUTION

$$M_c = 0.475 \text{ M} \qquad M_d = ?$$

$$V_c = 855 \text{ mL} = 0.855 \text{ L} \qquad V_d = 1.25 \text{ L}$$

$$M_d = \frac{0.475 \text{ M} \times 0.855 \text{ L}}{1.25 \text{ L}} = \underline{0.325 \text{ M}}$$

ANALYSIS

The key concept in working dilution problems is that while the volume of solution goes up, and the concentration of solution goes down, the overall number of moles of solute never changes. The same number of moles are present before the dilution and afterward. Though it goes beyond the scope of this text, this concept can be stretched to allow calculations when the dilution is not occurring with water but instead with a more dilute solution of the compound. For instance, if you wanted to turn your 1.0 M NH_3 into 3.0 M NH_3 by adding 9.0 M NH_3, determining the total number of moles before and after the dilution, and then converting into the $M \times V$ relationship, would allow you to solve the problem.

SYNTHESIS

Dilution is a process that occurs frequently in lab. Why so prevalent? Because manufacturers find that it is more convenient to produce and ship solutions that are concentrated. It makes a more convenient use of space and weight and costs less. Furthermore, end users can use the dilution formula to scale-down the concentrations to whatever value is needed for their particular use. We can always dilute, but without the original solute, we can't make commercial solutions more concentrated. As a result, acids are sold at very high concentrations (18 M for H_2SO_4, 12 M for HCl). Hence, there are numerous safety procedures in place whenever we deal with commercial bottles.

▶ **ASSESSING THE OBJECTIVE FOR SECTION 12-4**

EXERCISE 12-4(a) LEVEL 1: What is the concentration if 0.500 mol of NaOH is present in 0.250 L of solution?

EXERCISE 12-4(b) LEVEL 1: What is the concentration if 1.00 L of 0.400 M HCl is diluted to 2.00 L?

EXERCISE 12-4(c) LEVEL 2: What mass of KI is needed to make 500 mL of 0.0750 M solution?

EXERCISE 12-4(d) LEVEL 2: What volume of water must you *add* to 50.0 ml of 2.00 M $Cu(NO_3)_2$ solution to dilute it to 0.800 M? (Assume volumes are additive.)

EXERCISE 12-4(e) LEVEL 2: What volume of 6.00 M $H_2SO_4(aq)$ is needed to prepare 250 mL of 0.400 M $H_2SO_4(aq)$?

EXERCISE 12-4(f) LEVEL 3: Describe how you would make 250.0 mL of 0.500 M NaCl solution using the following:

(a) solid NaCl   **(b)** a 3.00 M solution of NaCl(aq)

EXERCISE 12-4(g) LEVEL3: What is the molarity of a 2.0% KCl solution? (Assume the density of the solution is 1.0 g/ml.)

For additional practice, work chapter problems 12-23, 12-27, and 12-30.

12-5 Stoichiometry Involving Solutions

▶ **OBJECTIVE FOR SECTION 12-5**
Perform calculations involving titrations and other solution stoichiometry.

LOOKING AHEAD! In Chapter 5, we related the moles of a substance to its molar mass and the number of molecules. The volume of a gas at a specific temperature and pressure also relates to moles, so gases could be included in the general stoichiometry scheme in Chapter 10. Since molarity also relates to moles, solutions can be now be added to our general stoichiometry scheme as discussed in this section. ■

FIGURE 12-8 General Procedure for Stoichiometry Problems

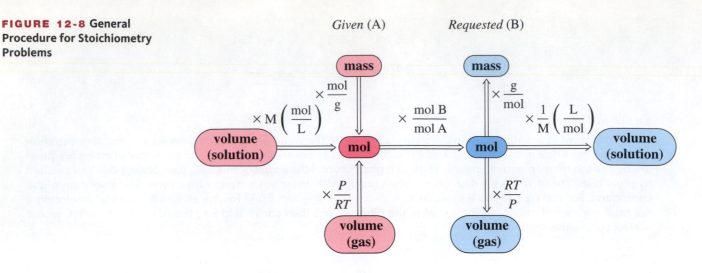

Given (A)   *Requested (B)*

12-5.1 Stoichiometry and Molarity

It is obvious by now that a large number of important chemical reactions take place in aqueous solution. With the use of molarity, the volume of a solution provides a direct path to the number of moles of a reactant or product. Conversion of the given information into moles has always been the first step in solving a stoichiometry problem. In Figure 12-8, the general scheme for working stoichiometry problems has been extended to include volumes of solutions. The following examples also illustrate the inclusion of solutions in stoichiometry problems.

EXAMPLE 12-8   **Calculating the Volume of a Reactant**

Given the balanced equation

$$3NaOH(aq) + H_3PO_4(aq) \rightarrow Na_3PO_4(aq) + 3H_2O(l)$$

what volume of 0.250 M NaOH is required to react completely with 4.90 g of H_3PO_4?

PROCEDURE

1. Convert mass of H_3PO_4 to moles using the molar mass of H_3PO_4.

2. Convert moles of H_3PO_4 to moles of NaOH using the mole ratio from the balanced equation.

3. Convert moles of NaOH to volume using the equation for molarity.

The complete unit map for this problem is shown below.

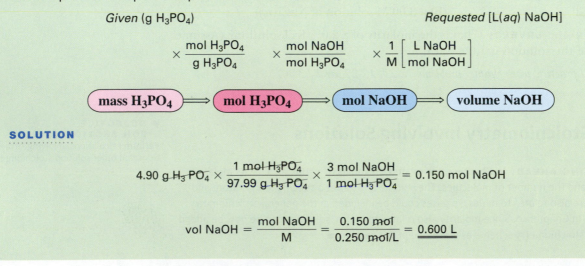

Given (g H₃PO₄)   *Requested [L(aq) NaOH]*

SOLUTION

$$4.90 \text{ g } H_3PO_4 \times \frac{1 \text{ mol } H_3PO_4}{97.99 \text{ g } H_3PO_4} \times \frac{3 \text{ mol NaOH}}{1 \text{ mol } H_3PO_4} = 0.150 \text{ mol NaOH}$$

$$\text{vol NaOH} = \frac{\text{mol NaOH}}{M} = \frac{0.150 \text{ mol}}{0.250 \text{ mol/L}} = \underline{0.600 \text{ L}}$$

ANALYSIS

In Section 7-1, we introduced the mass-to-mass stoichiometry problem. This problem is an example of a mass-to-volume (solution) calculation. The similarities are important. Compare the unit map in Example 7-3 in Section 7-1.3 to the one in this example. The only difference in units is in the last conversion factor, where molarity is used to calculate volume, rather than molar mass for the purpose of converting to grams. Otherwise, the procedure is identical. Substitution of a second molarity value for the second reactant would allow for a volume-to-volume calculation, negating the need to calculate a molar mass entirely. See Example 12-9.

SYNTHESIS

These problems are actually more common than the original stoichiometry problems presented in Chapter 7 simply because they represent reactions occurring in solution. The vast majority of chemical reactions are solution reactions rather than solid-phase reactions because of the mixing that occurs in solution. It's hard for solids to react together at all, and if they do, reaction times are generally slow. Solutions allow dissolved chemicals to mix thoroughly, bringing reactant molecule right up to reactant molecule and significantly increasing reaction rates.

EXAMPLE 12-9   Calculating the Volume of a Gas from a Solution Reaction

Given the balanced equation

$$2HCl(aq) + K_2S(aq) \rightarrow H_2S(g) + 2KCl(aq)$$

what volume of H_2S measured at STP would be evolved from 1.65 L of a 0.552 M HCl solution with excess K_2S present?

PROCEDURE

1. Convert volume and molarity of HCl to moles of HCl.

2. Convert moles of HCl to moles of H_2S using mole ratio from the balanced equation.

3. Convert moles of H_2S to volume of H_2S using the molar volume.

The complete unit map for the problem is shown below.

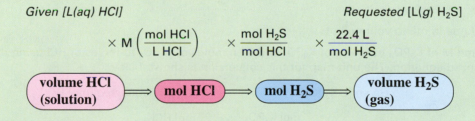

SOLUTION

$$V\,(\text{Solution}) \times M(\text{HCl}) = n\,(\text{HCl})$$

$$1.65\ \cancel{L} \times \frac{0.552\ \text{mol}}{\cancel{L}} = 0.911\ \text{mol HCl}$$

Since the volume of the gas is at STP, the molar volume relationship can be used rather than the ideal gas law.

$$0.911\ \cancel{\text{mol HCl}} \times \frac{1\ \text{mol H}_2\text{S}}{2\ \cancel{\text{mol HCl}}} \times \frac{22.4\ \text{L (STP)}}{\cancel{\text{mol H}_2\text{S}}} = \underline{10.2\ \text{L (STP)}}$$

ANALYSIS

This problem combines two different volume problems. One is a volume of solution and requires the use of molarity. The other is a volume of a gas and requires a gas law relationship, which was discussed in Chapter 10. Sometimes students get the two confused. Try to visualize the aqueous solution and the gas being evolved as the HCl solution is added. The solution to any stoichiometry problem lies in using the proper conversion factor: molar mass for grams, 22.4 L/mol for gases at STP, the ideal gas law for volumes at other conditions, and molarity for solutions.

SYNTHESIS

Which conversion factors would be necessary to determine the number of *molecules* of a product that could be produced from a given volume of reactant solution? Starting with the volume, we would need the molarity of the solution, the mole ratio between reactant and product, and finally Avogadro's number. The unit map for this calculation is

$$\frac{mol\ (reactant)}{L} \times \frac{mol\ (product)}{mol\ (reactant)} \times \frac{6.022 \times 10^{23}\ molecules}{mol\ (product)}$$

12-5.2 Titrations

Quite often, laboratory experiments require that we establish the molarity of a solution from stoichiometry data. For example, environmental scientists encounter this situation frequently when they sample river water. To know whether the water is safe to drink, the concentrations of Pb^{2+}, or Hg^{2+}, or some other heavy metal contaminant has to be established. The concentrations are determined through a stoichiometric procedure known as a *titration*.

Titrations *are usually volume-to-volume stoichiometry problems where a concentration is the unknown variable.* A measured volume of one reactant of known molarity is added to a known volume of the other reactant. The end of the reaction is generally noted by the addition of an *indicator*, which is a compound that provides a visual clue, such as a color change, that a reaction has been completed. When the indicator changes color, the titration is said to be at its *end point*. This occurs when a stoichiometric balance occurs between reactants. The molarity of the unknown solution can then be calculated. This is illustrated in Example 12-10.

EXAMPLE 12-10   Calculating an Unknown Concentration from Titration Data

Given the reaction: $3HCl(aq) + K_3PO_4(aq) \longrightarrow H_3PO_4(aq) + 3KCl(aq)$

What is the concentration of an HCl solution if 25.0 mL reacts with 38.5 mL of 0.145 M K_3PO_4 solution?

PROCEDURE

1. Convert volume in mL to volume in L.

2. Convert L and M of K_3PO_4 to moles of K_3PO_4 and then moles of K_3PO_4 to moles of HCl using the mole ratio from the balanced equation. The unit map for this conversion is shown below.

Given [L K_3PO_4]   *Requested* [mol HCl]

$$\times M\ \frac{mol\ K_3PO_4}{L\ K_3PO_4} \times \frac{mol\ HCl}{mol\ K_3PO_4}$$

volume K_3PO_4 (solution) $\rightarrow$ mol K_3PO_4 $\rightarrow$ mol HCl

3. Convert moles and volume of HCl to molarity of HCl.

SOLUTION

$$38.5\ mL \times \frac{1\ L}{10^3\ mL} = 0.0385\ L\ K_3PO_4$$

$$25.0\ mL \times \frac{1\ L}{10^3\ mL} = 0.250\ L\ HCl$$

$$0.0385\ L\ K_3PO_4 \times \frac{0.145\ mol\ K_3PO_4}{L\ K_3PO_4} \times \frac{3\ mol\ HCl}{1\ mol\ K_3PO_4} = 0.0167\ mol\ HCl$$

$$\text{molarity} = \frac{\text{mol}}{\text{L}} = \frac{0.0167 \text{ mol}}{0.250 \text{ L}} = \underline{\underline{0.668 \text{ M}}}$$

ANALYSIS

As a rule of thumb, concentrations above 1 M are considered fairly concentrated solutions. Concentrations below 0.1 M are considered dilute. The range between those two is what is frequently (though not exclusively) encountered. So when a value of 0.668 M is calculated for the molarity of a solution, it falls right in the range of common concentrations. The answer is very reasonable.

SYNTHESIS

Titrations find their biggest application in chemistry during acid–base reactions. There is generally a favorable reaction, and the accompanying change in the H^+ (aq) concentration (see Chapter 13) is the primary factor that causes the indicator to change color. Most indicators change color very close to the end point, making for a wide range of available indicators. The stoichiometry of the reaction is usually very straightforward and can be determined easily from the formulas of the acid and base.

▶ **ASSESSING THE OBJECTIVE FOR SECTION 12-5**

EXERCISE 12-5(a) LEVEL 1: What is the conversion factor needed to convert each of the following into moles?
(a) volume of a gas (at STP)   **(c)** mass of a solid
(b) volume of a solution   **(d)** number of molecules

EXERCISE 12-5(b) LEVEL 2: For the reaction
$$Ca(OH)_2(aq) + 2HNO_3(aq) \longrightarrow Ca(NO_3)_2(aq) + 2H_2O(l)$$
What volume of 0.125 M of $Ca(OH)_2$ reacts with 15.0 mL of 0.225 M HNO_3?

EXERCISE 12-5(c) LEVEL 2: For the reaction
$$Fe(OH)_3(s) + 3HCl(aq) \longrightarrow FeCl_3(aq) + 3H_2O(l)$$
What is the concentration of a solution of HCl if 33.5 mL reacts with 1.50 g of $Fe(OH)_3$?

EXERCISE 12-5(d) LEVEL 3: Sodium carbonate and hydrochloric acid react together to form carbon dioxide, water, and sodium chloride. If 50.0 mL of an HCl solution reacts with an excess of sodium carbonate to form 3.50 L of CO_2 at STP, what is the concentration of the HCl solution?

EXERCISE 12-5(e) LEVEL 3: What volume of 0.125 M phosphoric acid will react with 25.0 mL of 0.0850 M potassium hydroxide?

For additional practice, work chapter problems 12-43, 12-46, 12-49, and 12-51.

PART A SUMMARY

KEY TERMS

12-1.1	Two liquids are **miscible** if they form one phase and **immiscible** if they remain in two phases. p. 398
12-1.2	Ionic compounds dissolve in water as a result of **ion–dipole forces**. p. 398
12-1.3	Strong acids undergo **ionization** when dissolved in water. p. 400
12-2.1	**Solubility** is maximum amount of a solute that dissolves at a specific temperature. p. 401
12-2.1	Depending on the amount of solute dissolved and its solubility, the solution may be **unsaturated**, **saturated**, or, in certain circumstances, **supersaturated**. p. 401
12-2.1	The **concentration** of a solute refers to the amount present in a certain amount of solvent or solution. p. 401
12-2.2	**Recrystallization** takes advantage of the difference in solubility of a substance at different temperatures. p. 402

12-2.4 According to **Henry's law**, the solubility of gases in a liquid relates to the pressure above the liquid. p. 403

12-3 Concentration may be measured as **percent by mass**. p. 404

12-3.1 **Parts per million (ppm)** or **parts per billion (ppb)** are used for small concentrations. p. 405

12-4.1 **Molarity (M)** is a concentration unit that emphasizes the volume of the solution. p. 407

12-5.2 During a **titration**, an indicator can be used to determine the end point of the experiment. p. 414

SUMMARY CHART

Concentration Units

Concentration Unit	*Name*	*Relationship of solute*	*Use*
$\frac{\text{g solute}}{\text{100 g solvent}}$	———	Mass of solvent	Solubility tables
$\frac{\text{g solute}}{\text{g solution}} \times 100\%$	Percent by mass	Mass of solution	High concentrations (above 0.01%)
$\frac{\text{g solute}}{\text{g solution}} \times 10^6$ ppm	Parts per million (ppm)	Mass of solution	Low concentrations $(>10^{-4}\ \%)$
$\frac{\text{g solute}}{\text{g solution}} \times 10^9$ ppb	Parts per billion ppb	Mass of solution	Extremely low concentrations $(>10^{-7}\ \%)$
$\frac{\text{mole solute}}{\text{L solution}}$	Molarity (M)	Volume of solution	Measuring molar amount with volume and in stoichiometry problems

Part B

The Effects of Solutes on the Properties of Water

SETTING A GOAL

- You will learn how the physical properties of aqueous solutions differ from those of pure water.

OBJECTIVES

12-6 Explain the differences between nonelectrolytes, strong electrolytes, and weak electrolytes.

12-7 Calculate the boiling and melting points of aqueous solutions of electrolytes and nonelectrolytes.

▶ **OBJECTIVE FOR SECTION 12-6**
Explain the differences between nonelectrolytes, strong electrolytes, and weak electrolytes.

12-6 Electrical Properties of Solutions

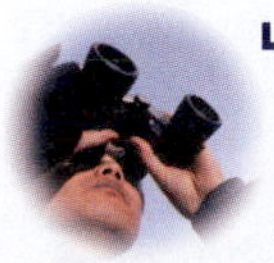

LOOKING AHEAD! The use of a hair dryer in the bathtub is extremely dangerous. Even though pure water is not a good conductor of electricity, ordinary tap water is because it contains a variety of dissolved compounds. It has long been known that the presence of a solute in water may affect its ability to conduct electricity. In this section, we will examine the electrical properties of water and aqueous solutions. ■

12-6.1 Conduction of Electricity

We have probably all been zapped at one time or another by touching a live wire. The culprit, naturally, was electricity. *Electricity is simply a flow of electrons through a substance called a* **conductor**. Metals are the most familiar conductors and, as such, find use in electrical wires. Because the outer electrons of metals are loosely held, they can be made to flow through a continuous length of wire. *Other substances resist the flow of electricity and are known as* **nonconductors** *or* **insulators**. Glass and wood are examples of nonconductors of electricity. When wires are attached to an automobile battery and then to a lightbulb, the light shines brightly. If the wire is cut, the light goes out because the circuit is broken. If the two ends of the cut wire are now immersed in pure water (distilled or rain), the light stays out, indicating that water does not conduct electricity under these circumstances. Now let's dissolve certain solutes in water and examine what happens. When compounds such as $CuSO_4$ or HCl are dissolved in water, the effect is obvious. The light immediately begins to shine, indicating that the solution is a good conductor of electricity. (See Figure 12-9b.) *Compounds whose aqueous solutions conduct electricity are known as* **electrolytes**. (Some ionic compounds have limited solubility in water, but if their molten state conducts electricity, they are also classified as electrolytes.)

We now understand that it is the presence of ions in the aqueous solution that allows the solution to conduct electricity. It is the movement of these ions that carries the electrical charge and completes the circuit. Soluble ionic compounds form ions in solution, and some polar covalent compounds dissociate upon dissolving to form ions. For example, both NaCl (ionic) and HCl (polar covalent) are classified as electrolytes because they form ions in aqueous solution. As you will recall, HCl is a molecular compound in the pure state. However, in water it is ionized as described earlier in this chapter.

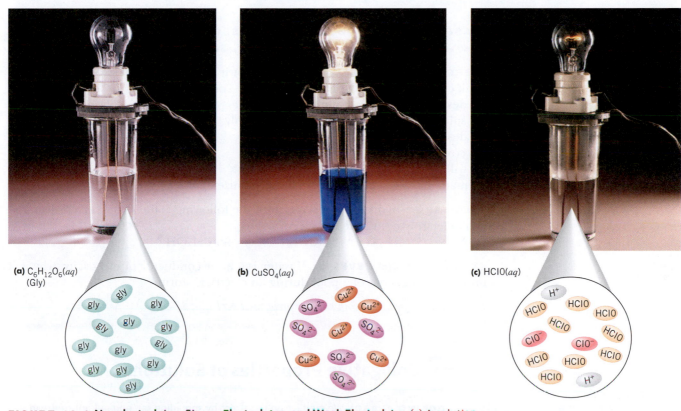

(a) $C_6H_{12}O_6(aq)$
(Gly)

(b) $CuSO_4(aq)$

(c) HClO(aq)

FIGURE 12-9 Nonelectrolytes, Strong Electrolytes, and Weak Electrolytes (a) A solution of glucose sugar ($C_6H_{12}O_6$) does not conduct electricity. (b) A $CuSO_4$ solution is a strong electrolyte and conducts electricity. (c) A hypochlorous acid solution [HClO(aq)] is a weak electrolyte and conducts a limited amount of electricity.

12-6.2 Nonelectrolytes and Electrolytes

Other compounds such as sucrose (table sugar), glucose (blood sugar), and alcohol dissolve in water, but their solutions do not conduct electricity. *Compounds whose aqueous solutions do not conduct electricity are known as* **nonelectrolytes**. Nonelectrolytes are molecular compounds that dissolve in water without formation of ions. (See Figure 12-9a.)

There are two classes of electrolytes: strong electrolytes and weak electrolytes. *Solutions of* **strong electrolytes** *are good conductors of electricity.* Almost all salts and strong acids are present only as ions in aqueous solution (e.g., NaCl, $CuSO_4$, and HCl) and are thus classified as strong electrolytes. (See Figure 12-9b.) *Solutions of* **weak electrolytes** *allow a limited amount of conduction.* When wires are immersed in solutions of weak electrolytes, the lightbulb glows, but very faintly. (See Figure 12-9c.) Even adding more of the solute does not cause increased conduction. Examples of weak electrolytes are ammonia (NH_3) and hypochlorous acid (HClO). As shown in Figure 12-9c, most of the HClO molecules in the solution remain as neutral molecules with only a small fraction present as ions. The ionization of HClO is an example of a reversible reaction that reaches a point of equilibrium and can be represented by the following equation. The equilibrium in this case lies far to the left.

$$HClO(aq) \rightleftharpoons H^+(aq) + ClO^-(aq)$$

The concept of equilibrium was discussed in relation to vapor pressure (Section 11-4.3) but will be discussed in more detail in Chapter 15. The identification of strong and weak acids will be explored in the next chapter.

(*Note*: Pure water actually does contain a very small concentration of ions, as we will discuss in the next chapter. The concentration of these ions in pure water is too low to detect by conduction of electrical current by the method described above.)

▶ **ASSESSING THE OBJECTIVE FOR SECTION 12-6**

EXERCISE 12-6(a) LEVEL 1: Are the following descriptions of strong electrolytes, weak electrolytes, or nonelectrolytes?
(a) describes all ionic compounds
(b) solution causing a lightbulb to glow dimly
(c) describes most molecular compounds
(d) conducts electricity well
(e) will not conduct electricity
(f) made from compounds that partially ionize in water

EXERCISE 12-6(b) LEVEL 2: Which of the following is the strong electrolyte, the weak electrolyte, and the nonelectrolyte?
(a) NH_3 **(b)** KNO_3 **(c)** CH_3OH

EXERCISE 12-6(c) LEVEL 3: If water is a poor conductor of electricity, why is it important to get out of a pool during an electrical storm?

For additional practice, work chapter problems 12-51 and 12-52.

▶ **OBJECTIVE FOR SECTION 12-7**
Calculate the boiling and melting points of aqueous solutions of electrolytes and nonelectrolytes.

12-7 Colligative Properties of Solutions

LOOKING AHEAD! We melt ice on the street by spreading salt and put antifreeze in our car's radiator—we take these actions because the presence of solutes alters some important physical properties of pure water. In this section we will examine why and how much the presence of solutes affects certain properties. ■

The presence of a solute in water may or may not affect its conductivity, depending on whether the solute is an electrolyte or a nonelectrolyte. There are other properties of water, however, that are always affected to some extent by the presence of a solute. Consider again what was mentioned in Chapter 3 as characteristic of a pure substance. Recall that a pure substance has a distinct and unvarying melting point and boiling point (at a specific atmospheric pressure). Mixtures, such as aqueous solutions, freeze and boil over a range of temperatures that are lower (for freezing) and higher (for boiling) than those of the pure solvent. The effect of these changes is to extend the liquid range for the solvent. The more solute, the more the melting and boiling points are affected. *A property that depends only on the relative amounts of solute in relation to the solvent is known as a* **colligative property**.

Colligative properties depend only on the amount or number of moles of particles present in solution, not their identity. The particles may be small molecules, large molecules, or ions—only the total number is relevant. This is like many of the gas laws, which, you may recall, also depend only on the total number of moles of gas present.

12-7.1 Vapor Pressure

The Dead Sea in Israel and the Great Salt Lake in the United States contain large concentrations of solutes. Since these bodies of water have no outlets to the ocean, dissolved substances have accumulated, forming saturated solutions. Even though both bodies of water exist in semiarid regions with high summer temperatures, they evaporate very slowly compared with a freshwater lake or even the ocean. If water in these lakes evaporated at the same rate as fresh water, both would nearly dry up in a matter of years. Why do they evaporate so slowly?

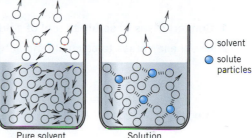

The presence of a nonvolatile solute * *in a solvent lowers the equilibrium vapor pressure from that of the pure solvent.* (See Section 11-4.3.) It is not hard to understand why this occurs based on what we have discussed about the interactions between solute and solvent. Because of various attractive forces, solvent molecules surround solute molecules or ions. Solute molecules or ions thus tend to tie up the solvent molecules, which in effect prevents them from escaping to the vapor. An increase in solute particles causes more solvent molecules to be held in the liquid phase. This results in a solution with a lower equilibrium vapor pressure than the pure solvent. (See Figure 12-10.) As we might predict from this model, the more solute particles present, the lower the vapor pressure of the solution. The resulting **vapor pressure lowering** affects the boiling point of the solution, as described next.

FIGURE 12-10 Vapor Pressure Lowering A nonvolatile solute reduces the number of solvent molecules escaping to the vapor.

12-7.2 Boiling Point

A direct effect of the lowered vapor pressure of a solution is a higher boiling point than that of the solvent. Recall that the normal boiling point of water is 100°C, which is the temperature at which its vapor pressure is equal to 760 torr. An aqueous salt solution, however, would have a vapor pressure lower than 760 torr. Thus, the solution would have to be heated above 100°C for the vapor pressure to reach 760 torr and begin boiling. Again, the more concentrated the solution, the lower is its vapor pressure and the higher is its boiling point.

The amount of **boiling-point elevation** is given by the equation

$$\Delta T_b = K_b m$$

*A nonvolatile solute is one that has essentially no vapor pressure at the relevant temperatures.

where

ΔT_b = the number of Celsius (or Kelvin) degrees that the boiling point is raised

K_b = a constant characteristic of the solvent [for water $K_b = 0.512$ (°C · kg)/mol; other values of K_b are given for particular solvents as needed in the problems]

m = a concentration unit called molality

For these calculations we need one more unit of concentration, which is known as molality. The definition of **molality** is shown below.

$$\text{molality } (m) = \frac{\text{moles of solute}}{\text{kg of solvent}}$$

This is a convenient concentration unit for this purpose since it emphasizes the relationship between the relative amounts of solute (expressed in moles) and mass of solvent (expressed in kg) present rather than the volume of the solution, as does molarity. Also, temperature does not affect the molality of a solution. Molarity, however, is affected by temperature as the volume of most liquids changes in response to the temperature.

EXAMPLE 12-11   Calculating Molality

What is the molality of methyl alcohol in a solution made by dissolving 18.5 g of methyl alcohol (CH_3OH) in 850 g of water?

PROCEDURE

1. Convert mass of solute to mol of solute and g of solvent to kg of solvent.

2. Use these values to calculate molality.

$$\text{molality} = \frac{\text{mol solute}}{\text{kg solvent}}$$

SOLUTION

$$\text{mol solute} = \frac{18.5 \text{ g}}{32.04 \text{ g/mol}} = 0.577 \text{ mol}$$

$$\text{kg solvent} = \frac{850 \text{ g}}{10^3 \text{ g/kg}} = 0.850 \text{ kg}$$

$$\text{molality} = \frac{0.577 \text{ mol}}{0.850 \text{ kg}} = \underline{0.679 \text{ m}}$$

ANALYSIS

When water is the solvent, the value of the molarity and the molality differ by just a fraction of a percent. That is because 1.0 L of aqueous solution has a mass of about 1.0 kg. Should they be closer together for concentrated or dilute solutions? In fact, the more dilute the solution, the closer the mass of 1 L of the solution is to 1.0 kg (the mass of 1.0 L of pure water). Thus, they are closer for dilute solutions. For example, the molarity of a 1.0% solution of LiCl is 0.237, while the molality is 0.238. However, the molarity of a 10% solution of LiCl is 2.49, while the molality is 2.62.

SYNTHESIS

So what if the solvent isn't water? Most liquids that routinely serve as solvents have densities significantly less than water's. For example, ethyl alcohol, a common solvent, has a density of 0.789 g/mL (0.789 kg/L). If we assume a dilute solution where 1 L of solution weighs about the same as 1 L of solvent, the number of moles of solute would be divided by 1 L for molarity but 0.789 kg for molality. That will cause the molality to be noticeably higher than the molarity. For the few solvents with densities higher than water [e.g., methylene chloride (1.33 g/mL)], the molality would be lower than the molarity.

EXAMPLE 12-12   Calculating the Boiling Point of a Solution

What is the boiling point of an aqueous solution containing 468 g of sucrose ($C_{12}H_{22}O_{11}$) in 350 g of water at 1 atm?

PROCEDURE

1. Convert g of solute (sucrose) to moles and g of solvent to kg.

2. Use these values and the value of K_b for water to calculate the temperature change.

$$\Delta T_b = K_b\, m = 0.512\ (°C \cdot kg)/mol \times \frac{mol\ solute}{kg\ solvent}$$

3. Calculate the actual boiling point from the temperature change.

SOLUTION

$$mol\ solute = \frac{468\ g}{342.5\ g/mol} = 1.37\ mol$$

$$kg\ solvent = \frac{350\ g}{10^3\ g/kg} = 0.350\ kg$$

$$\Delta T_b = 0.512\ (°C \cdot kg)/\ mol \times \frac{1.37\ mol}{0.350\ kg} = 2.0\ °C$$

$$normal\ boiling\ point\ of\ water = 100.0°C$$

$$boiling\ point\ of\ the\ solution = 100.0°C + \Delta T_b = 100.0 + 2.0 = \underline{102.0°C}$$

ANALYSIS

In this problem the mass of solute is greater than the mass of solvent, yet it created an increase in boiling point of only 2°C. Why so small a change? In this case, it's because the solute is made of comparatively heavy molecules. The molar mass of the sucrose is 342.5 g/mol. If our solute had a molar mass of only 50 g/mol or so, what would our boiling point have been? 50 is about 7 times less than 342.5, which means there would be seven times as many particles and seven times the rise in temperature. The new boiling point would be closer to 114°C, a significant increase.

SYNTHESIS

This process of boiling-point elevation (or freezing-point depression, discussed in the next section) represented one of the earliest ways chemists had to determine the molar mass of an unknown compound. First, we determine whether the compound is soluble in a polar or a nonpolar solvent. Knowing the proper solvent and its value of K_b, we can then do the experiment. It requires that we dissolve a measured mass of the compound in a measured mass of solvent. We then measure the boiling point of the pure solvent, the boiling point of the solution, and, from these two values, the boiling-point elevation. From this we solve for the molality. Since we know the mass of solute and the mass of solvent, the only variable that is left is the molar mass. We could try the experiment using the same compound in other solvents to confirm our values.

12-7.3 Freezing Point

The icy, cold winds of winter can be very hard on automobiles. If there is not enough antifreeze in the radiator, the coolant water may freeze. This could ruin the radiator and even crack the engine block, because water expands when it freezes. We illustrate the same principle as antifreeze in radiators when we spread salt on ice-covered streets or sidewalks. In both of these cases, we take advantage of the fact that solutions have lower freezing points than pure solvents. *The freezing point of a solution is lower just as the boiling point is higher than that of the pure solvent.*

The amount of **freezing-point lowering** is given by the equation

$$\Delta T_f = K_f m$$

where
ΔT_f = the number of Celsius (or Kelvin) degrees that the freezing point is lowered
K_f = a constant characteristic of the solvent [for water, $K_f = 1.86$ (°C · kg)/mol]
m = molality of the solution

EXAMPLE 12-13   Calculating the Freezing Point of a Solution

What is the freezing point of the solution in Example 12-12?

PROCEDURE

$$\Delta T_f = K_f\, m = 1.86 \text{ (°C · kg)/ mol} \times \frac{\text{mol solute}}{\text{kg solvent}}$$

SOLUTION

$$\Delta T_f = K_f\, m = 1.86 \text{ (°C · kg)/ mol} \times \frac{1.37 \text{ mol}}{0.350 \text{ kg}} = 7.3°C$$

freezing point of water = 0.0°C

freezing point of the solution = 0.0°C − ΔT_f = 0.0°C -7.3°C = $\underline{-7.3°C}$

ANALYSIS

This problem stresses the concept of a *colligative* property. In Examples 12-12 and 12-13 nothing changed about the solution, since there are the same number of solute particles in each case. Yet notice that the effect on the melting point is more pronounced than the effect on the boiling point. Since K_f of water (1.86) is over three times larger than the K_b of water (0.512), the freezing-point depression is over three times larger as well (7.3°C compared to 2°C). Because of the freezing-point *lowering* and boiling-point *elevation*, the liquid range in this example has been extended by over 9°C (i.e., −7.3°C to 102°C) for the solution compared to pure water. Permanent antifreeze not only protects the engine from freezing but in the summer it protects the water from boiling away. Until the late 1930s, methanol was used as antifreeze, but because it has a low boiling point, it would boil away in hot weather. It had to be removed and replaced with water when the weather warmed. Permanent antifreeze is now composed mostly of ethylene glycol ($C_2H_6O_2$). A 1:1 mixture protects down to −40°C (and up to about 110°C), although less may be used depending on the locality. Ethylene glycol, however, is very poisonous compared to propylene glycol ($C_3H_8O_2$), which is not. The latter is used in water-heating pipes in homes or other places where there is danger of accidental ingestion. (Because of its higher molar mass, however, it takes more propylene glycol to have the same effect as ethylene glycol.)

SYNTHESIS

Freezing-point depression is also used by chemists as a laboratory procedure to determine the molar mass of an unknown compound. See if you can calculate the molar mass of the compound in problem 12-91 in the chapter-end problems. Besides solubility, one looks for a solvent with the largest value for K_f. When you look at the equation for freezing-point depression, notice that the greater the value of K_f, the greater the freezing-point depression and the more precision (i.e., number of significant figures) in the measurement. For example, a measured depression of 12.2°C in one solvent is more precise than one of 1.3°C for the same solute in another solvent. Chloroform ($CHCl_3$) is very effective in this regard, since it has $K_f = 30$. One mole of solute in chloroform causes about 16 times as much freezing-point depression as in water. (This assumes that the solute is soluble in both solvents, which may not be the case.)

12-7.4 Osmotic Pressure

Food is preserved in salt water, drinking ocean water causes dehydration, tree and plant roots absorb water—these are all phenomena related to a colligative effect called osmosis. **Osmosis** *is the tendency for a solvent to move through a thin porous membrane from a dilute solution to a more concentrated solution.* The membrane is said to be *semipermeable,* which means that small solvent molecules can pass through but large hydrated solute species cannot. Figure 12-11 illustrates osmosis. On the right is a pure solvent and on the left, a solution. The two are separated by a semipermeable membrane. Solvent molecules can pass through the membrane in both directions, but the rate at which they diffuse to the right is lower because some of the water molecules on the left are held back by solute–solvent

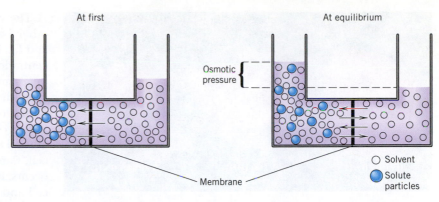

At first   **At equilibrium**

Osmotic pressure

Membrane

○ Solvent
● Solute particles

FIGURE 12-11 Osmotic Pressure Osmosis causes dilution of the more concentrated solution.

interactions. As a result of the uneven passage of water molecules, the water level rises on the left and drops on the right. This creates increased pressure on the left, which eventually counteracts the osmosis, and equilibrium is established. *The extra pressure required to establish this equilibrium is known as the* **osmotic pressure.** Like other colligative properties, it depends on the concentration of the solute. In Figure 12-11, the more concentrated the solution on the left (less solvent), the higher the osmotic pressure.

In Figure 12-11, if pressure greater than the osmotic pressure is applied on the left, reverse osmosis takes place and solvent molecules move from the solution to the pure solvent. This process is used in desalination plants that convert seawater (a solution) to drinkable water. This is important in areas of the world such as the Middle East, where there is a shortage of fresh water.

12-7.5 Electrolytes and Colligative Properties

Our final point concerns the effect of electrolytes on colligative properties. Electrolytes have a more pronounced effect on colligative properties than do nonelectrolytes. The reason is that these properties depend only on how many particles are present regardless of whether the particle is a neutral molecule, a cation, or an anion. For example, one mole of NaCl dissolves in water to produce two moles of particles, one mole of Na^+, and one mole of Cl^-.

Sodium chloride is used to melt ice from sidewalks, streets, and highways.

$$NaCl(s) \xrightarrow{\text{H}_2\text{O}} Na^+(aq) + Cl^-(aq)$$

Thus one mole of NaCl lowers the freezing point about twice as much as one mole of a nonelectrolyte. This effect is put to good use in the U.S. Snow Belt, where sodium chloride is spread on snow and ice to cause melting even though the temperature is below freezing. Even more effective in melting ice is calcium chloride ($CaCl_2$). This compound produces three moles of ions ($Ca^{2+} + 2Cl^-$) per mole of solute and therefore is three times as effective per mole as a nonelectrolyte in lowering the freezing point. Calcium chloride is occasionally used on roads when the temperature is too low for sodium chloride to be effective. Aqueous electrolyte solutions are quite corrosive toward metals because of their electrical conductivity. This is why they are not used in automobile radiators as an inexpensive antifreeze.

We see an example of the osmosis process whenever we leave our hands in water for extended periods of time. The oils that protect our skin eventually wash away, allowing water to come in contact with the skin, which is a semipermeable membrane. The movement of excess water molecules from the outside of our skin to the more concentrated fluid in our bodies causes puffy wrinkling. Pickles are wrinkled for the opposite reason. The cells of the cucumber have been dehydrated by the salty brine solution. In fact, brine solutions preserve many foods because the concentrated solution of salt removes water from the cells of bacteria, thus killing the bacteria. Trees and plants obtain water by absorbing water through the semipermeable membranes in their roots into the more concentrated solution inside the root cells.

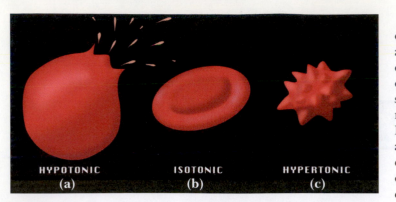

FIGURE 12-12 Blood Cells (a) At lower ion concentration, osmotic flow pumps water into the cell and may cause it to burst. (b) In isotonic solution, red blood cells are disk shaped. (c) At higher ion concentration, osmotic flow removes water from the cell interior, shrinking the cell.

The concentration of solutes in the blood and other bodily fluids is important in the hydration and health of cells in our bodies. When the concentration of ions and other solutes in the blood or plasma is the same as that inside of a cell, the solution is said to be *isotonic*. In this case, there is no net movement of water into or out of the cell. If the concentration of solutes in the solutions around the cells is greater in the blood than in the cell, water moves out of the cell and dehydration occurs. If the opposite occurs, water moves into the cell and expands it. (See Figure 12-12.) In either case, destruction of the cells may occur in extreme cases. Intravenous solutions, such as a glucose drip or injections in a saline solution, are isotonic with the blood so as not to dilute or concentrate the solution of the cells. As mentioned earlier in Example 12-1, a 0.89% (wt/v) saline solution is used as an isotonic solution.

MAKING IT REAL

Osmosis in a Diaper

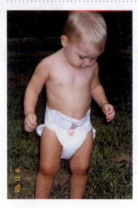

Except for the fact that a baby seems to be unusually heavy, one might not realize just how wet a diaper has become. In the old days, a quick touch would tell the condition. Modern superabsorbent diapers make use of a chemical that has a tremendous ability to absorb water. In fact, such diapers contain a white powdery material that can absorb 200 to 500 times its weight in water. Not only is such a diaper convenient for the diaper changer but it is also quite a bit more comfortable for the diaper wearer.

The compound used in these diapers is a polymeric material. The polymer used in diapers (sodium polyacrylate) is made of long chains of an ionic unit (called a monomer), sodium acrylate (shown below). Most familiar plastics such as Styrofoam cups, plastic bottles and bags, PVC pipes, and thousands of other everyday materials are made of various types of polymers. More detail about polymers is provided in Chapter 17.

The unique property of the diaper polymer is that its surface is semipermeable. Inside the surface, ions are present. The high concentration of ions inside causes water molecules to cross the membrane in the process of osmosis. The water then stays put inside the polymer. In fact, the diaper does not even feel wet, thus protecting the baby from nasty rashes. It is so effective that special "training diapers" can be used when it is time for baby to move on. These diapers are engineered to allow some wetness and discomfort as an incentive to become "trained."

Consider the absorbent ability of these diapers. A 1-g quantity of the polyacrylate can absorb up to 500 g of water. If one had an 8-oz glass of water, it would take less than 500 mg of superabsorbent to turn the glass of water into a wiggly, gelatinous mass. A 700-lb quantity of superabsorbent would be enough to turn a good-size swimming pool into something like Jell-O.

In 1999, world production of this polymer was 980 million tons. Not all of that went into diapers (fortunately). It is also used in agriculture, crafts, evaporative coolers, firefighting, toothpaste, and cosmetics.

An unpleasant (odiferous), drippy experience has been changed into something much more tolerable. Baby caregivers of the world are grateful for this chemical advance.

$$CH_2 = CH$$
$$|$$
$$CO_2^-Na^+$$
Sodium acrylate
(Monomer)

$$\{CH_2 - CH\}$$
$$|$$
$$CO_2^-Na^+$$
Sodium polyacrylate

▶ **ASSESSING THE OBJECTIVE FOR SECTION 12-7**

EXERCISE 12-7(a) LEVEL 1: Liquid A is a pure compound and liquid B contains a nonvolatile solute dissolved in liquid A.
(a) Which boils at the higher temperature?
(b) Which freezes at the higher temperature?
(c) Which has the higher osmotic pressure?
(d) Which has the higher vapor pressure?

EXERCISE 12-7(b) LEVEL 2: Calculate the molality if 15.0 g of $NaNO_3$ is dissolved in 50.0 g of H_2O.

EXERCISE 12-7(c) LEVEL 2: What is the freezing point of a solution if 0.750 mol of a nonelectrolyte solute is dissolved in 250.0 g of H_2O? What is the freezing point if it's 0.750 mol of $CaCl_2$ that's dissolved?

EXERCISE 12-7(d) LEVEL 3: Acetic acid $(HC_2H_3O_2)$ is a weak electrolyte. What would be the effect on the boiling point of water of adding one mole of acetic acid compared to 1 mole of some nonelectrolyte molecular solute? What would be the effect compared to adding 1 mole of NaCl?

EXERCISE 12-7(e) LEVEL 3: There is a limit as to how cold a salt (NaCl) solution can be lowered before it freezes. At this temperature the salt solution is saturated, which is at about 28 g/100 g H_2O. What is this temperature in Celsius and Fahrenheit degrees?

For additional practice, work chapter problems 12-56, 12-58, 12-60, 12-64, and 12-67.

PART B SUMMARY

KEY TERMS

12-6.1   A medium that conducts electricity is known as a **conductor**, whereas a medium that does not is an **insulator** (or **nonconductor**). p. 416

12-6.2   A solute that does not change water to a conductor is a **nonelectrolyte**, whereas an **electrolyte** allows the solution to become a conductor. p. 417

12-6.2   A **strong electrolyte** makes water a good conductor, whereas a **weak electrolyte** makes water a weak conductor. p. 418

12-7   A **colligative property** of a solution depends only on the amount of solute present. p. 419

12-7.1   A nonvolatile solute causes a **vapor pressure lowering** compared to the pure solvent. p. 419

12-7.2   A nonvolatile solute causes **boiling-point elevation**, the amount of which depends on the **molality** of the solution. p. 419

12-7.3   A solute causes a **freezing-point lowering**. p. 422

12-7.4   The presence of a solute allows **osmosis** to occur, which causes **osmotic pressure**. pp. 422–423

SUMMARY CHART

Colligative Properties		
Property of Solution	Effect	Result
Vapor pressure	Lowered	Solutions evaporate more slowly than pure solvents.
Boiling point	Raised	Solutions boil at higher temperatures than pure solvents.
Freezing point	Lowered	Solutions freeze at lower temperatures than pure solvents.
Osmotic pressure	Raised	Solvent from dilute solutions diffuses through a semipermeable membrane into concentrated solutions.

CHAPTER 12 SYNTHESIS PROBLEM

There are two ways we could melt the ice off of the front step on a nasty evening. We could either pour an alcoholic drink on it or spread some calcium chloride. Calcium chloride is more effective but a beverage may be more available in an emergency. In our study of chemistry, an understanding of how we measure concentrations, especially involving chemical reactions, is quite important. Also, how the physical proper- ties of solutions differ from pure solvents answers questions such as the melting of ice referred to previ- ously. In this problem, however, we will focus on these two very different water-soluble compounds to illustrate some of the many topics of this chapter.

PROBLEM	SOLUTION
a. What forces are present when ethyl alcohol (CH_3CH_2OH) and $CaCl_2$ dissolve in water?	**a.** Ethyl alcohol interacts with water much like methyl alcohol which was described earlier in this chapter. The alcohol has an O—H bond so it interacts with water with hydrogen bonding forces. It does not ionize in water, so the alcohol molecule remains intact in aqueous solution. Calcium chloride interacts with water with ion-dipole forces, which overcome the ion–ion forces in the solid crystal, thus causing solution. The solution process produces one Ca^{2+} and 2 Cl^- ions per formula unit of $CaCl_2$.
b. A 100.0-g quantity of alcohol and 100.0 g of calcium chloride are each dissolved in 275 mL of water. What is the mass percent of solute in each solution?	**b.** For water, 275 mL has a mass of 275 g since the density of water is nearly 1.00 g/mL. Therefore, both solutions have a mass of 100.0 g + 275 g = 375 g. $$\frac{100.0 \text{ g solute}}{375 \text{ g solution}} \times 100\% = \underline{26.7\% \text{ by mass solute}} \text{ in both solutions.}$$
c. A 100.0 g quantity of each of these two compounds is dissolved in enough water to make 275 mL of solu- tion. What is the molarity of these two solutions?	**c.** For alcohol: $100.0 \text{ g} \times \dfrac{1 \text{ mol}}{46.07 \text{ g}} = 2.17 \text{ mol}$ $$\frac{2.17 \text{ mol}}{0.275 \text{ L}} = \underline{7.89 \text{ M}}$$ For calcium chloride: $100.0 \text{ g} \times \dfrac{1 \text{ mol}}{111 \text{ g}} = 0.901 \text{ mol}$ $$\frac{0.901 \text{ mol}}{0.275 \text{ L}} = \underline{3.28 \text{ M}}$$
d. What is the molarity of the alcohol solution if 325 mL of water is added to the solution in part **c**? Assume that the volumes are additive.	**d.** Volume of the dilute = 275 mL + 325 mL = 600 mL $$M_d = \frac{M_c \times V_c}{V_d} = \frac{7.89 \text{ M} \times 275 \text{ mL}}{600 \text{ mL}} = \underline{3.62 \text{ M}}$$
e. What mass of CO_2 and volume at STP are produced from the complete combustion of 100.0 g of alcohol?	**e.** The balanced equation for the combustion reaction is: $$C_2H_5OH(l) + 3O_2(g) \longrightarrow 2CO_2(g) + 3H_2O(l)$$ $\boxed{\text{mol alcohol}} \longrightarrow \boxed{\text{mol } CO_2} \longrightarrow \boxed{\text{g } CO_2} \text{ (or) } \boxed{\text{Volume } CO_2}$ $2.17 \text{ mol } C_2H_5OH \times \dfrac{2 \text{ mol } CO_2}{\text{mol } C_2H_5OH} \times \dfrac{44.01 \text{ g } CO_2}{\text{mol } CO_2} = \underline{191 \text{ g } CO_2}$ $2.17 \text{ mol } C_2H_5OH \times \dfrac{2 \text{ mol } CO_2}{\text{mol } C_2H_5OH} \times \dfrac{22.4 \text{ L } CO_2}{\text{mol } CO_2} = \underline{97.2 \text{ L } CO_2}$
f. Describe the electrical conductivity of these two solutions.	**f.** Alcohol does not ionize in water so it is a <u>nonelectrolyte</u> and does not conduct electricity. Calcium chloride is ionized into $Ca^{2+}(aq)$ and $2Cl^-(aq)$ ions so it is a <u>strong electrolyte</u> and is a good conductor of electricity in solution.
g. What is the melting point of the alcohol solution and the melting point and boiling point of the calcium chloride solution? Use the information from part **b**.	**g.** For the aqueous solutions: $\Delta T_f = K_f m = 1.86 \text{ (°C} \cdot \text{kg)/mol} \times \dfrac{\text{mol solute}}{\text{kg solvent}}$ For alcohol: $m = \dfrac{2.17 \text{ mol solute}}{\dfrac{275 \text{ g solvent}}{1000 \text{ g solvent}}} = \underline{7.89 \text{ m}}$

PROBLEM	SOLUTION
	$\Delta T_f = 1.86 \times 7.89\ m = 14.7°C \qquad T_f = 0 - 14.7 = \underline{-14.7°C}$
	For calcium chloride, recall that there are three moles of ions per mole of compound.
	$m\ (\text{of ions}) = \dfrac{\dfrac{3 \times 0.901\ \text{mol solute}}{275\ \text{g solvent}}}{1000\ \text{g solvent}} = 9.83\ m$
	$\Delta T_f = 1.86 \times 9.83\ m = 18.3°C \qquad T_f = 0 - 18.3 = \underline{-18.3°C}$
	For the boiling point of the calcium chloride solution:
	$\Delta T_b = K_b m = 0.512\ (°C \cdot kg)\,/\,\text{mol} \times \dfrac{\text{mol solute}}{\text{kg solvent}}$
	$\Delta T_b = 3 \times 0.512 \times \dfrac{0.901\ \text{mol}}{0.275\ \text{kg solvent}} = 5.03°C$
	$T_b = 100.0 + 5.03 = \underline{105.0°C}.$

YOUR TURN

Consider two water-soluble compounds: acetone, H_3CCCH_3 (with a double-bonded O above the central C) and nitric acid, HNO_3.

a. What forces are present when acetone and nitric acid dissolve in water?

b. A 100.0-g quantity of acetone and 100.0 g of nitric acid are each present in 475 mL of water. What is the mass percent of solute in each solution?

c. A 100.0-g quantity of each of these two compounds is dissolved in enough water to make 475 mL of solution. What is the molarity of these two solutions?

d. What is the molarity of the acetone solution if it is diluted to a new volume of 1.25 L?

e. What mass of O_2 and volume at STP are required for the complete combustion of 100.0 g of acetone?

f. Describe the electrical conductivity of these two solutions.

g. What is the melting point of the acetone solution and the melting point and boiling point of the nitric acid solution? Use the information from part **b**.

Answers are on p 429.

CHAPTER SUMMARY

Water acts as an effective solvent, dispersing solutes into a homogeneous mixture known as a solution. When two liquids are **miscible**, they form a solution, but if they are **immiscible**, they remain a heterogeneous mixture. Some ionic compounds may dissolve in water because the ion–water forces **(ion–dipole)** overcome the forces holding the crystal together. Polar covalent molecular compounds may dissolve in water as discrete molecules, or they may undergo **ionization**. Although many ionic compounds are considered soluble, others are said to be insoluble since a very limited amount dissolves. The amount that dissolves—the **solubility** of a compound—is indicated by some convenient unit of **concentration**.

How much of a compound can dissolve at a certain temperature to make a **saturated** solution varies from compound to compound. **Unsaturated** solutions contain less than the maximum amount of a compound so that more of the compound may dissolve. **Supersaturated** solutions are unstable solutions containing more of a compound than the solubility would indicate. A precipitate often forms in such a solution. Solid compounds are generally more soluble at higher temperatures, whereas gaseous compounds are less soluble at higher temperatures. This property can be used to purify solids in a process called **recrystallization**. External pressure does not affect the solubility of solids but does affect gases, which illustrates **Henry's Law**.

Besides mass of solute per 100 g of solvent, which was used to illustrate comparative solubilities, other units of concentration are **percent by mass, molarity (M),** and **molality. Parts per million (ppm)** and **parts per billion (ppb)** are used for very small concentrations.

Since molarity relates volume of a solution to moles of solute, it can be incorporated into the general scheme for stoichiometry problems along with the mass of a compound (Chapter 9) and the volume of a gas (Chapter 10) (see Figure 12-6). **Titrations** are examples of solution stoichiometry.

We also studied the physical properties of solutions. In the first property mentioned, we found that certain solutes act as **nonelectrolytes** or as either **weak** or **strong electrolytes. Electrolytes** change water from a **nonconductor (insulator)** to a **conductor** of electricity. Electrolytes produce ions in solution, which allows water to become a conductor of electricity.

There are also four **colligative properties** of solutions. These are **vapor pressure lowering, boiling-point elevation,** and **freezing-point lowering.** The magnitude of the latter two properties relate to the **molality** of the solution. The process of **osmosis** leads to **osmotic pressure elevation,** the fourth colligative property.

OBJECTIVES

SECTION	YOU SHOULD BE ABLE TO...	EXAMPLES	EXERCISES	CHAPTER PROBLEMS
12-1	Describe the role of water in the formation of aqueous solutions of ionic compounds, strong acids, and polar molecules.		1a, 1b, 1c, 1d	1, 5, 6, 7
12-2	Describe the effects of temperature and pressure on the solubility of solids and gases in a liquid.		1d, 2a, 2b, 2c, 2d	2, 3, 9, 10, 11
12-3	Perform calculations of concentration involving percent by mass, ppm, and ppb.	12-1, 12-2	3a, 3b, 3c, 3d, 3e, 3f	14, 15, 16, 20, 22
12-4	Perform the calculation involving molarity and the dilution of solutions.	12-3, 12-4, 12-5, 12-6, 12-7	4a, 4b, 4c, 4d, 4e	23, 24, 25, 26, 27, 34, 35, 37, 40, 42
12-5	Perform calculations involving titrations and other solution stoichiometry.	12-8, 12-9, 12-10	5a, 5b, 5c	44, 45, 47, 49, 50, 51, 53
12-6	Explain the differences between nonelectrolytes, strong electrolytes, and weak electrolytes.		6a, 6b	56, 57
12-7	Calculate the boiling and melting points of aqueous solutions of electrolytes and nonelectrolytes.	12-11, 12-12, 12-13	7a, 7b, 7c, 7d, 7e	60, 62, 66, 67, 71, 72, 73, 74, 75, 80

▶ ANSWERS TO ASSESSING THE OBJECTIVES

PART A

EXERCISES

12-1(a) (a) yes (b) yes (c) no (d) yes (e) yes

12-1(b) (a) $2NH_4^+ (aq) + Cr_2O_7^{2-} (aq)$ (b) $H^+ (aq) + NO_3^- (aq)$ (c) $CH_3COCH_3 (aq)$ (d) $Zn^{2+} (aq) + 2Br^- (aq)$

12-1(c) (a) ion–dipole (b) hydrogen bonding (c) London force (the only force between nonpolar molecules)

12-1(d) The term *dissociates* here means "to break apart." The chemical is already ionic, so the water merely separates the positive ion from the negative ion. In the case of HNO_3, the compound was originally molecular in nature. But due to the strong interaction with water, the water was able to *ionize* the acid, or form ions out of the individual particles, so that what ends up dissolved is H^+ and NO_3^-.

12-2(a) If a soluble compound dissolves to the limit of its solubility at a given temperature, the solution is <u>saturated</u>; if the solute is not dissolved to the limit, the solution is <u>unsaturated</u>. If more solute is present than indicated by its solubility, the solution is <u>supersaturated</u> and the excess solute may eventually form a <u>precipitate</u>. Solids generally become <u>more</u> soluble at higher temperatures, whereas gases become <u>less</u> soluble. An increase in partial pressure of a gas above a liquid <u>increases</u> its solubility.

12-2(b) (a) Cl_2 is a gas. It is more soluble at low temperature and high pressures. (b) Na_2SO_4 is an ionic solid. It is more soluble at high temperatures but pressure has no effect.

12-2(c) (a) supersaturated (b) saturated (c) unsaturated

12-2(d) (a) supersaturated (b) unsaturated (c) saturated (d) unsaturated

12-2(e) Something that large should be ground up into smaller particles to increase the surface area. Since it's an ionic solid, the solvent can be gently heated to help dissolve the solid. Stirring or swirling the flask usually helps as well.

12-3(a) 8.0 g KBr/100 g solution, 8.0 g KBr/92 g H_2O, and 92 g H_2O/100 g solution

12-3(b) 15.0 g/60.0 g = 25.0% by mass

12-3(c) 154 g of solvent

12-3(d) 18 g

12-3(e) **(a)** 2.5×10^{-5}% **(b)** 0.25 ppm **(c)** 250 ppb

Based on the size of the numbers, the ppb gives the most appropriately sized value.

12-3(f) 0.75 mg lead

12-4(a) 2.00 M

12-4(b) 0.200 M

12-4(c) 6.23 g KI

12-4(d) 75.0 ml of H_2O must be added to bring the total volume to 125 ml.

12-4(e) 16.7 mL of 6.00 M H_2SO_4

12-4(f) **(a)** Using a 250-mL volumetric flask, measure out 7.31 g of NaCl. Transfer the solid into the flask. Fill the flask halfway with water and swirl until all the solid is dissolved. Fill the flask the rest of the way, using a dropper at the end to get the volume to the volumetric line. Cap and invert several times. **(b)** Using a buret, dispense 41.7 mL of 3.00 M NaCl solution into a 250-mL volumetric flask. Fill the rest of the way with water, swirling to ensure thorough mixing. Use a dropper at the end to get the volume to the volumetric line. Cap and invert several times.

12-4(g) 0.27 M

12-5(a) **(a)** molar volume (22.4 L/mol) **(b)** molarity **(c)** molar mass **(d)** Avogadro's number (6.022×10^{23} molecules/mol)

12-5(b) 13.5 mL of 0.125 M $Ca(OH)_2$

12-5(c) 1.26 M HCl

12-5(d) $Na_2CO_3 + 2HCl \longrightarrow 2NaCl + H_2O + CO_2$
6.25 M HCl

12-5(e) 5.67 mL

Part B

EXERCISES

12-6(a) **(a)** strong electrolytes **(b)** weak electrolytes **(c)** nonelectrolytes **(d)** strong electrolytes **(e)** nonelectrolytes **(f)** weak electrolytes

12-6(b) **(a)** weak electrolyte **(b)** strong electrolyte **(c)** nonelectrolytes

12-6(c) Swimming pool water has several compounds dissolved in it. Chlorine itself is a nonelectrolyte, but the forms in which it is delivered are mostly ionic strong electrolytes (such as calcium hypochlorite). Further, the water used in the pool naturally has several dissolved ions like Ca^{2+}, Mg^{2+} and Fe^{3+} in it already, which increase its conductivity. The compounds used to provide chlorine, though, are the primary conductive medium.

12-7(a) **(a)** liquid B **(b)** liquid A **(c)** liquid B **(d)** liquid A

12-7(b) 3.53 m

12-7(c) $-5.58°C$ for the nonelectrolyte solute; $-16.7°C$ for $CaCl_2$

12-7(d) The effect of acetic acid would be somewhere between a nonelectrolyte solute, which produces one particle per mole, and NaCl, which produces two particles per mole. Acetic acid is a weak electrolyte, which means only a small percent of the molecules are ionized. In fact, its effect would be only slightly greater than that of a nonelectrolyte.

12-7(e) $-17.8°C$. $0.0°F$. This is the method used to define zero on the Fahrenheit scale.

ANSWERS TO CHAPTER SYNTHESIS PROBLEM

a. Acetone is a polar compound that can interact with water molecules by hydrogen bonding between the O on acetone and the H's on water. It does not ionize when dissolved in water. Nitric acid is a strong acid and dissolves through ion-dipole interactions with ionization to form $H^+(aq)$ and $NO_3^-(aq)$ ions.

b. Both solutions are 17.4% solute by mass.

c. Acetone is 3.62 M; nitric acid is 3.35 M.

d. 1.38 M

e. 220 g O_2, 154 L O_2 (STP)

f. Acetone is a nonelectrolyte since it does ionize. Its solutions do not conduct electricity. Nitric acid is a strong electrolyte since strong acids are completely ionized in aqueous solution. Its solution is a good conductor of electricity.

g. Acetone; melting point = $-6.74°C$. For nitric acid, melting point = $-12.5°C$. The boiling point is $103.4°C$.

CHAPTER PROBLEMS

Throughout the text, answers to all exercises in color are given in Appendix E. The more difficult exercises are marked with an asterisk.

Aqueous Solutions (SECTION 12-1)

12-1. When an ionic compound dissolves in water, what forces in the crystal resist the solution process? What forces between water molecules and the crystal remove the ions from the lattice?

12-2. When a sample of KOH is placed in water, a homogeneous mixture of KOH is formed. Which is the solute, which is the solvent, and which is the solution?

12-3. Calcium bromide readily dissolves in water, but lead(II) bromide does not. Liquid benzene and water form a heterogeneous mixture, but liquid isopropyl alcohol and water mix thoroughly. Which of the above is said to be miscible, which immiscible, which insoluble, and which soluble?

12-4. Write equations illustrating the solution of each of the following ionic compounds in water.

(a) LiF   (b) $(NH_4)_3PO_4$ (c) Na_2CO_3 (d) $Ca(C_2H_3O_2)_2$

12-5. Write equations illustrating the solution of each of the following ionic compounds in water.

(a) $BaCl_2$ (b) $Al_2(SO_4)_3$ (c) $Cr(NO_3)_3$ (d) $Mg(ClO_4)_2$

12-6. Formaldehyde (H_2CO) dissolves in water without formation of ions. Write the Lewis structure of formaldehyde and show what types of interactions between solute and solvent are involved.

12-7. Nitric acid is a covalent compound that dissolves in water to form ions in the same manner as HCl. Write the equation illustrating the solution of nitric acid in water.

Temperature and Solubility (SECTION 12-2)

12-8. Referring to Figure 12-5, determine which of the following compounds is most soluble at 10°C: NaCl, KCl, or Li_2SO_4. Which is most soluble at 70°C?

12-9. Referring to Figure 12-5, determine what mass of each of the following dissolves in 250 g of H_2O at 60°C: KBr, KCl, and Li_2SO_4.

12-10. Referring to Figure 12-5, determine whether each of the following solutions is saturated, unsaturated, or supersaturated. (All are in 100 g of H_2O.)

(a) 40 g of KNO_3 at 40°C   (c) 75 g of KBr at 80°C

(b) 40 g of KNO_3 at 20°C   (d) 20 g of NaCl at 40°C

12-11. A 200-g sample of water is saturated with KNO_3 at 50°C. What mass of KNO_3 forms as a precipitate if the solution is cooled to the freezing point of water? (Refer to Figure 12-5.)

12-12. A 500-mL portion of water is saturated with Li_2SO_4 at 0°C. What happens if the solution is heated to 100°C? (Refer to Figure 12-5.)

Percent by Mass (SECTION 12-3)

12-13. What is the percent by mass of solute in a solution made by dissolving 9.85 g of $Ca(NO_3)_2$ in 650 g of water?

12-14. What is the percent by mass of solute if 14.15 g of NaI is mixed with 75.55 g of water?

12-15. A solution is 10.0% by mass NaOH. How many moles of NaOH are dissolved in 150 g of solution?

12-16. A solution contains 15.0 g of NH_4Cl in water and is 8.50% NH_4Cl. What is the mass of water present?

12-17. A solution is 23.2% by mass KNO_3. What mass of KNO_3 is present in each 100 g of H_2O?

12-18. A solution contains 1 mol of NaOH dissolved in 9 mol of ethyl alcohol (C_2H_5OH). What is the percent by mass NaOH?

12-19. Blood contains 10 mg of calcium ions in 100 g of blood serum (solution). What is this concentration in ppm?

12-20. A high concentration of mercury in fish is 0.5 ppm. What mass of mercury is present in each kilogram of fish? What is this concentration in ppb?

12-21. Seawater contains 1.2×10^{-2} ppb of gold ions. If all the gold could be extracted, what volume in liters of seawater is needed to produce 1.00 g of gold? (Assume the density of seawater is the same as that of pure water.)

12-22. The maximum allowable level of lead in drinking water is 50 ppb. What mass of lead in milligrams is allowed in 5000 gallons of water? (Assume that the density of the water is the same as that of pure water.)

Molarity (SECTION 12-4)

12-23. What is the molarity of a solution made by dissolving 2.44 mol of NaCl in enough water to make 4.50 L of solution?

12-24. Fill in the blanks.

Solute	M	Amount of Solute	Volume of Solution
(a) KI	_____	2.40 mol	2.75 L
(b) C_2H_5OH	_____	26.5 g	410 mL
(c) $NaC_2H_3O_2$	0.255	3.15 mol	_____ L
(d) $LiNO_2$	0.625	_____ g	1.25 L
(e) $BaCl_2$	_____	0.250 mol	850 mL
(f) Na_2SO_3	0.054	_____ mol	0.45 L
(g) K_2CO_3	0.345	14.7 g	_____ mL
(h) LiOH	1.24	_____ g	1650 mL
(i) H_2SO_4	0.905	0.178 g	_____ mL

12-25. What is the molarity of a solution of 345 g of Epsom salts $(MgSO_4 \cdot 7H_2O)$ in 7.50 L of solution?

12-26. What mass of $CaCl_2$ is in 2.58 L of a solution with a concentration of 0.0784 M?

12-27. What volume in liters of a 0.250 M solution contains 37.5 g of KOH?

12-28. What is the molarity of a solution made by dissolving 2.50×10^{-4} g of baking soda $(NaHCO_3)$ in enough water to make 2.54 mL of solution?

12-29. What are the molarities of the hydroxide ion and the barium ion if 13.5 g of $Ba(OH)_2$ is dissolved in enough water to make 475 mL of solution?

12-30. What is the molarity of each ion present if 25.0 g of $Al_2(SO_4)_3$ is present in 250 mL of solution?

*12-31. A solution is 25.0% by mass calcium nitrate and has a density of 1.21 g/mL. What is its molarity?

*12-32. A solution of concentrated NaOH is 16.4 M. If the density of the solution is 1.43 g/mL, what is the percent by mass NaOH?

*12-33. Concentrated nitric acid is 70.0% HNO_3 and 14.7 M. What is the density of the solution?

Dilution (SECTION 12-4)

12-34. What volume of 4.50 M H_2SO_4 should be diluted with water to form 2.50 L of 1.50 M acid?

12-35. If 450 mL of a certain solution is diluted to 950 mL with water to form a 0.600 M solution, what was the molarity of the original solution?

12-36. One liter of a 0.250 M solution of NaOH is needed. The only available solution of NaOH is a 0.800 M solution. Describe how to make the desired solution.

12-37. What is the volume in liters of a 0.440 M solution if it was made by dilution of 250 mL of a 1.25 M solution?

12-38. What is the molarity of a solution made by diluting 3.50 L of a 0.200 M solution to a volume of 5.00 L?

12-39. What volume of water in milliliters should be *added* to 1.25 L of 0.860 M HCl so that its molarity will be 0.545? Assume additive volumes.

12-40. What volume of water in milliliters should be *added* to 400 mL of a solution containing 35.0 g of KBr to make a 0.100 M KBr solution? Assume additive volumes.

*12-41. What volume in milliliters of *pure* acetic acid should be used to make 250 mL of 0.200 M $HC_2H_3O_2$? (The density of the pure acid is 1.05 g/mL.)

*12-42. What would be the molarity of a solution made by mixing 150 mL of 0.250 M HCl with 450 mL of 0.375 M HCl?

Stoichiometry and Titrations (SECTION 12-5)

12-43. Given the reaction

$$3KOH(aq) + CrCl_3(aq) \longrightarrow Cr(OH)_3(s) + 3KCl(aq)$$

what mass of $Cr(OH)_3$ would be produced if 500 mL of 0.250 M KOH were added to a solution containing excess $CrCl_3$?

12-44. Given the reaction

$$2KCl(aq) + Pb(NO_3)_2(aq) \longrightarrow PbCl_2(s) + 2KNO_3(aq)$$

what mass of $Pb(NO_3)_2$ is required to react with 1.25 L of 0.550 M KCl?

12-45. Given the reaction

$$Al_2(SO_4)_3(aq) + 3BaCl_2(aq) \longrightarrow 3BaSO_4(s) + 2AlCl_3(aq)$$

what mass of $BaSO_4$ is produced from 650 mL of 0.320 M $Al_2(SO_4)_3$?

12-46. Given the reaction

$$3Ba(OH)_2(aq) + 2Al(NO_3)_3(aq) \longrightarrow$$
$$2Al(OH)_3(s) + 3Ba(NO_3)_2(aq)$$

what volume of 1.25 M $Ba(OH)_2$ is required to produce 265 g of $Al(OH)_3$?

12-47. Given the reaction

$$2AgClO_4(aq) + Na_2CrO_4(aq) \longrightarrow Ag_2CrO_4(s) + 2NaClO_4(aq)$$

what volume of a 0.600 M solution of $AgClO_4$ is needed to produce 160 g of Ag_2CrO_4?

12-48. Given the reaction

$$3Ca(ClO_3)_2(aq) + 2Na_3PO_4(aq) \longrightarrow$$
$$Ca_3(PO_4)_2(s) + 6NaClO_3(aq)$$

what volume of a 2.22 M solution of Na_3PO_4 is needed to react with 580 mL of a 3.75 M solution of $Ca(ClO_3)_2$?

12-49. Consider the reaction

$$2HNO_3(aq) + 3H_2S(aq) \longrightarrow 2NO(g) + 3S(s) + 4H_2O(l)$$

(a) What volume of 0.350 M HNO_3 will completely react with 275 mL of 0.100 M H_2S?

(b) What volume of NO gas measured at 27°C and 720 torr will be produced from 650 mL of 0.100 M H_2S solution?

12-50. The concentration of acetic acid in vinegar can be determined by titration with sodium hydroxide. This reaction is represented by the following equation.

$$HC_2H_3O_2(aq) + NaOH(aq) \longrightarrow NaC_2H_3O_2(aq) + H_2O$$

What is the concentration of acetic acid in vinegar if 10.0 mL of vinegar takes 28.8 mL of 0.300 M NaOH to reach the endpoint?

12-51. A household antibacterial cleanser is made from a solution of sodium hydroxide. What is its concentration of NaOH if it takes 42.5 mL of 0.0500 M HCl to titrate 25.0 mL of the cleanser to the end point?

12-52. Dilute aqueous solutions of ammonia are used as cleansers, especially for grease stains. What is the concentration of the ammonia in the cleanser if 22.6 mL of 0.220 M sulfuric acid is needed to titrate 10.0 mL of the ammonia solution? The equation for neutralization is

$$2NH_3(aq) + H_2SO_4(aq) \longrightarrow 2NH_4^+(aq) + SO_4^{2-}(aq)$$

12-53. What is the molarity of 1.00 L of HNO_3 solution if it reacts completely with 25.0 g of $Ca(OH)_2$?

*12-54. Given the reaction

$$2NaOH(aq) + MgCl_2(aq) \longrightarrow Mg(OH)_2(s) + 2NaCl(aq)$$

what mass of $Mg(OH)_2$ would be produced by mixing 250 mL of 0.240 M NaOH with 400 mL of 0.100 M $MgCl_2$?

*12-55. Given the reaction

$$CO_2(g) + Ca(OH)_2(aq) \longrightarrow CaCO_3(s) + H_2O(l)$$

what is the molarity of a 1.00 L solution of $Ca(OH)_2$ that would completely react with 10.0 L of CO_2 measured at 25°C and 0.950 atm?

Properties of Solutions (SECTIONS 12-6 AND 12-7)

12-56. Three hypothetical binary compounds dissolve in water. AB is a strong electrolyte, AC is a weak electrolyte, and AD is a nonelectrolyte. Describe the extent to which each of these solutions conducts electricity and how each compound exists in solution.

12-57. Chlorous acid ($HClO_2$) is a weak electrolyte, and perchloric acid ($HClO_4$) is a strong electrolyte. Write equations illustrating the different behaviors of these two polar covalent molecules in water.

12-58. Explain the difference in the following three terms: 1 mole NaBr, 1 molar NaBr, and 1 molal NaBr.

12-59. What is the molality of a solution made by dissolving 25.0 g of NaOH in **(a)** 250 g of water and **(b)** 250 g of alcohol (C_2H_5OH)?

12-60. What is the molality of a solution made by dissolving 1.50 kg of KCl in 2.85 kg of water?

12-61. What mass of NaOH is in 550 g of water if the concentration is 0.720 m?

12-62. What mass of water is in a 0.430 m solution containing 2.58 g of CH_3OH?

12-63. What is the freezing point of a 0.20 m aqueous solution of a nonelectrolyte?

12-64. What is the boiling point of a 0.45 m aqueous solution of a nonelectrolyte?

12-65. When immersed in salty ocean water for an extended period, a person gets very thirsty. Explain.

12-66. Dehydrated fruit is wrinkled and shriveled up. When put in water, the fruit expands and becomes smooth again. Explain.

12-67. Explain how pure water can be obtained from a solution without boiling.

12-68. In industrial processes, it is often necessary to concentrate a dilute solution (much more difficult than diluting a concentrated solution). Explain how the principle of reverse osmosis can be applied.

*12-69. What is the molality of an aqueous solution that is 10.0% by mass $CaCl_2$?

*12-70. A 1.00 m KBr solution has a mass of 1.00 kg. What is the mass of the water?

*12-71. Ethylene glycol ($C_2H_6O_2$) is used as an antifreeze. What mass of ethylene glycol should be added to 5.00 kg of water to lower the freezing point to −5.0°C? (Ethylene glycol is a nonelectrolyte.)

*12-72. What is the boiling point of the solution in problem 12-71?

*12-73. Methyl alcohol can also be used as an antifreeze. What mass of methyl alcohol (CH_3OH) must be added to 5.00 kg of water to lower its freezing point to −5.0°C?

12-74. What is the molality of an aqueous solution that boils at 101.5°C?

12-75. What is the boiling point of a 0.15 m solution of a solute in liquid benzene? (For benzene, $K_b = 2.53$, and the boiling point of pure benzene is 80.1 °C.)

12-76. What is the boiling point of a solution of 75.0 g of naphthalene ($C_{10}H_8$) in 250 g of benzene? (See problem 12-75.)

12-77. What is the freezing point of a solution of 100 g of CH_3OH in 800 g of benzene? (For benzene, $K_f = 5.12$, and the freezing point of pure benzene is 5.5°C.)

12-78. What is the freezing point of a 10.0% by mass solution of CH_3OH in benzene? (See problem 12-77.)

12-79. A 1 m solution of HCl lowers the freezing point of water almost twice as much as a 1 m solution of HF. Explain.

12-80. What is the freezing point of automobile antifreeze if it is 40.0% by mass ethylene glycol ($C_2H_6O_2$) in water? (Ethylene glycol is a nonelectrolyte.)

12-81. In especially cold climates, methyl alcohol (CH_3OH) may be used as an automobile antifreeze. Would 40.0% by mass of an aqueous solution of methyl alcohol remain a liquid at −40°C? (Methyl alcohol is a nonelectrolyte.)

*12-82. Give the freezing point of each of the following in 100 g of water.

(a) 10.0 g of CH_3OH **(b)** 10.0 g of NaCl **(c)** 10.0 g of $CaCl_2$

General Problems

12-83. A mixture is composed of 10 g of KNO_3 and 50 g of KCl. What is the approximate amount of KCl that can be separated using the difference in solubility shown in Figure 12-5.

12-84. KBr and KNO_3 have equal solubilities at about 42°C. What is the composition of the precipitate if 100 g of H_2O saturated with these two salts at 42°C is then cooled to 0°C? (Refer to Figure 12-5.)

12-85. What is the percent composition by mass of a solution made by dissolving 10.0 g of sugar and 5.0 g of table salt in 150 mL of water?

12-86. What is the molarity of each ion in a solution that is 0.15 M $CaCl_2$, 0.22 M $Ca(ClO_4)_2$, and 0.18 M NaCl?

*12-87. 500 mL of 0.20 M $AgNO_3$ is mixed with 500 mL of 0.30 M NaCl. What is the concentration of Cl^- ion in the solution? The net ionic equation of the reaction that occurs is

$$Ag^+(aq) + Cl^-(aq) \longrightarrow AgCl(s)$$

*12-88. 400 mL of 0.15 M $Ca(NO_3)_2$ is mixed with 500 mL of 0.20 M Na_2SO_4. Write the net ionic equation of the precipitation reaction that occurs. What is its concentration of SO_4^{2-} remaining in solution?

*12-89. A certain metal (M) reacts with HCl according to the equation

$$M(s) + 2HCl(aq) \longrightarrow MCl_2(aq) + H_2(g)$$

1.44 g of the metal reacts with 225 mL of 0.196 M HCl. What is the metal?

*12-90. Another metal (Z) also reacts with HCl according to the equation

$$2Z(s) + 6HCl(aq) \longrightarrow 2ZCl_3(aq) + 3H_2(g)$$

24.0 g of Z reacts with 0.545 L of 2.54 M HCl. What is the metal? What volume of H_2 measured at STP is produced?

*12-91. A certain compound dissolves in a solvent, nitrobenzene. For nitrobenzene, $K_f = 8.10$. A solution with 3.07 g of the compound dissolved in 120 g of nitrobenzene freezes at 2.22°C. The freezing point of pure nitrobenzene is 5.67°C. Analysis of the compound shows it to be 40.0% C, 13.3% H, and 46.7% N. What is the formula of the compound?

12-92. Given 1.00 m aqueous solutions of **(a)** Na_3PO_4, **(b)** $CaCl_2$, **(c)** urea (a nonelectrolyte), **(d)** $Al_2(SO_4)_3$, and

(e) LiBr. Order these solutions from highest to lowest freezing points and explain.

12-93. Order the following solutions from lowest to highest boiling points.

(a) 0.30 m sugar (a nonelectrolyte) (d) 0.12 m KCl

(b) pure water (e) 0.05 m CrCl$_3$

(c) 0.05 m K$_2$CO$_3$

*12-94. One mole of an electrolyte dissolves in water to form three moles of ions. A 9.21-g quantity of this compound is dissolved in 175 g of water. The freezing point of this solution is −1.77°C. The compound is 47.1% K, 14.5% C, and 38.6% O. What is the formula of this compound?

*12-95. A sample of a metal reacts with water to form 487 mL of a 0.120-M solution of the metal hydroxide along with 1.10 L of hydrogen gas measured at 25°C and 0.650 atm. Is the metal Na or Ca?

12-96. What volume of NH$_3$ measured at 0.951 atm and 25°C is needed to form 250 mL of 0.450 M aqueous ammonia?

12-97. A 1.82-L volume of gaseous H$_2$S measured at 1.08 atm and 20°C is dissolved in water. What is the molarity of the aqueous H$_2$S if the volume of the solution is 2.00 L?

12-98. Sodium bicarbonate reacts with hydrochloric acid to form water, sodium chloride, and carbon dioxide gas. What volume of CO$_2$ measured at 35°C and 1.00 atm could be released by the reaction of 1.00 L of a 0.340 M solution of sodium bicarbonate with excess hydrochloric acid solution?

12-99. Aqueous calcium hydroxide solutions absorb gaseous carbon dioxide to form calcium bicarbonate solutions. What mass of calcium bicarbonate would be formed by reaction of 125.0 mL of 0.150 M calcium hydroxide with 450 mL of gaseous carbon dioxide measured at STP?

*12-100. The molecules of a compound are composed of one phosphorus and multiple chlorine atoms. A molecule of the compound is described as a trigonal pyramid. This gaseous compound dissolves in water to form a hydrochloric acid solution and phosphorus acid (H$_3$PO$_3$). What is the molarity of the hydrochloric acid if 750 mL of the gas, measured at STP, dissolves in 250 mL of water?

*12-101. A phosphorus-oxygen compound is 43.7% phosphorus. When the compound dissolves in water, it forms one compound, phosphoric acid. If 0.100 mol of the phosphorus-oxygen compound dissolves in 4.00 L of water to form a 0.100 M solution of phosphoric acid, what is the formula of the original compound?

*12-102. A 10.0-g quantity of a compound is dissolved in 100 g of water. The solution formed has a melting point of −7.14°C. Is the compound KCl, Na$_2$S, or CaCl$_2$?

12-103. An aqueous solution has a freezing point of −2.50°C. What is its boiling point?

*12-104. An aqueous solution of a nonelectrolyte is made by dissolving the solute in 1.00 L of water. The solution has a freezing point of −1.50°C. What volume of water must be added to change the freezing point to −1.15°C?

STUDENT WORKSHOP

Determining the Dipole of Planar Molecules

Purpose: To estimate the dipole moment of several simple molecules based on molecular shape and bond polarity. (Work in groups of three or four. Estimated time: 25 min.)

As we have seen in this chapter, polarity plays a major role in determining what compounds will dissolve in what solvents. "Like dissolves like" is a saying that indicates that polar solutes dissolve in polar solvents and nonpolar solutes dissolve in nonpolar solvents. Slightly polar solutes might dissolve in either, but certainly slightly polar solvents would be our first choice. The degree of polarity of a molecule can be estimated by the following activity. We will use only linear and planar molecules for simplicity. As an example, we will evaluate the polarity of formaldehyde, CH$_2$O.

1. Draw the Lewis structure of the molecule, attempting to be faithful to the bond angles predicted by VSEPR theory.

2. Draw arrows next to each bond, pointing in the direction of the more electronegative atom. The length of the arrow should be proportional to the difference in electronegativity between the two atoms. In this case, oxygen is significantly more electronegative than carbon, which is slightly more electronegative than hydrogen.

3. Next to the structure on the paper, redraw the arrows, tail to tip, in any order, maintaining their exact orientation.

4. Connect the tail of the first arrow to the tip of the last with another arrow. This last arrow represents the dipole of the molecule. The longer it is, the more polar the molecule, and it points toward the molecule's negative end. If the tail and tip are exactly in the same place, then there is no dipole, and the molecule is nonpolar.

Using the above procedure, evaluate the dipoles of the following molecules or ions:

- H$_2$O
- CO$_2$
- SO$_2$
- COCl$_2$
- HCN
- NO$_3^-$

Acids, Bases, and Salts

The lemon is known for its sour taste, which is caused by citric acid. Acids are a unique class of compounds that are discussed in this chapter.

SETTING THE STAGE

It can make your mouth pucker, your body shudder, and your eyes water. That's the reaction one gets from taking a bite of a fresh lemon. A taste of vinegar has the same effect. Even carbonated beverages produce a subtle sour taste that peps up the drink. All these substances have a similar effect because of the presence of a compound that produces a sour taste (but to different degrees). These compounds, known as acids, were characterized over 500 years ago, during the Middle Ages, in the chemical laboratories of alchemists. Substances were classified as acids because of their common properties (such as sourness) rather than a certain chemical composition. Acids are well known to the general population. They are very common in foods and drugs, such as citric acid in lemons, acetic acid in vinegar, lactic acid in sour milk, or acetylsalicylic acid in aspirin. Some must be handled with caution, such as sulfuric acid used in car batteries, or hydrochloric acid, used to clean concrete. We also relate the word *acid* to the serious environmental hazard of acid rain. When rain is acidic, reactions characteristic of acids lead to the degradation of stone used in buildings and the liberation of poisonous metal ions locked in soil.

We briefly introduced acids, bases, and the reaction between them in Section 6-6-2. At that time we restricted our discussion to strong acids and bases. We have covered quite a bit of chemistry since Chapter 6, so we will reexamine this important topic in more detail. We can expand on our discussion of these compounds by including weak acids and bases.

In Part A of this chapter, we will review and expand on the concept of acids and bases introduced in earlier chapters. How acidity is measured and reported is the emphasis of Part B. Finally, we expand the concept of acids and bases even further to include salts, buffers, and oxides in Part C.

Part A

Acids, Bases, and the Formation of Salts

SETTING A GOAL

- You will expand your knowledge of acids and bases with more general definitions that allow us to examine their comparative strengths.

OBJECTIVES

13-1 List the general properties of acids and bases.

13-2 Identify Brønsted acids and bases and conjugate acid–base pairs in a proton exchange reaction.

13-3 Calculate the hydronium ion concentration in a solution of a strong acid and a weak acid given the initial concentration of the acid and the percent ionization of the weak acid.

13-4 Write the molecular, total ionic, and net ionic equations for neutralization reactions.

▶ **OBJECTIVE FOR SECTION 13-1**

List the general properties of acids and bases.

FIGURE 13-1 Zinc and Limestone in Acid Zinc (top) reacts with acid to liberate hydrogen; limestone (CaCO₃, bottom) reacts with acid to liberate carbon dioxide.

13-1 Properties of Acids and Bases

LOOKING AHEAD! Historically, well before scientists knew much about their compositions, acids and bases were classified as such based on their common properties. These common properties relate to a specific "active ingredient." The nature of these properties and the active ingredients are the topics of this section. ■

The sour taste of acids accounts for the origin of the word itself. The word *acid* originates from the Latin *acidus,* meaning "sour," or the closely related Latin *acetum,* meaning "vinegar." This ancient class of compounds has several characteristic chemical properties. Acids are compounds that do the following:

1. Taste sour (of course, one *never* tastes laboratory chemicals)
2. React with certain metals (e.g., Zn and Fe), with the liberation of hydrogen gas (see Figure 13-1)
3. Cause certain organic dyes to change color (e.g., litmus paper turns from blue to red in acids)
4. React with limestone ($CaCO_3$), with the liberation of carbon dioxide gas (see Figure 13-1)
5. React with bases to form salts and water

Some familiar acids, their common names, and their formulas are shown below.

CHEMICAL NAME	COMMON NAME	FORMULA
hydrochloric acid	muriatic acid	HCl
sulfuric acid	oil of vitriol, battery acid	H_2SO_4
acetic acid	vinegar (sour ingredient)	$HC_2H_3O_2$
carbonic acid	carbonated water	H_2CO_3

The counterparts to acids are bases. Bases are compounds that do the following:

1. Taste bitter
2. Feel slippery or soapy
3. Dissolve oils and grease
4. Cause certain organic dyes to change color (e.g., litmus paper turns from red to blue in bases)
5. React with acids to form salts and water

13-1.1 Arrhenius Acids and Bases

The properties listed above relate to what acids and bases do, not to their chemical composition. It was not until 1884 that a Swedish chemist, Svante Arrhenius, suggested that the particular composition of these compounds determined their behavior. *He proposed that acids produced H^+ ions and bases produced OH^- ions in water.*

13-1.2 Strong Acids in Water

The ionization process of acids in water was illustrated in Figure 12-3 in the previous chapter. To illustrate the importance of water in the ionization process, the reaction of an acid with water can be represented as

$$HCl + H_2O \longrightarrow H_3O^+(aq) + Cl^-(aq)$$

Instead of H^+, the acid species is often represented as H_3O^+, which is known as the **hydronium ion**. The hydronium ion is simply a representation of the H^+ ion in a hydrated form. The acid species is represented as H_3O^+ rather than H^+ because it is somewhat closer to what is believed to be the actual species. In fact, the nature of H^+ in aqueous solution is even more complex than H_3O^+ (i.e., $H_5O_2^+$, $H_7O_3^+$, etc.). In any case, the acid species can be represented as $H^+(aq)$ or $H_3O^+(aq)$, depending on the convenience of the particular situation. *Just remember that both refer to the same species in aqueous solution.* If $H^+(aq)$ is used, it should be understood that it is not just a bare proton in aqueous solution but is associated with water molecules. (It is hydrated.)

It is the current practice to list on the label the active ingredient in medicines or drugs. In this regard, the active ingredient of acids is the $H^+(aq)$ ion. We can now see how this ion accounts for some of the behavior of acids listed previously. Equation 1 in the list below illustrates the reaction of an acid with a metal. In equation 2, the reaction of an acid with limestone is illustrated. And in equation 3, we show a neutralization reaction. The net ionic equations of these reactions are also shown, which emphasizes the role of the $H^+(aq)$ ion in each case.

1. Acids react with metals (e.g., Zn) and give off hydrogen gas.

$$Zn(s) + 2HCl(aq) \longrightarrow ZnCl_2(aq) + H_2(g)$$

$$Zn(s) + \underline{2H^+(aq)} \longrightarrow Zn^{2+}(aq) + H_2(g)$$

2. Acids react with limestone to give off carbon dioxide gas.

$$CaCO_3(s) + 2HNO_3(aq) \longrightarrow Ca(NO_3)_2(aq) + H_2O(l) + CO_2(g)$$

$$CaCO_3(s) + \underline{2H^+(aq)} \longrightarrow Ca^{2+}(aq) + H_2O(l) + CO_2(g)$$

3. Acids react with bases.

$$ACID + BASE \longrightarrow SALT + WATER$$

$$HClO_4(aq) + NaOH(aq) \longrightarrow NaClO_4(aq) + H_2O(l)$$

$$\underline{H^+(aq)} + OH^-(aq) \longrightarrow H_2O(l)$$

The last reaction is of prime importance and is discussed in more detail in Section 13-4.

13-1.3 Strong Bases in Water

Now let's turn our attention to bases. Bases are compounds that produce OH^- ions in water, forming what are known as basic solutions, sometimes referred to as *alkaline* or *caustic* solutions. Some of the commonly known bases are sodium hydroxide (also known as caustic soda, or lye), potassium hydroxide (caustic potash), calcium hydroxide (slaked lime), and ammonia. Except for ammonia, these compounds are all solid ionic compounds. Forming a solution in water simply releases the OH^- ion into the aqueous medium.

$$NaOH(s) \xrightarrow{H_2O} Na^+(aq) + OH^-(aq)$$

$$Ba(OH)_2(s) \xrightarrow{H_2O} Ba^{2+}(aq) + 2OH^-(aq)$$

The action of ammonia (NH_3) as a base is somewhat different from that of the ionic hydroxides and is better described by a more detailed look at acids and bases in the following section.

▶ **ASSESSING THE OBJECTIVE FOR SECTION 13-1**

EXERCISE 13-1(a) LEVEL 1: Fill in the blanks.
An acid is a compound that produces the _____ ion in solution, which is also written as the hydronium ion (_____). A base is a compound that produces an ion with the formula _____ and the name _____ ion.

EXERCISE 13-1(b) LEVEL 2: Write the formula and name of the acid or base formed from the following ions.
(a) ClO_4^- **(b)** Fe^{2+} **(c)** S^{2-} **(d)** Li^+

EXERCISE 13-1(c) LEVEL 2: Write an equation for the reaction that occurs when the following compounds are placed in water:
(a) HI   **(b)** LiOH

EXERCISE 13-1(d) LEVEL 3: Based on the list of properties of acids and bases, which of the two do the following most likely contain?
(a) drain cleaner   **(d)** dishwashing detergent
(b) orange juice   **(e)** bleach
(c) salad dressing   **(f)** antacid

For additional practice, work chapter problems 13-1, 13-3, 13-5, and 13-6

▶ **OBJECTIVE
FOR SECTION 13-2**
Identify Brønsted acids and bases and conjugate acid–base pairs in a proton exchange reaction.

13-2 Brønsted–Lowry Acids and Bases

LOOKING AHEAD! Besides neutral compounds that act as acids or bases in water, certain ions have acid or base behavior. However, to better describe the action of ions in water, we need a more inclusive definition of acids and bases. We do this in the following discussion. ∎

Limestone ($CaCO_3$) is quite a versatile compound. We can use it as solid rock in the construction of huge buildings or we can use it as a powder in chalk. It is also the major ingredient of many antacids, which are consumed to neutralize excess stomach acid. In this reaction, the carbonate ion (CO_3^{2-}) is the ingredient that reacts as a base (i.e., an antacid). We previously discussed this reaction as a *gas-forming neutralization reaction* in Section 6-6.3. However, from our previous definition of acids and bases, it is not immediately obvious how an anion such as CO_3^{2-} behaves as a base. In order to include anions as bases, we would be aided by a broader, more inclusive definition than that of Arrhenius, which focused mainly on molecular compounds. We will now focus on the role of the H^+ ion in solution. *In the* **Brønsted–Lowry** *definition, an* **acid** *is a proton (H^+) donor and a* **base** *is a proton acceptor.* To illustrate this definition, we again look at the reaction of HCl as an acid to form the H_3O^+ ion.

$$HCl(aq) + H_2O(l) \longrightarrow H_3O^+(aq) + Cl^-(aq)$$

13-2.1 Conjugate Acid–Base Pairs

HCl is an acid by the Arrhenius definition because it produces the H_3O^+ ion. It is also an acid by the Brønsted–Lowry definition because *it donates an H^+ to H_2O.* In this definition, however, the H_2O molecule also takes the role of a base because it accepts an H^+ from the HCl. The reaction of an acid and a base in water can be considered as an exchange of the proton. The two products formed could then conceivably act as an acid and a base in the reverse reaction. So, in our example, an acid (HCl) reacts with a base (H_2O) to form another acid (H_3O^+) and another base (Cl^-). *The base that remains when an acid donates a proton is known as the* **conjugate base** *of the acid. Likewise, the acid that is formed when the base accepts a proton is known as the* **conjugate acid** *of the base.* Thus HCl–Cl^- and H_3O^+–H_2O are known as *conjugate acid–base pairs.* The exchange of H^+ is illustrated below, where A_1 and B_1 refer to a specific conjugate acid–base pair and A_2 and B_2 refer to the other acid–base pair.

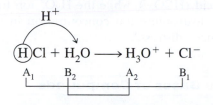

Now consider the reaction of ammonia (NH_3) in water. Ammonia is a base in the Arrhenius definition because it forms OH^- in aqueous solution even though the ammonia molecule itself does not contain the OH^- ion. If we examine its behavior in water as a Brønsted–Lowry base, however, it becomes more obvious how OH^- ions are produced.

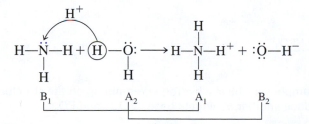

In the Brønsted–Lowry sense, the reaction can be viewed as simply an exchange of an H^+. When the base (NH_3) reacts, it adds H^+ to form its conjugate acid (NH_4^+). When the acid (H_2O) reacts, it loses an H^+ to form its conjugate base (OH^-). The NH_3, NH_4^+ (B_1 and A_1) and the H_2O, OH^- (A_2 and B_2) pairs are conjugate acid–base pairs.

13-2.2 Amphiprotic Ions

In the previous reaction, H_2O is an acid since it donates an H^+ to form NH_4^+. Recall that H_2O acts as a *base* when HCl is present. *A compound or ion that can either donate or accept H^+ ions is called* **amphiprotic**. Water is amphiprotic since it can accept H^+ ions when an acid is present or donate H^+ when a base is present. An amphiprotic substance has both a conjugate acid and a conjugate base. Examples of other amphiprotic substances include HS^- and $H_2PO_4^-$.

Before we look at other examples of Brønsted–Lowry acid–base reactions, we should emphasize the identification of conjugate acids and bases. *The conjugate base of a compound or ion results from removal of an H^+. The conjugate acid of a compound or ion results from the addition of an H^+.*

$$\text{conjugate acid} \underset{+H^+}{\overset{-H^+}{\rightleftarrows}} \text{conjugate base}$$

For example,

$$\underset{\text{acid}}{H_3PO_4} \underset{+H^+}{\overset{-H^+}{\rightleftarrows}} \underset{\text{base}}{H_2PO_4^-}$$

Notice that in the formation of a conjugate base, the base species ($H_2PO_4^-$) has one less hydrogen and the charge decreases by one from the acid (H_3PO_4). The reverse is true for formation of a conjugate acid from a base. That is, the acid has one additional hydrogen and the charge increases by one.

We are now ready to examine how the carbonate ion in calcium carbonate behaves as a base in antacid tablets. The CO_3^{2-} ion relieves acidic stomachs (containing excess H_3O^+) as illustrated by the following proton exchange reaction.

$$CO_3^{2-} + H_3O^+ \longrightarrow HCO_3^- + H_2O$$

Notice that the carbonate ion acts as a base by accepting the proton from H_3O^+ to form its conjugate acid (HCO_3^-), while the H_3O^+ ion forms its conjugate base (H_2O). Decreasing the hydronium ion concentration in the stomach is what is meant by "relief of stomach distress."

EXAMPLE 13-1   **Determining Conjugate Bases of Compounds**

What are the conjugate bases of **(a)** H_2SO_3 and **(b)** $H_2PO_4^-$?

PROCEDURE

Remove a hydrogen and subtract a positive charge from the remaining ion.

$$acid - H^+ = conjugate\ base$$

SOLUTION

(a) $H_2SO_3 - H^+ = \underline{HSO_3^-}$ **(b)** $H_2PO_4^- - H^+ = \underline{HPO_4^{2-}}$

ANALYSIS

Notice that the two conjugate bases in this example each have a hydrogen remaining on the structure. In these cases, the hydrogens can be removed to produce another conjugate base—SO_3^{2-} and PO_4^{3-}.

SYNTHESIS

It gets progressively more difficult to remove second and third hydrogen ions from acids. This makes sense, as you are now trying to pull something with a positive charge away from something with a negative charge. As a result, it takes a stronger and stronger base to remove the H^+. Said another way, each successive species becomes weaker and weaker.

EXAMPLE 13-2   **Determining Conjugate Acids of Compounds**

What are the conjugate acids of **(a)** CN^- and **(b)** $H_2PO_4^-$?

PROCEDURE

$$base + H^+ = conjugate\ acid$$

SOLUTION

(a) $CN^- + H^+ = \underline{HCN}$ **(b)** $H_2PO_4^- + H^+ = \underline{H_3PO_4}$

ANALYSIS

Notice that in this example the conjugate bases have negative charges and the conjugate acids are neutral. Another example of a conjugate acid–base pair is NH_4^+–NH_3, where the acid is positively charged and the base is neutral. This illustrates a fairly consistent pattern found in acids and bases. Acids are usually neutral or positively charged species. Bases are usually neutral or negatively charged species. In the case of amphiprotic anions, however, the anion can act as an acid because it contains an acidic hydrogen (e.g., $H_2PO_4^-$).

SYNTHESIS

Proteins consist of long chains of amino acids. Amino acids are interesting compounds in regard to acid–base pairs and amphiprotism. Just from the name, you might suspect that something is unique. *Amino* is similar to the name *ammonia,* which we know to be a base, and the term *acid* is unambiguous. Amino acids are well-known amphiprotic compounds. What makes them unique is that the acid–base reactions they undergo can occur within the same molecule! The neutral structure of a simple amino acid is

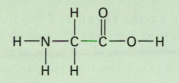

Under most natural conditions, the acid part on the right of the molecule donates a proton to the base part on the left. This produces the conjugate acid and base on each end of the molecule.

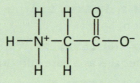

Amino acids thus have properties more typical of ionic compounds (i.e., high melting points). Can you identify the structural regions in the amino acid that form the conjugate acid–base pairs?

EXAMPLE 13-3   Writing Acid–Base Reactions

Write the equations illustrating the following Brønsted–Lowry acid–base reactions.
(a) H_2S as an acid with H_2O **(c)** $H_2PO_4^-$ as a base with H_3O^+
(b) $H_2PO_4^-$ as an acid with OH^- **(d)** CN^- as a base with H_2O

PROCEDURE

A Brønsted–Lowry acid–base reaction produces a conjugate acid and base. Remove an H^+ from the acid and transfer it to the base to form the products.

SOLUTION

Acid		Base		Conjugate Acid		Conjugate Base
(a) H_2S	+	H_2O	$\longrightarrow$	H_3O^+	+	HS^-
(b) $H_2PO_4^-$	+	OH^-	$\longrightarrow$	H_2O	+	HPO_4^{2-}
(c) H_3O^+	+	$H_2PO_4^-$	$\longrightarrow$	H_3PO_4	+	H_2O
(d) H_2O	+	CN^-	$\longrightarrow$	HCN	+	OH^-

ANALYSIS

Acid–base reactions always move from the stronger acid–base pair to the weaker pair. So we can examine reactions that work well to help rank acids and bases relative to each other. For instance, H_3O^+ reacts completely with $H_2PO_4^-$ to produce the conjugate acid H_3PO_4 [part (c)]. This tells us that the hydronium cation is a stronger acid than phosphoric acid. It's also not that surprising that a cation is more acidic than a neutral molecule. We will discuss the relative strengths of acids and bases in more detail in the next section.

SYNTHESIS

Note that equations (b) and (c) indicate that the $H_2PO_4^-$ is amphiprotic. Another question that could be asked, especially in equation (a), is what makes H_2S the acid and H_2O the base. Both are neutral, and they're very similar, with sulfur and oxygen both in the Group VIA column. Further, in (b), both species are negative. Why is either an acid? This becomes a question of relative acidic or basic strength of the two species involved, and is explored in the next section.

▶ **ASSESSING THE OBJECTIVE FOR SECTION 13-2**

EXERCISE 13-2(a) LEVEL 1: Fill in the blanks.
In the Brønsted–Lowry definition, acids are _____ donors and bases are _____ acceptors. A conjugate acid of a compound or ion results from the _____ of an _____ ion. A substance that has both a conjugate acid and a conjugate base is said to be _____.

EXERCISE 13-2(b) LEVEL 1: Use the Brønsted–Lowry definition to identify the acid and the base in the following reaction:

$$SO_3^{2-} + NH_4^+ \longrightarrow NH_3 + HSO_3^-$$

EXERCISE 13-2(c) LEVEL 2: Write the conjugate acid and the conjugate base of the $HC_2O_4^-$ ion.

EXERCISE 13-2(d) LEVEL 2: Write the reaction of the HCO_3^- ion with water **(a)** where water acts as a base and **(b)** where water acts as an acid.

EXERCISE 13-2(e) LEVEL 3: Complete the following reaction of a Brønsted–Lowry proton exchange. What are the two sets of conjugate acid–base pairs?

$$CH_3COOH + (CH_3)_3N: \longrightarrow$$

EXERCISE 13-2(f) LEVEL 3: The NH_2^- ion is a powerful base that can remove H^+ from many compounds. What is the Brønsted–Lowry acid–base reaction that NH_2^- undergoes with CH_3OH?

For additional practice, work chapter problems 13-7, 13-9, 13-11, and 13-13.

▶ **OBJECTIVE FOR SECTION 13-3**

Calculate the hydronium ion concentration in a solution of a strong acid and a weak acid given the initial concentration of the acid and the percent ionization of the weak acid.

13-3 Strengths of Acids and Bases

LOOKING AHEAD! The properties of acids and bases described in Section 13-1 are mostly associated with strong acids and bases. Other substances display these properties but in less dramatic fashion. For example, dilute acetic acid is sour but tame enough to use on a salad, and ammonia dissolves grease but is mild enough to clean oil stains from floors. There is a wide range of behavior that we regard as acidic or basic. This is the subject of this section. ■

Ammonia as a base would do a poor job of unclogging a stopped-up drain, yet we certainly wouldn't want to use lye to clean an oil spot from a carpet. In the former case, the base is too weak; in the latter, it is much too strong. Strong acids and bases are difficult and dangerous to handle and store. Weak acids and bases are quite easy and safe to have around. The large difference in acid or base behavior relates to the concentration of the active ingredient (H_3O^+ or OH^-) produced by the acid or base in water. This depends on its strength. First, we will consider the strength of acids.

13-3.1 The Strength of Acids

In Chapter 6, we indicated that strong acids were 100% dissociated into ions in solution. There are only six common strong acids. In addition to hydrochloric (HCl), which we have already discussed, there are two other binary acids, hydrobromic acid

(HBr) and hydroiodic acid (HI). The other strong acids are sulfuric acid (H_2SO_4), nitric acid (HNO_3), and perchloric acid ($HClO_4$). Sulfuric acid is a somewhat more complex case but will be considered shortly.

We have been using the 100% ionization criteria to describe a strong acid, but we still haven't formally answered the question "strong compared to what?" In the Brønsted–Lowry definition, the reaction of a molecular acid with water is considered a proton (H^+) exchange reaction. In fact, there is a competition between the proton-donating abilities of two acids (e.g., HCl and H_3O^+) and, like other competitions, the stronger prevails. *The stronger acid reacts with the stronger base to produce a weaker conjugate acid and conjugate base. In the case of a strong acid in water, the molecular acid (e.g., HCl) is a stronger proton donor than H_3O^+.* Therefore, the reaction proceeds essentially 100% to the right.

$$\underset{\text{stronger acid}}{HCl(aq)} + \underset{\text{stronger base}}{H_2O(l)} \longrightarrow \underset{\text{weaker acid}}{H_3O^+(aq)} + \underset{\text{weaker base}}{Cl^-(aq)}$$

Most acids that we may be familiar with, such as acetic acid, ascorbic acid (vitamin C), and citric acid, are all considered weak acids. The weak acids that we will discuss in this section are also neutral molecular acids. *A **weak molecular acid** is partially ionized (usually less than 5% at typical molar concentrations).*

The ionization of a weak acid is limited because it is a reversible reaction. Such a reaction was briefly mentioned in Section 12-6.1, since weak acids are examples of weak electrolytes. We will discuss the concept of reversible reactions and equilibrium in detail in Chapter 15, but a basic understanding of these concepts will help us understand the action of weak acids and bases in water. The ionization of HF, a weak acid, is as follows.

$$HF(aq) + H_2O(l) \rightleftharpoons H_3O^+(aq) + F^-(aq)$$

This is an example of a chemical reaction that reaches a state of equilibrium, which is represented by a double arrow ($\rightleftharpoons$). (A single arrow represents a reaction that essentially goes to completion.) *In a reaction at equilibrium, two reactions are occurring simultaneously.* In the ionization of HF, a forward reaction occurs to the right, producing ions (H_3O^+ and F^-), and a reverse reaction occurs to the left, producing the original reactants, which are the molecular compounds HF and H_2O.

$$\textit{Forward: } HF(aq) + H_2O(l) \longrightarrow H_3O^+(aq) + F^-(aq)$$

$$\textit{Reverse: } H_3O^+(aq) + F^-(aq) \longrightarrow HF(aq) + H_2O(l)$$

At equilibrium, the forward and reverse reactions occur at the same rate so the concentrations of all species remain constant. However, the identities of the individual molecules and ions do change. The reaction thus *appears* to have gone to a certain extent and then stopped. In fact, at equilibrium, a *dynamic (constantly changing)* situation exists in which two reactions going in opposite directions at the same rate keep the concentrations of all species constant.

For weak acids, the point of equilibrium lies far to the left side of the original ionization equation. This is because the H_3O^+ ion is a stronger proton donor than the HF molecular acid, opposite the case for strong acids. Thus, reactants (the molecular compounds on the left) are favored over the ionic products on the right. In other words, most of the fluorine is present in the form of molecular HF rather than fluoride ions. In a weak acid solution the concentration of the active ingredient, H_3O^+, and the anion is comparatively small. (See Figure 13-2.)

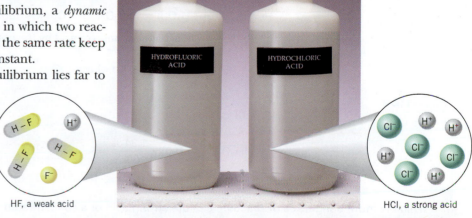

HF, a weak acid   HCl, a strong acid

FIGURE 13-2 Strong Acids and Weak Acids Strong acids are completely ionized in water; weak acids are only partially ionized.

In the following discussions and examples we will refer to the percent ionization of the weak acid. Like all percent problems, the actual amount or concentration of ions present is found by multiplying the total amount of acid initially present by the percent expressed in decimal form, which is obtained by dividing the percent ionization by 100%. Thus, if the original concentration of an acid is 0.20 M and it is 5.0% ionized, the concentration of each ion (the H_3O^+ cation and the specific anion) is

$$0.20 \text{ M} \times \frac{5.0 \text{ \%}}{100 \text{ \%}} = 0.20 \text{ M} \times 0.050 = 0.010 \text{ M}$$

The following examples illustrate the difference in acidity (the difference in H_3O^+ concentration) between a strong acid and a weak acid. In these examples, *the appearance of a species in brackets (e.g., $[H_3O^+]$) represents the numerical value of the concentration of that species in moles per liter (M).*

EXAMPLE 13-4   Calculating the H_3O^+ Concentration in a Strong Acid Solution

What is $[H_3O^+]$ in a 0.100 M HNO_3 solution?

PROCEDURE

HNO_3 is a strong acid, so the following reaction goes 100% to the right.

$$HNO_3(aq) + H_2O(l) \longrightarrow H_3O^+(aq) + NO_3^-(aq)$$

As in other stoichiometry problems involving complete reactions, the amount (or concentration) of a product is found from the amount (or concentration) of a reactant using a mole ratio conversion factor from the balanced equation.

SOLUTION

$$0.100 \text{ mol/L } HNO_3 \times \frac{1 \text{ mol/L } H_3O^+}{1 \text{ mol/L } HNO_3} = 0.100 \text{ mol/L } H_3O^+$$

$$[H_3O^+] = \underline{0.100 \text{ M}}$$

EXAMPLE 13-5   Calculating the H_3O^+ Concentration in a Weak Acid Solution

What is $[H_3O^+]$ in a 0.100 M $HC_2H_3O_2$ solution that is 1.34% ionized?

PROCEDURE

Calculate the concentration of H_3O^+ from the percent ionization and the initial concentration. Since $HC_2H_3O_2$ is a weak acid, the following ionization reaches equilibrium when 1.34% of the initial $HC_2H_3O_2$ is ionized.

$$HC_2H_3O_2(aq) + H_2O(l) \rightleftharpoons H_3O^+(aq) + C_2H_3O_2^-(aq)$$

The $[H_3O^+]$ is calculated by multiplying the original concentration of acid by the percent *expressed in fraction form.*

$$[H_3O^+] = [\text{original concentration of acid}] \times \frac{\% \text{ ionization}}{100\%}$$

SOLUTION

In this case,

$$[H_3O^+] = [0.100] \times \frac{1.34 \text{ \%}}{100 \text{ \%}} = \underline{1.34 \times 10^{-3} \text{ M}}$$

ANALYSIS

So which is more dangerous or reactive, a dilute strong acid or a concentrated weak acid? Of course the terms are vague, and we'd have to know the exact concentrations and the exact percent ionization of the weak acid. Actually, both factors play a role. We will have to come up with some convenient way of measuring the relative strength of such solutions, based on a common component of both. We'll take this up in Section 13-5.

SYNTHESIS

A 0.100-M solution seems to us to be relatively dilute. Furthermore, if it is less than 2% ionized, it would appear that the solution in question has very little acid in it at all. As it turns out, a little acid goes a long way. The solution described above is about as acidic as orange juice and more acidic than acid rain. Acid rain kills fish and trees, and eats away at marble buildings and structures. So, even at 1.34×10^{-3} M, there's still plenty of hydronium to do the damage associated with acids.

In the two preceding examples, we found that the concentration of H_3O^+ is about 100 times greater in the strong acid solution, although both were at the same original concentration. Only in the case of strong acids does the original concentration of the acid equal the concentration of H_3O^+ ions.

In Example 13-5 we used acetic acid ($HC_2H_3O_2$) as a typical weak acid. Perhaps one wonders why it is written that way, not as $H_4C_2O_3$. However, if we look at the Lewis representation of acetic acid, we notice that there are two types of hydrogen atoms. The one attached to the oxygen is polar and is potentially acidic. It is ionized (to a limited extent) by the water molecules, as illustrated in Figure 12-3 in the previous chapter. In the case of acetic acid, the three hydrogen atoms attached to the carbon are essentially nonpolar and do not interact with polar water molecules when placed in aqueous solution. Thus, the three hydrogen atoms attached to carbon are not affected by proton exchange and remain as part of the acetate ion. Except for the binary acids the acidic hydrogen in an acid is bonded to an oxygen.

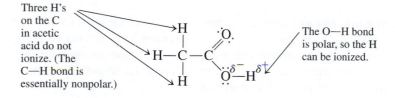

Three H's on the C in acetic acid do not ionize. (The C—H bond is essentially nonpolar.)

The O—H bond is polar, so the H can be ionized.

13-3.2 The Strength of Bases

Now we consider the case of bases. They also exhibit a range of strengths depending on the concentration of OH^- produced by the base. Strong bases are ionic compounds that dissolve in water to form OH^- anions. All alkali metal hydroxides are strong bases and are quite soluble in water. The alkaline earth hydroxides [except $Be(OH)_2$] also completely dissociate into ions in solution. However, $Mg(OH)_2$ has a very low solubility in water and so produces a very small concentration of aqueous OH^-. Because of its low solubility, it can be taken internally, as a solid suspension, to combat excess stomach acid (milk of magnesia).

The most familiar example of a weak molecular base and the one we will emphasize is ammonia (NH_3), whose reaction as a base was discussed in Section 13-2. *A* **weak molecular base** *is a base that is only partially converted into ions in solution.* The reaction of ammonia as a base is shown by the equation

$$NH_3(aq) + H_2O(l) \rightleftharpoons NH_4^+(aq) + OH^-(aq)$$

The tip-off that ammonia is a weak base is found in the equilibrium arrows rather than the single arrow that implies a complete reaction. As in the case of weak acids, the position of equilibrium lies far to the left. The vast majority of dissolved NH_3 molecules remain in the molecular form shown on the left of the double arrows rather than as ions, shown on the right. In the Brønsted–Lowry sense, we note that NH_4^+ is a stronger proton donor than H_2O and OH^- is a stronger proton acceptor than NH_3. Several other neutral nitrogen-containing compounds, such as methylamine (CH_3NH_2) and pyridine (C_5H_5N), also react with water in a similar manner to produce weakly basic solutions.

Milk of magnesia is a base taken for indigestion.

▶ **ASSESSING THE OBJECTIVE FOR SECTION 13-3**

EXERCISE 13-3(a) LEVEL 1: Fill in the blanks.
Strong acids are essentially _____% ionized in aqueous solution, whereas weak acids are _____ ionized. Sodium hydroxide is a strong base, but ammonia is a _____. When a weak acid reacts with water, the reaction is said to be at _____.

EXERCISE 13-3(b) LEVEL 2: A certain 0.10 M solution of an acid is 2.5% ionized. Is this a strong or a weak acid? What is the $[H_3O^+]$ in this solution?

EXERCISE 13-3(c) LEVEL 2: An acid with a concentration of 0.25 M contains a hydronium concentration of 0.0050 M. What percent of the acid is ionized? Is it strong or weak?

EXERCISE 13-3(d) LEVEL 3: Write the equation for each of the following reacting with water. Be sure to include the proper arrow ($\longrightarrow$ or $\rightleftharpoons$) as appropriate.
(a) HNO_3 **(b)** HNO_2 **(c)** $HClO_2$ **(d)** $HClO_4$

For additional practice, work chapter problems 13-18, 13-21, 13-23, 13-24, and 13-25.

▶ **OBJECTIVE FOR SECTION 13-4**
Write the molecular, total ionic, and net ionic equations for neutralization reactions.

13-4 Neutralization and the Formation of Salts

LOOKING AHEAD! Molecules or ions that act as acids in water have characteristic properties, as do molecules or ions that act as bases. When solutions of acids and bases are mixed in the proper amounts, the characteristic properties are destroyed or neutralized. The products of such a reaction are a salt and water. We will look at the interactions of solutions of acids with solutions of bases next. ■

In Chapter 6 we described a type of double-displacement reaction between acids and bases known as *neutralization*. If we mix the acid and base in stoichiometric amounts, the products are simply water and a salt. We will begin our discussion of neutralization reactions with a review of the reaction between a strong acid (hydrochloric acid) and a strong base (sodium hydroxide) as described in Chapter 6 and then move on to other cases. The molecular, total ionic, and net ionic equations are shown below. In this case, it is more convenient to represent the acid species as simply $H^+(aq)$ rather than H_3O^+.

$$\text{ACID} \quad + \quad \text{BASE} \quad \longrightarrow \text{SALT} + \text{WATER}$$

Molecular: $HCl(aq) + NaOH(aq) \longrightarrow NaCl(aq) + H_2O(l)$

Total ionic: $H^+(aq) + \cancel{Cl^-(aq)} + \cancel{Na^+(aq)} + OH^-(aq) \longrightarrow$
$$\cancel{Na^+(aq)} + \cancel{Cl^-(aq)} + H_2O(l)$$

Net ionic: $H^+(aq) + OH^-(aq) \longrightarrow H_2O(l)$

The key to what drives neutralization reactions is found in the net ionic equation. The active ingredient from the acid $[H^+(aq)]$ reacts with the active ingredient from the base $[OH^-(aq)]$ to form the molecular compound water. A salt is what is left over—usually present as spectator ions if the salt is soluble.

As a vital mineral needed to maintain good health, "salt" refers to just one substance, sodium chloride, as formed in the preceding reaction. Actually, salts can result from many different combinations of anions and cations from a variety of neutralizations. The following neutralization reactions, written in molecular form, illustrate the formation of some other salts.

	ACID		BASE		SALT		WATER
1.	$2HNO_3(aq)$	+	$Ca(OH)_2(aq)$	$\longrightarrow$	$Ca(NO_3)_2(aq)$	+	$2H_2O(l)$
2.	$HClO(aq)$	+	$LiOH(aq)$	$\longrightarrow$	$LiClO(aq)$	+	$H_2O(l)$
3.	$H_2SO_4(aq)$	+	$2NaOH(aq)$	$\longrightarrow$	$Na_2SO_4(aq)$	+	$2H_2O(l)$

13-4.1 Neutralization of a Strong Acid with a Strong Base

Each of these three neutralization reactions represents somewhat different situations, so we will look at these reactions one at a time in ionic form. Reaction 1 in the above list again represents the neutralization of a strong acid with a strong base. In this case, however, the base, $Ca(OH)_2$, dissolves in water to produce two OH^- ions per formula unit. Thus two moles of acid are needed per mole of base for complete neutralization. The total ionic and net ionic equations for reaction 1 are as follows.

$$2H^+(aq) + 2NO_3^-(aq) + Ca^{2+}(aq) + 2OH^-(aq) \longrightarrow Ca^{2+}(aq) + 2NO_3^-(aq) + 2H_2O(l)$$
$$H^+(aq) + OH^-(aq) \longrightarrow H_2O(l)$$

Notice that the net ionic equation is identical to the reaction illustrated at the beginning of this section.

13-4.2 Neutralization of a Weak Acid with a Strong Base

Reaction 2 illustrates a neutralization of hypochlorous acid ($HClO$), a weak acid, with a strong base. Recall that in the case of most weak acids, the overwhelming majority of molecules are present in solution in the molecular form [i.e., $HClO(aq)$] rather than as ions [i.e., $H^+(aq)$ and $ClO^-(aq)$]. Thus, when we write the total ionic and net ionic equations, the acid is displayed in the predominant molecular form. These two equations for reaction 2 are shown as follows.

Total ionic: $HClO(aq) + Li^+(aq) + OH^-(aq) \longrightarrow Li^+(aq) + ClO^-(aq) + H_2O(l)$

Net ionic: $HClO(aq) + OH^-(aq) \longrightarrow ClO^-(aq) + H_2O(l)$

13-4.3 Neutralization of a Polyprotic Acid with a Strong Base

Acids are sometimes designated by the number of H^+ ions that are available from each molecule. Thus, HCl, HNO_3, and HClO are known as **monoprotic acids**, *since only one H^+ is produced per molecule. Those acids that can produce more than one H^+ are known as* **polyprotic** *acids. More specifically, polyprotic acids may be* **diprotic** *(two H^+'s), such as H_2SO_4, or* **triprotic** *(three H^+'s) as in the case of H_3PO_4.* Except for sulfuric acid, all other polyprotic acids are weak acids.

Reaction 3 represents the neutralization of the strong diprotic acid H_2SO_4. In the case of neutralization of polyprotic acids, the neutralization takes place one acidic hydrogen at a time. Thus addition of one mole of NaOH to H_2SO_4 results in a partial neutralization, forming water and $NaHSO_4$ (sodium bisulfate).

$$H_2SO_4(aq) + NaOH(aq) \longrightarrow NaHSO_4(aq) + H_2O(l)$$

Sodium bisulfate (or sodium hydrogen sulfate) is an example of an acid salt. *An* **acid salt** *is an ionic compound containing an anion with one or more acidic hydrogens that can be neutralized by a base.*

A second mole of NaOH added to the $NaHSO_4$ solution completes the neutralization.

$$NaHSO_4(aq) + NaOH(aq) \longrightarrow Na_2SO_4(aq) + H_2O(l)$$

The sum of the two equations produces the overall, complete neutralization of H_2SO_4. (See Figure 13-3.)

$$H_2SO_4(aq) + 2NaOH(aq) \longrightarrow Na_2SO_4(aq) + 2H_2O(l)$$

FIGURE 13-3 Neutralization of H_2SO_4 The hydrogens of H_2SO_4 can be neutralized one at a time.

The total ionic and net ionic equations for the complete neutralization are represented below. For simplification, we will consider sulfuric acid as completely ionized to produce two H^+ ions in aqueous solution. In fact, the bisulfate ion in solution is not completely ionized.

$$2H^+(aq) + SO_4^{2-} + 2Na^+(aq) + 2OH^-(aq) \longrightarrow 2Na^+(aq) + SO_4^{2-}(aq) + 2H_2O(l)$$
$$H^+(aq) + OH^-(aq) \longrightarrow H_2O(l)$$

EXAMPLE 13-6 Writing a Complete Neutralization Reaction

Write the balanced equation in molecular form illustrating the complete neutralization of $Al(OH)_3$ with H_2SO_4.

PROCEDURE

Complete neutralization requires one H^+ for each OH^-. Since $Al(OH)_3$ has three available OH^- ions and H_2SO_4 can provide only two H^+ ions, the reaction requires two moles of $Al(OH)_3$ for three moles of H_2SO_4.

SOLUTION

$$2Al(OH)_3 + 3H_2SO_4 \longrightarrow Al_2(SO_4)_3 + 6H_2O$$

ANALYSIS

When the goal is complete neutralization, it is easy to determine the mole ratio between acid and base without having to write out the balanced reaction. Just use the procedure above to match up the H^+'s produced by the acid with the OH^-'s available from the base. What would the mole ratio for complete neutralization be for the reaction between NaOH and H_3PO_4? It would take three NaOH's for each acid.

SYNTHESIS

The experimental procedure used to neutralize an acid and a base is called a *titration*, as discussed in the previous chapter. Here, base solution of a known molarity is slowly added to a known amount of acid using a *buret*, an accurately calibrated glass tube with a stopcock at its base. A chemical indicator added to the original acid solution changes color when there is an excess of base present. Typically, one drop of indicator solution indicates visually when the solution changes from acidic to basic. A commonly used indicator is phenolphthalein. It is colorless when the solution is acidic but changes abruptly to pink when even a slight excess of base is present. The stoichiometric equivalence point is the point where the solution just begins to change color.

EXAMPLE 13-7 Writing a Partial Neutralization Reaction

Write the balanced molecular, total ionic, and net ionic equations illustrating the reaction of 1 mol of H_3PO_4 (a weak acid) with 1 mol of $Ca(OH)_2$.

PROCEDURE

Although 1 mol of H_3PO_4 has three available H^+ ions to neutralize, 1 mol of $Ca(OH)_2$ can react with only two of them. This would leave the HPO_4^{2-} ion in solution.

SOLUTION

Molecular: $H_3PO_4(aq) + Ca(OH)_2(aq) \longrightarrow CaHPO_4(aq) + 2H_2O(l)$

Total ionic: $H_3PO_4(aq) + Ca^{2+}(aq) + 2OH^-(aq) \longrightarrow Ca^{2+}(aq) + HPO_4^{2-} + 2H_2O(l)$

Net ionic: $H_3PO_4(aq) + 2OH^- \longrightarrow HPO_4^{2-}(aq) + 2H_2O(l)$

ANALYSIS

In most cases, the strength of the first acidic hydrogen in a polyprotic acid is several thousand times greater than the second acidic hydrogen, and if there is a third, the second is several thousand times greater than the third. As a result, when we neutralize acids with bases, we can do so one hydrogen at a time. It would not be the case that two moles of OH^- would form some PO_4^{3-}, and an equal amount of $H_2PO_4^-$ would not react. Instead, the reaction essentially takes place one step at a time.

SYNTHESIS

We can apply some common chemical sense to understand why it gets progressively harder to remove each successive H^+ from a polyprotic acid and its acid salt. Covalent bonds hold the hydrogens onto the molecule. These bonds must be broken, which requires energy. The first H^+ is removed from a neutral molecule. But the second H^+ must be removed from a negative ion. It is harder to remove a positive ion from a negative ion than from a neutral molecule. After it is removed, the resulting conjugate base is even more negative. Energetically, it takes progressively more and more energy to remove the next H^+. The weaker the acid, the harder it is to remove an H^+.

MAKING IT REAL

Forensic Chemistry: Salts and Fingerprint Imaging

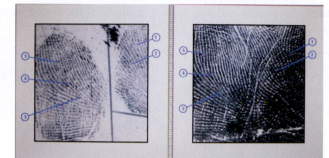

A latent fingerprint (left) is matched with one in a file (right).

The use of fingerprints to identify individuals has been in practice for over 2000 years. In the second century B.C., the Chinese used the imprint of a finger in a wax seal to identify the sender of a document. For over 400 years it has been known that the pattern of a fingerprint is unique to each individual. In modern times, fingerprints can be matched by comparing certain characteristics of the print called loops, whorls, and arches. There are about 20 to 30 characteristics of a fingerprint that are usually classified. Chances are less than one in a billion that two individuals would have more than eight of these characteristics the same.

There are three types of prints that are found at a crime scene. *Visible* prints are obvious when they are in a colored material such as blood, paint, or ink. *Plastic* prints are also obvious when they occur in wax, soap, or even dust. The most difficult to detect are *latent* prints, which are invisible and made by the transfer of perspiration or body oils to the material in question. Detection of latent prints is very difficult and requires special techniques depending on the surface where the print is suspected.

A crime scene that has latent prints is first "dusted for prints." Special powders are used that can be applied with a brush and will contrast with the surface.

The powder adheres to the perspiration or body oil of the print and results in an image. Charcoal and aluminum powder are often used in dusting.

The use of ultraviolet light is also used to detect prints without the use of powders. Ultraviolet light is directed at the surface and reflected back. A sensor detects the difference in reflected light from the surface and the print. The sensor also converts the ultraviolet light to visible light, which can then be recorded or made into a photographic print.

The use of chemicals is also important for some prints that are very hard to detect by other methods. A procedure usually performed in a laboratory in an enclosed container involves subjecting the object containing the print to vapors of an organic compound called cyanoacrylate ester (the major component of *superglue*). Fumes from the chemical adhere to the fingerprint, and eventually it becomes visible.

Lasers that emit an intense beam of light with a specific wavelength are also used to detect fingerprints. The laser causes compounds in the print to fluoresce, or give off light of a different specific wavelength. The investigator wears goggles, which protect the eyes from the laser light and transmit only the light from the print.

So what does all of this have to do with salts? Perspiration contains several salts, mainly sodium chloride and potassium chloride. Even after the perspiration dries, the solid salts are left behind. Recently, at an American Chemical Society Convention in 2005, scientists at Los Alamos National Laboratory reported that a type of X-rays called micro-X-rays cause the elements in the two salts to fluoresce. The light that is given off from the salts can be analyzed by a computer and converted into a print. It is hoped that this technique, combined with others, will provide yet another weapon in the identification of a hard-to-detect latent fingerprint that may be weeks or even months old.

▶ ASSESSING THE OBJECTIVE FOR SECTION 13-4

EXERCISE 13-4(a) LEVEL 1: Fill in the blanks.
The reaction between an acid and a base is known as a _____ reaction. The net ionic equation of this reaction always has _____ as a product. The spectator ions of the reaction form a _____. The partial neutralization of polyprotic acids produces an _____ _____ and water.

EXERCISE 13-4(b) LEVEL 2: Write the balanced molecular, total ionic, and net ionic equations illustrating the neutralization of HNO_3 with $Sr(OH)_2$.

EXERCISE 13-4(c) LEVEL 2: Write the balanced molecular, total ionic, and net ionic equations for the reaction between the weak base NH_3 and the strong acid HBr.

EXERCISE 13-4(d) LEVEL 3: Write the balanced molecular, total ionic, and net ionic equations illustrating the reaction of 1 mol of $H_2C_2O_4$ (oxalic acid) **(a)** with 1 mol of CsOH and then **(b)** with 2 mol of CsOH.

EXERCISE 13-4(e) LEVEL 3: What acid and base react together to form the following?
(a) K_2SO_4 **(b)** $KHSO_4$ **(c)** NH_4Cl

For additional practice, work chapter problems 13-35, 13-36, 13-41, and 13-45.

PART A SUMMARY

KEY TERMS

13-1.2	The **hydronium ion** represents the proton in hydrated form. p. 437
13-2	In the **Brønsted–Lowry** definition, an **acid** is a proton donor and a **base** is a proton acceptor. p. 438
13-2.1	A proton exchange reaction produces **conjugate acid** and **conjugate base** pairs. p. 439
13-2.2	An **amphiprotic** substance can act as either a proton acceptor or donor. p. 439
13-3.1	**Weak molecular acids** produce limited amounts of hydronium ion in solution. p. 443
13-3.2	**Weak molecular bases** produce limited amounts of hydroxide ion in solution. p. 445
13-4.3	A **monoprotic acid** can donate one proton. A **diprotic** (two protons) and **triprotic** (three protons) are **polyprotic** acids. p. 447
13-4.3	An **acid salt** is produced by the partial neutralization of a polyprotic acid. p. 447

SUMMARY CHART

Acids and Bases

Substance	Arrhenius	Brønsted–Lowry	Comments
Acid	Increases $H^+(aq)$ concentration	Proton donor	Strong—100% ionized Weak—usually less than 10% ionized
Base	Increases OH^- concentration	Proton acceptor	Strong—100% ionized Weak—usually less than 10% ionized

Neutralization and Partial Neutralization Reactions

Acid	+	Base	⟶	Salt	+	Water
1. $HX(aq)$	+	$M^+OH^-(aq)$	⟶	$M^+X^-(aq)$	+	$H_2O(l)$
2. $H_2Y(aq)$	+	$M^+OH^-(aq)$	⟶	$M^+HY^-(aq)$	+	$H_2O(l)$

Part B

The Measurement of Acid Strength

SETTING A GOAL

- You will learn about how the relative acidities of aqueous solutions are expressed.

OBJECTIVES

13-5 Using the ion product of water, relate the hydroxide ion and the hydronium ion concentrations.

13-6 Given the hydronium or hydroxide concentrations, calculate the pH and pOH and vice versa.

13-5 Equilibrium of Water

LOOKING AHEAD! We still need a convenient way to express the acidity of a solution, which relates to the concentration of H_3O^+ ions. To do this we need to take a closer look at water itself. As we will see in this section, there is more going on in pure water than we originally indicated. ■

▶ **OBJECTIVE FOR SECTION 13-5**
Using the ion product of water, relate the hydroxide ion and the hydronium ion concentrations.

13-5.1 Autoionization

In the previous chapter, pure water was classified as a nonconductor of electricity. This implied that no ions were present. This isn't exactly true, however. With more sensitive instruments, we find that there actually are very small and equal concentrations of hydronium and hydroxide ions in pure water. The presence of ions in pure water is due to a process known as autoionization. **Autoionization** *produces positive and negative ions from the dissociation of the molecules of the liquid.* For water, this is represented as follows, the double arrow again indicating that the reaction is reversible and reaches a state of equilibrium.

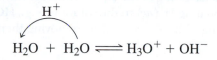

Although this equilibrium lies very far to the left, a small but important amount of H_3O^+ ions and OH^- ions coexist in pure water. It is this small concentration that we will focus on as a means of expressing acid and base behaviors and their relative strengths.

The concentration of each ion at 25°C has been found by experiment to be 1.0×10^{-7} mol/L. This means that only about one out of every 10 million water molecules is actually ionized at any one time. Other experimental results tell us that the product of the ion concentrations is a constant. This phenomenon will be explained in more detail in Chapter 15, but for now we accept it as fact. Therefore, at 25°C

$$[H_3O^+][OH^-] = K_w \text{ (a constant)}$$

Substituting the actual concentrations of the ions, we can now find the numerical value of the constant.

$$[1.0 \times 10^{-7}][1.0 \times 10^{-7}] = 1.0 \times 10^{-14}$$

13-5.2 The Ion Product of Water

$K_w (1.0 \times 10^{-14})$ *is known as the* **ion product** *of water* at 25°C. The importance of this constant is that it tells us the concentrations of H_3O^+ and OH^- not only in pure water but also in acidic and basic solutions. The following example illustrates this relationship. In Figure 13-4, the ion product of water is illustrated. Notice that there is an inverse relationship between the two ion concentrations. That is, the larger the concentration of H_3O^+, the smaller the concentration of OH^-. In pure water or a neutral solution, the concentrations are both equal to 1.0×10^{-7} M. If an acid is added to pure water, the balance is tipped toward the H_3O^+ side and the solution becomes acidic to some degree. This means that the concentration of H_3O^+ rises above 10^{-7} M, while the concentration of OH^- drops below 10^{-7} M. It is analogous to the action of a see-saw. As one side goes up, the other goes down.

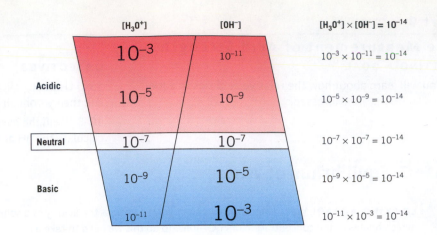

FIGURE 13-4 The Relationship Between H_3O^+ and OH^- in Water
A large concentration of H_3O^+ corresponds to a low concentration of OH^- in a solution, and vice versa.

In summary, an acidic, basic, or neutral solution can now be defined in terms of concentrations of ions.

$$Neutral:\ [H_3O^+] = [OH^-] = 1.0 \times 10^{-7}$$
$$Acidic:\ [H_3O^+] > 1.0 \times 10^{-7} \quad and \quad [OH^-] < 1.0 \times 10^{-7}$$
$$Basic:\ [H_3O^+] < 1.0 \times 10^{-7} \quad and \quad [OH^-] > 1.0 \times 10^{-7}$$

We should now incorporate this information into our understanding of acids and bases. In the Arrhenius definition, an acid is a substance that produces H_3O^+ ions in aqueous solution. But now we see that H_3O^+ is present in neutral and even basic solutions as well. A slight modification of the definition solves this problem. *An acid is any substance that increases $[H_3O^+]$ in water, and a base is any substance that increases $[OH^-]$ in water.* With our new understanding of the equilibrium, we can see that a substance can be an acid by directly donating $H^+(aq)$ to the solution (e.g., HCl, H_2S), or a substance can be an acid by reacting with OH^- ions, thus removing them from the solution.

EXAMPLE 13-8   Calculating $[OH^-]$ from $[H_3O^+]$

In a certain solution, $[H_3O^+] = 1.5 \times 10^{-2}$ M. What is $[OH^-]$ in this solution?

PROCEDURE

Use the relationship for K_w, $[H_3O^+][OH^-] = 1.0 \times 10^{-14}$, and solve for $[OH^-]$.

SOLUTION

$$[H_3O^+][OH^-] = 1.0 \times 10^{-14}$$

$$[OH^-] = \frac{1.0 \times 10^{-14}}{[H_3O^+]} = \frac{1.0 \times 10^{-14}}{1.5 \times 10^{-2}}$$

$$= 6.7 \times 10^{-13} \text{ M}$$

ANALYSIS

Notice that any concentration of acid or base above 1.0 M will lead to the other concentration being below 1.0×10^{-14}. This is perfectly acceptable, both mathematically and in practice. In the next section, we'll see that it leads to some calculations that many students initially assume are impossible.

SYNTHESIS

When the hydronium concentration is high, it requires large amounts of acid or base to change the relative values of the two. When the hydroxide concentration is high, the same situation occurs. However, in the middle range, where both concentrations are around 10^{-7}, it takes only small amounts of acid or base to dramatically alter the relative concentrations. What's the lesson here? In our typically neutral, or close-to-neutral, world, it doesn't take a lot of acidic or basic substances released into the environment to dramatically change things. Acid rain doesn't have to be very acidic to kill plants or fish. Small amounts will raise the concentration above a level that is toxic.

▶ ASSESSING THE OBJECTIVE FOR SECTION 13-5

EXERCISE 13-5(a) LEVEL 1: Fill in the blanks.
Pure water contains a small but important concentration of _____ and _____ ions. The product of the concentrations of these two ions, known as the _____ _____ of water, is symbolized by _____ and has a value of _____ at 25°C. As the concentration of _____ increases in water, the concentration of H_3O^+ _____.

EXERCISE 13-5(b) LEVEL 2: What is the $[H_3O^+]$ in a solution that has $[OH^-] = 7.2 \times 10^{-5}$?

EXERCISE 13-5(c) LEVEL 3: What are the hydronium and hydroxide concentrations of a 0.125-M solution of weak base that is 0.40% ionized?

EXERCISE 13-5(d) LEVEL 3: Is the solution in Exercise 13-5(b) acidic or basic? Is it strongly or weakly acidic or basic? What is an example of a compound that could produce a hydronium concentration at that level?

For additional practice, work chapter problems 13-51, 13-53, and 13-57.

13-6 The pH Scale

LOOKING AHEAD! The concentration of hydronium or hydroxide ions in aqueous solution is usually quite small. While scientific notation is a great help in expressing these very small numbers, it is still awkward in this case. There is another way. This involves expressing the numbers as logarithms, which then gives us a three- or four-digit number that tells us the same thing. The expression of these numbers in this manner is discussed in this section. ■

▶ OBJECTIVE
FOR SECTION 13-6
Given the hydronium or hydroxide concentrations, calculate the pH and pOH and vice versa.

13-6.1 The Definition of pH

The producers of commercial television advertising assume that the general population is aware not only of the importance of acidity but also of how it is scientifically expressed. For example, we often hear references to controlled pH in hair shampoo commercials and other products. In fact, pH is an important and convenient method for expressing acid strength. For example, in a typical acidic solution, assume $[H_3O^+]$ is equal to 1×10^{-5} M. Scientific notation is certainly better than using a string of nonsignificant zeros (i.e., 0.00001 M), but expressed as pH, the number is simply 5.0. The pH scale represents the negative exponent of 10 as a positive number. The exponent of 10 in a number is the number's common logarithm or, simply, log. **pH** *is a logarithmic expression of* $[H_3O^+]$.

$$pH = -\log [H_3O^+]$$

Therefore, a solution of pH = 1.00 has $[H_3O^+]$ equal to 1.0×10^{-1} M, and pure water has pH = 7.00 ($[H_3O^+] = 1.0 \times 10^{-7}$ M). In expressing pH, the number to the right of the decimal place should have the same number of significant figures as the original coefficient or number. That is,

$$\text{if } [H_3O^+] = \underset{\text{2 significant figures}}{\underline{1.0}} \times 10^{-4} \text{ M, then pH} = 4.\underset{\text{2 places to the right of the decimal}}{\underline{00}}$$

A much less popular but valid way of expressing $[OH^-]$ is **pOH**.

$$pOH = -\log [OH^-]$$

CONDITIONING
SHAMPOO
for normal/dry hair
SHAMPOOING REVITALISANT
pour les cheveux normaux/secs
CHAMPÚ ACONDICIONADOR
para cabello normal/seco

pH	MWS
4.5-5.5	150-2500

QUADRAMINE
COMPLEX

10.1 fl. oz. | 300 ml

This shampoo is supposedly desirable because it is "pH balanced."

A simple relationship between pH and pOH can be derived from the ion product of water.

$$[H_3O^+][OH^-] = 1.0 \times 10^{-14}$$

If we now take −log of both sides of the equation, then we have

$$-\log[H_3O^+][OH^-] = -\log(1.0 \times 10^{-14})$$

Since $\log(A \times B) = \log A + \log B$, the equation can be written as

$$-\log[H_3O^+] - \log[OH^-] = -\log 1.0 - \log 10^{-14}$$

Since $\log 1.0 = 0.00$ and $\log 10^{-14} = -14$, the equation is

$$pH + pOH = 14.00$$

Generally, pOH is not used extensively since pH relates to the OH⁻ concentration as well as the H_3O^+ concentration. The relationships among $[H_3O^+]$, $[OH^-]$, pH, and pOH are summarized as follows.

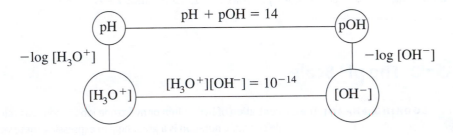

Although most of us are tempted to go straight to our calculators to change from scientific notation to logarithms, it is helpful to review the *meaning* of common logs and some of the rules of their use. You are encouraged to read Appendix C, which contains a brief discussion of common logarithms.

In Figure 13-5 we have included the pH and pOH in addition to the ion concentrations shown in Figure 13-4. Acidic, basic, and neutral solutions can now be defined in terms of pH and pOH.

Neutral: pH = pOH = 7.00

Acidic: pH < 7.00 and pOH > 7.00

Basic: pH > 7.00 and pOH < 7.00

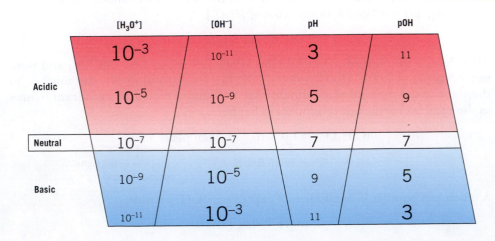

FIGURE 13-5 Concentrations of ions, pH, and pOH A low pH in a solution corresponds to a high pOH.

13-6.2 pH and the Acidity of Solutions

In the use of pH, one must remember that the *lower* the value for pH, the *higher* the concentration of H_3O^+. Also, *a change in one unit in the pH (e.g., from 4 to 3) corresponds to a 10-fold change in concentration (e.g., from 10^{-4} to $10^{-3}M$).* Another scientific scale that is logarithmic is the Richter scale for measuring earthquakes. This scale measures the amplitude of seismic waves set off by the tremor. An earthquake with a reading of 9.0 on this scale (such as the huge earthquake in Japan in 2011) indicates the waves are 10 times larger than an earthquake of magnitude 8.0. It is 100 times larger than one measuring 7.0, which is still considered strong.

In Table 13-1, we have listed the pH of some common chemicals, foods, or products. A 1.0-M HCL solution is the most acidic solution listed. Solutions of pH 1 or less are considered *strongly acidic*. Solutions with pH less than 7 but greater than 1 are considered *weakly acidic*. Even then, what we refer to as a weakly acidic solution still covers quite a range. A solution of pH 6 is 100,000 times less acidic than a solution of pH 1. On the other side of the scale, we have a 1-M solution of NaOH, which has a pH of 14. Solutions of pH 13 or greater are considered *strongly basic*. Solutions with pH greater than 7 but less than 13 are considered *weakly basic*.

TABLE 13-1

The pH Scale

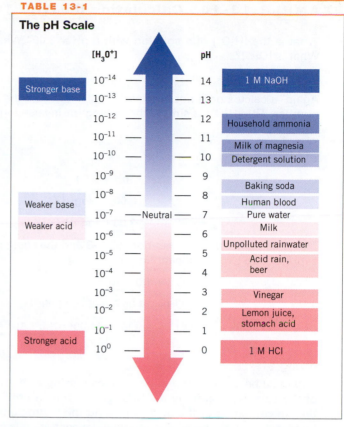

EXAMPLE 13-9   Expressing the pH and pOH of a Solution

What is the pH of a solution with $[H_3O^+] = 1.0 \times 10^{-8}$M? What is the pOH?

PROCEDURE

A calculator is not needed for calculations of pH where the coefficient is exactly 1. Understanding the definition is sufficient. The pH is simply the negative of the exponent of 10.

SOLUTION

$$[H_3O^+] = 1.0 \times 10^{-8} \text{ M}$$

$$pH = -\log(1.0 \times 10^{-8}) = \underline{8.00}$$

(Since there are two significant figures, there should be two places to the right of the decimal.)

$$pH + pOH = 14.00$$

$$pOH = 14.00 - pH$$

$$pOH = 14.00 - 8.00 = \underline{6.00}$$

ANALYSIS

The pOH could have been calculated in another way as well. Knowing the inverse relationship of H_3O^+ and OH^-, we determine that a 1.0×10^{-8} M hydronium solution would be $1.0 \times 10^{-14}/1.0 \times 10^{-8} = 1.0 \times 10^{-6}$ M hydroxide. The pOH is equal to $-\log(1.0 \times 10^{-6})$, which is 6.00.

SYNTHESIS

What is the pH of a 10.0-M solution of H_3O^+? 10.0 in scientific notation is 1.00×10^1. The pH equals −1.000, since the negative of the exponent in this case is −1. This solution has a pH of less than zero, as would any hydronium concentration greater than 1.0 M. Conversely, the pH scale can range higher than 14 as well, for particularly concentrated hydroxide solutions. They don't extend much beyond those values, though, as a pH of 15 would be 10 M OH^-, and a pH of 16 would have to be 100 M. Concentrations this high are unlikely.

EXAMPLE 13-10   Calculating the $[H_3O^+]$ from the pH

What is the $[H_3O^+]$ of a solution with a pH = 3.00? What is the pOH? What is $[OH^-]$?

PROCEDURE

Again, a calculator is not needed for calculations if the pH is a whole number. The $[H_3O^+]$ is simply the antilog (or inverse log) of the minus value of pH.

SOLUTION

$$pH = 3.00$$
$$3.00 = -\log[H_3O^+]$$
$$-3.00 = \log[H_3O^+]$$

This means that the exponent of 10 is −3 and the coefficient should be expressed to two significant figures.

$$[H_3O^+] = \underline{1.0 \times 10^{-3}}$$
$$pOH = 14.00 - 3.00 = \underline{11.00}$$
$$[OH^-] = antilog[-11.00] = \underline{1.0 \times 10^{-11}}$$

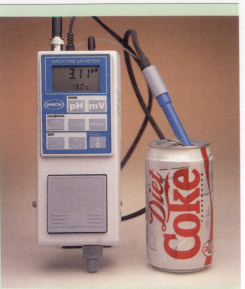

pH is read directly with this common laboratory instrument.

ANALYSIS

You must be careful of what you are expressing with pH. You are not necessarily calculating the concentration of the acid itself but only the H_3O^+ concentration produced by the specific acid. Whether this translates into the concentration of the acid producing the hydronium depends on whether the acid is strong or weak and therefore on how much of its original concentration is converted into hydronium.

SYNTHESIS

Occasionally advertisers will play on the general public's lack of scientific literacy by using terms associated with pH to mislead. Speaking of a substance's pH as being "higher" than normal has opposite meanings when the substance is an acid or a base. Acids with a higher pH are weaker, whereas bases with a higher pH are stronger. Not knowing which an advertiser is referring to can lead a consumer to believe that a chemical in question is more dangerous when in fact it is more benign. When advertisers say that the pH of a woman's perspiration is higher than a man's in order to sell a product, what they are really saying is that it is weaker, as perspiration is mildly acidic. But it sells the antiperspirant.

EXAMPLE 13-11   Calculating the pH and pOH of a Strong Acid Solution

What is the pH of a 1.5×10^{-2} M solution of $HClO_4$? What is the pOH of this solution?

PROCEDURE

$HClO_4$ is a strong acid, which means that it is 100% ionized in solution. Therefore, $[H_3O^+]$ is equal to the original $HClO_4$ concentration. Calculators are usually employed to calculate logarithms but there are several styles. The following instructions are valid for a majority of nongraphing calculators. If yours differs, you should consult your user's manual.

1. Enter the number. First enter "1.5," then push the exponent key, [EXP] or [EE] and enter "2". Push [+/−] the key to change the exponent to −2.

2. Push the [log] key.

3. The display reads −1.82, which is the log of the number you entered.

4. Change the reading to 1.82, since the pH is the negative of the log.

SOLUTION

$$[H_3O^+] = 1.5 \times 10^{-2} \text{ M}$$
$$\log (1.5 \times 10^{-2}) = -1.82$$
$$pH = -\log (1.5 \times 10^{-2}) = \underline{1.82}$$
$$pH + pOH = 14.00$$
$$pOH = 14.00 - 1.82 = \underline{12.18}$$

ANALYSIS

If this were a KOH solution of the same concentration, what would be the pH? The only difference in the solutions, as far as pH is concerned, is that the concentration is hydroxide. So it would be the pOH that would be calculated directly from the starting concentration and the pH that would be determined by subtracting from 14.00. pOH = 1.82 and pH = 12.18. This makes sense. KOH is a base, and therefore its solutions should have pH's greater than 7.0.

SYNTHESIS

The calculation above holds true only for a strong acid (or base, with the appropriate changes for pOH). If the acid in question were weak, you would have to know the % ionization to figure out what the concentration of H_3O^+ was. There are alternative methods that can be used, which are discussed in Chapter 15.

EXAMPLE 13-12 Converting pH to [H₃O⁺]

In a given weakly basic solution, pH = 9.62. What is $[H_3O^+]$?

PROCEDURE

In this case, it is necessary to take the antilog (or the inverse log on a calculator).

1. Enter the 9.62 on the calculator.

2. Change the sign ($\boxed{+/-}$ key).

3. Press the $\boxed{\text{inv}}$, $\boxed{\text{shift}}$, or $\boxed{\text{2}^{\text{nd}}}$ key, then the $\boxed{\text{log}}$ key. Consult the instructions for your calculator if these keys are different or not available.

4. The number displayed should be rounded off to two significant figures since there are two numbers to the right of the decimal place in the original number.

SOLUTION

$$-\log [H_3O^+] = 9.62 \qquad \log [H_3O^+] = -9.62$$
$$[H_3O^+] = \underline{2.4 \times 10^{-10} \text{ M}}$$

ANALYSIS

Remember to make sure that your answers make mathematical and chemical sense. Failure to convert the sign to negative gives a concentration of 4 billion molar. That number can be rejected immediately. Furthermore, if you forget the shift or 2nd key and simply take the log of a negative number, you'll get an error message. It is mathematically meaningless. Typically, acid concentrations will range from 1.0 M down to 10^{-5} M for chemically relevant acids. Those represent a pH range from 0 to 5. Bases have the same concentration range, and the pH generally ranges from 14 down to 9. A pH in the middle range, 6–8, tends to be very dilute, and the actual pH is affected by the dissociation of water itself, thus boosting the hydronium or hydroxide concentrations higher than you might expect.

SYNTHESIS

The more deeply you get into science, the more prevalent you'll find logarithmic scales. Along with the Richter scale, and, of course, pH, logarithmic scales are used to measure all sorts of phenomena whose values are spread out over several orders of magnitude. Stellar brightness is measured on a logarithmic scale (though not log 10, like pH), as is the decibel in acoustics. In spectroscopy, light absorbance, which is a direct function of concentration, is related logarithmically to the transmittance, or the percentage of light passing through the sample. The voltage of a battery depends on the log of the concentration of the chemicals in the battery. These are just a few examples, but they illustrate the importance of the use of logarithms in the study of science.

▶ **ASSESSING THE OBJECTIVE FOR SECTION 13-6**

EXERCISE 13-6(a) LEVEL 1: Fill in the blanks.
The pH of a solution is defined as _____. An acidic solution is one that has an $[H_3O^+]$ greater than _____ and a pH _____ than 7. A basic solution is one that has an $[OH^-]$ greater than _____ and a pH _____ than 7.

EXERCISE 13-6(b) LEVEL 2: What is the pH of the solution where $[OH^-] = 7.2 \times 10^{-5}$?

EXERCISE 13-6(c) LEVEL 2: What is the concentration of an LiOH solution if the pH is 13.40?

EXERCISE 13-6(d) LEVEL 2: For a solution of $HClO_4$ with a pH of 2.20, what are the following?
(a) pOH   **(b)** hydronium concentration   **(c)** hydroxide concentration

EXERCISE 13-6(e) LEVEL 3: For a solution of NaOH that is 0.075 M, what are the following?
(a) hydronium concentration   **(b)** hydroxide concentration   **(c)** pH   **(d)** pOH

EXERCISE 13-6(f) LEVEL 3: Which has the higher $[H_3O^+]$, a 0.005-M solution of HCl or a 0.2-M solution of $HC_2H_3O_2$, which is only 0.4% ionized?

For additional practice, work chapter problems, 13-61, 13-63, 13-69, and 13-71.

PART B SUMMARY

KEY TERMS

13-5.1 The ionization of the molecules of a solvent into positive and negative ions is known as **autoionization**. p. 451

13-5.2 The **ion product** of water (K_w) is the quantitative relationship between the hydronium and hydroxide ions. p. 451

13-6.1 **pH** is the logarithmic measurement of the hydronium ion concentration.
pOH is the same measurement of the hydroxide ion concentration. p. 453

SUMMARY CHART

Relative Acidity of Solutions				
Solution	$[H_3O^+]$	$[OH^-]$	pH	pOH
Strongly acidic	$> 10^{-1}$	$< 10^{-13}$	< 1.0	> 13.0
Weakly acidic	10^{-4}	10^{-10}	4.0	10.0
Neutral	10^{-7}	10^{-7}	7.0	7.0
Weakly basic	10^{-10}	10^{-4}	10.0	4.0
Strongly basic	$< 10^{-13}$	$> 10^{-1}$	> 13.0	< 1.0

Part C

Salts and Oxides as Acids and Bases

SETTING A GOAL

- You will learn how salts and oxides can also affect the pH of aqueous solutions.

OBJECTIVES

13-7 Write a hydrolysis reaction, if one occurs, for salt solutions to determine whether they are acidic or basic.

13-8 Write equations illustrating how a buffer solution can absorb either added acid or base.

13-9 Determine, if possible, whether a specific oxide is acidic or basic.

13-7 The Effect of Salts on pH—Hydrolysis

LOOKING AHEAD! Solutions of some salts are neutral, some basic, and some acidic. Why this is so is the subject of this section. ■

▶ OBJECTIVE
FOR SECTION 13-7
Write a hydrolysis reaction, if one occurs, for salt solutions to determine whether they are acidic or basic.

When a salt dissolves in water, both a cation and an anion are produced. These two ions may or may not react with water. The cation has the potential to act as an acid and the anion as a base. *The reactions of a cation as an acid or an anion as a base are known as* **hydrolysis reactions**. The general hydrolysis reaction of a hypothetical anion (e.g., X^-) is represented as follows:

$$X^-(aq) + H_2O(l) \rightleftharpoons HX(aq) + OH^-(aq)$$

The general hydrolysis reaction of a hypothetical cation (e.g., BH^+) is shown below.

$$BH^+(aq) + H_2O(l) \rightleftharpoons B(aq) + H_3O^+(aq)$$

We will consider four types of salt solutions. To understand why certain ions undergo hydrolysis we will first examine the ions that *do not* undergo this reaction.

13-7.1 Neutral Solutions of Salts

First, consider the Cl^- ion, which is the anion (conjugate base) of the strong acid HCl. Recall that in the ionization process of HCl in water, the forward reaction is 100% complete, so the reverse reaction does not occur (i.e., if it is 100% in one direction, it must be 0% in the other.) Thus we notice that the Cl^- ion, although formally known as a conjugate base, in fact does not exhibit proton-accepting ability in water to form an HCl molecule. *Therefore, the Cl^- ion does not undergo hydrolysis. Other anions of strong acids (i.e., NO_3^-, ClO_4^-, Br^-, and I^-) also do not undergo hydrolysis reactions so do not affect the pH of the solution.* (The HSO_4^-, an acid salt anion, is an exception and is moderately acidic in water.) The anions of strong bases (e.g., Na^+, K^+, Ca^{2+}, and Mg^{2+} and other alkali metal and alkaline earth metal ions) also *do not* undergo hydrolysis in water so do not affect the pH of the solution. These cations and anions exist simply as independent hydrated ion in aqueous solution. If neither the cation nor the anion undergoes a hydrolysis reaction, solutions of these salts do not affect the pH and remain the same as that of pure water (i.e., 7.0). We can identify these salts as having an anion originating from a strong acid and a cation originating from a strong base. [i.e., the alkali and alkaline earth metals (except Be^{2+})]. Examples of neutral salts are $NaCl$, $Ba(ClO_4)_2$, and KNO_3.

13-7.2 Salts Forming Basic Solutions: Anion Hydrolysis

Now consider the ionization equilibrium for the weak acid HClO.

$$HClO(aq) + H_2O(l) \rightleftharpoons H_3O^+(aq) + ClO^-(aq)$$

This reaction reaches a point of equilibrium that strongly favors the molecular reactants on the left. The reverse reaction shows that the ClO$^-$ *does* demonstrate some proton-accepting ability in water. Now consider what happens when this anion is added to water as part of a salt such as in NaClO (bleach). Since ClO$^-$ can accept a proton, it exhibits its weakly basic nature by producing a small concentration of hydroxide ion. (If you had a bleach solution on your fingers, you would notice that it is slippery, which is a property typical of bases.) The following equation represents the anion hydrolysis reaction of the hypochlorite ion.

$$ClO^-(aq) + H_2O(l) \rightleftharpoons HClO(aq) + OH^-(aq)$$

The extent of hydrolysis of anions is generally quite small, however, so again the equilibrium lies far to the left. We can now make a general statement. *The anions of weak acids (i.e., their conjugate bases) undergo hydrolysis in aqueous solution, producing weakly basic solutions.* We can make another generalization: the weaker the acid, the stronger is its conjugate base. There are a few anions (e.g., H$^-$) that are actually strong bases in water and, as mentioned previously, a few that are so weak (e.g., Cl$^-$) that they do not exhibit any basic behavior.

In summary, when the salt originates from the cation of a strong base and the anion of a weak acid, the cation does not affect the pH but the anion does. In the anion hydrolysis reaction, a small equilibrium concentration of OH$^-$ makes the solution basic. Examples of basic salts are KC$_2$H$_3$O$_2$, Ca(NO$_2$)$_2$, and NaF. The pH of solutions of these salts will be greater than 7.0.

13-7.3 Salts Forming Acidic Solutions: Cation Hydrolysis

Now consider the weak base ammonia. As indicated in Section 13-3, its reaction with water produces small concentrations of NH$_4^+$ and OH$^-$ ions. Since this is also an equilibrium situation, we can conclude that the NH$_4^+$ ion demonstrates proton-donating properties in the reverse reaction. When the ammonium ion is placed in water as part of a salt (e.g., NH$_4$Cl), it exhibits a weakly acidic nature by producing a small concentration of hydronium ion. The cation hydrolysis of the ammonium ion is illustrated by the following equation.

$$NH_4^+(aq) + H_2O(l) \rightleftharpoons H_3O^+(aq) + NH_3(aq)$$

We can generalize this behavior also. *Cations (conjugate acids) of weak bases undergo hydrolysis in aqueous solution, producing weakly acidic solutions.*

In summary, when a salt forms from the cation of a weak base and the anion of a strong acid, the anion does not affect the pH but the cation does. The primary example of a cation that undergoes hydrolysis is NH$_4^+$. Since the anions do not undergo hydrolysis, solutions of salts such as NH$_4$Br and NH$_4$NO$_3$ are weakly acidic. The pH of such solutions will be less than 7.0. Other examples of acidic cations include transition metal cations and Al^{3+}. The neutral hydroxides of these cations are also weak bases but generally insoluble in water. For example, solutions of FeCl$_3$ and Cu(NO$_3$)$_2$, which produce the Fe^{3+} and Cu^{2+} ions, respectively, form weakly acidic solutions.

13-7.4 Complex Cases

There are other salt solutions that are not easy to predict without quantitative data. For example, in a solution of the salt NH$_4$NO$_2$, both anion and cation undergo hydrolysis, but not to the same extent. In other cases, it is not immediately obvious

Solutions of ordinary bleach (NaClO) are weakly basic.

whether solutions of acid salts are acidic or basic. For example, the HCO_3^- ion is amphiprotic, which means that it is potentially an acid since it has a hydrogen that can react with a base. However, it is also the conjugate base of the weak acid H_2CO_3, so it can act as a base by undergoing hydrolysis. Sodium bicarbonate solutions are actually weakly basic and for that reason are sometimes used as an antacid to treat upset stomachs. Solutions of the acid salts $NaHSO_3$ and $NaHSO_4$, however, are acidic. Quantitative information, which is discussed in Chapter 15, would be needed, however, for us to have predicted these facts.

The discussion of the effect on pH when salts dissolve in water is summarized as follows.

ORIGIN OF CATION	ORIGIN OF ANION	PH OF SOLUTION	EXAMPLES
strong base	strong acid	7.0	KI, $Sr(ClO_4)_2$
strong base	weak acid	>7.0	KNO_2, BaS
weak base	strong acid	<7.0	NH_4ClO_4, $AlCl_3$
weak base	weak acid	more information needed	NH_4ClO, $N_2H_5NO_2$
strong base	acid salt	more information needed	KH_2PO_4, $Ca(HSO_3)_2$

EXAMPLE 13-13   Predicting the Acidity of Salt Solutions

Indicate whether the following solutions are acidic, basic, or neutral. If the solution is acidic or basic, write the equation illustrating the appropriate reaction.

(a) KCN   **(b)** $Ca(NO_3)_2$ **(c)** $(CH_3)_2NH_2^+Br^-$ [$(CH_3)_2NH$ is a weak base like NH_3.]

PROCEDURE

Identify the ions formed by the solution of each salt. Determine whether the anion originates from a weak or strong acid and the cation from a weak or strong base.

SOLUTION

(a) KCN: K^+ is the cation of the strong base KOH and does not hydrolyze. The CN^- ion, however, is the conjugate base of the weak acid HCN and hydrolyzes as follows.

$$CN^- + H_2O \rightleftharpoons HCN + OH^-$$

Since OH^- is formed in this solution, the solution is basic.

(b) $Ca(NO_3)_2$: Ca^{2+} is the cation of the strong base $Ca(OH)_2$ and does not hydrolyze. NO_3^- is the conjugate base of the strong acid HNO_3 and also does not hydrolyze. Since neither ion hydrolyzes, the solution is neutral.

(c) $(CH_3)_2NH_2^+Br^-$: $(CH_3)_2NH_2^+$ is the conjugate acid of the weak base $(CH_3)_2NH$. It undergoes hydrolysis according to the equation

$$(CH_3)_2NH_2^+ + H_2O \rightleftharpoons (CH_3)_2NH + H_3O^+$$

The Br^- ion is the conjugate base of the strong acid HBr and does not hydrolyze. Since only the cation undergoes hydrolysis, the solution is acidic.

ANALYSIS

Another way of evaluating the pH of a salt solution is to think about what compounds would have to be mixed together to make the salt. Consider KCN in **(a)**. It is formed from the neutralization of the strong base KOH with the weak acid HCN. Intuitive sense indicates that a mixture of a *strong* base and a *weak* acid would form a solution that is more basic than acidic, so we'd predict a basic salt. A salt like NH_4NO_3 works as the exact opposite. The *weak* base, NH_3, neutralized with the *strong* acid, HNO_3, should give an acidic salt. Note that this line of reasoning doesn't *explain* where the excess acid or base is coming from, unlike the equations above, but it is a quick way to arrive at the correct answer.

SYNTHESIS

A good example of how we take advantage of acidic and basic salts is by using them to alter soil pH. Typically, plants grow better in slightly acidic soils, although a few do better at a higher pH. Ammonium salts lower the pH of soils nicely, while also adding critical nitrogen. Aluminum sulfate is another product used to lower the pH of soils. Calcium, another critical mineral, tends to wash out of acidic soils, so basic calcium carbonate, or lime, is used to treat acidic soil and to increase the calcium content.

▶ **ASSESSING THE OBJECTIVE FOR SECTION 13-7**

EXERCISE 13-7(a) LEVEL 1: What type of salt (acidic, basic, or neutral) will the following produce?
(a) cation of a weak base with anion of a strong acid
(b) cation of a strong base with anion of a strong acid
(c) cation of a strong base with anion of a weak acid
(d) cation of a weak base with anion of a weak acid

EXERCISE 13-7(b) LEVEL 2: Identify the following as forming acidic, basic, or neutral solutions when dissolved in water.
(a) Li_3PO_4 **(b)** NH_4ClO_4 **(c)** $CaBr_2$ **(d)** K_2SO_3

EXERCISE 13-7(c) LEVEL 2: Write the hydrolysis reaction (if any) of the following salts.
(a) Na_2CO_3 **(b)** $BaBr_2$ **(c)** $N_2H_5NO_3$ (N_2H_4 is a weak base like ammonia.)

EXERCISE 13-7(d) LEVEL 3: What are two examples of compounds you could add to NaOH in order to form a basic salt?

EXERCISE 13-7(e) LEVEL 3: What are two examples of compounds you could add to ammonia in order to form an acidic salt?

EXERCISE 13-7(f) LEVEL 3: HSO_4^- can potentially undergo two reactions in water. Only one occurs. Which one and why?

For additional practice, work chapter problems 13-78, 13-79, 13-83, and 13-87.

▶ **OBJECTIVE FOR SECTION 13-8**
Write equations illustrating how a buffer solution can absorb either added acid or base.

13-8 Control of pH—Buffer Solutions

LOOKING AHEAD! Perhaps the most important chemical system involves the blood coursing through our veins. Our blood has a pH of about 7.4, but a variation of as little as 0.2 pH unit causes coma or even death. How is the pH of the blood so rigidly controlled despite all the acidic and basic substances we ingest? The control of pH is the job of buffers, which are discussed next. ∎

13-8.1 The Function of a Buffer

The word *buffer* is usually used in the context of "absorbing a shock." For example, a car bumper serves as a buffer for the passengers by absorbing the energy of an impact. A buffer solution has a similar effect on the pH of a solution. *A* **buffer** *solution resists changes in pH caused by the addition of limited amounts of a strong acid or a strong base.* Commercial products tell us that buffered solutions are important in items such as hair shampoo and aspirin. But to us it is most important because many chemical reactions in our bodies, including those that give us life's energy, must take place in a controlled pH environment.

A buffer works by having a substance in solution that is available to react with any added H_3O^+ or OH^-. The most logical candidates for this duty are weak bases and weak acids, respectively. Consider how a hypothetical monoprotic weak acid (HX) reacts with added OH^-.

$$HX(aq) + OH^-(aq) \longrightarrow X^-(aq) + H_2O(l)$$

In the previous section we found that the conjugate base of the weak acid (X^-) can act as a base in water. It reacts with H_3O^+ as follows:

$$X^-(aq) + H_3O^+(aq) \longrightarrow HX(aq) + H_2O(l)$$

Both of these reactions are essentially complete. So, if we have a solution that contains a significant concentration of both the weak acid and its conjugate weak base, we have a solution that reacts with either added H_3O^+ or OH^-. One may think that just the weak acid alone would be a buffer because the ionization produces some of its anion. However, the amount from ionization is very small, so we need additional anion concentration from another source such as a salt.

The buffer in this product controls the pH.

13-8.2 The Composition of a Buffer

Buffer solutions are made from (a) a solution of a weak acid that also contains a salt of its conjugate base or (b) a solution of a weak base that also contains a salt of its conjugate acid.

A typical buffer solution is put to use in swimming pools. To keep bacteria at bay, we use ordinary household bleach (NaClO) plus a limited amount of strong acid that converts some of the NaClO to HClO. This solution keeps the pH at a comfortable 7.5. If the pH gets too low (acidic), the water can sting the eyes and cause corrosion. If it gets too high (alkaline), the water can become cloudy and algae may thrive. The swimming pool has a significant concentration of both ClO^- and HClO, which together act as a buffer. The solution of these two compounds, one a weak acid and the other a strong electrolyte, is illustrated by the following two equations.

$$HClO(aq) + H_2O(l) \rightleftharpoons H_3O^+(aq) + ClO^-(aq)$$

$$NaClO(s) \longrightarrow Na^+(aq) + ClO^-(aq)$$

Notice that a mixture of these two compounds in water produces a solution containing three ions, H_3O^+, Na^+, and ClO^-, and one molecular compound, HClO. The Na^+ is the cation of a strong base and is a spectator ion so it can be neglected. (It does not interact with H_2O so does not affect the pH.) Our main focus, then, is on the significant concentration of molecular HClO and ClO^- ion, along with the small concentration of H_3O^+, which are shown in bold above. The relevant equilibrium is again illustrated as

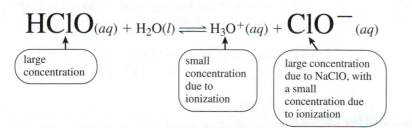

The HClO and the ClO^- ion (from the salt) are available to react with any H_3O^+ or OH^- added to the system. When OH^- is added to the system, the molecular HClO removes it as follows:

$$HClO(aq) + OH^-(aq) \longrightarrow ClO^-(aq) + H_2O(l)$$

Should some H_3O^+ be added to the system, the ClO^- is available to remove it.

$$ClO^-(aq) + H_3O^+(aq) \longrightarrow HClO(aq) + H_2O(l)$$

FIGURE 13-6 A Buffer Solution A buffer can react with added H_3O^+ or OH^-.

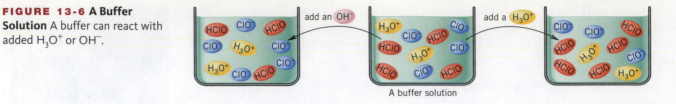

A buffer solution

The addition of a limited amount of acid or base is thus counteracted by species present in the buffer solution, and the pH changes very little. (See Figure 13-6.)

A strong acid and its salt (e.g., HCl and NaCl) cannot act as a buffer because the acid is not present in solution in molecular form—it is completely ionized in water. Therefore, there is no reservoir of HCl molecules that can react with added OH^- ions, nor can the very weak conjugate base (Cl^-) react with H_3O^+ ions as we indicated previously.

Consider the following analogy. One person has $20 in his pocket with no savings in the bank; another person has $20 in her pocket in addition to $100 in the bank. A $12 expense will change the first person's pocket money drastically, as it will decrease to $8. The second person, however, is "buffered" from this expense and can replenish the loss with the bank savings, maintaining the same amount of pocket money. The molecular HClO is analogous to "money in the bank." It is available to react with added OH^-. Thus the original concentration of H_3O^+, which is analogous to "money in the pocket," remains essentially unchanged. The person who keeps all his money in his pocket is analogous to a strong acid solution in which there is no molecular acid in reserve. In this case, since all the acid ionizes to H_3O^+, any added OH^- decreases the H_3O^+ concentration and thus increases the pH.

Solutions of weak bases and salts containing their conjugate acids (e.g., NH_3 and NH_4Cl) also serve as buffers. The two relevant reactions in this case are

Added acid: $NH_3 + H_3O^+ \longrightarrow NH_4^+ + H_2O$

Added base: $NH_4^+ + OH^- \longrightarrow H_2O + NH_3$

The pH of the buffer solution depends on the strength of the acid or base chosen. In Chapter 15, actual pH values of buffer solutions are calculated from quantitative information related to acid strength.

We began this section by mentioning the importance of the buffer system of our blood. The carbonic acid–bicarbonate buffer system is our main protection and is discussed in Making It Real, "The pH Balance in the Blood." However, another buffer system is at work in the cells of our bodies. This is the $H_2PO_4^- - HPO_4^{2-}$ buffer system. In this system the dihydrogen phosphate ion ($H_2PO_4^-$) acts as the acid species and the hydrogen phosphate ion (HPO_4^{2-}) acts as the base species. These two ions can remove H_3O^+ or OH^- ions produced in the cells as follows:

Added acid: $HPO_4^{2-} + H_3O^+ \longrightarrow H_2PO_4^- + H_2O$

Added base: $H_2PO_4^- + OH^- \longrightarrow HPO_4^{2-} + H_2O$

13-8.3 Buffer Capacity

There is a limit to how much a buffer system can resist change. If the added amount of OH^- exceeds the reserve of HClO (referred to as the *buffer capacity*), then the pH will rise. This is analogous to a $110 expense for the person with the $100 savings. It is more than she can cover with bank savings, so the amount in her pocket decreases. The buffer capacity also relates to the original amount of the salt (NaClO) present. The best buffers, then, are those that contain large and approximately equal concentrations of acid and conjugate base. They are capable of absorbing significant amounts of either hydronium or hydroxide as necessary.

MAKING IT REAL

The pH Balance in the Blood

The chemistry of the blood is a science unto itself. It is our river of life, delivering fuel and oxygen for the muscles, removing waste products, and keeping our body temperature steady. The pH of the blood is 7.40, just barely alkaline. As mentioned in an earlier Making It Real, enzymes run our body but they function only in a very narrow pH range. Deviations of only 0.2 pH unit either way can cause coma or even be fatal. If the pH drops below 7.20, a condition known as *acidosis* results. If the pH rises to above 7.60, the opposite condition is called *alkalosis*. Either situation seriously upsets our body chemistry.

Perhaps one of the most amazing properties of blood is its ability to maintain a nearly constant pH despite the fact that acids and bases find their way into the bloodstream. The pH of our blood is held constant by two buffering systems. We will look closely at the more important system, which is the carbonic acid–bicarbonate buffer. This system is represented by the equation

The pH of the blood is an important measurement.

$$H_2CO_3(aq) + H_2O \rightleftharpoons HCO_3^-(aq) + H_3O^+(aq)$$

The carbonic acid (H_2CO_3) part of the buffer reacts with any excess OH^- ion that enters our blood, and the HCO_3^- ion reacts with any excess H_3O^+. For example, exercise produces extra H_3O^+ from the lactic acid formed in the muscles. The H_3O^+ reacts with HCO_3^- to form carbonic acid. This is transported to the lungs where it decomposes to CO_2 gas and H_2O, which is then exhaled. Our bodies respond to exercise by faster breathing and faster release of carbon dioxide.

If extra OH^- enters the blood, it reacts with H_2CO_3 and more bicarbonate is formed, which is eventually excreted by the kidneys. Sometimes when people get nervous, they breathe too fast, which is called *hyperventilation*. This causes too much carbon dioxide to be exhaled, the blood becomes less acidic and the pH of the blood rises above 7.4. If the person faints, the breathing rate decreases and the pH quickly comes back into balance. Actually, the concentration of bicarbonate is about 10 times the concentration of carbonic acid, so we are more protected against extra acid than extra base, which works out well.

Health problems such as diabetes and kidney disease may disturb the buffers of the blood. Generally, however, we couldn't have a better system to maintain pH and with it the smooth functioning of our body.

▶ **ASSESSING THE OBJECTIVE FOR SECTION 13-8**

EXERCISE 13-8(a) LEVEL 1: Fill in the blanks.
A buffer solution consists of a weak acid and a salt of its _____ _____. Buffer solutions resist a change in _____ because any hydronium is neutralized with the _____ and any hydroxide is neutralized with the _____. The _____ _____ is a measure of how much strong acid or strong base the buffer can absorb.

EXERCISE 13-8(b) LEVEL 1: Which of these would form a buffer solution?
(a) $HNO_3 + KNO_3$ (c) $HF + NaF$
(b) $HNO_2 + KNO_2$ (d) $HBr + NaBr$

EXERCISE 13-8(c) LEVEL 2: What salt could you add in order to make a buffer solution from each of the following aqueous solutions?
(a) NH_3 (b) Na_2SO_3 (c) $HClO_2$

EXERCISE 13-8(d) LEVEL 3: A mixture of $H_2C_2O_4$ and $NaHC_2O_4$ can act as a buffer. (a) Write the reaction that occurs when H_3O^+ is added to the solution. (b) Write the reaction when OH^- is added to the solution.

EXERCISE 13-8(e) LEVEL 3: The addition of HCl to the weak base C_5H_5N can create a buffer solution if the number of moles of HCl added is less than the number of moles of C_5H_5N present. What two species make up that buffer?

For additional practice, work chapter problems 13-29, 13-95, and 13-97.

► OBJECTIVE
FOR SECTION 13-9
Determine, if possible, whether a
specific oxide is acidic or basic.

13-9 Oxides as Acids and Bases

LOOKING AHEAD! We have one more important question concerning acids. How do they get into the atmosphere so as to make rain acidic? The question of acid rain is a matter of international consequence. ∎

13-9.1 Acid Anhydrides

One of the more unfortunate consequences of the human race's progress is acid rain. Tall industrial smokestacks disperse exhaust gases high into the atmosphere, where prevailing winds may carry pollutants hundreds of miles before they return to the Earth in the form of acid rain. Since acid rain passes over the boundaries of many countries, it can be a touchy subject between nations. Its effect on lakes and forests can be devastating, and there is little doubt that the problem must be faced and solved. To understand the origin of acid rain, we will examine how the oxides of the elements also lead to acid–base behavior.

Most acid rain originates from the combustion of coal or other fossil fuels that contain sulfur as an impurity. Combustion of sulfur or sulfur compounds produces sulfur dioxide (SO_2). In the atmosphere, sulfur dioxide reacts with oxygen to form sulfur trioxide (SO_3).

$$2SO_2(g) + O_2(g) \longrightarrow 2SO_3(g)$$

The sulfur trioxide reacts with water in the atmosphere to form sulfuric acid.

$$SO_3(g) + H_2O(l) \longrightarrow H_2SO_4(aq)$$

The sulfuric acid solution that comes down in the rain is a strong acid, which is corrosive and destructive. As mentioned earlier, acids react with metals and lime-stone, which are both used externally in buildings. (See Figure 13-7.) In the above reaction, sulfur trioxide can be considered as simply the dehydrated form of sulfuric acid. It is thus known as an **acid anhydride**, *which means acid without water.* Many nonmetal oxides are acid anhydrides. When dissolved in water, they form acids. Three other reactions of nonmetal oxides to form acids follow.

$$CO_2(g) + H_2O(l) \rightleftharpoons H_2CO_3(aq)$$
carbonic acid

$$SO_2(g) + H_2O(l) \rightleftharpoons H_2SO_3(aq)$$
sulfurous acid

$$N_2O_5(l) + H_2O(l) \longrightarrow 2HNO_3(aq)$$
nitric acid

We have previously mentioned that carbon dioxide, the acid anhydride of carbonic acid, is responsible for the fizz and the tangy taste in carbonated soft drinks and beer. When all the carbon dioxide escapes, the beverage goes flat. Carbon dioxide is also present in the atmosphere and dissolves in rainwater to make rain naturally acidic. Carbon dioxide by itself lowers the pH of rain to only about 5.7. Oxides of sulfur and nitrogen, however, lower the pH to 4.0 or even lower.[*] The oxides of nitrogen originate mainly from engines in automobiles and other heat sources. The high temperature in the engine causes the two elements of air, nitrogen and oxygen, to combine to form nitrogen oxides. In eastern North America and western Europe, the acidity in rain is about two-thirds due to sulfuric acid and the remainder due to nitric acid. The amount of acid rain due to nitrogen oxides seems to be growing, however. There is also a small amount of hydrochloric acid in acid rain.

FIGURE 13-7 The Effect of Acid Rain The deterioration of the statue is due to acid rain.

---

[*] On April 10, 1974, a rain fell on Pitlochry, Scotland, that had a pH of 2.4, which is about the same as that of vinegar. This is the most acidic rain ever recorded.

13-9.2 Base Anhydrides

Ionic metal oxides dissolve in water to form bases and thus are known as **base anhydrides**. Some examples of these reactions are

$$Na_2O(s) + H_2O(l) \longrightarrow 2NaOH(aq)$$

$$CaO(s) + H_2O(l) \longrightarrow Ca(OH)_2(aq)$$

Salt is formed by the reaction between an acid anhydride and a base anhydride. For example, the following reaction and the neutralization of H_2SO_3 with $Ca(OH)_2$ in aqueous solution produce the same salt.

$$SO_2(g) + CaO(s) \longrightarrow CaSO_3(s)$$

$$H_2SO_3(aq) + Ca(OH)_2(aq) \longrightarrow CaSO_3(s) + 2H_2O(l)$$

The first reaction represents a way of removing SO_2 from the combustion products of an industrial plant so that some of our abundant high-sulfur coal can be used without harming the environment.

MAKING IT REAL

Acid Rain—The Price of Progress?

Acid rain has had a harmful effect on the environment.

A century ago, it was inconceivable that we could actually change our planet. The human population was less than 2 billion and the world seemed vast and the environment forgiving. Now we know better. Greenhouse gases, ozone holes, and polluted water are serious problems created by people. One problem that we are doing something about is *acid rain*. Acid rain results from the presence of nonmetal oxides such as SO_2 and NO_2 in the atmosphere as industrial or automobile by-products.

In Canada and Scandinavia some lakes have been seriously affected. The pH of these lakes has become progressively lower as a result of acid rain. Fish and any other aquatic creatures have completely disappeared. In these areas of the world, the soil and rocks contain little of the types of minerals that neutralize acids, such as limestone ($CaCO_3$, the active ingredient in many antacids).

In other situations, the vulnerability of $CaCO_3$ to acid rain causes problems. Marble statues and mortar that have survived for centuries are rapidly deteriorating in the industrialized world. Various attempts are being made to protect these treasures with coatings, but so far nothing completely effective has been found.

The fertility of the soil is being affected by acid rain. Metal oxides are an integral part of the soil. These oxides are insoluble in water but become more soluble at low pH, as illustrated by the following equation.

$$Al_2O_3(s) + 6H_3O^+(aq) \longrightarrow 2Al^{3+}(aq) + 9H_2O(l)$$

Some forests in the world seem to be seriously stressed. Acid rain is the suspected culprit. In addition to oxides, many exposed metals that are not affected by ordinary rain are corroded by acid rain.

There is progress, however. There are strict limits on sulfur emissions from power plants. Catalytic converters on automobiles are reducing the nitrogen emissions. Laws are on the books and more are coming as we attempt to reverse this serious problem. These solutions do not come cheaply, however. We are all paying the billions of dollars necessary to prevent these oxides from entering the atmosphere.

Another concern about the buildup of CO_2 in the atmosphere is its effect on the acidity of the oceans. As the partial pressure of CO_2 increases in the atmosphere, more dissolves in the ocean (recall Henry's law), decreasing its pH. In fact, it has recently been reported that in the past 200 years the pH of the surface of the ocean has decreased from 8.3 to 8.2. That may not seem like much, but considering that pH is a logarithmic scale, the change represents a 20% increase in H_3O^+ concentration (i.e., from 5×10^{-9} mol/L to 6×10^{-9} mol/L). It is projected that the pH could fall to 7.7 (a 400% increase) by 2100 if the CO_2 concentration in the atmosphere increases as projected. It is not possible, at this time, to predict how the increased acidity of the oceans will affect the ocean's ecology or Earth's temperature. But it is a matter of grave concern.

▶ ASSESSING THE OBJECTIVE FOR SECTION 13-9

EXERCISE 13-9(a) LEVEL 1: Fill in the blanks.
Most nonmetal oxides react with water to form _____ solutions, and ionic metal oxides react with water to form _____ solutions. Compounds like K_2O or Cl_2O are called base or acid _____ because they contain no water.

EXERCISE 13-9(b) LEVEL 1: Do you expect the following to be acidic or basic?
(a) P_4O_{10} **(b)** NO_2 **(c)** MgO **(d)** B_2O_3

EXERCISE 13-9(c) LEVEL 2: What acid could form directly from SO_2?

EXERCISE 13-9(d) LEVEL 3: What base could form directly from Li_2O?

EXERCISE 13-9(e) LEVEL 3: What is the problem with burning high-sulfur coal? Why would we want to do it? What can we do to remove the sulfur from the exhaust?

For additional practice, work chapter problems 13-99 and 13-101.

PART C   SUMMARY

KEY TERMS

13-7	The reaction of an ion as an acid or base in water is known as a **hydrolysis reaction**. p. 459	
13-8.1	A **buffer** is a solution that resists changes in pH and contains both a weak acid and a weak base. p. 466	
13-9.1	An **acid anhydride** is a nonmetal oxide that forms an acid in water. p. 466	
13-9.2	A **base anhydride** is a metal oxide that forms a base in water. p. 467	

SUMMARY CHART

The Identity of Acids and Bases

Type	Example	Reaction
ACIDS		
1. Molecular hydrogen compounds ($H^+ +$ ion)	$HClO_4$	$HClO_4 + H_2O \longrightarrow H_3O^+ + ClO_4^-$
2. Cations (conjugate acids of weak bases)	NH_4^+	$NH_4^+ + H_2O \rightleftharpoons H_3O^+ + NH_3$
3. Nonmetal oxides	SO_3	$SO_3 + H_2O \longrightarrow H_2SO_4$ $H_2SO_4 + H_2O \longrightarrow H_3O^+ + HSO_4^-$
BASES		
1. Ionic hydroxides	$Ca(OH)_2$	$Ca(OH)_2 \xrightarrow{H_2O} Ca^{2+} + 2OH^-$
2. Molecular nitrogen compounds	$(CH_3)_2NH$	$(CH_3)_2NH + H_2O \rightleftharpoons (CH_3)_2NH_2^+ + OH^-$
3. Anions (conjugate bases of weak acids)	CN^-	$CN^- + H_2O \rightleftharpoons HCN + OH^-$
4. Metal oxides	K_2O	$K_2O + H_2O \longrightarrow 2KOH$ $KOH \xrightarrow{H_2O} K^+ + OH^-$

CHAPTER 13 SYNTHESIS PROBLEM

The actions of acids and bases are critical topics in many, if not most, chemical processes including life and the environmental. This chapter has helped us sort out some of these important reactions. In the following problems we will ask questions about the reactions and measurements of two hypothetical acids, one strong and one weak, and two hypothetical bases, one strong and one weak.

PROBLEM	SOLUTION
A hypothetical strong acid has the formula HX and a hypothetical diprotic weak acid has the formula H_2A. (X^- and A^{2-} are hypothetical anions. H, as usual, is the symbol for hydrogen.) A hypothetical strong base has the formula LOH and a hypothetical weak base has the formula NR_3. (L^+ is a hypothetical cation and NR_3 is a nitrogen-type base like ammonia.) **a.** Write the reaction of all four of these compounds with water. Show equilibrium arrows where appropriate.	**a.** $HX(aq) + H_2O(l) \longrightarrow H_3O^+(aq) + X^-(aq)$ $H_2A(aq) + H_2O(l) \rightleftharpoons H_3O^+(aq) + HA^-(aq)$ $LOH(s) \xrightarrow{H_2O} L^+(aq) + OH^-(aq)$ $NR_3(aq) + H_2O(l) \rightleftharpoons NR_3H^+(aq) + OH^-(aq)$
b. In the reactions of HX and NR_3 with water, indicate the conjugate acid–base pairs.	**b.** acid (HX)–conjugate base X^-; base H_2O–conjugate acid H_3O^+ acid H_2O–conjugate base OH^-; base NR_3–conjugate acid NR_3H^+
c. Write four balanced neutralization reactions between the two acids and the two bases. Write the total ionic and the net ionic reaction between HX and LOH.	**c.** $HX(aq) + LOH(aq) \longrightarrow LX(aq) + H_2O(l)$ $HX(aq) + NR_3(aq) \longrightarrow NR_3H^+X^-(aq)$ $H_2A(aq) + 2LOH(aq) \longrightarrow L_2A(aq) + 2H_2O(l)$ $H_2A(aq) + 2NR_3(aq) \longrightarrow (NR_3H)_2A(aq)$ *Total ionic:* $H^+(aq) + X^-(aq) + L^+(aq) + OH^-(aq) \longrightarrow L^+(aq) + X^-(aq) + H_2O(l)$ *Net ionic:* $H^+(aq) + OH^-(aq) \longrightarrow H_2O(l)$
d. What is the $[H_3O^+]$ and the pH of 0.010-M solutions of each of the four compounds? The weak acid (H_2A) is 15% ionized. (Ignore the second ionization since it is very small.) The weak base is 2.5% ionized.	**d.** For HX: $[H_3O^+] = 0.010\ M = 1.0 \times 10^{-2}\ M$ pH $= -\log[1.0 \times 10^{-2}] = \underline{2.00}$ For H_2A: $[H_3O^+] = 0.010 \times 0.15 = 0.0015\ M = 1.5 \times 10^{-3}$ pH $= \underline{2.82}$ For LOH: $[OH^-] = 0.010 = 1.0 \times 10^{-2}$ $[H_3O^+] = \dfrac{K_w}{[OH^-]} = \dfrac{1.0 \times 10^{-14}}{1.0 \times 10^{-2}} = 1.0 \times 10^{-12};$ pH $= \underline{12.00}$ For NR_3: $[OH^-] = 0.010 \times 0.025 = 0.00025 = \underline{2.5 \times 10^{-4}}$ $[H_3O^+] = \dfrac{1.0 \times 10^{-14}}{2.5 \times 10^{-4}} = 4.0 \times 10^{-11};$ pH $= \underline{10.40}$
e. There are four salts formed from the reactions in part **c** above. Write the formulas of these salts and describe the acidity of aqueous solutions formed when these four salts are dissolved in water. Write any hydrolysis reactions that may occur involving these four ions.	**e.** L^+X^-: A solution of this salt is neutral. Neither ion undergoes hydrolysis. $NR_3H^+X^-(aq)$: This solution is slightly acidic due to cation hydrolysis. $\quad NR_3H^+(aq) + H_2O(l) \rightleftharpoons NR_3(aq) + H_3O^+(l)$ $L_2A(aq)$: This solution is slightly basic due to anion hydrolysis. $A^{2-}(aq) + H_2O(l) \rightleftharpoons HA^-(aq) + OH^-(aq)$ $(NR_3H)_2A(aq)$: Both ions undergo hydrolysis but more information is needed to determine the comparative extent of the hydrolysis reactions.
f. A solution of H_2A and LHA serves as a buffer solution. Illustrate with two equations how this buffer removes added H_3O^+ and OH^- from the solution. Name another combination of acid and conjugate base that would form a buffer.	**f.** *Added acid:* $HA^-(aq) + H_3O^+(l) \longrightarrow H_2A(aq) + H_2O(l)$ *Added base:* $H_2A(aq) + OH^-(aq) \longrightarrow HA^-(aq) + H_2O(l)$ Another buffer would be a solution of NR_3 containing the salt: $NR_3H^+X^-$.

YOUR TURN

In this case consider two real acids, one strong HBr and the other weak $HC_2H_3O_2$. The two real bases are $Ba(OH)_2$, which is strong, and CH_3NH_2, which is weak.

a. Write the reactions of all four species with water. Show equilibrium arrows when appropriate.

b. Write acid–base conjugate pairs for the reactions involving $HC_2H_3O_2$ and CH_3NH_2.

c. Write four neutralization reactions involving these four compounds. Write the total ionic and net ionic equations for the reaction between $HC_2H_3O_2$ and $Ba(OH)_2$.

d. What is the $[H_3O^+]$ and the pH of 0.10-M solutions of these four compounds? The weak acid is 1.3% ionized and the weak base is 6.6% ionized.

e. Write the formulas of the four salts formed in part **c** above and describe the acidity of solutions of these salts. Write any hydrolysis reactions that may occur involving these ions.

f. A solution of CH_3NH_2 and $CH_3NH_3^+Br^-$ is a buffer. Illustrate with two equations how this buffer removes added H_3O^+ and OH^- from the solution.

Answers are on p. 472.

CHAPTER SUMMARY

Compounds have been classified as acids or bases for hundreds of years on the basis of common sets of chemical characteristics. In the twentieth century, however, acid character was attributed to formation of H^+ [also represented as the **hydronium ion** (H_3O^+)] in aqueous solution. Base character is due to the formation of OH^- in solution.

Our understanding of acid–base behavior can be broadened somewhat by use of the **Brønsted–Lowry** definition, which defines **acids** as proton (H^+) donors and **bases** as proton acceptors. **Amphiprotic** substances can either donate or accept a proton. An acid–base reaction constitutes an H^+ exchange between an acid and a base to form a **conjugate base** and **conjugate acid**, respectively.

Acids and bases can also be classified according to strength. Strong acids and strong bases are 100% ionized in water, whereas **weak molecular acids** and **weak molecular bases** are only partially ionized. Partial ionization of a molecular acid occurs when a reaction reaches a point of equilibrium in which both molecules (on the left of the equation) and ions (on the right) are present. For weak molecular acids and bases, the point of equilibrium favors the left, or molecular, side of the equation. Therefore, the H_3O^+ concentration in a weak acid solution is considerably lower than in a strong acid solution at the same initial concentration of acid.

When acidic and basic solutions are mixed, the two active ions combine in a neutralization reaction to form water. Complete neutralization of a **monoprotic acid** results in a salt and water. Incomplete neutralization of a **polyprotic acid** (either **diprotic** or **triprotic**) produces an **acid salt** and water.

Even in pure water, there is a very small equilibrium concentration of H_3O^+ and OH^- (1.0×10^{-7} M) due to the **autoionization** of water. The product of these concentrations is a constant known as the **ion product** (K_w) of water. The ion product can be used to calculate the concentration of one ion from that of the other in any aqueous solution. A convenient method to express the H_3O^+ or OH^- concentrations of solutions involves the use of the logarithmic expressions **pH** and **pOH**.

The ions of a salt may or may not interact with water as acids or bases. If such a reaction of an ion does occur, it is known as **hydrolysis**. To predict the effect of the solution of a salt on the pH, possible hydrolysis reactions of both the cation and the anion must be examined. The resulting solution may be neutral, weakly basic, or weakly acidic, depending on the ion undergoing hydrolysis.

When a solution of a weak acid is mixed with a solution of a salt, providing its conjugate base, a **buffer** solution is formed. Buffer solutions resist changes in pH from addition of limited amounts of a strong acid or base. In a buffer, the reservoir of nonionized acid (e.g., $HC_2H_3O_2$) reacts with added OH^-, while the reservoir of the conjugate base (e.g., $C_2H_3O_2^-$) reacts with added H_3O^+. Weak bases and salts providing their conjugate acids also act as buffers (e.g., NH_3 and $NH_4^+Cl^-$).

Finally, the list of acids and bases was expanded to include oxides, which can be classified as **acid anhydrides** or **base anhydrides**.

OBJECTIVES

SECTION	YOU SHOULD BE ABLE TO...	EXAMPLES	EXERCISES	CHAPTER PROBLEMS
13-1	List the general properties of acids and bases.		1a, 1b, 1c, 1d	5, 6
13-2	Identify Brønsted acids and bases and conjugate acid–base pairs in a proton exchange reaction.	13-1, 13-2, 13-3	2a, 2b, 2c, 2d	1, 2, 7, 8, 9, 10, 11, 14
13-3	Calculate the hydronium ion concentration in a solution of a strong acid and a weak acid given the initial concentration of the acid and the percent ionization of the weak acid.	13-4, 13-5	3a, 3b, 3c	17, 23, 24, 26, 29
13-4	Write the molecular, total ionic, and net ionic equations for neutralization reactions.	13-6, 13-7	4a, 4b, 4c, 4d	18, 19, 21, 33, 34, 35, 36, 40, 42, 45
13-5	Using the ion product of water, relate the hydroxide ion and the hydronium ion concentrations.	13-8	5a, 5b, 5c	52, 53, 54, 56
13-6	Given the hydronium or hydroxide concentrations, calculate the pH and pOH and vice versa.	13-9, 13-10, 13-11, 13-12	6a, 6b, 6c, 6d, 6e, 6f	60, 62, 64, 66, 67, 70, 73, 76

13-7	Write a hydrolysis reaction, if one occurs, for salt solutions to determine whether they are acidic or basic.	13-13	7a, 7b, 7c, 7d, 7e, 7f	31, 79, 82, 84, 86, 87, 88
13-8	Write equations illustrating how a buffer solution can absorb either added acid or base.		8a, 8b, 8c, 8d, 8e	92, 94, 95, 97, 98
13-9	Determine, if possible, whether a specific oxide is acidic or basic.		9a, 9b, 9c, 9d, 9e	99, 100

▶ ANSWERS TO ASSESSING THE OBJECTIVES

Part A

EXERCISES

13-1(a) An acid is a compound that produces the <u>hydrogen</u> ion in solution, which is also written as the hydronium ion (H_3O^+). A base is a compound that produces an ion with the formula <u>OH$^-$</u> and the name <u>hydroxide</u> ion.

13-1(b) (a) $HClO_4$, perchloric acid (b) $Fe(OH)_2$, iron(II) hydroxide (c) H_2S, hydrosulfuric acid (d) $LiOH$, lithium hydroxide

13-1(c) (a) $HI + H_2O \longrightarrow H_3O^+ + I^-$

(b) $LiOH + H_2O \longrightarrow Li^+(aq) + OH^-(aq)$

13-1(d) (a) base (b) acid (c) acid (d) base (e) base (f) base

13-2(a) In the Brønsted–Lowry definition, acids are <u>proton</u> donors and bases are <u>proton</u> acceptors. A conjugate acid of a compound or ion results from the <u>exchange</u> of an H^+ ion. A substance that has both a conjugate acid and a conjugate base is said to be <u>amphiprotic</u>.

13-2(b) acid: NH_4^+; base: SO_3^{2-}

13-2(c) acid: $H_2C_2O_4$; base: $C_2O_4^{2-}$

13-2(d) (a) $HCO_3^- + H_2O \longrightarrow CO_3^{2-} + H_3O^+$

(b) $HCO_3^- + H_2O \longrightarrow H_2CO_3 + OH^-$

13-2(e) $CH_3COOH + (CH_3)_3N \longrightarrow CH_3COO^- + (CH_3)_3NH^+$.

The two sets of conjugate acid–base pairs are

CH_3COOH–CH_3COO^- and $(CH_3)_3NH^+$–$(CH_3)_3N$

13-2(f) $NH_2^- + CH_3OH \longrightarrow NH_3 + CH_3O^-$

13-3(a) Strong acids are essentially <u>100</u> % ionized in aqueous solution, whereas weak acids are <u>partially</u> ionized. Sodium hydroxide is a strong base, but ammonia is a <u>weak base</u>. When a weak acid reacts with water, the reaction is said to be at <u>equilibrium</u>.

13-3(b) This is a weak acid. $[H_3O^+] = 0.10 \times 0.025 = 0.0025$ M.

13-3(c) Since the hydronium concentration is less than the original acid concentration, this is a weak acid. The percent ionization $= 0.0050 \times 100/.25 = 2.0\%$.

13-3(d) (a) $HNO_3 + H_2O \longrightarrow H_3O^+ + NO_3^-$

(b) $HNO_2 + H_2O \rightleftharpoons H_3O^+ + NO_2^-$

(c) $HClO_2 + H_2O \rightleftharpoons H_3O^+ + ClO_2^-$

(d) $HClO_4 + H_2O \longrightarrow H_3O^+ + ClO_4^-$

13-4(a) The reaction between an acid and a base is known as a <u>neutralization</u> reaction. The net ionic equation of this reaction always has H_2O as a product. The spectator ions of the reaction form a <u>salt</u>. The partial neutralization of polyprotic acids produces an <u>acid salt</u> and water.

13-4(b)

Molecular: $2HNO_3(aq) + Sr(OH)_2(aq) \longrightarrow$
$$2H_2O(l) + Sr(NO_3)_2(aq)$$

Total Ionic: $2H^+(aq) + 2NO_3^-(aq) + Sr^{2+}(aq) + 2OH^-(aq) \longrightarrow$
$$2H_2O(l) + Sr^{2+}(aq) + 2NO_3^-(aq)$$

Net Ionic: $H^+(aq) + OH^-(aq) \longrightarrow H_2O(l)$

13-4(c)

Molecular: $NH_3(aq) + HCl(aq) \longrightarrow NH_4Cl(aq)$

Total Ionic: $NH_3(aq) + H^+(aq) + Cl^-(aq) \longrightarrow$
$$NH_4^+(aq) + Cl^-(aq)$$

Net Ionic: $NH_3(aq) + H^+(aq) \longrightarrow NH_4^+(aq)$

13-4(d) (a) *Molecular:* $H_2C_2O_4(aq) + CsOH(aq) \longrightarrow$
$$CsHC_2O_4(aq) + H_2O(l)$$

Total Ionic: $H_2C_2O_4(aq) + Cs^+(aq) + OH^-(aq) \longrightarrow$
$$Cs^+(aq) + HC_2O_4^-(aq) + H_2O(l)$$

Net Ionic: $H_2C_2O_4(aq) + OH^-(aq) \longrightarrow HC_2O_4^-(aq) + H_2O(l)$

(b) *Molecular:* $H_2C_2O_4(aq) + 2CsOH(aq) \longrightarrow$
$$Cs_2C_2O_4(aq) + 2H_2O(l)$$

Total Ionic: $H_2C_2O_4(aq) + 2Cs^+(aq) + 2OH^-(aq) \longrightarrow$
$$2Cs^+(aq) + C_2O_4^{2-}(aq) + 2H_2O(l)$$

Net Ionic: $H_2C_2O_4(aq) + 2OH^-(aq) \longrightarrow C_2O_4^{2-}(aq) + 2H_2O(l)$

13-4(e) (a) $H_2SO_4 + 2KOH$ (b) $H_2SO_4 + KOH$ (c) $NH_3 + HCl$

Part B

EXERCISES

13-5(a) Pure water contains a small but important concentration of H_3O^+ and OH^- ions. The product of the concentrations of these two ions, known as the <u>ion</u> <u>product</u> of water, is symbolized by K_w and has a value of 1×10^{-14} at 25°C. As the concentration of <u>OH$^-$</u> increases in water, the concentration of H_3O^+ <u>decreases</u>.

13-5(b) $(1.0 \times 10^{-14})/(7.2 \times 10^{-5}) = 1.4 \times 10^{-10}$ M

13-5(c) $[OH^-] = 0.00050$ M $= 5.0 \times 10^{-4}$ M $[H_3O^+] = 2.0 \times 10^{-11}$ M

13-5(d) $[OH^-]$ is larger than 1.0×10^{-7}, so the solution is basic. It would need to be close to 10^{-1} to be considered strongly basic, so this is weakly basic. A common weak base that could produce that solution would be NH_3.

13-6(a) The pH of a solution is defined as $-\log [H_3O^+]$. An acidic solution is one that has an $[H_3O^+]$ greater than 1×10^{-7} and a pH <u>less</u> than 7. A basic solution is one that has an $[OH^-]$ greater than 1×10^{-7} and a pH <u>greater</u> than 7.

13-6(b) pH $= 9.86$

13-6(c) $[LiOH] = 0.25$

13-6(d) (a) 11.80   (b) 0.0063 M H_3O^+ (c) 1.6×10^{-12} M OH^-

13-6(e) (a) 1.3×10^{-13} M H_3O^+ (b) 0.075 M OH^-
(c) $pH = 12.88$ (d) $pOH = 1.12$

13-6(f) The HCl is a strong acid and the H_3O^+ is determined directly from the concentration of acid. It is 0.005 M, and the pH is 2.3. The acetic acid's H_3O^+ comes only from the part that is ionized, or 0.0008 M, which is about six times less than the HCl solution. The pH of the second solution is 3.1.

Part C

EXERCISES

13-7(a) (a) acidic (b) neutral (c) basic (d) more information is needed

13-7(b) (a) basic (b) acidic (c) neutral (d) basic

13-7(c) (a) $CO_3^{2-} + H_2O \rightleftharpoons HCO_3^- + OH^-$
(b) No hydrolysis is occurring.
(c) $N_2H_5^+ + H_2O \rightleftharpoons N_2H_4 + H_3O^+$

13-7(d) Any weak acid would work, such as $HC_2H_3O_2$, $HClO_2$, HNO_2, or HF.

13-7(e) Any strong acid would work, such as HCl, HNO_3, or H_2SO_4.

13-7(f) $HSO_4^- + H_2O \rightleftharpoons H_3O^+ + SO_4^{2-}$

It can't produce hydroxide because the conjugate acid, sulfuric acid, is strong. Therefore, HSO_4^- has no basic properties.

13-8(a) A buffer solution consists of a weak acid and a salt of its conjugate base. Buffer solutions resist a change in pH because any hydronium is neutralized with the base and any hydroxide is neutralized with the acid. The buffer capacity is a measure of how much strong acid or strong base the buffer can absorb.

13-8(b) Possible buffers include (b) and (c).

13-8(c) (a) NH_4Cl (b) $NaHSO_3$ (c) $KClO_2$

13-8(d) (a) $H_3O^+ + HC_2O_4^- \longrightarrow H_2C_2O_4 + H_2O$
(b) $OH^- + H_2C_2O_4 \longrightarrow HC_2O_4^- + H_2O$

13-8(e) $C_5H_5NH^+Cl^-$ (the weakly acidic salt) and C_5H_5N (the weak base)

13-9(a) Most nonmetal oxides react with water to form acidic solutions, and ionic metal oxides react with water to form basic solutions. Compounds like K_2O or Cl_2O are called base or acid anhydrides because they contain no water.

13-9(b) (a) acidic (b) acidic (c) basic (d) acidic

13-9(c) H_2SO_3

13-9(d) LiOH

13-9(e) High-sulfur coal releases SO_2 into the environment, which can lead to the formation of acid rain. High-sulfur coal, though, is abundant and therefore costs less. To keep the SO_2 out of the environment, we can use scrubbers that react the acid anhydride SO_2 with the base anhydride CaO to form the insoluble salt $CaSO_3$.

ANSWERS TO CHAPTER SYNTHESIS PROBLEM

a. $HBr(aq) + H_2O(l) \longrightarrow H_3O^+(aq) + Br^-(aq)$
$HC_2H_3O_2(aq) + H_2O(l) \rightleftharpoons$
$\qquad H_3O^+(aq) + C_2H_3O_2^-(aq)$
$Ba(OH)_2(s) \xrightarrow{H_2O} Ba^{2+}(aq) + 2 OH^-(aq)$
$CH_3NH_2(aq) + H_2O(l) \rightleftharpoons$
$\qquad CH_3NH_3^+(aq) + OH^-(aq)$

b. acid ($HC_2H_3O_2$)–conjugate base $C_2H_3O_2^-$; base H_2O–conjugate acid H_3O^+ acid H_2O–conjugate base OH^-; base CH_3NH_2–conjugate acid $CH_3NH_3^+$

c. $2HBr(aq) + Ba(OH)_2(aq) \longrightarrow$
$\qquad BaBr_2(aq) + 2H_2O(l)$
$HBr(aq) + CH_3NH_2(aq) \longrightarrow$
$\qquad CH_3NH_3^+Br^-(aq)$
$2HC_2H_3O_2(aq) + Ba(OH)_2(aq) \longrightarrow$
$\qquad Ba(C_2H_3O_2)_2(aq) + 2H_2O(l)$
$HC_2H_3O_2(aq) + CH_3NH_2 \longrightarrow$
$\qquad CH_3NH_3^+C_2H_3O_2^-(aq)$
Total ionic:
$2HC_2H_3O_2(aq) + Ba^{2+}(aq) + 2OH^-(aq)$
$\longrightarrow Ba^{2+}(aq) + 2C_2H_3O_2^-(aq) +$
$\qquad 2H_2O(l)$
Net ionic: $2HC_2H_3O_2(aq) + 2OH^-(aq)$
$\qquad \longrightarrow 2C_2H_3O_2^-(aq) + 2H_2O(l)$

d. For HBr: $[H_3O^+] = 1.0 \times 10^{-1}$ M pH = 1.00
For $HC_2H_3O_2$: $[H_3O^+] = 1.3 \times 10^{-3}$ M pH = 2.89
For $Ba(OH)_2$: $[OH^-] = 0.10 \times 2 =$
$\qquad 2.0 \times 10^{-1}$ $[H_3O^+] = 5.0 \times 10^{-14}$;
$\qquad pH = 13.30$
For CH_3NH_2: $[OH^-] = 0.0066$ M $[H_3O^+] =$
$\qquad 1.5 \times 10^{-12}$ pH = 11.82

e. $BaBr_2$: A solution of this salt is neutral. Neither ion undergoes hydrolysis.
$CH_3NH_3^+Br^-(aq)$: This solution is slightly acidic due to cation hydrolysis.
$CH_3NH_3^+(aq) + H_2O(l) \rightleftharpoons$
$\qquad CH_3NH_2(aq) + H_3O^+(l)$
$Ba(C_2H_3O_2)_2(aq)$: This solution is slightly basic due to anion hydrolysis.
$C_2H_3O_2^-(aq) + H_2O(l) \rightleftharpoons$
$\qquad HC_2H_3O_2(aq) + OH^-(aq)$
$CH_3NH_3^+C_2H_3O_2^-(aq)$: Both ions undergo hydrolysis but more information is needed to determine the comparative extent of the hydrolysis reactions.

f. *Added acid:* $CH_3NH_2(aq) + H_3O^+(l)$
$\qquad \longrightarrow CH_3NH_3^+(aq) + H_2O(l)$
Added base: $CH_3NH_3^+(aq) + OH^-(aq)$
$\qquad \longrightarrow CH_3NH_2(aq) + H_2O(l)$

CHAPTER PROBLEMS

Throughout the text, answers to all exercises in color are given in Appendix E. The more difficult exercises are marked with an asterisk.

Acids and Bases (SECTION 13-1)

13-1. Give the formulas and names of the acid compounds derived from the following anions.

(a) NO_3^- (b) NO_2^- (c) ClO_3^- (d) SO_3^{2-}

13-2. Give the formulas and names of the acid compounds derived from the following anions.

(a) CN^- (b) CrO_4^{2-} (c) ClO_4^- (d) Br^-

13-3. Give the formulas and names of the base compounds derived from the following cations.

(a) Cs^+ (b) Sr^{2+} (c) Al^{3+} (d) Mn^{3+}

13-4. Give the formulas and names of the acid or base compounds derived from the following ions.

(a) Ba^{2+} (c) Pb^{2+} (e) $H_2PO_4^-$

(b) Se^{2-} (d) ClO^- (f) Fe^{3+}

13-5. Write reactions illustrating the acid or base behavior in water for the following.

(a) HNO_3 (b) $CsOH$ (c) $Ba(OH)_2$ (d) HBr

13-6. Write reactions illustrating the acid or base behavior in water for the following.

(a) HI (b) $Sr(OH)_2$ (c) $RbOH$ (d) $HClO_4$

Brønsted–Lowry Acids and Bases (SECTION 13-2)

13-7. What is the conjugate base of each of the following?

(a) HNO_3 (c) HPO_4^{2-} (e) H_2O

(b) H_2SO_4 (d) CH_4 (f) NH_3

13-8. What is the conjugate acid of each of the following?

(a) CH_3NH_2 (c) NO_3^- (e) H^+

(b) HPO_4^{2-} (d) O^{2-} (f) H_2O

13-9. Identify conjugate acid-base pairs in the following reactions.

(a) $HClO_4 + OH^- \longrightarrow H_2O + ClO_4^-$

(b) $HSO_4^- + ClO^- \longrightarrow HClO + SO_4^{2-}$

(c) $H_2O + NH_2^- \longrightarrow NH_3 + OH^-$

(d) $NH_4^+ + H_2O \longrightarrow NH_3 + H_3O^+$

13-10. Identify conjugate acid-base pairs in the following reactions.

(a) $HCN + H_2O \longrightarrow H_3O^+ + CN^-$

(b) $HClO_4 + NO_3^- \longrightarrow HNO_3 + ClO_4^-$

(c) $H_2S + NH_3 \longrightarrow NH_4^+ + HS^-$

(d) $H_3O^+ + HCO_3^- \longrightarrow H_2CO_3 + H_2O$

13-11. Write reactions indicating Brønsted–Lowry acid behavior with H_2O for the following. Indicate conjugate acid-base pairs.

(a) H_2SO_3 (c) HBr (e) H_2S

(b) $HClO$ (d) HSO_3^- (f) NH_4^+

13-12. Write reactions indicating Brønsted–Lowry base behavior with H_2O for the following. Indicate conjugate acid-base pairs.

(a) NH_3 (c) HS^- (e) F^-

(b) N_2H_4 (d) H^-

13-13. Write equations showing how HS^- can act as a Brønsted–Lowry base with H_3O^+ and as a Brønsted–Lowry acid with OH^-.

13-14. Bicarbonate of soda ($NaHCO_3$) acts as an antacid (base) in water. Write an equation illustrating how the HCO_3^- ion reacts with H_3O^+. Bicarbonate is amphiprotic. Write the reaction illustrating its behavior as an acid in water.

Strengths of Acids and Bases (SECTION 13-3)

13-15. Describe how a strong acid and a weak acid relate and differ.

13-16. When HBr ionizes in water, the reaction is 100% complete. Write the equation illustrating how HBr behaves as an acid. What does this tell us about the strength of the acid? An accepted observation is that *the stronger the acid, the weaker its conjugate base.* Compare the strength of HBr as a proton donor with Br^- as a proton acceptor.

13-17. Solutions of $HClO_2$ indicate that the ionization is very limited. Write the reaction illustrating how $HClO_2$ behaves as an acid. What does this tell us about the strength of the acid and the strength of its conjugate base?

13-18. Write equations illustrating the reactions with water of the acids formed in problem 13-1. Indicate strong acids with a single arrow and weak acids with equilibrium arrows.

13-19. Write equations illustrating the reactions with water of the acids formed in problem 13-2. Indicate strong acids with a single arrow and weak acids with equilibrium arrows.

13-20. Dimethylamine [$(CH_3)_2NH$] is a weak base that reacts in water like ammonia (NH_3). Write the equilibrium illustrating this reaction.

13-21. Pyridine (C_5H_5N) behaves as a weak base in water like ammonia. Write the equilibrium illustrating this reaction.

13-22. The concentration of a monoprotic acid (HX) in water is 0.10 M. The concentration of H_3O^+ ion in this solution is 0.010 M. Is HX a weak or a strong acid? What percent of the acid is ionized?

13-23. A 0.50-mol quantity of an acid is dissolved in 2.0 L of water. In the solution, $[H_3O^+] = 0.25$. Is this a strong or a weak acid? Explain.

13-24. What is $[H_3O^+]$ in a 0.55 M $HClO_4$ solution?

13-25. What is $[H_3O^+]$ in a 0.55 M solution of a weak acid, HX, that is 3.0% ionized?

13-26. What is $[OH^-]$ in a 1.45 M solution of NH_3 if the NH_3 is 0.95% ionized?

*13-27. What is $[H_3O^+]$ in a 0.354 M solution of H_2SO_4? Assume that the first ionization is complete but that the second is only 25% complete.

13-28. A 1.0 M solution of HF has $[H_3O^+] = 0.050$. What is the percent ionization of the acid?

13-29. A 0.10 M solution of pyridine (a weak base in water) has $[OH^-] = 4.4 \times 10^{-5}$. What is the percent ionization of the base?

13-30. The HSO_4^- ion is not amphiprotic in water. Which species cannot exist in water—its conjugate base or conjugate acid? Why not?

Neutralization and Salts (SECTION 13-4)

13-31. Identify each of the following as an acid, base, salt, or acid salt.

(a) H_2S (c) H_3AsO_4 (e) KNO_3

(b) $BaCl_2$ (d) $Ba(HSO_4)_2$ (f) $LiOH$

13-32. Explain why a strong acid is represented in aqueous solution as two ions but a weak acid is represented as one molecule.

13-33. Write the balanced molecular equations showing the complete neutralizations of the following.

(a) HNO_3 by $NaOH$ (c) $HClO_2$ by KOH

(b) $Ca(OH)_2$ by HI

13-34. Write the balanced molecular equations showing the complete neutralizations of the following.

(a) HNO_2 by $NaOH$ (c) H_2S by $Ba(OH)_2$

(b) H_2CO_3 by $CsOH$

13-35. Write the total ionic and net ionic equations for the reactions in problem 13-33.

13-36. Write the total ionic and net ionic equations for the reactions in problem 13-34.

13-37. Write the molecular, total ionic, and net ionic equations of the complete neutralization of $H_2C_2O_4$ with NH_3.

13-38. Write the molecular, total ionic, and net ionic equations for the complete neutralization of $HC_2H_3O_2$ with NH_3.

13-39. Write the formulas of the acid and the base that formed the following salts.

(a) $KClO_3$ (c) $Ba(NO_2)_2$

(b) $Al_2(SO_3)_3$ (d) NH_4NO_3

13-40. Write the formulas of the acid and the base that formed the following salts.

(a) Li_2CrO_4 (c) $Fe(ClO_4)_3$

(b) $NaCN$ (d) $Mg(HCO_3)_2$

13-41. Write balanced acid-base neutralization reactions that would lead to formation of the following salts or acid salts.

(a) $CaBr_2$ (c) $Ba(HS)_2$

(b) $Sr(ClO_2)_2$ (d) Li_2S

13-42. Write balanced acid-base neutralization reactions that would lead to formation of the following salts or acid salts.

(a) Na_2SO_3 (c) $Mg_3(PO_4)_2$

(b) AlI_3 (d) $NaHCO_3$

13-43. Write two equations illustrating the stepwise neutralization of H_2S with $LiOH$.

13-44. Write the two net ionic equations illustrating the two reactions in problem 13-43.

13-45. Write three equations illustrating the stepwise neutralization of H_3AsO_4 with $LiOH$. Write the total reaction.

13-46. Write the net ionic equations for the three reactions in problem 13-45.

13-47. Write the equation illustrating the reaction of 1 mol of H_2S with 1 mol of $NaOH$.

*13-48. Write the equation illustrating the reaction between 1 mol of $Ca(OH)_2$ and 2 mol of H_3PO_4.

Equilibrium of Water and K_w (SECTION 13-5)

13-49. If some ions are present in pure water, why isn't pure water considered to be an electrolyte?

13-50. Why can't $[H_3O^+] = [OH^-] = 1.0 \times 10^{-2}$ in water? What would happen if we tried to make such a solution by mixing 10^{-2} mol/L of KOH with 10^{-2} mol/L of HCl?

13-51. (a) What is $[H_3O^+]$ when $[OH^-] = 10^{-12}$ M?

(b) What is $[H_3O^+]$ when $[OH^-] = 10$ M?

(c) What is $[OH^-]$ when $[H_3O^+] = 2.0 \times 10^{-5}$ M?

13-52. (a) What is $[OH^-]$ when $[H_3O^+] = 1.50 \times 10^{-3}$ M?

(b) What is $[H_3O^+]$ when $[OH^-] = 2.58 \times 10^{-7}$ M?

(c) What is $[H_3O^+]$ when $[OH^-] = 5.69 \times 10^{-8}$ M?

13-53. When 0.250 mol of the strong acid $HClO_4$ is dissolved in 10.0 L of water, what is $[H_3O^+]$? What is $[OH^-]$?

13-54. Lye is a very strong base. What is $[H_3O^+]$ in a 2.55 M solution of $NaOH$? In the weakly basic household ammonia, $[OH^-] = 4.0 \times 10^{-3}$ M. What is $[H_3O^+]$?

13-55. Identify the solutions in problem 13-51 as acidic, basic, or neutral.

13-56. Identify the solutions in problem 13-52 as acidic, basic, or neutral.

13-57. Identify each of the following as an acidic, basic, or neutral solution.

(a) $[H_3O^+] = 6.5 \times 10^{-3}$ M   (c) $[OH^-] = 4.5 \times 10^{-8}$ M

(b) $[H_3O^+] = 5.5 \times 10^{-10}$ M   (d) $[OH^-] = 50 \times 10^{-8}$ M

13-58. Identify each of the following as an acidic, basic, or neutral solution.

(a) $[OH^-] = 8.1 \times 10^{-8}$ M   (c) $[H_3O^+] = 4.0 \times 10^{-3}$ M

(b) $[H_3O^+] = 10.0 \times 10^{-8}$ M   (d) $[OH^-] = 55 \times 10^{-8}$ M

pH and pOH (SECTION 13-6)

13-59. What is the pH of the following solutions?

(a) $[H_3O^+] = 1.0 \times 10^{-6}$ M   (d) $[OH^-] = 2.5 \times 10^{-5}$ M

(b) $[H_3O^+] = 1.0 \times 10^{-9}$ M   (e) $[H_3O^+] = 6.5 \times 10^{-11}$ M

(c) $[OH^-] = 1.0 \times 10^{-2}$ M

13-60. What is the pH of the following solutions?

(a) $[H_3O^+] = 1.0 \times 10^{-2}$ M   (d) $[OH^-] = 3.6 \times 10^{-9}$ M

(b) $[OH^-] = 1.0 \times 10^{-4}$ M   (e) $[OH^-] = 7.8 \times 10^{-4}$ M

(c) $[H_3O^+] = 1.0$ M   (f) $[H_3O^+] = 4.22 \times 10^{-4}$ M

13-61. What are the pH and pOH of the following?

(a) $[H_3O^+] = 0.0001$ (c) $[H_3O^+] = 0.020$

(b) $[OH^-] = 0.00001$ (d) $[OH^-] = 0.000320$

13-62. What are the pH and pOH of the following?

(a) $[H_3O^+] = 0.0000001$ (c) $[OH^-] = 0.0568$

(b) $[OH^-] = 0.0001$ (d) $[H_3O^+] = 0.00082$

13-63. What is $[H_3O^+]$ of the following?

(a) pH = 3.00   (d) pOH = 6.38

(b) pH = 3.54   (e) pH = 12.70

(c) pOH = 8.00

13-64. What is $[H_3O^+]$ of the following?

(a) pH = 9.0   (c) pH = 2.30

(b) pOH = 9.0   (d) pH = 8.90

13-65. Identify each of the solutions in problems 13-59 and 13-63 as acidic, basic, or neutral.

13-66. Identify each of the solutions in problems 13-60 and 13-64 as acidic, basic, or neutral.

13-67. A solution has pH = 3.0. What is the pH of a solution that is 100 times less acidic? What is the pH of a solution that is 10 times more acidic?

13-68. A solution has pOH = 4. What is the pOH of a solution that is 1000 times more acidic? What is the pOH of a solution that is 100 times more basic?

13-69. What is the pH of a 0.075 M solution of the strong acid HNO_3?

13-70. What is the pH of a 0.0034 M solution of the strong base KOH?

13-71. What is the pH of a 0.018 M solution of the strong base $Ca(OH)_2$?

13-72. A weak monoprotic acid is 10.0% ionized in solution. What is the pH of a 0.10 M solution of this acid?

13-73. A weak base is 5.0% ionized in solution. What is the pH of a 0.25 M solution of this base? (Assume one OH^- per formula unit.)

13-74. Identify each of the following solutions as strongly basic, weakly basic, neutral, weakly acidic, or strongly acidic.

(a) pH = 1.5   (d) pH = 13.0   (g) pOH = 7.5

(b) pOH = 13.0   (e) pOH = 7.0   (h) pH = −1.0

(c) pH = 5.8   (f) pH = 8.5

13-75. Arrange the following substances in order of increasing acidity.

(a) household ammonia, pH = 11.4

(b) vinegar, $[H_3O^+] = 2.5 \times 10^{-3}$ M

(c) grape juice, $[OH^-] = 1.0 \times 10^{-10}$ M

(d) sulfuric acid, pOH = 13.6

(e) eggs, pH = 7.8

(f) rainwater, $[H_3O^+] = 2.0 \times 10^{-6}$ M

13-76. Arrange the following substances in order of increasing acidity.

(a) lime juice, $[H_3O^+] = 6.0 \times 10^{-2}$ M

(b) antacid tablet in water, $[OH^-] = 2.5 \times 10^{-6}$ M

(c) coffee, pOH = 8.50

(d) stomach acid, pH = 1.8

(e) saliva, $[H_3O^+] = 2.2 \times 10^{-7}$ M

(f) a soap solution, pH = 8.3

(g) a solution of lye, pOH = 1.2

(h) a banana, $[OH^-] = 4.0 \times 10^{-10}$ M

*13-77. What is the pH of a 0.0010 M solution of H_2SO_4? (Assume that the first ionization is complete but the second is only 25% complete.)

Hydrolysis of Salts (SECTION 13-7)

13-78. Two of the following act as weak bases in water. Write the appropriate reactions illustrating the weak base behavior.

(a) ClO_4^- (b) $C_2H_3O_2^-$ (c) NH_3 (d) HF

13-79. Two of the following act as weak acids in water. Write the appropriate reactions illustrating the weak acid behavior.

(a) H_2CrO_4 (b) NH_4^+ (c) NH_2CH_3 (d) CrO_4^{2-}

13-80. Three of the following molecules or ions do not affect the pH of water. Which are they and why do they not affect the pH?

(a) K^+ (c) HCO_3^- (e) O_2

(b) NH_3 (d) NO_3^- (f) $N_2H_5^+$

13-81. Complete the following hydrolysis equilibria.

(a) $S^{2-} + H_2O \rightleftharpoons$ _____ $+ OH^-$

(b) $N_2H_5^+ + H_2O \rightleftharpoons N_2H_4 +$ _____

(c) $HPO_4^{2-} + H_2O \rightleftharpoons H_2PO_4^- +$ _____

(d) $(CH_3)_2NH_2^+ + H_2O \rightleftharpoons$ _____ $+ H_3O^+$

13-82. Complete the following hydrolysis equilibria.

(a) $CN^- + H_2O \rightleftharpoons HCN +$ _____

(b) $NH_4^+ + H_2O \rightleftharpoons$ _____ $+ H_3O^+$

(c) $B(OH)_4^- + H_2O \rightleftharpoons H_3BO_3 +$ _____

(d) $Al(H_2O)_6^{3+} + H_2O \rightleftharpoons Al(H_2O)_5(OH)^{2+} +$ _____

13-83. Write the hydrolysis equilibria (if any) for the following ions.

(a) F^- (d) HPO_4^{2-}

(b) SO_3^{2-} (e) CN^-

(c) $(CH_3)_2NH_2^+[(CH_3)_2NH]$ (f) Li^+

[is a weak base like ammonia.]

13-84. Write the hydrolysis equilibria (if any) for the following ions.

(a) Br^- (b) HS^- (c) ClO_4^- (d) H^- (e) Ca^{2+}

13-85. Calcium hypochlorite is used to purify water. When dissolved, it produces a slightly basic solution. Write the equation illustrating the solution of calcium hypochlorite in water and the equation illustrating its basic behavior.

13-86. Aqueous NaF solutions are slightly basic, whereas aqueous NaCl solutions are neutral. Write the appropriate equation that illustrates this. Why aren't NaCl solutions also basic?

13-87. Predict whether aqueous solutions of the following salts are acidic, neutral, or basic.

(a) $Ba(ClO_4)_2$ (d) KBr

(b) $N_2H_5^+NO_3^-$ (N_2H_4 is a weak base.)   (e) NH_4Cl

(c) $LiC_2H_3O_2$ (f) BaF_2

13-88. Predict whether aqueous solutions of the following salts are acidic, neutral, or basic.

(a) Na_2CO_3 (b) K_3PO_4 (c) NH_4ClO_4 (d) SrI_2

*13-89. Both C_2^{2-} and its conjugate acid HC_2^- hydrolyze 100% in water. From this information, complete the following equation.

$$CaC_2(s) + 2H_2O(l) \longrightarrow \underline{\quad}(g) + Ca^{2+}(aq) + 2\underline{\quad}(aq)$$

(The gas formed—acetylene—can be burned as it is produced. This reaction was once important for this purpose as a source of light in old miners' lamps.)

*13-90. Aqueous solutions of NH_4CN are basic. Write the two hydrolysis reactions and indicate which takes place to the greater extent.

13-91. Aqueous solutions of $NaHSO_3$ are acidic. Write the two equations (one hydrolysis and one ionization) and indicate which takes place to the greater extent.

Buffers (SECTION 13-9)

13-92. Identify which of the following form buffer solutions when 0.50 mol of each compound is dissolved in 1 L of water.

(a) HNO_2 and KNO_2 **(f)** HCN and $KClO$

(b) NH_4Cl and NH_3 **(g)** NH_3 and $BaBr_2$

(c) HNO_3 and KNO_2 **(h)** H_2S and $LiHS$

(d) HNO_3 and KNO_3 **(i)** KH_2PO_4 and K_2HPO_4

(e) $HClO$ and $Ca(ClO)_2$

13-93. A certain solution contains dissolved HCl and NaCl. Why can't this solution act as a buffer?

13-94. Write the equilibrium involved in the N_2H_4, N_2H_5Cl buffer system. (N_2H_4 is a weak base.) Write equations illustrating how this system reacts with added H_3O^+ and added OH^-.

13-95. Write the equilibrium involved in the HCO_3^-, CO_3^{2-} buffer system. Write equations illustrating how this system reacts with added H_3O^+ and added OH^-.

13-96. Write the equilibrium involved in the HPO_4^{2-}, PO_4^{3-} buffer system. Write equations illustrating how this system reacts with added H_3O^+ and added OH^-.

13-97. If 0.5 mol of KOH is added to a solution containing 1.0 mol of $HC_2H_3O_2$, the resulting solution is a buffer. Explain.

13-98. A solution contains 0.50 mol each of HClO and NaClO. If 0.60 mol of KOH is added, will the buffer prevent a significant change in pH? Explain.

Oxides as Acids and Bases (SECTION 13-9)

13-99. Write the formula of the acid or base formed when each of the following anhydrides is dissolved in water.

(a) SrO   **(c)** P_4O_{10} **(e)** N_2O_3

(b) SeO_3 **(d)** Cs_2O **(f)** Cl_2O_5

13-100. Write the formula of the acid or base formed when each of the following anhydrides is dissolved in water.

(a) BaO   **(b)** SeO_2 **(c)** Cl_2O **(d)** Br_2O **(e)** K_2O

13-101. Carbon dioxide is removed from the space shuttle by bubbling the air through a LiOH solution. Show the reaction and the product formed.

13-102. Complete the following equation.

$$Li_2O + N_2O_5(g) \longrightarrow \underline{\quad\quad} (s)$$

General Problems

13-103. Iron reacts with an acid, forming an aqueous solution of iron(II) iodide and a gas. Write the equation illustrating the reaction.

13-104. Aluminum reacts with perchloric acid. Write the equation illustrating this reaction.

13-105. Nitric acid reacts with sodium sulfite, forming sulfur dioxide gas, a salt, and water. Write the equation illustrating the reaction.

13-106. Perbromic acid reacts with sodium sulfide to form a pungent gas, hydrogen sulfide. Write the equation illustrating the reaction.

13-107. There are acid-base systems based on solvents other than H_2O. One is ammonia (NH_3), which is also amphiprotic. Write equations illustrating each of the following.

(a) the reaction of HCN with NH_3 acting as a base

(b) the reaction of H^- with NH_3 acting as an acid

(c) the reaction of HCO_3^- with NH_3 acting as a base

(d) the reaction between NH_4Cl and $NaNH_2$ in ammonia

13-108. The conjugate base of methyl alcohol (CH_3OH) is CH_3O^-. Its conjugate acid is $CH_3OH_2^+$. Write equations illustrating each of the following.

(a) the reaction of HCl with methyl alcohol acting as a base

(b) the reaction of NH_2^- with methyl alcohol acting as an acid

13-109. Sulfite ion (SO_3^{2-}) and sulfur trioxide (SO_3) look similar at first glance, but one forms a strongly acidic solution whereas the other is weakly basic. Write equations illustrating this behavior.

13-110. Tell whether each of the following compounds forms an acidic, basic, or neutral solution when added to pure water. Write the equation illustrating the acidic or basic behavior where appropriate.

(a) H_2S **(e)** $N_2H_5^+Br^-$ **(i)** H_2SO_3

(b) KClO   **(f)** $Ba(OH)_2$ **(j)** Cl_2O_3

(c) NaI   **(g)** $Sr(NO_3)_2$

(d) NH_3 **(h)** $LiNO_2$

13-111. Tell whether each of the following compounds forms an acidic, basic, or neutral solution when added to pure water. Write the equation illustrating the acidic or basic behavior where appropriate.

(a) HBrO   **(d)** N_2H_4 **(g)** RbBr

(b) CaO   **(e)** SO_2

(c) NH_4ClO_4 **(f)** $Ba(C_2H_3O_2)_2$

13-112. In a lab there are five different solutions with pH's of 1.0, 5.2, 7.0, 10.2, and 13.0. The solutions are LiOH, $SrBr_2$, KClO, NH_4Cl, and HI, all at the same concentration. Which pH corresponds to which compound? What must be the concentration of all of these solutions?

13-113. When one mixes a solution of baking soda ($NaHCO_3$) with vinegar ($HC_2H_3O_2$), bubbles of gas appear. Write equations for two reactions that indicate the identity of the gas.

*13-114. High-sulfur coal contains 5.0% iron pyrite (FeS_2). When the coal is burned, the iron pyrite also burns according to the equation

$$4FeS_2(s) + 11O_2(g) \longrightarrow 2Fe_2O_3(s) + 8SO_2(g)$$

What mass of sulfuric acid can eventually form from the combustion of 100 kg of coal? Sulfuric acid is formed according to the following equations.

$$2SO_2(g) + O_2(g) \longrightarrow 2SO_3(g)$$

$$SO_3(g) + H_2O(l) \longrightarrow H_2SO_4(aq)$$

13-115. A 2.50-g quantity of HCl is dissolved in 245 mL of water and then diluted to 890 mL. What is the pH of the concentrated and the dilute solution?

13-116. A 0.150-mole quantity of NaOH is dissolved in 2.50 L of water. In a separate container, 0.150 mole of HCl is present in 2.50 L of water. What is the pH of each solution? What is the pH of a solution made by mixing the two?

*13-117. A solution is prepared by mixing 10.0 g of HCl with 10.0 g of NaOH. What is the pH of the solution if the volume is 1.00 L?

*13-118. A solution is prepared by mixing 25.0 g of H_2SO_4 with 50.0 g of KOH. What is the pH of the solution if the volume is 500 mL?

*13-119. A solution is prepared by mixing 500 mL of 0.10 M HNO_3 with 500 mL of 0.10 M $Ca(OH)_2$. What is the pH of the solution after mixing?

STUDENT WORKSHOP

Plotting the Titration of a Strong Acid with a Strong Base

Purpose: To calculate pH, and create a graph which demonstrates how the pH of an acid solution changes as it is titrated with a base. (Work in groups of three or four. Estimated time: 25 min.)

In this activity, you will make a graph showing how the pH of a solution changes as you titrate a strong acid with a strong base. The graph paper (sideways) should be scaled from 0 to 14 on the y-axis (this will be the pH) and from 0 mL to 50 mL on the x-axis (the volume of added NaOH). The data for this experiment are for 25.0 mL of 0.10 M HNO_3 being titrated with 0.10 M NaOH.

At 25.0 mL of added NaOH, you will be at the equivalence point in the titration, and the pH will be 7.00. Prior to that, we expect acidic pH's, and past that point, pH's should be in the base range, as we add more base than there is acid to neutralize. The accompanying table records the acid or base concentration for each amount of added NaOH, after the reaction is complete. You should determine the pH at each point, plot the graph of pH versus added NaOH, and answer the following questions.

mL of NaOH added	mol/L [Acid]	mol/L [Base]
0.0	0.10	
5.0	0.067	
10.0	0.043	
15.0	0.025	
20.0	0.011	
24.0	0.0020	
25.0	0	0
26.0		0.0020
30.0		0.0091
35.0		0.017
40.0		0.023
50.0		0.033

1. Over what range(s) of added base is the pH changing the least?
2. Around what volume is the pH changing the most?
3. If the indicator changes color from pH 8 to pH 10, how much base does it take to cause the color to change?
4. The plot you have created is referred to as an *S curve*.
5. Using the Internet, can you find other areas where S-curve behavior applies?

Chapter

14

Oxidation-Reduction Reactions

Electricity generated by this large Tesla coil is dramatic and awesome. In fact, electricity has been essential to our lives for well over a century. In this chapter, we are concerned with how electricity is generated and put to use by chemical reactions.

SETTING THE STAGE

A distant rumble signals the ominous gathering of thunderstorms. We may cast a cautious eye toward the sky and think of shelter. The roll of thunder warns us about one force of nature for which we have a great respect, so we try to get out of its way. That, of course, is lightning. Lightning has no doubt caused fear as well as amazement in the human race since people first looked to the sky and wondered about its nature. But this force was not harnessed until modern times. The use of electricity (the same force as lightning) is so common to us now that it is taken for granted. Huge generating plants dot the rural landscape with towering smokestacks discharging smoke and steam. Not many decades ago, however, electricity was mainly a laboratory curiosity, until the experiments of inventors such as James Watt, Alexander Graham Bell, and Thomas Edison tapped its limitless potential. Even now, when we turn on a cell phone or an iPod or start a car, a flow of electrons (electricity) from a battery is put to immediate use. The electricity used by these devices originates from chemical reactions that involve an exchange of electrons between reactants.

Electron exchanges have not previously been defined as a specific type of reaction, but many of the classifications discussed in Chapter 6 fit into this broad category. Most combination, all combustion, and all single-replacement reactions can also be categorized as electron exchange reactions.

In Part A, we will examine the nature of these reactions and how the equations representing these reactions are balanced. The practical applications of redox reactions, such as the generation of electricity in batteries or to release metals from their ores, are among the topics discussed in Part B.

Part A

Redox Reactions—The Exchange of Electrons

SETTING A GOAL

■ You will learn of an important classification of chemical reactions that involve an exchange of electrons.

OBJECTIVES

14-1 Using oxidation states, determine the species oxidized, the species reduced, the oxidizing agent, and the reducing agent in an electron exchange reaction.

14-2 Balance redox reactions by the oxidation state (bridge) method.

14-3 Balance redox reactions by the ion-electron (half-reaction) method in both acidic and basic media.

▶ **OBJECTIVE FOR SECTION 14-1**

Using oxidation states, determine the species oxidized, the species reduced, the oxidizing agent, and the reducing agent in an electron exchange reaction.

14-1 The Nature of Oxidation and Reduction and Oxidation States

LOOKING AHEAD! There are two general types of reactions that involve the hydrogen atom. One involves the proton and the other involves the electron. In Chapter 13, we described the actions of acids and bases in water as an exchange of a proton. In this chapter, we describe reactions involving the electron. First, we examine this exchange and the terms used to describe it. ■

Sodium metal and chlorine gas react in a spectacular demonstration of chemical power. A small chunk of sodium placed in a flask filled with chlorine gas immediately glows white hot as the elements combine to form ordinary table salt. (See Figure 14-1.) We will examine the reaction of sodium and chlorine to illustrate the process of an electron exchange reaction. The equation for this reaction is as follows.

$$2Na(s) + Cl_2(g) \longrightarrow 2NaCl(s)$$

In Section 9-2, we briefly explained what was happening in the reaction. A sodium atom loses one electron, which is gained by a chlorine atom, resulting in the formation of two ions. The gain and loss of an electron were predicted as a logical consequence of the octet rule.

$$Na\!\odot \;\; + \;\; \ddot{\underset{..}{Cl}}\!: \longrightarrow Na^+ \; :\!\ddot{\underset{..}{Cl}}\!:^-$$

14-1.1 Half-Reactions

In the acid–base reactions discussed in the previous chapter, the atoms in the reactants keep their quota of electrons in changing to products. Such is not the case in the reaction shown above, however. We will now take a closer look at what happens as reactants change into products. An electron exchange reaction can be viewed as the sum of two half-reactions. *A* **half-reaction** *represents either the loss of electrons or the gain of electrons as a separate balanced equation.* Thus the half-reaction involving only sodium is

$$Na \longrightarrow Na^+ + e^-$$

In this half-reaction, notice that the neutral sodium atom has lost an electron to form a positive sodium ion. *A substance that loses electrons in a chemical reaction is said to be* **oxidized**.

Now consider what happens to the chlorine molecule in going from reactant to product.

$$2e^- + Cl_2 \longrightarrow 2Cl^-$$

FIGURE 14-1 Formation of NaCl An active metal reacts with a poisonous gas to form ordinary table salt (NaCl).

In this half-reaction, the neutral chlorine molecule has gained two electrons to form two chloride ions. *A substance that gains electrons in a chemical reaction is said to be **reduced***. A simple mnemonic helps us remember these two terms: OIL RIG can be used to recall **O**xidation **I**s **L**oss (of electrons)—**R**eduction **I**s **G**ain.

14-1.2 Redox Reactions

Obviously, the two processes (oxidation and reduction) complement each other similar to how acids and bases complement each other in a neutralization reaction. If there is an oxidation occurring, then there must also be a reduction, giving us the basis for this classification of chemical reaction. *Reactions involving an exchange of electrons are known as **oxidation–reduction**, or simply, **redox reactions***.

Now consider how the two half-reactions add together to make a complete, balanced equation. *An important principle of a redox reaction is that the electrons gained in the reduction process must equal the electrons lost in the oxidation process.* Note in our sample reaction that the reduction process involving Cl_2 requires two electrons. Therefore, the oxidation process must involve two Na's to provide these two electrons. The electrons on both sides of the equation must be equal so that they can be eliminated by subtraction when the two half-reactions are added.

Oxidation half-reaction: $\qquad 2Na \longrightarrow 2Na^+ + 2e^-$

Reduction half-reaction: $\quad 2e^- + Cl_2 \longrightarrow 2Cl^-$

Total reaction: $\qquad\qquad 2Na + Cl_2 \longrightarrow 2Na^+ + 2Cl^- \text{ (or 2NaCl)}$

Instead of identifying a reactant by what happened to it (i.e., it was oxidized or reduced), in this type of reaction it is sometimes more useful to emphasize what it does. *The substance that causes the oxidation (i.e., by accepting electrons) is called the **oxidizing agent**, and the substance that causes the reduction (i.e., by providing the electrons) is called the **reducing agent***. Thus, the substance reduced is the oxidizing agent and the substance oxidized is the reducing agent.

In the example above, Na is the *reducing agent* and is oxidized to Na^+. Cl_2 is the *oxidizing agent* and is reduced to Cl^-.

Many other familiar chemical changes are redox reactions. For example, the corrosion of iron to form rust involves an exchange of electrons. The formation of rust [iron(III) oxide] is illustrated below.

$$4Fe(s) + 3O_2(g) \longrightarrow 2Fe_2O_3(s)$$

Originally, the term *oxidation* referred specifically to reactions where a substance like iron adds oxygen. In the reverse reaction, oxygen is removed from iron(III) oxide, which "reduces" its mass. Thus, the removal of oxygen was known as "reduction." Now we define the process in terms of the exchange of electrons, so the terms are used regardless of whether oxygen is involved. In many redox reactions, however, the species undergoing electron exchange are not as obvious as the two examples that we have used in this section. So we need a tool to help us. This leads us to a concept known as oxidation states.

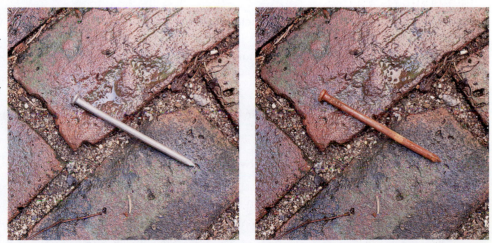

An iron nail soon forms a coating of rust when exposed to moisture and air.

14-1.3 Oxidation States

In Chapter 9 we presented Lewis structures for molecules and ions where electrons were shared between two atoms. With the concept of electronegativity (see Section 9-6), however, we found that electrons are not shared equally. The greater share of the electron pair in the bond is closer to the more electronegative atom, which gives it a partial negative charge. As a convenient method of keeping track of electrons in redox reactions, we will now assign *all* of the electrons in a bond to the more electronegative atom. *The **oxidation state** (or **oxidation number**) of an atom in a molecule or ion is the charge that atom would have if all electrons in its bonds were assigned to the more electronegative atom.* The most electronegative element is fluorine, oxygen is second, and the electronegativity of the elements then decreases diagonally to the left on the periodic table in the direction of the metal–nonmetal borderline in the periodic table. We actually don't need to know all the electronegativities of elements in order to assign oxidation states, however. A few general rules will suffice. These rules and some practice exercises follow.

1. The oxidation state of an element in its free (natural) state is zero [e.g., $Cu(s)$, $B(s)$]. This includes polyatomic elements [e.g., $Cl_2(g)$, $P_4(s)$].
2. The oxidation state of a monatomic ion is the same as the charge on that ion (e.g., $Na^+ = +1$ oxidation state, $O^{2-} = -2$, $Al^{3+} = +3$).
   a. Alkali metal ions are always $+1$ (same as the group number).
   b. Alkaline earth metal ions are always $+2$ (same as the group number).
3. The halogens are in a -1 oxidation state in binary (two-element) compounds, whether ionic or covalent, *when bound to a less electronegative element.*
4. Oxygen in a compound is usually -2. Certain compounds (which are rare) called peroxides or superoxides contain oxygen in a less negative oxidation state. Oxygen is positive only when bound to the more electronegative fluorine.
5. Hydrogen in a compound is usually $+1$. When combined with a less electronegative element (usually a metal), hydrogen has a -1 oxidation state (e.g., LiH).
6. The sum of the oxidation states of all the atoms in a neutral compound is zero. For a polyatomic ion, the sum of the oxidation states equals the charge on the ion.

EXAMPLE 14-1   Calculating Oxidation States

What is the oxidation state of the following?

(a) Fe in FeO   **(b)** N in N_2O_5 **(c)** S in H_2SO_3 **(d)** As in AsO_4^{3-}

PROCEDURE

An algebraic equation can be constructed from rule 6. For example, assume that we have a hypothetical compound M_2A_3. Then, from rule 6,

$$[2 \times (\text{oxidation state of M})] + [3 \times (\text{oxidation state of A})] = 0$$

or, to simplify,

$$2(\text{ox. state M}) + 3(\text{ox. state A}) = 0$$

If the formula represents a polyatomic ion, the quantity on the left is equal to the charge rather than zero.

SOLUTION

(a) FeO: The oxidation states of the two elements add to zero (rule 6).

$$(\text{ox. state Fe}) + (\text{ox. state O}) = 0$$

Since the oxidation of state of oxygen is -2 (rule 4),

$$\text{ox. state Fe} + (-2) = 0$$

$$\text{ox. state Fe} = \underline{\underline{+2}}$$

(b) N_2O_5: The oxidation states add to zero (rule 6), as shown by the equation

$$2(\text{ox. state N}) + 5(\text{ox. state O}) = 0$$

Since ox. state O is −2 (rule 4),

$$2(\text{ox. state N}) + 5(-2) = 0$$
$$2(\text{ox. state N}) = +10$$
$$\text{ox. state N} = \underline{\underline{+5}}$$

(c) H_2SO_3: The oxidation states add to zero (rule 6).

$$2(\text{ox. state H}) + (\text{ox. state S}) + 3(\text{ox. state O}) = 0$$

H is usually +1 and O is usually −2 (rules 4 and 5).

$$2(+1) + \text{ox. state S} + 3(-2) = 0$$
$$\text{ox. state S} = \underline{\underline{+4}}$$

(d) AsO_4^{3-}: The sum of the oxidation states of the atoms equals the charge on the ion (rule 6).

$$(\text{ox. state As}) + 4(\text{ox. state O}) = -3$$

Since O is −2 (rule 4),

$$\text{ox. state As} + 4(-2) = -3$$
$$\text{ox. state As} = \underline{\underline{+5}}$$

ANALYSIS

All charges on monatomic ions are oxidation states, though not all oxidation states are charges. The oxidation state of an atom in a molecule or ion would be the charge on the atom if all the other atoms were present as ions. For example, in the sulfate ion (SO_4^{2-}), if all atoms were ions, we would have $[S^{6+}4O^{2-}]^{2-}$. This is certainly not the case, since electrons are shared between the sulfur and oxygen atoms, as we illustrated in Chapter 9. However, even though the sulfur is not present as a +6 ion, the +6 assigned to the sulfur represents its oxidation state. We will see why we need this important information next.

SYNTHESIS

Sometimes, in assigning oxidation states, we have to choose between a rule that says "always" and one that says "usually." For example, in the compound magnesium hydride (MgH_2) all atoms cannot have positive oxidation states. However, magnesium is "always" +2, so the hydrogen, which is "usually" +1, must have a −1 oxidation state in this case. In the compound O_2F_2, all atoms could not have negative oxidation states. In this case, fluorine is the more electronegative, so it has a −1 oxidation state. This means that each oxygen must be in a +1 oxidation state. There are even a few cases where we find oxidation states that are not even numbers. What would the oxidation state of the iron atoms be in Fe_3O_4? If the oxygens are −2 each, then their total is −8. The three iron atoms must have oxidation states that add to +8, so each one is +2.67. (In fact, Fe_3O_4 is a one-to-one combination of FeO and Fe_2O_3.) Fortunately, all the odd exceptions are not enough to worry about, so our rules will hold in almost all cases that we will encounter.

14-1.4 Using Oxidation States in Redox Reactions

By noting the change of oxidation state of the same atom in going from a reactant to a product, we can trace the exchange of electrons in the reaction. We can now add to our definition of oxidation and reduction in terms of oxidation state. *Oxidation is a loss of electrons as indicated by an increase in the oxidation state. Reduction is a gain of electrons as indicated by a decrease (or reduction) in the oxidation state.* In the reaction of sodium with chlorine discussed in Section 14-1, notice that the oxidation state of the sodium increased from zero to +1, indicating oxidation, and that of chlorine decreased from zero to −1, indicating reduction.

In most compounds, usually only one of the elements undergoes a change in redox reactions. Thus, it is often necessary to calculate the oxidation states of all the elements in all compounds so that we can see which ones have undergone the change. With experience, however, the oxidized and reduced species are more easily recognized. In the following examples, we will find all the oxidation states so that we can identify the changes and label them appropriately.

EXAMPLE 14-2   Identifying Oxidized and Reduced Species

In the following unbalanced equations, indicate the reactant oxidized, the reactant reduced, the oxidizing agent, and the reducing agent. Indicate the products that contain the elements that were oxidized or reduced.

(a) $Al + HCl \longrightarrow AlCl_3 + H_2$

(b) $CH_4 + O_2 \longrightarrow CO_2 + H_2O$

(c) $MnO_2 + HCl \longrightarrow MnCl_2 + Cl_2 + H_2O$

(d) $K_2Cr_2O_7 + SnCl_2 + HCl \longrightarrow CrCl_3 + SnCl_4 + KCl + H_2O$

PROCEDURE

In the equations, we wish to identify the species that contain atoms of an element undergoing a change in oxidation state. At first, it may be necessary to calculate the oxidation state of every atom in the equation until you can recognize the changes by inspection.

SOLUTION

(a) Oxidation states of elements:

$$\overset{0}{Al} + \overset{+1\ -1}{HCl} \longrightarrow \overset{+3\ -1}{AlCl_3} + \overset{0}{H_2}$$

Reactant	Change	Agent	Product
Al	oxidized	reducing	$AlCl_3$
HCl	reduced	oxidizing	H_2

(b) Oxidation states of elements:

$$\overset{-4\ +1}{CH_4} + \overset{0}{O_2} \longrightarrow \overset{+4\ -2}{CO_2} + \overset{+1\ -2}{H_2O}$$

Reactant	Change	Agent	Product
CH_4	oxidized	reducing	CO_2
O_2	reduced	oxidizing	CO_2, H_2O

(c) Oxidation states of elements:

$$\overset{+4\ -2}{MnO_2} + \overset{+1\ -1}{HCl} \longrightarrow \overset{+2\ -1}{MnCl_2} + \overset{0}{Cl_2} + \overset{+1\ -2}{H_2O}$$

Reactant	Change	Agent	Product
HCl	oxidized	reducing	Cl_2
MnO_2	reduced	oxidizing	$MnCl_2$

(d) Oxidation states of elements:

$$\overset{+1\ +6\ -2}{K_2Cr_2O_7} + \overset{+2\ -1}{SnCl_2} + \overset{+1\ -1}{HCl} \longrightarrow \overset{+3\ -1}{CrCl_3} + \overset{+1\ -1}{KCl} + \overset{+4\ -1}{SnCl_4} + \overset{+1\ -2}{H_2O}$$

Reactant	Change	Agent	Product
$SnCl_2$	oxidized	reducing	$SnCl_4$
$K_2Cr_2O_7$	reduced	oxidizing	$CrCl_3$

ANALYSIS

You will notice that any substance present as a free element in a reaction is involved in either the oxidation or the reduction process. Also, hydrogen and oxygen are generally not oxidized or reduced if they remain part of compounds. They are involved only if present as free elements in either the reactants or products. Further note that none of the analysis done for this problem requires a balanced reaction. It is sufficient simply to know the identities of the reactants and their products.

SYNTHESIS

Consider the equation illustrating the decomposition of hydrogen peroxide:

$$2H_2O_2(aq) \longrightarrow 2H_2O(l) + O_2(g)$$

What has been oxidized and what has been reduced? In this case H_2O_2 plays both roles. In one molecule of H_2O_2, the oxidation state of the oxygen has decreased from -1 to -2 to form H_2O, so it has been reduced. In the other molecule of H_2O_2, the oxygen has increased from -1 to 0, so it has been oxidized. One molecule of hydrogen peroxide is the oxidizing agent and one is the reducing agent. In such a case, we say that the H_2O_2 has been *disproportionated*.

MAKING IT REAL

Lightning Bugs (Fireflies)—Nature's Little Night-Lights

Tiny little flashes of light on a summer night–those of us east of the Rockies know the sources as lightning bugs, also called fire-flies. They have been a source of curiosity since ancient times, but only recently have we known how these tiny creatures can light up.

Glowsticks—an example of chemiluminescence.

In photochemical reactions, such as photosynthesis, light energy *initiates* a chemical reaction. In the firefly, we have just the opposite situation. That is, light energy is *produced* by a chemical reaction. The production of light energy by a chemical reaction is known as *chemiluminescence*. If it is produced by living organisms, such as the firefly, it is known as *bioluminescence*.

We now have some understanding of what goes on in the firefly. The chemical reaction is an oxidation–reduction reaction, with O_2 from the air serving as the oxidizing agent. The other compounds are complex molecules, so we will simply refer to them by name (luciferin) or initials (ATP). Also involved is an enzyme (luciferase), which acts as a catalyst. This reaction is represented as follows:

luciferin + ATP + $O_2(g)$ $\xrightarrow{\text{luciferase}}$ products + hυ (cold light)

By studying the bioluminescent chemistry of the firefly, we also discovered that all fireflies have the same luciferin, but each species produces a different color of light. The color was found to be determined by the luciferase that is unique to each species of firefly. An amazing discovery about this reaction is that the production of light is incredibly efficient. Eighty out of 100 molecules that react produce light, thus giving it an 80% efficiency.

It wasn't until 1928 that a scientist, H. O. Albrecht, first described a nonbiological chemical reaction that could be conducted in a laboratory to generate chemiluminescence, but with only 0.1% efficiency. The reaction is similar to the one shown above except that nonbiological chemicals available to the scientist were used along with hydrogen peroxide (H_2O_2) as the oxidizing agent. In the early 1960s, other chemiluminescent reactions were discovered and patented. Today a substance called Cyalume™, a trademark product of American Cyanamid, is used to produce light with about 5% efficiency.

When you see children with glowsticks or other decorative jewelry that glows in the dark, thank the firefly for its amazing contribution to our world of chemistry.

▶ **ASSESSING THE OBJECTIVE FOR SECTION 14–1**

EXERCISE 14-1(a) LEVEL 1: Fill in the blanks.
If all the atoms in a compound were ions, the charge on the ions would be the same as their _____ _____. For oxygen in compounds, this is usually _____. Hydrogen is usually _____ in compounds. A substance oxidized undergoes a _____ of electrons and an _____ in oxidation state. This substance is also known as a _____ agent.

EXERCISE 14-1(b) LEVEL 1: What is the oxidation state of **(a)** B in H_3BO_3 and **(b)** S in $S_2O_3^{2-}$?

EXERCISE 14-1(c) LEVEL 1: What is the oxidation state of carbon in each of the following compounds?

(a) CO_3^{2-} **(b)** CH_3Cl **(c)** CF_4 **(d)** CH_2O

EXERCISE 14-1(d) LEVEL 2: In the following reaction, indicate the substance oxidized, the substance reduced, the oxidizing agent, and the reducing agent.

$$ClO_2 + H_2O_2 \longrightarrow O_2 + Cl^-$$

EXERCISE 14-1(e) LEVEL 2: Consider the following unbalanced equation:

$$CrI_3 + OH^- + Cl_2 \longrightarrow CrO_4^{2-} + IO_4^- + Cl^- + H_2O$$

What has been oxidized and what has been reduced?

EXERCISE 14-1(f) LEVEL 3: Is it possible to have a reduction without an oxidation or an oxidation without a reduction? Why or why not?

EXERCISE 14-1(g) LEVEL 3: Is the compound $KMnO_4$ more likely to be an oxidizing or reducing agent? Why?

For additional practice, work chapter problems 14-1, 14-6, 14-10, and 14-12.

▶ **OBJECTIVE FOR SECTION 14-2**
Balance redox reactions by the oxidation state (bridge) method.

14-2 Balancing Redox Equations: Oxidation State Method

LOOKING AHEAD! An important principle of redox reactions is that "electrons lost equal electrons gained." In the next two sections, we will put this concept to use as the key to balancing some rather complex reactions that would be quite difficult to balance by inspection, as we did in Chapter 6. There are two procedures for this endeavor. The oxidation state method, discussed in this section, is useful for balancing equations in molecular form. ∎

In a typical redox reaction, only two atoms undergo a change in their oxidation states. By identifying these two atoms and calculating the change in the oxidation state, we can arrive at a balanced equation. *The* **oxidation state**, *or* **bridge method**, *focuses on the atoms of the elements undergoing a change in oxidation state.*

The following reaction will be used to illustrate the procedures for balancing equations by the oxidation state method.

$$HNO_3(aq) + H_2S(aq) \longrightarrow NO(g) + S(s) + H_2O(l)$$

1. Identify the atoms whose oxidation states have changed.

$$\overset{+5}{H\underline{N}O_3} + \overset{-2}{H_2\underline{S}} \longrightarrow \overset{+2}{\underline{N}O} + \overset{0}{\underline{S}} + H_2O$$

2. Draw a bridge between the same atoms whose oxidation states have changed, indicating the electrons gained or lost. This is the change in oxidation state.

$$\overset{+3e^-}{\overset{\displaystyle\frown}{\underset{-2e^-}{\underset{\displaystyle\smile}{\overset{+5}{H}NO_3 + \overset{-2}{H_2}S \longrightarrow \overset{+2}{N}O + \overset{0}{S} + H_2}}}}$$

3. Multiply the two numbers (+3 and −2) by whole numbers that produce a common multiple. For 3 and 2 the common multiple is 6. (For example, $+3 \times \underline{2} = +6$; $-2 \times \underline{3} = -6$.) Use these multipliers as coefficients of the respective compounds or elements.
   Note that six electrons are lost (bottom) and six are gained (top).

$$\overset{+3e^- \times \enclose{circle}{2} = +6e^-}{\underset{-2e^- \times \enclose{circle}{3} = -6e^-}{2HNO_3 + 3H_2S \longrightarrow 2NO + 3S + H_2O}}$$

4. Balance the rest of the equation by inspection. Note that there are eight H's on the left, so *four* H₂O's are needed on the right. If the equation has been balanced correctly, the O's should balance. Note that they do.

$$2HNO_3 + 3H_2S \longrightarrow 2NO + 3S + 4H_2O$$

EXAMPLE 14-3   Balancing Equations by the Oxidation State Method

Balance the following equations by the oxidation state method.

(a) $Zn + AgNO_3 \longrightarrow Zn(NO_3)_2 + Ag$

(b) $Cu + HNO_3 \longrightarrow Cu(NO_3)_2 + H_2O + NO_2$

(c) $O_2 + HI \longrightarrow H_2O + I_2$

PROCEDURE (a)

(1) Determine the elements that have been oxidized and reduced. **(2)** From the oxidation state change, determine the common multiple and use the two multipliers as coefficients in the equation.

SOLUTION (a)

When a copper penny reacts with nitric acid, nitrogen dioxide gas (brown) is formed.

$$\overset{-2e^-}{\overset{\displaystyle\frown}{\underset{+1e^-}{\underset{\displaystyle\smile}{\overset{0}{Zn} + \overset{+1}{A}gNO_3 \longrightarrow \overset{+2}{Zn}(NO_3)_2 + \overset{0}{Ag}}}}}$$

The oxidation (top) should be multiplied by 1, and the reduction process (bottom) should be multiplied by 2.

$$\overset{-2e^- \times 1 = -2e^-}{\underset{+1e^- \times 2 = +2e^-}{Zn + 2AgNO_3 \longrightarrow Zn(NO_3)_2 + 2Ag}}$$

The final balanced equation is

$$Zn + 2AgNO_3 \longrightarrow Zn(NO_3)_2 + 2Ag$$

PROCEDURE (b)

Write a bridge between Cu and N, which have undergone oxidation state changes.

SOLUTION (b)

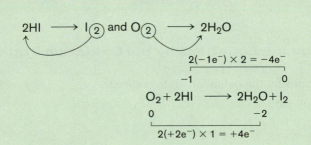

The equation, so far, is

$$Cu + 2HNO_3 \longrightarrow Cu(NO_3)_2 + H_2O + 2NO_2$$

Note, however, that four N's are present on the right, but only two are on the left. The addition of two more HNO_3's balances the N's, and the equation is completely balanced with two H_2O's on the right.

$$Cu + 4HNO_3 \longrightarrow Cu(NO_3)_2 + 2H_2O + 2NO_2$$

(In this aqueous reaction, HNO_3 serves two functions. Two HNO_3's are reduced to two NO_2's, and the other two HNO_3's provide anions for the Cu^{2+} ion. These latter NO_3^- ions are present in the solution as spectator ions. Spectator ions are not oxidized, reduced, or otherwise changed during the reaction.)

PROCEDURE (c)

The elements undergoing a change in oxidation state are oxygen and iodine. *If an atom that has changed is in a compound where it has a subscript other than 1, first balance these atoms by adding a temporary coefficient.*

SOLUTION (c)

$$2HI \longrightarrow I_{\textcircled{2}} \text{ and } O_{\textcircled{2}} \longrightarrow 2H_2O$$

$$
\begin{array}{c}
2(-1e^-) \times 2 = -4e^- \\
\overbrace{\quad} \\
-1 \qquad\qquad 0 \\
O_2 + 2HI \longrightarrow 2H_2O + I_2 \\
0 \qquad\qquad -2 \\
\underbrace{\quad} \\
2(+2e^-) \times 1 = +4e^-
\end{array}
$$

The final balanced equation is

$$O_2 + 4HI \longrightarrow 2H_2O + 2I_2$$

ANALYSIS

In (c), a common mistake is failing to ensure that the element undergoing a change is balanced first. If one neglected to do this, the change would be from 0 in O_2 to −2 in H_2O, for a loss of two electrons. Notice that there is actually a loss of four electrons, since the O_2 is reduced to two H_2O molecules. Likewise, there is a gain of two electrons when 2HI is oxidized to I_2.

SYNTHESIS

Consider the following reaction: $C_6H_{12}O_6 \longrightarrow C_2H_6O + CO_2$. This is the process of fermentation to form alcohol (C_2H_6O) from sugar ($C_6H_{12}O_6$), and it is significantly more complex than shown here. Based on this reaction, the carbon in $C_6H_{12}O_6$ would be assigned an oxidation state of zero. The carbons in the products would be −2 and +4, respectively. If you follow the procedure, you would predict that this equation balances thusly:

$$3\,C_6H_{12}O_6 \longrightarrow 6\,C_2H_6O + 6\,CO_2$$

or, simplified,

$$C_6H_{12}O_6 \longrightarrow 2\,C_2H_6O + 2\,CO_2$$

The point here, though, is that while the average oxidation states of the carbons were 0, −2, and +4, respectively, the actual oxidation states of the six carbons in sugar and the two carbons in alcohol vary. They merely average to those respective values. To be specific, in alcohol the oxidation states of the two carbons are actually −3 and −1, due to having different surrounding atoms. In the sugar, the oxidation states vary from −1 to +2, depending on the type of sugar. Be careful, whenever there are multiple atoms of the same type in a molecule, not to necessarily assume that all have the same oxidation state.

▶ ASSESSING THE OBJECTIVE FOR SECTION 14-2

EXERCISE 14-2(a) LEVEL 2: Balance the following equations by the bridge method.

(a) $Fe(s) + HCl(aq) \longrightarrow FeCl_3(aq) + H_2(g)$

(b) $HNO_3(aq) + HCl(aq) \longrightarrow NO_2(g) + Cl_2(g) + H_2O$

EXERCISE 14-2(b) LEVEL 3: In an alkaline dry cell (battery), manganese(IV) oxide reacts with water and zinc to produce manganese(III) oxide and zinc(II) hydroxide. Write the equation illustrating this reaction and balance the equation by the bridge method.

For additional practice, work chapter problems 14-16 and 14-17.

14-3 Balancing Redox Equations: Ion-Electron Method

▶ **OBJECTIVE FOR SECTION 14-3**
Balance redox reactions by the ion-electron (half-reaction) method in both acidic and basic media.

LOOKING AHEAD! One problem with the oxidation state method is that it may give the impression that some atoms exist as ions when they are actually *part* of a compound or ion (e.g., N^{5+}). The second method that we will discuss focuses on the entire ion or molecule containing the atom undergoing a change. This is a more realistic representation of the species involved. ■

While it is generally true that only two atoms actually undergo oxidation state changes, we can also consider the entire species that changes (e.g., NO_3^- rather than N^{5+}). *In the **ion-electron method** (also known as the half-reaction method), the total reaction is separated into half-reactions, which are balanced individually and then added back together.* The ion-electron method recognizes the complete change of an ion or molecule as it goes from reactant to products. As we will see later in this chapter, a balanced half-reaction is how we represent a specific change that occurs in a battery.

The rules for balancing equations are somewhat different in acidic solution [containing $H^+(aq)$ ion] than in basic solution [containing $OH^-(aq)$ ion]. The two solutions are considered separately, with acid solution reactions discussed first. To simplify the equations, only the net ionic equations are balanced.

14-3.1 Balancing Reactions in Acidic Solution

The balancing of an equation in aqueous acid solution is illustrated with the following unbalanced equation.

$$H^+(aq) + Cl^-(aq) + Cr_2O_7^{2-}(aq) \longrightarrow Cr^{3+}(aq) + Cl_2(g) + H_2O(l)$$

1. Separate out the molecule or ion that contains atoms of an element that has been oxidized or reduced and the product containing atoms of that element. If necessary, calculate the oxidation states of individual atoms until you are able to recognize the species that changes. In this method, it is actually not necessary to know the oxidation state. The reduction process is

$$Cr_2O_7^{2-} \longrightarrow Cr^{3+}$$

2. If a subscript of the atoms of the element undergoing a change in oxidation state is more than 1, balance those atoms with a temporary coefficient. In this case, it is the Cr.

$$Cr_2O_7^{2-} \longrightarrow 2Cr^{3+}$$

3. Balance the oxygens by adding H_2O on the side needing the oxygens (one H_2O for each O needed).

$$Cr_2O_7^{2-} \longrightarrow 2Cr^{3+} + 7H_2O$$

4. Balance the hydrogens by adding H^+ on the other side of the equation from the H_2O's ($2H^+$ for each H_2O added). Note that the H and O have not undergone a change in oxidation state.

$$14H^+ + Cr_2O_7^{2-} \longrightarrow 2Cr^{3+} + 7H_2O$$

5. The atoms in the half-reaction are now balanced. Check to make sure. Now comes the important step of balancing the charge on both sides of the equation. The charge is determined separately on each side of the equation. This is accomplished by multiplying the coefficient of any ion present times the charge on that ion. Neutral molecules are excluded from charge determination. Notice that there are 14 H^+ ions. The charge due to H^+ is 14 times the charge on the proton (+1), or +14. This is then added to the charge on the dichromate (−2), which makes a total charge on the left (i.e., the reactant side) of +12 [i.e., $(14 \times +1) + (-2) = +12$]. On the right (i.e., the product side), the charge on the Cr (+3) is multiplied by its coefficient (2), making a charge of +6 on the right. The H_2O is a neutral molecule, so it is not included. The total charge on the left is +12, and on the right it is +6. The charges on both sides of the reaction must now be balanced. To do this, add the appropriate number of *negative* electrons (−1 each) to the more *positive* side. Adding $6e^-$ on the left (the more positive side) balances the charges on both sides, and the half–reaction is balanced (i.e., $+12 - 6 = +6$).

$$6e^- + 14H^+ + Cr_2O_7^{2-} \longrightarrow 2Cr^{3+} + 7H_2O$$

6. Repeat the same procedure for the other half-reaction.

$$Cl^- \longrightarrow Cl_2$$
$$2Cl^- \longrightarrow Cl_2$$
$$2Cl^- \longrightarrow Cl_2 + 2e^-$$

7. Before the two half-reactions are added, we must make sure that electrons gained equal electrons lost. Sometimes the half-reactions must be multiplied by factors that give the same number of electrons. In this case, if the oxidation process is multiplied by 3 (and the reduction process by 1), there will be an exchange of $6e^-$. When these two half-reactions are added, the $6e^-$ can be subtracted from both sides of the equation.

$$3(2Cl^- \longrightarrow Cl_2 + 2e^-)$$
$$6Cl^- \longrightarrow 3Cl_2 + 6e^-$$

Addition produces the balanced net ionic equation.

$$6\cancel{e^-} + 14H^+ + Cr_2O_7^{2-} \longrightarrow 2Cr^{3+} + 7H_2O$$

$$6Cl^- \longrightarrow 3Cl_2 + 6\cancel{e^-}$$

$$\overline{14H^+(aq) + 6Cl^-(aq) + Cr_2O_7^{2-}(aq) \longrightarrow 2Cr^{3+}(aq) + 3Cl_2(g) + 7H_2O(l)}$$

An excellent way to check our answer is to make sure the net charges on both sides of the equation are the same. The net charge on the left side of the equation is $(14 \times +1) + (6 \times -1) + (-2) = \underline{+6}$. On the right the charge is $(2 \times +3) = \underline{+6}$. Recall that neutral molecules have no net charge.

EXAMPLE 14-4 **Balancing Redox Equations in Acidic Solution**

Balance the following equations for reactions occurring in acid solution by the ion-electron method.

(a) $MnO_4^-(aq) + SO_2(g) + H_2O(l) \longrightarrow Mn^{2+}(aq) + SO_4^{2-}(aq) + H^+(aq)$

(b) $Cu(s) + NO_3^-(aq) \longrightarrow Cu^{2+}(aq) + H_2O + NO(g)$

PROCEDURE

Balance each half-reaction in the following order: (1) the element that changes, (2) oxygens, (3) hydrogens, and (4) the charge. Add the two half-reactions together, and subtract out anything that appears on both sides.

SOLUTION (a)

$$\text{Reduction:} \qquad MnO_4^- \longrightarrow Mn^{2+}$$

$$H_2O: \qquad MnO_4^- \longrightarrow Mn^{2+} + 4H_2O$$

$$H^+: \qquad 8H^+ + MnO_4^- \longrightarrow Mn^{2+} + 4H_2O$$

$$e^-: \quad 5e^- + 8H^+ + MnO_4^- \longrightarrow Mn^{2+} + 4H_2O$$

$$\text{Oxidation:} \qquad SO_2 \longrightarrow SO_4^{2-}$$

$$H_2O: \qquad 2H_2O + SO_2 \longrightarrow SO_4^{2-}$$

$$H^+: \qquad 2H_2O + SO_2 \longrightarrow SO_4^{2-} + 4H^+$$

$$e^-: \qquad 2H_2O + SO_2 \longrightarrow SO_4^{2-} + 4H^+ + 2e^-$$

The reduction reaction is multiplied by 2 and the oxidation by 5 to produce 10 electrons for each process, as shown below.

$$2(5e^- + 8H^+ + MnO_4^- \longrightarrow Mn^{2+} + 4H_2O)$$

$$5(2H_2O + SO_2 \longrightarrow SO_4^{2-} + 4H^+ + 2e^-)$$

$$10e^- + 16H^+ + 2MnO_4^- \longrightarrow 2Mn^{2+} + 8H_2O$$

$$\underline{10H_2O + 5SO_2 \longrightarrow 5SO_4^{2-} + 20H^+ + 10e^-}$$

$$10H_2O + 16H^+ + 5SO_2 + 2MnO_4^- \longrightarrow 5SO_4^{2-} + 2Mn^{2+} + 8H_2O + 20H^+$$

Note that H_2O and H^+ are present on both sides of the equation. Therefore, $8H_2O$ and $16H^+$ can be subtracted from *both sides*, leaving the final balanced net ionic equation as

$$2MnO_4^-(aq) + 5SO_2(g) + 2H_2O(l) \longrightarrow 2Mn^{2+}(aq) + 5SO_4^{2-}(aq) + 4H^+(aq)$$

SOLUTION (b)

$$\text{Reduction:} \qquad NO_3^- \longrightarrow NO$$

$$H_2O: \qquad NO_3^- \longrightarrow NO + 2H_2O$$

$$H^+: \qquad 4H^+ + NO_3^- \longrightarrow NO + 2H_2O$$

$$e^-: \quad 3e^- + 4H^+ + NO_3^- \longrightarrow NO + 2H_2O$$

$$\text{Oxidation:} \qquad Cu \longrightarrow Cu^{2+}$$

$$e^-: \qquad Cu \longrightarrow Cu^{2+} + 2e^-$$

Multiply the reduction half-reaction by 2 and the oxidation half-reaction by 3, and then add the two half-reactions.

$$6e^- + 8H^+ + 2NO_3^- \longrightarrow 2NO + 4H_2O$$

$$\underline{3Cu \longrightarrow 3Cu^{2+} + 6e^-}$$

$$8H^+(aq) + 2NO_3^-(aq) + 3Cu(s) \longrightarrow 3Cu^{2+}(aq) + 2NO(g) + 4H_2O(l)$$

ANALYSIS

When checking to see that redox reactions are properly balanced, remember that both the number of atoms and the charge must be the same on both sides. In (a), there are 4 hydrogens, 20 oxygens, 5 sulfurs, and 2 manganese atoms on each side. In addition, the total charge on the left side is $2 \times (-1)$, or -2, and on the right side it is $[2 \times (+2)] + [5 \times (-2)] + [4 \times (+1)]$, or -2, as well. Both atoms and charges balance. What is the atom and electron inventory for the reaction in (b)? Does it balance?

SYNTHESIS

In the chapter problems, you will notice that some acids are represented as molecules and others as ions. Thus, nitric acid, HNO_3, appears as $H^+ + NO_3^-$ in ionic equations, but sulfurous acid, H_2SO_3, appears as H_2SO_3. Why is this so? Recall our previous discussion about the difference between strong acids and weak acids. Strong acids (e.g., HNO_3) do not exist as molecules in aqueous solution, since they are completely ionized. Weak acids (e.g., H_2SO_3), on the other hand, are present mostly as neutral molecules. A weak acid forms very limited amounts of H^+ in solution, so we represent it in its more prevalent form as a molecule.

14-3.2 Balancing Reactions in Basic Solution

In a basic solution, OH^-, rather than H^+, is in excess. Perhaps the simplest way to adjust to this condition is to follow the same procedure as in acid solution but change either each half-reaction or the total reaction to basic solution after it is otherwise balanced. We do this by adding one OH^- to *both sides* of the equation for each H^+ present in the equation. As we learned in the previous chapter, the H^+ ion combines with an OH^- ion to form H_2O [i.e., $H^+(aq) + OH^-(aq) \longrightarrow H_2O(l)$]. In effect, this procedure converts an H^+ ion to an H_2O on one side, leaving one OH^- ion on the opposite side of the equation. For example, consider the half-reaction shown below, which has been balanced in acid solution. We will adjust the equation to a basic solution as follows.

1. Reduction half-reaction balanced in acid solution:

$$2e^- + 2H^+ + ClO^- \longrightarrow Cl^- + H_2O$$

2. Add $2OH^-$ to both sides for the two H^+'s shown on the left:

$$2e^- + (2H^+ + \mathbf{2OH^-}) + ClO^- \longrightarrow Cl^- + H_2O + \mathbf{2OH^-}$$

3. Convert $2H^+ + 2OH^-$ to $2H_2O$:

$$2e^- + \mathbf{2H_2O} + ClO^- \longrightarrow Cl^- + H_2O + \mathbf{2OH^-}$$

4. Simplify by subtracting the fewer number of H_2O's (1) from both sides of the equation:

$$2e^- + H_2O + ClO^- \longrightarrow Cl^- + 2OH^-$$

Now consider a typical unbalanced equation representing a total redox reaction. The presence of OH^- ion in the equation tells us that this reaction occurs in basic solution.

$$MnO_4^-(aq) + C_2O_4^{2-}(aq) + OH^-(aq) \longrightarrow MnO_2(s) + CO_3^{2-}(aq) + H_2O(l)$$

Because tables of half-reactions are represented as either acidic (with H^+ ions) or basic (with OH^- ions), we will change each half-reaction to basic solution. (Alternatively, we could balance the total equation as if in acid solution and then change the final balanced equation to basic solution.)

1. Balance the reduction half-reaction involving MnO_4^- and MnO_2 in acid solution.

$$3e^- + 4H^+ + MnO_4^- \longrightarrow MnO_2 + 2H_2O$$

2. Change to basic solution by adding $4OH^-$ to each side.

$$3e^- + (4H^+ + 4OH^-) + MnO_4^- \longrightarrow MnO_2 + 2H_2O + 4OH^-$$

3. Combine $4H^+$ and $4OH^-$ to form $4H_2O$ and then subtract $2H_2O$ from both sides of the equation.

$$3e^- + 2H_2O + MnO_4^- \longrightarrow MnO_2 + 4OH^-$$

4. Balance the oxidation half-reaction involving $C_2O_4^{2-}$ and CO_3^{2-} in acid solution.

$$2H_2O + C_2O_4^{2-} \longrightarrow 2CO_3^{2-} + 4H^+ + 2e^-$$

5. Change to basic solution as before by adding $4OH^-$ to each side and simplify by subtracting out $2H_2O$ from each side.

$$4OH^- + C_2O_4^{2-} \longrightarrow 2CO_3^{2-} + 2H_2O + 2e^-$$

6. Multiply the reduction reaction by 2 and the oxidation half-reaction by 3 and add equations.

$$\cancel{6e^-} + \cancel{4H_2O} + \overset{4}{\cancel{12}}OH^- + 3C_2O_4^{2-} + 2MnO_4^- \longrightarrow$$

$$6CO_3^{2-} + \overset{2}{\cancel{6}}H_2O + 2MnO_2 + 8\cancel{OH^-} + \cancel{6e^-}$$

7. Simplify by subtracting out $6e^-$, $4H_2O$, and $8OH^-$.

$$4OH^- + 3C_2O_4{}^{2-} + 2MnO_4{}^- \longrightarrow 6CO_3{}^{2-} + 2H_2O + 2MnO_2$$

EXAMPLE 14-5   Balancing Redox Equations in Basic Solution

Balance the following equation in basic solution by the ion-electron method.

$$Bi_2O_3(s) + NO_3{}^-(aq) + OH^-(aq) \longrightarrow BiO_3{}^-(aq) + NO_2{}^-(aq) + H_2O(l)$$

PROCEDURE

(1) Identify and separate the reduced and oxidized species and their products. (2) Balance in acidic solution. (3) Change from acidic to basic solution. (4) Simplify.

SOLUTION

Reduction reaction:	$NO_3{}^- \longrightarrow NO_2{}^-$
Balance in acidic:	$2e^- + 2H^+ + NO_3{}^- \longrightarrow NO_2{}^- + H_2O$
Change to basic:	$2e^- + (2H^+ + 2OH^-) + NO_3{}^- \longrightarrow NO_2{}^- + H_2O + 2OH^-$
Simplify H_2O:	$2e^- + H_2O + NO_3{}^- \longrightarrow NO_2{}^- + 2OH^-$
Oxidation reaction:	$Bi_2O_3 \longrightarrow 2BiO_3{}^-$
Balance in acidic:	$3H_2O + Bi_2O_3 \longrightarrow 2BiO_3{}^- + 6H^+ + 4e^-$
Change to basic:	$6OH^- + 3H_2O + Bi_2O_3 \longrightarrow 2BiO_3{}^- + (6H^+ + 6OH^-) + 4e^-$
Simplify H_2O:	$6OH^- + Bi_2O_3 \longrightarrow 2BiO_3{}^- + 3H_2O + 4e^-$

Multiply the reduction reaction by 2 and add to the oxidation half-reaction.

$$4e^- + Bi_2O_3 + 2H_2O + 2NO_3{}^- + 6OH^- \longrightarrow 2BiO_3{}^- + 2NO_2{}^- + 4OH^- + 3H_2O + 4e^-$$

Simplify by subtracting $2H_2O$, $4OH^-$, and $4e^-$ from both sides of the equation.

$$Bi_2O_3(s) + 2NO_3{}^-(aq) + 2OH^-(aq) \longrightarrow 2BiO_3{}^-(aq) + 2NO_2{}^-(aq) + H_2O(l)$$

ANALYSIS

Balancing equations in basic solution requires a direct application of the material from the previous chapter. In an aqueous solution, you have at your disposal H_2O and excess H^+ or OH^-, depending on whether the solution is acidic or basic. By adding an OH^- ion to an H^+ ion, you are essentially changing the H^+ to an H_2O. This is the net ionic equation of a neutralization reaction between a strong acid and a strong base. Remember to add an OH^- to both sides of the equation, however.

SYNTHESIS

We make distinctions between acidic and basic solutions because some compounds produce reactions that are different in each case. For example, $MnO_4{}^-$ is reduced to Mn^{2+} in acid solution but to MnO_2 in basic solution. In another example, Fe^{3+} is present in acidic solution but forms insoluble $Fe(OH)_3$ in basic solution. Alternately, different forms of the compounds exist in different pH's. $HClO$ is present in acidic solutions and can be reduced to Cl_2. ClO^- is present in basic solutions and can be reduced to Cl^-. Furthermore, as we'll see in the next section, each one of these reactions has a different potential for occurring, so the amount of energy we can get out of the reaction varies in each case.

▶ ASSESSING THE OBJECTIVE FOR SECTION 14-3

EXERCISE 14-3(a) LEVEL 2: Balance the following equations by the ion-electron method.

(a) $NO_3{}^-(aq) + H_2SO_3(aq) \longrightarrow SO_4{}^{2-}(aq) + NO(g) + H^+(aq) + H_2O(l)$

(b) $H_2O(l) + MnO_4{}^-(aq) + S^{2-}(aq) \longrightarrow MnS(s) + S(s) + OH^-(aq)$

(Notice some sulfide ions change and others do not.)

EXERCISE 14-3(b) LEVEL 3: When elemental tin is placed in a nitric acid solution, a spontaneous redox reaction occurs, producing solid tin(IV) oxide and nitrogen dioxide gas. Write the equation illustrating this reaction and balance the equation by the ion-electron method. (Water is also a product.)

For additional practice, work chapter problems 14-18, 14-20, 14-22, and 14-24.

PART A SUMMARY

KEY TERMS

14-1.1	**Half-reactions** indicate either the **oxidation** or **reduction** process. pp. 480–481
14-1.2	An electron exchange reaction is known as an **oxidation-reduction**, or simply **redox**, reaction. p. 481
14-1.2	The species reduced is known as the **oxidizing agent**; the species oxidized is known as the **reducing agent**. p. 481
14-1.3	**Oxidation states** (or **oxidation numbers**) are used to identify the elements that change in redox reactions. p. 482
14-2	The **oxidation state** or **bridge method** balances equations by focusing on the oxidation state changes. p. 486
14-3	The **ion-electron method** is used to balance half-reactions separately before adding to the total reaction. p. 489

SUMMARY CHART

Oxidizing and Reducing Agents

$$\text{reactant oxidized} \xrightarrow{\text{by the}} \text{oxidizing agent}$$

is the / is the

$$\text{reactant reduced} \xrightarrow{\text{by the}} \text{reducing agent}$$

Balancing Redox Reactions in Aqueous Solution

Unbalanced core reaction: $X + AO^- \longrightarrow XO_2^- + A$

Acidic solution

Separate into two half-reactions

$$X \longrightarrow XO_2^- \qquad\qquad AO^- \longrightarrow A$$

Add H_2O

$$2H_2O + X \longrightarrow XO_2^- \qquad\qquad AO^- \longrightarrow A + H_2O$$

Add H^+

$$2H_2O + X \longrightarrow XO_2^- + \underline{4H^+} \qquad\qquad \underline{2H^+} + AO^- \longrightarrow A + H_2O$$

Add e^-

$$2H_2O + X \longrightarrow XO_2^- + 4H^+ + \underline{3e^-} \qquad\qquad \underline{e^-} + 2H^+ + AO^- \longrightarrow A + H_2O$$

Balance electron exchange

$$2H_2O + X \longrightarrow XO_2^- + 4H^+ + 3e^- \qquad\qquad \underline{3}e^- + \underline{6}H^+ + \underline{3}AO^- \longrightarrow \underline{3}A + \underline{3}H_2O$$

Add two reactions and simplify

$$\mathbf{2H^+ + X + 3AO^- \longrightarrow XO_2^- + 3A + H_2O}$$

To change to a basic solution

Add OH^- (on both sides) for each H^+

$$(2H^+ + \underline{2OH^-}) + X + 3AO^- \longrightarrow XO_2^- + 3A + H_2O + \underline{2OH^-}$$

Combine H^+ and OH^- to make H_2O and simplify

$$H_2O + X + 3AO^- \longrightarrow XO_2^- + 3A + 2OH^-$$

Part B

Spontaneous and Nonspontaneous Redox Reactions

SETTING A GOAL

■ You will understand the extensive practical applications of redox reactions.

OBJECTIVES

14-4 Using a table of relative strengths of oxidizing agents, determine whether a specific redox reaction is spontaneous.

14-5 Describe the structure and electricity-generating ability of voltaic cells and batteries.

14-6 Write the reactions that occur when a salt is electrolyzed.

14-4 Predicting Spontaneous Redox Reactions

LOOKING AHEAD! In the previous chapter, we found that favorable (spontaneous) reactions occur between a stronger acid and a stronger base, forming a weaker acid and a weaker base. Electron exchange reactions between a stronger oxidizing agent and a stronger reducing agent to produce a weaker oxidizing and reducing agent are also spontaneous. In addition, by observing the results of a few reactions, we can rank oxidizing and reducing agents and compile this information into a table. We can then use this table to predict the occurrence of a large number of reactions. ■

▶ **OBJECTIVE FOR SECTION 14-4**
Using a table of relative strengths of oxidizing agents, determine whether a specific redox reaction is spontaneous.

We are well aware that gold is not chemically reactive. Why is it so stable? In an earlier chapter, we found that it is hard to remove an electron from a gold atom because of its high ionization energy. Using our current language we would say that gold is very hard to oxidize. It is just the opposite for a metal such as sodium, which is very easily oxidized. In this section, we will compare and rank three metals (zinc, copper, and nickel) as to their ability to be oxidized. This requires that we observe the results of three experiments.

The first experiment involves comparing copper to zinc. We do this by immersing a strip of zinc metal in an aqueous solution containing Cu^{2+} ions. It is obvious that an oxidation–reduction reaction occurs spontaneously, as indicated by the coating of copper that forms on the zinc metal. (In Chapter 6, we referred to this same reaction as a single-replacement reaction.) The zinc metal is oxidized to Zn^{2+} and the Cu^{2+} ion is reduced to copper metal with the transfer of two electrons from the zinc metal to the copper ion. (See Figure 6-5, Section 6-4.) However, if we placed copper metal in a solution of Zn^{2+} ions, we would observe that no reaction occurs. The reverse reaction does not occur spontaneously, since reactions are spontaneous (favorable) in one direction only. (See Figure 6-6, Section 6-4.1.)

The net ionic equation for the spontaneous reaction is

$$Zn(s) + Cu^{2+}(aq) \longrightarrow Zn^{2+}(aq) + Cu(s)$$

This is a spontaneous reaction because a stronger oxidizing agent and a stronger reducing agent react to form a weaker oxidizing and reducing agent. Since the spontaneous reaction is from left to right, we can make the following conclusions about the relative strengths of the two oxidizing and reducing agents.

1. Cu^{2+} is a better oxidizing agent than Zn^{2+}.
2. Zn metal is a better reducing agent than Cu metal.

In the second experiment, we will compare copper to nickel. As in the first experiment, we find that copper metal also forms a coating on a strip of Ni immersed in a Cu^{2+} solution. This spontaneous reaction is illustrated by the following equation.

$$Ni(s) + Cu^{2+}(aq) \longrightarrow Ni^{2+}(aq) + Cu(s)$$

This result provides similar conclusions as the Zn–Cu experiments.

A coating of copper forms on a strip of zinc in a Cu^{2+} solution.

3. Cu^{2+} is a better oxidizing agent than Ni^{2+}.

4. Ni metal is a better reducing agent than Cu metal.

So far, we know that both zinc and nickel are better reducing agents than copper, but how do they compare to each other? In the third experiment we immerse a strip of zinc in a solution containing Ni^{2+} ions. A coating of nickel eventually appears on the zinc metal. (If we had first tried placing a strip of nickel in a Zn^{2+} solution, we would find that no reaction occurs.) The spontaneous reaction between Zn and Ni^{2+} is represented below.

$$Zn(s) + Ni^{2+}(aq) \longrightarrow Zn^{2+}(aq) + Ni(s)$$

Our conclusions from this experiment are:

5. Ni^{2+} is a better oxidizing agent than Zn^{2+}.

6. Zn metal is a better reducing agent than Ni metal.

We can now rank all three ions as oxidizing agents in order of decreasing strength as follows.

$$Cu^{2+} > Ni^{2+} > Zn^{2+}$$

Notice that the strength of the metals as reducing agents is inversely related to the strength of their ions as oxidizing agents. Because of the inverse relationship, the ranking of the three metals as reducing agents is in the reverse order.

$$Zn > Ni > Cu$$

These three metals were included in the *activity series* that was discussed in Chapter 6. In this chapter, we now explain the activity series by comparing the strengths of the metals as reducing agents and of their ions as oxidizing agents.

More experiments can provide additional ions and molecular species for our ranking. Eventually, we can construct a table of oxidizing agents ordered according to strength. Such a ranking is given in Table 14-1 and is an extension of the activity series of metals introduced in Chapter 6. In some cases, instruments are required to give quantitative values, known as *reduction potentials*, which indicate the comparative strength of a given oxidizing agent in relation to a defined standard. The strengths of the oxidizing agents are compared in these measurements at the same concentration for all ions involved (1.00 M) and at the same partial pressure of all gases involved (1.00 atm).

A redox reaction takes place between an oxidizing agent on the left and a reducing agent on the right. *A favorable or spontaneous reaction occurs between an element, ion, or compound on the left (an oxidizing agent) and a species on the right (a reducing agent) that lies below it in the table.* The strongest oxidizing agent (F_2) is at the top of Table 14-1, on the left. It will oxidize any reducing agent listed on the right. However, the reducing agents, shown on the right, become stronger *down* the table. Na is the strongest reducing agent listed and so will reduce any oxidizing agent listed on the left. The spontaneous reaction can be visualized as taking place in a clockwise direction, with the oxidizing agent forming a product to the right and the reducing agent below it forming a product in the opposite direction to the left. (See Figure 14-2.)

Table 14-1 is particularly useful in predicting the reactions of certain elements with water. (All the reactions shown in the table are assumed to occur in aqueous solution.) The boxed reaction near the top of the table represents the oxidation of water when read from right to left. Note that the gaseous elements F_2 and Cl_2 spontaneously oxidize water to produce oxygen gas and an acid. (The reaction of Cl_2 with water is quite slow, however.)

$$Cl_2 + 2e^- \rightleftharpoons 2Cl^-$$

$$O_2 + 4H^+ + 4e^- \rightleftharpoons 2H_2O$$

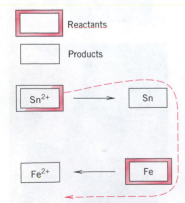

The spontaneous reaction is
$Sn^{2+} + Fe \longrightarrow Sn + Fe^{2+}$

Reactants

Products

$Sn^{2+} \longrightarrow Sn$

$Fe^{2+} \longleftarrow Fe$

FIGURE 14-2 Spontaneous Reaction The stronger oxidizing agent reacts with the stronger reducing agent.

TABLE 14-1

Oxidizing Agents and Reducing Agents

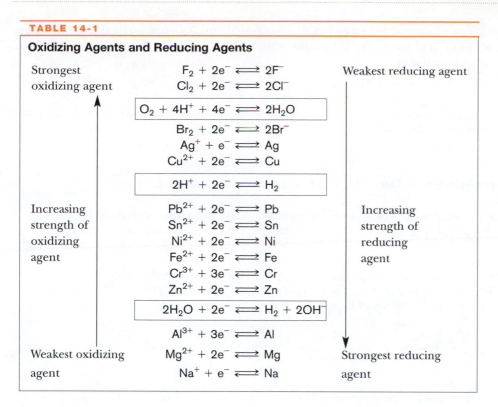

Strongest oxidizing agent

$F_2 + 2e^- \rightleftharpoons 2F^-$
$Cl_2 + 2e^- \rightleftharpoons 2Cl^-$

$O_2 + 4H^+ + 4e^- \rightleftharpoons 2H_2O$

$Br_2 + 2e^- \rightleftharpoons 2Br^-$
$Ag^+ + e^- \rightleftharpoons Ag$
$Cu^{2+} + 2e^- \rightleftharpoons Cu$

$2H^+ + 2e^- \rightleftharpoons H_2$

$Pb^{2+} + 2e^- \rightleftharpoons Pb$
$Sn^{2+} + 2e^- \rightleftharpoons Sn$
$Ni^{2+} + 2e^- \rightleftharpoons Ni$
$Fe^{2+} + 2e^- \rightleftharpoons Fe$
$Cr^{3+} + 3e^- \rightleftharpoons Cr$
$Zn^{2+} + 2e^- \rightleftharpoons Zn$

$2H_2O + 2e^- \rightleftharpoons H_2 + 2OH^-$

$Al^{3+} + 3e^- \rightleftharpoons Al$
$Mg^{2+} + 2e^- \rightleftharpoons Mg$
$Na^+ + e^- \rightleftharpoons Na$

Weakest reducing agent

Increasing strength of oxidizing agent

Increasing strength of reducing agent

Weakest oxidizing agent

Strongest reducing agent

The spontaneous reaction is

$$2Cl_2 + 2H_2O \longrightarrow O_2 + 4H^+ + 4Cl^-$$

The boxed reaction near the bottom of the table represents the reduction of water when read from left to right. Note that the metals Al, Mg, and Na spontaneously reduce water to produce hydrogen gas and a base. *(We can say that Na reduces water, or, conversely, we can say that water oxidizes Na.)*

$$2H_2O + 2e^- \rightleftharpoons H_2 + 2OH^-$$
$$Na^+ + e^- \rightleftharpoons Na$$

The spontaneous reaction is

$$2H_2O + 2Na \longrightarrow H_2 + 2Na^+ + 2OH^-$$

These metals are known as *active* metals because of their chemical reactivity with water.

The third boxed reaction, near the middle of the table, represents the reduction of aqueous acid solutions (1.00 M H^+) to form hydrogen gas. Note that metals such as Ni and Fe are not oxidized by water but are oxidized by strong acid solutions.

$$2H^+ + 2e^- \rightleftharpoons H_2$$
$$Fe^{2+} + 2e^- \rightleftharpoons Fe$$

The spontaneous reaction is

$$2H^+ + Fe \longrightarrow Fe^{2+} + H_2$$

Acid rain is so named because it contains a considerably higher H^+ concentration than ordinary rain. From this discussion, we can understand why metals such as iron and nickel are more likely to be corroded by acid rain.

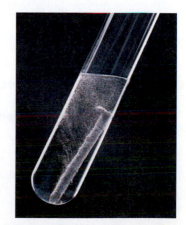

Iron is oxidized by a strong acid solution to form iron(II) ions and hydrogen gas.

Two of the metals shown in the table—Cu and Ag—are not oxidized by either water or acid solutions.* Thus, these metals are relatively unreactive and find use in jewelry and coins.

_____

*Copper is dissolved by nitric acid solutions, but the copper is oxidized by the nitrate ion, not the hydronium ion. See Example 14-3(b).

EXAMPLE 14-6   Determining the Direction of a Spontaneous Reaction

A strip of tin is placed in an aqueous solution of $Cr(NO_3)_3$ in one experiment. In another, a strip of chromium is placed in an aqueous solution of $Sn(NO_3)_2$. Write the net ionic equation illustrating the spontaneous reaction that occurs.

PROCEDURE

In Table 14-1, note that Sn^{2+} is a stronger oxidizing agent than Cr^{3+} and Cr is a stronger reducing agent than Sn. The reactants are, therefore, Sn^{2+} and Cr. A balanced equation is obtained by multiplying the Sn^{2+} half-reaction by 3 and the Cr half-reaction by 2 for an exchange of six electrons.

SOLUTION

$$3Sn^{2+}(aq) + 2Cr(s) \longrightarrow 3Sn(s) + 2Cr^{3+}(aq)$$

ANALYSIS

Notice that the spontaneous reaction takes place in a clockwise direction in the table. Sn^{2+} is in the higher position in the table as an oxidizing agent and reacts to the right, forming its product, Sn. Cr is below Sn so is the stronger reducing agent. It reacts to the left, forming its product, Cr^{3+}. Six electrons are exchanged to balance the equation. In using the table, remember that one species reacts to the right and one to the left.

SYNTHESIS

Nonmetal elements tend to be at the upper left of the reduction table and metal elements at the bottom right. These represent the reactive extremes of the table. As you move up the left side and down the right side, the elements become more reactive. It is not surprising that this order is also roughly close to the historical order in which these elements were discovered, isolated, and identified. The more reactive an element is, the harder it becomes to produce it in its elemental state. Metals at the upper right, like copper and silver, have been known since the beginning of civilization. So have iron, lead, mercury, and tin, which require only modest efforts to liberate from their ores. Metal ions of medium reduction potential, such as nickel, cobalt, and manganese, were isolated in the mid-1700s. The most reactive metals were finally isolated after electricity became a viable technique for promoting difficult chemical reactions in the nineteenth century. These include sodium, potassium, lithium, magnesium, and aluminum. Another factor, of course, is the relative abundance of the element, with rarer elements being more difficult to find. A very low reduction potential of the metal ions, though, was the primary roadblock to these elements' discovery.

EXAMPLE 14-7   Writing a Spontaneous Single-Replacement Reaction

A strip of tin metal is placed in an $AgNO_3$ solution. If a reaction takes place, write the equation illustrating the spontaneous reaction.

PROCEDURE

In Table 14-1, note that the oxidizing agent Ag^+ is above the reducing agent Sn. Therefore, a spontaneous reaction does occur. Write the balanced equation, noticing that two electrons are exchanged.

SOLUTION

$$2Ag^+(aq) + Sn(s) \longrightarrow Sn^{2+}(aq) + 2Ag(s)$$

ANALYSIS

Again, notice that the spontaneous reaction is in a clockwise direction. The Ag^+ reacts to the right and the Sn reacts to the left.

SYNTHESIS

A practical application of this reaction is to "silver coat" tin or some other metal, such as iron or nickel, that lies lower than Ag^+ in the table. Au^{3+} also ranks high in the table as an oxidizing agent like Ag^+. This means that a cheaper metal can be "gold plated" to give it the outer appearance of gold.

EXAMPLE 14-8 Writing a Spontaneous Reaction of a Metal with Water

A length of aluminum wire is placed in water. Does the aluminum react with water? What is the spontaneous reaction involving aluminum and water?

PROCEDURE

In Table 14-1, note that aluminum is an active metal and would react with water (as an oxidizing agent). The aluminum is oxidized and the water is reduced.

SOLUTION

$$6H_2O(l) + 2Al(s) \longrightarrow 2Al^{3+}(aq) + 6OH^-(aq) + 3H_2(g)$$

Since $Al(OH)_3$ is insoluble in water, however, the equation should be written as

$$6H_2O(l) + 2Al(s) \longrightarrow 2Al(OH)_3(s) + 3H_2(g)$$

ANALYSIS

How many electrons are being exchanged in this reaction? The oxidation of aluminum to Al^{3+} is a 3-electron half-reaction. In water, hydrogen is reduced from + 1 to 0 in H_2, but two hydrogen atoms require 2 electrons. The least common multiple of 3 and 2 is 6, so there are 6 electrons exchanging in this reaction.

SYNTHESIS

Theoretically, aluminum should dissolve (i.e., react) in water. How then can we have aluminum boats? Metallic aluminum reacts with oxygen in the air to form a coating of Al_2O_3, which protects the metal from coming into contact with water. In other words, it "self-paints." The reaction described above is actually limited to the outermost layer of aluminum. The only thing you would notice is that shiny aluminum forms a gray coating in water.

▶ **ASSESSING THE OBJECTIVE FOR SECTION 14-4**

EXERCISE 14-4(a) LEVEL 1: Fill in the blanks.
Spontaneous redox reactions occur between the _____ oxidizing and reducing agents to produce _____ oxidizing and reducing agents. In a table, the stronger oxidizing agent is located _____ in the table than the reducing agent.

EXERCISE 14-4(b) LEVEL 2: In which of the following would a spontaneous reaction occur? Write the balanced equation for that reaction.

(a) A strip of Pb is placed in a Cr^{3+} solution.
(b) A strip of Cr is placed in a Pb^{2+} solution.

EXERCISE 14-4(c) LEVEL 2: If zinc and iron were in electrical contact with one another, which would oxidize first?

EXERCISE 14-4(d) LEVEL 3: Given the following table of oxidizing and reducing agents, answer the questions below.

strongest ⟶
oxidizing
agent

$$I_2 \quad + 2e^- \rightleftharpoons 2I^-$$
$$Cr^{3+} + 3e^- \rightleftharpoons Cr$$
$$Mn^{2+} + 2e^- \rightleftharpoons Mn \qquad \text{strongest}$$
$$Ca^{2+} + 2e^- \rightleftharpoons Ca \leftarrow \text{reducing agent}$$

(a) Is the following reaction spontaneous or nonspontaneous?
$$Mn + Ca^{2+} \longrightarrow Ca + Mn^{2+}$$

(b) Write a balanced equation representing a spontaneous reaction involving the Cr, Cr^{3+} and the Mn, Mn^{2+} half-reactions.

(c) If an aqueous solution of I_2 (an antiseptic) is spilled on a chromium-coated automobile bumper, will a reaction occur? If so, write the balanced equation for the reaction.

(d) Which substance in the table will react with Mn^{2+}?

(e) Which substances in the table will not react with any other in the table?

EXERCISE 14-4(e) LEVEL 3: Write a balanced equation illustrating how a metal such as iron could be "gold plated" from an Au^{3+} solution.

For additional practice, work chapter problems 14-20, 14-27, 14-31, and 14-32.

▶ **OBJECTIVE FOR SECTION 14-5**
Describe the structure and electricity-generating ability of voltaic cells and batteries.

14-5 Voltaic Cells

LOOKING AHEAD! The electrons exchanged in a spontaneous redox reaction can be detoured through a wire and put to work. This is the principle used in all the different types of batteries that we depend on to run our calculators, remotes, cell phones, and even iPods and MP3 players. How this is accomplished is the topic of this section. ■

Releasing the brake on an automobile parked on the side of a hill brings no surprise. The automobile spontaneously rolls down the hill because the bottom of the hill represents a lower potential energy than the top. Chemical reactions occur spontaneously for the same reason. The products represent a position of lower potential energy than the reactants. In the case of the car, the difference in potential energy is due to position and the attraction of gravity; in the case of a chemical reaction, the difference in potential energy is due to the composition of the reactants and products. (See Figure 14-3.) When the car rolls down the hill, the difference in energy is transformed into the kinetic energy of the moving car and heat from friction. When a

FIGURE 14-3 Energy States
A chemical reaction proceeds in a certain direction for the same reason that a car rolls down a hill.

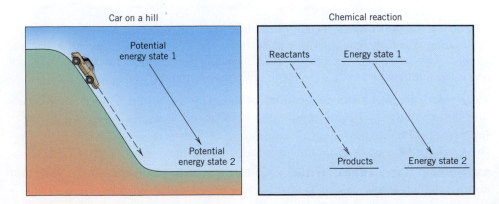

chemical reaction proceeds spontaneously from reactants to products, the difference is given off as heat, light, or electrical energy. In this section, we are concerned with the electrical energy provided by spontaneous redox reactions.

14-5.1 The Daniell Cell

A **voltaic cell** (*also called a galvanic cell*) *uses a favorable or spontaneous redox reaction to generate electrical energy through an external circuit.* One of the earliest voltaic cells put to use the spontaneous reaction discussed in the previous section, which is

$$Zn(s) + Cu^{2+}(aq) \longrightarrow Zn^{2+}(aq) + Cu(s)$$

This voltaic cell is known as the *Daniell cell* and was used to generate electrical current for the new telegraph and doorbells in homes.

In an earlier experiment discussed in the previous section, we found that a strip of zinc placed in a Cu^{2+} solution forms a coating of metallic copper on the zinc strip. To generate electricity, however, the oxidation reaction must be separated from the reduction reaction so that the electrons exchanged can flow in an external wire where they can be put to use. The Daniell cell is illustrated in Figure 14-4. A zinc strip is immersed in a Zn^{2+} solution, and, in a separate compartment, a copper strip is immersed in a Cu^{2+} solution. A wire connects the two metal strips. The two metal strips are called electrodes. *The **electrodes** are the surfaces in a cell at which the reactions take place. The electrode at which oxidation takes place is called the **anode**. Reduction takes place at the **cathode**.* In the compartment on the left, the strip of Zn serves as the anode, since the following reaction occurs when the circuit is connected.

$$Zn \longrightarrow Zn^{2+} + 2e^-$$

When the circuit is complete, the two electrons travel in the external wire to the Cu electrode, which serves as the cathode. The reduction reaction occurs at the cathode.

$$2e^- + Cu^{2+} \longrightarrow Cu$$

To maintain neutrality in the solution, some means must be provided for the movement of a SO_4^{2-} ion (or some other negative ion) from the right compartment, where a Cu^{2+} has been removed, to the left compartment, where a Zn^{2+} has been produced. The salt bridge is an aqueous gel that allows ions to migrate between compartments but does not allow the mixing of solutions. A porous plate separating the two solutions, as shown in Figure 14-4, also serves this function. (If Cu^{2+} ions mixed into the left compartment, they would form a coating of Cu on the Zn electrode, thus short-circuiting the cell.)

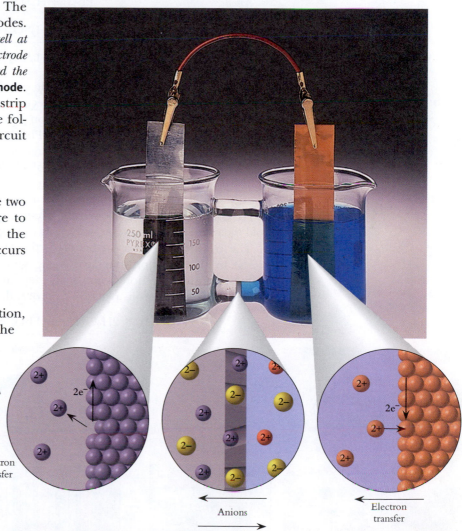

Electron transfer

Anions

Electron transfer

FIGURE 14-4 The Daniell Cell
This chemical reaction produced electricity for the first telegraphs.

The electrodes in this cell are called active electrodes because they are involved in the reaction. As the cell discharges (generates electrical energy), the Zn electrode becomes smaller but the Cu electrode becomes larger. The reaction can be stopped by interrupting the external circuit with a switch. If the circuit is open, the electrons can no longer flow, so no further reaction occurs until the switch is again closed. The energy of the cell can thus be stored between uses.

14-5.2 Lead-Acid Batteries

Two of the most common voltaic cells in use today are two of the oldest. (Both were invented in the late-nineteenth century.) These are the dry cell (flashlight battery) and the lead-acid cell (car battery). *(A **battery** is a collection of one or more separate cells joined together in one unit.)*

We still use good old lead–acid batteries to start our cars. Such a three-cell storage battery is illustrated in Figure 14-5. Each cell is composed of two grids separated by an inert spacer. One grid of a fully charged battery contains metallic lead. The other contains PbO_2, which is insoluble in H_2O. Both grids are immersed in a sulfuric acid solution (battery acid). When the battery is discharged by connecting the electrodes, the following half-reactions take place spontaneously.

Anode: $$Pb(s) + H_2SO_4(aq) \longrightarrow PbSO_4(s) + 2H^+(aq) + 2e^-$$

Cathode: $$2e^- + 2H^+(aq) + PbO_2(s) + H_2SO_4(aq) \longrightarrow PbSO_4(s) + 2H_2O(l)$$

Total reaction: $$Pb(s) + PbO_2(s) + 2H_2SO_4(aq) \longrightarrow 2PbSO_4(s) + 2H_2O(l)$$

The electrons released at the Pb anode travel through the external circuit to run a car's lights, starter, radio, or whatever is needed. The electrons return to the PbO_2 cathode to complete the circuit. As the reaction proceeds, both electrodes are converted to $PbSO_4$, and the H_2SO_4 is depleted. Since $PbSO_4$ is also insoluble, it remains attached to the grids as it forms. The degree of discharge of a battery can be determined by the density of the battery acid. Since the density of a fully discharged battery is 1.15 g/mL, the difference in density between this value and the density of a fully charged battery (1.35 g/mL) gives the amount of charge remaining in the battery. As the electrodes convert to $PbSO_4$, the battery loses power and eventually becomes "dead." Most lead batteries sold today are sealed, however, so density cannot be used to determine its condition.

The convenience of a car battery is that it can be recharged. After the engine starts, an alternator or generator is engaged to push electrons back into the cell in the opposite direction from which they came during discharge. This forces the reverse, nonspontaneous reaction to proceed.

$$2PbSO_4(s) + 2H_2O(l) \longrightarrow Pb(s) + PbO_2(s) + 2H_2SO_4(aq)$$

FIGURE 14-5 Lead Storage Battery The lead storage battery is rechargeable.

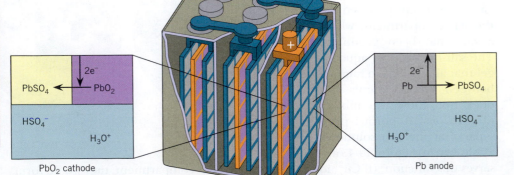

When the battery is fully recharged, the alternator shuts off, the circuit is open, and the battery is ready for the next start.

14-5.3 The Dry Cell

The dry cell (invented by Leclanché in 1866) is not rechargeable to any extent but is comparatively inexpensive and easily portable. (In contrast, the lead–acid battery is heavy and expensive and must be kept upright.) The dry cell illustrated in Figure 14-6 consists of a zinc anode, which is the outer rim, and an inert graphite electrode. (An *inert* or *passive* electrode provides a reaction surface but does not itself react.) In between is an aqueous paste containing NH_4Cl, MnO_2, and carbon. The reactions are as follows.

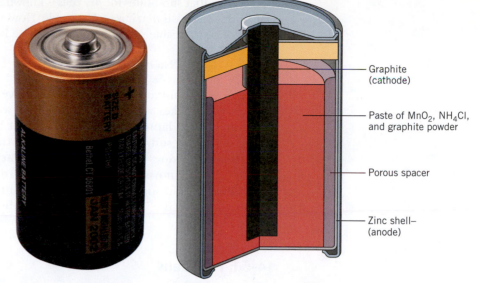

Graphite (cathode)

Paste of MnO_2, NH_4Cl, and graphite powder

Porous spacer

Zinc shell– (anode)

FIGURE 14-6 Dry Cell The dry cell (left) is comparatively inexpensive, light, and portable. Its contents are depicted on the right.

Anode: $Zn(s) \longrightarrow Zn^{2+}(aq) + 2e^-$

Cathode: $2NH_4^+(aq) + 2MnO_2(s) + 2e^- \longrightarrow Mn_2O_3(s) + 2NH_3(aq) + H_2O$

A disadvantage of the dry cell is that the NH_4^+ ion creates an acidic solution. (In the previous chapter, we discussed how this cation undergoes hydrolysis to form a weakly acidic solution.) The zinc electrode slowly reacts with the weakly acidic solution, so the shelf-life of these batteries is a matter of only a few months. In an *alkaline battery*, NaOH or KOH is substituted for the NH_4Cl, so the solution is basic and zinc reacts much more slowly. Alkaline batteries are more expensive, but they have much longer shelf-lives. The anode reaction in the alkaline battery is

$$Zn(s) + 2OH^-(aq) \longrightarrow ZnO(s) + H_2O(l) + 2e^-$$

14-5.4 Specialized Batteries

Other types of batteries that have been developed in modern times are more useful for calculators and wristwatches. These batteries are very small and deliver a small amount of current for a long time. Some batteries can produce current for three to five years. One is the *silver battery*, which uses a zinc anode and a silver(I) oxide cathode. Another, known as the *mercury battery*, also uses a zinc anode but a mercury(II) oxide cathode. (See Figure 14-7.) The two reactions in these batteries are

$$\text{Silver battery: } Zn(s) + Ag_2O(s) \longrightarrow ZnO(s) + 2Ag(s)$$

$$\text{Mercury battery: } Zn(s) + HgO(s) \longrightarrow ZnO(s) + Hg(l)$$

FIGURE 14-7 Mercury Battery These small batteries deliver electrical current for long periods.

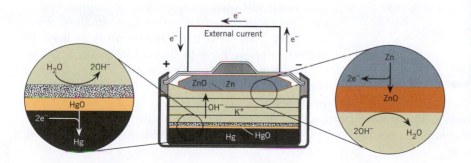

A battery that substitutes for dry cells is known as a *nickel-cadmium battery*. The advantage of this battery is that it is rechargeable. However, it is considerably more expensive initially. The reversible reaction that takes place in this cell is

$$Cd(s) + 2NiO(OH)(s) + 2H_2O(l) \longrightarrow 2Ni(OH)_2(s) + Cd(OH)_2(s)$$

There are many other batteries that have special uses, such as the *lithium battery*, which takes advantage of the light weight and strong reducing ability of lithium. Most laptop computers use what is known as a *lithium ion battery*. It provides a good source of power and is both rechargeable and comparatively light. It actually does not involve a redox reaction, however. Electrical energy is supplied by the movement of lithium ions from one electrode to another while electrons move through an external circuit. Lithium ion batteries are now under development for use in an electrical automobile to be produced by several major manufacturers.

14-5.5 Fuel Cells

Space travel requires a tremendous source of electrical energy. The requirements are that the source be continuous (no recharging necessary), lightweight, and dependable. Solar energy directly converts rays from the sun into electricity and is used in the international space station (as well as in an increasing number of homes). It was not practical for shorter runs such as the space shuttle. Although expensive, a source of power that fills the bill nicely is the fuel cell. A **fuel cell** *uses the direct reaction of hydrogen or a hydrogen compound and oxygen to produce electrical energy.* Figure 14-8 is an illustration of a fuel cell. Hydrogen and oxygen gases are fed into the cell, where they form water. As long as the gases enter the cell, power is generated. The water that is formed can be removed and used for other purposes in the spacecraft. Since reactants are supplied from external sources, the electrodes are not consumed, and the cell does not have to be shut down to be regenerated, as a car battery does. The best inert electrode surfaces at which the gases react are made of the extremely expensive metal platinum. The reactions that take place in a basic solution in the fuel cell are

Anode: $H_2(g) + 2OH^-(aq) \longrightarrow 2H_2O + 2e^-$

Cathode: $O_2(g) + 2H_2O + 4e^- \longrightarrow 4OH^-(aq)$

Overall: $2H_2(g) + O_2(g) \longrightarrow 2H_2O$

Fuel cells have had some large-scale application in power generation for commercial purposes. Currently, a great deal of research money and effort is being expended in fuel cell research. See Making It Real, "Fuel Cells—The Future Is (Almost) Here."

More efficient and durable batteries are the focus at the current time. For example, an efficient and relatively inexpensive electric automobile is the object of intensive efforts. A small electric car using lead–acid batteries requires at least 18 batteries. These have to be replaced every year or so, depending on use. Also, much of the power of these batteries must be used just to move the heavy batteries around, not including the car and passengers. There have been some encouraging possibilities for lighter, more durable batteries, and several types of electric automobiles are in the testing phase.

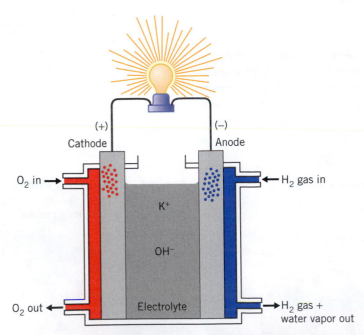

FIGURE 14-8 Fuel Cell The fuel cell can generate power without interruption for recharging.

MAKING IT REAL

Fuel Cells—The Future Is (Almost) Here

Special vehicles are currently testing fuel cells as a power source.

The simplest combustion reaction, hydrogen plus oxygen to form water, may be the key to the future. On a mass-of-fuel basis, it is by far the most energetic reaction. The combustion of two moles (4 g) of hydrogen produces 570 kJ of energy, much more than any other fuel. Since the combustion of hydrogen is a redox reaction, the energy can be released as electrical energy. This is the principle of the fuel cell, first demonstrated in the nineteenth century. It has long been used in the space program by NASA to continuously generate electricity without releasing toxic products. The two reactions that occur in the fuel cell and a diagram were illustrated in Section 14-5.5 and Figure 14-8.

The use of the fuel cell has been limited by the expense of inert electrodes that have traditionally contained rare platinum and/or palladium, which are more expensive than gold. Also, although hydrogen was previously discussed as the perfect fuel, it is hazardous to handle, it is not cheap in the pure state, and it is not easy to store in large quantities. Another problem involves the high operating temperatures necessary in a fuel cell.

All these problems, however, are being addressed with billions of dollars of investment and research. Much of the hope of making fuel cells practical in the short run rests in using fuels such as methane (natural gas), butane, methanol, ethanol, or even gasoline rather than pure hydrogen. These compounds are readily available and easy to transport. The problem with these fuels is that they tend to gunk up the electrodes and they produce carbon dioxide, which is a greenhouse gas. Still, it would be much easier to dispense natural gas or even butane at a service station rather than pure hydrogen. Hydrogen is ideal for the fuel cell, but its large-scale application may have to wait.

If you have not yet heard much about fuel cells, you will. Experimental cars and buses using fuel cells are already on the road. Reports come in almost daily about advancements in the research on fuel cells. The operating temperatures keep coming down; new and cheaper materials for the electrodes, along with more efficient use of fuels other than hydrogen, are being developed. Eventually, we will even have fuel cells generating electrical power scattered around a city, eliminating the need for extensive overhead transmission lines. Research is in progress to miniaturize the fuel cell so it can be used in cell phones and calculators. It would be smaller, cheaper (eventually), and easily refueled. (Some may even run on the alcohol in wine or beer.) Fuel cells are the power source of the future.

▶ **ASSESSING THE OBJECTIVE FOR SECTION 14-5**

EXERCISE 14-5(a) LEVEL 1: Fill in the blanks.
A spontaneous chemical reaction is used in a _____ cell. An example is the Daniell cell, where _____ _____ is oxidized at the _____ and _____ _____ is reduced at the _____. An electrolytic balance is maintained by means of a _____ bridge. In the lead–acid battery, _____ _____ is _____ at the anode and _____ is reduced at the _____.

EXERCISE 14-5(b) LEVEL 1: Refer to the text to identify the following cells:
(a) $2PbSO_4(s) + 2H_2O \longrightarrow Pb(s) + PbO_2(s) + 2H_2SO_4(aq)$
(b) $Cd(s) + 2NiO(OH)(s) + 2H_2O(l) \longrightarrow 2Ni(OH)_2(s) + Cd(OH)_2(s)$
(c) $Zn(s) + 2NH_4^+(aq) + 2MnO_2(s) \longrightarrow Zn^{2+}(aq) + Mn_2O_3(s) + 2NH_3(aq) + H_2O$
(d) $2H_2(g) + O_2(g) \longrightarrow 2H_2O$

EXERCISE 14-5(c) LEVEL 2: A cell is constructed of a Br_2, Br^- half-cell connected to an Fe^{2+}, Fe half-cell. Write a balanced equation representing the spontaneous reaction that occurs.

EXERCISE 14-5(d) LEVEL 3: If you were asked to design a battery to replace the lead–acid battery for an automobile, what balanced equation would you propose from the oxidizing and reducing agents listed in Table 14-1? Three things should be considered (in addition to cost, which we will ignore). (1) The mass of the reactants is important. (2) The farther apart the oxidizing and reducing agents are in the table, the more powerful the battery (i.e., the higher voltage). (3) The overall chemical reactivity of the reactants is important. For example, if a reactant explodes in air or water, we would not want to use it.

For additional practice, work chapter problems 14-36, 14-38, and 14-40.

▶ **OBJECTIVE**
FOR SECTION 14-6
Write the reactions that occur when a salt is electrolyzed.

14-6 Electrolytic Cells

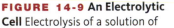

LOOKING AHEAD! Spontaneous reactions are like cars rolling down a hill. They go to lower energy states. Can a car or a chemical reaction go up a hill? Of course—but in order for this to happen, energy in the form of a push for a car or energy for a chemical reaction must be supplied. Nonspontaneous redox reactions can occur if sufficient electrical energy is supplied from an outside source. This is the final topic of this chapter. ■

Back in the good old days before inexpensive but strong plastics, automobiles were equipped with beautiful chrome bumpers and other chromium accessories. Chromium not only looks great but is also resistant to rust, so it protects the underlying iron (which does rust) from exposure to air and water. How is iron metal coated with a layer of chromium? According to Table 14-1, we would *not* predict that a coating of chromium would spontaneously form on iron immersed in a Cr^{3+} solution. But that doesn't mean we can't *make* it happen. Nonspontaneous redox reactions occur if enough electrical energy is supplied from an outside source. *Cells that convert electrical energy into chemical energy are called* **electrolytic cells**. They involve nonspontaneous redox reactions.

An example of an electrolytic cell is shown in Figure 14-9. When sufficient electrical energy is supplied to the electrodes from an outside source, the following non-spontaneous reaction occurs.

$$2H_2O(l) \longrightarrow 2H_2(g) + O_2(g)$$

The electrolysis of water to produce hydrogen using solar energy seems to be a logical solution to the energy crisis. The hydrogen produced could be used in a fuel cell. Unfortunately, the process using solar energy is currently very inefficient and expensive. At this time, the cheapest source of hydrogen is from a hydrocarbon such as methane (natural gas). However, much research is underway to improve the process.

For electrolysis of water to occur, an electrolyte such as K_2SO_4 must be present in solution. Pure water alone does not have a sufficient concentration of ions to allow conduction of electricity.

Another example of an electrolytic cell is the recharge cycle of the lead-acid battery described in Section 14-5.2. When energy from gasoline burning in the engine activates the alternator, electrical energy is supplied to the battery, and the nonspontaneous reaction occurs as an electrolysis reaction. This reaction re-forms the original reactants.

Electrolysis has many useful applications. For example, in addition to chromium, silver and gold can be electroplated onto cheaper metals. In

FIGURE 14-9 An Electrolytic Cell Electrolysis of a solution of potassium sulfate gives hydrogen gas and oxygen gas as products.

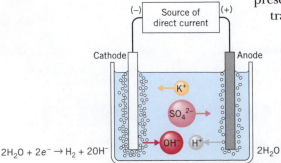

$2H_2O + 2e^- \rightarrow H_2 + 2OH^-$

$2H_2O \rightarrow O_2 + 4H^+ + 4e^-$

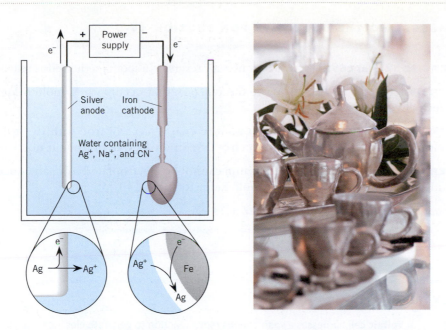

FIGURE 14-10 Electroplating With an input of energy, a spoon can be coated with silver. The service has been electroplated with silver.

Figure 14-10 the metal spoon is the cathode and the silver bar serves as the anode. When electricity is supplied, the Ag anode produces Ag^+ ions, and the spoon cathode reduces Ag^+ ions to form a layer of Ag. The silver-plated spoon can be polished and made to look as good as sterling silver, which is more expensive.

Electrolytic cells are used to free many elements from their compounds. Such cells are especially useful where metals are held in their compounds by strong chemical bonds. Examples are the metals aluminum, sodium, and magnesium. A little over 120 years ago, aluminum was produced through difficult chemical reactions involving elemental sodium as a reducing agent. As a result it was as expensive as gold. Producing molten Al_2O_3 (bauxite ore) for electrolysis was not practical because of its very high melting point (2000°C). In 1886, however, it was discovered that when bauxite was mixed with Na_3AlF_6 (a mineral called cryolite), the melting point of the mixture was much lower and electrolysis was possible. However, the process does require a large amount of energy. (Recycling requires only about 5% as much energy.) For that reason, recycling aluminum saves money and the environment. Commercial quantities of sodium and chlorine are also produced by electrolysis of molten sodium chloride. An apparatus used for the electrolysis of molten sodium chloride is illustrated in Figure 14-11. At the high temperature required to keep the NaCl in the liquid state, sodium forms as a liquid and is drained from the top of the cell.

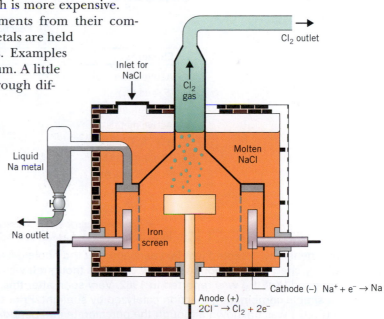

FIGURE 14-11 Electrolysis of NaCl Cross section of the Downs cell used for the electrolysis of molten sodium chloride. The cathode is a circular ring that surrounds the anode. The electrodes are separated from each other by an iron screen. During the operation of the cell, molten sodium collects at the top of the cathode compartment, from which it is periodically drained. The chlorine gas bubbles out of the anode compartment and is collected.

► **ASSESSING THE OBJECTIVE FOR SECTION 14-6**

EXERCISE 14-6(a) LEVEL 1: List four uses for reversing a spontaneous redox reaction.

EXERCISE 14-6(b) LEVEL 2: Write the reaction that occurs when molten $MgCl_2$ is electrolyzed.

EXERCISE 14-6(c) LEVEL 3: If you want to coat a piece of iron with Cr, it must be electroplated. Is it necessary to electroplate Sn on a piece of iron? If not, why?

EXERCISE 14-6(d) LEVEL 3: Aluminum cannot be electroplated from an aqueous solution containing Al^{3+} ions. Why not? Refer to Table 14-1.

For additional practice, work chapter problem 14-43.

PART B SUMMARY

KEY TERMS

14-5.1	A **voltaic cell** harnesses a spontaneous redox reaction to generate electricity. p. 501
14-5.1	A voltaic cell consists of two **electrodes**, an **anode** for oxidation and a **cathode** for reduction. p. 501
14-5.2	A **battery** consists of one or more voltaic cells arranged in a series. p. 502
14-5.5	A **fuel cell** generates electricity from the reaction of gases at inert electrodes. p. 504
14-6	Electrical energy is supplied to an **electrolytic cell** to cause a nonspontaneous redox reaction. p. 506

SUMMARY CHART

Reactions of Elements with Water

Elements	Reaction	Products
Active nonmetals (F_2 and Cl_2)	Oxidize water	O_2 and nonmetal acid (HF, HCl)
Other nonmetals	No reaction	
Active metals (e.g., K, Ca, Al)	Reduce water	H_2 and metal hydroxide [e.g., KOH, $Ca(OH)_2$]
Other metals (e.g., Fe, Cu, Au)	No reaction	

CHAPTER 14 SYNTHESIS PROBLEM

It was generally accepted just 50 years ago that the noble gases (also called the inert gases) could not form ordinary chemical compounds. A few chemists thought it was possible, however, and the first noble gas compound ($XePtF_6$) was reported in 1962. Very soon after this, xenon was found to form a fluoride [$XeF_2(g)$] by a simple combination reaction catalyzed by sunlight. Prior to the synthesis of XeF_2, the perbromate ion (BrO_4^-) was unknown although the perchlorate and periodate ions existed. (Chlorine is just above and iodine just below bromine on the periodic table.)

PROBLEM	SOLUTION
a. Write the unbalanced core equation illustrating the reaction between XeF_2 and bromate ion (BrO_3^-) to form Xe and the perbromate ion in aqueous solution.	**a.** $XeF_2(aq) + BrO_3^-(aq) \longrightarrow Xe(g) + BrO_4^-(aq)$

PROBLEM	SOLUTION
b. What is the substance oxidized and the substance reduced and their oxidation states in the reactants and products?	**b.** XeF_2 is reduced to Xe. The oxidation state of Xe in XeF_2 is +2 and is reduced to 0 in Xe. The oxidation state of the Br in BrO_3^- is +5 and is oxidized to +7 in BrO_4^-.
c. Balance the reaction in acid solution by the ion-electron method.	**c.** $2e^- + XeF_2 \longrightarrow Xe + 2F^-$ $H_2O + BrO_3^- \longrightarrow BrO_4^- + 2H^+ + 2e^-$ Add the two half-reactions so that the $2e^-$ cancel. balanced equation: $H_2O(l) + XeF_2(g) + BrO_3^-(aq) \longrightarrow$ $Xe(g) + BrO_4^-(aq) + 2HF(aq)$ (The HF is written in molecular form since it is a weak acid.)
d. Write the balanced equation as it would appear in basic solution.	**d.** $2OH^-(aq) + XeF_2(s) + BrO_3^-(aq) \longrightarrow$ $Xe(g) + BrO_4^-(aq) + 2F^-(aq) + H_2O(l)$ (Add two OH^- to each side. The two H^+ in HF neutralize two OH^- on the right to form $2H_2O$. Then cancel one H_2O on each side.
e. Elemental oxygen is a good oxidizing agent (hence the name) but XeF_2 is stronger. Write a balanced equation illustrating a spontaneous reaction involving the O_2–H_2O half-reaction and the XeF_2–Xe half-reaction.	**e.** $2XeF_2(g) + 2H_2O(l) \longrightarrow 2Xe(g) + O_2(g) + 4HF(aq)$

YOUR TURN

Bleach (hypochlorous acid, HClO) is a powerful oxidizing agent, which we use around the house to remove stains and to kill germs. We also use it in swimming pools to kill bacteria. We discussed the bromate ion in the problem above. It can be produced from bromide salts such as NaBr by reaction with hypochlorous acid, which forms chlorine gas on reaction.

    a. Write the unbalanced core equation illustrating the reaction between HOCl and the Br^- ion as described above.

    b. What is the substance oxidized and the substance reduced and their oxidation states in the reactants and products?

    c. Balance the reaction in acid solution by the ion-electron method.

    d. Write the balanced equation as it would appear in basic solution.

    e. Hypochlorous acid is a good oxidizing agent but hydrogen peroxide is stronger. (H_2O_2 is also used as a bleaching agent or antiseptic). Write a balanced equation in acid solution illustrating a spontaneous reaction involving the H_2O_2–H_2O half-reaction and the $HClO$–Cl_2 half-reaction.

Answers are on p. xxx.

CHAPTER SUMMARY

A common characteristic of a large number of chemical reactions is an exchange of electrons between reactants. These reactions are known as **oxidation–reduction reactions**, or simply **redox reactions**. In such reactions, the reactant that gives up or loses the electrons is **oxidized**, and the reactant that gains the electrons is **reduced**. The reactant oxidized is also known as a **reducing agent**, and the reactant reduced is known as an **oxidizing agent**.

To keep track of the electron exchange, we can follow the change in **oxidation states** (or **oxidation numbers**) of the elements in the compounds. The oxidation state increases in the substance oxidized and decreases in the substance reduced. We can use this understanding to balance equations by the **oxidation state (bridge) method**. In another method, a redox reaction can be divided into two **half-reactions**: an oxidation

and a reduction. These two processes take place so that all electrons lost in the oxidation process are gained in the reduction process. This fact is useful in balancing oxidation–reduction reactions by the **ion-electron method**. Most of these reactions would, at best, be very difficult to balance by inspection methods, as described in Chapter 6.

Each substance has its own inherent strength as either an oxidizing agent or a reducing agent. A table can be constructed in which oxidizing and reducing agents are ranked by strength as determined by observation or measurements with electrical instruments. From this table, a great many other spontaneous reactions can be predicted. In Table 14-1, stronger oxidizing agents are ranked higher on the left and stronger reducing agents are ranked lower on the right. Reactions can thus be predicted as shown in

Figure 14-2. Also, the reactions (or lack of reactions) of certain elements with water can be predicted.

Spontaneous chemical reactions occur because reactants are higher in chemical (potential) energy than products. We are familiar with many exothermic reactions, where the difference in energy is released as heat. The energy can also be released as electrical energy in a **voltaic cell**. In a voltaic cell, the two half-reactions are physically separated so that electrons travel in an external circuit or wire between **electrodes**. The **anode** is the electrode

at which oxidation takes place, and the **cathode** is the electrode at which reduction takes place. The Daniell cell, the car **battery**, the dry cell, and the **fuel cell** all involve spontaneous chemical reactions in which the chemical energy is converted directly into electrical energy.

Many reactions that are predicted to be unfavorable or nonspontaneous can be made to occur if electrical energy is supplied from an outside source. These are referred to as **electrolytic cells** and are useful in the commercial production of metals and in electroplating.

OBJECTIVES

SECTION	YOU SHOULD BE ABLE TO...	EXAMPLES	EXERCISES	CHAPTER PROBLEMS
14-1	Using oxidation states, determine the species oxidized, the species reduced, the oxidizing agent, and the reducing agent in an electron exchange reaction.	14-1, 14-2	1a, 1b, 1c, 1d, 1e, 1f, 1g	1, 2, 6, 8, 9, 11, 12, 14
14-2	Balance redox reactions by the oxidation state (bridge) method.	14-3	2a, 2b	16, 17
14-3	Balance redox reactions by the ion-electron (half-reaction) method in both acidic and basic media.	14-4, 14-5	3a, 3b	19, 21, 23, 25
14-4	Using a table or relative strengths of oxidizing agents, determine whether a specific redox reaction is spontaneous.	14-6, 14-7, 14-8	4a, 4b, 4c, 4d, 4e	27, 28, 30, 32
14-5	Describe the structure and electricity-generating ability of voltaic cells and batteries.		5a, 5c	35, 36, 37, 38, 39, 42
14-6	Write the reactions that occur when a salt is electrolyzed.		6a, 6b, 6c, 6d	43, 46, 47

▶ANSWERS TO ASSESSING THE OBJECTIVES

Part A

EXERCISES

14-1(a) If all the atoms in a compound were ions, the charge on the ions would be the same as their <u>oxidation</u> states. For oxygen in compounds, this is usually −2. Hydrogen is usually +1 in compounds. A substance oxidized undergoes a <u>loss</u> of electrons and an <u>increase</u> in oxidation state. This substance is also known as a <u>reducing</u> agent.

14-1(b) (a) +3 (b) +2

14-1(c) (a) +4 (b) −2 (c) +4 (d) 0

14-1(d) $ClO_2 \longrightarrow Cl^-$ ClO_2 is reduced and is the oxidizing agent. $H_2O_2 \longrightarrow O_2$ H_2O_2 is oxidized and is the reducing agent.

14-1(e) Both ions in CrI_3 are oxidized. The Cr is oxidized from +3 in Cr^{3+} to +6 in CrO_4^{2-}; the I is oxidized from −1 in I^- to 0 in I_2. The Cl is reduced from 0 in Cl_2 to −1 in Cl^-.

14-1(f) No, it is not. Reduction is the gain of electrons. They must come from somewhere. Oxidation is the loss of electrons. They must end up somewhere. The two processes always occur together. Certain reactions are referred to as being an oxidation or a reduction, but in these cases, it is because we focus on a molecule of interest. Other, less important molecules in these reactions are always undergoing the complementary process.

14-1(g) Mn is +7 (high of state). It would be an oxidizing agent.

14-2(a) (a) $2Fe(s) + 6HCl(aq) \longrightarrow 2FeCl_3(aq) + 3H_2(g)$
(b) $2HNO_3(aq) + 2HCl(aq) \longrightarrow 2NO_2(g) + Cl_2(g) + 2H_2O$

14-2(b) $2MnO_2(s) + Zn(s) + H_2O(l) \longrightarrow$
$$Mn_2O_3(s) + Zn(OH)_2(s)$$

14-3(a) (a) $2NO_3^-(aq) + 3H_2SO_3(aq) \longrightarrow$
$$3SO_4^{2-}(aq) + 2NO(g) + 4H^+(aq) + H_2O$$
(b) $8H_2O(l) + 2MnO_4^-(aq) + 7S^{2-}(aq) \longrightarrow$
$$2MnS(s) + 5S(s) + 16OH^-(aq)$$

14-3(b) $Sn(s) + 4HNO_3(aq) \longrightarrow SnO_2(s) + 4NO_2(g) + 2H_2O(l)$

Part B

EXERCISES

14-4(a) Spontaneous redox reactions occur between the <u>stronger</u> oxidizing and reducing agents to produce <u>weaker</u> oxidizing and reducing agents. In a table, the stronger oxidizing agent is located <u>higher</u> in the table than the reducing agent.

14-4(b) A spontaneous reaction would occur in (b).
$3Pb^{2+}(aq) + 2Cr(s) \longrightarrow 3Pb(s) + 2Cr^{3+}(aq)$

14-4(c) The zinc is the stronger reducing agent. It will be oxidized more easily.

14-4(d) (a) nonspontaneous (b) $2Cr^{3+} + 3Mn \longrightarrow 2Cr + 3Mn^{2+}$ (c) Yes. $3I_2 + 2Cr \longrightarrow 2Cr^{3+} + 6I^-$ (d) Only Ca (e) Ca^{2+} and I^-

14-4(e) $2Au^{3+}(aq) + 3Fe(s) \longrightarrow 3Fe^{2+}(aq) + 2Au(s)$

14-5(a) A spontaneous chemical reaction is used in a <u>voltaic</u> cell. An example is the Daniell cell, where <u>zinc metal</u> is oxidized at the <u>anode</u> and <u>copper ion</u> is reduced at the <u>cathode</u>. An electrolytic balance is maintained by means of a <u>salt bridge</u>. In the lead-acid battery, <u>lead–metal</u> is <u>oxidized</u> at the anode and PbO_2 is reduced at the <u>cathode</u>.

14-5(b) (a) lead storage battery (b) NiCad battery (c) dry cell (d) fuel cell

14-5(c) $Br_2(l) + Fe(s) \longrightarrow 2Br^-(aq) + Fe^{2+}(aq)$

14-5(d) There actually isn't one correct answer, since the factors involved sometimes conflict, such as mass of the reactants versus the voltage of the cell. The reaction shown above for Exercise 14-5(b) is one possibility because neither reactant reacts directly with water. However, bromine is very corrosive and causes severe burns. Although sodium metal is light and is the most powerful reducing agent, it is very difficult to use because it reacts violently

with water. Likewise, fluorine is almost impossible to use because of its high reactivity. It reacts with almost any container. This leaves magnesium, which is very light and stable, since it reacts with water only at high temperatures, and chlorine gas, which reacts very slowly with water. The reaction is

$$Mg(s) + Cl_2(aq) \longrightarrow Mg^{2+}(aq) + 2Cl^-(aq)$$

In fact, a workable battery has been made using Zn and Cl_2.

14-6(a) (1) electroplating metals like silver onto a surface (2) recharging batteries (3) producing fuels like hydrogen (4) producing reactive metals like aluminum and sodium

14-6(b) $MgCl_2(l) \longrightarrow Mg(l) + Cl_2(g)$

14-6(c) The formation of Sn from an Sn^{2+} solution onto a piece of iron is a spontaneous reaction, so energy does not need to be supplied. The spontaneous reaction is

$$Sn^{2+}(aq) + Fe \longrightarrow Sn(s) + Fe^{2+}(aq)$$

14-6(d) Aluminum or any other active metal that reacts with water, such as magnesium or sodium, cannot be electroplated from aqueous solution. All three of these metals are produced from electrolysis of their molten salts. Only metals above water (as a reducing agent) can be electroplated from solutions of their aqueous ions.

ANSWERS TO CHAPTER SYNTHESIS PROBLEM

a. $HClO(aq) + Br^-(aq) \longrightarrow$
$$BrO_3^-(aq) + Cl_2(g)$$

b. HClO is reduced to Cl_2. The oxidation state of Cl in HClO is +1 and is reduced to 0 in Cl_2. The oxidation state of the Br in Br^- is -1 and is oxidized to +5 in BrO_3^-.

c. $2e^- + 2H^+ + 2HClO \longrightarrow Cl_2 + 2H_2O$
$3H_2O + Br^- \longrightarrow BrO_3^- + 6H^+ + 6e^-$
Multiply the top reaction by three and add so that $6e^-$ cancel.
balanced equation:
$6HClO(aq) + Br^- \longrightarrow$
$$3Cl_2(g) + BrO_3^-(aq) + 3H_2O(l)$$

d. $3H_2O + 6ClO^-(aq) + Br^- \longrightarrow$
$$3Cl_2(g) + BrO_3^-(aq) + 6OH^-(aq)$$

e. $H_2O_2(aq) + Cl_2(aq) \longrightarrow 2HClO(aq)$

CHAPTER PROBLEMS

Throughout the text, answers to all exercises in color are given in Appendix E. The more difficult exercises are marked with an asterisk.

Oxidation States (SECTION 14-1)

14-1. Give the oxidation states of the elements in the following compounds.

(a) PbO_2 (c) C_2H_2 (e) LiH   (g) Rb_2Se

(b) P_4O_{10} (d) N_2H_4 (f) BCl_3 (h) Bi_2S_3

14-2. Give the oxidation states of the elements in the following compounds.

(a) ClO_2 (c) CO   (e) Mn_2O_3

(b) XeF_2 (d) O_2F_2 (f) Bi_2O_5

14-3. Which of the following elements form *only* the +1 oxidation state in compounds?

(a) Li   (c) Ca   (e) K   (g) Rb

(b) H   (d) Cl   (f) Al

14-4. Which of the following elements form *only* the +2 oxidation state in compounds?

(a) O   (c) Be   (e) Sc   (g) Ca

(b) B   (d) Sr   (f) Hg

14-5. What is the only oxidation state of Al in its compounds?

14-6. What is the oxidation state of each of the following?

(a) P in H_3PO_4 (e) S in SF_6

(b) C in $H_2C_2O_4$ (f) N in $CsNO_3$

(c) Cl in ClO_4^- (g) Mn in $KMnO_4$

(d) Cr in $CaCr_2O_7$

14-7. What is the oxidation state of the specified atom?

(a) S in SO_3 (b) Co in Co_2O_3

(c) U in UF_6

(d) N in HNO_3

(e) Cr in K_2CrO_4

(f) Mn in $CaMnO_4$

14-8. What is the oxidation state of each of the following?

(a) Se in SeO_3^{2-}

(b) I in H_5IO_6

(c) S in $Al_2(SO_3)_3$

(d) Cl in $HClO_2$

(e) N in $(NH_4)_2S$

Oxidation-Reduction (SECTION 14-1)

14-9. Which of the following reactions are oxidation-reduction reactions?

(a) $2H_2 + O_2 \longrightarrow 2H_2O$

(b) $CaCO_3 \longrightarrow CaO + CO_2$

(c) $2Na + 2H_2O \longrightarrow 2NaOH + H_2$

(d) $2HNO_3 + Ca(OH)_2 \longrightarrow Ca(NO_3)_2 + 2H_2O$

(e) $AgNO_3 + KCl \longrightarrow AgCl + KNO_3$

(f) $Zn + CuCl_2 \longrightarrow ZnCl_2 + Cu$

14-10. Identify each of the following half-reactions as either oxidation or reduction.

(a) $Na \longrightarrow Na^+ + e^-$

(b) $Zn^{2+} + 2e^- \longrightarrow Zn$

(c) $Fe^{2+} \longrightarrow Fe^{3+} + e^-$

(d) $O_2 + 4H^+ + 4e^- \longrightarrow 2H_2O$

(e) $S_2O_8^{2-} + 2e^- \longrightarrow 2SO_4^{2-}$

14-11. Identify each of the following changes as either oxidation or reduction.

(a) $P_4 \longrightarrow H_3PO_4$

(b) $NO_3^- \longrightarrow NH_4^+$

(c) $Fe_2O_3 \longrightarrow Fe^{2+}$

(d) $Al \longrightarrow Al(OH)_4^-$

(e) $S^{2-} \longrightarrow SO_4^{2-}$

14-12. For each of the following unbalanced equations, complete the table below.

(a) $MnO_2 + H^+ + Br^- \longrightarrow Mn^{2+} + Br_2 + H_2O$

(b) $CH_4 + O_2 \longrightarrow CO_2 + H_2O$

(c) $Fe^{2+} + MnO_4^- + H^+ \longrightarrow Fe^{3+} + Mn^{2+} + H_2O$

Reaction	Reactant Oxidized*	Product of Oxidation	Reactant Reduced	Product of Reduction	Oxidizing Agent	Reducing Agent
(a)						
(b)						
(c)						

*Element, molecule, or ion.

14-13. For the following two unbalanced equations, construct a table like that in Problem 14-12.

(a) $Al + H_2O \longrightarrow AlO_2^- + H_2 + H^+$

(b) $Mn^{2+} + Cr_2O_7^{2-} + H^+ \longrightarrow MnO_4^- + Cr^{3+} + H_2O$

14-14. For the following equations, identify the reactant oxidized, the reactant reduced, the oxidizing agent, and the reducing agent.

(a) $Sn + HNO_3 \longrightarrow SnO_2 + NO_2 + H_2O$

(b) $IO_3^- + SO_2 + H_2O \longrightarrow I_2 + SO_4^{2-} + H^+$

(c) $CrI_3 + OH^- + Cl_2 \longrightarrow CrO_4^{2-} + IO_4^- + Cl^- + H_2O$

(d) $I^- + H_2O_2 \longrightarrow I_2 + H_2O + OH^-$

14-15. Identify the product or products containing the elements oxidized and reduced in problem 14-14.

Balancing Equations by the Bridge Method (SECTION 14-2)

14-16. Balance each of the following equations by the oxidation state method.

(a) $NH_3 + O_2 \longrightarrow NO + H_2O$

(b) $Sn + HNO_3 \longrightarrow SnO_2 + NO_2 + H_2O$

(c) $Cr_2O_3 + Na_2CO_3 + KNO_3 \longrightarrow CO_2 + Na_2CrO_4 + KNO_2$

(d) $Se + BrO_3^- + H_2O \longrightarrow H_2SeO_3 + Br^-$

14-17. Balance the following equations by the oxidation state method.

(a) $I_2O_5 + CO \longrightarrow I_2 + CO_2$

(b) $Al + H_2O \longrightarrow AlO_2^- + H_2 + H^+$

(c) $HNO_3 + HCl \longrightarrow NO + Cl_2 + H_2O$

(d) $I_2 + Cl_2 + H_2O \longrightarrow HIO_3 + HCl$

Balancing Equations by the Ion-Electron Method (SECTION 14-3)

14-18. Balance the following half-reactions in acidic solution.

(a) $Sn^{2+} \longrightarrow SnO_2$

(b) $CH_4 \longrightarrow CO_2$

(c) $Fe^{3+} \longrightarrow Fe^{2+}$

(d) $I_2 \longrightarrow IO_3^-$

(e) $NO_3^- \longrightarrow NO_2$

14-19. Balance the following half-reactions in acidic solution.

(a) $P_4 \longrightarrow H_3PO_4$

(b) $ClO_3^- \longrightarrow Cl^-$

(c) $S_2O_3^{2-} \longrightarrow SO_4^{2-}$

(d) $NO_3^- \longrightarrow NH_4^+$

(e) $H_2O_2 \longrightarrow H_2O$

14-20. Balance each of the following by the ion-electron method. All are in acidic solution.

(a) $S^{2-} + NO_3^- + H^+ \longrightarrow S + NO + H_2O$

(b) $I_2 + S_2O_3^{2-} \longrightarrow S_4O_6^{2-} + I^-$

(c) $SO_3^{2-} + ClO_3^- \longrightarrow Cl^- + SO_4^{2-}$

(d) $Fe^{2+} + H_2O_2 + H^+ \longrightarrow Fe^{3+} + H_2O$

(e) $AsO_4^{3-} + I^- + H^+ \longrightarrow I_2 + AsO_3^{3-} + H_2O$

(f) $Zn + H^+ + NO_3^- \longrightarrow Zn^{2+} + NH_4^+ + H_2O$

14-21. Balance each of the following by the ion-electron method. All are in acidic solution.

(a) $Mn^{2+} + BiO_3^- + H^+ \longrightarrow MnO_4^- + Bi^{3+} + H_2O$

(b) $IO_3^- + SO_2 + H_2O \longrightarrow I_2 + SO_4^{2-} + H^+$

(c) $Se + BrO_3^- + H_2O \longrightarrow H_2SeO_3 + Br^-$

(d) $P_4 + HClO + H_2O \longrightarrow H_3PO_4 + Cl^- + H^+$

(e) $Al + Cr_2O_7^{2-} + H^+ \longrightarrow Al^{3+} + Cr^{3+} + H_2O$

(f) $ClO_3^- + I^- + H^+ \longrightarrow Cl^- + I_2 + H_2O$

(g) $As_2O_3 + NO_3^- + H_2O \longrightarrow AsO_4^{3-} + NO + H^+$

14-22. Balance the following half-reactions in basic solution.

(a) $SnO_2^{2-} \longrightarrow SnO_3^{2-}$ **(c)** $Si \longrightarrow SiO_3^{2-}$

(b) $ClO_2^- \longrightarrow Cl_2$ **(d)** $NO_3^- \longrightarrow NH_3$

14-23. Balance the following half-reactions in basic solution.

(a) $Al \longrightarrow Al(OH)_4^-$ **(c)** $N_2H_4 \longrightarrow NO_3^-$

(b) $S^{2-} \longrightarrow SO_4^{2-}$

14-24. Balance each of the following by the ion-electron method. All are in basic solution.

(a) $S^{2-} + OH^- + I_2 \longrightarrow SO_4^{2-} + I^- + H_2O$

(b) $MnO_4^- + OH^- + I^- \longrightarrow MnO_4^{2-} + IO_4^- + H_2O$

(c) $BiO_3^- + SnO_2^{2-} + H_2O \longrightarrow SnO_3^{2-} + OH^- + Bi(OH)_3$

(d) $CrI_3 + OH^- + Cl_2 \longrightarrow CrO_4^{2-} + IO_4^- + Cl^- + H_2O$

[Hint: In (d), two ions are oxidized; include both in one half-reaction.]

14-25. Balance each of the following by the ion-electron method. All are in basic solution.

(a) $ClO_2 + OH^- \longrightarrow ClO_2^- + ClO_3^- + H_2O$

(b) $OH^- + Cr_2O_3 + NO_3^- \longrightarrow CrO_4^{2-} + NO_2^- + H_2O$

(c) $Cr(OH)_4^- + BrO^- + OH^- \longrightarrow Br^- + CrO_4^{2-} + H_2O$

(d) $Mn^{2+} + H_2O_2 + OH^- \longrightarrow H_2O + MnO_2$

(e) $Ag_2O + Zn + H_2O \longrightarrow Zn(OH)_2 + Ag$

14-26. Balance the following two equations by the ion-electron method, first in acidic solution and then in basic solution.

(a) $H_2 + O_2 \longrightarrow H_2O$ **(b)** $H_2O_2 \longrightarrow O_2 + H_2O$

Predicting Redox Reactions (SECTION 14-4)

14-27. Using Table 14-1, predict whether the following reactions occur in aqueous solution. If not, write N.R. (no reaction).

(a) $2Na + 2H_2O \longrightarrow H_2 + 2NaOH$

(b) $Pb + Zn^{2+} \longrightarrow Pb^{2+} + Zn$

(c) $Fe + 2H^+ \longrightarrow Fe^{2+} + H_2$

(d) $Fe + 2H_2O \longrightarrow Fe^{2+} + 2OH^- + H_2$

(e) $Cu + 2Ag^+ \longrightarrow 2Ag + Cu^{2+}$

(f) $2Cl_2 + 2H_2O \longrightarrow 4Cl^- + O_2 + 4H^+$

(g) $3Zn^{2+} + 2Cr \longrightarrow 2Cr^{3+} + 3Zn$

14-28. Using Table 14-1, predict whether the following reactions occur in aqueous solution. If not, write N.R.

(a) $Sn^{2+} + Pb \longrightarrow Pb^{2+} + Sn$

(b) $Ni^{2+} + H_2 \longrightarrow 2H^+ + Ni$

(c) $Cu + F_2 \longrightarrow CuF_2$

(d) $Ni^{2+} + 2Br^- \longrightarrow Ni + Br_2$

(e) $3Ni^{2+} + 2Cr \longrightarrow 2Cr^{3+} + 3Ni$

(f) $2Br_2 + 2H_2O \longrightarrow 4Br^- + O_2 + 4H^+$

14-29. If a reaction occurs, write the balanced molecular equation.

(a) Nickel metal is placed in water.

(b) Bromine is dissolved in water that is in contact with tin metal.

(c) Silver metal is placed in an $HClO_4$ solution.

(d) Oxygen gas is bubbled into an HBr solution.

(e) Liquid bromine is placed on a sheet of aluminum.

14-30. If a reaction occurs, write the balanced molecular equation.

(a) Bromine is added to an HCl solution.

(b) Sodium metal is heated with solid aluminum chloride.

(c) Iron is placed in a $Pb(ClO_4)_2$ solution.

(d) Oxygen gas is bubbled into an HCl solution.

(e) Fluorine gas is added to water.

14-31. Which of the following elements react with water: **(a)** Pb, **(b)** Ag, **(c)** F_2, **(d)** Br_2, **(e)** Mg? Write the balanced equation for any reaction that occurs.

14-32. Which of the following species will be reduced by hydrogen gas in aqueous solution: **(a)** Br_2, **(b)** Cr, **(c)** Ag^+, **(d)** Ni^{2+}? Write the balanced equation for any reaction that occurs.

14-33. In Chapter 13, we mentioned the corrosiveness of acid rain. Why does rain containing a higher $H^+(aq)$ concentration cause more damage to iron exposed in bridges and buildings than pure H_2O? Write the reaction between Fe and $H^+(aq)$

14-34. Br_2 can be prepared from the reaction of Cl_2 with NaBr dissolved in seawater. Explain. Write the reaction. Can Cl_2 be used to prepare F_2 from NaF solutions?

Voltaic Cells (SECTION 14-5)

14-35. What is the function of the salt bridge in the voltaic cell?

14-36. In an alkaline battery, the following two half-reactions occur.

$$Zn(s) + 2OH^-(aq) \longrightarrow Zn(OH)_2(s) + 2e^-$$
$$2MnO_2(s) + 2H_2O(l) + 2e^- \longrightarrow 2MnO(OH)(s) + 2OH^-(aq)$$

Which reaction takes place at the anode and which at the cathode? What is the total reaction?

14-37. The following overall reaction takes place in a silver oxide battery.

$$Ag_2O(s) + H_2O(l) + Zn(s) \longrightarrow Zn(OH)_2(s) + 2Ag(s)$$

The reaction takes place in basic solution. Write the half-reaction that takes place at the anode and the half-reaction that takes place at the cathode.

14-38. The nickel–cadmium (NiCad) battery is used as a replacement for a dry cell because it is rechargeable. The overall reaction that takes place is

$$2NiO(OH)(s) + Cd(s) + 2H_2O(l) \longrightarrow$$
$$2Ni(OH)_2(s) + Cd(OH)_2(s)$$

Write the half-reactions that take place at the anode and the cathode.

14-39. Sketch a galvanic cell in which the following overall reaction occurs.

$$Ni^{2+}(aq) + Fe(s) \longrightarrow Fe^{2+}(aq) + Ni(s)$$

(a) What reactions take place at the anode and the cathode?

(b) In what direction do the electrons flow in the wire?

(c) In what direction do the anions flow in the salt bridge?

14-40. Describe how a voltaic cell could be constructed from a strip of iron, a strip of lead, an $Fe(NO_3)_2$ solution, and a $Pb(NO_3)_2$ solution. Write the anode reaction, the cathode reaction, and the total reaction.

14-41. Judging from the relative difference in the strengths of the oxidizing agents (Fe^{2+} vs. Pb^{2+}) and (Zn^{2+} vs. Cu^{2+}), which do you think would be the more powerful cell, the one in problem or the Daniell cell? Why?

*14-42. The power of a cell depends on the strength of both the oxidizing and the reducing agents. Write the equation illustrating the most powerful redox reaction possible between an oxidizing agent and a reducing agent *in aqueous solution.* Consider only the species shown in Table 14-1.

Electrolytic Cells (SECTION 14-6)

14-43. Chrome plating is an electrolytic process. Write the reaction that occurs when an iron bumper is electroplated using a $CrCl_3$ solution. Are there any metals shown in Table 14-1 on which a chromium layer would spontaneously form?

14-44. Why can't elemental sodium be formed in the electrolysis of an aqueous NaCl solution? Write the reaction that does occur at the cathode. How is elemental sodium produced by electrolysis?

14-45. Why can't elemental fluorine be formed by electrolysis of an aqueous NaF solution? Write the reaction that does occur at the anode. How is elemental fluorine produced?

14-46. A "tin can" is made by forming a layer of tin on a sheet of iron. Is electrolysis necessary for such a process or does it occur spontaneously? Write the equation for this reaction. Is electrolysis necessary to form a layer of tin on a sheet of lead? Write the relevant equation.

14-47. Certain metals can be purified by electrolysis. For example, a mixture of Ag, Zn, and Fe can be dissolved so that their metal ions are present in aqueous solution. If a solution containing these ions is electrolyzed, which metal ion would be reduced to the metal first?

General Problems

14-48. Nitrogen exists in nine oxidation states. Arrange the following compounds in order of increasing oxidation state of N: K_3N, N_2O_4, N_2, NH_2OH, N_2O, $Ca(NO_3)_2$, N_2H_4, N_2O_3, NO.

14-49. Given the following information concerning metal strips immersed in certain solutions, write the net ionic equations representing the reactions that occur.

Metal Strip	Solution	Reaction
Cd	$NiCl_2$	Ni coating formed
Cd	$FeCl_2$	no reaction
Zn	$CdCl_2$	Cd coating formed
Fe	$CdCl_2$	no reaction

Where does Cd^{2+} rank as an oxidizing agent in Table 14-1?

14-50. A hypothetical metal (M) forms a coating of Sn when placed in an $SnCl_2$ solution. However, when a strip of Ni is placed in an MCl_2 solution, a coating of the metal M forms on the nickel. Write the net ionic equations representing the reactions that occur. Where does M^{2+} rank as an oxidizing agent in Table 14-1?

14-51. A solution of gold ions (Au^{3+}) reacts spontaneously with water to form metallic gold. Metallic gold does not react with chlorine but does react with fluorine. Write the equations representing the two spontaneous reactions, and locate Au^{3+} in Table 14-1 as an oxidizing agent.

14-52. Given the following *unbalanced* equation

$$H^+(aq) + Zn(s) + NO_3^-(aq) \longrightarrow$$
$$Zn^{2+}(aq) + N_2(g) + H_2O(l)$$

what mass of Zn is required to produce 0.658 g of N_2?

14-53. Given the following *unbalanced* equation

$$MnO_2(s) + HBr(aq) \longrightarrow MnBr_2(aq) + Br_2(l) + H_2O(l)$$

what mass of MnO_2 reacts with 228 mL of 0.560 M HBr?

14-54. Given the following *unbalanced* equation

$$H^+(aq) + NO_3^-(aq) + Cu_2O(s) \longrightarrow$$
$$Cu^{2+}(aq) + NO(g) + H_2O(l)$$

what volume of NO gas measured at STP is produced by the complete reaction of 10.0 g of Cu_2O?

*14-55. Given the following *unbalanced* equation in acid solution

$$H_2O(l) + HClO_3(aq) + As(s) \longrightarrow H_3AsO_3(aq) + HClO(aq)$$

If 200 g of As reacts with 200 g of $HClO_3$, what mass of H_3AsO_3 is produced? *(Hint: Calculate the limiting reactant.)*

*14-56. Given the following *unbalanced* equation in basic solution

$$Zn(s) + NO_3^-(aq) \longrightarrow NH_3(g) + Zn(OH)_4^{2-}(aq)$$

what volume of NH_3 is produced by 6.54 g of Zn? The NH_3 is measured at 27.0°C and 1.25 atm pressure.

14-57. Solutions of potassium permanganate are a deep-purple color. Permanganate is a strong oxidizing agent that forms the Mn^{2+} ion in acid solution when it is reduced. When these purple solutions are added to a reducing agent, the purple color disappears until all of the reducing agent reacts. So we have a very convenient way to know when the reducing agent is used up—the solution suddenly turns purple. Balance the following net ionic equations involving permanganate in acidic solution.

(a) $MnO_4^-(aq) + Fe^{2+}(aq) \longrightarrow Fe^{3+}(aq) + Mn^{2+}(aq)$

(b) $MnO_4^-(aq) + Br^- \longrightarrow Br_2(l) + Mn^{2+}(aq)$

(c) $MnO_4^-(aq) + C_2O_4^{2-}(aq) \longrightarrow CO_2(g) + Mn^{2+}(aq)$

14-58. Using the balanced equation from problem 14-57, determine the volume (in mL) of 0.220 M $KMnO_4$ needed to completely react with 25.0 g of $FeCl_2$ dissolved in water.

14-59. Using the balanced equation from problem 14-57, determine the volume (in mL) of 0.450 M KBr needed to completely react with 125 mL of 0.220 M $KMnO_4$.

14-60. The active ingredient in household bleach is sodium hypochlorite, which is a strong oxidizing agent. We can tell how much of the active ingredient is present in a two-step analysis. First the hypochlorite oxidizes excess iodide to elemental iodine according to the unbalanced equation that occurs in basic solution.

$$ClO^-(aq) + I^-(aq) \longrightarrow I_2(s) + Cl^-(aq)$$

Next, a solution of sodium thiosulfate ($Na_2S_2O_3$) is added to react with the iodine, as illustrated by the equation

$$I_2(s) + S_2O_3^{2-}(aq) \longrightarrow I^-(aq) + S_4O_6^{2-}(aq)$$

Balance the two equations.

*14-61. Household bleach is 5.00% by weight sodium hypochlorite. Using the balanced equations from problem 14-60, determine the volume (in mL) of 0.358 M $Na_2S_2O_3$ solution needed to react with all the sodium hypochlorite in 100 mL of household bleach. Consider the density of bleach to be the same as water.

STUDENT WORKSHOP

Balancing Redox Reactions

Purpose: To practice balancing a complex redox reaction by the ion-electric method. (Work in groups of three or four. Estimated time: 10 min.)

This is a challenging problem about balancing redox reactions; it is best done as a group effort to make sure that no atoms or charges are missed. Balance the following reaction in a basic solution, using the ion-electron method:

$$Ce^{4+} + Fe(CN)_6^{4-} \longrightarrow Ce(OH)_3 + Fe(OH)_3 + CO_3^{2-} + NO_3^-$$

Once completed, do an atom and charge inventory to confirm that it is balanced. Then identify the oxidizing and reducing agents for the reaction.

Chapter
15

Reaction Rates and Equilibrium

Because water enters this small pond at the same rate that it leaves, the level of the pond stays the same. This is similar to the equilibrium that occurs in certain chemical reactions.

SETTING THE STAGE

Many of us have probably neglected to put some food back into the refrigerator because it was hidden behind something or other. Unfortunately, in a matter of a day or two, mold or some bad-smelling, green life-form starts to grow on the food. This should lead to its quick disposal. In fact, the decay and formation of mold that we are witnessing on natural food products are chemical reactions that occur rapidly at room temperature. Had we remembered to put the food item in the refrigerator, however, these nasty chemical reactions could have been stopped or at least delayed for weeks. Decay, growth of mold or other unwanted life-forms, and, in fact, all chemical reactions slow down as the temperature is lowered. An understanding of how chemical reactions occur, as we will discuss in this chapter, can help us understand why temperature affects the rate of a reaction.

Some of the chemical reactions that are discussed in this chapter reach a point of "balance" between reactants and products. This is like the balance that many of us strive for between school or job and the need to relax and have fun. Or perhaps we seek the balance of having money come in at least as fast as it goes out. In a chemical sense, balance means an equilibrium between two competing reactions, one forming products and the other, in the reverse direction, reforming the original reactants.

In Part A in this chapter, we will examine the path of chemical reactions from reactants to products and then discuss reversible reactions that reach a point of equilibrium. In Part B, we will emphasize the quantitative calculations associated with the distribution between reactants and products in reactions that reach a point of equilibrium.

Part A

Collisions of Molecules and Reactions at Equilibrium

SETTING A GOAL

- You will learn how chemical reactions occur as well as the factors that affect how fast reactions occur and how much product they produce.

OBJECTIVES

15-1 Describe how a chemical reaction takes place on a molecular level by using an energy diagram.

15-2 List the factors that affect the rate of a chemical reaction.

15-3 Using Le Châtelier's principle, predict what effect changing certain conditions will have on the equilibrium point.

▶ **OBJECTIVE**
FOR SECTION 15-1
Describe how a chemical reaction takes place on a molecular level by using an energy diagram.

15-1 How Reactions Take Place

LOOKING AHEAD! Chemical reactions don't just happen. Somehow, reactant molecules or ions must find a way to come together so that the atoms can be reshuffled into product molecules or ions. The concept of how this happens is surprisingly straightforward and is the topic of this section. ∎

15-1.1 Collision Theory

Taking a trip from Los Angeles to Chicago raises an obvious question. How are you going to get there? One could drive, fly, or take a train. In any case, we can't get from point A to point B without some means of transportation. Likewise, chemical reactions must have some way for reactant molecules to be transformed into product molecules. The chemist's view of how chemical reactions occur is actually quite simple. Since molecules have kinetic energy, they are all in constant motion in one way or another. In the gaseous and liquid states, this motion leads to frequent collisions between molecules. These collisions may lead to chemical changes. *The assumption that chemical reactions are due to the collisions of molecules is known as* **collision theory**. In some collisions chemical bonds are broken in the colliding molecules. If the molecular or atomic fragments created from the collision reform in a different arrangement, a chemical reaction results. We will illustrate collision theory by means of a hypothetical chemical reaction that can be custom-designed to clearly illustrate the basic principles involved. Real-life reactions are generally more complicated, and explanations are required for each complication or exception, so we will keep it simple.

Our reaction involves the combination of two hypothetical diatomic gaseous elements, A_2 and B_2, to form two molecules of a gaseous product, AB. The reaction is illustrated by the equation

$$A_2(g) + B_2(g) \longrightarrow 2AB(g)$$

A reasonable assumption here is that the products form from the collision of an A_2 molecule with a B_2 molecule. If all collisions led to products, however, this and all other reactions would be essentially instantaneous because of the large number of collisions per second. However, two conditions are necessary for a collision to lead to the formation of products. First, *the collision between two reactant molecules must take place in the right geometric orientation*. As illustrated in Figure 15-1, for new bonds to form, the two reactant molecules must meet in a side-to-side manner rather than end to end or side to end. This condition, by itself, severely limits the number of collisions that can lead to a reaction. A second condition is that *the collision must occur with enough energy to break the bonds in the reactants so that new bonds can form in the products.* The second condition even more severely limits the number of collisions that lead to product formation. If the two conditions are not met, the colliding molecules simply recoil from each other unchanged. (See Figure 15-1.)

15-1.2 Activation Energy

Imagine that you are trying to roll a bowling ball over a small incline. When you swing the ball behind you and then release it, you are imparting motion (kinetic energy) to the ball. As it starts up the hill, however, it begins to slow. It slows because the kinetic energy is being converted into potential energy as a result of the increased elevation of the ball. If you swing the ball hard enough, it may have enough kinetic energy to overcome the potential energy barrier of the hill. In that case, it will make it to the top of the hill and go down the other side. If not, the ball will stop before reaching the top when all the kinetic energy has been converted into potential energy; the ball will then roll back at you as the potential energy converts back to the kinetic energy. (See Figure 15-2.)

Chemical reactions also have an energy barrier. *The minimum kinetic energy needed for the reaction to occur is known as the* **activation energy**. If properly colliding molecules have kinetic energy equal to or greater than the activation energy (which represents the potential energy barrier), a reaction occurs; if not, they recoil unchanged.

15-1.3 The Course of a Reaction

To illustrate what happens in a chemical reaction, we will follow the fate of the two hypothetical diatomic elements A_2 and B_2 as they move from being far apart, to colliding with each other in a proper orientation that has enough energy for the reaction, and finally to forming products. Recall that energy is conserved, meaning that the sum of the kinetic energy and potential energy of the molecules remains the same throughout

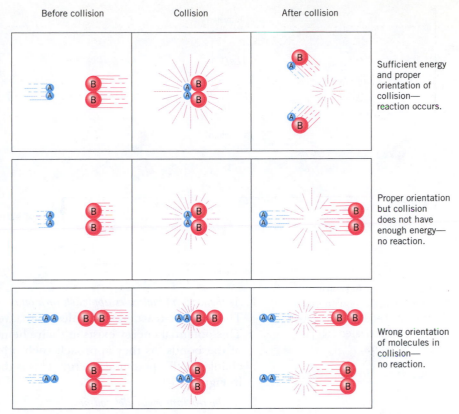

Before collision | Collision | After collision

Sufficient energy and proper orientation of collision—reaction occurs.

Proper orientation but collision does not have enough energy—no reaction.

Wrong orientation of molecules in collision—no reaction.

FIGURE 15-1 Reaction of A_2 and B_2 Reactions occur only when colliding molecules have the minimum amount of energy and the right orientation.

FIGURE 15-2 The Potential Energy Barrier On the left, the ball does not have enough kinetic energy to overcome the potential energy barrier. On the right, the ball has enough kinetic energy to make it over.

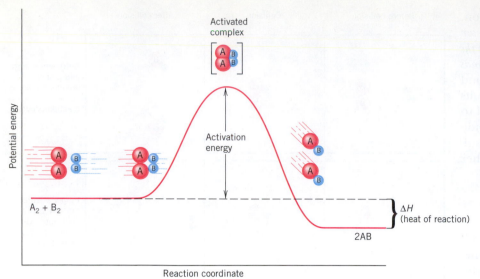

FIGURE 15-3 Activation
Energy The activation energy is
the potential energy difference
between the activated complex
and the reactants.

the reaction. In Figure 15-3, we
attempt to illustrate the progress
of the reaction as the molecules
collide. The curve represents the
potential energy (on the vertical
axis) of the species involved in the
reaction. This horizontal axis is
referred to as the *reaction coordinate*
(also called the *path of the reaction*)
and can be viewed as a time axis.
We will follow the reaction in four
stages: (1) as the reactant molecules approach each other, (2)
as the reactant molecules collide,
(3) as the reaction occurs, and
finally, (4) as the product molecules recoil.

1. *As the reactant molecules approach each other*
   The separate reactant molecules each have a specific amount of potential energy.
   This potential energy exists in their chemical bonds and depends on the nature
   of the bonds. As they approach each other but before they collide there is no
   change in this potential energy. This is the flat part of the curve on the left side
   in Figure 15-3.

2. *As the reactant molecules collide*
   The potential energy of the molecules does not change until they make contact
   and begin to compress together. As they collide, the molecules lose velocity,
   which means they lose kinetic energy. This kinetic energy changes to potential
   energy as the molecules compress, so the potential energy of the compressing
   molecules rises as shown by the ascending curve on the left. The collision is analogous to what happens when we hit a tennis ball with a racquet. As the ball makes
   contact with the racquet, it slows down and compresses. The kinetic energy of
   motion is converted to potential energy in the compressed ball.

3. *As the reaction occurs*
   At some point during the collision, the compression reaches a maximum where
   the potential energy reaches its highest point and the kinetic energy is zero (i.e.,
   the motion has stopped). In the tennis ball analogy, this is the instant when the
   tennis ball has no motion but has maximum potential energy from compression.
   At maximum impact, both molecules are compressed together, forming what is known as an
   activated complex. Since we are assuming a reaction takes place, the activated complex represents the state that is intermediate between reactants and products. That
   is, old bonds are partially broken and new ones are partially formed. The activation energy for the forward reaction is represented by the difference in potential
   energy of the activated complex (at the peak of the curve) and the reactants.

4. *As the product molecules recoil*
   After new bonds have formed, product molecules recoil and the potential energy
   decreases as it is converted back into the kinetic energy of the recoiling molecules. In our tennis ball analogy, as the ball recoils, the potential energy of the
   compressed ball reconverts to kinetic energy of motion and the ball speeds away
   from the racquet.

15-1.4 The Heat of the Reaction

Notice in Figure 15-3 that the product molecules (AB) in this example end up
with lower potential energy than the original reactants. This results from some of

At the instant the tennis ball is not
moving, all the energy is in the form
of potential energy.

the potential energy in the original chemical bonds (i.e., in A_2 and B_2) being released as extra kinetic energy to the product molecules (i.e., 2AB). Since the product molecules have more kinetic energy than the original reactants, the product molecules move faster and are "hotter."

The difference between the potential energy of the products and that of the reactants represents the *heat of the reaction, or enthalpy,* and is given the symbol Δ*H. In an exothermic process, the potential energy of the products is lower than the potential energy of the reactants, the difference being the heat that is released.* In our tennis ball analogy, we wish to have the ball recoil at a faster rate by giving the ball some "heat."

When the potential energy of the products is higher than that of the reactants, the reaction is endothermic and the product molecules are "cooler." Heat is required for the reaction to occur and is extracted from the environment. An actual endothermic reaction between N_2 and O_2 to form NO is represented in Figure 15-4.

Most spontaneous reactions are exothermic. For example, blood sugar or glucose ($C_6H_{12}O_6$) undergoes combustion in the body as shown by the following equation.

$$C_6H_{12}O_6 + 6O_2 \longrightarrow 6CO_2 + 6H_2O \quad \Delta H = -2529 \text{ kJ}$$

The sum of all of the bond energies in the glucose and molecular oxygen reactants is less than the sum of the bond energies in the CO_2 and H_2O products. In other words, the bonds in the products are stronger (in total) than in the reactants. The difference in the total bond energies of reactants and products is the source of the enthalpy (Δ*H*) of the reaction. In the process of photosynthesis, chlorophyll from plants converts this same amount of light energy from the sun, CO_2 from the air, and H_2O from the rain back into glucose or other carbohydrates. The reaction also recycles oxygen back into the air.

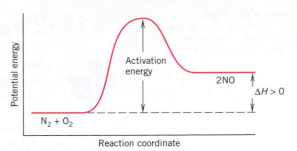

FIGURE 15-4 An Endothermic Reaction In an endothermic reaction, the potential energy of the reactants is less than that of the products.

▶ **ASSESSING THE OBJECTIVE FOR SECTION 15-1**

EXERCISE 15-1(a) LEVEL 1: Fill in the blanks.
Chemical reactions occur because of _____ between reactant molecules. In order for a reaction to take place, molecules must _____ in the proper orientation and with a _____ amount of energy, known as the _____ energy. An _____ forms at the point of maximum potential energy.

EXERCISE 15-1(b) LEVEL 2: In the reaction between SO_3 and H_2O to form H_2SO_4, the product molecule has less potential energy than the reactant molecules. Draw the energy diagram for this reaction.

EXERCISE 15-1(c) LEVEL 3: A hypothetical reaction is endothermic by 25 kJ/mol, and has an activation energy of 50 kJ/mol. Sketch the energy diagram, and label these values.

EXERCISE 15-1(d) LEVEL 3: SO_3 molecules collide with H_2O molecules to produce an activated complex that eventually leads to sulfuric acid. Draw the Lewis structures of SO_2 and H_2O. Show what the activated complex from collision would most likely look like. What rearrangements must the activated complex undergo to form the final product molecule?

For additional practice, work chapter problems 15-1 and 15-2.

▶ **OBJECTIVE
FOR SECTION 15-2**
List the factors that affect the
rate of a chemical reaction.

15-2 Rates of Chemical Reactions

LOOKING AHEAD! Some reactions are essentially instantaneous, such as in an explosion, whereas others may take months, such as the rusting of a nail. Why the big range? We are now ready to examine the factors that affect the rate of a reaction. ■

In the previous section, we asked how one might travel from Los Angeles to Chicago. The next obvious question involves how fast you want to get there. If you fly, you may average 500 mi/hr, but if you drive you may average only 50 mi/hr. We measure the rate of travel by distance per unit of time, which in this case is miles or kilometers per hour. Chemical reactions also take place at specific rates. As any chemical reaction occurs, reactants disappear and products appear. *The* **rate of a reaction** *measures the increase in concentration of a product or the decrease in concentration of a reactant per unit of time.* There are several factors that affect the rate of a reaction. These include the magnitude of the activation energy, the temperature at which the reaction takes place, the concentration of reactants, the size of the particles of solid reactants, and the presence of a catalyst. We will discuss these factors individually.

15-2.1 The Magnitude of the Activation Energy

Perhaps one of the most significant characteristics of any chemical reaction is its activation energy, which is determined by the nature of the reactants. Consider, for example, the reactions of two nonmetals with oxygen. An allotrope of phosphorus known as white phosphorus reacts almost instantly with the oxygen in the air, even at room temperature, producing an extremely hot flame. As a result, this form of phosphorus must be stored under water to prevent exposure to the air. Hydrogen also reacts with oxygen to form water. This is also a highly exothermic reaction, one that is used in space rockets. A mixture of hydrogen and oxygen, however, does not react at room temperature. In fact, no appreciable reaction occurs unless the temperature is raised to at least 400°C or the mixture is ignited with a spark. The difference in the rates of combustion of these two elements at room temperature lies in the activation energies of the two reactions. Figure 15-5 illustrates the activation energies (labeled E_a) for the combustion reactions of hydrogen and phosphorus. Note that the phosphorus reaction has a much lower activation energy than the hydrogen reaction. This explains why collisions between phosphorus molecules and oxygen have enough energy for reaction at room temperature, whereas collisions of the same energy between hydrogen and oxygen do not have the required energy to overcome the activation energy barrier. If one is rolling a ball over a hill, it is obviously easier to roll it over a small hill than a large hill.

White phosphorus ignites instantly in air.

FIGURE 15-5 Combustion P₄ and H₂ P_4 undergoes combustion at room temperature because of its lower activation energy.

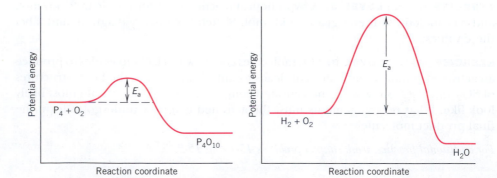

15-2.2 The Temperature

We have to be very careful about the chicken salad on a hot summer day. Food spoilage occurs amazingly fast at a warm temperature. In fact, all chemical reactions occur faster on hot summer days. Previously, we established that the rate of a reaction depends on the kinetic energy of the reactant molecules.

r(rate) $\propto$ energy of colliding molecules

In Chapter 10 we discussed the distribution of kinetic energies of molecules at a given temperature. This is again illustrated in Figure 15-6. Notice that, since temperature is related to average kinetic energy, the higher the temperature, the higher the average kinetic energy of the molecules. In the figure, the dashed line represents the activation energy (E_a). As mentioned, this is the minimum amount of kinetic energy that colliding molecules must have for products to form. Note that at the higher temperature (T_2) the curve intersects the dashed line at a higher point than does T_1, meaning that a greater fraction of molecules can overcome the activation energy in a collision. As a result, the rates of chemical reactions increase as the temperature increases.

There is another reason why an increase in temperature increases the rate of a reaction. In addition to the energy of the collisions, the rate of a reaction depends on the frequency of collisions (the number of collisions per second).

r(rate) $\propto$ frequency of collisions

Recall that the kinetic energy (K.E.) of a moving object is given by the relationship

$$\text{K.E.} = \frac{1}{2}mv^2 \ (m = \text{mass}, v = \text{velocity})$$

This indicates that the more kinetic energy a moving object has, the faster it is moving.

As the temperature increases, the average velocity of the molecules increases. This means that the frequency of collisions also increases. As an analogy, imagine a box containing red and blue Ping-Pong balls, as shown in Figure 15-7. If we jiggle the box, the balls move around and collide. If we jiggle the balls faster (analogous to a higher temperature), the balls move around and there are more frequent collisions. The increased noise we hear tells us that, indeed, collisions are not only more energetic but also more frequent. Most of the increase in the rate of a chemical reaction is due to the increased energy of the collisions, with a lesser contribution from the increased rate of collisions. A good rule of thumb for many typical chemical reactions is that the rate of reaction approximately doubles for each 10°C rise in temperature.

15-2.3 The Concentrations of Reactants

Increasing the rate at which we shake a box of Ping-Pong balls obviously increases the rate of collisions between the balls. Another way of increasing the rate of collisions is to increase the number of balls in the box. In Figure 15-8, we have two situations. In the box on top, there are four red and four blue balls. In the box at the bottom, there are four red balls, but the number of blue balls has been increased to eight. The concentration (number of balls per unit volume) has been increased. Increasing the number of balls of either color increases the number of blue–red collisions, which is signaled by the more intense noise even though the box is jiggled at the same rate (analogous to the same temperature). In the reaction of A_2 and B_2, the same phenomenon applies. The

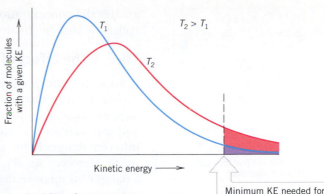

Minimum KE needed for reaction to occur

FIGURE 15-6 Kinetic Energy Distributions for a Reaction Mixture at Two Different Temperatures The sizes of the shaded areas under the curves are proportional to the total fractions of the molecules that possess the minimum activation energy.

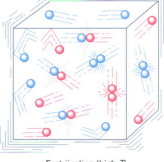

Slow jiggling (low *T*)

Fast jiggling (high *T*)

FIGURE 15-7 Effect of Velocity on Collisions Collisions occur more frequently and with more force as the velocity of the balls increases.

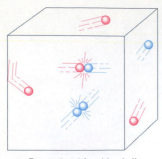

Four red and four blue balls
(low concentration)

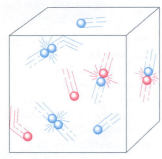

Four red and eight blue balls
(high concentration)

FIGURE 15-8 Effect of Concentration on Collisions Collisions occur more frequently when there are more balls in the box.

greater the concentration of either or both of the reactants, the more frequent the collisions and the greater the rate of the reaction.

15-2.4 The Effect of Particle Size

If we wish to burn an old dead tree, there is a slow way and there is a quick way. Trying to burn it as one big log could take days. If we cut the tree up into smaller logs, we could burn it in hours. In fact, if we changed the tree into a pile of sawdust and spread it around, we might burn it in a matter of minutes. Related to this is the inherent danger in the storage of grain in large silos. Normally, a pile of grain burns slowly. However, small particles of grain dust can become suspended in air, forming a dangerous mixture that can actually detonate. Such explosions have occurred. In the reaction between a solid and a gas, a solid and a liquid, or a liquid and a gas, the surface area of the solid or liquid obviously affects the rate of the reaction. The more area that is exposed, the faster the rate of the reaction.

15-2.5 The Presence of Catalysts

One of the principles of road building in the mountains is that if it's too difficult to go over the mountain, go through it. The mountain is like the activation energy of a reaction, and the tunnel represents a pathway or mechanism with a lower activation energy. *A* **catalyst** *is a substance that provides an alternative reaction mechanism with a lower activation energy.* (See Figure 15-9.) As a result, *a catalyst increases the rate of a reaction, but since the products are identical to the original reaction, the catalyst is not consumed in the reaction.*

An environmentally important example of how catalysts work is provided by the catalytic destruction of ozone (O_3) in the stratosphere. (This problem is entirely different from the concern over global warming and the greenhouse effect.) The following reaction is very slow in the absence of catalysts.

$$O_3(g) + O(g) \longrightarrow 2O_2(g)$$

The atoms of oxygen (O) originate from the dissociation of normal oxygen molecules (O_2) by solar radiation. Chlorofluorocarbons (e.g., CF_2Cl_2), used primarily as refrigerants, diffuse into the stratosphere and are also dissociated by solar radiation into Cl atoms (not Cl_2 molecules). When an atom of chlorine is present, two reactions take place that lead to the rapid conversion of ozone to normal oxygen (O_2).

$$Cl + O_3 \longrightarrow ClO + O_2$$

$$\frac{ClO + O \longrightarrow Cl + O_2}{Cl + ClO + O_3 + O \longrightarrow Cl + ClO + 2O_2}$$

Net reaction: $O_3 + O \longrightarrow 2O_2$

There are two significant facts about these two reactions. First, the presence of chlorine atoms dramatically increases the rate of conversion of ozone to oxygen. Of more importance is the fact that a chlorine atom is a reactant in the first reaction but is regenerated as a product in the second. This is how it acts as a catalyst. Because it is not consumed in the reaction, one chlorine atom can destroy thousands of ozone molecules. In fact, the series of reactions leading to the destruction of ozone in the stratosphere appears to be more complex than that shown above. Studies are currently under way to help us understand exactly how chlorine atoms interact with ozone. In the meantime, the use of chlorofluorocarbons has been phased out by international agreement and the situation seems to be slowly improving.

In the reactions just described, the catalyst is intimately mixed with reactants and is actually involved in the reaction. Catalysts of this nature are referred to as *homogeneous catalysts.* In other cases, a catalyst may simply provide a surface on which reactions take place (*heterogeneous catalyst*).

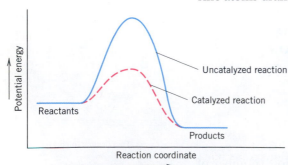

FIGURE 15-9 Activation Energy and Catalysis A catalyst speeds a reaction by lowering the activation energy.

Perhaps the most familiar application of heterogeneous catalysts is in the catalytic converter of an automobile. (See Figure 15-10.) The exhaust from the engine contains poisonous carbon monoxide, unburned fuel (mainly C_8H_{18}), and nitric oxide (NO). All contribute to air pollution, which is still a serious environmental concern in many localities. The catalytic converter, which is attached to the exhaust pipe, contains finely divided platinum and/or palladium. These metals provide a surface for the following reactions, which normally occur only at a very high temperature in the absence of a catalyst.

$$2CO(g) + O_2(g) \longrightarrow 2CO_2(g)$$
$$2C_8H_{18}(g) + 25O_2(g) \longrightarrow 16CO_2(g) + 18H_2O(g)$$
$$2NO(g) \longrightarrow N_2(g) + O_2(g)$$

All three reactions are exothermic, which explains why the catalytic converter becomes quite hot when the engine is running.

FIGURE 15-10 **The Catalytic Converter** This device on the automobile helps to reduce air pollution.

▶ **ASSESSING THE OBJECTIVE FOR SECTION 15-2**

EXERCISE 15-2(a) LEVEL 1: Fill in the blanks.
The rate of appearance of a product as a function of _____ is a measure of the _____ of the reaction. At _____ temperatures, more molecules can overcome this energy barrier, so the rate of the reaction _____. Increasing the concentration of colliding molecules _____ the rate of the reaction. The presence of a _____ does not change the distribution of products, but it _____ the rate of the reaction.

EXERCISE 15-2(b) LEVEL 2: Will doing the following speed up or slow down a reaction?
(a) Heating it   **(d)** Crushing a solid reactant
(b) Stirring it   **(e)** Adding a catalyst
(c) Diluting the reactants

EXERCISE 15-2(c) LEVEL 3: Explain the following.
(a) In certain types of surgery the body is cooled, which allows more time for the procedure.
(b) In acetylene torches pure oxygen rather than compressed air is used to provide a hotter flame.
(c) Hydrogen peroxide (H_2O_2) decomposes to water and oxygen very slowly. When solid manganese dioxide is added, however, oxygen evolves rapidly.
(d) People who are lactose intolerant cannot break down milk sugar (lactose). They lack the enzyme lactase.

For additional practice, work chapter problem 15-3.

15-3 **Equilibrium and Le Châtelier's Principle**

▶ **OBJECTIVE FOR SECTION 15-3**
Using Le Châtelier's principle, predict what effect changing certain conditions will have on the equilibrium point.

LOOKING AHEAD! Some chemical reactions or processes do not appear to go to completion. This is because these processes are reversible and reach what is called a point of equilibrium. A system at equilibrium has a remarkable ability to adapt to external changes. Equilibrium and how it is affected by conditions is the topic of this section. ∎

15-3.1 **Reversible Reactions**

Monetary equilibrium for a typical student is very desirable. We desperately need as much money coming in as going out. Chemical reactions may also reach a point of equilibrium where product molecules are formed at the same rate as they re-form reactants. To illustrate chemical equilibrium, let's return to the hypothetical reaction discussed in Section 15-2 and illustrated in Figure 15-1. We will assume that this reaction is reversible. *A*

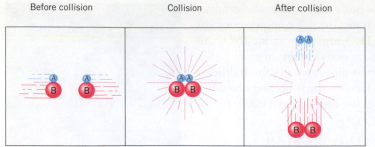

FIGURE 15-11 Reaction of 2AB
Collisions between AB molecules having the minimum energy and correct orientation lead to the formation of reactants.

reversible reaction *is one where both a forward reaction (forming products) and a reverse reaction (re-forming reactants) can occur.* Reversible reactions where both reactions occur simultaneously reach a point of equilibrium. *The* **point of equilibrium** *in a reversible process is when both the forward and reverse processes proceed at the same rate, so the concentrations of reactants and products remain constant.*

In our hypothetical reaction, the reaction mixture initially contains only A_2 and B_2. As the reaction proceeds, the concentrations of these two molecules begin to decrease as the concentration of the product, AB, increases. The rate of buildup of AB begins to decrease until, eventually, the concentration of AB no longer changes despite the presence of excess reactant molecules. To understand this, we turn our attention to the product molecule AB. If we had started with pure AB, we would find that AB slowly decomposes to form A_2 and B_2, just the reverse of the original reaction. This reaction occurs when two AB molecules collide with the proper orientation and sufficient energy to form A_2 and B_2. (See Figure 15-11.)

Now we put these two reactions together. If we again start with pure A_2 and B_2, at first only the forward reaction occurs, leading to the formation of AB. As the concentration of AB increases, however, the reverse reaction begins to occur. Eventually, the rate of formation of products (the forward reaction) is exactly offset by the rate of formation of reactants (the reverse reaction). That is, for every A_2 and B_2 that react to form two AB's, two AB molecules react to re-form one A_2 and one B_2. At this point the concentrations of all species (A_2, B_2, and AB) remain constant. However, if we could follow the fate of one atom of A, we would find that part of the time it existed in a molecule of A_2 and at other times it existed as part of the molecule AB. This phenomenon is referred to as a **dynamic equilibrium**, *which emphasizes the changing identities of reactants and products despite the fact that the total amounts of each do not change.* In Figure 15-12, the reaction and the point of equilibrium are graphically illustrated.

A reversible reaction that reaches a point of equilibrium is indicated by the use of double arrows instead of a single arrow, which implies a complete reaction. Thus our hypothetical reaction is represented as

$$A_2(g) + B_2(g) \rightleftharpoons 2AB(g)$$

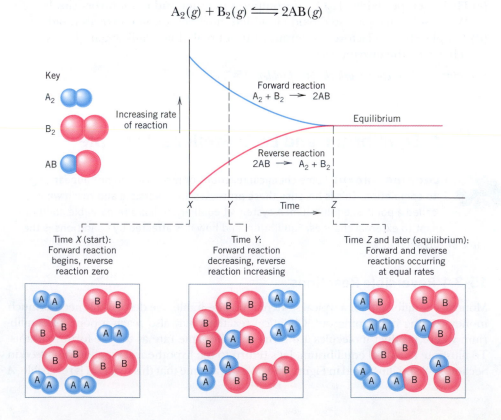

FIGURE 15-12 A_2, B_2, AB Equilibrium Equilibrium is achieved when the rates of the forward and reverse reactions are equal.

15-3.2 Systems at Equilibrium

In earlier parts of this text, we mentioned chemical and physical systems that reach a point of equilibrium. Our first mention of this phenomenon was in Chapter 6, where we discussed the formation of ammonia from its elements. This is a classic example of a system at equilibrium. Under the reaction conditions leading to the formation of ammonia, it also decomposes back to its elements. This is just the reverse of the forward reaction.

$$N_2(g) + 3H_2(g) \rightleftharpoons 2NH_3(g)$$

We encountered a system in Chapter 11 involving equilibrium between two physical states of water. When water is placed in a closed container at a certain temperature, it establishes an equilibrium vapor pressure. At the point of equilibrium, the vapor pressure is constant, which means that the rate of condensation equals the rate of evaporation. The equation for this equilibrium is

$$H_2O(l) \rightleftharpoons H_2O(g)$$

The equilibrium reactions discussed most recently were in Chapter 13 and involved the weak acids and bases. Although the ionization of a weak acid in water is usually very limited, it does occur to a small but measurable extent. At equilibrium, most molecules are present in water in the molecular rather than the ionized state, as represented by the following equilibrium equation.

$$HF(aq) + H_2O(l) \rightleftharpoons H_3O^+(aq) + F^-(aq)$$

The quantitative aspects of the reactions of weak acids and bases in water are discussed later in this chapter.

15-3.3 Le Châtelier's Principle

Life, at times, seems to be a continual attempt to maintain equilibrium despite all the stresses that come our way. If we spend too much money on clothes, we have to counteract that loss by cutting back on expenses somewhere else. If we stay up too late one night (studying, naturally), we must compensate the next night by getting to bed early or, better yet, sleeping in the next morning. Chemical reactions are analogous. There are several external changes of conditions (stresses) that may affect the point of equilibrium and cause the system to compensate.

How a system at equilibrium reacts to a change in conditions is summarized by **Le Châtelier's principle**, which states: *When stress is applied to a system at equilibrium, the system reacts in such a way to counteract the stress.* The changes in a chemical system that affect the point of equilibrium include the following:

1. A change in concentration of a reactant or product
2. A change in the pressure on a gaseous system
3. A change in the temperature

We will discuss these three changes individually. To illustrate how these conditions affect the point of equilibrium, we will use the following reaction.

$$N_2(g) + 3H_2(g) \rightleftharpoons 2NH_3(g)$$

The synthetic conversion of elemental nitrogen to ammonia was possibly the most important scientific discovery of the twentieth century. As mentioned previously, plants and people need nitrogen compounds, particularly as part of proteins, to sustain life. Nitrogen is plentiful in the atmosphere, but breaking the N_2 bond to get nitrogen into compounds is not easy, so nature provides a very limited supply. We will consider the above reaction as a way to change elemental nitrogen into ammonia. Unfortunately, this reaction has a high activation energy, so it must be run at a very high temperature (over 1000°C) in order for it to occur at a reasonable rate. However, little ammonia is produced at this high temperature, so it is not a feasible way to "fix" nitrogen without adjustments.

This large complex manufactures ammonia from nitrogen and hydrogen.

In 1914, a German scientist found a way to make this reaction economical. Unfortunately, the ammonia produced was first used to make explosives, which were used by Germany in World War I. Now, however, the reaction (known as the *Haber process*, after its discoverer, Fritz Haber) is used mainly to produce ammonia for fertilizer or other nitrogen compounds used for the same purpose. Without this process at work today, it is estimated that only about 40% of the world's population could be fed. In other words, three out of five of us alive today could not be here without the Haber process. What follows is how the conditions of this reaction can be manipulated to maximize the yield of ammonia.

15-3.4 Changing Concentrations

If the system is at equilibrium and then an additional portion of a reactant is introduced, the system shifts to produce more products. We can rationalize this by consideration of collision theory. The increase in concentration of reactants causes more collisions between reactant molecules, which leads to the formation of more products. As the concentration of products increases, the rate of the reverse reaction increases. Eventually, equilibrium is reestablished, but now with a higher concentration of products.

We can also rationalize this observation with Le Châtelier's principle. The additional N_2 can be considered as a stress on a system initially at equilibrium. The stress can be relieved by removal of some of the additional N_2 to form more product ammonia. We can generalize this observation. *An increase in the concentration of a reactant or product ultimately leads to an increase in the concentration of the species on the other side of the equation.* Conversely, a decrease in the concentration of a reactant or product ultimately leads to formation of more of that reactant or product.

In reactions where a precipitate forms or a gas is evolved, that product is, in effect, entirely removed from the system. The system shifts to replace this loss, but if the loss is complete, the reaction continues to completion. This is known as an *irreversible reaction*, since products cannot re-form reactants.

15-3.5 Changing Pressure

In a gaseous system, a change in the volume of the reaction container or the pressure may also affect the point of equilibrium. Boyle's law tells us that *a decrease in volume of a quantity of gas is caused by an increase in pressure.* Increasing the pressure on a gaseous mixture of reactants and products at equilibrium (or decreasing the volume of the container) is another example of stress on a system. The system can counteract the stress of increased pressure by shifting to a lower volume if possible. In our sample system, there are four moles of gaseous reactants ($N_2 + 3H_2$) but only two moles of the gaseous product ($2NH_3$). Therefore, an increase in pressure on this system shifts the point of equilibrium to the right, increasing the concentration of NH_3 until equilibrium is eventually reestablished. *In general, an increase in pressure on a gaseous system shifts the point of equilibrium in the direction of the smaller number of moles of gas.* Alternatively, if there is the same number of moles of gas on each side of the equation, a change in pressure does not shift the point of equilibrium. When gases are not involved in a reaction at equilibrium, such as one in aqueous solution, pressure changes have negligible effect, since solids and liquids are essentially incompressible.

15-3.6 Changing Temperature

The reaction of N_2 with H_2 is an exothermic reaction, which means that heat energy is produced along with NH_3. The equation showing heat as a product can be written as follows.

$$N_2(g) + 3H_2(g) \rightleftharpoons 2NH_3(g) + 92 \text{ kJ}$$

Since heat can be considered a component of the reaction, it can be treated in the same way as any other component, as discussed above, according to Le Châtelier's principle. That is, if we add heat (increase the temperature), the reaction shifts in a direction to remove that heat. If we remove heat by cooling the reaction mixture, the equilibrium shifts in a direction to replace that heat loss. *For an exothermic reaction, cooling the reaction mixture increases the concentration of products when equilibrium is reestablished. For an endothermic reaction, where heat can be considered as a reactant, heating the reaction mixture increases the concentration of products at equilibrium.* In our system, an increased concentration of NH_3 at equilibrium is favored at low temperatures. The system attempts to compensate for the loss of heat by formation of more NH_3 and, with it, more heat.

15-3.7 Adding a Catalyst

In our example system, we now have a problem. When we lower the temperature to increase the yield of NH_3, the reaction slows down. Why? Recall that the rate of the reaction decreases as the temperature decreases. Unfortunately, if we lower the temperature enough to get a reasonable amount of NH_3 at equilibrium, the reaction essentially stops. Fritz Haber discovered a heterogeneous catalyst (certain

---

MAKING IT REAL

The Lake That Exploded

The explosion of carbon dioxide from this lake caused a disaster.

In 1986, Lake Nyos in Cameroon, a country in Africa, literally blew its top. What happened in this huge lake is the same thing that happens when one opens a well-shaken can of soda. As we all know, gas (carbon dioxide) comes out of solution and suddenly propels the liquid at whatever or whomever the opening of the can is pointed. In the case of Lake Nyos, tons of carbon dioxide gas suddenly erupted from deep in the lake, forming a thick cloud. Since carbon dioxide is denser than air, the cloud hugged the ground, flowed over the rim of the crater, and began to creep down into the surrounding valleys. As the cloud moved, it pushed the air, with its life-sustaining oxygen, out of the way. Normal air is 0.03% CO_2. If the concentration rises to 10%, it can be fatal. Eventually, 1700 people and many thousands of cattle were asphyxiated. Two years earlier, the same thing had happened at nearby Lake Monoun, causing 37 casualties.

What happened to cause such a calamity and how does this relate to equilibrium? The lake is what is known as a volcanic crater lake. From the hot magma, 50 miles down in the Earth, carbon dioxide rises through vents underneath the lake and enters the cold water near the bottom. Recall that gases are more soluble at lower temperatures. Also, in the equilibrium between the gaseous state and dissolved state, the high pressure near the bottom of the lake favors keeping the gas in solution. In this equilibrium, higher pressure favors the side of the

equation with the lower volume, which is to the right with the CO_2 in the dissolved state (i.e., Le Châtelier's principle).

$$CO_2(g) \rightleftharpoons CO_2(aq)$$

As long as it is not disturbed, the carbon dioxide stays in solution in the cold, lower layers of the lake. A small earthquake, an underwater landslide, or even a violent storm may have upset the layers. Once the layer of water containing the carbon dioxide came closer to the surface, the reduced pressure allowed the equilibrium to shift to the gaseous state, suddenly forming huge volumes of gas. Once it started, it continued violently, like when opening a shaken bottle of soda. The cloud of gas expanded for over 20 miles from the lake until it dissipated.

Crater Lake in Oregon is much like Lake Nyos, but it is deeper and has much more CO_2 injected into its lower depths. Can it explode? Probably not. In winter, the cold surface water sinks to the bottom because it is denser than warm water. The lake thus "turns over." As the bottom water moves toward the surface, the dissolved CO_2 is gradually released each year. In equatorial Africa, it is hot all year long, so the colder bottom water is trapped and gas can accumulate over the centuries until it is suddenly released.

Efforts are underway by scientists and engineers to prevent this from happening again in Lake Nyos and two other lakes like it. In one attempt, pumps extend to the deep parts of the lakes, bringing some of the lower layer to the surface in pipes. In this way, the carbon dioxide can be released gradually before it again builds to dangerous levels. Still, much needs to be done or this tragedy could happen again.

transition metal oxides) that increased the rate of the reaction. In the presence of the catalyst, the reaction can be run at around 400°C instead of over 1000°C without a catalyst. At the lower temperature, the reaction proceeds smoothly with a significant yield of NH$_3$. The discovery of a catalyst was the key that made the process economically feasible and eventually changed the world.

Catalysts play a very significant role in many chemical reactions with industrial significance. *A catalyst in a system that reaches a point of equilibrium increases the rate of both the forward and reverse reactions proportionally but does not change the point of equilibrium.* The function of a catalyst is simply to reach the point of equilibrium in less time. However, with a catalyst, exothermic reactions can be run at considerably lower temperatures, thus increasing the yield of products.

▶ **ASSESSING THE OBJECTIVE FOR SECTION 15-3**

EXERCISE 15-3(a) LEVEL 1: Fill in the blanks.
Reactions that are reversible may reach a _____ where the concentrations of reactants and products remain _____. The point of equilibrium may be affected by a change of _____ of reactants, a change of _____, and a change of _____ (in a gaseous system).

EXERCISE 15-3(b) LEVEL 1: Which of the following is not the role of a catalyst?
(a) increase the rate of the reaction
(b) increase the amount of products formed
(c) lower the activation energy for the forward reaction
(d) lower the activation energy of the reverse reaction
(e) increase the heat of the reaction

EXERCISE 15-3(c) LEVEL 2: Given the equilibrium illustrated by the following equation
$$2H_2O(g) \rightleftharpoons 2H_2(g) + O_2(g) \quad \Delta H = +484$$
Will the concentration of H$_2$ at equilibrium be increased, decreased, or not affected by the following?
(a) an increase in concentration of H$_2$O
(b) an increase in concentration of O$_2$
(c) an increase in temperature
(d) an increase in pressure

EXERCISE 15-3(d) LEVEL 3: If more product forms at lower temperature for a exothermic reaction, why not run that type of reaction at the lowest feasible temperature?

EXERCISE 15-3(e) LEVEL 3: If we could observe one molecule of product of a reaction that had reached equilibrium, would we ever see anything happen to it?

For additional practice, work chapter problems 15-6, 15-8, 15-10, and 15-12.

PART A SUMMARY

KEY TERMS

15-1.1   **Collision theory** describes how reactants are transformed into products. p. 518
15-1.2   The **activation energy** is the energy required to form this complex. p. 519
15-1.3   An **activated complex** is formed at the instant of transformation. p. 520
15-2   The **rate of the reaction** is a measure of the time required to transform reactants into products. p. 522
15-2.5   A **catalyst** affects the rate of the reaction but does not change the eventual amount of reactants and products. p. 524
15-3.1   A **reversible reaction** reaches a **point of equilibrium**. It is referred to as a **dynamic equilibrium**, since the actual reactants and products are constantly changing. pp. 525–526
15-3.3   **Le Châtelier's** predicts the effect of changing conditions on the eventual distribution of reactants and products. p. 527

SUMMARY CHART

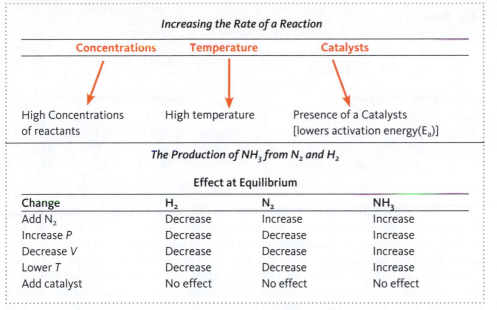

Increasing the Rate of a Reaction

Concentrations	Temperature	Catalysts

High Concentrations of reactants	High temperature	Presence of a Catalysts [lowers activation energy(E_a)]

The Production of NH_3 from N_2 and H_2

Effect at Equilibrium

Change	H_2	N_2	NH_3
Add N_2	Decrease	Increase	Increase
Increase P	Decrease	Decrease	Increase
Decrease V	Decrease	Decrease	Increase
Lower T	Decrease	Decrease	Increase
Add catalyst	No effect	No effect	No effect

Part B

The Quantitative Aspects of Reactions at Equilibrium

SETTING A GOAL

- You will learn how to mathematically manipulate reactions that reach a state of equilibrium.

OBJECTIVES

15-4 Perform calculations involving the equilibrium constant and equilibrium concentrations.

15-5 Perform calculations involving equilibrium constants and pH for acid–base reactions.

15-6 Perform calculations involving solubility constants and molar solubility.

15-4 The Equilibrium Constant

▶ **OBJECTIVE FOR SECTION 15-4**
Perform calculations involving the equilibrium constant and equilibrium concentrations.

LOOKING AHEAD! Long before chemists knew anything about collision theory, reaction rates, or activation energy, they realized that the concentrations of reactants and products were distributed in a predictable manner when a reaction reached the point of equilibrium. Our next subject concerns these quantitative aspects of the point of equilibrium. ■

15-4.1 The Equilibrium Constant Expression

To illustrate how the concentrations of reactants and products relate, we will consider a hypothetical reversible reaction involving *a* moles of reactant A combining with *b* moles of reactant B to form *c* moles of product C and *d* moles of product D.

$$aA + bB \rightleftharpoons cC + dD$$

For a reaction at equilibrium, the distribution of reactants and products is given by the following relationship.

$$K_{eq} = \frac{[C]^c\,[D]^d}{[A]^a\,[B]^b}$$

TABLE 15-1

The Value of K_{eq}

For the reaction $H_2(g) + I_2(g) \rightleftharpoons 2HI(g)$ at 450°C,

$$K_{eq} = \frac{[HI]^2}{[H_2][I_2]} \qquad [X] = \text{concentration in mol/L}$$

	INITIAL CONCENTRATION			EQUILIBRIUM CONCENTRATION			
Expt.	$[H_2]$	$[I_2]$	$[HI]$	$[H_2]$	$[I_2]$	$[HI]$	K_{eq}
1	2.000	2.000	0	0.428	0.428	3.144	54.0
2	0	0	2.000	0.214	0.214	1.572	54.0
3	1.000	1.000	1.000	0.321	0.321	2.358	54.0

Note that the coefficients (a, b, c, and d) of the substances (A, B, C, and D) become the exponents of the molar concentrations (mol/L) of the same substances. The products are written in the numerator, and the reactants are written in the denominator. This relationship for a particular reaction consists of two parts on each side of the equal sign. K_{eq} is called the **equilibrium constant** *and has a definite numerical value at a given temperature. The ratio to the right of the equality is called the* **equilibrium constant expression**. We will apply this relationship to gas-phase reactions in this section and to certain reactions that occur in aqueous solution in the next section.

15-4.2 The Value of the Equilibrium Constant

Using the equilibrium relationship, the distribution of reactants and products at equilibrium can be predicted regardless of initial concentrations. This is illustrated in Table 15-1. In a series of three experiments, we start with an initial concentration of reactants only (expt. 1), an initial concentration of products only (expt. 2), and an initial concentration of both reactants and products (expt. 3). By substituting the final equilibrium concentrations of all species in the equilibrium constant expression, a consistent value for the equilibrium constant is obtained for all three cases.

In our first example, we will write the equilibrium constant expressions for three reversible reactions that occur in the gaseous phase.

EXAMPLE 15-1 Writing Equilibrium Constant Expressions

Write the equilibrium constant expression for each of the following reactions.

(a) $N_2(g) + 3H_2(g) \rightleftharpoons 2NH_3(g)$

(b) $4NH_3(g) + 5O_2(g) \rightleftharpoons 4NO(g) + 6H_2O(g)$

(c) $PCl_3(g) + Cl_2(g) \rightleftharpoons PCl_5(g)$

PROCEDURE

Put the concentrations of products in the numerator and reactants in the denominator. Raise each concentration to the power of the coefficient in the balanced equation.

SOLUTIONS

(a) $K_{eq} = \dfrac{[NH_3]^2}{[N_2][H_2]^3}$ **(b)** $K_{eq} = \dfrac{[NO]^4[H_2O]^6}{[NH_3]^4[O_2]^5}$ **(c)** $K_{eq} = \dfrac{[PCl_5]}{[PCl_3][Cl_2]}$

ANALYSIS

What can we predict about how far a reaction progresses based on the value of K_{eq}? A large value for K_{eq} indicates that the numerator is much larger than the denominator in the equilibrium constant expression. This means that the concentrations of products are larger than those of reactants. For example, consider the simple equilibrium

$$A \Longleftrightarrow B \qquad K_{eq} = 10^2$$

$$K_{eq} = \frac{[B]}{[A]} = 10^2 \qquad \text{thus} \qquad [B] = 10^2[A] = 100[A]$$

The large value of the equilibrium constant signifies that, for this reaction at the point of equilibrium, the concentration of the product, [B], is 100 times the concentration of the reactant, [A]. That is, the equilibrium lies far to the right. However, if the value of K_{eq} were small (e.g.,10^{-2}), then the concentration of the reactant, [A], would be 100 times the concentration of the product, [B], and this equilibrium would lie far to the left.

SYNTHESIS

Another useful quantity is called the reaction quotient, given the symbol Q. It has an expression written the same way as the equilibrium constant expression, except that the values used are the starting concentrations, not the equilibrium concentrations. By comparing the values of Q and K_{eq}, we can determine which way the reaction will move to reach equilibrium. When $Q < K_{eq}$, the product concentrations are too low and the reactant concentrations are too high to be in a state of equilibrium. As a result, the system will produce more product at the expense of reactants. If $Q > K_{eq}$, then the opposite situation occurs, and the system will move to produce reactant at the expense of product. Should $Q = K_{eq}$, then the system is already at equilibrium and we would expect no relative shifting of concentrations.

The equilibrium constant is calculated from the experimental determination of the distribution of reactants and products at equilibrium. The following two examples illustrate such calculations.

EXAMPLE 15-2   Calculating the Value of K_{eq}

What is the value of K_{eq} for the following system at equilibrium?

$$2NO(g) + O_2(g) \Longleftrightarrow 2NO_2(g)$$

At a given temperature, the equilibrium concentrations of the gases are [NO] $= 0.890$, [O$_2$] $= 0.250$, and [NO$_2$] $= 0.0320$.

PROCEDURE

Substitute the concentrations into the equilibrium constant expression and raise the concentration to the appropriate power.

SOLUTION

For this reaction,

$$K_{eq} = \frac{[NO_2]^2}{[NO]^2[O_2]}$$

$$= \frac{[0.0320]^2}{[0.890]^2[0.250]} = \frac{1.024 \times 10^{-3}}{0.198} = \underline{\underline{5.17 \times 10^{-3}}}$$

ANALYSIS

Equilibrium problems are great places to practice your ability to estimate whether the answer you got is reasonable. In this problem we see that the equilibrium concentrations of the reactants are considerably higher than the product. This means that the equilibrium position favors reactants. Equilibrium constants for reactant-favored reactions are typically quite small, so our answer is very reasonable.

SYNTHESIS

What would happen to each of the values, qualitatively, if more oxygen is now added to the system? According to Le Châtelier's principle, the system would shift toward formation of more products. Therefore, the amount of NO$_2$ would rise, and the amount of NO would fall. The oxygen concentration would fall somewhere in between where it was after the initial equilibrium and where it was after the excess had been added. Though the calculation would be somewhat involved, it is doable once we know the value of the equilibrium constant.

EXAMPLE 15-3 **Calculating Equilibrium Concentrations and K_{eq}**

For the equilibrium

$$N_2(g) + 3H_2(g) \rightleftharpoons 2NH_3(g)$$

Complete the table and compute the value of the equilibrium constant.

Initial Concentrations			Equilibrium Concentrations		
[H_2]	[N_2]	[NH_3]	[H_2]	[N_2]	[NH_3]
0.200	0.200	0	?	?	0.0450

PROCEDURE

1. In a stoichiometry calculation, find the hydrogen and nitrogen concentrations needed to form the 0.0450 mol/L of NH_3. (See Section 7-1.)

2. Find the [H_2] and [N_2] remaining at equilibrium by subtracting the concentration that reacted from the initial concentration; that is,

$$[N_2]_{eq} = [N_2]_{initial} - [N_2]_{reacted}$$

3. Substitute the concentrations of all compounds present at equilibrium into the equilibrium constant expression and solve to find the value of K_{eq}.

SOLUTION

1. If 0.0450 mol of NH_3 is formed, calculate the number of moles of N_2 that reacted (per liter).

$$0.0450 \text{ mol NH}_3 \times \frac{1 \text{ mol N}_2}{2 \text{ mol NH}_3} = 0.0225 \text{ mol N}_2 \text{ reacted (per liter)}$$

The number of moles of H_2 that reacted is

$$0.0450 \text{ mol NH}_3 \times \frac{3 \text{ mol H}_2}{2 \text{ mol NH}_3} = 0.0675 \text{ mol H}_2 \text{ reacted (per liter)}$$

2. At equilibrium

$$[N_2]_{eq} = 0.200 - 0.0225 = 0.178$$

$$[H_2]_{eq} = 0.200 - 0.0675 = 0.132$$

3. $$K_{eq} = \frac{[NH_3]^2}{[N_2][H_2]^3} = \frac{(0.0450)^2}{(0.178)(0.132)^3} = \underline{\underline{4.95}}$$

ANALYSIS

This problem may at first seem confusing until we realize that it requires typical stoichiometry calculations. The first part of the problem is much like similar problems we worked in Chapter 7, where we converted the amount of a product to the amount of reactant required. That is, we are given the concentration of NH_3 and we need to find out the concentrations of reactants that formed the given concentration of NH_3. The concentration present at equilibrium of the reactants is simply the concentration left over from the amount that reacted.

SYNTHESIS

It is helpful to know what direction the equilibrium is headed so we know whether to subtract reacted amounts from the reactants or products. Again, the Q values can aid us. If any reactant or product is absent from the reaction, then its concentration is zero. For reactants, this means a Q value of infinity, and a shift toward reactants (subtract the reacted amount from products). In the above, the product concentration is zero. Therefore, the Q value is 0, and the reaction must make more product. Subtract the reacted amount from the reactants.

EXAMPLE 15-4   Calculating Equilibrium Concentrations from K_{eq}

In the preceding equilibrium, what is the concentration of NH_3 at equilibrium if the equilibrium concentrations of N_2 and H_2 are 0.22 and 0.14 mol/L, respectively?

PROCEDURE

In this example, use the value of K_{eq} found in the previous example. The concentration of NH_3 can be found by substituting the concentrations of the species given and solving for the one unknown.

SOLUTION

$$K_{eq} = \frac{[NH_3]^2}{[N_2][H_2]^3} \qquad \begin{aligned} [N_2] &= 0.22 \\ [H_2] &= 0.14 \end{aligned} \qquad K_{eq} = 4.95$$

$$K_{eq} = \frac{[NH_3]^2}{[0.22][0.14]^3} = 4.95$$

$$[NH_3]^2 = 2.99 \times 10^{-3}$$

$$[NH_3] = 5.5 \times 10^{-2} = 0.055$$

Thus the concentration of $NH_3 = \underline{0.055 \text{ mol/L}}$.

ANALYSIS

This problem requires the use of some simple algebra. That is, we isolate the unknown (i.e., NH_3) on the left side of the equation and all the known quantities on the right (i.e, K_{eq}, $[N_2]$, and $[H_2]$). Notice in this case that the unknown is squared, so the solution requires that we take the square root of the calculated quantity on the right.

SYNTHESIS

This type of problem could also be solved knowing the initial values of the starting materials, rather than their equilibrium values. In a realistic setting, it is those values that are most readily available. However, solving that sort of problem introduces complications that require some sophisticated mathematical assumptions and techniques that are beyond the scope of this text. Nevertheless, it's important to know what's possible.

▶ ASSESSING THE OBJECTIVE FOR SECTION 15-4

EXERCISE 15-4(a) LEVEL 1: Fill in the blanks.
A gaseous equilibrium can be expressed as a number known as the _____ _____ and a ratio of reactants and products known as the _____ _____ _____. A small value for K_{eq} (e.g., 10^{-5}) indicates that _____ are favored at equilibrium.

EXERCISE 15-4(b) LEVEL 1: Write the equilibrium constant expression for the following equilibria.

(a) $CO(g) + 3H_2(g) \rightleftharpoons CH_4(g) + H_2O(g)$

(b) $2NO(g) + Br_2(g) \rightleftharpoons 2NOBr(g)$

EXERCISE 15-4(c) LEVEL 2: What is the value of K_{eq} in part (a) of Exercise 15-4 (b) if there are 2.0×10^{-3} mol of CO, 1.0×10^{-2} mol of H_2, 3.5×10^{-4} mol of CH_4, and 5.0×10^{-5} mol of H_2O present in a 10.0-L container at equilibrium?

EXERCISE 15-4(d) LEVEL 2:

(a) What is the value of K_{eq} in part (b) of Exercise 15-4(b) if we start with 4.00 mol of NOBr only in a 20.0-L container, and, at equilibrium, it is found that $[NOBr] = 0.10$?

(b) Using K_{eq} from part (a) above, what is the concentration of $[NOBr]$ if $[NO] = 0.45$ and $[Br_2] = 0.22$ at equilibrium?

EXERCISE 15-4(e) LEVEL 3: Heterogeneous equilibria involve a solid and a gas. An example of such an equilibrium is

$$CaCO_3(s) \rightleftharpoons CaO(s) + CO_2(g)$$

The equilibrium constant expression is $K_{eq} = [CO_2]$. Why don't the concentrations of calcium carbonate and calcium oxide appear in the expression?

For additional practice, work chapter problems 15-15, 15-20, 15-25, and 15-30

▶ **OBJECTIVE**
FOR SECTION 15-5
Perform calculations involving equilibrium constants and pH for acid–base reactions.

15-5 Equilibria of Weak Acids and Weak Bases in Water

LOOKING AHEAD! We now return to weak acids, weak bases, and buffer solutions, which we first discussed in Chapter 13. Aqueous solutions of these substances are in equilibria, so they also have an equilibrium constant expression. In these cases the equilibrium occurs in aqueous solution, so these concentrations are the same as the molarity of the species. ■

If one is in a leaky boat and no one bails—down you go. No equilibrium is established between input and output, and the boat fills with water. The reaction of a strong acid in water is analogous to this. That is, no equilibrium occurs, since there is no reverse reaction. As a result, acids such as HNO_3 are completely ionized in water, which means that the molecular form of the acid does not exist in aqueous solution. In this case, a single arrow indicates a complete reaction.

$$HNO_3(aq) + H_2O(l) \longrightarrow H_3O^+(aq) + NO_3^-(aq)$$

15-5.1 Equilibrium Constant Expressions for Weak Acids and Weak Bases

The difference between a strong acid and a weak acid lies in the concentration of H_3O^+ produced by the acid. A weak acid such as hypochlorous acid (HClO) produces only a small concentration of ions because the ionization is a reversible reaction in which the equilibrium actually lies far to the left. The ionization of HClO is represented by the following equation.

$$HClO(aq) + H_2O(l) \rightleftharpoons H_3O^+(aq) + ClO^-(aq)$$

The equilibrium constant expression can be written as follows, where the concentrations now represent molarities.

$$K_{eq} = \frac{[H_3O^+][ClO^-]}{[H_2O][HClO]}$$

In this equilibrium, the concentrations of H_3O^+, ClO^-, and HClO in aqueous solution can all be varied. H_2O is the solvent, however, and is present in a very large excess compared with the other species. Since the amount of H_2O actually reacting is very small compared with the total amount present, the concentration of H_2O is essentially a constant. The $[H_2O]$ can therefore be included with the other constant, K_{eq}, to produce another constant labeled $\mathbf{K_a}$, *which is known as an* **acid ionization constant**.

$$K_{eq}[H_2O] = K_a = \frac{[H_3O^+][ClO^-]}{[HClO]}$$

In a weak acid solution we essentially have three species (besides water). There are two ions, H_3O^+ and the anion of the acid (i.e., ClO^-) and the acid in molecular

Vinegar contains a weak acid and ammonia, a weak base.

form (i.e., HClO). It is this limited amount of H_3O^+ produced that classifies hypochlorous acid as a weak acid.

Ionization of the weak base NH_3 is represented as

$$NH_3(aq) + H_2O(l) \rightleftharpoons NH_4^+(aq) + OH^-(aq)$$

In this case, the three species present are the cation (i.e., NH_4^+), the OH^- ion, and molecular NH_3. As in weak acids, H_2O does not appear in the equilibrium constant expression. For this reaction, *the equilibrium constant is labeled K_b, which is the* **base ionization constant**.

$$K_b = \frac{[NH_4^+][OH^-]}{[NH_3]}$$

15-5.2 Determining the Value of K_a or K_b

The values of K_a and K_b for weak acids and bases are determined experimentally. To calculate K_a or K_b, we need to know, or be able to determine from other data, the concentrations of both ions and the nonionized (i.e., molecular) form of the acid or base. In the following two examples, the needed concentrations are not directly given but must be inferred from the amount that ionizes (Example 15-5) or the pH of the solution (Example 15-6). After we illustrate how these constants are calculated, we will discuss their significance and how they are used.

EXAMPLE 15-5   **Calculating K_a from Equilibrium Concentrations**

In a 0.20-M solution of HNO_2, it is found that 0.009 mol/L of the HNO_2 ionizes. What are the concentrations of H_3O^+, NO_2^-, and HNO_2 at equilibrium, and what is the value of K_a?

PROCEDURE

1. Write the equilibrium equation for K_a. Remember that H_2O is not included in the expression.

2. Calculate the concentration of nonionized (molecular) HNO_2 present at equilibrium. Remember that the initial concentration given (0.20 M) represents the concentration of nonionized HNO_2 present *plus* the concentration that ionizes. Therefore, $[HNO_2]$ at equilibrium is the initial concentration minus the concentration that ionizes.

$$[HNO_2]_{eq} = [HNO_2]_{initial} - [HNO_2]_{ionized}$$

3. Note that one H_3O^+ and one NO_2^- are formed for each HNO_2 that ionizes (from the equation stoichiometry).

$$[H_3O^+]_{eq} = [NO_2^-]_{eq} = [HNO_2]_{ionized}$$

4. Calculate K_a.

SOLUTION

1. $$HNO_2(aq) + H_2O(l) \rightleftharpoons H_3O^+(aq) + NO_2^-(aq)$$

2. $$[HNO_2]_{initial} = 0.20 \qquad [HNO_2]_{ionized} = 0.009$$

3. $$[HNO_2]_{eq} = 0.20 - 0.009 = \underline{0.19}$$

3. $$[H_3O^+]_{eq} = [NO_2^-]_{eq} = 0.009$$

4. $$K_a = \frac{[H_3O^+][NO_2^-]}{[HNO_2]} = \frac{(0.009)(0.009)}{0.19}$$

$$= \underline{4 \times 10^{-4}}$$

ANALYSIS

It is important to note that one molecule of acid ionizes to produce one H_3O^+ ion and one anion, which in this case is NO_2^- The concentration of the two ions will thus be equal. If you know one, you know the other. It's like having a dozen pairs of shoes. If they are separated, you know there will be one dozen left shoes and one dozen right shoes. Note also that the acidity of the solution is determined only by the concentration of H_3O^+, not the concentration of the molecular form of the acid.

SYNTHESIS

Instead of being given the actual concentrations of the ions, the percent ionization of the acid is sometimes stated. In these cases, the concentrations of the ions are calculated like any other percent problem—by multiplying the percent in decimal form times the original concentration of acid. For example, if a 0.10-M acid is 18% ionized, the concentrations of both ions are 0.10 M $\times$ 0.18 = 0.018 M, and the concentration of unionized acid is 0.10 − 0.018 = 0.08 M. The value of K_a equals $[0.018][0.018]/0.08 = 4 \times 10^{-3}$.

EXAMPLE 15-6   Calculating K_a from pH

A 0.25-M solution of HCN has a pH of 5.00. What is K_a?

PROCEDURE

1. Write the equilibrium reaction.

2. Convert pH to $[H_3O^+]$.

3. Note that $[CN^-] = [H_3O^+]$ at equilibrium.

4. Calculate [HCN] at equilibrium, which is

$$[HCN]_{eq} = [HCN]_{initial} - [HCN]_{ionized}$$

$[HCN]_{ionized} = [H_3O^+]_{eq}$, since one H_3O^+ is produced at equilibrium for every HCN that ionizes.

5. Use these values to calculate K_a.

SOLUTION

1.
$$HCN(aq) + H_2O(l) \rightleftharpoons H_3O^+(aq) + CN^-(aq)$$

2.
$$[H_3O^+] = \text{antilog} \, (- 5.00)$$
$$[H_3O^+] = 1.0 \times 10^{-5} = [CN^-]$$
$$[HCN]_{eq} = 0.25 - (1.0 \times 10^{-5}) \approx 0.25$$

$$K_a = \frac{[H_3O^+] \, [CN^-]}{[HCN]}$$

$$K_a = \frac{(1.0 \times 10^{-5}) \, (1.0 \times 10^{-5})}{0.25}$$

$$= \underline{4.0 \times 10^{-10}}$$

ANALYSIS

Recall that pH is simply a way to express the H_3O^+ concentration. If we know the pH, we can calculate $[H_3O^+]$. If the pH is an integer, it becomes the negative exponent of 10. If it is not an integer, then we need to use a calculator. Refer to Example 13-9 in Chapter 13 for examples of these calculations. Notice also that the concentration of HCN is essentially unchanged. How can some of it ionize yet the original concentration not change? In fact, the amount of HCN that ionizes (10^{-5} M) is negligible compared with the initial concentration of HCN (0.25 M). Therefore, $[HCN]_{eq} \approx [HCN]_{initial}$. ($\approx$ means approximately equal). It is like a crowd of about 10,000 people. Ten people more or less in this crowd does not change the overall estimate. For the purposes of these calculations (two significant figures), a number is considered negligible compared with another if it is less than 10% of the larger number. In this case, $10^{-3}/0.25 \times 100\% = 0.4\%$ ionized.

SYNTHESIS

Make a quick prediction based on principles and mathematics. 0.25 M HCN with a pH of 5.0 has a $K_a = 4 \times 10^{-10}$. About what would another acid's K_a be if 0.25 M had a pH equal to 4.0? The same concentration produces a lower pH, indicating more molecules ionized. It is therefore a stronger acid and should have a larger equilibrium constant. Because both the concentration of the H_3O^+ and the acid anion are affected at the same time, one power of 10 in pH becomes two powers of 10 for an equilibrium constant. The value is around 4×10^{-8}.

15-5.3 pK_a and pK_b

The value for K_a and K_b can also be expressed as a negative log in same manner as pH. That is

$$pK_a = -\log K_a \quad \text{and} \quad pK_b = -\log K_b$$

Notice that the weaker the acid, the larger the value of pK_a. For example, if a weak acid had $K_a = 10^{-4}$, p$K_a = 4.0$. If an even weaker acid had $K_a = 10^{-9}$, p$K_a = 9.0$. pK_a is often used in organic and biochemistry to indicate relative acid strength and is used in the discussion of buffer solutions that follows.

15-5.4 Using K_a or K_b to Determine pH

Once a value for K_a or K_b for a specific weak acid or base is known, we have the ability to calculate the pH of a solution of known concentration of a weak acid (Example 15-7) or a weak base (Example 15-8).

TABLE 15-2

K_a and pK_a for Some Weak Acids

$K_a = \dfrac{[H_3O^+][X^-]}{[HX]}$ (HX symbolizes a weak acid, X⁻ its conjugate base)

ACID	FORMULA	K_a	pK_a	
hydrofluoric	HF	6.7×10^{-4}	3.17	
nitrous	HNO_2	4.5×10^{-4}	3.35	
formic	$HCHO_2$	1.8×10^{-4}	3.74	Decreasing
acetic	$HC_2H_3O_2$	1.8×10^{-5}	4.74	acid
hypochlorous	HCIO	3.2×10^{-8}	7.49	strength
hypobromous	HBrO	2.1×10^{-9}	8.68	
hydrocyanic	HCN	4.0×10^{-10}	9.40	

TABLE 15-3

K_b and K_b for some Weak Bases

$K_b = \dfrac{[HB^+][OH^-]}{[B]}$ (B symbolizes a weak bases, HB⁺ its conjugate acid)

BASE	FORMULA	K_b	pK_b	
dimethylamine	$(CH_3)_2NH$	7.4×10^{-4}	3.13	Decreasing
ammonia	NH_3	1.8×10^{-5}	4.74	base
hydrazine	N_2H_4	9.8×10^{-7}	6.01	strength

EXAMPLE 15-7 **Calculating the pH of a Solution of a Weak Acid**

What is the pH of a 0.155-M solution of HClO?

PROCEDURE

1. Write the equilibrium involved.

2. Write the appropriate equilibrium constant expression.

3. Let $x = [H_3O^+]$ at equilibrium; since $[H_3O^+] = [ClO^-]$, $x = [ClO^-]$.

4. At equilibrium, $[HClO]_{eq} = [HClO]_{initial} - [HClO]_{ionized}$.

5. Using the value for K_a in Table 15-2, solve for x.

6. Convert x to pH.

In summary:

	HClO	**$[H_3O^+]$**	**$[ClO^-]$**
initial	0.155	0	0
equilibrium	$0.155 - x$	x	x

SOLUTION

1.
$$HClO(aq) + H_2O(l) \rightleftharpoons H_3O^+(aq) + ClO^-(aq)$$

2.
$$K_a = \frac{[H_3O^+][ClO^-]}{[HClO]} = 3.2 \times 10^{-8}$$

3. At equilibrium, $[H_3O^+] = [ClO^-] = x$.

4. At equilibrium, $[HClO] = 0.155 - x$.

5.
$$\frac{(x)(x)}{0.155 - x} = 3.2 \times 10^{-8}$$

The solution of this equation appears to require the quadratic equation. However, a simplification of this calculation is possible. Note that K_a is a small number, indicating that the degree of ionization is small (the equilibrium lies far to the left). This means that x is a very small number. Since very small numbers make little or no difference when added to or subtracted from large numbers, they can be ignored with little or no error.

$$0.155 - x \approx 0.155$$

Therefore the expression can now be simplified.

$$\frac{(x)(x)}{0.155 - x} = \frac{x^2}{0.155} = 3.2 \times 10^{-8}$$

$$x^2 = 5.0 \times 10^{-9}$$

To solve for x, take the square root of each side of the equation.

$$\sqrt{x^2} = \sqrt{5.0 \times 10^{-9}}$$

$$x = [H_3O^+] = 7.1 \times 10^{-5}$$

(Note that the x is indeed much smaller than 0.155.)

6.
$$pH = -\log[H_3O^+] = -\log(7.1 \times 10^{-5}) = \underline{4.15}$$

ANALYSIS

Finding the pH of a weak acid solution given K_a is somewhat more mathematically involved than the earlier problems. Students often neglect the important mathematical simplification and assume that the quadratic equation is needed. In fact, it is rarely needed because almost all weak acids are weak enough that we can assume that very little acid or base ionizes. This occurs for K_a or K_b values less than about 10^{-3}, which covers most common weak acids and bases. The problem then simply requires the taking of a square root. Sometimes students get confused about which x to ignore. When x stands alone, as when it represents a quantity such as H_3O^+, it is *not* ignored. When it is subtracted from a much large number, it can be ignored.

How would one calculate the pH of a 0.01-M (10^{-2} M) solution of chlorous acid ($HClO_2$, $K_a = 10^{-2}$)? If we solved for x as above, $x = 10^{-2}$ M. That means it is 100% ionized, which means that the assumption that x is small compared to 10^{-2} is obviously incorrect. If we use the quadratic, $x = 0.008$ M, which means that it is 80% ionized. That is more reasonable. See if you can follow the reasoning and calculations.

EXAMPLE 15-8   Calculating the pH of a Solution of a Weak Base

What is the pH of a 0.245-M solution of N_2H_4?

PROCEDURE

1. Write the equilibrium involved.

2. Write the appropriate equilibrium constant expression.

3. Let $x = [OH^-]$ at equilibrium; since $[OH^-] = [N_2H_5^+]$, $[N_2H_5^+] = x$, too.

4. At equilibrium, $[N_2H_4]_{eq} = [N_2H_4]_{initial} - [N_2H_2]_{ionized}$

5. Using the value for K_b in Table 15-3, solve for x.

6. Convert x to pOH. Convert pOH to pH.

SOLUTION

1.
$$N_2H_4(aq) + H_2O(l) \rightleftharpoons N_2H_5^+(aq) + OH^-(aq)$$

2.
$$K_b = \frac{[N_2H_5^+][OH^-]}{[N_2H_4]} = 9.8 \times 10^{-7} \quad \text{(from Table 14-3)}$$

3. Let $x = [OH^-] = [N_2H_5^+]$ (at equilibrium)

4. $[N_2H_4] = 0.245 - x$ (at equilibrium). Since K_b is very small, x is very small. Therefore,
$$0.245 - x \approx 0.245$$

5.
$$\frac{(x)(x)}{0.245} = 9.8 \times 10^{-7}$$
$$x^2 = 2.4 \times 10^{-7}$$
$$x^2 = 4.9 \times 10^{-4} = [OH^-]$$
$$pOH = -\log[OH^-] = -\log(4.9 \times 10^{-4}) = 3.31$$

6.
$$pOH = 14.00 - pOH = 14.00 - 3.31 = \underline{10.69}$$

ANALYSIS

This is the same procedure as in Example 15-7, with the appropriate changes made for a base as opposed to an acid. This includes the use of K_b instead of K_a and the fact that the x stands for the hydroxide concentration instead of the hydronium concentration. Therefore, it is pOH that is being calculated directly. A quick conversion yields pH. Certainly, the results seem reasonable. This is a modest concentration of a weak base. A calculated pH value of 10.69 seems to be appropriate for this system.

SYNTHESIS

How do these acids and bases compare to those we studied in Chapter 13, when we weren't concerned with equilibrium constants? Consider a comparison between 0.245 M N_2H_4, the *weak* base in this problem, with 0.245 M NaOH, a *strong* base from Chapter 13. The pH for the N_2H_4 solution is 10.69. The pH for the NaOH solution is 13.39, or nearly a thousand times more concentrated. Why the difference? The pH of the strong base is determined directly from the concentration, so an equilibrium constant is not needed. Any equilibrium lies so far toward producing hydroxide that we assume the two are equal. The pH of the weak base requires the equilibrium expression, where the small value of K_b makes the pH lower as well.

15-5.5 The pH of Buffer Solutions

As mentioned in Chapter 13, buffer solutions are those that resist changes in pH and are made from mixtures of a weak acid (or base) and a salt containing the conjugate base (or acid). They have many important applications, especially in life processes such as maintaining the nearly constant pH of the blood. We can also calculate the pH of buffers using a form of the equation for K_a outlined previously. In fact, calculations of the pH of buffers are comparatively simple if we rearrange the equilibrium constant expression, as illustrated here with HClO.

$$[H_3O^+] = K_a \times \frac{[HClO]}{[ClO^-]}$$

By taking $-\log$ of both sides of the equation, we now have

$$-\log[H_3O^+] = -\log K_a - \log \frac{[HClO]}{[ClO^-]}$$

Now we substitute pH for $-\log[H_3O^+]$ and pK_a for $-\log K_a$ and invert the ratio of the concentrations, which changes the sign of the log.

$$pH = pK_a + \log \frac{[ClO^-]}{[HClO]}$$

Notice that the ClO^- is the base species in the equilibrium and HClO is the acid species. We can thus make this a general equation for calculating the pH of buffers by substituting [acid] for [HClO] and [base] for [ClO⁻]. *This general equation is known as the* **Henderson–Hasselbalch equation.**

$$pH = pK_a + \log \frac{[base]}{[acid]} \qquad pOH = pK_b + \log \frac{[acid]}{[base]}$$

Since weak acids and bases are used to make buffers, we are quite safe in assuming that concentrations of the acid and base species present in solution are essentially the same as the originally measured amounts. Also, the conjugate acid or base is added in the form of a salt where one of the ions is not involved in the equilibrium (i.e., the Na^+ in NaClO). The most effective buffers are made from solutions containing equal molar amounts of acid and base species. In such cases, the ratio of base to acid equals unity. Since log 1 = 0, pH equals pK_a. Thus, pK_a *is sometimes used as a measure of the acid strength and as the pH of the most effective buffer formed from this weak acid.* The pHs of other ratios of acid and base species can be calculated by simply substituting the amounts in moles or the concentrations into the appropriate Henderson–Hasselbalch equation. *Note that this equation is useful only for buffers, not for solutions of weak acids or weak bases alone.*

EXAMPLE 15-9   **Calculating the pH of a Buffer Solution**

What is the pH of a buffer solution that is made by dissolving 0.30 mol of $HC_2H_3O_2$ and 0.50 mol of $NaC_2H_3O_2$ in enough water to make 1.00 L of solution?

PROCEDURE

1. Obtain the value of pK_a from Table 15-2.

2. Employ the Henderson–Hasselbalch equation to solve for pH.

SOLUTION

In this solution, the concentration of the acid species is $[HC_2H_3O_2] = 0.30$ mol/L, and the concentration of the base species is $[C_2H_3O_2^-] = 0.50$ mol/L.

$$pK_a = -\log K_a = -\log(1.8 \times 10^{-5}) = 4.74$$

$$pH = 4.74 + \log\frac{[0.50]}{[0.30]} = 4.74 + 0.22 = \underline{4.96}$$

ANALYSIS

The buffer capacity of a system relates to the original concentration of the acid and base species in the solution. In this case the buffer capacity of the base is 0.50 mol/L. If any more than 0.50 mol/L of H_3O^+ is added to the system, the base of the buffer is used up and the pH decreases rapidly. If any more than 0.30 mol/L of OH^- is added to the system, the buffering capacity of the acid is used up and the pH rises dramatically. For example, if we added 0.60 mol/L of HCl to the system, the buffer could react with 0.50 mol/L of that amount. The 0.10 mol/L of excess H_3O^+ would lower the pH to 1.0.

SYNTHESIS

Reread Making It Real, "The pH Balance in the Blood," in Chapter 13. This is a very important biological concept because metabolism occurs in a very narrow pH range, so the blood is very heavily buffered. The normal pH of blood is 7.4, yet the pK_a of carbonic acid is 6.4. Using the Henderson–Hasselbalch equation, we can see what this means. The pH of this buffer would be 7.4 if the concentration of bicarbonate ion is 10 times the concentration of carbonic acid, as shown below. Mathematically, we are saying $[HCO_3^-] = 10[H_2CO_3]$:

$$pH = pKa + \log\frac{B}{A} = 6.4 + \log\frac{[HCO_3^-]}{[H_2CO_3]} = 6.4 + \log\frac{10[H_2CO_3]}{[H_2CO_3]} = 6.4 + \log 10 = 6.4 + 1.0 = 7.4$$

What this tells us is that our blood is 10 times more buffered against the addition of acid than base. In fact, more acid (e.g., lactic acid from exercise) enters our blood than base, so this works out well.

▶ **ASSESSING THE OBJECTIVE FOR SECTION 15-5**

EXERCISE 15-5(a) LEVEL 1: Write the equilibrium involved and the equilibrium constant expression of an aqueous solution of formic acid ($HCHO_2$). Write the equilibrium involved and the equilibrium constant expression of an aqueous solution of the hypothetical weak nitrogen base NX_3.

EXERCISE 15-5(b) LEVEL 1: The concentration of H_3O^+ in a 0.10-M solution of a weak acid is 1.3×10^{-3}M. What are the concentrations of each species in the equilibrium expression? What is the weak acid from those listed in Table 15-2?

EXERCISE 15-5(c) LEVEL 1: A buffer is made by dissolving an equal number of moles of the weak base $(CH_3)_2NH$ and $(CH_3)_2NH_2^+Br^-$. If $K_b = 7.4 \times 10^{-4}$, what is the pH of the solution?

EXERCISE 15-5(d) LEVEL 2: What is the pH of a 0.10-M solution of $HCHO_2$? (Refer to Table 15-2.)

EXERCISE 15-5(e) LEVEL 2: What is the pH of a buffer made by mixing 0.068 mol of NH_3 and 0.049 mol of NH_4Cl in 2.0 L of solution? (Refer to Table 15-3.)

EXERCISE 15-5(f) LEVEL 3: A 0.25-M solution of a hypothetical weak nitrogen base (NX_3) is 1.0% ionized. What is K_b?

EXERCISE 15-5(g) LEVEL 3: If you wished to construct a buffer solution of pH = 7.0 using compounds from Table 15-2, what would be the best choice? What ratio of base to acid would produce the correct pH?

For additional practice, work chapter problems 15-35, 15-39, 15-45, 15-51, and 15-56.

MAKING IT REAL

Buffers and Swimming Pool Chemistry

Control of pH in a swimming pool is essential.

When we take a refreshing dive into a swimming pool, we may not appreciate the fact that maintenance of this pleasure requires application of some serious chemistry. The balance of chemicals in a pool must be closely monitored on a daily basis by some lucky person.

First, the pool must have the proper content of chlorine, which is usually in the form of hypochlorous acid (HOCl), which destroys bacteria and keeps the pool safe. The most common source of chlorine is sodium hypochlorite (bleach) added as a 10% solution. Sodium hypochlorite dissolves to produce the hypochlorite ion, which undergoes basic hydrolysis (reaction **1**). By itself this solution would produce a swimming pool with a pH of around 10 because of hydrolysis of the ClO^- ion. This is way too alkaline. The pH is lowered to about 7.5 by adding hydrochloric acid or sodium bisulfate (reaction **2**). At pH = 7.5, pH = pK_a for this system, which means that equal molar amounts of HOCl and OCl^- are present (reaction **3**). Recall that this is a buffer solution, so the pH is stabilized around 7.5 even if some acid or base enters the pool. However, if the pH does drop below 7.2, the water can sting the

eyes and cause corrosion of metal pipes. Sodium carbonate is added until the pH returns to 7.5 (reaction **4**). If it rises above 7.8, the pool may become cloudy and scaling (i.e., deposits of $CaCO_3$) may occur. The pool manager adds more sodium bisulfate to bring the pH down. The reactions that we have discussed are illustrated by the equations below.

1. Add NaOCl:

$$NaOCl(aq) \xrightarrow{H_2O} Na^+(aq) + OCl^-(aq)$$
$$OCl^- + H_2O \rightleftharpoons HOCl + OH^- \qquad (pH = 10)$$

2. Lower pH:

$$NaHSO_4(s) \xrightarrow{H_2O} Na^+(aq) + HSO_4^-(aq)$$
$$HSO_4^-(aq) + OH^- \longrightarrow SO_4^{2-}(aq) + H_2O$$

3. Buffer equilibrium (pH = pK_a = 7.5):

$$HOCl(aq) + H_2O \rightleftharpoons OCl^-(aq) + H_3O^+(aq)$$

4. Raise pH:

$$Na_2CO_3(s) \xrightarrow{H_2O} 2Na^+(aq) + CO_3^{2-}(aq)$$
$$CO_3^{2-}(aq) + H_3O^+(aq) \longrightarrow HCO_3^-(aq) + H_2O$$

A good clean pool needs even more help. The presence of humans in the pool and their fluids (perspiration, etc.) produce ammonia. Ammonia reacts with HOCl to produce chloramines (i.e., NH_2Cl). If "the pool smells like chlorine," the smell is actually due to chloramines. It is removed by "shocking" the pool by adding another special chemical. If algae form, the pool becomes green and slimy, but another chemical clears it up.

Next time you are invited to take a dip in a nice swimming pool, remember how chemistry, work, and lots of money make it all possible as well as pleasant.

▶ **OBJECTIVE
FOR SECTION 15-6**
Perform calculations involving solubility constants and molar solubility.

15-6 Solubility Equilibria

LOOKING AHEAD! We mentioned in Chapters 6 and 12 that all salts have a specific solubility in water. In fact, saturated solutions of salts demonstrate a dynamic equilibrium between the solid salt and the ions in solution. This is reexamined in this section. ■

Passing a kidney stone is an event that one does not easily forget. It can be as painful as childbirth. The most common type of kidney stone is composed of calcium

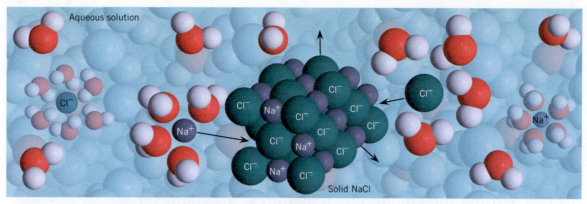

Aqueous solution

Solid NaCl

FIGURE 15-13 Solubility and Equilibrium At equilibrium, the rate at which ions are dissolving equals the rate at which ions are forming the solid.

oxalate (CaC_2O_4), which can be deposited as a solid in the kidney in a typical precipitation reaction when Ca^{2+} ions combine with $C_2O_4^{2-}$ ions. Fortunately, a little knowledge of the equilibrium involved in the formation of kidney stones and an appropriate diet can help prevent their formation in the first place. We will return to kidney stones shortly.

First we will review what happens when a salt such as NaCl is added to water. As discussed in Chapter 12, if enough solid is added to the water, the solution eventually becomes saturated. At this point it appears that the solution process has stopped. Instead, we find that the rate of solution of the salt is equal to the rate of deposition of the solid from the ions in solution and we have a dynamic equilibrium situation. (See Figure 15-13). This is illustrated by the equilibrium shown below.

$$NaCl(s) \rightleftharpoons Na^+(aq) + Cl^-(aq)$$

The principle of equilibrium is demonstrated by a common but interesting chemical phenomenon that is often displayed at science fairs. This is the science (and art) of crystal growing. In this experiment, a comparatively large crystal of a compound such as blue $CuSO_4$ is suspended in a saturated solution of that compound with excess small crystals at the bottom of the container. The larger crystal grows slowly at the expense of the smaller crystals at the bottom. During this time, the concentrations of Cu^{2+} and SO_4^{2-} in solution remain constant, as does the total mass of solid crystals. The changing size of the crystals, however, indicates that an equilibrium exists in which the identities of ions in the solid and dissolved states are changing. (See Figure 15-4.) This phenomenon occurs because small crystals, with their greater surface area, dissolve faster than large crystals. Thus, although the total mass of the crystals remains constant, the unequal rate of solution favors solution of small crystals and precipitation on the large crystal.

15-6.1 The Solubility Product

Now consider the equilibrium between solid calcium oxalate and its ions in solution.

$$CaC_2O_4(s) \rightleftharpoons Ca^{2+}(aq) + C_2O_4^{2-}(aq)$$

The extent of the solubility of this compound can be represented by an equilibrium constant known as the **solubility product constant (K_{sp})** or simply the **solubility product**. In this case the constant expression is

$$K_{sp} = [Ca^{2+}][C_2O_4^{2-}]$$

It is set up like any other equilibrium constant expression except that the solid (on the left) is not included. The reason for this is that the concentration of a

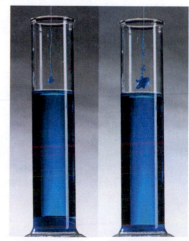

FIGURE 15-14 Crystal Growth After the $CuSO_4$ crystal on the left was suspended in the solution for a few days, it grew larger.

solid is a constant. *So, as long as at least some solid is present, how much solid is present is not relevant to the concentration of ions in solution.* Another example of a solubility product involves the solution of lead(II) chloride.

$$PbCl_2(s) \rightleftharpoons Pb^{2+}(aq) + 2Cl^-(aq) \qquad K_{sp} = [Pb^{2+}][Cl^-]^2$$

Notice that the $[Cl^-]$ concentration is squared because there are two chlorides in the equilibrium equation.

The equilibrium constant expressions can be written for other salts, as illustrated in Example 15-10.

EXAMPLE 15-10 **Writing the Equilibrium Constant Expressions for K_{sp}**

Write the equilibrium constant expressions (K_{sp}) for a solution of **(a)** Ag_2CrO_4, **(b)** $Fe(OH)_3$, and **(c)** $Ca_3(PO_4)_2$.

PROCEDURE

Write the equations for the solution of the salts in water. The ionic solid is the reactant, and the aqueous ions are the products. It may be necessary to review Chapter 4 and specifically Table 4-2 to determine the identity of the ions formed from the compounds.

(a) $Ag_2CrO_4(s) \rightleftharpoons 2Ag^+(aq) + CrO_4^{2-}(aq)$

(b) $Fe(OH)_3(s) \rightleftharpoons Fe^{3+}(aq) + 3OH^-(aq)$

(c) $Ca_3(PO_4)_2(s) \rightleftharpoons 3Ca^{2+}(aq) + 2PO_4^{3-}(aq)$

SOLUTION

(a) $K_{sp} = [Ag^+]^2[CrO_4^{2-}]$

(b) $K_{sp} = [Fe^{3+}][OH^-]^3$

(c) $K_{sp} = [Ca^{2+}]^3[PO_4^{3-}]^2$

ANALYSIS

As ionic compound formulas involve cations and anions with charges of 3 or less, there are relatively few forms the solubility product can take:

Salt Formula	Solubility Product
MX	$K_{sp} = [M^+][X^-]$
MX_2 or M_2X	$K_{sp} = [M^{2+}][X^-]^2$ or $K_{sp} = [M^+]^2[X^{2-}]$
MX_3 or M_3X	$K_{sp} = [M^{3+}][X^-]^3$ or $K_{sp} = [M^+]^3[X^{3-}]$
M_2X_3 or M_3X_2	$K_{sp} = [M^{3+}]^2[X^{2-}]^3$ or $K_{sp} = [M^{2+}]^3[X^{3-}]^2$

Any of the compounds you'll run into will have one of these forms.

SYNTHESIS

For even the most difficult problems that chemists are likely to encounter, writing the reaction that is occurring is typically the first step. But since many reactions are equilibrium reactions, what you write as the reactant and what you write as the product are not always clear-cut. Any one of the reactions above could be written in the opposite direction and still be a valid chemical equation. The equilibrium expression would then simply be the reverse of the one we've already written, and the equilibrium constant would be the inverse of the reported value. Because there is no natural restriction on how we need to write a reaction, we have to agree to a convention that all chemists will follow. The convention for solubility products is to write the solid compound first, dissolving into its component ions.

15-6.2 The Value of K_{sp}

The values for K_{sp} are determined from molar solubility data, as illustrated in Examples 15-11 and 15-12.

EXAMPLE 15-11 **Determining K_{sp} from Solubility Data**

The molar solubility of AgCl is 1.3×10^{-5}. What is the value of K_{sp} for AgCl?

PROCEDURE

Write the equation for the solution of AgCl and the expression for K_{sp}.

$$AgCl(s) \rightleftharpoons Ag^+(aq) + Cl^-(aq) \quad K_{sp} = [Ag^+][Cl^-]$$

Notice from the stoichiometry of the equilibrium that for every mole of AgCl that dissolves, one mole of Ag^+ and one mole of Cl^- are formed. Therefore, the concentration of each ion is equal to the amount that dissolves, or the molar solubility.

SOLUTION

$$\text{solubility} = 1.3 \times 10^{-5} \text{ mol/L} = [Ag^+] = [Cl^-]$$

$$[1.3 \times 10^{-5}][1.3 \times 10^{-5}] = \underline{1.7 \times 10^{-10}}$$

EXAMPLE 15-12 **Determining K_{sp} from Solubility Data**

What is the value of K_{sp} for $PbCl_2$ if the molar solubility of this compound is 3.9×10^{-2} mol/L?

PROCEDURE

Write the solution equilibrium and the expression for K_{sp}.

$$PbCl_2 \rightleftharpoons Pb^{2+}(aq) + 2Cl^-(aq) \quad K_{sp} = [Pb^{2+}][Cl^-]^2$$

Notice from the stoichiometry of the solution that for each mole of $PbCl_2$ that dissolves, it produces one mole of Pb^{2+} and *two* moles of Cl^-. Therefore, the concentration of Pb^{2+} is equal to the actual moles/L that dissolve (i.e., the solubility), but the concentration of Cl^- is twice the solubility.

SOLUTION

$$[Pb^{2+}] = 3.9 \times 10^{-2} \quad [Cl^-] = 2 \times 3.9 \times 10^{-2} = 7.8 \times 10^{-2}$$

$$K_{sp} = [3.9 \times 10^{-2}] [7.8 \times 10^{-2}]^2 = (3.9 \times 10^{-2}) (6.1 \times 10^{-3}) = \underline{2.4 \times 10^{-4}}$$

ANALYSIS

Notice that the 2 in front of Cl^- is used twice—once to multiply the concentration by 2 and again to raise the concentration to the second power. What would it be for a hypothetical chemical, M_2X_3, whose molar solubility is 0.10 mol/L? The 2 and the 3 would each be used twice: $(2 \times 0.10)^2 \times (3 \times 0.10)^3 = 1.1 \times 10^{-3}$.

SYNTHESIS

Unlike K_a for acids, the solubility of two different salts cannot be compared directly by simply observing the value of K_{sp}. In setting up the expression for K_a, the equilibrium constant expressions for weak acids were all the same. The equilibrium constant expressions for salts, however, take on various forms depending on the stoichiometry of the salts. For example, two salts, MX and MY_2, may have the same solubility (e.g., 10^{-2} mol/L) but very different values for K_{sp}. For MX, $K_{sp} = 1 \times 10^{-4}$ and for MY_2, $K_{sp} = 4 \times 10^{-6}$.

EXAMPLE 15-13 **Determining Molar Solubility from K_{sp}**

What is the molar solubility of CaC_2O_4? For CaC_2O_4, $K_{sp} = 2.1 \times 10^{-9}$.

PROCEDURE

Write the solution equilibrium for CaC_2O_4 and the expression for K_{sp}.

$$CaC_2O_4(s) \rightleftharpoons Ca^{2+}(aq) + C_2O_4^{2-}(aq) \quad K_{sp} = [Ca^{2+}] [C_2O_4^{2-}]$$

Notice that the number of moles of Ca^{2+} in solution is equal to the number of moles of $C_2O_4{}^{2-}$ and both are equal to the molar solubility. We will let $x =$ the molar solubility.

SOLUTION

If $x =$ the molar solubility, then at equilibrium $[Ca^{2+}] = [C_2O_4{}^{2-}] = x$.
Thus $[x][x] = K_{sp} = 2.1 \times 10^{-9}$.
   Taking the square root of K_{sp}, the molar solubility is 4.6 x 10^{-5} mol/L.

ANALYSIS

When one dozen pairs of shoes dissociate, they form one dozen right shoes and one dozen left shoes. One dozen pairs produces two dozen shoes. Likewise, one mole of CaC_2O_4 produces one mole of Ca^{2+} and one mole of $C_2O_4{}^{2-}$. To extend this to the abstract, x moles of CaC_2O_4 dissociate into x moles of Ca^{2+} and x moles of $C_2O_4{}^{2-}$.

SYNTHESIS

Consider a more complex case where you are asked to calculate the molar solubility of a salt such as PbI_2. In this case, x would equal the concentration of Pb^{2+}, as in this example, but the concentration of I^- would be $2x$. The solution has $4x^3$ equal to K_{sp}. See if you can set this up.

15-6.3 K_{sp} and the Formation of Precipitates

Most kidney stones are formed from the precipitation of calcium oxalate.

Now we return to the formation of calcium oxalate kidney stones as described at the beginning of this section. Why does calcium oxalate precipitate from an aqueous solution such as the bloodstream? In a complex system such as in our blood, Ca^{2+} ion and $C_2O_4{}^{2-}$ ion originate from many sources, so they are not exactly equal. However, whenever the product of the concentrations of the two ions is equal to K_{sp}, the solution is saturated with that salt and at equilibrium. In Example 15-14 we have calculated the concentration of oxalate that is present in a saturated solution with a specific calcium ion concentration. The answer represents the maximum amount of that ion that can be present without formation of solid calcium oxalate. If the concentration of Ca^{2+} and $C_2O_4{}^{2-}$ is such that the product of their concentrations is greater than the value for K_{sp}, the system is not at equilibrium. *To reestablish equilibrium, the concentration of the ions must decrease by coming together to form a precipitate.* So how does one use this information to avoid calcium oxalate kidney stones? A stone (precipitate) will not form if the product of the two ion concentrations is equal to or less than K_{sp}. There are three ways to accomplish this. (1) Lower the concentration of both ions by dilution. In other words, drink plenty of water. This is always a good idea and probably the most important. (2) Lower the concentration of Ca^{2+} ion. This is not a good idea. Ca^{2+} is also incorporated into teeth and bones. If we lower the Ca^{2+} ion concentration too much, the solubility equilibrium involving bones may be affected and cause bone tissue to dissolve (osteoporosis). We don't want to avoid kidney stones by having brittle bones. Lowering calcium ion concentration can have an effect opposite to what we desire. (3) Lower the concentration of oxalate ion. Many foods such as celery, spinach, and beets have relatively high amounts of oxalate salts. If one is susceptible to these types of kidney stones, the best advice is to drink plenty of water and limit high-oxalate foods (which also include chocolate, unfortunately).

EXAMPLE 15-14   Determining the Concentrations of Ions in a Saturated Solution

What is the maximum oxalate ion concentration in a solution where $[Ca^{2+}] = 4.4 \times 10^{-4}$ mol/L?

PROCEDURE

Solve for $[C_2O_4{}^{2-}]$ in the expression for K_{sp}.

SOLUTION

$$K_{sp} = [Ca^{2+}][C_2O_4^{2-}]$$

$$[C_2O_4^{2-}] = K_{sp}/[Ca^{2+}] = 2.1 \times 10^{-9}/4.4 \times 10^{-4} = \underline{4.8 \times 10^{-6} \text{ M}}$$

If the concentration of oxalate were any higher than the calculated value, a precipitate of CaC_2O_4 would form.

ANALYSIS

It is important to note that the concentrations of ions do not have to be the same for a precipitate to form. The critical point is that if the ion product, set up like the equilibrium constant expression, exceeds the value of K_{sp}, a precipitate forms. In fact, as you can see from the expression, the higher the concentration of oxalate ion, the lower the concentration of calcium ion that can be tolerated before a solid forms. Also, it does not matter what other ions may be present. For example, the Ca^{2+} may be in solution from a variety of compounds and the oxalate may have entered the blood in solution as sodium oxalate.

SYNTHESIS

This concept can also be used in conjunction with pH. Most hydroxides of metals have very low solubilities. Since one of the ions produced in solution is the OH^- ion, addition of acid to the solution will affect the solubility. This is because the added H^+ reacts with OH^- to form water (neutralization). According to Le Châtelier's principle, removing the OH^- ion in the equilibrium leads to a shift of the equilibrium to the right. Therefore, most metal hydroxides are more soluble at lower pH's than they are at higher pH's. Water companies make use of this fact to treat hard water by increasing the pH of the water at the processing plant. Increasing the pH, along with adding "flocculants," causes metal ions such as Fe^{3+}, Ca^{2+}, and Mg^{2+}, which cause hard water in many locales, to precipitate out as solid hydroxides. This solid is then allowed to settle, and is removed.

▶ **ASSESSING THE OBJECTIVE FOR SECTION 15-6**

EXERCISE 15-6(a) LEVEL 1: Write the solubility constant expression for
(a) $Mg_3(PO_4)_2$ **(b)** $Al(OH)_3$ **(c)** $CaCO_3$

EXERCISE 15-6(b) LEVEL 2: The solubility of $Mg(OH)_2$ is 1.3×10^{-4} mol/L. What is the value of K_{sp}?

EXERCISE 15-6(c) LEVEL 2: What is the molar solubility of barium carbonate? For $BaCO_3$, $K_{sp} = 8.1 \times 10^{-9}$.

EXERCISE 15-6(d) LEVEL 2: What is the Mg^{2+} concentration in a solution with pH = 10.00? Use K_{sp} for $Mg(OH)_2$ from Exercise 15-6(b).

EXERCISE 15-6(e) LEVEL 3: Which compound is more soluble in mol/L, CuI ($K_{sp} = 1.1 \times 10^{-12}$) or $Ni(OH)_2$ ($K_{sp} = 2.0 \times 10^{-15}$)? Would the addition of acid to either of these solutions affect the solubility?

For additional practice, work chapter problems 15-62 and 15-66.

PART B SUMMARY

KEY TERMS

15-4.1   The **equilibrium constant (K_{eq})** has a specific value. The **equilibrium constant expression** is a fraction related to the equilibrium involved. p. 532

15-5.1   The equilibrium constant expression for a weak acid is known as the **acid ionization constant (K_a)** and for a base is known as the **base ionization constant (K_b)**. p. 536

15-5.5   The **Henderson–Hasselbalch equation** is used to calculate the pH of a buffer solution. p. 542

15-6.1   The solubility of an ionic compound in water can be expressed as a **solubility product constant (K_{sp})** or simply **solubility product**. p. 545

SUMMARY CHART

Equilibrium Constant Expressions

Gases	Weak Acid in Water	Ionic Compound in Water
$A_2(g) + B_2(g) \rightleftharpoons 2AB(g)$	$HX(aq) + H_2O \rightleftharpoons H_3O^+ + A^-(aq)$	$MX_2 \rightleftharpoons M^{2+}(aq) + 2X^-(aq)$
$K_{eq} = \dfrac{[AB]^2}{[A_2][B_2]}$	$K_a = \dfrac{[H_3O^+][A^-]}{[HA]}$	$K_{sp} = [M^{2+}][X^-]^2$

CHAPTER 15 SYNTHESIS PROBLEM

Hydrogen sulfide (H_2S) is a poisonous gas, which results from the decay of foods. Fortunately, it is very pungent so the nose tells us that it is present well before we inhale toxic amounts. It is also responsible for the tarnishing of silver and corrosion of copper pipes. H_2S is produced as a by-product from the production of carbon disulfide (CS_2), which is a common solvent. This is a gas-phase reaction that takes place at a high temperature. We then look into the action of H_2S as a weak acid in water.

PROBLEM	SOLUTION
a. The following equilibrium reaction takes place at 600°C, where sulfur is in the gaseous phase. It is an exothermic reaction. Sulfur is the more expensive reactant. $CH_4(g) + 4S(g) \rightleftharpoons CS_2(g) + 2H_2S(g)$ List three ways the rate of the reaction can be increased.	**a.** Increase the concentration of CH_4, increase the temperature, and add a catalyst.
b. List three ways the concentration of H_2S can be increased at equilibrium. Why is the reaction run at such a high temperature as 600°C ?	**b.** Increase the concentration of CH_4, or S, increase the pressure, and lower the temperature. The temperature must be high enough so that the rate of formation of H_2S is faster although the concentration of H_2S will be lower. A catalyst is used in this reaction to minimize the temperature.
c. Write the equilibrium constant expression for this reaction.	**c.** $K_{eq} = \dfrac{[CS_2][H_2S]^2}{[CH_4][S]^4}$
d. Given the following equilibrium concentrations: $[CS_2] = 0.25$, $[H_2S] = 0.30$, $[CH_4] = 0.015$, and $[S] = 0.10$, what is the value of K_{eq}?	**d.** $K_{eq} = \dfrac{(0.25)(0.30)^2}{(0.015)(0.10)^4} = 1.5 \times 10^4$
e. H_2S is a weak acid in water. Write the reactions of H_2S in water.	**e.** H_2S is a diprotic acid. $H_2S(aq) + H_2O \rightleftharpoons H_3O^+(aq) + HS^-(aq)$ $HS^-(aq) + H_2O \rightleftharpoons H_3O^+(aq) + S^{2-}(aq)$
f. The $[H_3O^+]$ concentration comes almost entirely from the first ionization so the second ionization can be ignored. Write the expression for K_a for the first ionization and calculate its value if the pH = 4.00 in a 0.10-M solution of H_2S.	**f.** $K_a = \dfrac{[H_3O^+][HS^-]}{[H_2S]}$ pH = 4.00, $[H_3O^+] = 1.0 \times 10^{-4} = [HS^-]$ $H_2S = 0.10 - 0.001 \approx 0.10$ $K_a = \dfrac{(1.0 \times 10^{-4})(1.0 \times 10^{-4})}{0.10} = 1.0 \times 10^{-7}$
g. What is the pH of a buffer made by dissolving 0.20 mol of NaHS in 1.00 L of a 0.10-M solution of H_2S? Write two reactions illustrating how this buffer solution reacts with added H_3O^+ and added OH^-.	**g.** pH = $pK_a + \log\dfrac{[base]}{[acid]}$ $pK_a = -\log K_a = 7.00$. pH = $7.00 + \log\dfrac{(0.20)}{(0.10)}$ pH = $7.00 + 0.30 = \underline{7.30}$ added acid: $H_3O^+ + HS^-(aq) \longrightarrow H_2S(aq) + H_2O$ added base: $OH^- + H_2S(aq) \longrightarrow HS^-(aq) + H_2O$

PROBLEM	SOLUTION
h. In a 0.10-M H_2S solution, $[S^{2-}] = 1 \times 10^{-19}$, which is very low. Does a precipitate of FeS form in a 0.010-M solution of $[Fe^{2+}]$? K_{sp} for FeS $= 6 \times 10^{-19}$.	**h.** The equation for the solubility of FeS is $$FeS(s) \rightleftharpoons Fe^{2+}(aq) + S^{2-}(aq) \quad K_{sp} = [Fe^{2+}][S^{2-}]$$ In the solution, $(0.010)(1 \times 10^{-19}) = 1 \times 10^{-21}$. This is a smaller value than K_{sp} so the solution is not saturated with FeS and a precipitate does not form.

YOUR TURN

Hydrogen fluoride (HF) is also a toxic gas. In aqueous solution it is a weak acid but very dangerous to handle since it causes severe burns and is absorbed into the bloodstream. Its aqueous solutions are used to etch glass since HF reacts with SiO_2 (a component of glass).

a. The following reaction involving HF takes place in the gaseous phase.

$$CF_4\,(g) + 2H_2O(g) \rightleftharpoons 4HF(g) + CO_2(g) \quad \Delta H = 42\ kJ$$

List three ways the rate of the forward reaction can be increased. CF_4 is the more expensive reactant.

b. List three ways the concentration of CF_4 can be increased at equilibrium.

c. Write the equilibrium constant expression for this reaction.

d. The value of K_{eq} at this temperature is 4.0×10^{-6} and $[HF] = 1.0 \times 10^{-1}$, $[CO_2] = 2.0 \times 10^{-3}$, and $[H_2O] = 0.50$. What is $[CF_4]$?

e. Write the reaction of HF in water.

f. Write the K_a expression for the ionization of HF and calculate the pH of a 0.20-M solution. For HF, $K_a = 6.7 \times 10^{-4}$.

g. What is the pH of a buffer made by dissolving 0.10 mol of NaF in 1.00 L of a 0.20-M solution of HF? Write two reactions illustrating how this buffer solution reacts with added H_3O^+ and added OH^-.

h. Using the value for the concentration of F^- from part **f** above, determine whether a precipitate of CaF_2 forms in a solution that has $[Ca^{2+}] = 0.010$. For CaF_2, $K_{sp} = 4.0 \times 10^{-11}$.

Answers are on p. 554.

CHAPTER SUMMARY

This chapter concerns several important chemistry topics, such as how reactions take place, the rates at which they occur, and the point of equilibrium in reversible reactions. **Collision theory** tells us that reactions take place through collisions of molecules. The **rate of a reaction** is determined by the frequency of these collisions and the **activation energy** of the reaction. There are several things that influence these two factors and thus affect the rate of the reaction. These are: (1) The magnitude of the *activation energy*. Each set of potential reactants requires a different amount of energy to form the **activated complex** that leads to the formation of products. If collisions do not have the necessary kinetic energy, no products form. (2) The *temperature*. (3) The *concentration of reactants*. (4) The *particle size* in the case of heterogeneous reactions. (5) A **catalyst**.

Reversible reactions often reach a **point of equilibrium** where measurable concentrations of reactants and products coexist. In fact, a **dynamic equilibrium** exists because both the forward and reverse reactions are occurring at the same rate. The point of equilibrium can be influenced by reaction conditions. However, changes in these reaction conditions have a predictable effect on the point of equilibrium, which is correlated by **Le Châtelier's principle**. The factors that can shift the point of equilibrium include (1) *concentration*, (2) *pressure*, and (3) *temperature*.

For exothermic reactions, an increase in temperature leads to a decrease in the proportion of products. Just the opposite occurs for an endothermic reaction. Many important industrial reactions do not take place readily, or even at all, without a catalyst present. A catalyst brings a reaction mixture to the point of equilibrium faster but does not change the actual point of equilibrium.

The distribution of concentrations of reactants and products can be predicted. This is done using a numerical value (K_{eq}) called the **equilibrium constant**, which is equal to a ratio of reactants and products constructed from the balanced equation and known as the **equilibrium constant expression**.

The value of the equilibrium constant at a given temperature is obtained by experimental measurements of the distribution of reactants and products at equilibrium. The constant and the initial concentrations can then be used to calculate the concentration of a given gas or ion at equilibrium.

The equilibrium of weak acids and bases can also be expressed in terms of an equilibrium constant expression. The **acid ionization constants** for weak acids (K_a) and the **base ionization constants** for weak bases (K_b) are useful for calculation of the pH of these solutions. We are also able to use equilibrium constants to calculate the pH of buffers. The **Henderson–Hasselbalch equation** can be used to calculate the pH of buffer solutions. The solubility of salts in water is determined by the equilibrium constant K_{sp}, which is known as the **solubility product constant** or simply the **solubility product**. K_{sp} can be calculated from solubility data, and vice versa.

OBJECTIVES

SECTION	YOU SHOULD BE ABLE TO...	EXAMPLES	EXERCISES	CHAPTER PROBLEMS
15-1	Describe how a chemical reaction takes place on a molecular level by using an energy diagram.		1a, 1 b, 1c	2, 4, 8, 9
15-2	Describe the types and relative strengths of intermolecular forces that can occur between two molecules.		2a, 2b	
15-3	Using Le Châtelier's principle, predict what effect changing certain conditions will have on the equilibrium point.		3a, 3b, 3c, 3d	11, 13, 18
15-4	Perform calculations involving the equilibrium constant and equilibrium concentrations.	15-1, 15-2, 15-3, 15-4	4a, 4b, 4c, 4d, 4e	15, 16, 20, 22, 24, 26, 27, 28, 29, 33, 34
15-5	Perform calculations involving equilibrium constants and pH for acid–base reactions.	15-5, 15-6, 15-7, 15-8, 15-9	5a, 5b, 5c, 5d, 5e, 5f, 5g	35, 36, 37, 41, 42, 46, 48, 50, 52, 57
15-6	Perform calculations involving solubility constants and molar solubility.	15-10, 15-11, 15-12, 15-13, 15-14	6a, 6b, 6c, 6d, 6e	60, 61, 63, 64, 65, 67, 68, 69

▶ ANSWERS TO ASSESSING THE OBJECTIVES

Part A

EXERCISES

15-1(a) Chemical reactions occur because of <u>collisions</u> between reactant molecules. In order for a reaction to take place, molecules must <u>collide</u> in the proper orientation and with a <u>minimum</u> amount of energy, known as the <u>activation</u> energy. An <u>activated complex</u> forms at the point of maximum potential energy.

15-1(b)

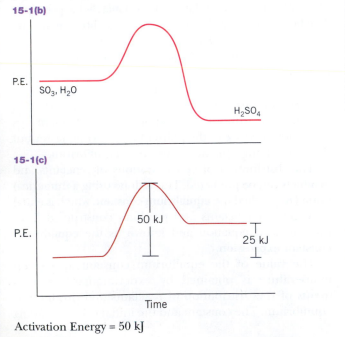

15-1(c)

Activation Energy = 50 kJ

ΔH = 25 kJ

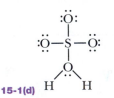

15-1(d)

To form the product molecule, a hydrogen must transfer from the oxygen in the water molecule to another oxygen. Sulfuric acid has two —O—H bonds.

15-2(a) The rate of appearance of a product as a function of <u>time</u> is a measure of the <u>rate</u> of the reaction. At <u>higher</u> temperatures, more molecules can overcome this energy barrier, so the rate of the reaction <u>increases</u>. Increasing the concentration of colliding molecules <u>increases</u> the rate of the reaction. The presence of a <u>catalyst</u> does not change the distribution of products, but it <u>increases</u> the rate of the reaction.

15-2(b) **(a)** Speed it up   **(b)** Speed it up
(c) Slow it down   **(d)** Speed it up   **(e)** Speed it up

15-2(c) **(a)** Cooling the body slows the rate of metabolism, which means organs like the brain can go longer without oxygen. This increases the time that blood flow can be interrupted before damage occurs. **(b)** Pure oxygen increases the concentration of a reactant of combustion. This increases the rate of the reaction and the amount of heat generated. **(c)** Manganese dioxide acts as a catalyst, which increases the rate of the reaction. **(d)** Enzymes are biological catalysts. They are necessary to break down sugars or to cause thousands of other chemical reactions in the human body.

15-3(a) Reactions that are reversible may reach a <u>point of equilibrium</u> where the concentrations of reactants and products remain <u>constant</u>. The point of equilibrium may be affected by a change of <u>concentrations</u> of reactants, a change of <u>temperature</u>, and a change of <u>pressure</u> (in a gaseous system).

15-3(b) A catalyst does not increase the amount of product formed (**b**). It just produces it faster. It also does not change the heat of the reaction (**e**).

15-3(c) (**a**) increased (**b**) decreased (**c**) increased (**d**) decreased

15-3(d) Recall that all reactions slow as the temperature decreases. We reach a point where the reaction stops as the temperature is lowered. The temperature must be raised for the reaction to proceed, even though the yield of product is reduced.

15-3(e) Yes, we would. Equilibria in chemical reactions are dynamic, which means that over time, that product molecule will be converted back into reactant, and later into product again. It will continue to move back and forth until it is involved in a different reaction.

Part B

EXERCISES

15-4(a) A gaseous equilibrium can be expressed as a number known as the <u>equilibrium constant</u> and a ratio of reactants and products known as the <u>equilibrium</u> constant <u>expression</u>. A small value for K_{eq} (e.g., 10^{-5}) indicates that <u>reactants</u> are favored at equilibrium.

15-4(b)

(**a**) $K_{eq} = \dfrac{[CH_4][H_2O]}{[CO][H_2]^3}$

(**b**) $K_{eq} = \dfrac{[NOBr]^2}{[NO]^2[Br_2]}$

15-4(c)

$$K_{eq} = \frac{\left(\dfrac{3.5 \times 10^{-4}\,mol}{10.0\,L}\right)\left(\dfrac{5.0 \times 10^{-5}\,mol}{10.0\,L}\right)}{\left(\dfrac{2.0 \times 10^{-3}\,mol}{10.0\,L}\right)\left(\dfrac{1.0 \times 10^{-2}\,mol}{10.0\,L}\right)^3}$$

$$= \frac{17.5 \times 10^{-11}}{2.0 \times 10^{-13}} = \underline{\underline{880}}$$

15-4(d) (**a**) $K_{eq} = 20$ (**b**) $[NOBr] = 0.94$ mol/L

15-4(e) The concentration (mol/L) of a pure solid or liquid is essentially a constant, just like its density (g/mL). Pressure and temperature have little effect on the volume of a liquid or solid. However, the concentration of a gas is a variable very much dependent on the temperature and pressure. Only variables appear in the equilibrium constant expression. As long as $CaCO_3$ and CaO are present, the distribution of reactants and products at equilibrium depends only on the concentration of CO_2.

15-5(a) $HCHO_2(aq) + H_2O(l) \rightleftharpoons H_3O^+(aq) + CHO_2^-(aq)$

$$K_a = \frac{[H_3O^+][CHO_2^-]}{[HCHO_2]}$$

$NX_3(aq) + H_2O(l) \rightleftharpoons NX_3H^+(aq) + OH^-(aq)$

$$K_b = \frac{[NX_3H^+][OH^-]}{[NX_3]}$$

15-5(b) $[H_3O^+] = [anion] = 1.3 \times 10^{-3}$ $[Acid] = 0.10 - 0.0013 = 0.10$

$$K_a = \frac{(1.3 \times 10^{-3})(1.3 \times 10^{-3})}{0.10} = 1.7 \times 10^{-5}$$

This K_a is closest to acetic acid ($HC_2H_3O_2$).

15-5(c)

$$pK_b = 3.13 \quad pOH = pK_b + \log\frac{[acid]}{[base]} = pK_b + \log 1$$

$$pOH = pK_b = 3.13 \quad pH = 14.00 - 3.13 = \underline{10.87}$$

15-5(d) From Table 15-2 for $HCHO_2$, $K_a = 1.8 \times 10^{-4}$. Let $x = [H_3O^+] = [CHO_2^-]$ at equilibrium. Then $[HCHO_2] = 0.10 - x \approx 0.10$ at equilibrium.

$$\frac{x^2}{0.10} = 1.8 \times 10^{-4} \quad x^2 = 1.8 \times 10^{-5} \quad x = 4.2 \times 10^{-3}$$

$$pH = -\log(4.2 \times 10^{-3}) = \underline{2.38}$$

15-5(e) The base species is NH_3: [base] = 0.068 mol/2.0 L.

The acid species is NH_4^+: [acid] = 0.049 mol/2.0 L.

(The Cl^- is a spectator ion and is not part of the equilibrium.)

$$pOH = 4.74 + \log\frac{(0.049\,mol/2.0\,L)}{(0.068\,mol/2.0\,L)} = 4.74 - 0.14 = 4.60$$

$$pH = 14.00 - 4.60 = \underline{9.40}$$

15-5(f) $[OH^-] = [NX_3H^+] = 0.0025$, $[NX_3] = 0.25$ $K_b = 2.5 \times 10^{-5}$

15-5(g) Try to match the pH to the pK_a. The closest is hypochlorous acid, with $pK_a = 7.5$. The two compounds should therefore be HClO and its salt, NaClO (as an example). Use the Henderson–Hasselbalch equation to determine the ratio of the two. $7.0 = 7.5 + \log[ClO^-]/[HClO]$:

$$\log\frac{[ClO^-]}{[HClO]} = -0.5$$

$$\frac{[ClO^-]}{[HClO]} \approx \frac{1}{3}$$

A 3:1 mole ratio of HClO to NaClO will buffer a solution at pH = 7.0

15-6(a) (**a**) $K_{sp} = [Mg^{2+}]^3[PO_4^{3-}]^2$ (**b**) $K_{sp} = [Al^{3+}][OH^-]^3$ (**c**) $K_{sp} = [Ca^{2+}][CO_3^{2-}]$

15-6(b) $K_{sp} = [1.3 \times 10^{-4}][2.6 \times 10^{-4}]^2 = 8.8 \times 10^{-12}$

15-6(c) $x^2 = 8.1 \times 10^{-9}$ $x = 9.0 \times 10^{-5}$

15-6(d) $[Mg^{2+}][OH^-]^2 = 8.8 \times 10^{-12}$ $[Mg^{2+}] = \dfrac{K_{sp}}{[OH^-]^2}$

If pH = 10.00, pOH = 4.00, $[OH^-] = 1.0 \times 10^{-4}$

$$[Mg^{2+}] = \frac{8.8 \times 10^{-12}}{(1.0 \times 10^{-4})^2} = 8.8 \times 10^{-4}\,mol/L$$

15-6(e) You cannot compare K_{sp} values directly. Calculate solubilities.

CuI: $K_{sp} = [Cu^+][I^-]$ ⠀⠀⠀ $1.1 \times 10^{-12} = x^2$

⠀⠀⠀⠀⠀ $x = 1.0 \times 10^{-6}$ mol/L

Ni(OH)$_2$: $K_{sp} = [Ni^{2+}][OH^-]^2$ ⠀ $2.0 \times 10^{-15} = 4x^3$

⠀⠀⠀⠀⠀ $x = 7.9 \times 10^{-6}$ mol/L.

Ni(OH)$_2$ is marginally more soluble despite the smaller K_{sp}. However, because it contains OH^-, it is susceptible to pH. Decreasing the pH decreases the OH^- concentration and increases the solubility of the compound.

ANSWERS TO CHAPTER SYNTHESIS PROBLEM

a. Add H_2O, increase the temperature, and add a catalyst.

b. Increase the concentration of HF, or CO_2, increase the pressure, and lower the temperature.

c. $K_{eq} = \dfrac{[CO_2][HF]^4}{[CF_4][H_2O]^2}$

d. $[CF_4] = 0.20$

e. $HF(aq) + H_2O \rightleftharpoons$
⠀⠀⠀ $H_3O^+(aq) + F^-(aq)$

f. $K_a = \dfrac{[H_3O^+][F^-]}{[HF]}$

Let $x = [H_3O^+] = [F^-]$ $[HF] = 0.20 - x \approx 0.20$

$\dfrac{x^2}{0.20} = 6.7 \times 10^{-4}$

$x^2 = 1.3 \times 10^{-4}$ ⠀ $x = 0.011$

pH $= -\log(0.011) = 1.96$

g. pH $= pK_a + \log \dfrac{[base]}{[acid]}$

$pK_a = -\log K_a = 3.17$ pH $=$

$3.17 + \log\left(\dfrac{0.10}{0.20}\right)$

pH $= 3.17 - 0.30 = 2.87$

added acid: $H_3O^+ + F^-(aq)$
⠀⠀ $\longrightarrow HF(aq) + H_2O$
added base: $OH^- + HF(aq)$
⠀⠀ $\longrightarrow F^-(aq) + H_2O$

h. $K_{sp} =$
$[Ca^{2+}][F^-]^2 (0.010)(0.011)^2 =$
1.2×10^{-6}
This is a larger value than K_{sp} so a precipitate of CaF$_2$ will form.

CHAPTER PROBLEMS

Throughout the text, answers to all exercises in color are given in Appendix E. The more difficult exercises are marked with an asterisk.

Collision Theory and Reaction Rates

(SECTIONS 15-1 AND 15-2)

15-1. Explain why all collisions between reactant molecules do not lead to products.

15-2. The equation

$$NO_2(g) + CO(g) \rightleftharpoons NO(g) + CO_2(g)$$

represents a reaction in which the mechanism of the forward reaction involves a collision between the two reactant molecules. Describe the probable orientation of the molecules in this collision.

15-3. Explain the following facts from your knowledge of collision theory and the factors that affect the rates of reactions.

(a) The rates of chemical reactions approximately double for each 10°C rise in temperature.

(b) Eggs cook more slowly in boiling water at higher altitudes where water boils at temperatures less than 100°C.

(c) H_2 and O_2 do not start to react to form H_2O except at very high temperatures (unless initiated by a spark)

(d) Wood burns explosively in pure O_2 but slowly in air, which is about 20% O_2.

(e) Coal dust burns faster than a single lump of coal.

(f) Milk sours if left out for a day or two but will last two weeks in the refrigerator.

(g) H_2 and O_2 react smoothly at room temperature in the presence of finely divided platinum.

15-4. Explain the following facts from your knowledge of collision theory and the factors that affect the rates of reactions.

(a) A wasp is lethargic at temperatures below 65°F.

(b) Charcoal burns faster if you blow on the coals.

(c) A 0.10 M boric acid solution ($[H_3O^+] = 7.8 \times 10^{-6}$) can be used for eyewash, but a 0.10 M hydrochloric acid solution ($[H_3O^+] = 0.10$) would cause severe damage.

(d) To keep apple juice from fermenting to apple cider, the apple juice must be kept cold.

(e) Coal does not spontaneously ignite at room temperature, but white phosphorus does.

(f) Potassium chlorate decomposes at a much lower temperature when mixed with MnO_2.

Equilibrium and Le Châtelier's Principle (SECTION 15-3)

15-5. Explain the point of equilibrium in terms of reaction rates.

15-6. In the hypothetical reaction between A_2 and B_2, when was the rate of the forward reaction at a maximum? In the same reaction, when was the rate of the reverse reaction at a maximum? (Refer to Figure 15-12.)

15-7. Give a reason, besides reaching a point of equilibrium, why certain reactions do not go to completion (100% to the right).

15-8. Compare the activation energy for the reverse reaction with that for the forward reaction in Figure 15-3. Which are easier to form, reactants from products or products from reactants? Which system would come to equilibrium faster, a reaction starting with pure products or one starting with pure reactants?

15-9. Consider an endothermic reaction that comes to a point of equilibrium such as shown in Figure 15-4. Which is greater, the activation energy for the forward or for the reverse reaction? Which system would come to equilibrium faster, a reaction starting with pure products or one starting with pure reactants?

15-10. The following equilibrium represents an important industrial process to produce sulfur trioxide, which is then converted to sulfuric acid. More sulfuric acid is manufactured on a worldwide bases by this process than any other chemical. The production of sulfur trixoide from sulfur dioxide is aided by a catalyst. Sulfur dioxide is produced from the combustion of elemental sulfur or sulfur compounds. The reaction is run at about 400°C.

$$2SO_2(g) + O_2(g) \rightleftharpoons 2SO_3(g) + \text{heat}$$

Determine the direction in which the equilibrium will be shifted by the following changes.

(a) increasing the concentration of O_2

(b) increasing the concentration of SO_3

(c) decreasing the concentration of SO_2

(d) decreasing the concentration of SO_3

(e) compressing the reaction mixture

(f) removing the catalyst

(g) How will the yield of SO_3 be affected by raising the temperature to 750°C?

(h) How will the yield of SO_3 be affected by lowering the temperature to 0°C? How will this affect the rate of formation of SO_3?

15-11. Consider the equilibrium

$$4NH_3(g) + 5O_2(g) \rightleftharpoons 4NO(g) + 6H_2O(g) + \text{heat energy}$$

How will this system at equilibrium be affected by the following changes?

(a) increasing the concentration of O_2

(b) removing all of the H_2O as it is formed

(c) increasing the concentration of NO

(d) compressing the reaction mixture

(e) increasing the volume of the reaction container

(f) increasing the temperature

(g) decreasing the concentration of O_2

(h) adding a catalyst

15-12. The following equilibrium takes place at high temperatures.

$$N_2(g) + 2H_2O(g) + \text{heat energy} \rightleftharpoons 2NO(g) + 2H_2(g)$$

How will the concentration of NO at equilibrium be affected by these changes?

(a) increasing $[N_2]$

(b) decreasing $[H_2]$

(c) compressing the reaction mixture

(d) decreasing the volume of the reaction container

(e) decreasing the temperature

(f) adding a catalyst

15-13. Consider the equilibrium

$$2SO_3(g) + CO_2(g) + \text{heat energy} \rightleftharpoons CS_2(g) + 4O_2(g)$$

How will the concentration of CS_2 at equilibrium be affected by these changes?

(a) decreasing the volume of the reaction vessel

(b) adding a catalyst

(c) increasing the temperature

(d) increasing the original concentration of CO_2

(e) removing some O_2 as it forms

15-14. Fortunately for us, the major components of air, N_2 and O_2, do not react under ordinary conditions. The reaction shown is endothermic.

$$N_2(g) + O_2(g) \rightleftharpoons 2NO(g)$$

Would the formation of NO in an automobile be affected by these changes?

(a) a lower pressure

(b) a lower temperature

(c) a lower concentration of O_2

The Equilibrium Constant Expression (SECTION 15-4)

15-15. Write the equilibrium constant expression for each of the following equilibria.

(a) $CO(g) + Cl_2(g) \rightleftharpoons COCl_2(g)$

(b) $CH_4(g) + 2H_2O(g) \rightleftharpoons CO_2(g) + 4H_2(g)$

(c) $4HCl(g) + O_2(g) \rightleftharpoons 2Cl_2(g) + 2H_2O(g)$

(d) $CH_4(g) + Cl_2(g) \rightleftharpoons CH_3Cl(g) + HCl(g)$

15-16. Write the equilibrium constant expression for each of the following equilibria.

(a) $3O_2(g) \rightleftharpoons 2O_3(g)$

(b) $N_2(g) + 2O_2(g) \rightleftharpoons 2NO_2(g)$

(c) $C_2H_2(g) + 2H_2(g) \rightleftharpoons C_2H_6(g)$

(d) $4H_2(g) + CS_2(g) \rightleftharpoons CH_4(g) + 2H_2S(g)$

15-17. For the hypothetical reaction $A_2 + B_2 \rightleftharpoons 2AB$, $K_{eq} = 1.0 \times 10^8$, are reactants or products favored at equilibrium?

15-18. For the hypothetical reaction $2C + 3B \rightleftharpoons 2D + F$, $K_{eq} = 5 \times 10^{-7}$, are reactants or products favored at equilibrium?

15-19. For the reaction $H_2(g) + I_2(g) \rightleftharpoons 2HI(g)$, $K_{eq} = 45$ at a given temperature, are reactants or products favored at equilibrium?

15-20. Given the system

$$3O_2(g) \rightleftharpoons 2O_3(g)$$

at equilibrium, $[O_2] = 0.35$ and $[O_3] = 0.12$. What is K_{eq} for the reaction under these conditions?

15-21. Given the system

$$N_2(g) + 2O_2(g) \rightleftharpoons 2NO_2(g)$$

at a given temperature, there are 1.25×10^{-3} mol of N_2, 2.50×10^{-3} mol of O_2, and 6.20×10^{-4} mol of NO_2 in a 1.00-L container. What is K_{eq} for this reaction at this temperature?

15-22. Given the system

$$2SO_3(g) + CO_2(g) \rightleftharpoons CS_2(g) + 4O_2(g)$$

what is K_{eq} if, at equilibrium, $[SO_3] = 2.0 \times 10^{-2}$ mol/L, $[CO_2] = 4.5 \times 10^{-3}$ mol/L, $[CS_2] = 6.2 \times 10^{-4}$ mol/L, and $[O_2] = 1.0 \times 10^{-4}$ mol/L

15-23. Given the system

$$CH_4(g) + 2H_2O(g) \rightleftharpoons CO_2(g) + 4H_2(g)$$

at equilibrium, we find 2.20 mol of CO_2, 4.00 mol of H_2, 6.20 mol of CH_4, and 3.00 mol of H_2O in a 30.0-L container. What is K_{eq} for the reaction? *(Hint:* Convert amount and volume to concentration.)

15-24. Given the system

$$C_2H_2(g) + 2H_2(g) \rightleftharpoons C_2H_6(g)$$

at equilibrium, we find 296 g of C_2H_6 present along with 3.50 g of H_2 and 21.0 g of C_2H_2 in a 400-mL container. What is K_{eq}? *(Hint:* Convert amount and volume to concentration.)

*15-25. Consider the following system:

$$2HI(g) \rightleftharpoons H_2(g) + I_2(g)$$

(a) If we start with $[HI] = 0.60$, what would $[H_2]$ and $[I_2]$ be if all the HI reacts?

(b) If we start with $[HI] = 0.60$ and $I_2 = 0.20$, what would $[H_2]$ and $[I_2]$ be if all the HI reacts?

(c) If we start with only $[HI] = 0.60$ and 0.20 mol/L of HI reacts, what are $[HI]$, $[I_2]$, and $[H_2]$ at equilibrium?

(d) From the information in (c), calculate K_{eq}.

(e) What is the K_{eq} for the reverse reaction? How does this value differ from the value given in Table 15-1?

*15-26. In the reaction

$$N_2(g) + 3H_2(g) \rightleftharpoons 2NH_3(g)$$

initially 1.00 mol of N_2 and 1.00 mol of H_2 are mixed in a 1.00-L container. At equilibrium, it is found that $[NH_3] = 0.20$.

(a) What are the concentrations of N_2 and H_2 at equilibrium?

(b) What is the K_{eq} for the system at this temperature?

*15-27. At a given temperature, N_2, H_2, and NH_3 are mixed. The initial concentration of each is 0.50 mol/L. At equilibrium, the concentration of N_2 is 0.40 mol/L.

(a) Calculate the concentrations of H_2 and NH_3 at equilibrium.

(b) What is the K_{eq} at this temperature?

*15-28. At the start of the reaction

$$4NH_3(g) + 5O_2 \rightleftharpoons 4NO(g) + 6H_2O(g)$$

$[NH_3] = [O_2] = 1.00$. At equilibrium, it is found that 0.25 mol/L of NH_3 has reacted.

(a) What concentration of O_2 reacts?

(b) What are the concentrations of all species at equilibrium?

(c) Write the equilibrium constant expression and substitute the proper values for the concentrations of reactants and products.

*15-29. At the start of the reaction

$$CO(g) + Cl_2(g) \rightleftharpoons COCl_2(g)$$

$[CO] = 0.650$ and $[Cl_2] = 0.435$. At equilibrium, it is found that 10.0% of the CO has reacted. What are $[CO]$, $[Cl_2]$, and $[COCl_2]$ at equilibrium? What is K_{eq}?

Calculations Involving K_{eq} (SECTION 15-4)

15-30. For the reaction

$$PCl_3(g) + Cl_2(g) \rightleftharpoons PCl_5(g)$$

$K_{eq} = 0.95$ at a given temperature. If $[PCl_3] = 0.75$ and $[Cl_2] = 0.40$ at equilibrium, what is the concentration of PCl_5 at equilibrium?

15-31. At a given temperature, $K_{eq} = 46.0$ for the reaction

$$4HCl(g) + O_2(g) \rightleftharpoons 2Cl_2(g) + 2H_2O(g)$$

At equilibrium, $[HCl] = 0.100$, $[O_2] = 0.455$, and $[H_2O] = 0.675$. What is $[Cl_2]$ at equilibrium?

15-32. Using the value for K_{eq} calculated in problem 15-23, find the concentration of H_2O at equilibrium if, at equilibrium, $[CH_4] = 0.50$, $[CO_2] = 0.24$ and $[H_2] = 0.20$.

*15-33. For the following equilibrium, $K_{eq} = 56$ at a given temperature.

$$CH_4(g) + Cl_2(g) \rightleftharpoons CH_3Cl(g) + HCl(g)$$

At equilibrium, $[CH_4] = 0.20$ and $[Cl_2] = 0.40$. What are the equilibrium concentrations of CH_3Cl and HCl if they are equal?

*15-34. Using the equilibrium in problem 15-20, calculate the equilibrium concentration of O_3 if the equilibrium concentration of O_2 is 0.80 mol/L.

Equilibria of Weak Acids and Weak Bases
(SECTION 15-5)

15-35. Write the expression for K_a or K_b for each of the following equilibria. Where necessary, complete the equilibrium.

(a) $HBrO + H_2O \rightleftharpoons H_3O^+ + BrO^-$

(b) $NH_3 + H_2O \rightleftharpoons NH_4^+ + OH^-$

(c) $H_2SO_3 + H_2O \rightleftharpoons H_3O^+ + HSO_3^-$

(d) $HSO_3^- + H_2O \rightleftharpoons \rule{1cm}{0.4pt} + SO_3^{2-}$

(e) $H_3PO_4 + H_2O \rightleftharpoons H_3O^+ + \rule{1cm}{0.4pt}$

(f) $(CH_3)_2NH + H_2O \rightleftharpoons (CH_3)_2NH_2^+ + \rule{1cm}{0.4pt}$

15-36. Write the expression for K_a or K_b for each of the following equilibria. Where necessary, complete the equilibrium.

(a) $N_2H_4 + H_2O \rightleftharpoons N_2H_5^+ + OH^-$

(b) $HCN + H_2O \rightleftharpoons H_3O^+ + CN^-$

(c) $H_2C_2O_4 + H_2O \rightleftharpoons H_3O^+ + \rule{1cm}{0.4pt}$

(d) $H_2PO_4^- + H_2O \rightleftharpoons \rule{1cm}{0.4pt} + HPO_4^{2-}$

15-37. A 0.10 M solution of a weak acid HX has a pH of 5.0. A 0.10 M solution of another weak acid HB has a pH of 5.8. Which is the weaker acid? Which has the larger value for K_a?

15-38. A hypothetical weak acid HZ has a K_a of 4.5×10^{-4}. Rank the following 0.10 M solutions in order of increasing pH: HZ, $HC_2H_3O_2$, and HClO.

15-39. In a 0.20 M solution of cyanic acid, HOCN, $[H_3O^+] = [OCN^-] = 6.2 \times 10^{-3}$.

(a) What is [HOCN] at equilibrium?

(b) What is K_a?

(c) What is the pH?

15-40. A 0.58 M solution of a weak acid (HX) is 10.0% dissociated.

(a) Write the equilibrium equation and the equilibrium constant expression.

(b) What are $[H_3O^+], [X^-]$, and [HX] at equilibrium?

(c) What is K_a?

(d) What is the pH?

15-41. Nicotine (Nc) is a nitrogen base in water (similar to ammonia). Write the equilibrium equation illustrating this base behavior. In a 0.44 M solution of nicotine, $[OH^-] = [NcH^+] = 5.5 \times 10^{-4}$. What is K_b for nicotine? What is the pH of the solution?

15-42. In a 0.085 M solution of carbolic acid, HC_6H_5O, the pH = 5.48. What is K_a?

15-43. Novocaine (Nv) is a nitrogen base in water (similar to ammonia). Write the equilibrium equation. In a 1.25 M solution of novocaine, pH = 11.46. What is K_b?

15-44. In a 0.300 M solution of chloroacetic acid $HC_2Cl_3O_2$, $[HC_2Cl_3O_2] = 0.277$ M at equilibrium. What is the pH? What is K_a?

15-45. What is the pH of a 0.65 M solution of HBrO?

15-46. What is the pH of a 1.50 M solution of HNO_2?

15-47. What is $[OH^-]$ in a 0.55 M solution of NH_3?

15-48. What is $[H_3O^+]$ in a 0.25 M solution of $HC_2H_3O_2$?

15-49. What is the pH of a 1.00 M solution of $(CH_3)_2NH$?

15-50. What is the pH of a 0.567 M solution of N_2H_4?

15-51. What is the pH of a buffer made by mixing 0.45 mol of NaCN and 0.45 mol of HCN in 2.50 L of solution?

15-52. What is the pH of a buffer made by dissolving 1.20 mol of NH_3 and 1.20 mol of NH_4ClO_4 in 13.5 L of solution?

15-53. What is the pH of a buffer made by dissolving 0.20 mol of KBrO and 0.60 mol of HBrO in 850 mL of solution? What is the pH if the solution is diluted to 2.00 L?

15-54. What is the pH of a buffer solution made by mixing 0.044 mol of $HCHO_2$ with 0.064 mol of $KCHO_2$ in 1.00 L of solution?

15-55. What is the pH of a buffer solution made by mixing 0.058 mol of HClO and 5.50 g of $Ca(ClO)_2$ in a certain amount of water?

15-56. What is the pH of a buffer made by mixing 1.50 g of N_2H_4 with 1.97 g of $N_2H_5^+Cl^-$ in 2.00 L of solution?

15-57. What is the pH of a buffer that contains 150 g of HNO_2 and 150 g of $LiNO_2$?

*15-58. A buffer of pH = 7.50 is desired. Which two of the following compounds should be dissolved in water in equimolar amounts to provide this buffer solution: HNO_2, HClO, HCN, $NaNO_2$, NH_3, KCN, KClO, NH_4Cl?

*15-59. A buffer of pH = 9.25 is desired. Which two of the following compounds should be dissolved in water in equimolar amounts to provide this buffer solution: $HC_2H_3O_2$, HClO, NH_3, N_2H_4, $KC_2H_3O_2$, NH_4Cl, NaClO, N_2H_5Br?

Solubility Equilibria (SECTION 15-6)

15-60. Write the expression of K_{sp} for the following salts: **(a)** FeS, **(b)** Ag_2S, **(c)** $Zn(OH)_2$.

15-61. Write the expression for K_{sp} for the following salts: **(a)** $PbCO_3$, **(b)** Bi_2S_3, **(c)** PbI_2.

15-62. What is the value of K_{sp} for copper(I) iodide if the molar solubility is 2.2×10^{-6} mol/L?

15-63. What is the value of K_{sp} for magnesium carbonate $(MgCO_3)$ if a saturated solution contains 0.287 g/L?

15-64. What is the value of K_{sp} for $Ca(OH)_2$ if the molar solubility is 1.3×10^{-2} mol/L?

*15-65. What is the value of K_{sp} for $Cr(OH)_3$ if the molar solubility is 3×10^{-8} mol/L?

15-66. What is the molar solubility of AgI? For AgI, $K_{sp} = 8.3 \times 10^{-17}$.

*15-67. Lead(II) fluoride is present in the earth as the mineral fluorite. It is considered insoluble in water. What is the molar solubility of lead(II) fluoride? For PbF_2, $K_{sp} = 4.1 \times 10^{-8}$.

15-68. The concentration of Ag^+ ion in a solution is 7×10^{-6} M, and the concentration of Br^- is 8×10^{-7} M. Will a precipitate of AgBr form? For AgBr, $K_{sp} = 7.7 \times 10^{-13}$.

15-69. The concentration of Ag^+ in a solution is 0.012 M, and the concentration of SO_4^{2-} is 0.050 M. Will a precipitate of form? For Ag_2SO_4, $K_{sp} = 1.4 \times 10^{-5}$.

General Problems

15-70. In a given gaseous equilibrium, chlorine reacts with ammonia to produce nitrogen trichloride and hydrogen chloride.

(a) Write the balanced equation illustrating this equilibrium.

(b) Write the expression for K_{eq}.

(c) How is the equilibrium concentration of ammonia affected by an increase in pressure?

(d) How is the equilibrium concentration of ammonia affected by the addition of some chlorine gas?

(e) The value of K_{eq} for this reaction is 2.4×10^{-9}. Are there more products or reactants present at equilibrium?

(f) If the equilibrium concentrations of chlorine = 0.10 M, nitrogen trichloride = 2.0×10^{-4} M, and hydrogen chloride = 1.0×10^{-3} M, what is the equilibrium concentration of ammonia?

15-71. In a given gaseous equilibrium, acetylene (C_2H_2) reacts with HCl to form $C_2H_4Cl_2$.

(a) Write the balanced equation illustrating this equilibrium.

(b) Write the expression for K_{eq}.

(c) How does an increase in pressure affect the equilibrium concentration of $C_2H_4Cl_2$?

(d) The reaction is exothermic. How does an increase in temperature affect the value of K_{eq}? How does it affect the equilibrium concentration of HCl?

(e) At a given temperature at equilibrium, $[C_2H_2] = 0.030$, $[HCl] = 0.010$, and $[C_2H_4Cl_2] = 0.60$. What is the value of K_{eq}?

(f) What does the value of K_{eq} tell us about the point of equilibrium at this temperature?

15-72. Dinitrogen monoxide (nitrous oxide) decomposes to its elements in a gaseous equilibrium reaction.

(a) Write the balanced equation illustrating this equilibrium.

(b) Write the expression for K_{eq}.

(c) How does an increase in pressure affect the equilibrium concentration of nitrogen?

(d) If the reaction is exothermic, how does a decrease in temperature affect the equilibrium concentration of dinitrogen monoxide?

(e) At a given temperature, 0.10 mol of dinitrogen monoxide is placed in a 1.00-L container. At equilibrium, it is found that 1.5% of the dinitrogen oxide has decomposed. What is the value of K_{eq}?

15-73. Nitrogen dioxide is in a gaseous equilibrium with dinitrogen tetroxide.

(a) Write the balanced equation illustrating this reaction. Show nitrogen dioxide as the reactant and dinitrogen tetroxide as the product.

(b) Write the expression for K_{eq}.

(c) How does an increase in pressure affect the concentration of nitrogen dioxide at equilibrium?

(d) At equilibrium at a given temperature, it is found that there are 10.0 g of nitrogen dioxide and 12.0 g of dinitrogen tetroxide in a 2.50-L container. What is the value of K_{eq}?

15-74. The bicarbonate ion (HCO_3^-) can act as either an acid or a base in water. For the acid reaction

$$HCO_3^- + H_2O \rightleftharpoons H_3O^+ + CO_3^{2-} \quad K_a = 4.7 \times 10^{-11}$$

For the base reaction

$$HCO_3^- + H_2O \rightleftharpoons OH^- + H_2CO_3 \quad K_b = 2.2 \times 10^{-8}$$

(a) Based on the values of K_a and K_b, is a solution of $NaHCO_3$ acidic or basic?

(b) How is bicarbonate of soda used medically?

(c) Write the reactions that occur when sodium bicarbonate reacts with stomach acid (HCl).

(d) Can sodium bisulfate act in the same way toward acids as sodium bicarbonate does?

*15-75. Given 0.10 M solutions of the compounds KOH, NaCl, HNO_2, NH_3, HNO_3, $HCHO_2$, and N_2H_2, rank them in order of increasing pH.

15-76. What is the pH of a buffer that is made by mixing 265 mL of 0.22 M $HC_2H_3O_2$ with 375 mL of 0.12 M $Ba(C_2H_3O_2)_2$?

*15-77. What is the pH of a buffer made by adding 0.20 mol of NaOH to 1.00 L of a 1.00 M solution of $HC_2H_3O_2$? Assume no volume change. *(Hint:* First consider the partial neutralization of by $HC_2H_3O_2$ by NaOH.)

*15-78. What is the value of K_{sp} for $Mg(OH)_2$ if the pH of a saturated solution is equal to 10.50?

*15-79. What is the pH of a saturated solution of $Fe(OH)_2$? For $Fe(OH)_2$, $K_{sp} = 1.6 \times 10^{-14}$.

15-80. Aqueous solutions of barium salts are very poisonous, yet we ingest barium sulfate as a contrasting agent for X-rays of the intestinal tract. What is the Ba^{2+} concentration in a saturated solution of $BaSO_4$? What is the mass of $BaSO_4$ present in one liter of solution? For $BaSO_4$, $K_{sp} = 1.1 \times 10^{-10}$.

STUDENT WORKSHOP

Equilibrium and Le Châtelier's Principle

Purpose: To examine the effect of changes in concentration on the amounts of reactants and products of a reaction at equilibrium. (Work in groups of three or four. Estimated time: 25 min.)

In order to illustrate the interplay of chemicals involved in establishing equilibrium, play this little game. You'll require two dozen interlocking blocks (such as Lego©) in two colors. You'll assemble "molecules" based on the equation:

$$A + B \longrightarrow AB$$

(In other words, the "reaction" is a block of each color combining together to form a product.)

The equilibrium condition is that there must be the same number of AB molecules as the total number of A's and B's. (Real equilibria work on the mathematical relationship between the concentrations of all reactants and products (i.e., K_a, but this serves as an easy way to illustrate the point.)

1. Begin with 12 A's and 12 B's. How many AB's must you produce to establish equilibrium?

2. Perturb the system by adding an additional 6 A's from the reserve. How must things alter to reestablish equilibrium?

3. Now take away 3 AB's. What has to happen?

4. Add 9 B's. How does the equilibrium shift?

5. Add 6 AB's. What are the new values for each chemical species?

Fill in the following chart for each of the changes:

Initial			Shift ←, →	Equilibrium		
A	B	AB	#	A	B	AB
12	12	0	8−	4	4	8
___	___	___	___	___	___	___
___	___	___	___	___	___	___
___	___	___	___	___	___	___
___	___	___	___	___	___	___

- In what other ways can you alter the equilibrium position?
- How would the dynamics change if the equation were $2A + B \rightarrow A_2B$?

Nuclear Chemistry

Deep in the interior of the sun, nuclear reactions are occurring that liberate vast amounts of energy. This energy is transmitted to us as light, which makes life on this planet possible.

SETTING THE STAGE

In July 1945, the most powerful explosion yet produced by the human race shook the New Mexico desert. The first nuclear bomb had been detonated, and our world would never be the same—the nuclear age was now upon us. But the dawn of this age actually began quite innocently 49 years earlier. At that time, there was certainly no intent to launch the human race in a radical new direction. In 1896, a French scientist named Henri Becquerel was investigating how sunlight interacted with a uranium mineral called *pitchblende*. He suspected that sunlight caused pitchblende to give off the mysterious "X-rays" that had been discovered the previous year by Wilhelm Roentgen. X-rays, a high-energy form of light, were known to penetrate the covering of photographic plates, thus exposing the film. One cloudy day, Becquerel was unable to do an experiment, but he left the uranium on top of unexposed film in a photographic plate. Several days later, he developed the film anyway and discovered that the part underneath the sample of pitchblende was exposed. Obviously, the uranium had spontaneously emitted some form of radiation that had nothing to do with sunlight. It was this discovery that marked the true beginning of the nuclear age.

Since we first discussed the atom in Chapters 1 and 2, we have addressed the results of interactions of the electrons that reside outside a tiny but comparatively massive nucleus. Interactions of electrons account for the way atoms of an element combine chemically with other atoms. Now we turn our attention to the nucleus, which is composed of two types of particles, neutrons and protons, both called *nucleons*. Also recall that elements are composed of isotopes. Isotopes of a specific element are those atoms that have the same number of protons but different numbers of neutrons.

As we will see, most natural nuclear processes transform one element into another. Deep in castle dungeons around 600 years ago, alchemists thought, or at least hoped, that they could accomplish this feat. Actually, it wasn't just any element they wanted to produce—it was gold. The synthesis of one element from another never happened at that time, but it wasn't long into the nuclear age that this was actually accomplished. Beginning with a simple experiment by Lord Rutherford in the early 1900s and extending to the use of huge accelerators today, humans have been transforming the elements.

In Part A, we examine in more detail the changes that can occur in the nucleus of a particular isotope as a result of the radiation that Becquerel first observed. In Part B, we will see how human intervention has produced results that range from the fear of nuclear bombs to the hope of nuclear medicine.

Part A

Naturally Occurring Radioactivity

SETTING A GOAL

■ You will learn of the different types of radiation and how it affects surrounding matter.

OBJECTIVES

16-1 Write balanced nuclear equations for the five types of radiation.

16-2 Using half-lives, estimate the time needed for various amounts of radioactive decay to occur.

16-3 Describe how the three natural types of radiation affect matter.

16-4 Describe the methods and units used to detect and measure radiation.

▶ **OBJECTIVE FOR SECTION 16-1**
Write balanced nuclear equations for the five types of radiation.

16-1 Radioactivity

LOOKING AHEAD! It was a monumental discovery that one element could change into another through natural processes. We will look closely at the nature of those changes in this section. ■

The radiation process discovered by Becquerel was more complex than originally thought. The term *radioactive* was first suggested by Marie Curie to describe elements that spontaneously emit **radiation** (*particles or energy*). Marie Curie and her husband, Pierre Curie, pursued the observation of Becquerel and discovered several more naturally **radioactive isotopes** of elements besides uranium (thorium, polonium, and radium). Lord Rutherford later identified three types of natural radiation. We will look at these three processes individually, as well as two additional processes since discovered.

16-1.1 Alpha (α) Particles

The first nuclear change that was identified involved the emission of a helium nucleus from the nucleus of a heavy isotope. *The emitted helium nucleus, composed of two protons and two neutrons but no electrons, is referred to as an* **alpha (α) particle**. This process is conveniently illustrated by a nuclear equation. *A* **nuclear equation** *shows the initial nucleus to the left of the reaction arrow and the product nuclei or particles to the right of the arrow.* Just as chemical equations are balanced, so are nuclear equations.

Nuclear equations are balanced by having the same totals of positive charges (the numbers that are subscripts on the left of the symbol) and mass numbers (the numbers that are superscripts on the left of the symbol) on both sides of the equation. The emission of an alpha particle by ^{238}U is illustrated in Figure 16-1 and represented by the following nuclear equation.

$$^{238}_{92}U \longrightarrow {}^{234}_{90}Th + {}^{4}_{2}He$$

Note that the loss of four nucleons from the original (parent) nucleus leaves the remaining (daughter) nucleus with 234 nucleons ($238 - 4 = 234$). The loss of two protons leaves 90 protons in the daughter nucleus ($92 - 2 = 90$). Because a nucleus with 90 protons is an isotope of thorium, one element has indeed changed into another.

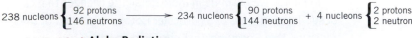

$$^{238}_{92}U \longrightarrow {}^{234}_{90}Th + {}^{4}_{2}He \text{ (alpha particle)}$$

238 nucleons $\begin{cases} 92 \text{ protons} \\ 146 \text{ neutrons} \end{cases} \longrightarrow$ 234 nucleons $\begin{cases} 90 \text{ protons} \\ 144 \text{ neutrons} \end{cases} +$ 4 nucleons $\begin{cases} 2 \text{ protons} \\ 2 \text{ neutrons} \end{cases}$

FIGURE 16-1 Alpha Radiation
An alpha particle is a helium nucleus emitted from a larger nucleus.

16-1.2 Beta (β) Particles

A second form of radiation involves the emission of an electron from the nucleus. *An electron emitted from the nucleus is known as a* **beta (β) particle**. *In effect, beta particle emission changes a neutron in a nucleus into a proton.* This causes the atomic number of the isotope to increase by 1 while the mass number remains constant. This type of radiation is illustrated in Figure 16-2 and represented by the following nuclear equation.

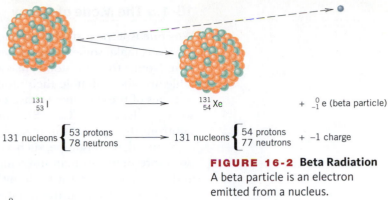

FIGURE 16-2 Beta Radiation
A beta particle is an electron emitted from a nucleus.

$$^{131}_{53}\text{I} \longrightarrow \,^{131}_{54}\text{Xe} + \,^{0}_{-1}\text{e}$$

Notice that, as with alpha radiation, the sum of the atomic masses and the atomic numbers on both sides of the arrow are the same.

16-1.3 Gamma (γ) Rays

A third type of radioactive decay involves the emission of a high-energy form of light called a **gamma (γ) ray**. Like all light, gamma rays travel at 3.0×10^8 m/sec (186,000 mi/sec). This type of radiation may occur alone or in combination with alpha or beta radiation. The following equation illustrates gamma ray emission.

$$^{60m-}_{27}\text{Co} \longrightarrow \,^{60}_{27}\text{Co} + \gamma$$

Gamma radiation alone does not result in a change in either the mass number or the atomic number of the isotope. In the equation shown above, the *m* stands for *metastable*, which means that the isotope is in a high-energy state. A nucleus in a metastable, or high-energy, state may emit energy in the form of gamma radiation. The product nucleus has the same identity but is then in a lower energy state.

16-1.4 Positron Particles and Electron Capture

This type of radiation is not among the three types of naturally occurring radiation but is found in a few artificially produced isotopes. *A* **positron ($_{+1}^{0}$e) particle** *is an electron with a positive charge rather than a negative charge. In effect, positron emission changes a proton into a neutron.* As a result, the product nucleus has the same mass number but a lower atomic number. A positron is known as a form of *antimatter* because, when it comes in contact with normal matter (e.g., a negatively charged electron), the two particles annihilate each other, producing two gamma rays that travel in exactly opposite directions.

$$^{13}_{7}\text{N} \longrightarrow \,^{13}_{6}\text{C} + \,^{0}_{+1}\text{e}$$
$$^{0}_{+1}\text{e} + \,^{0}_{-1}\text{e} \longrightarrow 2\gamma$$

Another nuclear process, known as **electron capture**, involves *the capture of an inner-shell electron by the nucleus.* This process is very rare in nature but is more common in artificially produced elements. When an isotope captures an orbital electron, the effect is the same as positron emission. *That is, a proton changes into a neutron.* Electron capture is detected by the X-rays that are emitted as outer electrons fall into the inner shell, where there is a vacancy due to the captured electron. In most cases, gamma rays are also emitted by the product nuclei. Electron capture is illustrated by the following nuclear equation.

$$^{50}_{23}\text{V} + \,^{0}_{-1}\text{e} \longrightarrow \,^{50}_{22}\text{Ti} + \text{X-rays} + \gamma$$

16-1.5 The Mode of Decay

Why one isotope is stable and another is not is a somewhat involved topic. We can, however, make some simple observations that help us predict some types of radiation. Notice that the atomic masses of most of the lighter elements in the periodic table are about double their atomic numbers. That is because the most common isotopes of these elements have equal numbers of protons and neutrons, or nearly so (e.g., ^{4}He, ^{16}O, ^{40}Ca, ^{32}S). The heavier elements, however, tend to have more neutrons than protons in their most common and stable isotopes.

Notice that an isotope, such as, $^{14}_{6}C$ has eight neutrons and six protons, so it has two more neutrons than its common stable isotope $^{12}_{6}C$. This isotope is radioactive, and we might expect it to decay by a process that would increase the number of protons and decrease the number of neutrons. This would be accomplished if a neutron turned into a proton. That process is beta particle emission. The result of this radiation is a change to a more stable neutron/proton ratio (e.g., $^{14}_{7}N + ^{0}_{-1}e$). Isotopes that have excess protons (e.g., $^{10}_{6}C$) also tend to be radioactive but decay by positron emission (or electron capture). The result of this radiation is a change of a proton into a neutron and, again, a more stable nucleus with a better neutron/proton ratio (e.g., $^{10}_{5}B + ^{0}_{+1}e$). A good example of predicting radioactive decay concerns the four known isotopes of nitrogen—^{12}N, ^{14}N, ^{15}N, and ^{16}N. ^{12}N decays by electron capture to form the stable isotope ^{12}C. ^{14}N and ^{15}N are stable. ^{16}N decays by beta particle emission to form the stable isotope ^{16}O.

$$^{12}N \longrightarrow ^{12}C + ^{0}_{+1}e$$

$$^{16}N \longrightarrow ^{16}O + ^{0}_{-1}e$$

Since the heaviest stable element is bismuth, isotopes heavier than bismuth must all lose both mass and charge to become stable. This is alpha particle emission.

The five types of radiation that have been discussed are summarized in Table 16-1.

TABLE 16-1

Types of Radiation

RADIATION	MASS NUMBER	CHARGE	IDENTITY
alpha (α)	4	+2	helium nucleus
beta (β)	0	−1	electron (out of nucleus)
gamma (γ)	0	0	light energy
positron	0	+1	positive electron
electron capture	0	−1	electron (into nucleus)

EXAMPLE 16-1 Writing Equations for Nuclear Decay

Complete the following nuclear equations:

(a) $^{218}_{84}Po \longrightarrow$ _____ $+ ^{4}_{2}He$ **(b)** $^{210}_{81}Tl \longrightarrow ^{210}_{82}Pb +$ _____ **(c)** _____ $\longrightarrow ^{8}_{4}Be + ^{0}_{+1}e$

PROCEDURE

For the missing isotope or particle:

1. Find the total number of nucleons; it is the same on both sides of the equation.

2. Find the total charge or atomic number; it is also the same on both sides of the equation.

3. If what's missing is the isotope of an element, find the symbol of the element that matches the atomic number from the list of elements inside the front cover.

SOLUTION

(a) Nucleons: $218 = x + 4$, so $x = 214$
Atomic number: $84 = y + 2$, so $y = 82$
From the inside cover, the element having an atomic number of 82 is Pb. The isotope is

$$^{214}_{82}\text{Pb}$$

(b) Nucleons: $210 = 210 + x$, so $x = 0$
Atomic number: $81 = 82 + y$, so $y = -1$
An electron or beta particle has negligible mass (compared with a nucleon) and a charge of -1.

(c) Nucleons: $x = 8 + 0$, so $x = 8$
Atomic number: $y = 4 + 1$, so $y = 5$
From the inside cover, the element with an atomic number of 5 is B. The isotope is

$$^{8}_{5}\text{B}$$

ANALYSIS

Radioactive particles are either emitted (alpha particles, beta particles, positrons) and are therefore found on the product side of a nuclear equation, or absorbed (electrons by electron capture) and therefore appear on the reactant side of the equation. Each one has a symbol and a specific charge (lower left) and mass (upper left) associated with that symbol. Referring to Table 16-1, can you write the symbol for each? $^{4}_{2}\alpha$ for alpha, $^{0}_{-1}\beta$ for beta, $^{0}_{+1}e$ for a positron, and $^{0}_{-1}e$ for an electron being captured. ($^{0}_{0}\gamma$ for gamma radiation can be included, but it doesn't affect the mathematics.) Once the equation is completed, confirm that the superscripts on both sides of the arrow add up to the same value and that the subscripts do likewise.

SYNTHESIS

Recall that we can predict what type of emission may occur based on the ratio of protons to neutrons in an isotope. Each element on the periodic table with an atomic number less than bismuth has at least one stable isotope (except Pm and Tc). Furthermore, the stable isotopes have atomic masses very close to their average atomic masses listed on the periodic table. If the mass number of a particular isotope is very much above this number, we would predict beta particle emission. If, however, the mass number is very much below average, we would predict positron emission (or electron capture). For very large isotopes (ones larger than bismuth), alpha emission, which lowers the size of the nucleus, is the most common form of radiation.

▶ **ASSESSING THE OBJECTIVE FOR SECTION 16-1**

EXERCISE 16-1(a) LEVEL 1: Fill in the blanks.
There are five types of _____.
1. An alpha particle is a _____ nucleus.
2. A beta particle is an _____.
3. A gamma ray is a high-energy form of _____.
4. A positron is a _____.
5. Electron capture has the same effect on a nucleus as _____ emission.

EXERCISE 16-1(b) LEVEL 2: Write the radioactive isotope that leads to the indicated nucleus and particle or ray.

(a) _____ $^{87}_{38}\text{Sr}$ + beta particle

(b) _____ $^{223}_{87}\text{Fr}$ + alpha particle

(c) _____ $^{54}_{26}\text{Fe}$ + positron

(d) _____ + electron capture $\longrightarrow$ $^{123}_{51}\text{Sb}$

(e) _____ $^{137}_{55}\text{Cs}$ + gamma ray

EXERCISE 16-1(c) LEVEL 3: Predict the mode of decay (beta particle or positron particle) if appropriate for the following isotopes: (a) ^{14}O, (b) ^{45}Ca, and (c) ^{32}S.

For additional practice, work chapter problems, 16-1, 16-16, and 16-18.

16-2 Rates of Decay of Radioactive Isotopes

LOOKING AHEAD! Uranium is a naturally occurring element, but it is radioactive. Why hasn't it all decayed by now? Why some radioactive elements are still present on this ancient Earth is the subject of this section. ■

16-2.1 Half-Life

The heaviest element in the periodic table with at least one stable isotope is bismuth (#83). All elements heavier than bismuth have no stable isotopes (i.e., all isotopes are radioactive). Some heavy elements, such as uranium, still exist in the Earth in significant amounts, however, because they have very long "half-lives." *The **half-life (t$_{1/2}$)** of a radioactive isotope is the time required for one-half of a given sample to decay.* There are many isotopes lighter than bismuth that are also radioactive. Each radioactive isotope has a specific and constant half-life. Unlike the rate of chemical reactions, the half-life of an isotope is independent of conditions such as temperature, pressure, and whether the element is in the free state or part of a compound. The half-lives of radioactive isotopes vary from billions of years to fractions of a second. The half-lives and modes of decay of several radioactive isotopes are listed in Table 16-2.

As an example consider ^{131}I listed in Table 16-2. It is a product of the fission that takes place in nuclear power plants such as those that were damaged in the 2011 earthquake in Japan. (Fission will be discussed later in this chapter.) If this isotope is released into the environment, it can be absorbed by animals where it accumulates in the thyroid gland. This may lead to cancer of that gland. The good news, if there is any in this tragedy, is that this isotope has a comparatively short half-life of 8 days. If one started with 10.00 μg of ^{131}I, there would be 5.00 μg left after 8 days. After 16 days (two half-lives), half of the amount present after 8 days would be left, which is 2.50 μg (1/4 of the original amount). After three half-lives, one-half of that would remain, which is 1.25 μg (1/8 of the original amount). A similar calculation is shown in Example 16-2.

The decay of an isotope is illustrated graphically in Figure 16-3.

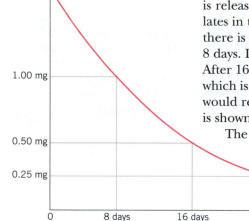

FIGURE 16-3 Radioactive Decay The graph illustrates the progress of the decay of iodine-131.

16-2.2 Radioactive Decay Series

$^{238}_{92}$U has a half-life of 4.5×10^9 years. Since that is roughly the age of Earth, about one-half of the $^{238}_{92}$U originally present when Earth formed from hot gases and dust is still present. (Half of what is now present will decay in another 4.5 billion years.)

TABLE 16-2

Half-Lives

ISOTOPE	$t_{1/2}$	MODE OF DECAY	PRODUCT
$^{238}_{92}$U	4.5×10^9 years	α, γ	$^{234}_{90}$Th
$^{242}_{94}$Pu	3.76×10^5 years	α	$^{238}_{92}$U
$^{234}_{90}$Th	24.1 days	β	$^{234}_{91}$Pa
$^{226}_{88}$Ra	1620 years	α	$^{222}_{86}$Rn
$^{14}_{6}$C	5760 years	β	$^{14}_{7}$N
$^{131}_{53}$I	8.0 days	β, γ	$^{131}_{54}$Xe
$^{218}_{85}$At	1.3 sec	α	$^{214}_{83}$Bi

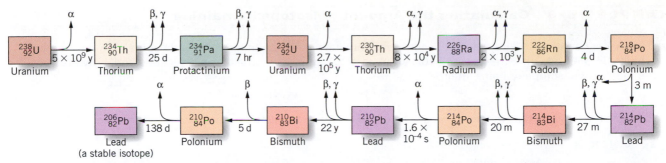

FIGURE 16-4 Radioactive Decay Series This series starts with ²³⁸U and ends with ²⁰⁶Pb.

Note in Table 16-2 that when $^{238}_{92}U$ decays, it forms an isotope ($^{234}_{90}Th$) that also decays (very rapidly compared with $^{238}_{92}U$). The $^{234}_{91}Pa$ formed from the decay of $^{234}_{90}Th$ also decays, and so forth, until finally the stable isotope $^{206}_{82}Pb$ is formed. This is known as the $^{238}_{92}U$ radioactive decay series. A **radioactive decay series** *starts with a naturally occurring radioactive isotope with a half-life near the age of Earth.* (If it was very much shorter, there wouldn't be any left.) *The series ends with a stable isotope.* (See Figure 16-4.) There are at least two other naturally occurring decay series: the $^{235}_{92}U$ series, which ends with $^{207}_{82}Pb$ and the $^{232}_{90}Th$ series, which ends with $^{208}_{82}Pb$. As a result of the $^{238}_{92}U$ decay series, where uranium is found in rocks, we also find other radioactive isotopes as well as lead. In fact, by examining the ratio of $^{238}_{92}U$ to $^{206}_{82}Pb$ in a rock, its age can be determined. For example, a rock from the moon showed that about half of the original $^{238}_{92}U$ had decayed to $^{206}_{82}Pb$. This meant that the rock was formed about 4.5×10^9 years ago.

EXAMPLE 16-2   Calculating the Time to Decay

If we started with 4.0 mg of ^{14}C, how long would it take until only 0.50 mg remained?

PROCEDURE
Find the half-life of ^{14}C from Table 16-2 and calculate the amount remaining after each half-life.

SOLUTION
After each half-life, the amount is reduced by one-half.

After 5760 years, $\frac{1}{2} \times 4.0$ mg = 2.0 mg remaining

After another 5760 years, $\frac{1}{2} \times 2.0$ mg = 1.0 mg remaining

After another 5760 years, $\frac{1}{2} \times 1.0$ mg = 0.50 mg remaining

Total time = 17,280 years

Therefore, in 17,280 years, 0.50 mg remains.

ANALYSIS
It is often wrongly assumed that the half of a sample decays in the first half-life and the second half in the second half-life. In other words, it would all be gone in two half-lives. Note that each half-life starts with the amount remaining from the previous decay. If we started with one mole of particles (6.02×10^{23}), it would, theoretically, take 79 half-lives before every single particle decayed. In practice, our ability to detect isotopes fails long before the concentration falls that low.

SYNTHESIS
It is calculations like these that allow us to determine the age of objects containing radioactive particles. The natural amount of an isotope is established, and then, by detecting how much is left in a sample, we can calculate how many half-lives the sample has undergone. By working down, we can estimate the amount of time since the object was formed. We will give examples of such calculations later in this chapter.

EXAMPLE 16-3   Calculating the Amount of Isotope Remaining

^{103}Pd has a half-life of 17 days and is used to treat prostate cancer. A titanium capsule containing the isotope remains in the body after treatment. If 10.0 μg (1 μg = 10^{-6} g) is inserted into the prostate gland, how much of the radioactive isotope remains after 68 days?

PROCEDURE

Calculate how much would remain after each half-life. Determine the number of half-lives in 68 days (i.e., 68/17 = 4).

SOLUTION

After one half-life (17 days), $\frac{1}{2} \times$ 10.0 μg = 5.00 μg

After two half-lives (34 days), $\frac{1}{2} \times$ 5.00 μg = 2.50 μg

After three half-lives (51 days), $\frac{1}{2} \times$ 2.50 μg = 1.25 μg

After four half-lives (68 days), $\frac{1}{2} \times$ 1.25 μg = 0.625 μg

ANALYSIS

After four half-lives, 0.625 μg remains, which means that in a little over two months only 6.25% of the radioactive isotope is still present. Even if the time period doesn't work out to exactly round numbers of half-lives, the amount can be estimated. (There are also specific equations that will allow us to calculate it directly, but they are beyond our scope.)

SYNTHESIS

In most cases, chemical reactions occur because two individual molecules collide. In other cases, there may be just one molecule involved in the reaction but some external energy source, such as heat or light, initiates the reaction. But a nuclear decay reaction is unique in that it involves just a single type of atom, and the impetus for the reaction comes internally from the atom itself. This is why the rate at which the reaction occurs, measured by the half-life, is not dependent on concentration, temperature, pressure, or any other external factor. Half-life is an innate function of the isotope itself.

▶ ASSESSING THE OBJECTIVE FOR SECTION 16-2

EXERCISE 16-2(a) LEVEL 1: Which isotopes decay faster: ones with long half-lives or ones with short half-lives?

EXERCISE 16-2(b) LEVEL 2: What percent of an isotope has decayed after three half-lives?

EXERCISE 16-2(c) LEVEL 2: After 36.0 minutes, it was found that 4.0 μg remains of a sample of a radioactive isotope that originally weighed 64 μg. What is the half-life of the isotope?

EXERCISE 16-2(d) LEVEL 3: Why is it not possible to isolate a pure sample of a radioactive substance?

For additional practice, work chapter problems 16-20 and 16-22.

▶ OBJECTIVE
FOR SECTION 16-3
Describe how the three natural types of radiation affect matter.

16-3 The Effects of Radiation

LOOKING AHEAD! We tend to fear the effects of radiation. Certainly, it may cause cancer and sickness, but this same radiation can save lives when used in medical diagnosis or cancer treatment. How radiation interacts with matter is our next subject. ∎

A generation grew up with a deep fear of radiation as a result of a possible nuclear holocaust. Although the threat of nuclear bombs remains, we now see radiation in a more positive light because of all the medical applications that are in common use.

The same radiation that we fear so much is also used to shrink cancerous tumors, treat other diseases, and diagnose illnesses that would otherwise go undetected.

16-3.1 Generation of Heat

Nuclear decay is a process in which a nucleus spontaneously gives off high-energy particles (alpha or beta) or high-energy light (gamma). The interaction of these fast-moving particles or light waves with surrounding matter eventually results in increased velocity for all molecules in the area. In other words, the temperature increases. In fact, much of the heat generated in the interior of Earth and other planets is a direct result of natural radiation from decaying isotopes. The presence of radioactive isotopes in waste from nuclear power plants makes these materials extremely hot. In most cases the wastes must be stored in water continuously so that they do not melt.

16-3.2 Formation of Ions and Free Radicals

Radiation also interacts with matter to cause ionization. *Besides producing heat, collisions of radioactive particles or rays leave a trail of ions along their path, so this type of radiation is sometimes referred to as* **ionizing radiation.** The formation of ions is caused by the removal of an electron from a neutral molecule.

If high-energy particles or rays collide with an H_2O molecule, an electron may be ejected from the molecule, leaving an H_2O^+ ion behind. The ion is chemically reactive and can be very destructive to biological molecules involved in life processes. If the molecule that is ionized is a large, complex molecule that is part of a cell in a living system, the ionization causes damage, mutation, or even eventual destruction of the cell.

Another important effect of radiation is the creation of free radicals. **Free radicals** *are neutral atoms or neutral parts of molecules that have unpaired electrons.* For example, the H_2O^+ ion is formed from radiation. The H_2O^+ ion can break into an H^+ ion and a neutral OH free radical as shown below.

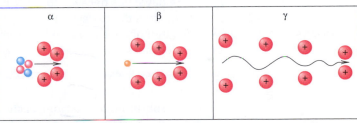

$$H_2O^+ \longrightarrow H^+ + OH$$

Free radicals such as OH are particularly destructive to living cells.

High ionization along path but low penetration   Moderate ionization along path but low penetration   Low ionization along path but very high penetration

FIGURE 16-5 Ionization and Penetration Gamma radiation has the most penetrating power of the three types of radiation.

16-3.3 Ionization and Types of Radiation

As shown in Figure 16-5, an alpha particle causes the most ionization and is the most destructive along its path. However, these particles do not travel more than a few centimeters in air before they acquire two electrons to become an ordinary helium atom. They can be stopped by a piece of paper. The danger of alpha emitters (such as uranium and plutonium) is that they can be ingested through food or inhaled into the lungs, and eventually accumulate in bones. There, in intimate contact with the blood-producing cells of the bone marrow, they slowly damage the surrounding tissue at the molecular level. Ultimately, the radiation can cause certain cells to become abnormal. Rapid reproduction of these altered cells in the bone marrow causes the type of cancer commonly referred to as leukemia.

Beta radiation is less ionizing along its path than alpha radiation but is more penetrating. Still, a thin sheet of metal such as aluminum will stop most beta radiation. This type of radiation can cause damage to surface tissue such as skin and eyes but does not reach internal organs unless ingested. In the past, countries tested nuclear devices in the atmosphere. This produced large quantities of ^{90}Sr (a beta emitter with a half-life of 28 years). Strontium is an alkaline earth metal like calcium, a major component of bones. Strontium can

Lead aprons are worn to protect workers from X-rays and gamma rays.

MAKING IT REAL

Radioactivity and Smoke Detectors

Smoke detectors make use of a synthetic element.

We don't usually think of the presence in our home of radioactive isotopes as a good thing. However, an isotope of Americium (^{241}Am) in smoke detectors has saved many lives. Americium is a rather new element on the periodic table. ^{243}Am ($t_{1/2} =$ 7370 years) was first discovered in 1945 in airborne dust particles after the first atomic bombs. The isotope used in smoke detectors is a byproduct of nuclear reactors. It is formed from the decay of ^{241}Pu, which is extracted from spent reactor elements. It all starts in the reactor with ^{238}U, the most abundant isotope of uranium. The neutrons produced in the reactor form ^{241}Pu in the following steps.

$$^{238}U + \,^{1}_{0}n \longrightarrow \,^{239}U \xrightarrow{\beta^-} \,^{239}Np \xrightarrow{\beta^-}$$
$$^{239}Pu + \,^{1}_{0}n \longrightarrow \,^{240}Pu + \,^{1}_{0}n \longrightarrow \,^{241}Pu$$

Most of the ^{241}Pu ($t_{1/2} = 14$ days) has decayed to ^{241}Am ($t_{1/2} = 432$ years) by the time the extraction occurs. It is quite expensive, though. It is produced by the U.S. Atomic Energy Commission, where it sells for about $1,500 per gram. However, since only 0.2 mg of ^{241}Am (as the compound AmO_2) is used in each unit, 5000 smoke detectors can be made from one gram of the isotope. That amount of Americium produces about 3×10^4 decays/second.

Smoke detectors work on simple principles. The ^{241}Am decays by alpha particle emission. The alpha particles (^{4}He) ionize the major molecules of air (N_2 and O_2) to N_2^+ form and O_2^+ ions. These ions form between plates of two electrodes that are connected to a battery or house current. The ions serve as a conduction pathway between the two electrodes and complete a circuit. When smoke drifts in between the two electrodes, it absorbs the alpha particles, so fewer ions are formed. When the detector senses the circuit is disrupted, the alarm sounds.

Are smoke detectors safe? The answer is yes. Alpha particles have a short range of a few centimeters at most, so they stay within the confines of the detector. The detectors themselves do not have enough radioactivity to pose a problem to the environment. The current necessary for continual operation is very small, so a 9-volt battery should last at least six months.

Modern smoke detectors are life-saving devices that are the direct product of the nuclear age. We are glad to have them.

---

substitute for calcium, accumulate in the bones, and eventually cause leukemia and bone cancer. Good sense finally prevailed on the major atomic powers, and atmospheric testing of nuclear weapons has not occurred to any extent for many years.

The radiation from each radioactive isotope also has a specific energy. The energy of the emitted particles or rays is also important in the amount of ionization or free radical formation that occurs.

▶ **ASSESSING THE OBJECTIVE FOR SECTION 16-3**

EXERCISE 16-3(a) LEVEL 1: Fill in the blanks.
Radiation causes _____ as well as the formation of _____ along its path. The most ionizing type of radiation is _____ _____, whereas the most penetrating is _____ _____.

EXERCISE 16-3(b) LEVEL 2: Which of the following is an ion that may be formed from radiation and which is a free radical?
$$Cl^-, H_2O^+, OH^-, OH, H_2O, Ar$$

EXERCISE 16-3(c) LEVEL 3: The heat in Earth's interior that forms the lava in volcanoes is due to radiation. Explain how radiation causes heating.

For additional practice, work chapter problems 16-26 and 16-27.

16-4 The Detection and Measurement of Radiation

LOOKING AHEAD! How do we know how much radiation we are absorbing? Since we can't see it or feel it, we must have some method of measuring radiation. The detection and measurement of radiation are obviously important in understanding the danger or the benefit. ■

16-4.1 Film Badges and Radon Detectors

Becquerel first discovered radiation by noticing that it exposed photographic film. This is still an important method of both detecting and measuring radiation. Film badges are worn by anyone who works near a source of radiation. (See Figure 16-6.) When the film is developed, the degree of darkening is proportional to the amount of radiation absorbed. Different filters are used so that the amount of each type of radiation received (alpha, beta, or gamma) can be measured.

An application of the film badge is used in radon detectors. ^{222}Rn is a radioactive isotope formed as part of the natural decay series starting with ^{238}U. Because radon is a noble gas, it can escape from the ground into the air. In certain areas of the United States where there is a significant amount of uranium in the soil, radon may accumulate in basements or other enclosed areas through foundation cracks. When inhaled, radon is usually exhaled unchanged, causing no harm. Its half-life, however, is only 10.82 days, which means that some actually does undergo radioactive decay when it is inhaled. If so, it decays to ^{218}Po, which stays in the lungs because polonium is a solid. This is also a highly radioactive isotope that then decays in the lungs and may cause damage. The longer one is exposed to radon, the greater the chances that it will cause cancer. When the detector is left in a basement, it measures the number of radon disintegrations. After a specified time, the detector is sent to a laboratory, where the film is developed and the amount of radiation from radon is determined.

FIGURE 16-6 Film Badges and Radon Detector These devices are used to detect radiation.

16-4.2 Geiger and Scintillation Counters

As mentioned in the previous section, radiation causes ionization. Ionization is also used to measure radiation in the **Geiger counter.** The Geiger counter is composed of a long metal tube with a thin membrane at one end that allows all types of radiation to penetrate. The metal tube is filled with a gas such as argon at low pressure. A metal wire in the center of the tube is positively charged. When a radioactive particle enters the tube, it causes ionization. The electrons from this ionization move quickly to the positive electrode, causing a pulse of electricity to flow. This pulse is amplified and causes a light to flash, produces an audible "beep," or registers on a meter in some unit such as "counts per minute." (See Figure 16-7.)

A third type of radiation detector is a **scintillation counter**. Certain compounds such as zinc sulfide are known as *phosphors*. When radiation strikes a phosphor, a tiny flash of light is emitted in a process known as *fluorescence*. The flash of light is converted into a pulse of electricity that can be counted in the same manner as in the Geiger counter.

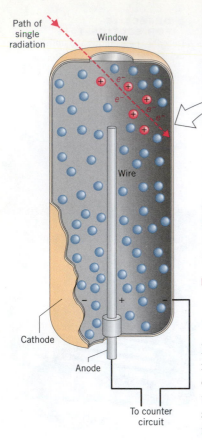

Path of single radiation

Window

Wire

Cathode

Anode

To counter circuit

FIGURE 16-7 The Geiger Counter This device measures radiation by the ionization it causes.

16-4.3 Radiation and the Human Body

How much radiation a person receives and how it affects the human body are matters of great importance. As a result, various units of measurement have been devised. The first measure of radiation simply relates to the number of disintegrations (counts) that occur in a specified time interval. This is known as the sample's activity. This unit is the **curie** (named after Madame Curie). It is defined as the decay rate of one gram of radium. The milli- and microcuries refer to smaller and more meaningful measures of radiation.

$$1 \text{ curie (Ci)} = 1.0 \times 10^{10} \text{ counts/s}$$

$$1 \text{ millicurie (mCi)} = 1.0 \times 10^{7} \text{ counts/s}$$

$$1 \text{ microcurie (μCi)} = 1.0 \times 10^{4} \text{ counts/s}$$

The **becquerel** (Bq) is 1 disintegration per second and is used to measure very low levels of activity.

An important unit of radiation specifically refers to how different types of radiation affect the human body. The unit that describes the effect on the human body is the **rem** (roentgen equivalent in man). Other units used for this measure are **millirems (mrems)** ($1 \text{ mrem} = 10^{-3}$ rems) and **sieverts** (Sv) ($1 \text{ Sv} = 100$ rems). Sieverts and **millisieverts** (10^{-3} Sv) are the SI units and are now used in most standard publications and internationally. Radiation levels from the damaged nuclear reactors in Japan in 2011 were generally reported in millisieverts **(mSv)**.

When one is exposed to various types of radiation, the total amount of radiation in rems or sieverts becomes a critical number. Our natural environment provides what is referred to as background radiation. It comes from radon in the air; radiation from the sun and outer space, known as cosmic rays; and radioactive elements in rocks and in concrete. Even TV and smoking provide a small amount of radiation exposure. Although the amount varies geographically, on the average we absorb about 300 mrems (3 mSv) annually from background radiation. Modern X-rays vary from about 20 mrems (0.2 mSv) for dental to over 1000 mrems (10 mSv) for certain extensive diagnostic X-rays of the intestines.

The amount of radiation the average person absorbs in one year is a concern. Those who live in higher altitudes or fly frequently in airplanes receive more cosmic radiation than those who stay at sea level. People who work in the nuclear industry, such as hospitals or nuclear power plants, receive extra radiation and must wear film badges to measure the rems of radiation received. The federal government puts the health limit at 5000 mrems (50 mSv) as the maximum amount one should receive in one year.

Although the amount of cumulative radiation can be dangerous, it is far more dangerous to receive a large dose of radiation at one time. Over a period of time, the body can recover to a certain extent. If one receives a dose of 100 rems (1000 mSv) at one exposure, radiation sickness occurs. A sharp reduction of white blood cells occurs along with nausea, vomiting, and fatigue. Above 300 rems (3000 mSv), hair loss and severe infection may occur because of the absence of white blood cells.

Doses above 500 rems (5000 mSv) cause death in 50% of those who have received that amount of radiation. This is known as the **LD$_{50}$**, *the dose that is fatal to one-half of those receiving that amount.* A dose above 600 rems (6000 mSv) is considered fatal to all humans within a few weeks.

▶ **ASSESSING THE OBJECTIVE FOR SECTION 16-4**

EXERCISE 16-4(a) LEVEL 1: Fill in the blanks.
Three major ways in which radiation is measured are the _____ badge, the _____ counter, and the _____ counter. The major unit used to measure the total amount of radiation is the _____. The major units that measure the amount of radiation received by a person are the _____ and the _____. The amount of radiation that is fatal to half of those who receive it is known as the _____.

EXERCISE 16-4(b) LEVEL 3: Radiation causes fluorescence. Historically, this was put to use in a common item that almost everyone wore. Do you know what it was?

For additional practice, work chapter problems 16-28 and 16-29.

PART A SUMMARY

KEY TERMS

16-1	**Radiation** refers to the particles or light energy. **Radioactive isotopes** spontaneously emit radiation. p. 562
16-1.1	**Alpha particles (α)** are helium nuclei emitted from a nucleus. A nuclear equation illustrates changes in the nucleus. p. 562
16-1.2	**Beta particles (β)** are electrons emitted from a nucleus. p. 563
16-1.3	**Gamma rays (γ)** are composed of high-energy light from a nucleus. p. 563
16-1.4	**Positrons** are positively charged electrons. Positron emission and **electron capture** both decrease the nuclear charge of the emitting nucleus by 1. p. 563
16-2.1	The **half-life ($t_{1/2}$)** of an isotope is the time required for one-half of a sample to decay. p. 566
16-2.2	A **radioactive decay series** relates to the isotopes formed between a long-lived isotope and a stable isotope of lead. p. 567
16-3.2	**Ionizing radiation** is responsible for the production of ions. **Free radicals** are also formed as a result of radiation. p. 569
16-4.2	**Geiger counters** and **scintillation counters** are used to detect radiation. p. 571
16-4.3	The **curie** and **becquerel** are units that relate to the number of disintegrations. The **rem**, **mrem**, **sievert**, and **mSv** are units that relate to the effect of radiation on humans. p. 572
16-4.3	The **LD$_{50}$** relates to the amount of radiation fatal to one-half of victims. p. 573

SUMMARY CHART

Effects of Type of Radiation on an Isotope

	Alpha (α)	Beta (β)	Gamma (γ)	Positron (or Electron capture)
Identity	$^{4}_{2}He^{2+}$	$^{0}_{-1}e$	light energy	$^{0}_{+1}e$
Effect on	mass −4	mass 0	mass 0	mass 0
Isotope	charge −2	charge +1	charge 0	charge −1

Part B

Induced Nuclear Changes and their Uses

SETTING A GOAL

- You will learn how we have applied synthetic nuclear reactions to develop a wide range of nuclear processes.

OBJECTIVES

16-5 Write balanced nuclear equations for the synthesis of specific isotopes by nuclear reactions.

16-6 List and describe the beneficial uses of radioactivity.

16-7 Identify the differences between natural radioactivity, nuclear fission, and nuclear fusion.

▶ **OBJECTIVE FOR SECTION 16-5**
Write balanced nuclear equations for the synthesis of specific isotopes by nuclear reactions.

16.5 Nuclear Reactions

LOOKING AHEAD! Elements spontaneously change into other elements by emitting alpha or beta particles. But can scientists do the same thing? Our next topic involves the accomplishment of changing one element into another. ■

16-5.1 Transmutation

Transmutation *of an element is the conversion of that element into another.* The first example of transmutation was discovered by Ernest Rutherford in 1919; earlier, he had proposed the nuclear model of the atom. Rutherford found that bombarding ^{14}N atoms with alpha particles caused a nuclear reaction that produced ^{17}O and a proton. This nuclear reaction is illustrated by the nuclear equation

$$^{14}_{7}N + {}^{4}_{2}He \longrightarrow {}^{17}_{8}O + {}^{1}_{1}H$$

Note that the total number of nucleons is conserved by the reaction ($14 + 4$ on the left $= 17 + 1$ on the right). The total charge is also conserved during the reaction ($7 + 2$ on the left $= 8 + 1$ on the right). Both the number of nucleons and the total charge must be balanced (the same on both sides of the equation) in a nuclear reaction.

Over a decade later, Iréne Curie, daughter of Pierre and Marie Curie, and her husband, Frédéric Joliot, produced an isotope of phosphorus that was radioactive. This was the first artificially induced radioactivity, for which Curie and Joliot received the 1935 Nobel Prize in chemistry. The radioactive isotope formed was ^{30}P, which decays to ^{30}Si as illustrated by the following nuclear equations.

$$^{27}_{13}Al + {}^{4}_{2}He \longrightarrow {}^{30}_{15}P + {}^{1}_{0}n$$

$$^{30}_{15}P \longrightarrow {}^{30}_{15}Si + {}^{0}_{+1}e$$

16-5.2 Particle Accelerators

Because the nucleus has a positive charge and many of the bombarding nuclei (except neutrons) also have a positive charge (e.g., $^{4}_{2}He$, $^{1}_{1}H$, $^{2}_{1}H$), the particles must have a high energy (velocity) to overcome the natural repulsion between these two like charges. From the collision of the nucleus and the particle, a high-energy nucleus is formed. As a result, another particle (usually a neutron) is then emitted by the nucleus to carry away excess energy from the collision. This is analogous to a head-on collision of two cars, where fenders, doors, and bumpers come flying off after impact. The nucleus–particle collision is illustrated in Figure 16-8.

FIGURE 16-8 Collision of an Alpha Particle with a Nucleus
Alpha particles must have sufficient energy to overcome the repulsion of the target nucleus for a collision to take place.

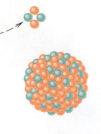

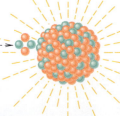

Alpha particle does not have enough energy, thus no collision

Collision occurs

High-energy nucleus gives off particle

The invention of particle accelerators, which increase the velocity and energy of nuclei and subatomic particles, opened the door to vast possibilities for artificial nuclear reactions. The larger accelerators also allow studies of collisions so powerful that the colliding nuclei are blasted apart. From these experiments, scientists have glimpsed the most basic composition of matter. These accelerators are large and expensive. Completed in the early 1970s, the Fermi National Accelerator in Batavia, Illinois, has a circumference of about 6 miles and cost $245 million. A huge accelerator, known as the *Superconducting Super Collider* (*SSC*), was about one-third completed in Texas and would have had a circumference of 52 miles. The cost of about $12 billion was considered too much, however, so it was canceled by the U.S. Congress. CERN, the largest accelerator, is in Switzerland and is known as the *Large Hydron Collider* (LHC). It is hoped it will create conditions that have not occurred since the beginning of the universe known as the "big bang." (See Figure 16-9.)

FIGURE 16-9 CERN Accelerator This is the largest accelerator in use today.

16-5.3 Synthetic Heavy Elements

The use of particle accelerators has made possible the extension of the periodic table past element number 94, plutonium. Elements number 95 through number 118 have been synthesized by means of large particle accelerators. Element numbers 113 through 116 were synthesized at The *Joint Institute for Nuclear Research* in Dubna, Russia. In late 2006, the synthesis of three atoms of element number 118 was reported and in 2009 the synthesis of element 114 was confirmed. Finally, the most recent element synthesized was element number 117, reported in early 2010. This element completed the seventh period. Before the most recently discovered elements officially take their place on the periodic table, however, the experiments must be duplicated at other facilities. The latest official addition (February 2010) to the periodic table is element number 112, which is named copernicium (symbol Cn).

The following nuclear reactions illustrate the formation of two heavy elements.

$$^{238}_{92}\text{U} + ^{12}_{6}\text{C} \longrightarrow ^{244}_{98}\text{Cf} + 6^{1}_{0}\text{n}$$

$$^{238}_{92}\text{U} + ^{16}_{8}\text{O} \longrightarrow ^{250}_{100}\text{Fm} + 4^{1}_{0}\text{n}$$

EXAMPLE 16-4   Writing Equations for Nuclear Reactions

Complete the following nuclear equations.

(a) $^{9}_{4}\text{Be} + ^{2}_{1}\text{H} \longrightarrow$ _____ $+ ^{1}_{0}\text{n}$ **(b)** $^{252}_{98}\text{Cf} + ^{10}_{5}\text{B} \longrightarrow$ _____ $+ 5^{1}_{0}\text{n}$

PROCEDURE

Complex nuclear equations are balanced in the same manner as discussed in Example 16-1. Determine the missing isotope or particle by recalling that the total mass and the total charge are the same on both sides of the equation.

SOLUTION

(a) Nucleons: $9 + 2 = x + 1$, so $x = 10$
    Atomic number: $4 + 1 = y + 0$, so $y = 5$
From the list of elements, we find that the element is boron (atomic number 5). The isotope is

$$^{10}_{5}\text{B}$$

(b) Nucleons: $252 + 10 = x + (5 \times 1)$, so $x = 257$.
    Atomic number: $98 + 5 = y + (5 \times 0)$, so $y = 103$
From the list of elements, the isotope is

$$^{257}_{103}\text{Lr}$$

ANALYSIS

The new nuclear symbol for this set of exercises is the neutron, $_0^1n$, which can be emitted when two particles collide. Notice that the elemental species that are involved in the reactions have a degree of redundancy to them. Along with the symbol, the atomic number is reported in the lower-left corner. These values can be determined by the symbols themselves, so their inclusion is not critical but is convenient.

SYNTHESIS

Notice that the product of each of these reactions, along with the daughter particle, is one or more neutrons. These high-energy particles are themselves potential reactants in nuclear transmutations [see Exercise 16-5(b) in Assessing the Objective for Section 16-5]. Neutrons, which have no charge, are not affected by repulsion of a nucleus as are alpha particles or other nuclei. Therefore, because it takes less energy for a neutron to collide with a nucleus, reactions initiated by neutrons can take place under less extreme conditions than those that require a particle accelerator.

▶ **ASSESSING THE OBJECTIVE FOR SECTION 16-5**

EXERCISE 16-5(a) LEVEL 1: When completing a nuclear reaction, what must be balanced?

EXERCISE 16-5(b) LEVEL 2: Fill in the blanks in the following equations.

(a) $_{23}^{51}V + _1^2H \longrightarrow$ _____ $+ _1^1H$

(b) $_{92}^{235}U + _0^1n \longrightarrow _{38}^{90}Sr +$ _____ $+ 2\,_0^1n$

(c) $_{96}^{246}Cm +$ _____ $\longrightarrow _{102}^{254}No + 4\,_0^1n$

EXERCISE 16-5(c) LEVEL 3: The formation of element 118 was first reported in 1999. It was a fraudulent claim, however, so the report was withdrawn. In late 2006, the formation of three atoms of element 118 was again reported by reliable scientists who had previously reported the discovery of elements 113, 114, 115, and 116. Element 118 was made by bombarding ^{249}Cf with ^{48}Ca. The new element had 176 neutrons. It decayed almost immediately by giving off three alpha particles one at a time.

(a) Write the nuclear equation showing the formation of element 118 and the three products of the alpha particle decays.

(b) If enough of element 118 could be produced to study, what would we predict to be its physical state at room temperature?

For additional practice, work chapter problems 16-30 and 16-32.

16-6 Applications of Radioactivity

LOOKING AHEAD! Despite their horrible reputation, radioactive isotopes have a huge number of beneficial applications. How we use radioactivity to date artifacts, make our food healthier, and help diagnose and even cure our illnesses is the subject of this section. ∎

The benefits of the nuclear age are awesome. Heart scans tell us whether bypass surgery is indicated. Radiation treatments have saved thousands of lives and prolonged many thousands of others. The shelf-life of perishable food has been extended by weeks. These and other uses have far outweighed the dangers from the misuse of radioactivity. We will begin our discussion of how radioactivity improves our lives with how we use a naturally produced radioactive isotope to date artifacts.

16-6.1 Carbon Dating

Perhaps one of the most interesting applications of radioactive isotopes is the dating of wood or other carbon-containing substances that were once alive. ^{14}C is a radioactive isotope with a half-life of 5760 years. It is produced in the stratosphere by a nuclear reaction involving *cosmic rays* (a variety of radiation and particles) from the sun. ^{14}C dating is effective in dating artifacts that are from about 1000 to about 50,000 years old. If the sample is much older, the level of radiation become too low to detect.

$$^{14}_{7}N + ^{1}_{0}n \longrightarrow ^{14}_{6}C + ^{1}_{1}H$$

The ^{14}C then mixes with the normal and stable isotopes of carbon in the form of carbon dioxide. Carbon dioxide is taken up by living systems and, through photosynthesis, becomes part of the carbon structure of the organism. As long as the carbon-based system is alive, the ratio of ^{14}C to normal carbon is the same as in the atmosphere. When the system dies, the amount of ^{14}C in the organism begins to decrease. By comparing the amount of radiation from ^{14}C in the artifact with that in living systems, the age can be determined. (See Example 16-2.)

The procedure can analyze carbon-containing objects like wood, cloth, and bone. Some famous examples include dating the Shroud of Turin to around A.D. 1300 and the "Ice-Man," found near the Italian–Austrian border in 1991, to around 3300 B.C.

Besides ^{14}C dating, the decay of other isotopes (e.g., ^{40}K) can be studied to determine the age of rocks that were formed from molten material millions of years ago.

The Shroud of Turin was dated to the fourteenth century.

16-6.2 Neutron Activation Analysis

Most naturally occurring isotopes are not radioactive. **Neutron activation analysis** *is a process whereby stable isotopes are made radioactive by absorption of neutrons in a nuclear reactor.* An example is the production of ^{60}Co, which is used in cancer therapy, from a stable isotope of cobalt.

$$^{59}_{27}Co + ^{1}_{0}n \longrightarrow ^{60m}_{27}Co$$

Another application of neutron activation concerns the analysis of arsenic in human hair. Arsenic compounds can be used as a slow poison, but arsenic accumulates in human hair in minute amounts. By subjecting the human hair to neutron bombardment, the stable isotope of arsenic is changed to a metastable nucleus.

$$^{75}_{33}As + ^{1}_{0}n \longrightarrow ^{76m}_{33}As \longrightarrow ^{76}_{33}As + \gamma$$

The amount of gamma radiation from the metastable arsenic can be measured and is proportional to the amount of arsenic present. The method is very sensitive to even trace amounts of arsenic.

16-6.3 Food Preservation

In a very simple procedure, many types of food can be irradiated with gamma radiation, which kills bacteria and other microorganisms that cause food spoilage without changing the taste or appearance of the food. This increases the shelf-life of food before decay sets in. For example, mold begins to form on strawberries, even when refrigerated, in just a few days. After irradiation, the same strawberries can be stored for two weeks without decay. (See Figure 16-10.) There has been some buyer resistance to irradiated foods. The public worries that irradiation somehow may make the food radioactive. The procedure is perfectly safe, however, and as public apprehension declines, we will probably see more and more irradiated foods at the grocery store.

FIGURE 16-10 Irradiated Strawberries The strawberries at the top are moldy after a few days. Those on the bottom have been irradiated and are still fresh after two weeks.

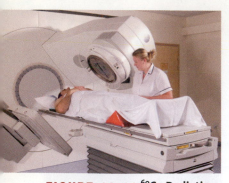

FIGURE 16-11 **⁶⁰Co Radiation Treatment** A beam of gamma rays can be used to destroy cancerous tissue.

16-6.4 Medical Therapy

⁶⁰Co gamma radiation can be focused into a narrow beam. Although gamma radiation destroys healthy as well as malignant cells, healthy cells recover faster. By focusing the beam at different angles on tumors located within the body, the gamma rays can be concentrated on the tumor. (See Figure 16-11.) This treatment has many unpleasant side effects because of the destruction of normal, healthy cells.

A procedure known as *brachytherapy* is used increasingly in the treatment of prostate cancer. Tiny "seeds" of ¹⁰³Pd ($t_{1/2} = 17$ days) or ¹²⁵I ($t_{1/2} = 8$ days) inside a titanium capsule are implanted in the cancerous gland. As these isotopes decay by electron capture, they release low-energy X-rays, which destroy prostate tissue but penetrate only about 1 cm into the surrounding tissue. Since the radiation has a localized effect, nearby organs are minimally affected. The radiation lasts only a few months because of the short half-lives, so the seeds are left permanently in place. Although there are several other radioactive isotopes used in therapies, the greatest application of these isotopes is in diagnosis.

16-6.5 Medical Diagnosis

Today, hardly any large hospital could be without its nuclear medicine division. We have come to rely on the use of radioactive isotopes to help diagnose many diseases and conditions. Various radioactive isotopes can be injected into the body, and their movement through the body or where they accumulate can be detected outside the body with radiation detectors. ⁵⁹Fe is used to measure the rate of formation and lifetime of red blood cells. ⁹⁹ᵐTe is used to image the brain, heart, and other organs. ¹³¹I is used to detect thyroid malfunction and can also be used for the treatment of thyroid tumors. (If large amounts are absorbed, however, it can cause cancer.) ²⁴Na is used to study the circulatory system.

One of the more useful (and expensive) tests is known as a *PET (positron emission tomography)* scan, which produces an image of a two-dimensional slice through a portion of the body. The body is injected with a compound (e.g., glucose) containing a radioactive isotope such as ¹¹C. The glucose containing this isotope of carbon is metabolized along with glucose produced by the body containing the stable isotope ¹²C. The parts of the brain that are particularly active in metabolism of glucose will display increased radioactivity. Abnormal glucose metabolism in the brain can then be detected, which may indicate a tumor or Alzheimer's disease. This diagnostic procedure is considered noninvasive compared to surgery. (See Figure 16-12.)

The ¹¹C decays by positron emission.

$$\ce{^{11}_6C -> ^{11}_5B + ^0_{+1}e}$$

$$\ce{^0_{+1}e + ^0_{-1}e -> 2\gamma}$$

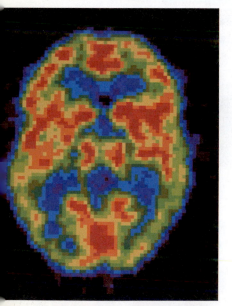

FIGURE 16-12 PET Scan This PET scan of the brain reveals areas of activity during sleep.

Almost immediate annihilation of the positron by an electron leads to two gamma rays that exit the body in exactly opposite directions. Scintillation counters are positioned around the body to detect these two gamma rays and ignore others from background radiation. Computers are then used to translate the density and location of the gamma rays into two-dimensional images.

▶ **ASSESSING THE OBJECTIVE FOR SECTION 16-6**

EXERCISE 16-6(a) LEVEL 1: Fill in the blanks.
Both naturally occurring and synthetic radioactive isotopes find many uses. The decay of carbon-_____ can be used to date many carbon-containing fossils. Medically useful radioactive isotopes can be prepared by _____ activation. Food can be preserved by radiation with _____ _____. PET scans make use of isotopes that emit _____.

EXERCISE 16-6(b) LEVEL 2: The most commonly used isotope in PET scans is ^{18}F, with a half-life of 110 min compared to ^{11}C, with a half-life of 20 min. These isotopes must be synthesized in a nearby cyclotron because of their short half-lives. What isotope would ^{16}O be bombarded with to produce ^{18}F? What isotope would be used to produce ^{11}C if it was bombarded with protons?

EXERCISE 16-6(c) LEVEL 3:

(a) Why can't a fossil that is over 50,000 years old be dated by normal carbon dating?

(b) Why does adding a neutron to a stable isotope often produce a radioactive isotope?

For additional practice, work chapter problems 16-25 and 16-42.

16-7 Nuclear Fission and Fusion

► OBJECTIVE FOR SECTION 16-7
Identify the differences between natural radioactivity, nuclear fission, and nuclear fusion.

LOOKING AHEAD! In 1938, a discovery was made that at first suggested massive destruction and, later, on the positive side, inexpensive energy. The discovery was the atomic bomb and then the control of its process in nuclear power plants. The nuclear process that produces this power is the next topic. ∎

16-7.1 The Atomic Bomb

In 1934, two Italian physicists, Enrico Fermi and Emilio Segré, attempted to expand the periodic table by bombarding ^{238}U with neutrons to produce neutron-rich isotopes that would decay by beta particle emission to form elements with atomic numbers greater than 92 (the last element on the periodic table at the time).

$$^{238}_{92}U + {}^{1}_{0}n \longrightarrow {}^{239}_{92}U \longrightarrow {}^{239}_{93}X + {}^{0}_{-1}e$$

The experiment seemed to work, but they were perplexed by the presence of several radioactive isotopes produced in addition to the presumed element number 93. In 1938, these experiments were repeated by two German scientists, Otto Hahn and Fritz Strassman. They were able to identify the excess radioactivity as coming from isotopes such as barium, lanthanum, and cerium, which had about half the mass of the uranium isotope. Two other German physicists, Lise Meitner and Otto Frisch, were able to show that the rare isotope of uranium (^{235}U, 0.7% of naturally occurring uranium) was undergoing fission into roughly two equal parts after absorbing a neutron. **Fission** *is the splitting of a large nucleus into two smaller nuclei of similar size.* It should be noted that there are a variety of products from the fission of ^{235}U in addition to those represented by the following equation. (See Figure 16-13.)

$$^{235}_{92}U + {}^{1}_{0}n \longrightarrow {}^{139}_{56}Ba + {}^{94}_{36}Kr + 3{}^{1}_{0}n$$

Two points about this reaction had monumental consequences for the world, and scientists in Europe and America were quick to grasp their meaning in a world about to go to war.

1. Comparison of the masses of the product nuclei with that of the original nucleus indicated that a significant amount of mass was lost in the reaction. According to Einstein's equation ($E = mc^2$), the lost mass must be converted to a tremendous amount of energy. Fission of a few kilograms of ^{235}U all at once could produce energy equivalent to tens of thousands of tons of the conventional explosive TNT.

2. What made the rapid fission of a large sample of ^{235}U feasible was the potential for a chain reaction. *A nuclear chain reaction is a reaction that is*

self-sustaining. The reaction generates the means to trigger additional reactions. Note in Figure 16-13 that the reaction of one original neutron caused three to be released. These three neutrons cause the release of nine neutrons and so forth. If a given densely packed "critical mass" of ^{235}U is present, the whole mass of uranium can undergo fission in an instant, with a quick release of energy in the form of radiation.

Unfortunately, the world thus entered the nuclear age in pursuit of a bomb. In the early 1940s, a massive secret effort—the Manhattan Project—began research into producing "the atomic bomb." The first nuclear bomb was exploded over Alamogordo Flats in New Mexico on July 16, 1945. This bomb and the bomb exploded over Nagasaki, Japan, were made of ^{239}Pu, which is a synthetic fissionable isotope. The bomb exploded over Hiroshima, Japan, was made of ^{235}U.

16-7.2 Nuclear Power

This enormously destructive device, however, can be tamed. The chain reaction can be controlled by absorbing excess neutrons with cadmium bars. A typical nuclear reactor is illustrated in Figure 16-14. In a reactor core, uranium in the form of pellets is encased in long rods called fuel elements. Cadmium bars are raised and lowered among the fuel elements to control the rate of fission by absorbing excess neutrons. If the cadmium bars are lowered all the way, the fission process can be halted altogether. In normal operation, the bars are raised just enough so that the fission reaction occurs at the desired level. Energy released by the fission and by the decay of radioactive products formed from the fission is used to heat water, which circulates among the fuel elements. This water, which is called the primary coolant, becomes very hot (about 300°C) without boiling because it is under high pressure. The water heats a second source of water, changing it to steam. The steam from the secondary coolant is cycled outside of the containment building, where it is used to turn a turbine that generates electricity.

FIGURE 16-13 Fission and Chain Reactions In fission, one neutron produces two fragments of the original atom and an average of three neutrons. These three neutrons can cause additional fission reactions and lead to a chain reaction.

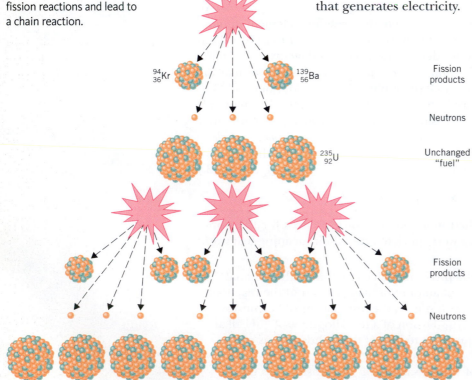

Perhaps the greatest advantage of nuclear energy is that it does not pollute the air with oxides of sulfur and carbon as conventional power plants do. Sulfur oxides have been implicated as a major cause of acid rain. It also seems likely that large amounts of carbon dioxide in the atmosphere are leading to significant changes in the weather as a result of the greenhouse effect. Another advantage is that nuclear power does not deplete the limited supply of fossil fuels (oil, coal, and natural gas), which are also used to make plastics and fertilizers. Until the tragedy in Chernobyl, Ukraine, in April 1986, there had been no loss of human life from the use of nuclear energy to generate electricity. At the current time, about 28 countries

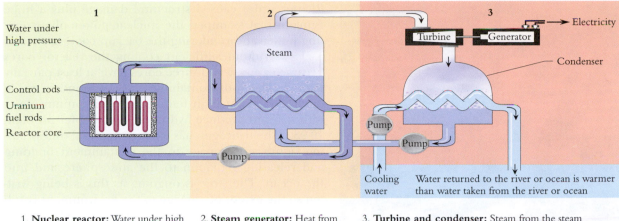

1. **Nuclear reactor:** Water under high pressure carries heat generated in the nuclear reactor core to the steam generator.

2. **Steam generator:** Heat from the reactor vaporizes water in the steam generator, creating steam.

3. **Turbine and condenser:** Steam from the steam generator powers a turbine, producing useable electricity. The condenser uses cooling water from a river or ocean to recondense the steam from the turbine.

FIGURE 16-14 A Nuclear Reactor Nuclear energy is converted to heat energy, the heat energy to mechanical energy, and the mechanical energy to electrical energy.

produce electricity through nuclear power. The United States currently generates about 20% and France about 80% of their electricity from nuclear power. (See Figure 16-15.) Proponents of nuclear power feel that if the proper systems were used with adequate safeguards and backups, catastrophic accidents would be avoided. In fact, new designs of nuclear power plants solve many of the legitimate concerns of citizens. A safe reactor would be one that shuts itself down in case of an emergency and would not overheat or explode. (Such an explosion would be a conventional explosion, not an atomic explosion.) Experts from several countries are also currently planning a relatively safe reactor that will use as a fuel the ^{239}Pu salvaged from dismantled nuclear bombs of the United States and of the former Soviet Union. This reactor would be highly efficient, converting about 50% of the heat generated into electricity, compared with an efficiency of about 33% for a typical commercial nuclear power plant.

The disadvantages were made painfully obvious at Three Mile Island in Pennsylvania in 1979, the catastrophic explosion in Chernobyl, Ukraine, in 1986, and most recently, at the Fukushima nuclear complex in Japan on March 11, 2011. The accident at Three Mile Island took more than 14 years to clean up at a cost greater than $1 billion. Fortunately, there was very little release of radioactivity and no loss of life in this accident.

The catastrophic explosion (which was conventional, not nuclear) and subsequent fire in reactor 4 in Chernobyl, Ukraine, dispersed more dangerous radiation into the environment than all the atmospheric bomb tests by all nations put together. [See Figure 16-16(a).] The surroundings for at least 30 mi around the plant, and even farther downwind, are still uninhabitable. The reactor itself has been entombed in concrete and must remain so for hundreds of years. The extent of the damage is still being assessed. The number of actual deaths is also a matter of controversy. Estimates range from less than 100 to several thousand. At a minimum, many lives will probably be shortened by exposure to the radiation.

The world's most recent nuclear tragedy occurred in Japan on March 11, 2011, at the Fukushima nuclear complex. [See Figure 16-16(b).] A large earthquake (9.0 on the

FIGURE 16-15 A Nuclear Power Plant In such plants, the nuclear energy of uranium is converted into electrical energy.

(a)

(b)

FIGURE 16-16 (a) The Chernobyl Reactor and (b) The Fukushima Reactor The Nuclear Reactors at (a) Chernobyl in 1986 and in (b) Japan in 2011 suffered severe damage.

Richter scale) caused structural damage plus a huge tsunami that swamped the nuclear power plants cutting power and rendering the backup cooling systems useless. Without sufficient cooling the reactor cores became exposed and three cores have at least partially melted. The rods containing the uranium are made of zirconium. At high temperatures zirconium reacts with water to produce hydrogen (see chapter problem 10-93.) The hydrogen has exploded several times causing significant damage to the containment building and releasing radiation to the atmosphere and into the ocean. The story is ongoing as this is being written so the extent of damage to the environment is still unknown.

A meltdown occurs if the temperature of the reactor core exceeds 1130°C, the melting point of uranium. Theoretically, the mass of molten uranium, together with all the highly radioactive decay products, could accumulate on the floor of the reactor, melt through the many feet of protective concrete, and eventually reach groundwater. If this were to happen, vast amounts of deadly radioactive wastes could be spread through the groundwater into streams and lakes.

The type of accident that occurred in Ukraine is unlikely in most of the world's commercial reactors. Unlike reactors in the former Soviet Union, these reactors are surrounded by an extremely strong structure, called a containment building, made of reinforced concrete. These buildings should contain an explosion such as occurred at Chernobyl. The reactors in the United States are built to withstand the collision of a jet airliner and 300 mi/hr winds. Also, most reactors do not contain graphite, which is highly combustible and led to the prolonged fire in Chernobyl that spread so much radioactivity into the atmosphere. It should be noted that bomb-grade uranium is at least 90% ^{235}U, while that used in nuclear reactors is 2–3% ^{235}U. The uranium used in nuclear reactors could not be used for a bomb.

A growing problem in the use of nuclear energy involves used, or spent, fuel. Originally, used fuel was to be processed at designated centers where the unused fuel could be separated and reused and the radioactive wastes disposed of. However, problems in transportation of this radioactive material and the danger of theft of the ^{239}Pu that is produced in a reactor have hindered solution of the problem. (^{239}Pu is also fissionable and could be used in a nuclear bomb.) Currently, spent fuel elements, which must be replaced every four years, are being stored at the reactor sites. These fuel rods are not only highly radioactive but remain extremely hot and must be continually cooled. It is estimated that they will remain dangerously radioactive for at least 1000 years. This is a problem that must soon be resolved. At one time, it was planned to store the waste from U.S. reactors in Yucca Flats in Nevada. Although over $14 billion has been spent preparing the site, this plan has been abandoned.

Because of the accidents, new regulations regarding safeguards have made the cost of a new nuclear power plant considerably higher than that of a conventional power plant using fossil fuels. As a result, there haven't been any new plants in the United States for several decades. Recently, a few permits have been granted for construction of new nuclear power plants in the near future in the United States. The recent event in Japan may again slow the process, however.

16-7.3 Nuclear Fusion

In the late 1930s, another source of nuclear energy was proposed by Hans Bethe. This involved fusing small nuclei together rather than splitting large ones. *This reaction involves the* **fusion**, *or bringing together, of two small nuclei.* An example of a fusion reaction is

$$\ce{^{3}_{1}H} + \ce{^{2}_{1}H} \longrightarrow \ce{^{4}_{2}He} + \ce{^{1}_{0}n}$$

($\ce{^{3}_{1}H}$ is called tritium; it is a radioactive isotope of hydrogen.)

As in the fission process, a significant amount of mass is converted into energy. In fact, fusion energy is the origin of almost all our energy, since it powers the sun. Millions of tons of matter are converted to energy in the sun every second. Because of its large mass, however, the sun contains enough hydrogen to "burn" for additional billions of years. It is amazing that science did not know how or why the sun gave off energy until a little over 70 years ago.

The principle of fusion was first demonstrated on this planet with a tremendously destructive device called the hydrogen bomb, which was first exploded in 1952. This bomb can be more than 1000 times as powerful as the atomic bomb, which uses the fission process.

Fission is controlled in nuclear power plants. However, controlling fusion is technically extremely difficult. Temperatures on the order of 50 million degrees Celsius are needed so that the colliding nuclei have enough kinetic energy to overcome their mutual repulsions and cause fusion. The necessary temperature is hotter than the interior of the sun. No known materials can withstand these temperatures, so alternative containment procedures must be used. Producing electricity from a fusion reactor is perhaps the greatest technical challenge yet faced by the human race. The effort has been underway for many years. The design and construction of a large fusion reactor (International Thermonuclear Experimental Reactor, or ITER) is being financed and staffed by an international consortium of nations including the United States, Japan, Russia, and the European Union. It is hoped that eventually this reactor will exceed the *breakeven point*, where the same amount of energy is produced by the fusion as is needed to bring about the fusion. In about 20 years, if things go well, the first fusion reactor designed to produce electricity could be built. It is still a long-shot, but fusion could solve the planet's need for clean energy.

The advantages of controlled fusion power are impressive.

1. It should be relatively clean. Few radioactive products are formed.
2. Fuel is inexhaustible. On one hand, the oceans of the world contain enough deuterium, one of the reactants, to provide the world's energy needs for a trillion years. On the other hand, there is a very limited supply of fossil fuels and uranium.
3. There is no possibility of the reaction going out of control and causing a meltdown. Fusion will occur in power plants in short bursts of energy that can be stopped easily in case of mechanical problems.

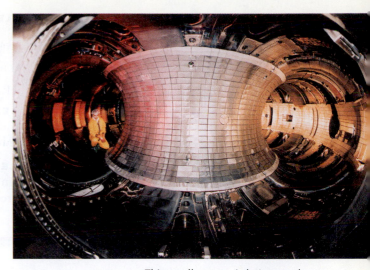

This small reactor is being used to study the feasibility of the generation of fusion power.

▶ **ASSESSING THE OBJECTIVE FOR SECTION 16-7**

EXERCISE 16-7(a) LEVEL 1: Fill in the blanks.
A nuclear reaction known as fission can occur as a nuclear _____ reaction. The process whereby small nuclei combine to form a larger nucleus and a small particle is known as _____.

EXERCISE 16-7(b) LEVEL 2: Identify the nuclear process occurring in each of these equations.

(a) $^3\text{He} + {}^3\text{He} \longrightarrow {}^6\text{Be}$

(b) $^{137}\text{Cs} \longrightarrow {}^{137}\text{Ba} + {}_{-1}\text{e}$

(c) $^{241}\text{Am} \longrightarrow {}^{237}\text{Np} + {}^4\text{He}$

(d) $^{239}\text{Pu} + {}^1\text{n} \longrightarrow {}^{141}\text{Ba} + {}^{97}\text{Sr} + 2\ {}^1\text{n}$

(e) $^{179}\text{Ta} + {}_{-1}\text{e} \longrightarrow {}^{179}\text{Hf}$

EXERCISE 16-7(c) LEVEL 3: Much of the free world opposes countries such as Iran and North Korea in their efforts to use nuclear power. How can a nuclear power plant be used to produce material for an atomic bomb?

EXERCISE 16-7(d) LEVEL 3: Fusion is putting two atoms together. Fission is breaking one atom apart. In this section, we see that *both* of these processes release energy. How can opposite processes both produce energy?

For additional practice, work chapter problems 16-36 and 16-43.

MAKING IT REAL

Revisiting the Origin of the Elements

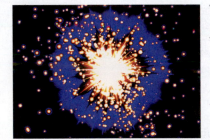

Elements are created in the interiors of stars.

We are literally made of stardust. This sounds rather melodramatic, but it is indeed true. Other than hydrogen, all other atoms in our bodies originated in the violent interiors of ancient stars or from their catastrophic explosions. We can now revisit this topic with some of the information presented in this chapter.

The universe itself began violently about 14 billion years ago. All the matter and energy in the universe originated from a single point called the *singularity*. This point expanded in what is called the Big Bang. The fundamental particles—electrons, neutrons, and protons—were formed within minutes. Hydrogen was the primeval element from which other elements would form.

Sometime after the Big Bang, clouds of hydrogen gas began to clump together and contract because of the attraction of gravity. As contraction continued, the hydrogen cloud became denser and hotter (i.e., Charles's law). Eventually, the nuclei of hydrogen had enough energy to undergo fusion to produce helium. This process liberates huge amounts of energy, so the first stars began to shine brightly and the universe began to light up. After a few million years, the early stars had used up their supply of

hydrogen and the helium nuclei underwent fusion to form even heavier elements like carbon and oxygen, as shown here.

$$_2^4\text{He} + {}_2^4\text{He} \longrightarrow {}_4^8\text{Be} + {}_2^4\text{He} \longrightarrow {}_6^{12}\text{C} + {}_2^4\text{He} \longrightarrow {}_8^{16}\text{O}$$

The fusion reactions continued in stages up to the formation of elements as heavy as iron. At this point, further fusion reactions forming heavier nuclei do not liberate energy and thus do not occur.

So how did heavier elements like lead and gold form? When a large star exhausted its supply of elements that can liberate energy by fusion, the nuclear reactions suddenly stopped. Without the energy needed to counteract gravity, the star immediately collapsed and then exploded, ejecting its elements into outer space along with a huge cloud of neutrons. This violent event is called a *supernova*. The elements from the star absorbed the neutrons singly or in batches. As mentioned in this chapter, isotopes with excess neutrons decay by beta emission, forming the next higher element on the periodic table. Absorption of neutrons increased the mass number, and beta emission increased the atomic number of the isotope. An example is shown here.

$$_{39}^{99}\text{Y} + {}_0^1\text{n} \longrightarrow {}_{39}^{91}\text{Y} \longrightarrow {}_{40}^{91}\text{Zr} + {}_{-1}^0\text{e}$$

These elements drifted through space as dust and gases. About 4.5 billion years ago, a cloud of hydrogen containing the dust from previous stars and supernovas contracted and formed our sun and its planets. We are the inhabitants of the third planet out from this ordinary star.

PART B   SUMMARY

KEY TERMS

16-5.1   **Transmutation** is the process of changing one element into another by nuclear reactions. p. 574

16-6.2   **Neutron activation analysis** is used to change stable isotopes into radioactive isotopes for analysis. p. 577

16-7.1   Specific heavy nuclei undergo **fission** when bombarded with neutrons to produce lighter nuclei and release energy. p. 579

16-7.3   **Fusion** is a nuclear process whereby light elements combine to form heavier elements and release energy. p. 583

SUMMARY CHART

Three Types of Nuclear Reactions

Transmutation

$$^{253}_{99}\text{Es} + ^{4}_{2}\text{He} \longrightarrow ^{256}_{101}\text{Md} + ^{1}_{0}\text{n}$$

Fission

$$^{235}_{92}\text{U} + ^{1}_{0}\text{n} \longrightarrow ^{94}_{38}\text{Sr} + ^{139}_{54}\text{Xe} + 3\,^{1}_{0}\text{n}$$

Fusion

$$^{2}_{1}\text{H} + ^{2}_{1}\text{H} \longrightarrow ^{3}_{2}\text{He} + ^{1}_{0}\text{n}$$

CHAPTER 16 SYNTHESIS PROBLEM

The discovery of element number 117 was reported in 2010. It was given the temporary name of *ununseptium* with the symbol Uus. Two isotopes were synthesized—mass numbers 294 and 293. The heavier isotope had a half-life of 78 milliseconds (78×10^{-3} seconds). This was actually long by superheavy-element standards. The discovery of this element completed the seventh period in the periodic table and fit between elements number 116 and 118, which had been previously discovered.

PROBLEM	SOLUTION
a. Uus-294 was made by bombarding Bk-249 with Ca-48 in a particle accelerator. This accomplished by an international team in Dubna, Russia. Write the balanced nuclear equation illustrating this reaction.	**a.** $^{249}_{97}\text{Bk} + ^{48}_{20}\text{Ca} \longrightarrow ^{294}_{117}\text{Uus} + 3\,^{1}_{0}\text{n}$ Three neutrons must have been released to balance the equation.
b. Uus-294 decays by several alpha and beta emissions to Db-262. How many alpha particles and how many beta particles were emitted in this overall process?	**b.** An alpha particle decreases the mass number by four and the positive charge by two. The overall mass change is $294 - 262 = 32$. This corresponds to the loss of <u>eight</u> alpha particles. Eight alpha particles would decrease the charge by 16 (i.e., 8×2). This would bring the charge down to $117 - 16 = 101$. A beta particle increases the charge by one, so <u>four</u> beta particles must have been emitted in the process to bring the charge from 101 up to 105.
c. Ca-48 is probably radioactive. What type of decay would be predicted and how many times would it occur to form a stable isotope?	**c.** Ca-48 is neutron rich, so it would probably decay by beta emission to form the next higher element. If it occurred twice, it would produce a stable isotope of titanium. $^{48}_{20}\text{Ca} \longrightarrow ^{48}_{21}\text{Sc} + ^{0}_{-1}\text{e} \longrightarrow ^{48}_{22}\text{Ti} + ^{0}_{-1}\text{e}$
d. Bk-249 is also radioactive with a half-life of 330 days. What mass of the 22.0 mg that was used in this nuclear reaction would be left after 990 days?	**d.** After 330 days, 11.0 mg would remain. After 660 days, 5.5 mg would remain. After 990 days, <u>2.75 mg</u> would remain.
e. Besides alpha and beta particles, Bk-249 undergoes spontaneous fission. What does this mean?	**e.** The isotope divides into elements that are about half its mass plus a few neutrons. Since the process occurs spontaneously, no neutrons are needed to initiate the process.
f. Although just a few atoms of Uus-294 have existed for only a fraction of a second, what chemical and physical properties would you predict for the element if there was ever enough of it to study?	**f.** It would be in Group VIIA and behave as a halogen. The element should be chemically reactive and form diatomic molecules. It would probably be a metalloid solid in elemental form.

YOUR TURN

Element number 112 was the latest element to be officially named by IUPAC (the International Union of Pure and Applied Chemistry). This was announced in February 2010. It was named copernicium (Cn) after Nicolaus Copernicus, the Renaissance scholar who first proposed that the Earth revolved around the sun and not vice versa. The element was first reported in 1996 from the synthesis of one atom. It took 13 more years for the discovery to be confirmed so that it could be officially named.

a. Cn was prepared by bombarding Pb-208 with Zn-70 nuclei. Besides Cn, one neutron is emitted in the reaction. Write the total balanced nuclear equation.

b. The Cn isotope decays by two alpha particle emissions. Write equations illustrating these emissions.

c. Zn-64 is the lightest, stable zinc isotope while Zn-60 is radioactive. What type of decay would you expect for Zn-60? In fact, it undergoes these modes of decay to form Ni-60. Write the two nuclear equations.

d. If one starts with 10.0 mg of Zn-60, only 2.50 mg remains after 4.80 minutes. What is the half-life of Zn-60?

e. Predict the physical and chemical properties of Cn based on its position in the periodic table.

Answers are on p. 588.

CHAPTER SUMMARY

The nuclei of **radioactive isotopes** are unstable and emit **radiation**. Although radioactive isotopes exist for each element, all isotopes with an atomic number greater than 83 are unstable. Originally, three types of radiation were discovered, **alpha (α) particles**, **beta (β) particles**, and **gamma (γ) rays**. Since then, two other types of radiation have been characterized that occur rarely in nature but more commonly in artificially produced radioactive isotopes. These are **positron** ($_{+1}^{0}e$) **particles** and **electron capture**. These five modes of decay are illustrated by the following **nuclear equations**.

Alpha particle: $\quad _{96}^{240}Cm \longrightarrow _{94}^{236}Pu + _{2}^{4}He$

Beta particle: $\quad _{30}^{71}Zn \longrightarrow _{31}^{71}Ga + _{-1}^{0}e$

Gamma radiation: $\quad _{43}^{99m}Tc \longrightarrow _{43}^{99}Tc + \gamma$

Positron particle: $\quad _{10}^{19}Ne \longrightarrow _{9}^{19}F + _{+1}^{0}e$

Electron capture: $\quad _{4}^{7}Be + _{-1}^{0}e \longrightarrow _{3}^{7}Li$

Radioactive isotopes decay at widely different rates. A measure of the rate of decay is called the **half-life** ($t_{1/2}$) of the isotope. Half-lives vary from billions of years to fractions of a second. A long-lived isotope is ^{238}U, which begins a **radioactive decay series** and ends with a stable isotope of lead after a series of alpha and beta decays.

Ionizing radiation not only forms ions in surrounding matter, but causes heating and **free radical** formation. In living matter, the ions or free radicals may lead to destruction of the cells. Alpha and beta emitters are most dangerous when ingested, since they cause a high degree of ionization in close proximity to cells. Gamma rays are less ionizing but very penetrating; thus, they are also very dangerous. Radiation is detected and measured by means of film badges, **Geiger counters**, and **scintillation counters**.

The **curie** and **becquerel** are units that measure radiation activity while the **rem** and **sievert** are units that measure how radiation affects living tissue. A dose of 500 rems is the **LD$_{50}$** for humans. Nuclear reactions, including **neutron activation analysis**, occur when nuclei are bombarded by particles or other nuclei. This leads to the artificial **transmutation** of one element into another. **Fission** and **fusion** are also nuclear reactions. Fission is the splitting of one heavy nucleus into two more or less equal fragments and neutrons. Fusion is the joining of two light nuclei to form a heavier nucleus and a particle.

OBJECTIVES

SECTION	YOU SHOULD BE ABLE TO...	EXAMPLES	EXERCISES	CHAPTER PROBLEMS
16-1	Write balanced nuclear equations for the five types of radiation.	16-1	1a, 1b, 1c	2, 3, 5, 7, 9, 12, 15, 17, 19
16-2	Using half-lives, estimate the time needed for various amounts of radioactive decay to occur.	16-2, 16-3	2a, 2b, 2c, 2d	21, 22, 24, 25
16-3	Describe how the three natural types of radiation affect matter.		3a, 3b, 3c	10, 13

16-4	Describe the methods and units used to detect and measure radiation.		4a, 4b	26, 27, 28, 29
16-5	Write balanced nuclear equations for the synthesis of specific isotopes by nuclear reactions.	16-4	5a, 5b, 5c	30, 31, 33, 35
16-6	List and describe the beneficial uses of radioactivity.		6a, 6b, 6c	
16-7	Identify the differences between natural radioactivity, nuclear fission, and nuclear fusion.		7a, 7b,7c, 7d	36, 37

▶ ANSWERS TO ASSESSING THE OBJECTIVES

Part A
EXERCISES

16-1(a) There are five types of <u>radiation</u>.

1. An alpha particle is a <u>helium</u> nucleus.

2. A beta particle is an <u>electron</u>.

3. A gamma ray is a high-energy form of <u>light</u>.

4. A positron is a <u>positively charged electron</u>.

5. Electron capture has the same effect on a nucleus as <u>positron</u> emission.

16-1(b)

(a) $^{87}_{37}\text{Rb}$ **(b)** $^{227}_{89}\text{Ac}$ **(c)** $^{54}_{27}\text{Co}$ **(d)** $^{123}_{52}\text{Te}$ **(e)** $^{137m}_{55}\text{Cs}$

16-1(c) (a) Positron emission, since this is a proton-rich isotope. (The product would be the stable isotope, ^{14}N.)
(b) Beta particle emission, since this is a neutron-rich isotope. (The product would be the stable isotope ^{45}Sc.)
(c) This is not a radioactive isotope as the neutron/proton ratio is 1:1.

16-2(a) The shorter the half-life, the faster an isotope decays.

16-2(b) After one half-life, 50% has decayed; after two half-lives, 75% has decayed; and after three half-lives, 87.5% has decayed.

16-2(c) After four half-lives, 4.0 μg of the original sample is left. $t_{1/2} = 36/4 = 9.0$ min (i.e., after 9.0 min, 32 μg remains; after 18 min, 16 μg; after 27 min, 8.0 μg; and after 36 min, 4.0 μg).

16-2(d) We can never have a pure sample of a radioactive isotope because it is constantly decaying into different isotopes. There will always be a percentage, constantly changing, of other isotopes.

16-3(a) Radiation causes <u>heating</u> as well as the formation of <u>ions</u> along its path. The most ionizing type of radiation is <u>alpha radiation</u>, whereas the most penetrating is <u>gamma radiation</u>.

16-3(b) An ion formed is H_2O^+ and a free radical is OH. The ion is a normal molecule minus an electron, and the free radical is a neutral molecule or part of a molecule with an unpaired electron.

16-3(c) All the particles and rays come out of a nucleus with high velocity. Collisions with the surrounding molecules or atoms in Earth's interior transmit kinetic energy to the surrounding atoms, thus increasing the heat.

16-4(a) Three major ways in which radiation is measured are the <u>film</u> badge, the <u>Geiger</u> counter, and the <u>scintilla</u>-tion counter. The major unit used to measure the total amount of radiation is the <u>curie</u>. The major units that measure the amount of radiation received by a person are the rem and the <u>sievert</u>. The amount of radiation that is fatal to half of those who receive it is known as the LD_{50}.

16-4(b) The first watch dials that glowed in the dark were coated with paint that contained zinc sulfide and radium. The steady emission of alpha particles from the radium caused the dial to glow. Unfortunately, workers who painted the dial began to die of radiation poisoning, so this process was discontinued. Luminescent dials now glow in the dark from a small current of electricity that causes material called phosphors to glow. The clock by your bed probably has a liquid crystal display or a light-emitting diode (LED).

Part B
EXERCISES

16-5(a) Balanced nuclear reactions have both a balanced mass and a balanced nuclear charge.

16-5(b) (a) $^{52}_{23}\text{V}$ **(b)** $^{144}_{54}\text{Xe}$ **(c)** $^{12}_{6}\text{C}$

16-5(c)

(a) $^{249}_{98}\text{Cf} + {}^{48}_{20}\text{Ca} \longrightarrow {}^{294}_{118}\text{Uuo} + 3{}^{1}_{0}\text{n}$

$^{294}\text{Uuo} \longrightarrow {}^{290}\text{Uuh} + \alpha \longrightarrow {}^{286}\text{Uuq} + \alpha \longrightarrow {}^{282}_{112}\text{Cn} + \alpha$

(b) Element 118 would be a noble gas.

16-6(a) Both naturally occurring and synthetic radioactive isotopes find many uses. The decay of carbon-<u>14</u> can be used to date many carbon-containing fossils. Medically useful radioactive isotopes can be prepared by <u>neutron</u> activation. Food can be preserved by radiation with <u>gamma radiation</u>. PET scans make use of isotopes that emit <u>positrons</u>.

16-6(b) $^{16}_{8}\text{O} + {}^{2}_{1}\text{H} \longrightarrow {}^{18}_{9}\text{F}$ $^{10}_{5}\textbf{B} + {}^{1}_{1}\text{H} \longrightarrow {}^{11}_{6}\text{C}$

16-6(c) (a) After 50,000 years (about nine half-lives), less than 0.2% of the original ^{14}C remains. That would be very difficult to detect and measure accurately with a radiation counter.
(b) The new isotope becomes "neutron rich," which means that it likely will be radioactive and decay by beta emission.

16-7(a) A nuclear reaction known as fission can occur as a nuclear <u>chain</u> reaction. The process whereby small nuclei combine to form a larger nucleus and a small particle is known as <u>fusion</u>.

16-7(b) (a) fusion (b) beta emission (c) alpha emission (d) fission (e) electron capture

16-7(c) Some of the neutrons produced from the fission process combine with ^{238}U, the more plentiful isotope. The ^{239}U produced undergoes two beta decays to form ^{239}Pu. This can be extracted (though not easily) from the control rods and made into atomic bombs.

16-7(d) The determining factor is the size of the nuclei. Small nuclei can undergo fusion, with the release of energy. A few large nuclei can undergo fission, with the release of energy. The breakeven point on the periodic table is iron. This is why the nuclear fusion reactions in stars end with the production of iron.

ANSWERS TO CHAPTER SYNTHESIS PROBLEM

a. $^{208}_{82}Pb + {}^{70}_{30}Zn \longrightarrow {}^{277}_{112}Cn + {}^{1}_{0}n$

The isotope formed is Cn-277

b. $^{277}_{112}Cn \longrightarrow {}^{273}_{110}Ds + {}^{4}_{2}He \longrightarrow {}^{269}_{108}Hs + {}^{4}_{2}He$

c. Zn-60 is proton rich so it would undergo either positron emission or electron capture. Both would provide the same outcome but it actually undergoes two electron captures.

$^{60}_{30}Zn + {}^{0}_{-1}e \longrightarrow {}^{60}_{29}Cu + {}^{0}_{-1}e \longrightarrow {}^{60}_{28}Ni$

d. This would be two half-lives, so the half-life is 4.80/2 or 2.40 minutes.

e. Cn would be in Group IIB under mercury. It may be a liquid metal like mercury with oxidation states of +2 and possibly +1.

CHAPTER PROBLEMS

Throughout the text, answers to all exercises in color are given in Appendix E. The more difficult exercises are marked with an asterisk.

Nuclear Radiation (SECTION 16-1)

16-1. Write isotope symbols in the form $^{A}_{Z}M$ for isotopes of the following compositions.

(a) 84 protons and 126 neutrons

(b) 46 neutrons and a mass number of 84

(c) 100 protons and a mass number of 257

(d) lead-206

(e) uranium-233

16-2. Write isotope symbols in the form $^{A}_{Z}M$ for isotopes of the following compositions.

(a) 86 protons and 134 neutrons

(b) 6 protons and a mass number of 13

(c) 22 neutrons and a mass number of 40

(d) potassium-41

(e) americium-243

16-3. Give the symbols, including mass number and charge, for the following.

(a) an alpha particle

(b) a beta particle

(c) a neutron

16-4. A deuteron has a mass number of 2 and a positive charge of 1. Write its isotope symbol.

16-5. A triton has a mass number of 3 and a positive charge of 1. Write its isotope symbol.

16-6. Write the isotope symbol, including mass number, for the isotope that results when each of the following emits an alpha particle.

(a) $^{210}_{84}Po$

(b) ^{152}Gd

(c) fermium-252

(d) Mt-266

16-7. Write the isotope symbol, including mass number, for the isotope that results when each of the following emits an alpha particle.

(a) $^{234}_{92}U$

(b) $^{222}_{88}Ra$

(c) ^{210}Bi

(d) thorium-229

16-8. Write the isotope symbol, including mass number, for the isotope that results when each of the following emits a beta particle.

(a) $^{3}_{1}H$

(b) ^{153}Gd

(c) iron-59

(d) sodium-24

16-9. Write the isotope symbol, including mass number, for the isotope that results when each of the following emits a beta particle.

(a) $^{131}_{53}I$

(b) ^{234}Pa

(c) lead-210

(d) nitrogen-16

16-10. Give the symbol of a positron. What effect does the emission of a positron have on an isotope? How are positrons detected?

16-11. Manganese-51 undergoes positron emission. Write the nuclear equation illustrating this reaction.

16-12. A certain isotope undergoes positron emission to form ^{23}Na. Write the nuclear equation illustrating this reaction.

16-13. What happens to an isotope that undergoes electron capture? How is electron capture detected?

16-14. Germanium-68 undergoes electron capture. Write the nuclear equation illustrating this reaction.

16-15. A certain isotope undergoes electron capture to form ^{55}Mn. Write the nuclear equation illustrating this reaction.

16-16. Complete the following nuclear equations.

(a) $^{214}_{83}\text{Bi} \longrightarrow {}^{214}_{84}\text{Po} + \underline{\hphantom{xx}}$

(b) $^{90}_{37}\text{Rb} \longrightarrow \underline{\hphantom{xx}} + {}^{0}_{-1}\text{e}$

(c) $^{26}_{14}\text{Si} \longrightarrow \underline{\hphantom{xx}} + {}^{0}_{+1}\text{e}$

(d) $^{235}_{92}\text{U} \longrightarrow \underline{\hphantom{xx}} + {}^{4}_{2}\text{He}$

(e) $^{179}_{73}\text{Ta} + {}^{0}_{-1}\text{e} \longrightarrow \underline{\hphantom{xx}}$

(f) $\underline{\hphantom{xx}} \longrightarrow {}^{41}_{21}\text{Sc} + {}^{0}_{-1}\text{e}$

(g) $\underline{\hphantom{xx}} \longrightarrow {}^{210}_{82}\text{Pb} + {}^{0}_{-1}\text{e}$

16-17. Complete the following nuclear equations.

(a) $^{239}_{93}\text{Np} \longrightarrow \underline{\hphantom{xx}} + {}^{0}_{-1}\text{e}$

(b) $\underline{\hphantom{xx}} \longrightarrow {}^{93}_{44}\text{Ru} + {}^{0}_{+1}\text{e}$

(c) $^{226}_{88}\text{Ra} \longrightarrow {}^{222}_{86}\text{Rn} + \underline{\hphantom{xx}}$

(d) $\underline{\hphantom{xx}} \longrightarrow {}^{235}_{92}\text{U} + {}^{4}_{2}\text{He}$

(e) $^{80}_{37}\text{Rb} + \underline{\hphantom{xx}} \longrightarrow {}^{80}_{36}\text{Kr}$

(f) $^{32}_{15}\text{P} \longrightarrow {}^{32}_{16}\text{S} + \underline{\hphantom{xx}}$

16-18. From the following information, write nuclear equations that include all isotopes and particles.

(a) $^{230}_{90}\text{Th}$ decays to $^{226}_{86}\text{Ra}$

(b) $^{214}_{84}\text{Po}$ emits an alpha particle.

(c) $^{210}_{84}\text{Po}$ emits a beta particle.

(d) An isotope emits an alpha particle and forms lead-214.

(e) $^{14}_{6}\text{C}$ decays to form $^{14}_{7}\text{N}$.

(f) Chromium-50 is formed by positron emission.

(g) Argon-37 captures an electron.

*16-19. The decay series of $^{238}_{92}\text{U}$ to $^{210}_{82}\text{Pb}$ involves alpha and beta emissions in the following sequence: α, β, β, α, α, α, α, α, β, α, β, β, β, α. Identify all isotopes formed in the series.

Nuclear Decay and Half-Life (SECTION 16-2)

16-20. What fraction of a radioactive isotope remains after four half-lives?

16-21. What percent of a radioactive isotope remains after five half-lives?

16-22. If one starts with 20 mg of a radioactive isotope with a half-life of 2.0 days, how much remains after each interval?

(a) four days

(b) eight days

(c) four half-lives

16-23. The half-life of a given isotope is 10 years. If we start with a 10.0-g sample of the isotope, how much is left after 20 years?

16-24. Start with 12.0 g of a given radioactive isotope. After 11 years, only 3.0 g is left. What is the half-life of the isotope?

16-25. The isotope $^{14}_{6}\text{C}$ is used to date fossils of formerly living systems, such as prehistoric animal bones. If the radioactivity of the carbon in a sample of bone from a mammoth is one-fourth of the radioactivity of the current level, how old is the fossil? ($t_{1/2} = 5760$ years)

Effects of Radiation and Its Detection
(SECTIONS 16-3 AND 16-4)

16-26. Of the three types of radiation, which is the most ionizing? Which is the most penetrating? How does each type of radiation cause damage to cells?

16-27. How does radiation cause ionization? How does ionization cause damage to living tissues?

16-28. How does a film badge work? How can a film badge tell which type of radiation is being absorbed?

16-29. How does a Geiger counter work? How does a scintillation counter work?

Nuclear Reactions (SECTION 16-5)

16-30. Complete the following nuclear equations.

(a) $^{35}_{17}\text{Cl} + {}^{1}_{0}\text{n} \longrightarrow \underline{\hphantom{xx}} + {}^{1}_{1}\text{H}$

(b) $^{27}_{13}\text{Al} + \underline{\hphantom{xx}} \longrightarrow {}^{25}_{12}\text{Mg} + {}^{4}_{2}\text{He}$

(c) $^{27}_{13}\text{Al} + {}^{4}_{2}\text{He} \longrightarrow \underline{\hphantom{xx}} + {}^{1}_{0}\text{n}$

(d) $^{238}_{92}\text{U} + 15{}^{1}_{0}\text{n} \longrightarrow {}^{253}_{100}\text{Fm} + \underline{\hphantom{xx}}$

(e) $^{244}_{96}\text{Cm} + {}^{12}_{6}\text{C} \longrightarrow \underline{\hphantom{xx}} + 2{}^{1}_{0}\text{n}$

(f) $\underline{\hphantom{xx}} + {}^{2}_{1}\text{H} \longrightarrow {}^{238}_{93}\text{Np} + {}^{1}_{0}\text{n}$

(g) $^{242}_{94}\text{Pu} + {}^{22}_{10}\text{Ne} \longrightarrow {}^{260}_{104}\underline{\hphantom{xx}} + \underline{\hphantom{xx}}$

16-31. Complete the following nuclear equations.

(a) $^{249}_{96}\text{Cm} + \underline{\hphantom{xx}} \longrightarrow {}^{260}_{103}\text{Lr} + 4{}^{1}_{0}\text{n}$

(b) $^{15}_{7}\text{N} + {}^{1}_{1}\text{H} \longrightarrow \underline{\hphantom{xx}} + {}^{4}_{2}\text{He}$

(c) $^{10}_{5}\text{B} + {}^{4}_{2}\text{He} \longrightarrow \underline{\hphantom{xx}} + {}^{1}_{1}\text{H}$

(d) $^{249}_{98}\text{Cf} + \underline{\hphantom{xx}} \longrightarrow {}^{263}_{106}\underline{\hphantom{xx}} + 4{}^{1}_{0}\text{n}$

16-32. An element on the periodic table was prepared by the following nuclear reaction.

$$^{209}_{83}\text{Bi} + {}^{58}_{26}\text{Fe} \longrightarrow {}^{266}_{109}\text{Mt}$$

How many neutrons were emitted in the reaction?

16-33. Dubnium-260 plus four neutrons is prepared by bombarding Californium-249 with what isotope?

16-34. Hassium-265 plus one neutron is prepared by bombarding an isotope of lead with iron-58. What is the isotope of lead?

16-35. Bismuth-209 can be bombarded with chromium-54, producing one neutron and what heavy element isotope?

Fission and Fusion (SECTION 16-7)

16-36. When $^{235}_{92}U$ and $^{239}_{94}Pu$ undergo fission, a variety of reactions take place. Complete the following.

(a) $^{235}_{92}U + ^{1}_{0}n \longrightarrow \underline{\qquad} + ^{146}_{58}Ce + 3^{1}_{0}n$

(b) $^{239}_{94}Pu + ^{1}_{0}n \longrightarrow ^{141}_{56}Ba + \underline{\qquad} + 2^{1}_{0}n$

16-37. What is the difference between fission and fusion? What is the source of energy for these two processes?

*16-38. In 1989, some scientists reported that they had achieved "cold fusion." This was supposedly accomplished by electrolyzing heavy water (deuterium oxide) with palladium electrodes. It was suggested that the deuterium was absorbed into the electrodes and somehow two deuterium nuclei could overcome the strong repulsions at a low temperature and fuse. Evidence was presented to indicate that more energy came out of the reaction than could be accounted for by a chemical process. There has been little support for these initial experiments, but, for a while, they generated nightly reports on the national news. The fusion of two deuterium nuclei should produce helium-4, which, according to theory, would have too much energy to be stable and so would decompose to either helium-3 or tritium. What other two particles would be produced by this decomposition? Write the appropriate nuclear equations.

General Problems

16-39. Write equations illustrating each of the following nuclear processes.

(a) Palladium-106 absorbs an alpha particle to produce an isotope and a proton.

(b) Meitnerium-266 emits an alpha particle to form an isotope that also emits an alpha particle.

(c) Bismuth-212 emits a beta particle.

(d) An isotope emits a positron to form copper-60.

(e) Plutonium-239 absorbs a neutron to produce cesium-140, another isotope, and three neutrons.

(f) Lead-206 is bombarded by an isotope to produce Seaborgium-257 and three neutrons.

(g) An isotope captures an electron to form niobium-93.

16-40. Write equations illustrating each of the following nuclear processes.

(a) An isotope of a heavy element is bombarded with boron-11 to form lawrencium-257 and four neutrons.

(b) An isotope emits an alpha particle to form actinium-231.

(c) Oxygen-14 emits a particle to form nitrogen-14.

(d) Sulfur-31 captures an electron to form an isotope.

(e) Uranium-235 absorbs a neutron to form rubidium-90, cesium-142, and several neutrons.

(f) Two helium nuclei fuse (in the sun) to form lithium-7 and a particle.

(g) An isotope emits a beta particle to form lead-208.

(h) Plutonium-239 is bombarded with an isotope to form curium-242 and a neutron

16-41. The radioactive isotope $^{90}_{38}Sr$ can accumulate in bones, where it replaces calcium. It emits a high-energy beta particle, which eventually can cause cancer.

(a) What is the product of the decay of $^{90}_{38}Sr$?

(b) How long would it take for a 0.10-mg sample of $^{90}_{38}Sr$ to decay to where only 2.5×10^{-2} mg was left? (The half-life of $^{90}_{38}Sr$ is 25 years.)

16-42. The radioactive isotope $^{131}_{53}I$ accumulates in the thyroid gland. On the one hand, this can be useful in detecting diseases of the thyroid and even in treating cancer at that location. On the other hand, exposure to excessive amounts of this isotope, such as from a nuclear power plant, can cause cancer of the thyroid. $^{131}_{53}I$ emits a beta particle with a half-life of 8.0 days. What is the product of the decay of $^{131}_{53}I$? If one started with 8.0×10^{-6} g of $^{131}_{53}I$, how much would be left after 32 days?

16-43. The fissionable isotope $^{239}_{94}Pu$ is made from the abundant isotope of uranium, $^{238}_{92}U$, in nuclear reactors. When $^{238}_{92}U$ absorbs a neutron from the fission process, $^{239}_{94}Pu$ eventually forms. This is the principle of the breeder reactor, although $^{239}_{94}Pu$ is formed in all reactors. Complete the following reaction.

$$^{238}_{92}U + ^{1}_{0}n \longrightarrow \underline{\qquad} + ^{0}_{-1}e$$

$$\longrightarrow ^{239}_{94}Pu + \underline{\qquad}$$

16-44. An isotope of hydrogen, known as tritium (^3H), has a half-life of 12 years. If a sample of tritium was prepared 60 years ago, what was its original mass if its current mass is 0.42 μg?

16-45. A particular isotope has a half-life of 10.0 days. What percent of the original sample is left after 30.0 days?

16-46. A new element was created in 1994. What is the atomic number and mass number of the isotope if the nuclear reaction involved bombardment of ^{209}Bi with ^{64}Ni to form the new element plus one neutron?

16-47. The isotope ^{269}Ds was synthesized in 1994. The nuclear reaction involved bombardment of an isotope with ^{62}Ni and produced the new element and one neutron. What is the isotope involved in the reaction?

16-48. A team in Germany reported the formation of one atom of ^{277}Cn in 1996 by bombarding ^{208}Pb with a lighter nucleus. If one neutron is also produced, what is the isotopic notation for the lighter nucleus?

16-49. A team in Germany reported element ^{282}Uuq, which, according to theory, should be more stable than other super-heavy elements. The reaction involves bombarding ^{208}Pb with a lighter isotope. If two neutrons are also produced, what is the isotopic notation of the lighter nucleus?

Foreword to the Appendices

The successful athlete must be "in shape" physically. The successful chemistry student must be "in shape" mathematically.

WHY do some (one or two, anyway) students seem so self-assured in the study of chemistry and yet others (all the rest) seem so worried? Most likely, it has a lot to do with preparation. Preparation in this case probably does not mean a prior course in chemistry, but it does mean having a solid mathematical background. Most of the students using this text probably are a little rusty on at least some aspects of basic arithmetic, algebra, and scientific notation. There are several reasons for this. Some have not had a good secondary school background in math courses, and others have been away from their high school or college math courses for a number of years. It makes no difference—most students need access to a few reminders, hints, and review exercises to get in shape. The sooner students admit that they have forgotten some math, the faster they do something about it and start to enjoy chemistry. It is really difficult to appreciate the study of this science if math deficiencies get in the way.

Appendix A reviews some of the basic arithmetic concepts such as manipulation of fractions, expressing decimal fractions, and, very importantly, the expression and use of percent. **Appendix B** reviews the manipulation and solution of simple algebra equations, which are so important in the quantitative aspects of chemistry. **Appendix C** supplements the discussion of scientific notation in Chapter 1 with more examples and exercises. Also included in this appendix is a discussion of the concept of logarithms, which is a convenient way to express exponential numbers in certain situations. **Appendix D** is a glossary of terms, and **Appendix E** contains answers to more than half of the exercises at the ends of the chapters.

Basic Mathematics

The following is a very quick refresher of fundamentals of math. This may be sufficient to aid you if you are just a little rusty on some of the basic concepts. For more thorough explanations and practice, however, you are urged to use a more comprehensive math review workbook or consult with your instructor.

One may ask, "Why not just use a calculator?" The answer is that serious science students need a "feeling" for the numbers they use. This can be accomplished only by understanding the calculations involved. Therefore, it is well worth the time to go through this appendix *without a calculator*. Being able to do these calculations on your own will certainly pay off. Someday the battery on the calculator may die during an exam.

A-1 Addition and Subtraction

Since most calculations in this text use numbers expressed in decimal form, we will emphasize the manipulation of this type of number. In addition and subtraction, it is important to line up the decimal point carefully before doing the math.

Subtraction is simply the addition of a negative number. Remember that subtraction of a negative number changes the sign to a plus (two negatives make a positive). For example, $4 - 7 = -3$, but $4 - (-7) = 4 + 7 = 11$.

EXAMPLE A-1   Addition and Subtraction

Carry out the following calculations.

(a) $16.75 + 13.31 + 175.67$

$$
\begin{array}{r}
16.75 \\
13.31 \\
175.67 \\
\hline
205.73
\end{array}
$$

(b) $11.8 + 13.1 - 6.1$

$$
\begin{array}{r}
11.8 \\
+13.1 \\
\hline
24.9
\end{array}
\qquad
\begin{array}{r}
24.9 \\
-6.1 \\
\hline
18.8
\end{array}
$$

(c) $47.82 - 111.18 - (-12.17)$

This is the same as $47.82 - 111.18 + 12.17$.

$$
\begin{array}{r}
47.82 \\
+12.17 \\
\hline
59.99
\end{array}
\qquad
\begin{array}{r}
-111.18 \\
+59.99 \\
\hline
-51.19
\end{array}
$$

EXERCISES

A-1. Carry out the following calculations.

(a) $47 + 1672$

(e) $0.897 + 1.310 - 0.063$

(b) $11.15 + 190.25$

(f) $-0.377 - (-0.101) + 0.975$

(c) $114 + 26 - 37$

(g) $17.489 - 318.112 - (0.315) + (-3.330)$

(d) $-97 + 16 - 118$

Answers: **(a)** 1719   **(b)** 201.40   **(c)** 103   **(d)** -199 **(e)** 2.144   **(f)** 0.699
(g) -303.638

A-2 Multiplication

Multiplication is expressed in various ways as follows:

$$13.7 \times 115.35 = 13.7 \cdot 115.35 = (13.7)\ (115.35) = 13.7\ (115.35)$$

If it is necessary to carry out the multiplication in longhand, you must be careful to place the decimal point correctly in the answer. Count the *total* number of digits to the right of the decimal point in both multipliers (three in this example). The answer has that number of digits to the right of the decimal point in the answer. Finally, round off to the proper number of significant figures, which is three in this case.

$$13.7 \times 2.15 = 29455 = 29.455 = 29.5^{*}$$

When a number (called a *base*) is multiplied by itself one or more times, it is said to be raised to a *power*. The power (called the *exponent*) indicates the number of bases multiplied. For example, the exact values of the following numbers raised to a power are

$$4^2 = 4 \times 4 = 16 \text{ (“four squared”)}$$

$$2^2 = 2 \times 2 \times 2 \times 2 = 16 \text{ (“two to the fourth power”)}$$

$$4^3 = 4 \times 4 \times 4 = 64 \text{ (“four cubed”)}$$

$$(14.1)^2 = 14.1 \times 14.1 = 198.81 = 199^{*}$$

In the calculations used in this book, most numbers have specific units. In multiplication, the units as well as the numbers are multiplied. For example,

$$3.7 \text{ cm} \times 4.61 \text{ cm} = 17 \text{ (cm} \times \text{cm)} = 17 \text{ cm}^2$$

$$(4.5 \text{ in.})^3 = 91 \text{ in.}^3$$

In the multiplication of a series of numbers, grouping is possible.

$$(a \times b) \times c = a \times (b \times c)$$

$$3.0 \text{ cm} \times 148 \text{ cm} \times 3.0 \text{ cm} = (3.0 \times 3.0) \times 148 \times (\text{cm} \times \text{cm} \times \text{cm})$$

$$= \underline{1300 \text{ cm}^3}$$

When multiplying signs, remember:

$$(+) \times (-) = - \qquad (+) \times (+) = + \qquad (-) \times (-) = +$$

For example, $(-3) \times 2 = -6$; $(-9) \times (-8) = +72$.

---

*Rounded off to three significant figures. See Section 1-2.

EXERCISES

A-2. Carry out the following calculations. For (a) through (d) carry out the multiplications completely. For (e) through (h) round off the answer to the proper number of significant figures and include units.

(a) $16.2 \times (-118)$

(d) $(-47.8) \times (-9.6)$

(b) $(4 \times 2) \times 879$

(e) $3.0 \text{ ft} \times 18 \text{ lb}$

(c) $(-8) \times (-2) \times (-37)$

(f) $17.7 \text{ in.} \times (13.2 \text{ in.} \times 25.0 \text{ in.})$

(g) What is the area of a circle where the radius is 2.2 cm? (Area $= \pi r^2$. $\pi = 3.14$.)

(h) What is the volume of a cylinder 5.0 in. high with a cross-sectional radius of 0.82 in.? (Volume = area of cross section $\times$ height.)

Answers: **(a)** -1911.6 **(b)** 7032 **(c)** -592 **(d)** 458.88 **(e)** $54 \text{ ft} \cdot \text{lb}$
(f) 5840 in.^3 **(g)** 15 cm^2 **(h)** 11 in.^3

A-3 Roots of Numbers

A root of a number is a fractional exponent. It is expressed as

$$\sqrt[x]{a} = a^{1/x}$$

If x is not shown (on the left), it is assumed to be 2 and is known as the *square root*. The square root is the number that when multiplied by itself gives the base a. For example,

$$\sqrt{4} = 2 \qquad (2 \times 2 = 4)$$
$$\sqrt{9} = 3 \qquad (3 \times 3 = 9)$$

The square root of a number may have either a positive or a negative sign. Generally, however, we are interested only in the positive root in chemistry calculations.

If the square root of a number is not a whole number, it may be computed on a calculator or found in a table. Without these tools available, an educated approximation can come close to the answer. For example, the square root of 54 lies between 7 ($7^2 = 49$) and 8 ($8^2 = 64$) but closer to 7. An educated guess of 7.3 would be excellent.

The cube root of a number is expressed as

$$\sqrt[3]{b} = b^{1/3}$$

It is the number multiplied by itself two times that gives b. For example,

$$\sqrt[3]{27} = 3.0 \qquad (3 \times 3 \times 3 = 27)$$
$$\sqrt[3]{64} = 4.0 \qquad (4 \times 4 \times 4 = 64)$$

A hand calculator is the most convenient source of roots of numbers.

EXERCISES

A-3. Find the following roots. If necessary, first approximate the answer, then check with a calculator.

(a) $\sqrt{25}$ (b) $\sqrt{36 \text{ cm}^2}$ (c) $\sqrt{144 \text{ ft}^4}$ (d) $\sqrt{40}$

(e) $\sqrt{7.0}$ (f) $110^{1/2}$ (g) $100^{1/3}$ (h) $\sqrt[3]{50}$

(i) What is the radius of a circle that has an area of 150 ft^2? (Area $= \pi r^2$)

(j) What is the radius of the cross section of a cylinder that has a volume of 320 m^3 and a height of 6.0 m? (Volume $= \pi r^2 \times$ height)

Answers: **(a)** 5.0 **(b)** 6.0 cm **(c)** 12.0 ft^2 **(d)** 6.3 **(e)** 2.6 **(f)** 10.5 **(g)** 4.64
(h) 3.7 **(i)** 6.91 ft **(j)** 4.1 m

A-4 Division, Fractions, and Decimal Numbers

Common fractions express ratios or portions of numbers. A *proper fraction* is less than one so it has a larger denominator than numerator. An *improper fraction* is greater than one so it has a smaller denominator than numerator. In chemistry most fractions are expressed as decimal numbers, which show proper fractions less than 1.0 and improper fractions greater than 1.0. To convert a common fraction to a decimal number, the numerator is divided by the denominator. Consider the two fractions shown below where the decimal is shown to three digits.

proper fraction	improper fraction
$7/8 = 7 \div 8 = 0.675$	$11/9 = 11 \div 9 = 1.22$
common decimal	common decimal
fraction fraction	fraction fraction

Many divisions can be simplified by cancellation, which is the elimination of common factors in the numerator and denominator. This is possible because a number divided by itself is equal to unity (e.g., $25/25 = 1$). As in multiplication, all units also must be divided. If identical units appear in both numerator and denominator, they also can be canceled. This is the basis of dimensional analysis for problem solving, introduced in Chapter 1.

$$\frac{a \times c}{b \times c} = \frac{a}{b}$$

$$\frac{190 \times 4 \text{ torr}}{190 \text{ torr}} = 4$$

$$\frac{2500 \text{ cm}^3}{150 \text{ cm}} = \frac{50 \times 50 \text{ cm} \times \text{cm} \times \text{cm}}{50 \times 3 \text{ cm}} = \frac{50 \text{ cm}^2}{3} = 17 \text{ cm}^2$$

$$\frac{2800 \text{ mi}}{45 \text{ hr}} = \frac{5 \times 560 \text{ mi}}{5 \times 9 \text{ hr}} = \frac{62 \text{ mi}}{1 \text{ hr}} = 62 \text{ mi/hr}$$

This is read as 62 miles "per" one hour or simply 62 miles per hour. The word per implies a fraction or a ratio with the unit after per in the denominator. If a number is not written or read in the denominator with a unit, it is assumed that the number is unity and is known to as many significant figures as the number in the numerator (i.e., 62 miles per 1.0 hr).

EXERCISES

A-4. Express the following in decimal form. Express (a)–(c) to three digits.

(a) $3/7$ (d) $892 \text{ mi} \div 41 \text{ hr}$ (g) $\dfrac{67.5 \text{ g}}{15.2 \text{ mL}}$ (j) $\dfrac{0.8772 \text{ ft}^3}{0.0023 \text{ ft}^2}$

(b) $14/19$ (e) $982.6 \div 0.250$ (h) $\dfrac{1890 \text{ cm}^3}{66 \text{ cm}}$ (k) $\dfrac{37.50 \text{ ft}}{0.455 \text{ sec}}$

(c) $19/14$ (f) $195 \div 2650$ (i) $\dfrac{146 \text{ ft} \cdot \text{hr}}{0.68 \text{ ft}}$

Answers: **(a)** 0.429 **(b)** 0.737 **(c)** 1.36 **(d)** 22 mi/hr **(e)** 3930 **(f)** 0.0736 **(g)** 4.44 g/mL **(h)** 29 cm^2 **(i)** 210 hr **(j)** 380 ft **(k)** 82.4 ft/sec

A-5 Multiplication and Division of Fractions

When two or more fractions are multiplied, all numbers *and units* in both numerator and denominator can be combined into one fraction.

The division of one fraction by another is the same as the multiplication of the numerator by the *reciprocal* of the denominator. The reciprocal of a fraction is simply the fraction in an inverted form (e.g., 3/5 is the reciprocal of 5/3).

$$\frac{a}{b/c} = a \times \frac{c}{b} \qquad \frac{a/b}{c/d} = \frac{a}{b} \times \frac{d}{c} = \frac{a \times d}{b \times c}$$

EXAMPLE A-2   Multiplication and Division

Carry out the following calculations. Round off the answer to two digits.

(a) $\dfrac{3}{5} \times \dfrac{75}{4} \times \dfrac{16}{7} = \dfrac{3 \times 75 \times 16}{5 \times 4 \times 7} = \dfrac{3 \times \overset{15}{\cancel{75}} \times \overset{4}{\cancel{16}}}{\cancel{5} \times \cancel{4} \times 7} = \dfrac{180}{7} = \underline{\underline{26}}$

(b) $\dfrac{42 \text{ mi}}{\text{hr}} \times \dfrac{3}{7} \text{ hr} \times \dfrac{5280 \text{ ft}}{\text{mi}} = \dfrac{\overset{6}{\cancel{42}} \times 3 \times 5280 \text{ mi} \times \text{hf} \times \text{ft}}{\underset{1}{7} \text{ hr} \times \text{mi}} = \underline{\underline{95{,}000 \text{ ft}}}$

(c) $\dfrac{3}{4} \text{ mol} \times \dfrac{0.75 \text{ g}}{\text{mol}} \times \dfrac{1 \text{ mL}}{19.3 \text{ g}} = \dfrac{3 \times 0.75 \times 1 \times \text{mol} \times \text{g} \times \text{mL}}{4 \times 1 \times 19.3 \times \text{mol} \times \text{g}} = \underline{\underline{0.029 \text{ mL}}}$

(d) $\dfrac{1650}{3/5} = 1650 \times \dfrac{5}{3} = \underline{\underline{2800}}$

(e) $\dfrac{145 \text{ g}}{7.5 \text{ g/mL}} = 145 \text{ g} \times \dfrac{1 \text{ mL}}{7.5 \text{ g}} = \underline{\underline{19 \text{ mL}}}$

EXERCISES

A-5. Express the following answers in decimal form. If units are not used, round off the answer to three digits. If units are included, round off to the proper number of significant figures and include units in the answer.

(a) $\dfrac{3}{8} \times \dfrac{4}{7} \times \dfrac{21}{20}$

(b) $\dfrac{250}{273} \times \dfrac{175}{300} \times (-6)$

(c) $\dfrac{4}{9} \times \left(-\dfrac{5}{8}\right) \times \left(-\dfrac{3}{4}\right)$

(d) $195 \text{ g/mL} \times 47.5 \text{ mL}$

(e) $0.75 \text{ mol} \times 17.3 \text{ g/mol}$

(f) $(3.57 \text{ in.})^2 \times 0.85 \text{ in.} \times \dfrac{16.4 \text{ cm}^3}{\text{in.}^3}$

(g) $\dfrac{\dfrac{150}{350}}{\dfrac{25}{42}}$

(h) $\dfrac{\left(-\dfrac{3}{7}\right)}{\left(-\dfrac{4}{9}\right)}$

(i) $\dfrac{\left(-\dfrac{17}{3}\right)}{\dfrac{8}{9}}$

(j) $\dfrac{\dfrac{16}{9} \times \dfrac{10}{14}}{\dfrac{5}{6}}$

(k) $\dfrac{75.2 \text{ torr}}{760 \text{ torr/atm}}$

(m) $\dfrac{305 \text{ K} \times 62.4 \dfrac{\text{L} \cdot \text{torr}}{\text{K} \cdot \text{torr}} \times 0.25 \text{ mol}}{650 \text{ torr}}$

(l) $\dfrac{(55.0 \text{ mi/hr}) \times (5280 \text{ ft/mi}) \times (1 \text{ hr/60 min})}{60 \text{ sec/min}}$

Answers: **(a)** 0.225 **(b)** −3.21 **(c)** 0.208 **(d)** 9260 g **(e)** 13 g **(f)** 180 cm^3
(g) 0.720 **(h)** 0.964 **(i)** −6.38 **(j)** 1.52 **(k)** 0.0989 atm **(l)** 80.7 ft/sec **(m)** 7.3 L

A-6 Decimal Numbers and Percent

In the examples of fractions thus far, we have seen that the units of the numerator can be profoundly different from those of the denominator (e.g., miles/hr, g/mL, etc.). In other problems in chemistry we use fractions to express a component part in the numerator to the total in the denominator. In most cases such fractions are expressed without units and in decimal form.

EXAMPLE A-3   Decimal Numbers

(a) A box of nails contains 985 nails; 415 of these are 6-in. nails, 375 are 3-in. nails, and the rest are roofing nails. What is the fraction of roofing nails in decimal form?

SOLUTION

$$\text{Roofing nails} = \text{total} - \text{others} = 985 - (415 + 375) = 195$$

$$\frac{\text{component}}{\text{total}} = \frac{195}{375 + 415 + 195} = \underline{\underline{0.198}}$$

(b) A mixture contains 4.25 mol of N_2, 2.76 mol of O_2, and 1.75 mol of CO_2. What is the fraction of moles of O_2 present in the mixture? (This fraction is known, not surprisingly, as "the mole fraction." The mole is a unit of quantity, like dozen.)

SOLUTION

$$\frac{\text{component}}{\text{total}} = \frac{2.76}{4.25 + 2.76 + 1.75} = \underline{\underline{0.315}}$$

EXERCISES

A-6. A grocer has 195 dozen boxes of fruit; 74 dozen boxes are apples, 62 dozen boxes are peaches, and the rest are oranges. What is the fraction of the boxes that are oranges?

A-7. A mixture contains 9.85 mol of gas. A 3.18-mol quantity of the gas is N_2, 4.69 mol is O_2, and the rest is He. What is the mole fraction of He in the mixture?

A-8. The total pressure of a mixture of two gases, N_2 and O_2, is 0.72 atm. The pressure due to O_2 is 0.41 atm. What is the fraction of the pressure due to O_2?

Answers: **A-6** 0.30   **A-7** 0.201   **A-8** 0.57

The decimal numbers that have just been discussed are frequently expressed as percentages. *Percent* simply means *parts per 100*. Percent is obtained by multiplying a fraction in decimal form by 100%.

EXAMPLE A-4   Expressing Percent

If 57 out of 180 people at a party are women, what is the percent women?

SOLUTION

The fraction of women in decimal form is

$$\frac{57}{180} = 0.317$$

The percent women is

$$0.317 \times 100\% = \underline{31.7\% \text{ women}}$$

The general method used to obtain percent is

$$\frac{\text{component}}{\text{total}} \times 100\% = \underline{\hspace{2cm}}\% \text{ of component}$$

To change from percent back to a decimal number, divide the percent by 100%, which moves the decimal to the left two places.

$$86.2\% = \frac{86.2\%}{100\%} \times 0.862 \,(\text{fraction in decimal form})$$

EXERCISES

A-9. Express the following fractions or decimal numbers as percents: $\dfrac{1}{4}, \dfrac{3}{8}, \dfrac{9}{8}, \dfrac{55}{25},$ 0.67, 0.13, 1.75, 0.098.

A-10. A bushel holds 198 apples, 27 of which are green. What is the percent of green apples?

A-11. A basket contains 75 pears, 8 apples, 15 oranges, and 51 grapefruit. What is the percent of each?

Answers: **A-9** $\frac{1}{4} = 25\%$, $\frac{3}{8} = 37.5\%$, $\frac{9}{8} = 112.5\%$, $\frac{55}{25} = 220\%$, $0.67 = 67\%$, $0.13 = 13\%$, $1.75 = 175\%$, $0.098 = 9.8\%$ **A-10** 13.6%   **A-11** 50.3% pears, 5.4% apples, 10.1% oranges, 34.2% grapefruit

We have seen how the percent is calculated from the total and the component part. We now consider problems where percent is given and we calculate either the component part as in Example A-5 (a) or the total as in Example A-5 (b). Such problems can be solved in two ways. The method we employ here uses the percent as a conversion factor, and the problems are solved by dimensional analysis. (See Section 1-5.) They also can be solved algebraically as is done in Appendix B.

EXAMPLE A-5   Using Percent in Calculations

(a) A crowd at a rock concert was composed of about 87% teenagers. If the crowd totaled 586 people, how many were teenagers?

PROCEDURE

Remember that percent means "per 100." In this case it means 87 teenagers per 100 people or, in fraction form,

$$\frac{87 \text{ teenagers}}{100 \text{ people}}$$

If this fraction is then multiplied by the number of people, the result is the component part or the number of teenagers.

$$586 \text{ people} \times \frac{87 \text{ teenagers}}{100 \text{ people}} = (586 \times 0.87) = \underline{510 \text{ teenagers}}$$

(b) A professional baseball player got a hit 28.7% of the times he batted. If he got 246 hits, how many times did he bat?

PROCEDURE

The percent can be written in fraction form and then inverted. It thus relates the total at bats to the number of hits:

$$28.7\% = \frac{28.7 \text{ hits}}{100 \text{ at bats}} \quad \text{or} \quad \frac{100 \text{ at bats}}{28.7 \text{ hits}}$$

If this is now multiplied by the number of hits, the result is the total number of at bats.

$$246 \text{ hits} \times \frac{100 \text{ at bats}}{28.7 \text{ hits}} = \frac{246}{0.287} = \underline{857 \text{ at bats}}$$

EXERCISES

A-12. In a certain audience, 45.9% were men. If there were 196 people in the audience, how many women were present?

A-13. In the alcohol molecule, 34.8% of the mass is due to oxygen. What is the mass of oxygen in 497 g of alcohol?

A-14. The cost of a hamburger in 2003 is 216% of the cost in 1970. If hamburgers cost $0.75 each in 1970, what do they cost in 2003?

A-15. In a certain audience, 46.0% are men. If there are 195 men in the audience, how large is the audience?

A-16. If a solution is 23.3% by mass HCl and it contains 14.8 g of HCl, what is the total mass of the solution?

A-17. An unstable isotope has a mass of 131 amu. This is 104% of the mass of a stable isotope. What is the mass of the stable isotope?

Answers: **A-12** 106 women   **A-13** 173 g   **A-14** $1.62   **A-15** 424 people   **A-16** 63.5 g
A-17 126 amu

Basic Algebra

There are two aspects of the use of algebra that affect chemistry students. First is the actual skill and application of basic concepts and second is the ability to translate words or quantitative concepts into a proper algebraic relationship. In the first section, we will concentrate on the algebra equation itself and how it can be manipulated. In the two sections following this we will concentrate on how we can express quantitative concepts as equations.

B-1 Operations of Basic Equations

Many of the quantitative problems of chemistry require the use of basic algebra. As an example of a simple algebra equation we use

$$x = y + 8$$

In any algebraic equation the equality remains valid when identical operations are performed on both sides of the equation. The following operations illustrate this principle.

1. A quantity may be added to or subtracted from both sides of the equation.

$$\text{(add 8)} \qquad x \underline{+ 8} = y + 8 \underline{+ 8} \quad x + 8 = y + 16$$

$$\text{(subtract 8)} \quad x \underline{- 8} = y + 8 \underline{- 8} \quad x - 8 = y$$

2. Both sides of the equation may be multiplied or divided by the same quantity.

$$\text{(multiple by 4)} \quad 4x = 4(y + 8) = 4y + 32$$

$$\text{(divide by 2)} \qquad \frac{x}{2} = \frac{(y + 8)}{2} \quad \frac{x}{2} = \frac{y}{2} + 4$$

3. Both sides of the equation may be raised to a power, or a root of both sides of an equation may be taken.

$$\text{(equation squared)} \quad x^2 = (y + 8)^2$$

$$\text{(square root taken)} \quad \sqrt{x} = \sqrt{y + 8}$$

4. Both sides of an equation may be inverted.

$$\frac{1}{x} = \frac{1}{y + 8}$$

In addition to operation on both sides of an equation, two other points must be recalled.

1. As in any fraction, identical factors in the numerator and the denominator in an algebraic equation may be canceled.

$$\frac{\cancel{4}x}{\cancel{4}} = x = y + 8 \quad \text{or} \quad x = \frac{\cancel{t}(y + 8)}{\cancel{t}} = y + 8$$

2. Quantities equal to the same quantity are equal to each other. Thus substitutions for equalities may be made in algebraic equations.

$$x = y + 8$$
$$x = 27$$

Therefore, since $x = x$,

$$y + 8 = 27$$

We can use these basic rules to solve algebraic equations. Usually, we need to isolate one variable on the left-hand side of the equation, with all other numbers and variables on the right-hand side of the equations. The operations previously listed can be simplified for this purpose in two ways.

In practice, a number or a variable may be moved to the other side of an equation with a change of sign. For example, if

$$x + z = y$$

then subtracting z from both sides, in effect, gives us

$$x = y - z$$

Also, the numerator of a fraction on the left becomes the denominator on the right. The denominator of a fraction on the left becomes the numerator on the right. For example, consider the following two cases.

If $xz = y$

then dividing both sides by z, in effect, gives us

$$x(z) = y$$
$$x = \frac{y}{z}$$

If $\dfrac{x}{k + 5} = B$,

then multiplying both sides by $k + 5$

$$\frac{x}{k + 5} = B$$
$$x = B(k + 5)$$

The following examples illustrate the isolation of one variable (x) on the left-hand side of the equation.

EXAMPLE B-1   Solving Algebra Equations for a Variable

(a) Solve for x in $x + y + 8 = z + 6$.

SOLUTION

Move $+y$ and $+8$ to the right by changing signs.

$$x = z + 6 - y - 8$$
$$= z - y + 6 - 8 = z - y - 2$$

(b) Solve for **x** in

$$\frac{x + 8}{y} = z$$

SOLUTION

First, move y to the right by multiplying both sides by y.

$$y \cdot \frac{x + 8}{y} = z \cdot y$$

This leaves

$$x + 8 = zy$$

Subtract 8 from both sides to obtain the final answer.

$$x = zy - 8$$

(c) Solve for x in

$$\frac{4x + 2}{3 + x} = 7$$

SOLUTION

First, multiply both sides by $(3 + x)$ to clear the fraction.

$$(3 + x) \cdot \frac{4x + 2}{(3 + x)} = 7(3 + x)$$

This leaves

$$4x + 2 = 21 + 7x$$

To move integers to the right and the x variable to the left, subtract $7x$ and 2 from both sides of the equation. This leaves

$$-3x = 19$$

Finally, divide both sides by -3 to move the -3 to the right.

$$x = -\frac{19}{3} = \underline{\underline{-6.33}}$$

(d) Solve for T_2 in

$$\frac{P_1 V_1}{T_1} = \frac{P_2 V_2}{T_2}$$

SOLUTION

To move T_2 to the left, multiply both sides by T_2.

$$T_2 \cdot \frac{P_1 V_1}{T_1} = T_2 \cdot \frac{P_2 V_2}{T_2} = P_2 V_2$$

Move $P_1 V_1$ to the right by dividing by $P_1 V_1$.

$$\frac{T_2}{P_1 V_1} \cdot \frac{P_1 V_1}{T_1} = \frac{P_2 V_2}{P_1 V_1}$$

Finally, move T_1 to the right by multiplying both sides by T_1.

$$T_2 = \frac{T_1 P_2 V_2}{P_1 V_1}$$

EXERCISES

B-1. Solve for x in $17x = y - 87$.

B-2. Solve for x in

$$\frac{y}{x} + 8 = z + 16$$

B-3. Solve for T in $PV = (mass/MM) RT$.

B-4. Solve for x in

$$\frac{7x - 3}{6 + 2x} = 3r$$

B-5. Solve for x in $18x - 27 = 2x + 4y - 35$. If $y = 3x$, what is the value of x?

B-6. Solve for x in

$$\frac{x}{4y} + 18 = y + 2$$

B-7. Solve for x in $5x^2 + 12 = x^2 + 37$.

B-8. Solve for r in

$$\frac{80}{2r} + \frac{y}{r} = 11$$

What is the value of r if $y = 14$?

Answers: **B-1** $x = (y - 87)/17$ **B-2** $x = y/(8 + z)$ **B-3** $T = PV \cdot MM/\text{mass} \cdot R$
B-4 $x = 3(6r + 1)/(7 - 6r)$ **B-5** $x = (y - 2)/4$. When $y = 3x$, $x = -2$.
B-6 $x = 4y(y - 16)$ **B-7** $x = \pm 2.5$ **B-8** $r = (40 + y)/11$. When $y = 14$, $r = \dfrac{54}{11}$

B-2 Word Problems and Algebra Equations

Eventually, a necessary skill in chemistry is the ability to translate word problems into algebra equations and then solve. The key is to assign a variable (usually x) to be equal to a certain quantity and then to treat the variable consistently throughout the equation. Again, examples are the best way to illustrate the problems.

EXAMPLE B-2 Solving Abstract Word Equations

Translate each of the following to an equation.

(a) A number x is equal to a number that is 4 larger than y.

$$\underline{x = y + 4}$$

(b) A number z is equal to three-fourths of u.

$$\underline{z = \frac{3}{4}u}$$

(c) The square of a number r is equal to 16.9% of the value of w.

$$\underline{r^2 = 0.169w} \text{ (change percent to a decimal number)}$$

(d) A number t is equal to 12 plus the square root of q.

$$\underline{t = 12 + \sqrt{q}}$$

EXERCISES

B-9. Write algebraic equations for the following:

(a) A number n is equal to a number that is 85 smaller than m.

(b) A number y is equal to one-fourth of z.

(c) Fifteen percent of a number k is equal to the square of another number d.

(d) A number x is equal to 14 more than the square root of v.

(e) Four times the sum of two numbers, q and w, is equal to 68.

(f) Five times the product of two variables, s and t, is equal to 16 less than the square of s.

(g) Five-ninths of a number C is equal to 32 less than a number F.

Answers: **(a)** $n = m - 85$ **(b)** $y = z/4$ **(c)** $0.15k = d^2$ **(d)** $x = \sqrt{v} + 14$
(e) $4(q + w) = 68$ **(f)** $5st = s^2 - 16$ **(g)** $\dfrac{5}{9}C = F - 32$

We now move from the abstract to the real. In the following examples it is necessary to translate the problem into an algebraic expression, as in the previous examples. There are two types of examples that we will use. The first you will certainly recognize, but the second type may be unfamiliar, especially if you have just begun the study of chemistry. However, it is *not* important that you understand the units of chemistry problems at this time. What *is* important is for you to notice that the problems are worked in the same manner regardless of the units.

EXAMPLE B-3   Solving Concrete Word Equations

(a) John is 2 years more than twice as old as Mary. The sum of their ages is 86. How old is each?

SOLUTION

Let x = age of Mary. Then $2x + 2$ = age of John.

$$x + (2x + 2) = 86$$
$$3x = 84$$
$$x = \underline{28} \text{ (age of Mary)}$$
$$2(28) + 2 = \underline{58} \text{ (age of John)}$$

(b) One mole of SF_6 has a mass 30.0 g less than four times the mass of 1 mol of CO_2. The mass of 1 mol of SF_6 plus the mass of 1 mol of CO_2 is equal to 190 g. What is the mass of 1 mol of each?

SOLUTION

Let x = mass of 1 mol of CO_2. Then $4x - 30$ = mass of 1 mol of SF_6.

$$x + (4x - 30) = 190$$
$$x = \underline{44 \text{ g}} \text{ (mass of 1 mol of } CO_2)$$
$$4(44) - 30 = \underline{146 \text{ g}} \text{ (mass of 1 mol of } SF_6)$$

(c) Two students took the same test, and their percent scores differed by 10%. If there were 200 points on the test and the total of their point scores was 260 points, what was each student's percent score?

PROCEDURE

Set up an equation relating each person's percent scores to their total points (260).

Let x = percent score of higher test.

Then $x - 10$ = percent score of lower test.

The points that each person scores is the percent in fraction form multiplied by the points on the test.

$$\frac{\% \text{ grade}}{100\%} \times (\text{points on test}) = \text{points scored}$$

SOLUTION

$$\left[\frac{x}{100} (200 \text{ points}) \right] + \left[\frac{x - 10}{100} (200 \text{ points}) \right] = 260 \text{ points}$$
$$200x + 200x - 2000 = 26{,}000$$
$$400x = 28{,}000$$
$$x = 70$$

higher score = $\underline{70\%}$ lower score = $70 - 10 = \underline{60\%}$

(d) If an 8.75-g quantity of sugar represents 65.7% of a solution, what is the mass of the solution?

SOLUTION

Let x = mass of the solution. Then

$$\frac{65.7}{100} x = 0.657x = 8.75$$
$$x = \frac{8.75}{0.657} = \underline{13.3 \text{ g}}$$

(e) A used car dealer has Fords, Chevrolets, and Hondas. There are 120 Fords, 152 Chevrolets, and the rest are Hondas. If the fraction of Fords is 0.310, how many Hondas are on the lot?

SOLUTION

Let x = number of Hondas.

$$\text{fraction of Fords} = \frac{\text{number of Fords}}{\text{total number of cars}} = 0.310$$
$$\frac{120}{120 + 152 + x} = 0.310$$
$$120 = 0.310(272 + x)$$
$$120 = 84.3 + 0.310x$$
$$x = \underline{115 \text{ Hondas}}$$

In the following exercises, a problem concerning an everyday situation is followed by one or more closely analogous problems concerning a chemistry situation. In both cases the mechanics of the solution are similar. Only the units differ.

EXERCISES

B-10. The total length of two boards is 18.4 ft. If one board is 4.0 ft longer than the other, what is the length of each board?

B-11. An isotope of iodine has a mass 10 amu less than two-thirds the mass of an isotope of thallium. The total mass of the two isotopes is 340 amu. What is the mass of each isotope?

B-12. An isotope of gallium has a mass 22 amu more than one-fourth the mass of an isotope of osmium. The difference in the two masses is 122 amu. What is the mass of each?

B-13. An oil refinery held 175 barrels of oil. When refined, each barrel yields 24 gallons of gasoline. If 3120 gallons of gasoline were produced, what percentage of the original barrels of oil was refined?

B-14. A solution contained 0.856 mol of a substance A_2X. In solution some of the A_2Xs break up into As and Xs. (Note that each mole of A_2X yields 2 mol of A.) If 0.224 mol of A is present in the solution, what percentage of the moles of A_2X dissociated (broke apart)?

B-15. In Las Vegas, a dealer starts with 264 decks of cards. If 42.8% of the decks were used in an evening, how many jacks (four per deck) were used?

B-16. A solution originally contains a 1.45-mol quantity of a compound A_3X_2. If 31.5% of the A_3X_2 dissociates (three As and two Xs per A_3X_2), how many moles of A are formed? How many moles of X? How many moles of undissociated A_3X_2 remain? How many moles of particles (As, Xs, and A_3X_2s) are present in the solution?

B-17. The fraction of kerosene that can be recovered from a barrel of crude oil is 0.200. After a certain amount of oil was refined, 8.90 gal of kerosene, some gasoline, and 18.6 gal of other products were produced. How many gallons of gasoline were produced?

B-18. The fraction of moles (mole fraction) of gas A in a mixture is 0.261. If the mixture contains 0.375 mol of gas B and 0.175 mol of gas C as well as gas A, how many moles of gas A are present?

Answers: **B-10** 7.2 ft, 11.2 ft **B-11** thallium, 210 amu; iodine, 130 amu **B-12** gallium, 70 amu; osmium, 192 amu **B-13** 74.3% **B-14** 13.1% **B-15** 452 jacks **B-16** 1.37 mol of A, 0.914 mol of X, 0.99 mol of A_3X_2, 3.27 mol total **B-17** 17.0 gal **B-18** 0.195 mol

B-3 Direct and Inverse Proportionalities

There is one other point that should be included in a review on algebra—direct and inverse proportionalities. We use these often in chemistry.

When a quantity is directly proportional to another, it means that an increase in one variable will cause a corresponding increase of the same percent in the other variable. A direct proportionality is shown as

$$A \propto B \quad (\propto \text{ is the proportionality symbol})$$

which is read "*A* is directly proportional to *B*." A proportionality can be easily converted to an algebraic equation by the introduction of a constant (in our examples designated *k*), called a constant of proportionality. Thus the proportion becomes

$$A = kB$$

or, rearranging,

$$\frac{A}{B} = k$$

Note that k is not a variable but has a certain numerical value that does not change as do A and B under experimental conditions.

A common, direct proportionality that we will study relates Kelvin temperature T and volume V of a gas at constant pressure. This is written as

$$V \propto T \qquad V = kT \qquad \frac{V}{T} = k$$

(This is known as Charles's law.)

When a quantity is inversely proportional to another quantity, an increase in one brings about a corresponding *decrease* in the other. An inverse proportionality between A and B is written as

$$A \propto \frac{1}{B}$$

As before, the proportionality can be written as an equality by the introduction of a constant (which has a value different from the example above).

$$A = \frac{k}{B} \qquad \text{or} \qquad AB = k$$

A common inverse proportionality that we use relates the volume V of a gas to the pressure P at a constant temperature. This is written as

$$V \propto \frac{1}{P} \qquad V = \frac{k}{P} \qquad PV = k$$

(This is known as Boyle's law.)

When one variable (e.g., x) is directly proportional to two other variables (e.g., y and z), the proportionality can be written as the product of the two.

$$\text{If } x \propto y \text{ and } x \propto z, \text{ then}$$

$$x \propto yz$$

When one variable (e.g., a) is directly proportional to one variable, (e.g., b) and inversely proportional to another (e.g., c,), the proportionality can be written as the ratio of the two.

$$\text{If } a \propto b \text{ and } a \propto \frac{1}{c}, \text{ then}$$

$$a \propto \frac{b}{c}$$

Quantities can be directly or inversely proportional to the square, square root, or any other function of another variable or number, as illustrated by the examples that follow.

EXAMPLE B-4   Solving Equations with Proportionalities

(a) A quantity C is directly proportional to the square of D. Write an equality for this statement and explain how a change in D affects the value of C.

SOLUTION

The equation is

$$\underline{C = kD^2}$$

Note that a change in D will have a significant effect on the value of C. For example,

$$\text{If } D = 1, \text{ then } C = k$$

If $D = 2$, then $C = 4k$

If $D = 3$, then $C = 9k$

Note that when the value of D is doubled, the value of C is increased *fourfold*.

(b) A variable X is directly proportional to the square of the variable Y *and* inversely proportional to the square of another variable Z. Write an equality for this statement.

SOLUTION

This can be written as two separate equations if it is assumed that Y is constant when Z varies and vice versa.

$$X = k_1 Y^2 \qquad (Z \text{ constant})$$
$$X = k_2 / Z^2 \qquad (Y \text{ constant})$$

k_1 and k_2 are different constants. This relationship can be combined into one equation when both Y and Z are variables.

$$X = \frac{k_3 Y^2}{Z^2}$$

k_3 is a third constant that is a combination of k_1 and k_2.

EXERCISES

B-19. Write equalities for the following relations.

(a) X is inversely proportional to $Y + Z$.

(b) $[H_3O^+]$ is inversely proportional to $[OH^-]$.

(c) $[H_2]$ is directly proportional to the square root of r.

(d) B is directly proportional to the square of y and the cube of z.

(e) The pressure P of a gas is directly proportional to the number of moles n and the temperature T, and inversely proportional to the volume V.

Answers: **(a)** $X = k/(Y + Z)$ **(b)** $[H_3O^+] = k/[OH^-]$ **(c)** $[H_2] = k\sqrt{r}$ **(d)** $B = ky^2 z^3$
(e) $P = knT/V$

Scientific Notation

Although this topic was first introduced in Section 1-3 in this text, we will focus on a review of the mathematical manipulation of numbers expressed in scientific notation in this appendix. Specifically, addition, multiplication, division, and taking the roots of numbers expressed in scientific notation are covered. We conclude this section with how we can simplify the expression of numbers in scientific notation with the use of logarithms. As mentioned in Chapter 1, scientific notation makes use of powers of 10 to express awkward numbers that employ more than two or three zeros that are not significant figures. The exponent of 10 simply indicates how many times we should multiply or divide a number (called the coefficient) by 10 to produce the actual number. For example, $8.9 \times 10^3 = 8.9$ (the coefficient) multiplied by 10 *three* times, or

$$8.9 \times 10 \times 10 \times 10 = 8900$$

Also, $4.7 \times 10^{-3} = 4.7$ (the coefficient) divided by 10 *three* times, or

$$\frac{4.7}{10 \times 10 \times 10} = 0.0047$$

C-1 Review of Scientific Notation

The method for expressing numbers in scientific notation was explained in Section 1-3. However, to simplify a number or to express it in the standard form with one digit to the left of the decimal point in the coefficient, it is often necessary to change a number already expressed in scientific notation. If this is done in a hurry, errors may result. Thus it is worthwhile to practice moving the decimal point of numbers expressed in scientific notation.

EXAMPLE C-1   Changing Normal Numbers to Scientific Notation

Change the following numbers to the standard form in scientific notation.

(a) 489×10^4 **(b)** 0.00489×10^8

PROCEDURE

All you need to remember is to raise the power of 10 one unit for each place the decimal point is moved to the left, and lower the power of 10 one unit for each place that the decimal point is moved to the right in the coefficient.

SOLUTION

(a) $489 \times 10^4 = (4.89) \times 10^4 = 4.89 \times 10^{4+2} = \underline{4.89 \times 10^6}$

(b) $0.00489 \times 10^8 = (0.00489) \times 10^8 = 4.89 \times 10^{8-3} = 4.89 \times 10^5$

As an aid to remembering whether you should raise or lower the exponent as you move the decimal point, it is suggested that you write (or at least imagine) the coefficient on a slant. For each place that you move the decimal point *up*, add one to the exponent. For each place that you move the decimal point *down*, subtract one from the exponent. Note that the exponent moves up or down with the decimal point. It may be easier to recall "up or down" rather than "right or left."

EXAMPLE C-2   Changing the Decimal Point in the Coefficient

Change the following numbers to the standard form in scientific notation.

(a) 4223×10^{-7} **(b)** 0.00076×10^{18}

SOLUTION

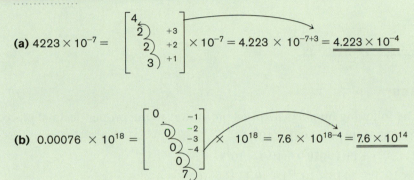

(a) $4223 \times 10^{-7} = 4.223 \times 10^{-7+3} = \underline{\underline{4.223 \times 10^{-4}}}$

(b) $0.00076 \times 10^{18} = 7.6 \times 10^{18-4} = \underline{\underline{7.6 \times 10^{14}}}$

EXERCISES

C-1. Change the following numbers to standard scientific notation with one digit to the left of the decimal point in the coefficient.

(a) 787×10^{-6} **(c)** 0.015×10^{-16} **(e)** 49.3×10^{15}

(b) 43.8×10^{-1} **(d)** 0.0037×10^{9} **(f)** 6678×10^{-16}

C-2. Change the following numbers to a number with two digits to the left of the decimal point in the coefficient.

(a) 9554×10^{4} **(c)** 1×10^{6} **(e)** 0.023×10^{-1}

(b) 1.6×10^{-5} **(d)** 116.5×10^{4} **(f)** 0.005×10^{23}

Answers: **C-1:** **(a)** 7.87×10^{-4} **(b)** 4.38 **(c)** 1.5×10^{-18} **(d)** 3.7×10^{6} **(e)** 4.93×10^{16}
(f) 6.678×10^{-13} **C-2:** **(a)** 95.54×10^{6} **(b)** 16×10^{-6} **(c)** 10×10^{5} **(d)** 11.65×10^{5}
(e) 23×10^{-4} **(f)** 50×10^{19}

C-2 Addition and Subtraction

Addition or subtraction of numbers in scientific notation can be accomplished only when all coefficients have the same exponent of 10. When all the exponents are the same, the coefficients are added and then multiplied by the power of 10. The correct number of places to the right of the decimal point, as discussed in Section 1-2, must be shown.

EXAMPLE C-3   Addition of Numbers in Scientific Notation

(a) Add the following numbers: 3.67×10^{-4}, 4.879×10^{-4}, and 18.2×10^{-4}.

SOLUTION

$$
\begin{array}{l}
3.67 \ \times 10^{-4} \\
4.879 \times 10^{-4} \\
\underline{18.2 \ \ \times 10^{-4}} \\
26.749 \times 10^{-4} = 26.7 \times 10^{-4} \ = \ 26.7 \times 10^{-3}
\end{array}
$$

(b) Add the following numbers: 320.4×10^3, 1.2×10^5, and 0.0615×10^7.

SOLUTION

Before adding, change all three numbers to the same exponent of 10.

$$
\begin{array}{rcl}
320.4 \times 10^3 &=& 3.204 \times 10^5 \\
1.2 \times 10^5 &=& 1.2 \ \ \times 10^5 \\
0.0615 \times 10^7 &=& \underline{6.15 \ \times 10^5} \\
&& 10.554 \times 10^5 = \underline{10.6 \times 10^5} = \underline{1.06 \times 10^6}
\end{array}
$$

EXERCISES

C-3. Add the following numbers. Express the answer to the proper decimal place.

(a) $152 + (8.635 \times 10^2) + (0.021 \times 10^3)$

(b) $(10.32 \times 10^5) + (1.1 \times 10^5) + (0.4 \times 10^5)$

(c) $(1.007 \times 10^{-8}) + (118 \times 10^{-11}) + (0.1141 \times 10^{-6})$

(d) $(0.0082) + (2.6 \times 10^{-4}) + (159 \times 10^{-4})$

C-4. Carry out the following calculations. Express your answer to the proper decimal place.

(a) $(18.75 \times 10^{-6}) - (13.8 \times 10^{-8}) + (1.0 \times 10^{-5})$

(b) $(1.52 \times 10^{-11}) + (17.7 \times 10^{-12}) - (7.5 \times 10^{-15})$

(c) $(481 \times 10^6) - (0.113 \times 10^9) + (8.5 \times 10^5)$

(d) $(0.363 \times 10^{-6}) + (71.2 \times 10^{-9}) + (519 \times 10^{-12})$

Answers: **C-3:** **(a)** 1.037×10^3 **(b)** 1.18×10^6 **(c)** 1.254×10^{-7} **(d)** 2.44×10^{-2}
C-4: **(a)** 2.9×10^{-5} **(b)** 3.29×10^{-11} **(c)** 3.69×10^8 **(d)** 4.35×10^{-7}

C-3 Multiplication and Division

When numbers expressed in scientific notation are multiplied, the exponents of 10 are *added*. When the numbers are divided, the exponent of 10 in the denominator (the divisor) is subtracted from the exponent of 10 in the numerator (the dividend).

EXAMPLE C-4   Multiplication and Division

(a) Carry out the following calculation.

$$(4.75 \times 10^6) \times (3.2 \times 10^5)$$

SOLUTION

In the first step, group the coefficients and the powers of 10. Carry out each step separately.

$$
\begin{array}{rcl}
(4.75 \times 3.2) \times (10^6 \times 10^5) &=& 15.200 \times 10^{6+5} \\
&=& 15 \times 10^{11} = \underline{1.5 \times 10^{12}}
\end{array}
$$

(b) Carry out the following calculation.

$$(1.62 \times 10^{-8}) \div (8.55 \times 10^{-3})$$

SOLUTION

$$\frac{1.62 \times 10^{-8}}{8.55 \times 10^{-3}} = \frac{1.62}{8.55} \times \frac{10^{-8}}{10^{-3}} = 0.189 \times 10^{-8-(-3)}$$

$$= 0.189 \times 10^{-5} = \underline{1.89 \times 10^{-6}}$$

EXERCISES

C-5. Carry out the following calculations. Express your answer to the proper number of significant figures with one digit to the left of the decimal point.

(a) $(7.8 \times 10^{-6}) \times (1.12 \times 10^{-2})$

(b) $(0.511 \times 10^{-3}) \times (891 \times 10^{-8})$

(c) $(156 \times 10^{-12}) \times (0.010 \times 10^{4})$

(d) $(16 \times 10^{9}) \times (0.112 \times 10^{-3})$

(e) $(2.35 \times 10^{3}) \times (0.3 \times 10^{5}) \times (3.75 \times 10^{2})$

(f) $(6.02 \times 10^{23}) \times (0.0100)$

C-6. Follow the instructions in Problem C-5.

(a) $(14.6 \times 10^{8}) \div (2.2 \times 10^{8})$

(b) $(6.02 \times 10^{23}) \div (3.01 \times 10^{20})$

(c) $(0.885 \times 10^{-7}) \div (16.5 \times 10^{3})$

(d) $(0.0221 \times 10^{3}) \div (0.57 \times 10^{18})$

(e) $238 \div (6.02 \times 10^{23})$

C-7. Follow the instructions in Problem C-5.

(a) $[(8.70 \times 10^{6}) \times (3.1 \times 10^{8})] \div (5 \times 10^{-3})$

(b) $(47.9 \times 10^{-6}) \div [(0.87 \times 10^{6}) \times (1.4 \times 10^{2})]$

(c) $1 \div [(3 \times 10^{6}) \times (4 \times 10^{10})]$

(d) $1.00 \times 10^{-14} \div [(6.5 \times 10^{5}) \times (0.32 \times 10^{-5})]$

(e) $[(147 \times 10^{-6}) \div (154 \times 10^{-6})] \div (3.0 \times 10^{12})$

Answers: **C-5:**   **(a)** 8.7×10^{-8} **(b)** 4.55×10^{-9} **(c)** 1.6×10^{-8} **(d)** 1.8×10^{6}
(e) 3×10^{10} **(f)** 6.02×10^{21} **C-6:** **(a)** 6.6 **(b)** 2.00×10^{3} **(c)** 5.36×10^{-12}
(d) 3.9×10^{-17} **(e)** 3.95×10^{-22} **C-7:** **(a)** 5×10^{17} **(b)** 3.9×10^{-13} **(c)** 8×10^{-18}
(d) 4.8×10^{-15} **(e)** 3.2×10^{-13}

C-4 Powers and Roots

When a number expressed in scientific notation is raised to a power, the coefficient is raised to the power and the exponent of 10 is *multiplied* by the power.

For a number expressed in scientific notation, we take the root of the coefficient and *divide* the exponent by the root. (A square root is the same as raising the number to the 1/2 power, a cube root to the 1/3 power, etc.) In the interest of easy viewing in the exercises, we will adjust the number so that division of the exponent by the root produces a whole number. It is not necessary to adjust the number when using a calculator.

EXAMPLE C-5   Powers and Roots

(a) Carry out the following calculation.

$$(3.2 \times 10^3)^2$$

SOLUTION

$$(3.2 \times 10^3)^2 = (3.2)^2 \times 10^{3 \times 2}$$
$$= 10.24 \times 10^6 = \underline{1.0 \times 10^7}$$
$$[(10^3)^2 = 10^3 \times 10^3 = 10 \times 10 \times 10 \times 10 \times 10 \times 10 = 10^6]$$

(b) Carry out the following calculation.

$$(1.5 \times 10^{-3})^3$$

SOLUTION

$$(1.5 \times 10^{-3})^3 = (1.5)^3 \times 10^{-3 \times 3}$$
$$= \underline{3.4 \times 10^{-9}}$$

(c) Carry out the following calculation.

$$\sqrt{2.9 \times 10^5}$$

SOLUTION

It is easier to see and do (even when using a calculator) if you first adjust the number so that the exponent of 10 is divisible by 2.

$$\sqrt{2.9 \times 10^5} = \sqrt{29 \times 10^4} = \sqrt{29} \times \sqrt{10^4} = \sqrt{29} \times 10^{4/2}$$
$$= 5.4 \times 10^2$$

(d) Carry out the following calculation.

$$\sqrt[3]{6.9 \times 10^{-8}}$$

SOLUTION

Adjust the number so that the exponent of 10 is divisible by 3.

$$\sqrt[3]{6.9 \times 10^{-8}} = \sqrt[3]{69 \times 10^{-9}} = \sqrt[3]{69} \times \sqrt[3]{10^{-9}}$$
$$= 4.1 \times 10^{-9/3} = \underline{4.1 \times 10^{-3}}$$

EXERCISES

C-8. Carry out the following operations.

(a) $(6.6 \times 10^4)^2$ **(d)** $(0.035 \times 10^{-3})^3$

(b) $(0.7 \times 10^6)^3$ **(e)** $(0.7 \times 10^7)^4$

(c) $(1200 \times 10^{-5})^2$

(It will be easier to square if you change the number to 1.2×10^5 first.)

C-9. Take the following roots. Approximate the answer if necessary.

(a) $\sqrt{36 \times 10^4}$ **(c)** $\sqrt{64 \times 10^9}$ **(e)** $\sqrt{81 \times 10^{-7}}$

(b) $\sqrt[3]{27 \times 10^{12}}$ **(d)** $\sqrt[3]{1.6 \times 10^5}$ **(f)** $\sqrt{180 \times 10^{10}}$

Answers: **C-8:** **(a)** 4.4×10^9 **(b)** 3×10^{17} **(c)** 1.4×10^{-4} **(d)** 4.3×10^{-14} **(e)** 2×10^{27}
C-9: **(a)** 6.0×10^2 **(b)** 3.0×10^4 **(c)** 2.5×10^5 **(d)** 54 **(e)** 2.8×10^{-3} **(f)** 1.3×10^6

C-5 Logarithms

Scientific notation is particularly useful in expressing very large or very small numbers. In certain areas of chemistry, however, such as in the expression of H_3O^+ concentration, even the repeated use of scientific notation becomes tedious. In this situation, it is convenient to express the concentration as simply the *exponent of 10. The exponent to which 10 must be raised to give a certain number is called its* **common logarithm**. With common logarithms (or just logs) it is possible to express both the coefficient and the exponent of 10 as one number.

Since logarithms are simply exponents of 10, logs of exact multiples of 10 such as 100 can be easily determined. Note that 100 can be expressed as 10^2, so that the log of 100 is exactly 2. Other examples of simple logs of numbers (that we assume to be exact) are

$1 = 10^0$	$\log 1 = 0$	$0.1 = 10^{-1}$	$\log 0.1 = -1$
$10 = 10^1$	$\log 10 = 1$	$0.01 = 10^{-2}$	$\log 0.01 = -2$
$100 = 10^2$	$\log 100 = 2$	$0.001 = 10^{-3}$	$\log 0.001 = -3$
$1000 = 10^3$	$\log 1000 = 3$	$0.0001 = 10^{-4}$	$\log 0.0001 = -4$

There are two general rules regarding logarithms.

1. $\log (A \times B) = \log A + \log B$

2. $\log (A/B) = \log A - \log B$

We can see how these rules apply when we multiply and divide multiples of ten.

	Exponents	**Logarithms**
multiplication	$10^4 \times 10^3 = 10^{4+3}$ $= 10^7$	$\log(10^4 \times 10^3) = \log 10^4 + \log 10^3$ $= 4 + 3 = 7$
division	$\dfrac{10^{10}}{10^4} = 10^{10-4} = 10^6$	$\log \dfrac{10^{10}}{10^4} = \log 10^{10} - \log 10^4$ $= 10 - 4 = 6$

Although we will use the calculator to determine the logs of numbers that are not simple multiples of ten, such as those above, it is helpful to have a sense of how the log of a specific number will appear. The log of a number has two parts. *The number to the left of the decimal is known as the* **characteristic** *and represents the exact exponent of 10 in the exponential number. The number to the right of the decimal is known as the* **mantissa** *and is the log of the coefficient in the exponential number.* For example, consider the log of the following exponential number.

Exponential number $\quad 5.7 \times 10^6$

$$(\log 10^6 = 6) \quad + \quad (\log 5.7 = 0.76)$$

$$\log (5.7 \times 10^6) \quad = \quad \boxed{6.\,|76|}$$

characteristic $\qquad\qquad$ mantissa

Notice that the mantissa should be expressed with the same number of significant figures as the coefficient of the original exponential number. In this case, that is two significant figures.

Although the calculator will easily provide logs and antilogs (inverse logs), we should be aware of some of the meanings of these numbers. For example, logs of numbers between one and ten are positive numbers where the characteristic is a zero (i.e., $5.8 = 5.8 \times 10^0$). For example,

$$\log 5.8 = \underline{0.76}$$

Logs of numbers greater than ten have a characteristic greater than zero. For example,

$$\log 4.7 \times 10^3 = \underline{3.67}$$

Logs of numbers that are less than one have a negative value since the exponent is a negative value. For example,

number $\quad 0.66 = 6.6 \times 10^{-1}$ $\qquad\qquad 7.3 \times 10^{-4}$

$\qquad\qquad \log 6.6 + \log 10^{-1}$ $\qquad\qquad \log 7.3 + \log 10^{-4}$

$\qquad \log 0.66 = 0.82 + (-1) = \underline{-0.18} \quad \log (7.3 \times 10^{-4}) = 0.86 - 4 = \underline{-3.14}$

Now consider how the *antilog* or *inverse log* of a number will appear. The antilog is the opposite of a logarithm. It is the number whose log has a certain value. For example, consider the antilog (x) of 2.

$$\log x = 2 \quad x = 10^2 \text{ since } \log 10^2 = 2$$

Since the log of a number between one and ten is a positive number with zero as the characteristic, then the antilog of a number with zero as a characteristic is between one and ten. If the characteristic of a positive number is greater than zero, that number is the exponent of ten and the antilog of the mantissa is the coefficient. For example,

$$\text{antilog } 0.62 = \underline{4.2} \quad \text{antilog } 6.62 = \underline{4.2 \times 10^6}$$

The antilog of a negative number is somewhat different. In this case, the calculator actually computes the positive value of the characteristic. To change a characteristic with a negative value to one with a positive value requires a mathematical manipulation. The negative number is separated into its two parts, the characteristic and the mantissa. A value of 1 is added to the characteristic (which makes it a positive number) and a value of 1 is subtracted from the mantissa. For example, consider the antilogs of −3.28. This number can be also expressed as 0.74 − 4. This calculation is shown below.

$$-3.28 = -0.28 - 3$$

$$(-0.28 + 1) - (3 - 1) = 0.72 - 4$$

$$\text{antilog } -3.28 = \text{antilog } 0.72 + \text{antilog } -4 = \underline{5.2 \times 10^{-4}}.$$

EXAMPLE C-6 Logs

Give the logarithms of the following numbers.

(a) 5.8	number between one and ten	$\log = \underline{0.76}$
(b) 4.7×10^3	number greater than ten	$\log = \underline{3.67}$
(c) 0.085	number less than one expressed in scientific notation as 8.5×10^{-2}	
	$\log 8.5 + \log 10^{-2} = +0.yy - 2 = -1.xx$	$\log = \underline{-1.07}$
(d) 8.7×10^{-7}	number less than one	$\log = \underline{-6.06}$

EXAMPLE C-7 Antilogs

Give the number whose log is the following.

(a) 0.84	antilog between zero and one	exponent = 0	$6.9 \times 10^0 = \underline{6.9}$
(b) 4.65	antilog greater than one	exponent = 4	$\underline{4.5 \times 10^{-4}}$
(c) −0.020	antilog less than zero (antilog 0.980 −1)	exponent = −1	$\underline{0.95}$
(d) −4.54	antilog less than zero (antilog 0.46 −5)	exponent = −5	$\underline{2.9 \times 10^{-5}}$

EXERCISES

C-10. Give the value of the logarithm for each of the following numbers.

(a) 7.4   (b) 0.087   (c) 1700   (d) 7.3×10^4 (e) 32×10^{-5} (f) 32×10^5

C-11. Give the number whose log is the following.

(a) 0.34   (b) −5.48   (c) −0.070   (d) 8.40   (e) 10.94   (f) −2.60

Answers: C-10: (a) 0.86   (b) −1.06   (c) 3.23   (d) 4.86   (e) −3.49   (f) 6.51
C-11: (a) 2.19   (b) 3.3×10^{-6} (c) 0.85   (d) 2.5×10^8 (e) 8.7×10^{10} (f) 2.5×10^{-3}

APPENDIX D

Glossary

The numbers in parentheses at the end of each entry refer to the chapter in which the entry is first discussed plus any chapter in which the topic is discussed in detail.

A

absolute zero. Theoretically, the lowest possible temperature. The temperature at which translational motion ceases. Defined as zero on the Kelvin scale or −273°C. (1, 10)

accuracy. How close a measurement is to the true value. (1)

acid. A compound that increases the H^+ concentration in aqueous solution. In the Brønsted definition, an acid is a proton donor. (4, 6, 13, 15)

acid anhydride. A molecular oxide that dissolves in water to form an oxyacid. (13)

acid ionization constant (K_a). The equilibrium constant specifically for the ionization of a weak acid in water. The magnitude of the constant relates to the strength of the acid. (15)

acid salt. An ionic compound containing one or more acidic hydrogens on the anion. (13)

acidic solution. A solution with pH < 7 or $[H_3O^+] > 10^{-7}$ M. (13)

actinide. One of the 14 elements between Ac and Rf. An element whose $5f$ subshell is filling. (4, 8)

activated complex. The transition state formed at the instant of maximum impact between reacting molecules. (15)

activation energy. The minimum energy needed by reactant molecules to form an activated complex so that a reaction may occur. It is the difference in potential energy between the reactants and the activated complex. (15)

activity series. The ability of metals to replace other metal ions in a single-replacement reaction and ranked in a series. (6)

actual yield. The experimentally measured amount of product in a chemical reaction. (7)

alkali metal. An element in Group IA (1) in the periodic table (except H). Elements with the electron configuration $[NG]ns^1$. (4, 8)

alkaline earth metal. An element in Group IIA (2) in the periodic table. Elements with the electron configuration $[NG]ns^2$. (4, 8)

allotropes. Different forms of the same element. (9)

alloy. A homogeneous mixture of metallic elements in one solid phase. (3)

alpha (α) particle. A helium nucleus $^4_2He^{2+}$ emitted from a radioactive nucleus. (2, 16)

amorphous solid. A solid without a defined shape. The molecules in such a solid do not occupy regular, symmetrical positions. (11)

amphiprotic. Refers to a molecule or ion that has both a conjugate acid and a conjugate base. (13)

anion. A negatively charged ion. (2)

anode. The electrode at which oxidation takes place. (14)

atmosphere. The sea of gases above the surface of earth. Also, the average pressure of the atmosphere at sea level, which is defined as the standard pressure and abbreviated atm. (10)

atom. The smallest fundamental particle of an element that has the properties of that element. (1, 2)

atomic mass. The weighted average of the isotopic masses of all of the naturally occurring isotopes of an element. The mass of an "average" atom of a naturally occurring element compared to ^{12}C. (1, 2)

atomic mass unit (amu). A mass that is exactly one-twelfth of the mass of an atom of ^{12}C, which is defined as exactly 12 amu. (2)

atomic number. The number of protons (positive charge) in the nuclei of the isotopes of a particular element. (2)

atomic radius. The distance from the nucleus of an atom to the outermost electron. (8)

atomic theory. A theory first proposed by John Dalton in 1803 that holds that the basic components of matter are atoms. (2)

Aufbau principle. A rule that states that electrons fill the lowest available energy level first. (8)

autoionization. The dissociation of a solvent to produce positive and negative ions. (13)

Avogadro's law. A law that states that the volume of a gas is directly proportional to the number of moles of gas present at constant temperature and pressure (i.e., $V = kn$). (10)

Avogadro's number. The number of objects or particles in one mole, which is 6.022×10^{23}. (5)

B

balanced equation. A chemical equation that has the same number and types of atoms on both sides of the equation. (6)

barometer. A device that measures the atmospheric pressure. (10)

base. A compound that increases the OH^- concentration in aqueous solution. In the Brønsted definition, a base is a proton acceptor. (6, 13)

base anhydride. An ionic oxide that dissolves in water to form an ionic hydroxide. (13)

base ionization constant (K_b). An equilibrium constant specifically for the ionization of a weak base in water. The magnitude of the constant relates to the strength of the base. (15)

basic solution. A solution with $pH > 7$ or $[OH^-] > 10^{-7}$ M. (13)

battery. One or more voltaic cells joined together as a single unit. (14)

beta (β) particle. A high-energy electron emitted from a radioactive nucleus. (16)

binary acid. An acid composed of hydrogen and one other element. (4)

Bohr model. A model of the atom in which electrons orbit around a central nucleus in discrete energy levels. (8)

boiling point. The temperature at which the vapor pressure of a liquid equals the restraining pressure. (3, 11, 12)

box diagram. See *orbital diagram.*

Boyle's law. A law that states that the volume of a gas is inversely proportional to the pressure at constant temperature (i.e., $PV = k$). (10)

Brønsted-Lowry definition. A definition of acids as proton donors and bases as proton acceptors. (13)

buffer solution. A solution that resists changes in pH from the addition of limited amounts of strong acid or base. Made from (a) a solution of a weak acid and a salt containing its conjugate base, or (b) a weak base and a salt containing its conjugate acid. (13, 15)

C

calorie. The amount of heat required to raise the temperature of one gram of water one degree Celsius. Equal to exactly 4.184 joules. (3)

catalyst. A substance that is not consumed in a reaction but whose presence increases the rate of the reaction. A catalyst lowers the activation energy. (15)

cathode. The electrode at which reduction takes place. (14)

cation. A positively charged ion. (2)

Celsius. A temperature scale with 100 equal divisions between the freezing and boiling points of water at average sea level pressure, with exactly zero assigned to the freezing point of water. (1)

chain reaction. A self-sustaining reaction. In a nuclear chain reaction, one reacting neutron produces between two and three other neutrons that in turn cause reactions. (16)

Charles's law. A law that states that the volume of a gas is directly proportional to the Kelvin temperature at constant pressure (i.e., $V = kT$). (10)

chemical change. A change in a substance to another substance or substances. (3)

chemical equation. The representation of a chemical reaction using the symbols of the elements and the formulas of compounds. (6)

chemical property. The property of a substance relating to its tendency to undergo chemical changes. (3)

chemical thermodynamics. The study of heat and its relationship to chemical changes. (7)

chemistry. The branch of science dealing with the nature, composition, and structure of matter and the changes it undergoes. (1,3)

coefficient. The number in scientific notation that is raised to a power of ten. (1) The number before an element or compound in a balanced chemical equation indicating the number of molecules or moles of that substance. (1, 7)

colligative property. A property that depends only on the relative amounts of solute and solvent present, not on their identities. (12)

collision theory. A theory that states that chemical reactions are brought about by collisions of molecules. (15)

combination reaction. A chemical reaction whereby one compound is formed from two elements and/or compounds. (6)

combined gas law. A law that relates the pressure, volume, and temperature of a gas (i.e., $PV = kT$). (10)

combustion reaction. A chemical reaction whereby an element or compound reacts with elemental oxygen. (6)

compound. A pure substance composed of two or more elements that are chemically combined in fixed proportions. (1, 2)

concentrated solution. A solution containing a relatively large amount of a specified solute per unit volume. (12)

concentration. The amount of solute present in a specified amount of solvent or solution. (3, 12)

condensation point. The temperature at which a pure substance changes from the gas to the liquid or solid state. (3, 11)

conductor. A substance that allows a flow of electricity. (12)

conjugate acid. An acid formed by addition of one H^+ to a base. (13)

conjugate base. A base formed by removal of one H^+ from an acid. (13)

conservation of energy. A law that states that energy is neither created nor destroyed but can be transformed from one form to another. (3)

conservation of mass. A law that states that matter is neither created nor destroyed in a chemical reaction. (3)

continuous spectrum. A spectrum where one color blends gradually into another. The visible spectrum containing all wavelengths of visible light. (8)

conversion factor. A relationship between two units or quantities expressed in fractional form. (1)

Coulomb's law A law that states that forces of attraction increase as the charges increase and decrease as the distance between the charges increases. (8)

covalent bond. The force that bonds two atoms together by a shared pair or pairs of electrons. (2, 9)

crystal lattice. See *lattice*.

crystalline solid. A solid with a regular, symmetrical shape where the molecules or ions occupy set positions in the crystal lattice. (11)

D

Dalton's law. A law that states that the total pressure of a gas in a system is the sum of the partial pressures of all component gases. (10)

Daniell cell. A voltaic cell made up of zinc and copper electrodes immersed in solutions of their respective ions that are connected by a salt bridge. (14)

decomposition reaction. A chemical reaction whereby one compound decomposes to two or more elements and/or compounds. (6)

density. The ratio of the mass (usually in grams) to the volume (usually in milliliters or liters) of a substance. (3, 10)

diffusion. The mixing of one gas or liquid into others. (10)

dilute solution. A solution containing a relatively small amount of a specified solute per unit volume. (12)

dilution. The preparation of a dilute solution from a concentrated solution by the addition of more solvent. (12)

dimensional analysis. A problem-solving technique that converts from one unit to another by use of conversion factors. (1)

dipole. Two poles—one positive and one negative—that may exist in a bond or molecule. (9)

dipole–dipole attractions. The force of attraction between a dipole on one polar molecule and a dipole on another polar molecule. (11)

diprotic acid. An acid that can produce two H^+ ions per molecule. (13)

discrete spectrum. A spectrum containing specific wavelengths of light originating from the hot gaseous atoms of an element. (8)

distillation. A laboratory procedure where a solution is separated into its components by boiling the mixture and condensing the vapor to form a liquid. (3)

double bond. The sharing of two pairs of electrons between two atoms. (9)

double-replacement reaction. A chemical reaction whereby the cations and anions in two compounds exchange, leading to formation of either a precipitate or a molecular compound such as water. (6)

dry cell. A voltaic cell composed of zinc and graphite electrodes immersed in an aqueous paste of NH_4Cl and MnO_2. (14)

dynamic equilibrium. See *point of equilibrium*.

E

effusion. The movement of gases through an opening or hole. (10)

electricity. A flow of electrons through a conductor. (14)

electrode. A surface in a cell where the oxidation or reduction reaction takes place. (14)

electrolysis. The process of forcing electrical energy through an electrolytic cell, thereby causing a chemical reaction. (14)

electrolyte. A solute whose aqueous solution or molten state conducts electricity. (12)

electrolytic cell. A voltaic cell that converts electrical energy into chemical energy by means of a nonspontaneous reaction. (14)

electron. A negatively charged particle in an atom, with a comparatively very small mass. (2)

electron capture. The capture of an orbital electron by the nucleus of a radioactive isotope. (16)

electron configuration. The designation of all of the electrons in an atom into specific shells and subshells. (8)

electronegativity. The ability of the atoms of an element to attract electrons in a covalent bond. (7)

electrostatic forces. The forces of attraction between unlike charges and of repulsion by like charges. (2)

element. A pure substance that cannot be broken down into simpler substances. The most basic form of matter existing under ordinary conditions. (1, 2)

empirical formula. The simplest whole-number ratio of atoms in a compound. (5)

endothermic reaction. A chemical reaction that absorbs heat from the surroundings. (3, 7, 15)

energy. The capacity or the ability to do work. It has several different forms (e.g., light and heat) and two types (kinetic and potential). (3)

equilibrium. See *point of equilibrium*.

equilibrium constant. A number that defines the position of equilibrium for a particular reaction at a specified temperature. (15)

equilibrium constant expression. The ratio of the concentrations of products to reactants, each raised to the power corresponding to its coefficient in the balanced equation. (15)

equilibrium vapor pressure. The pressure exerted by a vapor above a liquid at a certain temperature. (11)

evaporation. The vaporization of a liquid below its boiling point. (11)

exact number. A number that results from a definition or an actual count. (1)

excited state. Occupation by an electron of higher energy level than the lowest available energy level. (8)

exothermic reaction. A chemical reaction that occurs with the evolution of heat to the surroundings. (3, 7, 15)

exponent. In scientific notation, it is the power to which ten is raised. (1)

F

Fahrenheit. A temperature scale with 180 divisions between the freezing and boiling points of water, with exactly 32 assigned to the freezing point of water. (1)

family. See *group.*

filtration. A laboratory procedure where solids are removed from liquids by passing the heterogeneous mixture through a filter. (3)

fission. The splitting of a large, unstable nucleus into two smaller nuclei of similar size, resulting in the production of energy. (16)

formula. The symbols of the elements and the number of atoms of each element that make up a compound. (2)

formula unit. The simplest whole-number ratio of ions in an ionic compound. (2)

formula weight. The mass of a compound (in amu), which is determined from the number of atoms and the atomic masses of the elements indicated by the formula. (5)

freezing point. The temperature at which a pure substance changes from the liquid to the solid state. (3, 12)

fuel cell. A voltaic cell that can generate a continuous flow of electricity from the reaction of hydrogen and oxygen to produce water. (14)

fusion (nuclear). The combination of two small nuclei to form a larger nucleus, resulting in the production of energy. (16)

G

gamma (γ) ray. A high-energy form of light emitted from a radioactive nucleus. (16)

gas. A physical state that has neither a definite volume nor a definite shape and fills a container uniformly. (3, 10)

gas constant. The constant of proportionality (R) in the ideal gas law. (10)

gas law. A law governing the behavior of gases that is consistent with the kinetic molecular theory as applied to gases. (10)

Gay-Lussac's law. A law that states that the pressure of a gas is directly proportional to the Kelvin temperature at constant volume (i.e., $P = kT$). (10)

Graham's law. A law that states that the rates of diffusion of gases are inversely proportional to the square root of their molar masses. (10)

ground state. Occupation of an electron of the lowest available energy level in an atom. (8)

group. A vertical column of elements in the periodic table. (4)

H

half-life. The time required for one-half of a given sample of an isotope to undergo radioactive decay. (16)

half-reaction. The oxidation or reduction process in a redox reaction written separately. (14)

halogen. An element in Group VIIA (17) in the periodic table. Elements with the electron configuration [NG] ns^2np^5. (4, 8)

heat of fusion. The amount of heat in calories or joules required to melt one gram of a substance. (11)

heat of reaction. The amount of heat energy absorbed or evolved in a specified chemical reaction. (7)

heat of vaporization. The amount of heat in calories or joules required to vaporize one gram of the substance. (11)

heating curve. The graphical representation of the temperature as a solid is heated through two phase changes, plotted as a function of the time of heating. (11)

Henderson-Hasselbalch equation. Equations that are used to calculate the pH of buffer solutions. (13)

$$pH = pK_a + \log \frac{[base]}{[acid]} \qquad pOH = pK_b + \log \frac{[acid]}{[base]}$$

heterogeneous mixture. A nonuniform mixture containing two or more phases with definite boundaries between phases. (3)

homogeneous mixture. A mixture that is the same throughout and contains only one phase. (3)

Hund's rule. A rule that states that electrons occupy separate orbitals of the same energy with parallel spins if possible. (8)

hydrogen bonding. A force of attraction between a lone pair of electrons on an N, O, or F atom on one molecule and a hydrogen bonded to an N, O, or F atom on another. (11)

hydrolysis reaction. The reaction of an anion as a base or a cation as an acid with water. (13)

hydronium ion. A representation of the hydrogen ion in aqueous solution (H_3O^+). (13)

hypothesis. A tentative explanation of related data. It can be used to predict results of more experiments. (Prologue)

I

ideal gas. A hypothetical gas whose molecules are considered to have no volume or interactions with each other. An ideal gas would obey the ideal gas law under all conditions. (10)

ideal gas law. A relationship between the pressure, volume, temperature, and number of moles of gas (i.e., $PV = nRT$) (10)

immiscible liquids. Two liquids that do not mix and thus form a heterogeneous mixture. (12)

improper fraction. A fraction whose numerator is larger than the denominator and thus has a value greater than one. (Appendix A)

infrared light. Light with wavelengths somewhat longer than those of red light in the visible spectrum. (8)

inner transition element. Either a lanthanide, where the $4f$ subshell is filling, or an actinide, where the $5f$ subshell is filling. (4, 8)

insoluble compound. A compound that does not dissolve to any appreciable extent in a solvent. (6, 15)

insulator. See *nonconductor.*

intermolecular forces. The attractive forces between molecules. (11)

ion. An atom or group of covalently bonded atoms that has a net electrical charge. (2)

ion product (K_w). The equilibrium expression of the anion and the cation of water (i.e., $[H_3O^+][OH^-] = K_w$). (13)

ion–dipole force. The force between an ion and the dipoles of a polar molecule. (12)

ion-electron method. A method of balancing oxidation–reduction reactions where two half-reactions are balanced separately and then added so that electrons gained equal electrons lost. (14)

ionic bond. The electrostatic force holding the positive and negative ions together in an ionic compound. (2, 9)

ionic compound. Compounds containing positive and negative ions. (2, 9)

ionic solid. A solid where the crystal lattice positions are occupied by ions. (11)

ionization. The process of forming an ion or ions from a molecule or atom. (8, 12, 13, 16)

ionization energy. The energy required to remove an electron from a gaseous atom or ion. (8)

isoelectronic Two species having the same number of electrons and having the same electron configuration. (8)

isotopes. Atoms of the same element but having different numbers of neutrons. (2)

isotopic mass. The mass of an isotope compared to ^{12}C, which is defined as having a mass of exactly 12 amu. (2)

J

Joule. The SI unit for measurement of heat energy. (3)

K

Kelvin scale. A temperature scale in which 0 K is the lowest possible temperature. $T(K) = T(°C) + 273$. (1, 10)

kinetic energy. Energy as a result of motion; equal to $\frac{1}{2}mv^2$ (3, 11)

kinetic molecular theory. A theory advanced in the late 1800s to explain the nature of gases. (10)

L

lanthanide. One of 14 elements between La and Hf. Elements whose $4f$ subshell is filling. (4, 8)

lattice. A three-dimensional array of ions or molecules in a solid crystal. (9)

law. A concise statement or mathematical relationship that describes some behavior of matter. (Prologue)

law of conservation of energy. A law that states that energy cannot be created or destroyed but only transformed from one form to another. (3)

law of conservation of mass. A law that states that matter is neither created nor destroyed in a chemical reaction. (3)

lead–acid battery. A rechargeable voltaic battery composed of lead and lead dioxide electrodes in a sulfuric acid solution. (14)

Le Châtelier's principle. A principle that states that when stress is applied to a system at equilibrium, the system reacts in such a way as to counteract the stress. (15)

Lewis dot symbols. The representation of an element by its symbol with its valence electrons as dots. (9)

Lewis structure. The representation of a molecule or ion showing the order and arrangement of the atoms as well as the bonded pairs and unshared electrons of all the atoms. (9)

limiting reactant. The reactant that produces the least amount of product when that reactant is completely consumed. (7)

liquid. A physical state that has a definite volume but not a definite shape. Liquids take the shape of the lower part of the container. (3, 11)

London forces. The instantaneous dipole–induced dipole forces between molecules caused by an instantaneous imbalance of electrical charge in a molecule. The force is roughly dependent on the size of the molecule. (11)

M

main group element. See *representative element.*

mass. The quantity of matter (usually in grams or kilograms) in a sample. (1)

mass number. The number of nucleons (neutrons and protons) in a nucleus. (2)

matter. Anything that has mass and occupies space. (2)

measurement. The quantity, dimensions, or extent of something, usually in comparison to a specific unit. (1)

melting point. The temperature at which a pure substance changes from the solid to the liquid state. (3, 11)

metal. An element with a comparatively low ionization energy that forms positive ions in compounds. Generally, metals are hard, lustrous elements that are ductile and malleable. (4, 8)

metallic solid. A solid made of metals where positive metal ions occupy regular positions in the crystal lattice with the valence electrons moving freely among these positive ions. (11)

metalloid. Elements with properties intermediate between metals and nonmetals. Many of the elements on the metal–nonmetal borderline in the periodic table. (4)

metallurgy. The conversion of metal ores into metals. (Prologue)

metric system. A system of measurement based on multiples of 10. (1)

miscible liquids. Two liquids that mix or dissolve in each other to form a solution. (12)

model. A description or analogy used to help visualize a phenomenon. (8)

molality. A temperature-independent unit of concentration that relates the moles of solute to the mass (kg) of solvent. (12)

molar mass. The atomic mass of an element or the formula weight of a compound expressed in grams. (5)

molar volume. The volume of one mole of a gas at STP, which is 22.4 L. (10)

molarity. A unit of concentration that relates moles of solute to volume (in liters) of solution. (12)

mole. A unit of 6.022×10^{23} atoms, molecules, or formula units. It is the same number of particles as there are atoms in exactly 12 grams of ^{12}C. It also represents the atomic mass of an element or the formula weight of a compound expressed in grams. (5)

mole ratio. The ratios of moles from a balanced equation that serve as conversion factors in stoichiometric calculations. (7)

molecular compound. A compound composed of discrete molecules. (2)

molecular dipole. The combined or net effect of all of the bond dipoles in a molecule as determined by the molecular geometry. (9)

molecular equation. A chemical equation showing all reactants and products as neutral compounds. (6)

molecular formula. See *formula*.

molecular geometry. The geometry of a molecule or ion described by the bonded atoms. It does not include the unshared pairs of electrons. (9)

molecular solid. A solid where the individual molecules in the crystal lattice are held together by London forces, dipole–dipole attractions, or hydrogen bonding. (11)

molecular weight. The formula weight of a molecular compound. (8)

molecule. The basic unit of a molecular compound, which is two or more atoms held together by covalent bonds. (1, 2)

monoprotic acid. An acid that can produce one H^+ ion per molecule. (13)

N

net ionic equation. A chemical equation shown in ionic form with spectator ions eliminated. (6)

network solid. A solid where the atoms are covalently bonded to each other throughout the entire crystal. (11)

neutral. Pure water or a solution with pH = 7. (13)

neutralization reaction. A reaction whereby an acid reacts with a base to form a salt and water. The reaction of $H^+(aq)$ with $OH^-(aq)$. (6, 13)

neutron. A particle in the nucleus with a mass of about 1 amu and no charge. (2)

noble gas. An element with a full outer s and p subshell. Group VIIIA (17) in the periodic table. (2, 4, 8)

nonconductor. A substance that does not conduct electricity. (12)

nonelectrolyte. A solute whose aqueous solution or molten state does not conduct electricity. (12)

nonmetal. Elements to the right in the periodic table. These elements generally lack metallic properties. They have relatively high ionization energies. (4, 8, 9)

nonpolar bond. A covalent bond in which electrons are shared equally. (9)

normal boiling point. The temperature at which the vapor pressure of a liquid is equal to exactly one atmosphere pressure. (11)

nuclear equation. A symbolic representation of the changes of a nucleus or nuclei into other nuclei and particles. (16)

nuclear reactor. A device that can maintain a controlled nuclear fission reaction. Used either for research or generation of electrical power. (16)

nucleons. The protons and neutrons that make up the nucleus of the atom. (2, 16)

nucleus. The core of the atom containing neutrons, protons, and most of the mass. (2, 16)

O

octet rule. A rule that states that atoms of representative elements form bonds so as to have access to eight electrons either through bonds or unshared pairs of electrons. (9)

orbital. A region of space where there is the highest probability of finding a particular electron. There are four types of orbitals; each has a characteristic shape. (8)

orbital diagram. The representation of specific orbitals of a subshell as boxes and the electrons as arrows in the boxes. (8)

osmosis. The tendency of a solvent to move through a semipermeable membrane from a region of low concentration to a region of high concentration of solute. (12)

osmotic pressure. The pressure needed to counteract the movement of solvent through a semipermeable membrane from a region of low concentration of solute to a region of high concentration. (12)

oxidation. The loss of electrons as indicated by an increase in oxidation state. (14)

oxidation–reduction reaction. A chemical reaction involving an exchange of electrons. (14)

oxidation state (number). The charge on an atom in a compound if all atoms were present as monatomic ions. The electrons in bonds are assigned to the more electronegative atom. (14)

oxidation state method. A method of balancing oxidation–reduction reactions that focuses on the atoms of the elements undergoing a change in oxidation state. (14)

oxidizing agent. The element, compound, or ion that oxidizes another reactant. It is reduced. (14)

oxyacid. An acid composed of hydrogen and an oxyanion. (4)

oxyanion. An anion composed of oxygen and one other element. (4)

P

partial pressure. The pressure of one component in a mixture of gases. (10)

parts per billion (ppb). A unit of concentration obtained by multiplying the ratio of the mass of solute to the mass of solution by 10^9 ppb. (12)

parts per million (ppm). A unit of concentration obtained by multiplying the ratio of the mass of solute to the mass of solution by 10^6 ppm. (12)

Pauli exclusion principle. A rule that states that no two electrons can have the same spin in the same orbital. (8)

percent by mass. The mass of solute expressed as a percent of the mass of solution. (3, 12)

percent composition. The mass of each element expressed per 100 mass units of the compound. (5)

percent yield. The actual yield in grams or moles divided by the theoretical yield in grams or moles and multiplied by 100%. (7)

period. A horizontal row of elements between noble gases in the periodic table. (4)

periodic law. A law that states that the properties of elements are periodic functions of their atomic numbers. (4)

periodic table. An arrangement of elements in order of increasing atomic number. Elements with the same number of outer electrons are arranged in vertical columns. (4, 8)

pH. The negative of the common logarithm of the H_3O^+ concentration. (13)

phase. A homogeneous state (solid, liquid, or gas) with distinct boundaries and uniform properties. (3)

physical change. A change in physical state or dimensions of a substance that does not involve a change in composition. (3)

physical properties. Properties that can be observed without changing the composition of a substance. (3)

physical states. The physical condition of matter—solid, liquid, or gas. (3)

pOH. The negative of the common logarithm of the OH^- concentration. (13)

point of equilibrium. The point at which the forward and reverse processes in a reversible process occur at the same rate so that the concentrations of all species remain constant. (11, 13, 15)

polar covalent bond. A covalent bond that has a partial separation of charge due to the unequal sharing of electrons. (9)

polyatomic ion. A group of atoms covalently bonded to each other that have a net electrical charge. (2, 4)

polyprotic acid. An acid that can produce more than one H^+ ion per molecule. (13)

positron. An antimatter particle with the same mass as an electron but positively charged. Emitted from certain radioactive isotopes. (16)

potential energy. Energy as a result of position or composition. (3, 15)

precipitate. A solid compound formed in a solution. (6)

precipitation reaction. A type of double-replacement reaction in which an insoluble ionic compound is formed by an exchange of ions in the reactants. (6)

precision. The reproducibility of a measurement as indicated by the number of significant figures expressed. (1)

pressure. The force per unit area. (10)

principal quantum number (n). A number that corresponds to a particular shell occupied by the electrons in an atom. (8)

product. An element or compound in an equation that is formed as a result of a chemical reaction. (6)

proper fraction. A frac... than the denominato... one. (Appendix A) ...umerator is smaller ...a value less than

property. A particular charact... (2, 3)

proton. A particle in the nucleus of a substance. ...amu and a charge of +1. (2) ...of about 1

pure substance. A substance that has ...tion with definite and unchanging p... ments and compounds). (3) ...composi- ...e., ele-

Q

quantized energy level. An energy level with a de... measurable energy. (8)

R

radiation. Particles or high-energy light rays that are emi... ted by an atom or a nucleus of an atom. (8, 16)

radioactive decay series. A series of elements formed from the successive emission of alpha and beta particles starting from a long-lived isotope and ending with a stable isotope. (16)

radioactivity. The emission of energy or particles from an unstable nucleus. (16)

rate of reaction. A measure of the increase in concentration of a product or the decrease in concentration of a reactant per unit time. (15)

reactant. An element or compound in an equation that undergoes a chemical reaction. (6)

recrystallization. A laboratory procedure whereby a solid compound is purified by saturating a solution at a high temperature and then forming a precipitate at a lower temperature. (12)

redox reaction. See *oxidation-reduction reaction.*

reducing agent. An element, compound, or ion that reduces another reactant. It is oxidized. (14)

reduction. The gain of electrons as indicated by a decrease in the oxidation state. (14)

representative element. Elements whose outer *s* and *p* subshells are filling. The A Group elements (i.e., groups 1, 2, 13–17) in the periodic table [except for VIIIA (18)]. (4, 8)

resonance hybrid. The actual structure of the molecule as implied by the separate resonance structures. (9)

resonance structure. A Lewis structure showing one of two or more possible Lewis structures. (9)

reversible reaction. A reaction where both a forward reaction (forming products) and a reverse reaction (reforming reactants) can occur. (15)

room temperature. The standard reference temperature for physical state, which is usually defined as 25°C. (4)

S

salt. An ionic compound formed by the combination of most cations and anions. Also, the compound that forms from the cation of a base and the anion of an acid. (4, 6, 13)

...l that allows anions to migrate ...s in a voltaic cell. (14)

salt bridge. ...solution containing the maximum ...betwee...d solute at a specific temperature. (12)

saturated ...he method whereby modern scientists ...amo...ehavior of nature with hypotheses and

scienti...describe the behavior of nature with laws. ...ex...

...**ation.** A number expressed with one nonzero ...the left of the decimal point multiplied by 10 ...d to a given power. (1, Appendix C)

...**etal.** See *metalloid*.

...l. The principal energy level that contains one or more subshells. (8)

SI units. An international system of units of measurement. (1)

significant figure. A digit or number in a measurement that either is reliably known or is estimated. (1)

single-replacement reaction. A chemical reaction whereby a free element substitutes for another element in a compound. (6)

solid. A physical state with both a definite shape and a definite volume. (3, 11)

solubility. The maximum amount of a solute that dissolves in a specific amount of solvent at a certain temperature. (6, 12)

solubility product constant (K_{sp}). The equilibrium constant associated with the solution of ionic compounds. (15)

soluble compound. A compound that dissolves to an appreciable extent in a solvent. (6)

solute. A substance that dissolves in a solvent. (6, 12)

solution. A homogeneous mixture with one phase. It is composed of a solute dissolved in a solvent. (3, 6, 12)

solvent. A medium, usually a liquid, that disperses a solute to form a solution. (6, 12)

specific gravity. The ratio of the mass of a substance to the mass of an equal volume of water at the same temperature. (3)

specific heat. The amount of heat required to raise the temperature of one gram of a substance one degree Celsius. (3)

spectator ion. An ion that is in an identical state on both sides of an equation and not specifically involved in a reaction. (6)

spectrum. The separate color components of a beam of light. (8)

standard temperature and pressure (STP). The defined standard conditions for a gas, which are exactly 0°C and one atmosphere pressure. (10)

Stock method. A method used to name metal–nonmetal or metal–polyatomic ion compounds where the charge on the metal is indicated by Roman numerals enclosed in parentheses. (4)

stoichiometry. The quantitative relationship among reactants and products. (7)

strong acid (base). An acid (or base) that is completely ionized in aqueous solution. (6, 13)

structural formula. Formulas written so that the order and arrangement of specific atoms are shown. (2, 9)

sublimation. The vaporization of a solid. (11)

subshell. The orbitals of the same type within a shell. The subshells are named for the types of orbitals, that is, *s, p, d,* or *f.* (8)

substance. A form of matter. Usually thought of as either an element or a compound. (3)

supersaturated solution. A solution containing more than the maximum amount of solute indicated by the compound's solubility at that temperature. (12)

surface tension. The forces of attraction between molecules that cause a liquid surface to contract. (11)

symbol. One or two letters from an element's English or, in some cases, Latin name. (2)

T

temperature. A measure of the intensity of heat of a substance. It relates to the average kinetic energy of the substance. (1, 10)

theoretical yield. The calculated amount of product that would be obtained if all of a reactant were converted to a certain product. (7)

theory. A hypothesis that withstands the test of time and experiments designed to test the hypothesis. (Prologue)

thermochemical equation. A balanced equation that includes the amount of heat energy. (7)

thermometer. A device that measures temperature. (1)

torr. A unit of gas pressure equivalent to the height of one millimeter of mercury. (10)

total ionic equation. A chemical equation showing all soluble compounds that exist primarily as ions in aqueous solution as separate ions. (6)

transition element. Elements whose outer *s* and *d* subshells are filling. The B Group elements (i.e., groups 3–12) in the periodic table. (4, 8)

transmutation. The changing of one element into another by a nuclear reaction. (16)

triple bond. The sharing of three pairs of electrons in a bond between two atoms. (7)

triprotic acid. An acid that can produce three H^+ ions per molecule. (13)

U

ultraviolet light. Light with wavelengths somewhat shorter than those of violet light. (8)

unit. A definite quantity adapted as a standard of measurement. (1)

unit factor. A fractional expression that relates a quantity in a certain unit to "one" of another unit. (1)

unit map. A shorthand representation of the procedure for solving a problem that indicates the conversion of units in one or more steps. (1, 5, 7)

unsaturated solution. A solution that contains less than the maximum amount of solute indicated by the compound's solubility at that temperature. (12)

V

valence electron. An outer *s* or *p* electron in the atom of a representative element. (8)

valence shell electron-pair repulsion theory (VSEPR). A theory that predicts that electron pairs either unshared or in a bond repel each other to the maximum extent. (7)

vapor pressure. See *equilibrium vapor pressure.*

viscosity. A measure of the resistance of a liquid to flow. (11)

volatile. Refers to a liquid or solid with a significant vapor pressure. (11)

voltaic cell. A spontaneous oxidation–reduction reaction that can be used to produce electrical energy. (14)

volume. The space that a certain quantity of matter occupies. (1)

W

wavelength (λ). The distance between two adjacent peaks in a wave. (8)

wave mechanics. A complex mathematical approach to the electrons in an atom that considers the electron as having both a particle and a wave nature. (8)

weak acid (base). An acid (or base) that is only partially ionized in aqueous solution. (13)

weak electrolyte. A solute whose aqueous solution allows only a limited amount of electrical conduction. (12)

weight. A measure of the attraction of gravity for a sample of matter. (1)

Answers to Chapter Problems

CHAPTER 1

1-1. (b) 74.212 gal (the most significant figures)

1-2. A device used to produce a measurement may provide a reproducible answer to several significant figures, but if the device itself is inaccurate (such as a ruler with the tip broken off) the measurement is inaccurate.

1-4. (a) three (b) two (c) three (d) one (e) four
f) two (g) two (h) three

1-6. (a) ±10 (b) ±0.1 (c) ±0.01 (d) ±0.01
(e) ±1 (f) ±0.001 (g) ±100 (h) ±0.00001

1-8. (a) 16.0 (b) 1.01 (c) 0.665 (d) 489
(e) 87,600 (f) 0.0272 (g) 301

1-10. (a) 0.250 (b) 0.800 (c) 1.67 (d) 1.17

1-12. (a) ±0.1 (b) ±1000 (c) ±1 (d) ±0.01

1-14. (a) 188 (b) 12.90 (c) 2300 (d) 48 (e) 0.84

1-16. 37.9 qt

1-18. (a) 7.0 (b) 137 (c) 192 (d) 0.445 (e) 3.20 (f) 2.9

1-20. (a) two (b) three (c) two (d) one

1-23. (a) 6.07 (b) 0.08 (c) 8.624 (d) 24 (e) 0.220
(f) 0.52

1-26. (a) $(63) + 75.0 = \underline{138}$
(b) $(45) \times 25.6 = \underline{1200}$
(c) $(2.7) \times (10.52) = \underline{28}$

1-28. 13%

1-30. 0.11%

1-32. (a) 1.57×10^2 (b) 1.57×10^{-1} (c) 3.00×10^{-2}
(d) 4.0×10^7 (e) 3.49×10^{-2} (f) 3.2×10^4
(g) 3.2×10^{10} (h) 7.71×10^{-4} (i) 2.34×10^3

1-34. (a) 9×10^7 (b) 8.7×10^7 (c) 8.70×10^7

1-36. (a) 0.000476 (b) 6550 (c) 0.0078 (d) 48,900
(e) 4.75 (f) 0.0000034

1-38. (a) 4.89×10^{-4} (b) 4.56×10^{-5} (c) 7.8×10^3
(d) 5.71×10^{-2} (e) 4.975×10^8 (f) 3.0×10^{-4}

1-40. (b) < (f) < (g) < (d) < (a) < (e) < (c)

1-42. (a) 1.597×10^{-3} (b) 2.30×10^7 (c) 3.5×10^{-5}
(d) 2.0×10^{14}

1-44. (a) 10^7 (b) $10^0 = 1$ (c) 10^{29} (d) 10^9

1-46. (a) 3.1×10^{10} (b) 2×10^9 (c) 4×10^{13} (d) 14
(e) 2.56×10^{-14}

1-48. (a) 2.0×10^{12} (b) 3.7×10^{16} (c) 6.0×10^2
(d) 2×10^{-12} (e) 1.9×10^8

1-49. (a) 1.225×10^7 (b) 9.00×10^{-12} (c) 3.0×10^{-24}
(d) 9×10^4 (e) 1×10^{10}

1-51. (a) milliliter (mL) (b) hectogram (hg)
(c) nanojoule (nJ) (d) centimeter (cm)
(e) microgram (μg) (f) decipascal (dPa)

1-53. (a) 720 cm, 7.2 m, 7.2×10^{-3} km
(b) 5.64×10^4 mm, 5640 cm, 0.0564 km
(c) 2.50×10^5 mm, 2.50×10^4 cm, 250 m

1-54. (a) 8.9 g, 8.9×10^{-3} kg
(b) 2.57×10^4 mg, 0.0257 kg
(c) 1.25×10^6 mg, 1250 g

1-56. (a) 12 = 1 doz (c) 3 ft = 1 yd (e) 10^3 m = 1 km

1-58. (a) $\dfrac{1\text{ g}}{10^3\text{ mg}}$ (b) $\dfrac{1\text{ km}}{10^3\text{ m}}$ (c) $\dfrac{1\text{ L}}{100\text{ cL}}$
(d) $\dfrac{1\text{ m}}{10^3\text{ mm}}, \dfrac{1\text{ km}}{10^3\text{ m}}$

1-60. (a) $\dfrac{1\text{ ft}}{12\text{ in.}}$ (b) $\dfrac{2.54\text{ cm}}{\text{in.}}$ (c) $\dfrac{5280\text{ ft}}{\text{mi}}$
(d) $\dfrac{1.057\text{ qt}}{\text{L}}$ (e) $\dfrac{1\text{ qt}}{2\text{ pt}}, \dfrac{1\text{ L}}{1.057\text{ qt}}$

1-62. (a) 47 L (b) 98 cm (c) 1.85 mi (d) 51.56 yd
(e) 92 m (f) 10.27 bbl (g) 32 Gg

1-64. (a) 7.8 km, 4.8 mi, 2.5×10^4 ft
(b) 2380 ft, 0.724 km, 724 m
(c) 1.70 mi, 2.74 km, 2740 m
(d) 4.21 mi, 2.22×10^4 ft, 6780 m

1-65. (a) 25.7 L, 27.2 qt (b) 630 L, 170 gal
(c) 8.12×10^3 qt, 2.03×10^3 gal

1-67. 55.3 kg

1-69. $28.0\text{ m} \times \dfrac{10^2\text{ cm}}{\text{m}} \times \dfrac{1\text{ in.}}{2.54\text{ cm}} \times \dfrac{1\text{ ft}}{12\text{ in.}} \times \dfrac{1\text{ yd}}{3\text{ ft}}$
$= \underline{30.6\text{ yd}}$ (New punter is needed.)

1-71. 0.355 L

1-73. 6 ft 10 1/2 in. = 82.5 in.
$82.5\text{ in.} \times \dfrac{2.54\text{ cm}}{\text{in.}} \times \dfrac{1\text{ m}}{10^2\text{ cm}} = \underline{2.10\text{ m}}$

$$212 \text{ lb} \times \frac{1 \text{ kg}}{2.205 \text{ lb}} = \underline{96.1 \text{ kg}}$$

1-74. 14.5 gal

1-76. $0.200 \text{ gal} \times \frac{4 \text{ qt}}{\text{gal}} = \underline{0.800 \text{ qt}}$

$$0.800 \text{ qt} \times \frac{1 \text{ L}}{1.057 \text{ qt}} \times \frac{1 \text{ mL}}{10^{-3} \text{ L}} = \underline{757 \text{ mL}}$$

There is slightly more in a "fifth" than in 750 mL.

1-78. 105 km/hr

1-82. $\frac{\$0.899}{\text{gal}} \times \frac{1 \text{ gal}}{4 \text{ qt}} \times \frac{1.057 \text{ qt}}{\text{L}} = \underline{\$0.238/\text{L}}$

$19.04 (2001), $58.31 (2010)

1-83. $72.39 (551 mi), $39.36 (482 km)

1-84. 674 km

1-88. $28.20

1-89. $38.21 (the hybrid); $79.53 (the SUV)

1-92. 3.31 hr

1-93. **(a)** 7.28 euro **(b)** $11.29 **(c)** $19.79

1-94. 0.824 pound/euro, 21,100 pound

1-97. 2100 s, 0.58 hr

1-98. 572°F

1-99. 24°C

1-101. −38°F

1-103. 95.0°F

1-104. **(a)** −98°C **(b)** 22°C **(c)** 27°C **(d)** −48°C **(e)** 600°C

1-105. **(a)** 320 K **(b)** 296 K **(c)** 200 K **(d)** 261 K
(e) 291 K **(f)** 244 K

1-107. Since $T \, (°C) = T \, (°F)$ substitute $T \, (°C)$ for $T \, (°F)$ and set the two equations equal.

$$[T \, (°C) \times 1.8] + 32 = \frac{T \, (°C) - 32}{1.8}$$

$$(1.8)^2 T \, (°C) - T \, (°C) = -32 - 32(1.8)$$

$$T \, (°C) = -40°C$$

1-108. **(a)** 3×10^2 **(b)** 8.26 g · cm **(c)** 5.24 g/mL
(d) 19.1

1-109. **(a)** $\dfrac{1 \text{ g}}{10^3 \text{ mg}}, \dfrac{1 \text{ lb}}{453.6 \text{ g}}$ **(b)** $\dfrac{1.057 \text{ qt}}{\text{L}}, \dfrac{2 \text{ pt}}{\text{qt}}$

(c) $\dfrac{1 \text{ km}}{10 \text{ hm}}, \dfrac{1 \text{ mi}}{1.609 \text{ km}}$ **(d)** $\dfrac{1 \text{ in.}}{2.54 \text{ cm}}, \dfrac{1 \text{ ft}}{12 \text{ in.}}$

1-110. $5.34 \times 10^{10} \text{ ng} \times \dfrac{10^{-9} \text{ g}}{\text{ng}} \times \dfrac{1 \text{ lb}}{453.6 \text{ g}} = \underline{0.118 \text{ lb}}$

1-112. $40,182

1-113. $\dfrac{247 \text{ lb}}{82.3 \text{ doz}} = \underline{3.00 \text{ lb/doz}} \quad \dfrac{82.3 \text{ doz}}{247 \text{ lb}} = \underline{0.333 \text{ doz/lb}}$

1-114. $12.0 \text{ fur} \times \dfrac{1 \text{ mi}}{8 \text{ fur}} \times \dfrac{5280 \text{ ft}}{\text{mi}} \times \dfrac{12 \text{ in.}}{\text{ft}} \times \dfrac{1 \text{ hand}}{4 \text{ in.}}$

$$= \underline{2.38 \times 10^4 \text{ hands}}$$

1-116. 1030 pkgs, 1.41 years

1-117. 5.02 L

1-119. $5.4 \times 10^7 °F \quad 3.0 \times 10^7 °C + 273 = 3.0 \times 10^7 \text{ K}$

1-121. 9,300 lb, 4.65 ton,

CHAPTER 2

2-2. cadmium-Cd, calcium-Ca, californium-Cf, carbon-C, cerium-Ce, cesium-Cs, chlorine-Cl, chromium-Cr, cobalt-Co, copper-Cu, curium-Cm

2-5. **(a)** Ba **(b)** Ne **(c)** Cs **(d)** Pt **(e)** Mn **(f)** W

2-7. **(a)** boron **(b)** bismuth **(c)** germanium
(d) uranium **(e)** cobalt **(f)** mercury **(g)** beryllium
(h) arsenic

2-8. **(b)** and **(e)**

2-9. **(c)**

2-12. **(a)** 21 p, 21 e, 24 n **(b)** 90 p, 90 e, 142 n **(c)** 87 p, 87 e, 136 n **(d)** 38 p, 38 e, 52 n

2-14.

Isotope Name	Isotope Notation	Atomic Number	Mass Number	p	n	e
(a) silver-108	$^{108}_{47}\text{Ag}$	47	108	47	61	47
(b) silicon-28	$^{28}_{14}\text{Si}$	14	28	14	14	14
(c) potassium-39	$^{39}_{19}\text{K}$	19	39	19	20	19
(d) cerium-140	$^{140}_{58}\text{Ce}$	58	140	58	82	58
(e) iron-56	$^{56}_{26}\text{Fe}$	26	56	26	30	26
(f) tin-110	$^{110}_{50}\text{Sn}$	50	110	50	60	50
(g) iodine-118	$^{118}_{53}\text{I}$	53	118	53	65	53
(h) mercury-196	$^{196}_{80}\text{Hg}$	80	196	80	116	80

2-16. ^{59}Co

2-18. **(a)** The identity of a specific element is determined by its atomic number. An element is a basic form of matter and its atomic number relates to the number of protons in the nuclei of its atoms.

(b) Both relate to the particles in the nucleus of the atoms of an element. The atomic mass is the mass of an average atom since elements are usually composed of more than one isotope. The atomic number is the number of protons.

(c) Both relate to the number of particles in a nucleus. The mass number relates to the total number of protons and neutrons in an isotope of an element while the atomic number is the number of protons in the atoms of a specific element.

(d) All isotopes of an element have the same atomic number.

(e) Different isotopes of a specific element have the same number of protons but different numbers of neutrons or mass numbers.

2-19. (a) Re: at. no. 75, at. wt. 186.2   (b) Co: at. no. 27, at. wt. 58.9332   (c) Br: at. no. 35, at. wt. 79.904
(d) Si: at. no. 14, at. wt. 28.086

2-20. copper (Cu)

2-22. O: at. no. 8, mass no. 16

N: at. no. 7, mass no. 14

Si: at. no. 14, mass no. 28

Ca: at. no. 20, mass no. 40

2-24. $5.81 \times 12.00 = 69.7$ amu. The element is Ga.

2-26. 79.9 amu

2-27. ^{28}Si: $0.9221 \times 27.98 = 25.80$

^{29}Si: $0.0470 \times 28.98 = 1.362$

^{30}Si: $0.0309 \times 29.97 = \underline{0.926}$

$$28.088 \ = \ \underline{28.09 \text{ amu}}$$

2-29. Let x = decimal fraction of ^{35}Cl and y = decimal fraction of ^{37}Cl. Since there are two isotopes present, $x + y = 1$, $y = 1 - x$.

$(x \times 35) + (y \times 37) = 35.5$

$(x \times 35) + [(1 - x) \times 37] = 35.5$

$x = 0.75$ (75% ^{35}Cl) $y = 0.25$ (25% ^{37}Cl)

2-31. (a) They are both basic units of matter. Most elements are composed of individual atoms and many compounds are composed of individual molecules. Molecules are composed of atoms chemically bonded together.

(b) A compound is a pure form of matter. It is composed of individual units called molecules.

(c) They are both pure forms of matter. Compounds, however, are composed of two or more elements chemically combined.

(d) Most elements are composed of individual atoms. Some elements, however, are composed of molecules, which in most cases, contain two atoms.

2-32. (b) Br_2 (d) S_8 (f) P_4

2-33. P, phosphorus   O, oxygen   Br, bromine   F, fluorine   S, sulfur   Mg, magnesium

2-34. Hf is the symbol of the element hafnium. HF is the formula of a compound composed of one atom of hydrogen and one atom of fluorine.

2-36. (b) CO, diatomic compound   (e) N_2, diatomic element

2-39. (a) six carbons, four hydrogens, two chlorines

(b) two carbons, six hydrogens, one oxygen

(c) one copper, one sulfur, 18 hydrogens, 13 oxygens

(d) nine carbons, eight hydrogens, four oxygens

(e) two aluminums, three sulfurs, 12 oxygens

(f) two nitrogens, eight hydrogens, one carbon, three oxygens

2-40. (a) 12   (b) 9   (c) 33   (d) 21   (e) 17   (f) 14

2-41. (a) 8   (b) 7   (c) 4   (d) 3

2-42. (a) SO_2 (b) CO_2 (c) H_2SO_4 (d) C_2H_2

2-44. (a) They are both basic forms of matter. Atoms are neutral but ions are atoms that have acquired an electrical charge. Positive and negative ions are always found together.

(b) They are both basic forms of matter containing more than one atom. Molecules are neutral but polyatomic ions have acquired an electrical charge.

(c) Both have an electrical charge. Cations have a positive charge and anions have a negative charge.

(d) Both are classified as compounds, which are composed of the atoms of two or more elements. The basic unit of a molecular compound is a neutral molecule but the basic units of ionic compounds are cations and anions.

(e) Both are the basic units of compounds. Molecules are the basic entities of molecular compounds, and an ionic formula unit represents the smallest whole number of cations and anions representing a net charge of zero.

2-46. (a) $Ca(ClO_4)_2$ (b) $(NH_4)_3PO_4$ (c) $FeSO_4$

2-47. (a) one calcium, two chlorines, and eight oxygens
(b) three nitrogens, 12 hydrogens, one phosphorus, and four oxygens   (c) one iron, one sulfur, and four oxygens

2-50. (c) S^{2-}

2-52. (d) Li^+

2-54. FeS, Li_2SO_3

2-56. SO_3 represents the formula of a compound. It could be a gas. SO_3^{2-} is an anion and does not exist independently. It is part of an ionic compound with the other part being a cation.

2-57. (a) K^+: 19 p, 18 e   (b) Br^-: 35 p, 36 e
(c) S^{2-}: 16 p, 18 e   (d) NO_2^-: $7 + 16 = 23$ p, 24 e
(e) Al^{3+}: 13 p, 10 e   (f) NH_4^+: $7 + 4 = 11$ p, 10 e

2-59. (a) Ca^{2+} (b) Te^{2-} (c) PO_3^{3-} (d) NO_2^+

2-61. This is the Br^- ion. It is part of an ionic compound.

2-64. (a) $^{90}_{38}Sr^{2+}$ (b) $^{52}_{24}Cr^{3+}$ (c) $^{79}_{34}Se^{2-}$ (d) $^{14}_{7}N^{3-}$ (e) $^{139}_{57}La^{3+}$

2-66. (a) Na: 11 p, 12 n, and 11 e; Na^+ has 10 electrons.

(b) Ca: 20 p, 20 n, and 20 e; Ca^{2+} has 18 electrons.

(c) F: 9 p, 10 n, and 9 e; F^- has 10 electrons.

(d) Sc: 21 p, 24 n, and 21 e; Sc^{3+} has 18 electrons.

2-68. Let x = mass no. of I and y = mass no. of Tl
Then (1) $x + y = 340$ or $x = 340 - y$

$(2)\, x = \dfrac{2}{3}y - 10$

Substituting for x from (1) and solving for y

$y = 210$ amu (Tl) and $x = 340 - 210 = 130$ amu (I)

2-70. 121.8 (Sb) Sb^{3+} has $51 - 3 = 48$ electrons. ^{121}Sb has $121 - 51 = 70$ neutrons. ^{123}Sb has $123 - 51 = 72$ neutrons. ^{121}Sb, 57.9% due to neutrons; ^{121}Sb, 58.5% due to neutrons

2-71. 118 neutrons and 78 protons [platinum (Pt)]

$78 - 2 = 76$ electrons for Pt^{2+}

2-73. Mass of other atom = 16 (oxygen)
$NO^+ = (7 + 8) - 1 = 14$ electrons

2-75. (a) H: $\dfrac{1.008}{12.00} \times 8.000 = 0.672$

(b) N: $\dfrac{14.01}{12.00} \times 8.000 = 9.34$

(c) Na: $\dfrac{22.99}{12.00} \times 8.000 = 15.3$

(d) Ca: $\dfrac{40.08}{12.00} \times 8.000 = 26.7$

2-76. $\dfrac{43.3}{10.0} = 4.33$ times as heavy as ^{12}C

4.33×12.0 amu = 52.0 amu. The element is Cr.

CHAPTER 3

3-1. (c) It has a definite volume but not a definite shape.

3-2. The gaseous state is compressible because the basic particles are very far apart and thus the volume of a gas is mostly empty space.

3-4. (a) physical **(b)** chemical **(c)** physical **(d)** chemical **(e)** chemical **(f)** physical **(g)** physical **(h)** chemical **(i)** physical

3-6. (a) chemical **(b)** physical **(c)** physical **(d)** chemical **(e)** physical

3-9. Physical property: melts at 660°C; Physical change: melting; Chemical property: burns in oxygen; Chemical change: formation of aluminum oxide from aluminum and oxygen.

3-10. Original substance: green, solid (physical); can be decomposed (chemical). Substance is a compound. Gas: gas, colorless (physical); can be decomposed (chemical). Substance is a compound since it can be decomposed. Solid: shiny, solid (physical); cannot be decomposed (chemical). Substance is an element since it cannot be decomposed.

3-12. 2.60 g/mL

3-14. 1064 g/657 mL = 1.62 g/mL (carbon tetrachloride)

3-17. 1450 g

3-18. 670 g

3-19. 5.6 lb

3-21. 625 mL

3-22. 1.74 g/mL (magnesium)

3-23. 0.476 g/mL; Yes, it floats.

3-24. 0.951 g/mL, 4790 mL

Pumice floats in water but sinks in alcohol.

3-27. 2080 g

3-28. 111.0 g/125 mL = 0.888 g/mL

3-30. Water: 1000 g; Gasoline: 670 g

One liter of water has a greater mass.

3-31. 160 g/8.3 mL = 19 g/mL. It's gold.

3-34. One needs a conversion factor between mL (cm³) and ft³.

$\left(\dfrac{2.54\text{ cm}}{\text{in.}}\right)^3 = \dfrac{16.4\text{ cm}^3}{\text{in.}^3} = \dfrac{16.4\text{ mL}}{\text{in.}^3}\left(\dfrac{12\text{ in.}}{\text{ft}}\right)^3 = \dfrac{1728\text{ in.}^3}{\text{ft}^3}$

$\dfrac{1.00\text{ g}}{\text{mL}} \times \dfrac{1\text{ lb}}{453.6\text{ g}} \times \dfrac{16.4\text{ mL}}{\text{in.}^3} \times \dfrac{1728\text{ in.}^3}{\text{ft}^3} = 62.5\text{ lb/ft}^3$

3-35. 2.0×10^5 lb (100 tons)

3-36. Carbon dioxide is a compound composed of carbon and oxygen. It can be prepared from a mixture of carbon and oxygen, but the compound is no longer a mixture of the two elements.

3-38. Ocean water is the least pure because it contains a large amount of dissolved compounds. That is why it is not drinkable and cannot be used for crop irrigation. Drinking water also contains chlorine and some dissolved compounds but not nearly as much as ocean water. Rainwater is most pure but still contains some dissolved gases from the air.

3-40. (a) liquid only

3-41. (a) homogeneous **(b)** heterogeneous **(c)** heterogeneous **(d)** homogeneous **(e)** homogeneous **(f)** homogeneous solution **(g)** heterogeneous **(h)** homogeneous

3-43. (a) liquid **(b)** various solid phases **(c)** gas and liquid **(d)** liquid **(e)** solid **(f)** liquid **(g)** liquid and gas **(h)** gas

3-45. A mixture of all three would have carbon tetrachloride on the bottom, water in the middle, and kerosene on top. Water and kerosene float on carbon tetrachloride; kerosene floats on water.

3-47. Ice is less dense than water. An ice–water mixture is pure but heterogeneous.

3-48. (a) solution (a solid dissolved in a liquid) **(b)** heterogeneous mixture (probably a solid suspended in a liquid such as dirty water) **(c)** element **(d)** compound **(e)** solution (two liquids)

3-50. Mass of mixture = 22.6 + 855 = 878 g

$\dfrac{22.6}{878} \times 100\% = 2.57\%$ NaCl $(100\% - 2.57 = 97.4\%$ water)

3-52. 255 kg solution $\times \dfrac{25\text{ g solute}}{100\text{ kg solution}} = 64$ kg solute

3-54. 122 lb iron $\times \dfrac{100\text{ lb duriron}}{86\text{ lb iron}} = 140$ lb duriron

3-56. (a) exothermic **(b)** endothermic **(c)** endothermic **(d)** exothermic **(e)** exothermic

3-57. Gasoline is converted into heat energy when it burns. The heat energy causes the pistons to move, which is mechanical energy. The mechanical energy turns the alternator, which generates electrical energy. The electrical energy is converted into chemical energy in the battery.

3-59. (a) potential **(b)** kinetic **(c)** potential (It is stored because of its composition.) **(d)** kinetic **(e)** kinetic

3-62. Kinetic energy is at a maximum nearest the ground when the swing is moving the fastest. Potential energy is at a maximum when the swing has momentarily stopped at the highest point. Assuming no gain or loss of energy, the total of the two energies is constant.

3-63. $0.853 \, J/(g \cdot °C)$

3-64. $\dfrac{56.6 \text{ cal}}{365 \text{ g} \cdot 5.0 \text{ °C}} = 0.031 \text{ cal}/(g \cdot °C) \text{ (gold)}$

3-66.

$$°C = \frac{\text{cal}}{\text{sp. heat} \times g} = \frac{150 \text{ cal}}{0.092 \dfrac{\text{cal}}{g \cdot °C} \times 50.0 \text{ g}} = 33°C \text{ rise}$$

$$T°C = 25 + 33 = \underline{58°C}$$

This compares to a 3.0°C rise in temperature for 50.0 g of water.

3-68. 506 J

3-69. $58 - 25 = 33°C$ rise in temperature

$$g = \frac{\text{cal}}{\text{sp. heat} \cdot °C} = \frac{16.0 \text{ cal}}{0.106 \dfrac{\text{cal}}{g \cdot °C} \cdot 33°C} = \underline{4.6 \text{ g}}$$

3-70. The copper skillet, because it has a lower specific heat. The same amount of applied heat will heat the copper skillet more than the iron.

3-72. Iron: 9.38°C rise Gold: 32°C rise Water: 0.997°C rise

3-74. 45°C

3-76. 2910 J

3-80. 860 g

3-81. heat lost by metal = heat gained by water
$100.0 \text{ g} \times 68.7°C \times \text{specific heat} = 100.0 \text{ g} \times 6.3°C \times 4.184 \, J/g \cdot °C$

Specific heat = $0.38 \, J/g \cdot °C$. The metal is copper.

3-82. Density of A = 0.86 g/mL; density of B = 0.89 g/mL
Liquid A floats on liquid B.

3-84. 15 g of sugar

3-85. 9.27 g/mL

3-87. 3.17% salt

3-89. $50.0 \text{ mL gold} \times \dfrac{19.3 \text{ g}}{\text{mL gold}} = 965 \text{ g gold}$

$50.0 \text{ mL alum.} \times \dfrac{2.70 \text{ g}}{\text{mL alum.}} = 135 \text{ g alum.}$

$\dfrac{965 \text{ g gold}}{(965 + 135) \text{ g alloy}} \times 100\% = \underline{87.7\% \text{ gold}}$

3-91. specific heat = $0.13 \dfrac{J}{g \cdot °C}$ (gold)

$25.0 \text{ g gold} \times \dfrac{1 \text{ mL}}{19.3 \text{ g gold}} = \underline{1.30 \text{ mL}}$

3-94. When a log burns, most of the compounds formed in the combustion are gases and dissipate into the atmosphere. Only some solid residue (ashes) is left. When zinc and sulfur (both solids) combine, the only product is a solid so there is no weight change. When iron burns, however, its only product is a solid. It weighs more than the original iron because the iron has combined with the oxygen gas from the air.

CHAPTER 4

4-1. An active metal reacts with water and air. A noble metal is not affected by air, water, or most acids.

4-2. 32

4-3. (c) I_2, and **(g)** Br_2

4-5. (a) Fe, and **(d)** La—transition elements **(b)** Te, **(f)** H, and **(g)** In—representative elements **(e)** Xe—noble gas **(c)** Pm—inner transition element

4-7. (b) Ti, **(e)** Pd, and **(g)** Ag

4-8. The most common physical state is a solid, and metals are more common than nonmetals.

4-10. (a) Ne, **(d)** Cl, and **(f)** N

4-12. (a) Ru, **(b)** Sn, **(c)** Hf, and **(h)** W

4-13. (d) Te and **(f)** B

4-14. (a) Ar **(b)** Hg **(c)** N_2 **(d)** Be **(e)** Po

4-16. Element 118 is in Group VIIIA. It should be a noble gas. In fact, it should be the last nonmetal on the periodic table.

4-18. (a) lithium fluoride **(b)** barium telluride **(c)** strontium nitride **(d)** barium hydride **(e)** aluminum chloride

4-20. (a) Rb_2Se **(b)** SrH_2 **(c)** RaO **(d)** Al_2Te_3 **(e)** BeF_2

4-22. (a) bismuth (V) oxide **(b)** tin (II) sulfide **(c)** tin (IV) sulfide **(d)** copper(I) telluride **(e)** titanium(IV) oxide

4-24. (a) Cu_2S **(b)** V_2O_3 **(c)** AuBr **(d)** Ni_3P_2 **(e)** CrO_3

4-26. In 4-22: **(a)** Bi_2O_5, **(c)** SnS_2, and **(e)** TiO_2; In 4-24: **(e)** CrO_3

4-28. (c) ClO_3^-

4-30. ammonium, NH_4^+

4-31. (b) permanganate (MnO_4^-), **(c)** perchlorate (ClO_4^-), **(e)** phosphate (PO_4^{3-}), and **(f)** oxalate ($C_2O_4^{2-}$)

4-32. (a) chromium(II) sulfate **(b)** aluminum sulfite **(c)** iron(II) cyanide **(d)** rubidium hydrogen carbonate **(e)** ammonium carbonate **(f)** ammonium nitrate **(g)** bismuth(III) hydroxide

4-34. (a) $Mg(MnO_4)_2$ **(b)** $Co(CN)_2$ **(c)** $Sr(OH)_2$ **(d)** Tl_2SO_3 **(e)** $Fe_2(C_2O_4)_3$ **(f)** $(NH_4)_2Cr_2O_7$ **(g)** $Hg_2(C_2H_3O_2)_2$

4-36.

	HSO_3^-	Te^{2-}	PO_4^{3-}
NH_4^+	NH_4HSO_3 ammonium bisulfite	$(NH_4)_2Te$ ammonium telluride	$(NH_4)_3PO_4$ ammonium phosphate
Co^{2+}	$Co(HSO_3)_2$ cobalt(ll) bisulfite	$CoTe$ cobalt(ll) telluride	$Co_3(PO_4)_2$ cobalt(ll) phosphate
Al^{3+}	$Al(HSO_3)_3$ aluminum bisulfite	Al_2Te_3 aluminum telluride	$AlPO_4$ *aluminum phosphate*

4-38. (a) sodium chloride   (b) sodium hydrogen carbonate   (c) calcium carbonate   (d) sodium hydroxide   (e) sodium nitrate   (f) ammonium chloride   (g) aluminum oxide   (h) calcium hydroxide   (i) potassium hydroxide

4-39. (a) Ca_2XeO_6 (b) K_4XeO_6 (c) $Al_4(XeO_6)_3$

4-40. (a) phosphonium fluoride   (b) potassium hypobromite   (c) cobalt(III) iodate   (d) calcium silicate (actual name is calcium metasilicate)   (e) aluminum phosphite   (f) chromium(II) molybdate

4-41. (a) Si   (b) I   (c) H   (d) Kr   (e) H   (f) As

4-43. (a) carbon disulfide   (b) boron trifluoride   (c) tetraphosphorus decoxide   (d) dibromine trioxide   (e) methane   (f) dichlorine oxide or dichlorine monoxide   (g) phosphorus pentachloride   (h) sulfur hexafluoride

4-45. (a) P_4O_6 (b) CCl_4 (c) IF_3 (d) C_6H_{14} (e) SF_6 (f) XeO_2

4-47. (a) hydrochloric acid   (b) nitric acid   (c) hypochlorous acid   (d) permanganic acid   (e) periodic acid   (f) hydrobromic acid

4-48. (a) HCN   (b) H_2Se (c) $HClO_2$ (d) H_2CO_3 (e) HI   (f) $HC_2H_3O_2$

4-50. (a) hypobromous acid   (b) iodic acid   (c) phosphorous acid   (d) molybdic acid   (e) perxenic acid

4-52. ClO_2 chlorine dioxide

4-53. A = F   X = Br BrF_5 (Br is more metallic.) bromine pentafluoride

4-55. Gas = N_2; Al forms only +3. Thus the formula is AlN (N^{3-}), for aluminum nitride. Ti_3N_2, titanium(II) nitride

4-57. Co^{2+} and Br^-: $CoBr_2$ cobalt(II) bromide

4-58. Metal = Mg, nonmetal = S: MgH_2, magnesium hydride; H_2S, hydrogen sulfide or hydrosulfuric acid

4-60. NiI_2, nickel(II) iodide; H_3PO_4, phosphoric acid; $Sr(ClO_3)_2$, strontium chlorate; H_2Te, hydrogen telluride or hydrotelluric acid; As_2O_3, diarsenic trioxide; Sb_2O_3, antimony(III) oxide; SnC_2O_4, tin(II) oxalate

4-62. tin(II) hypochlorite, $Sn(ClO)_2$; chromic acid, H_2CrO_4; xenon hexafluoride, XeF_6; barium nitride, Ba_3N_2; hydrofluoric acid, HF; iron(III) telluride, Fe_2Te_3; lithium phosphate, Li_3PO_4

4-63. (e) $Rb_2C_2O_4$

4-65. Rb_2O_2, MgO_2, $Al_2(O_2)_3$, $Ti(O_2)_2$. H_2O_2, hydrogen peroxide, hydroperoxic acid

4-67. (a) Na_2CO_3 (b) $CaCl_2$ (c) $KClO_4$ (d) $Al(NO_3)_3$ (e) $Ca(OH)_2$ (f) NH_4Cl

4-69. N^{3-}, nitride; NO_2^-, nitrite; NO_3^-, nitrate; NH_4^+, ammonium; CN^-, cyanide

4-71. (e) chromium(III) carbonate

4-73. (c) barium chlorite

CHAPTER 5

5-1. 6.09 lb of pennies

5-3. $145 \; \cancel{g \, Au} \times \dfrac{108 \; g \, Ag}{197.0 \; \cancel{g \, Au}} = \underline{79.5 \; g \, Ag}$

5-4. 94.4 lb C

5-6. 71.5 g Cu

5-8. $25.0 \; \cancel{g \, C} \times \dfrac{x \; g}{12.01 \; \cancel{g \, C}} = 33.3 \; g \quad x = 16.0 \; g(O)$

The compound is CO.

5-10. 40.1 lb S

5-11. $\dfrac{9.548 \times 10^{15}}{6.8 \times 10^9} = \underline{1.4 \times 10^6 \; \text{years} \; (1.4 \; \text{million})}$

5-13. 6.022×10^{26} (if mass in kg); 6.022×10^{20} (if mass in mg)

5-14. (a) 0.468 mol P, 2.82×10^{23} atoms P

(b) 150 g Rb, 1.05×10^{24} atoms

(c) Al: 27.0 g, 1.00 mol,

(d) 5.00 mol X element is Ge

(e) 1.66×10^{-24} mol, 7.95×10^{-23} g

5-16. (a) 63.5 g Cu   (b) 16 g S   (c) 40.1 g Ca

5-18. (a) 1.93×10^{25} atoms   (b) 6.03×10^{23} atoms   (c) 1.20×10^{24} atoms

5-20. $50.0 \; \cancel{g \, Al} \times \dfrac{1 \; mol \, Al}{26.98 \; \cancel{g \, Al}} = 1.85 \; mol \, Al$

$= 0.895 \; mol \, Fe$

There are more moles of atoms (more atoms) in 50.0 g of Al.

5-21. $20.0 \; \cancel{g \, Ni} \times \dfrac{1 \; mol \, Ni}{58.69 \; \cancel{g \, Ni}} = 0.341 \; mol \, Ni$

$2.85 \times 10^{23} \; \cancel{atoms} \times \dfrac{1 \; mol \, Ni}{6.022 \times 10^{23} \; \cancel{atoms}}$

$= 0.473 \; mol \, Ni$

The 2.85×10^{23} atoms of Ni contain more atoms than 20.0 g.

5-23. 1.40×10^{21} atoms $= 2.32 \times 10^{-3}$ mol

$0.251 \; g / (2.32 \times 10^{-3} \; mol) = 108 \; g/mol$ (silver)

5-25. (a) 106.6 amu   (b) 80.07 amu   (c) 108.0 amu   (d) 98.09 amu   (e) 106.0 amu   (f) 60.05 amu   (g) 459.7 amu

5-27. $Cr_2(SO_4)_3$ $(2 \times 52.00) + (3 \times 32.07) + (12 \times 16.00)$

$$= \underline{392.2 \text{ amu}}$$

5-29. **(a)** 189 g H_2O, 6.32×10^{24} molecules

(b) 5.00×10^{-3} mol BF_3, 0.339 g BF_3

(c) 0.219 mol SO_2, 1.32×10^{23} molecules

(d) 0.0209 g K_2SO_4, 7.23×10^{19} formula units

(e) 7.47 mol SO_3, 598 g SO_3

(f) 7.61×10^{-3} mol, 4.58×10^{21} molecules

5-31. 21.5 g/0.0684 mol = $\underline{314 \text{ g/mol}}$

5-33. 161 g/mol

5-35. 5.10 mol C, 15.3 mol H, 2.55 mol O: Total = 23.0 mol of atoms; 61.3 g C, 15.4 g H, 40.8 g O: Total mass = $\underline{117.5 \text{ g}}$

5-36. 0.135 mol $Ca(ClO_3)_2$, 0.135 mol Ca, 0.270 mol Cl, 0.810 mol O Total = $\underline{1.215 \text{ mol of atoms}}$

5-38. $1.50 \text{ mol } H_2SO_3 \times \dfrac{2 \text{ mol H}}{\text{mol } H_2SO_4} \times \dfrac{1.008 \text{ g H}}{\text{mol H}}$

$$= \underline{3.02 \text{ g H}}$$

48.1 g S, 72.0 g O

5-40. $1.20 \times 10^{22} \text{ molecules} \times \dfrac{1 \text{ mol } O_2}{6.022 \times 10^{23} \text{ molecules}}$

$$= \underline{0.0199 \text{ mol } O_2}$$

$0.0199 \text{ mol } O_2 \times \dfrac{2 \text{ mol O atoms}}{\text{mol } O_2}$

$$= \underline{0.0398 \text{ mol O atoms}}$$

$0.0199 \text{ mol } O_2 \times \dfrac{32.00 \text{ g } O_2}{\text{mol } O_2}$

$$= \underline{0.637 \text{ g } O_2} \text{ The mass is the same.}$$

5-42. Total mass of compound = 1.375 + 3.935 = 5.310 g, 25.89% N, 74.11% O

5-43. 46.7% Si, 53.3% O

5-45. **(a)** C_2H_6O 52.14% C, 13.13% H, 34.73% O

(b) C_3H_6 85.62% C, 14.38% H

(c) C_9H_{18} 85.66% C, 14.34% H

(b) and **(c)** are actually the same. The difference comes from rounding off.

(d) Na_2SO_4 32.36% Na, 22.57% S, 45.07% O

(e) $(NH_4)_2CO_3$ 29.16% N, 8.392% H, 12.50% C, 49.95% O

5-47. 12.06% Na, 11.34% B, 71.31% O, 5.286% H

5-49. $C_7H_5SNO_3$. Formula weight = $(7 \times 12.01) + (5 \times 1.008) + 32.07 + 14.01 + (3 \times 16.00) = 183.2$ amu

C: $\dfrac{84.07 \text{ amu}}{183.2 \text{ amu}} \times 100\% = \underline{45.89\% \text{ C}}$

H: $\dfrac{5.040 \text{ amu}}{183.2 \text{ amu}} \times 100\% = \underline{2.751\% \text{ H}}$

S: $\dfrac{32.07 \text{ amu}}{183.2 \text{ amu}} \times 100\% = \underline{17.51\% \text{ S}}$

N: $\dfrac{14.01 \text{ amu}}{183.2 \text{ amu}} \times 100\% = \underline{7.647\% \text{ C}}$

O: $100\% - (45.89 + 2.751 + 17.51 + 7.647)$

$$= \underline{26.20\% \text{ O}}$$

5-51. $Na_2C_2O_4$ Formula weight = $(2 \times 22.99) + (2 \times 12.01) + (4 \times 16.00) = 134.0$ amu

There is 24.02 g (2×12.01) of C in 134.0 g of compound.

$125 \text{ g } Na_2C_2O_4 \times \dfrac{24.02 \text{ g C}}{134.0 \text{ g } Na_2C_2O_4} = \underline{22.4 \text{ g C}}$

5-52. 4.72 lb P

5-54. 1.40×10^3 lb Fe

5-55. **(a)** N_2O_4 and **(d)** $H_2C_2O_4$

5-56. **(a)** FeS   **(b)** SrI_2 **(c)** $KClO_3$ **(d)** I_2O_5

(e) $Fe_2O_{2.66} = Fe_3O_4$ **(f)** $C_3H_5Cl_3$

5-58. N_2O_3

5-60. KO_2

5-62. MgC_2O_4

5-63. CH_2Cl

5-65. $N_2H_8SO_3$

5-66. $C_{8/3}H_{8/3}O = C_8H_8O_3$

5-68. $C_3H_4Cl_4$ (empirical formula) $C_9H_{12}Cl_{12}$ (molecular formula)

5-70. $B_2C_2H_6O_4$ (molecular formula)

5-71. Empirical formula = $KC_2NH_3O_2$ Empirical mass = 112.2 g/emp. unit

$\dfrac{224 \text{ g/mol}}{112.2 \text{ g/emp. unit}} = 2$ emp. units/mol

$$K_2C_4N_2H_6O_4 \text{ (molecular formula)}$$

5-73. I_6C_6

5-75. 7.5×10^{-10} mol pennies

5-76. $0.443 \text{ g N} \times \dfrac{1 \text{ mol N}}{14.01 \text{ g N}} = 0.0316$ mol N

Thus 1.420 g of M also equals 0.0316 mol M since M and N are present in equimolar amounts.

1.420 g/0.0316 mol = $\underline{44.9 \text{ g/mol [scandium (Sc)]}}$

5-79. 2.78×10^{-3} mol P_4

$2.78 \times 10^{-3} \text{ mol } P_4 \times \dfrac{4 \text{ mol P}}{\text{mol } P_4}$

$\times \dfrac{6.022 \times 10^{23} \text{ atoms P}}{\text{mol P}} = 6.70 \times 10^{21}$ atoms P

5-80. 100 mol H_2 = 202 g H_2 therefore

100 H atoms < 100 H_2 molecules < 100 g H_2

$$< 100 \text{ mol } H_2$$

5-82. 120 g/mol of compound $120 - 55.8 = 64$ g of S

$\dfrac{64 \text{ g S}}{32.07 \text{ g S /mol}} = 2$ mol S

Formula = FeS_2

5-83. (a) $2Na^+$ and $S_4O_6^{2-}$ **(b)** 27.9 g S   **(c)** NaS_2O_3
(d) 270.3 g/mol   **(e)** 0.0925 mol $Na_2S_4O_6$, 5.57×10^{22}
formula units   **(f)** 35.5% oxygen

5-85. $\dfrac{2N}{2N + xO} = 0.368$ $\dfrac{28.02}{28.05 + 16.00x} = 0.368$

$x = 3$ N_2O_3 dinitrogen trioxide

5-87. $C_{12}H_4Cl_4O_2$ (molecular formula)

5-89. Empirical formula $CrCl_3O_{12}$ Actual formula =
$Cr(ClO_4)_3$ chromium(III) perchlorate

5-90. Assume exactly 100 g of compound. There is then
51.1 g H_2O and 48.9 g $MgSO_4$.

$MgSO_4$:48.9 g $MgSO_4 \times \dfrac{1\ mol}{120.4\ g\ MgSO_4}$

$= 0.406$ mol $MgSO_4$

2.94 mol H_2O/0.406 mol $MgSO_4$

$= 7.0$ mol H_2O/mol $MgSO_4$

The formula is $MgSO_4 \cdot 7H_2O$

5-93. $1.20\ \overline{g\ CO_2} \times \dfrac{1\ \overline{mol\ CO_2}}{44.01\ \overline{g\ CO_2}} \times \dfrac{1\ mol\ C}{\overline{mol\ CO_2}}$

$= 0.0273$ mol C

$0.489\ \overline{g\ H_2O} \times \dfrac{1\ \overline{mol\ H_2O}}{18.02\ \overline{g\ H_2O}} \times \dfrac{2\ mol\ H}{\overline{mol\ H_2O}}$

$= 0.0543$ mol H

C: $\dfrac{0.0273}{0.0273} = 1.0$ H: $\dfrac{0.0543}{0.0273} = 2.0$ CH_2

CHAPTER 6

6-1. (a) $Cl_2(g)$ **(b)** $C(s)$ **(c)** $K_2SO_4(s)$ **(d)** $H_2O(l)$
(e) $P_4(s)$ **(f)** $H_2(g)$ **(g)** $Br_2(l)$ **(h)** $NaBr(s)$
(i) $S_8(s)$ **(j)** $Na(s)$ **(k)** $Hg(l)$ **(l)** $CO_2(g)$

6-2. (a) $CaCO_3 \longrightarrow CaO + CO_2$

(b) $4\,Na + O_2 \longrightarrow 2Na_2O$

(c) $H_2SO_4 + 2NaOH \longrightarrow Na_2SO_4 + 2H_2O$

(d) $2H_2O_2 \longrightarrow 2H_2O + O_2$

6-4. (a) $2Al + 2H_3PO_4 \longrightarrow 2AlPO_4 + 3H_2$

(b) $Ca(OH)_2 + 2HCl \longrightarrow CaCl_2 + 2H_2O$

(c) $3Mg + N_2 \longrightarrow Mg_3N_2$

(d) $2C_2H_6 + 7O_2 \longrightarrow 4CO_2 + 6H_2O$

6-6. (a) $Mg_3N_2 + 6H_2O \longrightarrow 3Mg(OH)_2 + 2NH_3$

(b) $2H_2S + O_2 \longrightarrow 2S + 2H_2O$

(c) $Si_2H_6 + 8H_2O \longrightarrow 2Si(OH)_4 + 7H_2$

(d) $C_2H_6 + 5Cl_2 \longrightarrow C_2HCl_5 + 5HCl$

6-8. (a) $2B_4H_{10} + 11O_2 \longrightarrow 4B_2O_3 + 10H_2O$

(b) $SF_6 + 2SO_3 \longrightarrow 3O_2SF_2$

(c) $CS_2 + 3O_2 \longrightarrow CO_2 + 2SO_2$

(d) $2BF_3 + 6NaH \longrightarrow B_2H_6 + 6NaF$

6-10. (a) $2Na(s) + 2H_2O(l) \longrightarrow H_2(g) + 2NaOH(aq)$

(b) $2KClO_3(s) \longrightarrow 2KCl(s) + 3O_2(g)$

(c) $NaCl(aq) + AgNO_3(aq) \longrightarrow AgCl(s) + NaNO_3(aq)$

(d) $2H_3PO_4(aq) + 3Ca(OH)_2(aq) \longrightarrow$
$Ca_3(PO_4)_2(s) + 6H_2O(l)$

6-12. $Ni(s) + 2N_2O_4(l) \longrightarrow Ni(NO_3)_2(s) + 2NO(g)$

6-14. In exercise 6-2, **(a)** and **(d)** are decomposition reactions and **(b)** is a combination and combustion reaction. In exercise 6-4, **(c)** is a combination reaction and **(d)** is a combustion reaction.

6-16. (a) $2C_7H_{14}(l) + 21O_2(g) \longrightarrow 14CO_2(g) + 14H_2O(l)$

(b) $2LiCH_3(s) + 4O_2(g) \longrightarrow$
$Li_2O(s) + 2CO_2(g) + 3H_2O(l)$

(c) $C_4H_{10}O(l) + 6O_2(g) \longrightarrow 4CO_2(g) + 5H_2O(l)$

(d) $2C_2H_5SH(g) + 9O_2(g) \longrightarrow$
$2SO_2(g) + 4CO_2(g) + 6H_2O(l)$

6-18. (a) $Ba(s) + H_2(g) \longrightarrow BaH_2(s)$

(b) $8Ba(s) + S_8(s) \longrightarrow 8BaS(s)$

(c) $Ba(s) + Br_2(l) \longrightarrow BaBr_2(s)$

(d) $3Ba(s) + N_2(g) \longrightarrow Ba_3N_2(s)$

6-20. (a) $Ca(HCO_3)_2(s) \longrightarrow CaO(s) + 2CO_2(g) + H_2O(l)$

(b) $2Ag_2O(s) \longrightarrow 4Ag(s) + O_2(g)$

(c) $N_2O_3(g) \longrightarrow NO_2(g) + NO(g)$

6-22. (a) $2K(s) + Cl_2(g) \longrightarrow 2KCl(s)$

(b) $2C_6H_6(l) + 15O_2(g) \longrightarrow 12CO_2(g) + 6H_2O(l)$

(c) $2Au_2O_3(s) \longrightarrow 4Au(s) + 3O_2(g)$

(d) $2C_3H_8O(l) + 9O_2(g) \longrightarrow 6CO_2(g) + 8H_2O(l)$

(e) $P_4(s) + 10F_2(g) \longrightarrow 4PF_5(s)$

6-24. (a) $Na_2S \longrightarrow 2Na^+(aq) + S^{2-}(aq)$

(b) $Li_2SO_4 \longrightarrow 2Li^+(aq) + SO_4^{2-}(aq)$

(c) $K_2Cr_2O_7 \longrightarrow 2K^+(aq) + Cr_2O_7^{2-}(aq)$

(d) $CaS \longrightarrow Ca^{2+}(aq) + S^{2-}(aq)$

(e) $(NH_4)_2S \longrightarrow 2NH_4^+(aq) + S^{2-}(aq)$

(f) $Ba(OH)_2 \longrightarrow Ba^{2+}(aq) + 2OH^-(aq)$

6-26. (a) $HNO_3(aq) \longrightarrow H^+(aq) + NO_3^-(aq)$

(b) $Sr(OH)_2(s) \longrightarrow Sr^{2+}(aq) + 2OH^-(aq)$

6-28. (a) no reaction

(b) $Fe + 2H^+ \longrightarrow Fe^{2+} + H_2$

(c) $Cu + 2Ag^+ \longrightarrow Cu^{2+} + 2Ag$

(d) no reaction

6-30. (a) $CuCl_2(aq) + Fe(s) \longrightarrow FeCl_2(aq) + Cu(s)$

$Cu^{2+}(aq) + 2Cl^-(aq) + Fe(s) \longrightarrow$
$Fe^{2+}(aq) + 2Cl^-(aq) + Cu(s)$

$Cu^{2+}(aq) + Fe(s) \longrightarrow Fe^{2+}(aq) + Cu(s)$

(b) and **(c)** no reaction

(d) $3Zn(s) + 2Cr(NO_3)_3(aq) \longrightarrow$
$$3Zn(NO_3)_2(aq) + 2Cr(s)$$
$3Zn(s) + 2Cr^{3+}(aq) + 6NO_3^-(aq) \longrightarrow$
$$3Zn^{2+}(aq) + 6NO_3^-(aq) + 2Cr(s)$$
$3Zn(s) + 2Cr^{3+}(aq) \longrightarrow 3Zn^{2+}(aq) + 2Cr(s)$

6-32. $6Na(l) + Cr_2O_3(s) \longrightarrow 2Cr(s) + 3Na_2O(s)$
$3Na + Cr^{3+} \longrightarrow Cr + 3Na^+$

6-34. Insoluble compounds are **(b)** $PbSO_4$, and **(d)** Ag_2S

6-36. (a) $AgBr$ **(b)** Ag_2CO_3 **(c)** Ag_3PO_4

6-38. (a) $CuCO_3$ **(b)** $CdCO_3$ **(c)** $Cr_2(CO_3)_3$

6-40. Hg_2Cl_2

6-42. (b) $Ca_3(PO_4)_2$

6-44. (a) $2KI(aq) + Pb(C_2H_3O_2)_2(aq) \longrightarrow$
$$PbI_2(s) + 2KC_2H_3O_2(aq)$$

(b) and **(c)** no reaction occurs

(d) $BaS(aq) + Hg_2(NO_3)_2(aq) \longrightarrow$
$$Hg_2S(s) + Ba(NO_3)_2(aq)$$

(e) $FeCl_3(aq) + 3KOH(aq) \longrightarrow$
$$Fe(OH)_3(s) + 3KCl(aq)$$

6-46. (a) $2K^+(aq) + 2I^-(aq) + Pb^{2+}(aq) + 2C_2H_3O_2^-(aq)$
$\longrightarrow PbI_2(s) + 2K^+(aq) + 2C_2H_3O_2^-(aq)$
$Pb^{2+}(aq) + 2I^-(aq) \longrightarrow PbI_2(s)$

(d) $Ba^{2+}(aq) + S^{2-}(aq) + Hg_2^{2+}(aq) + 2NO_3^-(aq) \longrightarrow$
$$Hg_2S(s) + Ba^{2+}(aq) + 2NO_3^-(aq)$$
$Hg_2^{2+}(aq) + S^{2-}(aq) \longrightarrow Hg_2S(s)$

(e) $Fe^{3+}(aq) + 3Cl^-(aq) + 3K^+(aq) + 3OH^-(aq)$
$\longrightarrow Fe(OH)_3(s) + 3K^+(aq) + 3Cl^-(aq)$
$Fe^{3+}(aq) + 3OH^-(aq) \longrightarrow Fe(OH)_3(s)$

6-48. (a) $2K^+(aq) + S^{2-}(aq) + Pb^{2+}(aq) + 2NO_3^-(aq)$
$\longrightarrow PbS(s) + 2K^+(aq) + 2NO_3^-(aq)$
$S^{2-}(aq) + Pb^{2+}(aq) \longrightarrow PbS(s)$

(b) $2NH_4^+(aq) + CO_3^{2-}(aq) + Ca^{2+}(aq) + 2Cl^-(aq)$
$\longrightarrow CaCO_3(s) + 2NH_4^+(aq) + 2Cl^-(aq)$
$CO_3^{2-}(aq) + Ca^{2+}(aq) \longrightarrow CaCO_3(s)$

(c) $2Ag^+(aq) + 2ClO_4^-(aq) + 2Na^+(aq) + CrO_4^{2-}(aq)$
$\longrightarrow Ag_2CrO_4(s) + 2Na^+(aq) + 2ClO_4^-(aq)$
$2Ag^+(aq) + CrO_4^{2-}(aq) \longrightarrow Ag_2CrO_4(s)$

6-50. (a) $CuCl_2(aq) + Na_2CO_3(aq) \longrightarrow$
$$CuCO_3(s) + 2NaCl(aq)$$
Filter the solid $CuCO_3$.

(b) $(NH_4)_2SO_4(aq) + Pb(NO_3)_2(aq) \longrightarrow$
$$PbSO_4(s) + 2NH_4NO_3(aq)$$
Filter the solid $PbSO_4$.

(c) $2KI(aq) + Hg_2(NO_3)_2(aq) \longrightarrow$
$$Hg_2I_2(s) + 2KNO_3(aq)$$
Filter the solid Hg_2I_2.

(d) $NH_4Cl(aq) + AgNO_3(aq) \longrightarrow$
$$AgCl(s) + NH_4NO_3(aq)$$
Filter the solid AgCl; the desired product remains after water is removed by boiling.

(e) $Ca(C_2H_3O_2)_2(aq) + K_2CO_3(aq) \longrightarrow$
$$CaCO_3(s) + 2KC_2H_3O_2(aq)$$
Filter the solid $CaCO_3$; the desired product remains after the water is removed by boiling.

6-51. (b) HF

6-53. (c) $Al(OH)_3$

6-55. (a) $HI(aq) + CsOH(aq) \longrightarrow CsI(aq) + H_2O(l)$

(b) $2HNO_3(aq) + Ca(OH)_2(aq) \longrightarrow$
$$Ca(NO_3)_2(aq) + 2H_2O(l)$$

(c) $H_2SO_4(aq) + Sr(OH)_2(aq) \longrightarrow$
$$SrSO_4(s) + 2H_2O(l)$$

(d) $HNO_3(aq) + NaHCO_3(s) \longrightarrow$
$$NaNO_3(aq) + CO_2(g) + H_2O(l)$$

6-57. (a) $H^+(aq) + I^-(aq) + Cs^+(aq) + OH^-(aq) \longrightarrow$
$$Cs^+(aq) + I^-(aq) + H_2O(l)$$
$H^+(aq) + OH^-(aq) \longrightarrow H_2O(l)$

(b) $2H^+(aq) + 2NO_3^-(aq) + Ca^{2+}(aq) + 2OH^-(aq) \longrightarrow$
$$Ca^{2+}(aq) + 2NO_3^-(aq) + 2H_2O(l)$$
$H^+(aq) + OH^-(aq) \longrightarrow H_2O(l)$

(c) $2H^+(aq) + SO_4^{2-}(aq) + Sr^{2+}(aq) + 2OH^-(aq) \longrightarrow$
$$SrSO_4(s) + 2H_2O(l)$$
$H^+(aq) + OH^-(aq) \longrightarrow H_2O(l)$

(d) $H^+(aq) + NO_3^-(aq) + Na^+(aq) + HCO_3^-(s) \longrightarrow$
$$Na^+(aq) + NO_3^-(aq) + CO_2(g) + H_2O(l)$$
$H^+(aq) + HCO_3^-(s) \longrightarrow CO_2(g) + H_2O(l)$

Net ionic equation is the same as the total ionic equation since $SrSO_4$ precipitates.

6-59. $Mg(OH)_2(s) + 2HCl(aq) \longrightarrow MgCl_2(aq) + 2H_2O(l)$
$Mg(OH)_2(s) + 2H^+(aq) + 2Cl^-(aq) \longrightarrow$
$$Mg^{2+}(aq) + 2Cl^-(aq) + 2H_2O(l)$$
$Mg(OH)_2(s) + 2H^+(aq) \longrightarrow Mg^{2+}(aq) + 2H_2O(l)$

6-61. $C_3H_8(g) + 5O_2(g) \longrightarrow 3CO_2(g) + 4H_2O(l)$
$2C_4H_{10}(g) + 13O_2(g) \longrightarrow 8CO_2(g) + 10H_2O(l)$
$2C_8H_{18}(l) + 25O_2(g) \longrightarrow 16CO_2(g) + 18H_2O(l)$
$C_2H_5OH(l) + 3O_2(g) \longrightarrow 2CO_2(g) + 3H_2O(l)$

6-62. In both of these reactions, the reactants change from neutral atoms to cations and anions in the products. Ions are formed from neutral atoms from the loss or gain of electrons.

6-64. (a) $Ba(s) + I_2(s) \longrightarrow BaI_2(s)$

(b) $HBr(aq) + RbOH(aq) \longrightarrow RbBr(aq) + H_2O(l)$

(c) $Ca(s) + 2HNO_3(aq) \longrightarrow Ca(NO_3)_2(aq) + H_2(g)$

(d) $C_{10}H_8(s) + 12O_2(g) \longrightarrow 10CO_2(g) + 4H_2O(l)$

(e) $(NH_4)_2CrO_4(aq) + BaBr_2(aq) \longrightarrow$
$$BaCrO_4(s) + 2NH_4Br(aq)$$

(f) $2Al(OH)_3(s) \longrightarrow Al_2O_3(s) + 3H_2O(g)$

6-65. (b) $H^+(aq) + Br^-(aq) + Rb^+(aq) + OH^-(aq) \longrightarrow$
$$Rb^+(aq) + Br^-(aq) + H_2O(l)$$
$$H^+(aq) + OH^-(aq) \longrightarrow H_2O(l)$$
(c) $Ca(s) + 2H^+(aq) + 2NO_3^-(aq) \longrightarrow$
$$Ca^{2+}(aq) + 2NO_3^-(aq) + H_2(g)$$
$$Ca(s) + 2H^+(aq) \longrightarrow Ca^{2+}(aq) + H_2(g)$$
(e) $2NH_4^+(aq) + CrO_4^{2-}(aq) + Ba^{2+}(aq) + 2Br^-(aq)$
$$\longrightarrow BaCrO_4(s) + 2NH_4^+ + 2Br^-(aq)$$
$$CrO_4^{2-}(aq) + Ba^{2+}(aq) \longrightarrow BaCrO_4(s)$$

6-68. $H^+(aq) + OH^-(aq) \longrightarrow H_2O(l)$
$$Ba^{2+}(aq) + SO_4^{2-}(aq) \longrightarrow BaSO_4(s)$$

6-70. $Fe^{3+}(aq) + PO_4^{3-}(aq) \longrightarrow FePO_4(s)$
$$2Fe^{3+}(aq) + 3S^{2-}(aq) \longrightarrow Fe_2S_3(s)$$
$$Pb^{2+}(aq) + 2I^-(aq) \longrightarrow PbI_2(s)$$
$$3Pb^{2+}(aq) + 2PO_4^{3-}(aq) \longrightarrow Pb_3(PO_4)_2(s)$$
$$Pb^{2+}(aq) + S^{2-}(aq) \longrightarrow PbS(s)$$

CHAPTER 7

7-1. (a) $\dfrac{1 \text{ mol } H_2}{1 \text{ mol } Mg}$ **(b)** $\dfrac{2 \text{ mol } HCl}{1 \text{ mol } Mg}$ **(c)** $\dfrac{1 \text{ mol } H_2}{2 \text{ mol } HCl}$

(d) $\dfrac{2 \text{ mol } HCl}{1 \text{ mol } MgCl_2}$

7-2. (a) $\dfrac{2 \text{ mol } C_4H_{10}}{8 \text{ mol } CO_2}$ **(b)** $\dfrac{2 \text{ mol } C_4H_{10}}{13 \text{ mol } O_2}$

(c) $\dfrac{13 \text{ mol } O_2}{8 \text{ mol } CO_2}$ **(d)** $\dfrac{10 \text{ mol } H_2O}{13 \text{ mol } O_2}$

7-4. (a) 3.33 mol Al_2O_3, 3.33 mol $AlCl_3$, 10.0 mol NO, 20.0 mol H_2O **(b)** 1.00 mol Al_2O_3, 1.00 mol $AlCl_3$, 3.00 mol NO, 6.00 mol H_2O

7-6. (a) 15.0 mol O_2 and 10.0 mol CH_4 react
(b) 6.67 mol HCN and 20.0 mol H_2O produced

7-8. (a) 126 g SiF_4 and **(b)** 43.8 g H_2O produced
(c) 73.0 g SiO_2 reacts

7-9. (a) mol H_2O $\longrightarrow$ mol H_2
$$0.400 \text{ mol } H_2O \times \frac{2 \text{ mol } H_2}{2 \text{ mol } H_2O} = \underline{\underline{0.400 \text{ mol } H2}}$$

(b) g O_2 $\longrightarrow$ mol O_2 $\longrightarrow$ mol H_2O $\longrightarrow$ g H_2O
$$0.640 \text{ g } O_2 \times \frac{1 \text{ mol } O_2}{32.00 \text{ g } O_2} \times \frac{2 \text{ mol } H_2O}{1 \text{ mol } O_2} \times \frac{18.02 \text{ g } H_2O}{\text{mol } H_2O}$$
$$= \underline{\underline{0.721 \text{ g } H_2O}}$$

(c) g O_2 $\longrightarrow$ mol O_2 $\longrightarrow$ mol H_2 $\longrightarrow$ g H_2
$$0.032 \text{ g } O_2 \times \frac{1 \text{ mol } O_2}{32.00 \text{ g } O_2} \times \frac{2 \text{ mol } H_2}{1 \text{ mol } O_2} \times \frac{2.016 \text{ g } H_2}{\text{mol } H_2}$$
$$= \underline{\underline{0.0040 \text{ g } H_2}}$$

(d) g H_2 $\longrightarrow$ mol H_2 $\longrightarrow$ mol H_2O $\longrightarrow$ g H_2O
$$0.400 \text{ g } H_2 \times \frac{1 \text{ mol } H_2}{2.016 \text{ g } H_2} \times \frac{2 \text{ mol } H_2O}{2 \text{ mol } H_2} \times \frac{18.02 \text{ g } H_2O}{\text{mol } H_2O}$$
$$= \underline{\underline{3.58 \text{ g } H_2O}}$$

7-10. (a) 1.35 mol CO_2, 1.80 mol H_2O, 2.25 mol O_2
(b) 4.81 g H_2O **(c)** 1.10 g C_3H_8 **(d)** 44.1 g C_3H_8
(e) 5.26 g CO_2 **(f)** 0.0996 mol H_2O

7-12. 47.2 g N_2

7-14. 0.728 g HCl

7-16. mol FeS_2 $\longrightarrow$ mol H_2S $\longrightarrow$ molecules H_2S
$$0.520 \text{ mol } FeS_2 \times \frac{1 \text{ mol } H_2S}{1 \text{ mol } FeS_2}$$
$$\times \frac{6.022 \times 10^{23} \text{ molecules}}{\text{mol } H_2S} = \underline{\underline{3.13 \times 10^{23} \text{ molecules}}}$$

7-17. 16,900 g (16.9 kg) HNO_3

7-19. 2.30×10^3 g (2.30 kg) C_2H_5OH

7-20. 8140 g (8.14 kg) CO

7-21. (a) CuO limiting reactant producing 1.00 mol N_2
(b) stoichiometric mixture producing 1.00 mol N_2
(c) NH_3 limiting reactant producing 0.500 mol N_2
(d) CuO limiting reactant producing 0.209 mol N_2
(e) NH_3 limiting reactant producing 1.75 mol N_2

7-22. (a) $3.00 \text{ mol } CuO \times \dfrac{2 \text{ mol } NH_3}{3 \text{ mol } CuO}$
$$= 2.00 \text{ mol } NH_3 \text{ used}$$
$$3.00 - 2.00 = 1.00 \text{ mol } NH_3 \text{ in excess}$$
(c) 1.50 mol CuO in excess

7-25. H_2SO_4 is the limiting reactant and the yield of H_2 is 1.00 mole.
$$1.00 \text{ mol } H_2SO_4 \times \frac{2 \text{ mol } Al}{3 \text{ mol } H_2SO_4}$$
$$= 0.667 \text{ mol of Al used}$$
$$0.800 - 0.667 = \underline{0.133 \text{ mol Al remaining}}$$

7-26. $3.44 \text{ mol } C_5H_6 \times \dfrac{10 \text{ mol } CO_2}{2 \text{ mol } C_5H_6} = 17.2 \text{ mol } CO_2$
$$20.6 \text{ mol } O_2 \times \frac{10 \text{ mol } CO_2}{13 \text{ mol } O_2} = 15.8 \text{ mol } CO_2$$
Since O_2 is the limiting reactant:
$$15.8 \text{ mol } CO_2 \times \frac{44.01 \text{ g } CO_2}{\text{mol } CO_2} = \underline{\underline{695 \text{ g } CO_2}}$$

7-28. Since NH_3 produces the least N_2, it is the limiting reactant and the yield of N_2 is 0.750 mol.

7-29. $20.0 \text{ g } AgNO_3 \times \dfrac{1 \text{ mol } AgNO_3}{169.9 \text{ g } AgNO_3} \times \dfrac{2 \text{ mol } AgCl}{2 \text{ mol } AgNO_3}$
$$= 0.118 \text{ mol } AgCl$$
$$10.0 \text{ g } CaCl_2 \times \frac{1 \text{ mol } CaCl_2}{111.0 \text{ g } CaCl_2} \times \frac{2 \text{ mol } AgCl}{1 \text{ mol } CaCl_2}$$
$$= 0.180 \text{ mol } AgCl$$
Since $AgNO_3$ produces the least AgCl, it is the limiting reactant.
$$0.118 \text{ mol } AgCl \times \frac{143.4 \text{ g } AgCl}{\text{mol } AgCl} = \underline{\underline{16.9 \text{ g } AgCl}}$$
Convert moles of AgCl (the limiting reactant) to grams of $CaCl_2$ used.
$$0.118 \text{ mol } AgCl \times \frac{1 \text{ mol } CaCl_2}{2 \text{ mol } AgCl} \times \frac{111.0 \text{ g } CaCl_2}{\text{mol } CaCl_2}$$
$$= \underline{\underline{6.55 \text{ g } CaCl_2 \text{ used}}}$$
$$10.0 \text{ g} - 6.55 \text{ g} = \underline{3.5 \text{ g } CaCl_2 \text{ remaining}}$$

7-31. Products: 3.53 g H_2O, 4.71 g S, 2.94 g NO
Reactants remaining: remaining: 3.8 g HNO_3 remaining

7-32. 30.0 g SO_3 (theoretical yield) and 70.7% yield

7-35. <u>86.4%</u>

7-36. If 86.4% is converted to CO_2, the remainder
(13.6%) is converted to CO. Thus, 0.136×57.0 g = 7.75 g
of C_8H_{18} is converted to CO. Notice that 1 mole of C_8H_{18}
forms 8 moles of CO (because of the eight carbons in
C_8H_{18}). Thus

$$7.75 \text{ g } C_8H_{18} \times \frac{1 \text{ mol } C_8H_{18}}{114.2 \text{ g } C_8H_{18}} \times \frac{8 \text{ mol } CO}{1 \text{ mol } C_8H_{18}}$$
$$\times \frac{28.01 \text{ g CO}}{\text{mol } CO} = 15.2 \text{ g CO}$$

7-37. Theoretical yield $\times 0.700 = 250$ g (actual yield)
Theoretical yield = 250 g/0.700 = 357 g N_2

g $N_2 \longrightarrow$ mol $N_2 \longrightarrow$ mol $H_2 \longrightarrow$ g H_2
$$357 \text{ g } Na_2 \times \frac{1 \text{ mol } N_2}{28.02 \text{ g } N_2} \times \frac{4 \text{ mol } H_2}{1 \text{ mol } N_2} \times \frac{2.016 \text{ g } H_2}{\text{mol } H_2}$$
$$= 103 \text{ g } H_2$$

7-40. $2Mg(s) + O_2 \longrightarrow 2MgO(s) + 1204 \text{ kJ}$
$2Mg(s) + O_2(g) \longrightarrow 2MgO(s)$ $\Delta H = -1204 \text{ kJ}$

7-42. $CaCO_3(s) + 176 \text{ kJ} \longrightarrow CaO(s) + CO_2(g)$
$CaCO_3(s) \longrightarrow CaO(s) + CO_2(g)$ $\Delta H = 176 \text{ kJ}$

7-43. $1.00 \text{ g } C_8H_{18} \times \frac{1 \text{ mol } C_8H_{18}}{114.2 \text{ g } C_8H_{18}} \times \frac{5840 \text{ kJ}}{\text{mol } C_8H_{18}}$
$$= 48.0 \text{ kJ}$$

$1.00 \text{ g } CH_4 \times \frac{1 \text{ mol } CH_4}{16.04 \text{ g } CH_4} \times \frac{890 \text{ kJ}}{\text{mol } CH_4} = 55.6 \text{ kJ}$

7-45. **kJ $\longrightarrow$ mol Al $\longrightarrow$ g Al**
$$35.8 \text{ kJ} \times \frac{2 \text{ mol Al}}{850 \text{ kJ}} \times \frac{26.98 \text{ g Al}}{\text{mol Al}} = \underline{2.27 \text{ g Al}}$$

7-46. 69.7 g $C_6H_{12}O_6$

7-48. $125 \text{ g } Fe_2O_3 \times \frac{1 \text{ mol } Fe_2O_3}{159.7 \text{ g } Fe_2O_3} \times \frac{2 \text{ mol } Fe_3O_4}{3 \text{ mol } Fe_2O_3}$
$$\times \frac{3 \text{ mol } FeO}{1 \text{ mol } Fe_3O_4} \times \frac{1 \text{ mol } Fe}{1 \text{ mol } FeO} \times \frac{55.85 \text{ g } Fe}{\text{mol } Fe}$$
$$= 87.4 \text{ g } Fe$$

7-49. $2KClO_3 \longrightarrow 2KCl + 3O_2$

Find the mass of $KClO_3$ needed to produce 12.0 g O_2.

g $O_2 \longrightarrow$ mol $O_2 \longrightarrow$ mol $KClO_3 \longrightarrow$ g $KClO_3$
$$12.0 \text{ g } O_2 \times \frac{1 \text{ mol } O_2}{32.00 \text{ g } O_2} \times \frac{2 \text{ mol } KClO_3}{3 \text{ mol } O_2} \times \frac{122.6 \text{ g } KClO_3}{\text{mol } KClO_3}$$
$$= 30.7 \text{ g } KClO_3$$
$$\text{percent purity} = \frac{30.7 \text{ g}}{50.0 \text{ g}} \times 100\% = \underline{61.4\%}$$

7-50. <u>4.49% FeS_2</u>

7-52. **(1)** NH_3 is the limiting reactant.
   (2) 141 g NO (theoretical yield), <u>28.4% yield</u>

7-54. H_2O is the limiting reactant producing 11.0 g
$CaCl_2 \cdot 6H_2O$

7-55. Molecular formula = CH_2Cl_2
$CH_4(g) + 2Cl_2(g) \longrightarrow CH_2Cl_2(l) + 2HCl(g)$
Cl_2 is the limiting reactant producing 9.00 g CH_2Cl_2

CHAPTER 8

8-1. Ultraviolet light has shorter wavelengths but higher
energy than visible light. Because of this high energy,
ultraviolet light can damage living cells in tissues, thus
causing a burn.

8-2. Since these two shells are close in energy, transitions
of electrons from these two levels to the $n = 1$ shell have
similar energy. Thus, the wavelengths of light from the
two transitions are very close together.

8-3. Since these two shells are comparatively far apart in
energy, transitions from these two levels to the $n = 1$ shell
have comparatively different energies. Thus, the wavelengths of light from the two transitions are quite different.
(The $n = 3$ to $n = 1$ transition has a shorter wavelength and
higher energy than the $n = 2$ to the $n = 1$ transition.)

8-5. $1p$ and $3f$

8-8. A $4p$ orbital is shaped roughly like a two-sided baseball bat with two "lobes" lying along one of the three axes.
This shape represents the region of highest probability of
finding the $4p$ electrons.

8-9. The $3s$ orbital is spherical in shape. There is an equal
probability of finding the electron regardless of the
orientation from the nucleus. (In fact, the probability lies
in three concentric spheres with the highest probability
in the sphere farthest from the nucleus.) The highest
probability of finding the electron lies farther from the
nucleus in the $3s$ than in the $2s$.

8-11. One $3s$, three $3p$, and five $3d$, for a total of nine.

8-13. **(a)** $3p$, three   **(b)** $4d$, five   **(c)** $6s$, one

8-15. $2n^2$: $2(3)^2 = 18$ and $2(2)^2 = 8$

8-18. $2(5)^2 = 50$

8-19. The first four subshells in the fifth shell (s, p, d, and f)
hold 32 electrons. The g subshell holds $50 - 32 = 18$
electrons.

8-20. Since each orbital holds two electrons, there are
nine orbitals in this subshell.

8-21. The $1s$ subshell always fills first.

8-24. **(a)** $6s$ **(b)** $5p$ **(c)** $4p$ **(d)** $4d$

8-26. $4s, 4p, 5s, 4d, 5p, 6s, 4f$

8-27. **(a)** Mg: $1s^2 2s^2 2p^6 3s^2$
   (b) Ge: $1s^2 2s^2 2p^6 3s^2 3p^6 4s^2 3d^{10} 4p^2$
   (c) Cd: $1s^2 2s^2 2p^6 3s^2 3p^6 4s^2 3d^{10} 4p^6 5s^2 4d^{10}$
   (d) Si: $1s^2 2s^2 2p^6 3s^2 3p^2$

8-29. $1s^2 2s^2 2p^6 3s^2 3p^6$

8-31. **(a)** S: $[Ne]3s^2 3p^4$ **(b)** Zn: $[Ar]4s^2 3d^{10}$
   (c) Hf: $[Xe]6s^2 5d^2 4f^1$ **(d)** I: $[Kr]5s^2 4d^{10} 5p^5$

8-33. (a) In   **(b)** Y   **(c)** Ce   **(d)** Ar

8-35. (a) $3p$ **(b)** $5p$ **(c)** $5d$ **(d)** $5s$

8-37. Both have three valence electrons. The outer electron in IIIA is in a p subshell; in IIIB it is in a d subshell.

8-39. (a) F   **(b)** Ga   **(c)** Ba   **(d)** Gd   **(e)** Cu

8-41. (a) VIIA   **(b)** IIIA   **(c)** IIA   **(e)** IB

8-42. VA: [NG] $ns^2 np^3$ VB: [NG] $ns^2(n-1)d^3$

8-44. (b) and **(e)** belong to Group IVA

8-46. (a) IVA   **(b)** VIIIA   **(c)** IB   **(d)** Pr and Pa

8-47. (a) [NG] ns^2 **(b)** [NG] $ns^2 (n-1)d^{10}$

   (c) [NG] $ns^2 np^4$ or [NG] $ns^2 (n-1)d^{10} np^4$

   (d) [NG] $ns^2 (n-1)d^2$

8-48. Helium does not have a filled p subshell. (There is no $1p$ subshell.)

8-50. [Ne] $3s^2 3p^6$

8-52. The theoretical order of filling is $6d$, $7p$, $8s$, $5g$. The $6d$ is completed at element number 112. The $7p$ fills at element number 118, and the $8s$ fills at element number 120. Thus element number 121 would theoretically begin the filling of the $5g$ subshell. This assumes the normal order of filling.

8-54. $7s^2 6d^{10} 5f^{14} 7p^6$ Total 32

8-55. $8s^2 5g^{18} 7d^{10} 6f^{14} 8p^6$ Total 50

8-56. (a) transition   **(b)** representative   **(c)** noble gas   **(d)** inner transition

8-58. element number 118 (under Rn)

8-60. (a) This is excluded by Hund's rule since electrons are not shown in separate orbitals of the same subshell with parallel spins,   **(b)** This is correct,   **(c)** This is excluded by the Aufbau principle because the 2s subshell fills before the 2p. **(d)** This is excluded by the Pauli exclusion principle since the two electrons in the 2s orbital cannot have the same spin.

8-61. (a) S:

(b) V:

(c) Br:

(d) Pm:

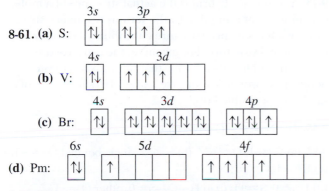

8-63. IIB, none; VB, three; VIA, two; VIIA, one; $[ns^1(n-1)d^5]$, six; Pm, five

8-65. IVA and VIA

8-67. (a) As   **(b)** Ru   **(c)** Ba   **(d)** I

8-69. Cr, 117 pm; Nb, 134 pm

8-71. (a) V   **(b)** Cl   **(c)** Mg   **(d)** Fe   **(e)** B

8-72. In, 558 kJ/mol; Ge, 762 kJ/mol

8-73. Te, 869 kJ/mol; Br, 1140 kJ/mol

8-75. (a) Cs^+ easiest, **(b)** Rb^{2+} hardest

8-77. (d) I

8-79. (a) Rb

8-81. (a) Mg^{2+} **(b)** K^+ **(c)** S   **(d)** Mg   **(e)** S^{2-} **(f)** Se^{2-}

8-83. The outer electron in Hf is in a shell higher in energy than that in Zr. This alone would make Hf a larger atom. However, in between Zr and Hf lie several subshells including the long $4f$ subshell (Ce through Lu). The filling of these subshells, especially the $4f$, causes a gradual contraction that offsets the higher shell for Hf.

8-84. C^+ (1086 kJ/mol), C^{2+} (3439 kJ/mol), C^{3+} (8059 kJ/mol), C^{4+} (14,282 kJ/mol), C^{5+} (52,112 kJ/mol). Notice that the energy required to form C^+ (1086 kj) is about twice the energy required to form Ga^+ (579 kJ), which is a metal. Thus it is apparent that metal cations form more easily than nonmetal cations.

8-85. (a) B   **(b)** Kr   **(c)** K, Cr, and Cu   **(d)** Ga   **(e)** Hf

8-87. (a) B   **(b)** Br   **(c)** Sb   **(d)** Si

8-89. (a) Sr, metal, representative element, IIA

   (b) Pt, metal, transition metal, VIIIB

   (c) Br, nonmetal, representative element, VIIA

8-91. P

8-93. Z = Sn, X = Zr

8-97. (a) Ca   **(b)** Br^- **(c)** S   **(d)** S^{2-} **(e)** Na^+

8-99.

s^1		d^1	d^2	d^3	d^4	d^5	d^6	d^7	d^8				p^4
1	s^2									p^1	p^2	p^3	2
3	4									5	6	7	8
9	10									11	12	13	14
15	16	17	18	19	20	21	22	23	24	25	26	27	28
29	30	31	32	33	34	35	36	37	38	39	40	41	42
43	44	45*											

*46–50 would be in a $4f$ subshell

(a) six in second period, 14 in fourth

(b) third period, #14; fourth period, #28

(c) first inner transition element is #46 (assuming an order of filling like that on Earth)

(d) Elements #11 and #17 are most likely to be metals.

(e) Element #12 would have the larger radius in both cases.

(f) Element #7 would have the higher ionization energy in all three cases.

(g) The ions that would be reasonable are 16^{2+} (metal cation), 13^- (nonmetal anion), 15^+ (metal cation), and 1^- (nonmetal anion). The 9^{2+} ion is not likely because the second electron would come from a filled inner subshell. The 7^+ ion is not likely because it would be a nonmetal cation. The 17^{4+} ion is not likely because #17 has only three electrons in the outer shell.

CHAPTER 9

9-1. (a) Ca· **(b)** ·Sb· **(c)** ·Sn· **(d)** ·Ï:

(e) :Ne: **(f)** ·Bi· **(g)** ·VIA:

9-2. (a) Group IIIA **(b)** Group VA **(c)** Group IIA

9-4. The electrons from filled inner subshells are not involved in bonding.

9-6. (b) Sr and S **(c)** H and K **(d)** Al and F

9-8. (b) S⁻ **(c)** Cr²⁺ **(e)** In⁺ **(f)** Pb²⁺ **(h)** Tl³⁺

9-10. They have the noble gas configuration of He, which requires only two electrons.

9-11. (a) K⁺ **(b)** ·Ö:⁻ **(c)** ·Ï:⁻ **(d)** :P:³⁻ **(e)** Ba⁺

(f) ·Xe:⁺ **(g)** Sc³⁺

9-13. (b) O⁻ **(e)** Ba⁺ **(f)** Xe⁺

9-14. (a) Mg²⁺ **(b)** Ga³⁺ (pseudo-noble gas configuration.)
(c) Br⁻ **(d)** S²⁻ **(e)** P³⁻

9-17. Se²⁻, Br⁻, Rb⁺, Sr²⁺, Y³⁺

9-18. Tl³⁺ has a full 5d subshell. This is a pseudo-noble gas configuration.

9-19. Cs₂S, Cs₃N, BaBr₂, BaS, Ba₃N₂, InBr₃, In₂S₃, InN

9-21. (a) CaI₂ **(b)** CaO **(c)** Ca₃N₂ **(d)** CaTe **(e)** CaF₂

9-23. (a) Cr³⁺ **(b)** Fe³⁺ **(c)** Mn²⁺ **(d)** Co²⁺ **(e)** Ni²⁺
(f) V³⁺

9-26. (a) H₂Se **(b)** GeH₄ **(c)** ClF **(d)** Cl₂O **(e)** NCl₃
(f) CBr₄

9-28. (b) Cl₃ **(d)** NBr₄ **(e)** H₃O

9-30. (a) SO₄²⁻ **(b)** IO₃⁻ **(c)** SeO₄²⁻ **(d)** H₂PO₃⁻
(e) S₂O₃²⁻

9-32.

(a) H—C—C—H (with H H above and H H below, C–C chain)   **(b)** H—Ö—Ö—H

(c) :F—N—F: (with :F: below)   **(d)** :Cl—S—Cl:

(e) H—C—Ö—C—H   H—C—C—Ö—H
(each with H's above and below)

9-34. (a) :C≡O: **(b)** ·O.   .O· with S below and :O: (SO₂ structure) **(c)** K⁺[:C≡N:]⁻

(d) H—Ö:
      :S—Ö:
      :Ö—H

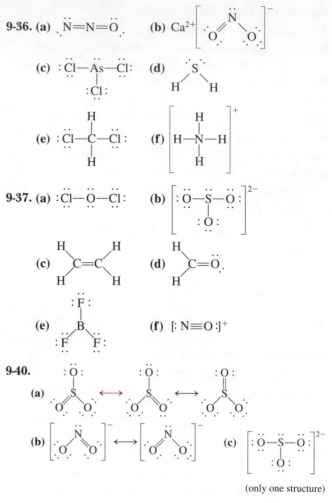

9-36. (a) Ṅ=N=Ö. **(b)** Ca²⁺ [O—N—O]⁻ (nitrite)

(c) :Cl—As—Cl: (with :Cl: below) **(d)** S with H H below

(e) :Cl—C—Cl: (with H above and H below) **(f)** [H—N—H]⁺ (with H above and H below)

9-37. (a) :Cl—Ö—Cl: **(b)** [:Ö—S—Ö:]²⁻ (with :O: below)

(c) C=C (ethylene, H H on each carbon) **(d)** C=Ö. (with H H)

(e) B (with :F: above and :F: :F: below) **(f)** [:N≡O:]⁺

9-40.

(a) three resonance structures of SO₃ connected by ↔

(b) [O—N—O]⁻ ↔ [O—N—O]⁻ (nitrite resonance)   **(c)** [:Ö—S—Ö:]²⁻ (with :O: below)

(only one structure)

9-42.

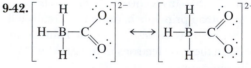

[H—B—C with O and O]²⁻ ↔ [H—B—C with O]²⁻ (with H's on B)

9-43. A resonance hybrid is the actual structure of a molecule or ion that is implied by the various resonance structures. Each resonance structure contributes a portion of the actual structure. For example, the two structures shown in problem 9–42 imply that both carbon-oxygen bonds have properties that are halfway between those of a single and a double bond.

9-45. :Ö—N̈=Ö. (−1 +1)   :Ö=N̈—Ö: (+1)

9-47. (a) C = −1, O = +1 **(b)** S = +2, single bonded O's = −1
(c) K = +1; C = −1; N = 0

(d) H's = 0; S = +1; O in H—O—S = 0, other O = −1

9-49. :Ö—Cl̈—Ö: (−1 +2 −1)   :Ö=Cl̈—Ö.
      :O:⁻¹   :O:⁻¹

9-51. :N̈=N=Ö. (−1 +1) ↔ :N≡N—Ö: (+1 −1)   :N̈=O=N̈: (−1 +2 −1)

Both resonance structures for NNO have much less formal charge than the NON structure.

9-52. Cs, Ba, Be, B, C, Cl, O, F

9-53. (a) $\overset{\delta-}{N} \underset{\rightleftarrows}{} \overset{\delta+}{H}$ **(b)** $\overset{\delta+}{B} \underset{\rightleftarrows}{} \overset{\delta-}{H}$ **(c)** $\overset{\delta+}{Li} \underset{\rightleftarrows}{} \overset{\delta-}{H}$
(d) $\overset{\delta-}{F} \underset{\rightleftarrows}{} \overset{\delta+}{O}$ **(e)** $\overset{\delta-}{O} \underset{\rightleftarrows}{} \overset{\delta+}{Cl}$ **(f)** $\overset{\delta-}{S} \underset{\rightleftarrows}{} \overset{\delta-}{Se}$
(g) $\overset{\delta-}{C} \underset{\rightleftarrows}{} \overset{\delta+}{B}$ **(h)** $\overset{\delta+}{Cs} \underset{\rightarrow}{} \overset{\delta-}{N}$
(i) C—S (very low polarity)

9-54. (i) nonpolar, **(b)** = **(f)**, **(d)** = **(e)** = **(g)**, **(a)**, **(c)**, **(h)**

9-55. (a) CF polar covalent **(b)** AlF ionic **(c)** FF nonpolar

9-56. Only **(d)** Al — F is predicted to be ionic on this basis since the electronegativity difference is 2.5.

9-58. (b) I—I nonpolar, **(c)** C—H (nearly nonpolar)

9-59. If the molecule has a symmetrical geometry and all bonds are the same, the bond dipoles cancel and the molecule is nonpolar.

9-60. (a) $\overset{..}{S}$ is V-shaped **(b)** $\overset{..}{S}=C=\overset{..}{S}$ is linear
    $FF:$

(c) $:\overset{..}{Cl}:$ is tetrahedral **(d)** $\overset{..}{O}\overset{..}{Cl}:$ is V-shaped
    $ClCF$
    $:F:$

(e) $\overset{..}{O}$ is V-shaped
    $ClCl$

9-62. (a) No angle between two points **(b)** 180° **(c)** 120°

9-64. (a) H
    $N—\overset{..}{O}—H$ N–trigonal pyramid, O–V-shaped
    H

(b) $H—\overset{..}{O}—C\equiv N:$ O–V-shaped, C–linear

(c) $H—\overset{H}{\underset{H}{C}}—C\equiv N:$ C–tetrahedral, C–linear

(d) See problem 9-42. B– tetrahedral, C– trigonal planar

9-66. If the molecule has a symmetrical geometry and all bonds are the same, the bond dipoles cancel and the molecule is nonpolar.

9-67. (a) polar **(b)** nonpolar **(c)** polar **(d)** polar
(e) polar

9-69. $:\overset{..}{O}:$
    S
    $\overset{..}{O}\overset{..}{Cl}:$ The molecule is polar.
    $:\overset{..}{Cl}:$

9-70. Since the H — S bond is much less polar than the H — O bond, the resultant molecular dipole is much less. The H_2S molecule is less polar than the H_2O molecule.

9-72. The CHF_3 molecule is more polar than the $CHCl_3$ molecule. The C — F bond is more polar than the C — Cl bond, which means that the resultant molecular dipole is larger for CHF_3.

9-74. $\overset{..}{O}$
    $||$
    S All bond dipoles in SO_3 cancel.
    $\overset{..}{O}\overset{..}{O}$

S Bond dipoles in SO_2 do not cancel.
$\overset{..}{O}\overset{..}{O}:$

9-76. (a) e.g., RbCl **(b)** e.g., SrO **(c)** e.g., AlN

9-77. $\overset{..}{Xe}$ is trigonal pyramid; the molecule is polar.
    $\overset{..}{O}|\overset{..}{O}$
    $:\overset{..}{O}:$

9-79. H
    $|$ No resonance structures.
    $H—B—C\equiv O:$ Geometry around B is tetrahedral.
    $|$ Geometry around C is linear.
    H

9-81. $N—N$ Geometry is V-shaped
    $\overset{..}{O}\overset{..}{O}$ around N's.

Other resonance structures:

$:\overset{..}{O}—N\equiv N—\overset{..}{O}: \longleftrightarrow :O\equiv N—\overset{..}{N}—\overset{..}{O}: \longleftrightarrow :\overset{..}{O}—\overset{..}{N}—N\equiv O:$

9-84. Valence electrons for neutral $PO_3 = 5(P) + (3 \times 6)$
(O's) = 23. Since the species has 26 electrons it must be an anion with a – 3 charge (i.e., PO_3^{3-})

$:\overset{..}{O}—\overset{..}{P}—\overset{..}{O}:^{3-}$
$|$
$:\overset{..}{O}:$

The geometry of the anion is trigonal pyramid with a O—P—O angle of about 109°.

9-85. Valence electrons for neutral $ClI_2 = 3 \times 7 = 21$.
Since the species has 20 valence electrons it must be a cation with a + 1 charge (i.e. ClI_2^+).

$\overset{..}{Cl}^+$
$\overset{..}{I}\overset{..}{I}$

The geometry of the cation is V-shaped with an angle of about 109°.

9-87. K_3N, potassium nitride $KN_3 = K^+N_3^-$ Resonance structures:

$\overset{..}{N}=N=\overset{..}{N} \longleftrightarrow :\overset{..}{N}—N\equiv N: \longleftrightarrow :N\equiv N—\overset{..}{N}:$

In all resonance structures, the geometry around N is linear.

9-90. $H\diagdown$
    $\overset{..}{O}—C\equiv N:$

The angle of 105° indicates that the H—O—C angle is V-shaped with an angle of about 109°. This would involve two bonds and two pairs of electrons on the O. The geometry around the C is linear.

9-91. $H\diagdown$
    $\overset{..}{S}=C=\overset{..}{N}:$

The angle of 116° indicates that the H—S—C angle is V-shaped with an angle of about 120°. This would involve three bonds and one pair of electrons on the S. The geometry around the C is linear.

9-92. $:\ddot{F}—\ddot{N}=\ddot{N}—\ddot{F}:$

9-94. Oxygen difluoride, OF_2. Lewis structure: $:\ddot{F}\quad\overset{\cdot\cdot\overset{\cdot\cdot}{O}}{}\quad\ddot{F}:$

Dioxygen difluoride, O_2F_2. Lewis structure: $:\ddot{F}:\quad\overset{\ddot{O}—\ddot{O}}{}\quad:\ddot{F}:$

Oxygen is less electronegative (more metallic) than fluorine and so is named first. Both molecules are angular; thus they are polar.

9-96. Formula $= NaHCO_3 = Na^+HCO_3^-$

$$H—\ddot{\underset{\cdot\cdot}{O}}—C\overset{\overset{\cdot\cdot}{\ddot{O}}:^-}{\underset{\ddot{O}:}{\diagup}}$$

The geometry around the C is trigonal planar with the approximate H—O—C angle of 120°.

9-96. Formula $= CaH_2S_2O_6 = Ca(HSO_3)_2$ calcium bisulfite or calcium hydrogen sulfite

$$H—\ddot{\underset{\cdot\cdot}{O}}—\underset{\underset{:\ddot{O}:}{|}}{S}—\ddot{O}:^-$$

The geometry around the S is trigonal pyramid. The H—O—S angle is about 109°.

9-99. (1) (a) 17   (b) 31   (c) 51_3 (d) 97   (e) 7(13)

(f) $10(13)_2$ (g) 67_2 (h) 3_26

(2) On Zerk, six electrons fill the outer s and p orbitals to make a noble gas configuration. Therefore, we have a "sextet" rule on Zerk.

(a) $1—\ddot{7}:$ (b) $3^+1:^-$ (ionic)   (c) $\overset{}{\underset{\underset{1}{|}}{1—5—1}}$

(d) $9^+:\ddot{7}:^-$ (ionic)   (e) $:\ddot{7}—1\ddot{3}:$ (f) $10^{2+}(:1\ddot{3}:^-)_2$ (ionic)

(g) $:\ddot{7}—\ddot{6}—\ddot{7}:$ (h) $(3^+)_2:\ddot{6}:^{2-}$ (ionic)

CHAPTER 10

10-1. The molecules of water are closely packed together and thus offer much more resistance. The molecules in a gas are dispersed into what is mostly empty space.

10-3. Since a gas is mostly empty space, more molecules can be added. In a liquid, the space is mostly occupied by the molecules so no more can be added.

10-5. Gas molecules are in rapid but random motion. When gas molecules collide with a light dust particle suspended in the air, they impart a random motion to the particle.

10-6. 1260 mi/hr

10-7. The molecule with the largest formula weight travels the slowest. SF_6(146.1 amu) $< SO_2$ (64.07 amu)

$< N_2O$ (44.02 amu) $< CO_2$ (44.01 amu)

$< N_2$ (28.02 amu) $< H_2$ (2.016 amu)

10-9. When the pressure is high, the gas molecules are forced close together. In a highly compressed gas the molecules can occupy an appreciable part of the total volume. When the temperature is low, molecules have a lower average velocity. If there is some attraction, they can momentarily stick together when moving slowly.

10-10. (a) 2.17 atm   (b) 0.0266 torr   (c) 9560 torr

(d) 0.0558 atm   (e) 3.68 lb/in.^{2} (f) 11 kPa

10-11. (a) 768 torr   (b) 2.54×10^4 atm   (c) 8.40 lb/in.^{2}

(d) 19 torr

10-13. 0.0102 atm

10-15. Assume a column of Hg has a cross-section of 1 cm^2 and is 76.0 cm high. Weight of Hg $= 76.0$ cm

$\times$ 1cm^2 $\times$ 13.6 g/cm^2 $= 1030$ g. If water is substituted, 1030 g of water in the column is required. Height $\times 1$ cm^2 $\times 1.00$ g/cm$^3 = 1030$ g height $= 1030$ cm

$$1030 \text{ cm} \times \frac{1 \text{ in.}}{2.54 \text{ cm}} \times \frac{1 \text{ ft}}{12 \text{ in.}} = \underline{33.8 \text{ ft}}$$

If a well is 40 ft deep, the water cannot be raised in one stage by suction since 33.8 ft is the theoretical maximum height that is supported by the atmosphere.

10-16. 10.2 L

10-18. 978 mL

10-19. 67.9 torr

10-22. V_{final} (V_f) $= 15$ $V_{initial}$ (V_i) $\dfrac{V_f}{V_i} = \dfrac{P_i}{P_f}$;

$$\frac{15 V_i}{V_i} = \frac{0.950 \text{ atm}}{P_f} \quad P_f = 0.950 \text{ atm} \times 15 = 14.3 \text{ atm}$$

10-23.

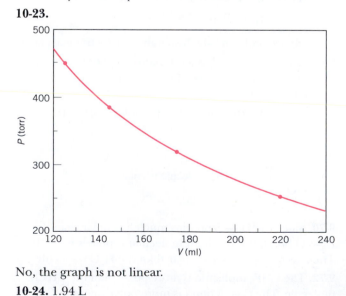

No, the graph is not linear.

10-24. 1.94 L

10-26. 77°C

10-28. 2.60×10^4 L

10-29. 341 K (68°C)

10-31. 2.94 atm

10-32. 191 K − 273 = −82°C

10-34. 596 K (323°C)

10-35. 30.2 lb/in.^{2}

10-38. (a) and (d)

10-40. Exp. 1, T increases; Exp. 2, V decreases; Exp. 3, P increases; Exp. 4, T increases

10-42. 1.24 atm

10-43. 76.8 L

10-45. 88.5 K = −185°C

10-48. 258 K(−15°C)

10-49. 409 mL

10-50. 1.70 L

10-51. Let x = total moles needed in the expanded balloon.

$$188 \text{ L} \times \frac{x}{8.40 \text{ mol}} = 275 \text{ L} \quad x = 12.3 \text{ mol}$$

12.3 − 8.4 = 3.9 mol must be added.

10-53.

$$n_2 = 2.50 \times 10^{-3} \text{mol} \times \frac{164 \text{ mL}}{75.0 \text{ mL}} = 5.47 \times 10^{-3} \text{mol}$$

$$(5.47 \times 10^{-3}) - (2.50 \times 10^{-3}) = 2.97 \times 10^{-3} \text{ mol}$$

$$2.97 \times 10^{-3} \text{ mol N}_2 \times \frac{28.02 \text{ g N}_2}{\text{mol N}_2} = 0.0832 \text{ g N}_2$$

10-55. 98°C

10-57. 13.0 g

10-58. 589 torr

10-60. 3.35 g

10-62. $n = 1.0 \times 10^6$ mol of gas in the balloon. 8800 lb He or 64,000 lb air

Lifting power with He = 64,000 − 8800 = 55,000 lb

Lifting power with H$_2$ = 64,000 − 4000 = 60,000 lb

Helium is a noncombustible gas whereas hydrogen forms an explosive mixture with O$_2$.

10-64. 762 torr

10-66. 6.8 torr

10-69. P_{N_2} = 756 torr; P_{O_2} = 84.0 torr; P_{SO_2} = 210 torr

10-70. 731 torr

10-72. P_{N_2} = 300 torr; P_{O_2} = 85 torr × $\dfrac{4.00 \text{ L}}{2.00 \text{ L}}$ = 170 torr;

P_{CO_2} = 225 torr; P_{tot} = 695 torr;

10-73. P_A = 0.550 atm; P_B = 0.300 atm;

10-75. 7.63 L

10-77. 112 L

10-79. 1.39×10^{-3} g

10-80. 1.24 g/L

10-82. 34.0g/mol

10-84. $1.00 \text{ L} \times \dfrac{273 \text{ K}}{298 \text{ K}} \times \dfrac{1.20 \text{ atm}}{1.00 \text{ atm}} = 1.10 \text{ L (STP)}$

$\dfrac{3.60 \text{ g}}{1.10 \text{ L}} = 3.27 \text{ g/L (STP)}$

10-85. Find moles of N$_2$ in 1 L at 500 torr and 22°C using the ideal gas law. $n = 0.272$ mol N$_2$, mass = 0.762 g N$_2$ Density = 0.762 g/L (500 torr and 22°C)

10-87. 25.7 L CO$_2$ (STP)

10-88. Vol. O$_2$ ⟶ mol O$_2$ ⟶ mol Mg ⟶ g Mg

$$5.80 \text{ L O}_2 \times \frac{1 \text{ mol O}_2}{22.4 \text{ L O}_2} \times \frac{2 \text{ mol Mg}}{1 \text{ mol O}_2} \times \frac{24.31 \text{ g Mg}}{\text{mol Mg}}$$

$$= 12.6 \text{ g Mg}$$

10-90. 3.47 L

10-92. (a) g C$_4$H$_{10}$ ⟶ mol C$_4$H$_{10}$ ⟶ mol CO$_2$ ⟶ Vol CO$_2$

$$85.0 \text{ g C}_4\text{H}_{10} \times \frac{1 \text{ mol C}_4\text{H}_{10}}{58.12 \text{ g C}_4\text{H}_{10}} \times \frac{8 \text{ mol CO}_2}{2 \text{ mol C}_4\text{H}_{10}} \times$$

$$\frac{22.4 \text{ L}}{\text{mol CO}_2} = 131 \text{ L}$$

(b) 96.1 L O$_2$ (c) 124 L CO$_2$

10-93. 2080 kg Zr (2.29 tons)

10-95. n_{O_2} = 1.46 mol O$_2$ $1.46 \text{ mol O}_2 \times \dfrac{1 \text{ mol CO}_2}{2 \text{ mol O}_2} =$

0.730 mol CO$_2$ V = 12.0 L CO$_2$

10-96. Force = $12.0 \text{ cm}^2 \times 15.0 \text{ cm} \times \dfrac{13.6 \text{ g}}{\text{cm}^3}$ = 2450 g

$$P = \frac{2450 \text{ g}}{12.0 \text{ cm}^2} = 204 \text{ g/cm}^2$$

$$1 \text{ atm} = 76.0 \text{ cm} \times \frac{13.6 \text{ g}}{\text{cm}^3} = 1030 \text{ g/cm}^2$$

$$204 \text{ g/cm}^2 \times \frac{1 \text{ atm}}{1030 \text{ g/cm}^2} = 0.198 \text{ atm}$$

10-98. $n = 0.0573$ mol $\dfrac{8.37 \text{ g}}{0.0573 \text{ mol}} = 146 \text{ g/mol}$

10-99. Empirical formula = CH$_2$ molar mass = 41.9 g/mol

molecular formula = C$_3$H$_6$

10-101. n_{tot} = 0.265 mol O$_2$ + 0.353 mol N$_2$ +

0.160 mol CO$_2$ = 0.778 mol of gas

V = 6.92 L

$$P_{O_2} = \frac{0.265}{0.778} \times 2.86 \text{ atm} = 0.974 \text{ atm}$$

P_{N_2} = 1.30 atm P_{CO_2} = 0.59 atm

10-102. $n = 5 \times 10^{-20}$ mol $P = 4 \times 10^{-17}$ atm

10-103. 19.6 L/mol density of CO$_2$ = 2.25 g/L

10-104. Density = $\dfrac{\text{mass}}{V} = \dfrac{P \times MM}{RT}$ = 0.525 g/L (hot air)

density at STP = 1.29 g/L

0.525/1.29 = 0.41 (Hot air is less than half as dense as air at STP.)

10-106. 0.0103 mol of H$_3$BCO produces 0.0412 mol of gaseous products; V = 1.12 L (products at 25°C and 0.900 atm)

10-108. $2Al(s) + 3F_2(g) \longrightarrow 2AlF_3(s)$

original $\underline{F_2 = 0.310 \text{ mol } F_2}$

leftover $F_2 = 0.0921$ mol

$0.310 - 0.092 = 0.218$ mol F_2 reacts forming $\underline{12.2 \text{ g } AlF_3}$

10-110. 135 L

10-112. 0.763 mol H_2O in the ice. V (of vapor) $= \underline{645 \text{ L}}$

10-113. 9.81×10^{-4} mol NH_3 produces 11.0 mL N_2H_4

10-115. Empirical formula $= NO_2$

(2) Find molar mass of product compound.

$$n = \frac{PV}{RT} = \frac{\dfrac{715 \text{ torr}}{760 \text{ torr/atm}} \times 1.05 \text{ L}}{0.0821 \dfrac{\text{L} \cdot \text{atm}}{\text{K} \cdot \text{mol}} \times 273 \text{ K}} = 0.0441 \text{ mol}$$

$$\text{Molar mass} = \frac{2.03 \text{ g}}{0.0441 \text{ mol}} = 46.0 \text{ g/mol}$$

(3) Since one compound decomposes to one other compound, the reactant compound must have the same empirical formula as the product compound. Since the empirical mass of $NO_2 = 46.01$ g/emp unit, then the product must be $NO_2 (MM = 46.0$ g/mol). Since 0.0220 mol of reactant form 0.0441 mol of product (1:2 ratio), the reaction must be

$$N_2O_4(l) \longrightarrow 2NO_2(g)$$

10-117. $6Li(s) + N_2(g) \longrightarrow 2Li_3N(s)$

$3Mg(s) + N_2(g) \longrightarrow Mg_3N_2(s)$

10.8 mol N_2 reacts with $\underline{450 \text{ g Li}}$ or $\underline{788 \text{ g Mg}}$

CHAPTER 11

11-1. Since gas molecules are far apart, they travel a comparatively large distance between collisions. Liquid molecules, on the other hand, are close together so do not travel far between collisions. The farther molecules travel, the faster they mix.

11-3. Both the liquid molecules and food coloring molecules are in motion. Through constant motion and collisions the food coloring molecules eventually become dispersed.

11-5. Generally, solids have greater densities than liquids. (Ice and water are notable exceptions.) Since the molecules of a solid are held in fixed positions, more of them usually fit into the same volume compared to the liquid state. This is similar to being able to get more people into a room if they are standing still than if they are moving around.

11-6. $CH_4 < CCl_4 < GeCl_4$

11-8. All are nonpolar molecules with only London forces between molecules. The higher the molar mass, the greater the London forces and the more likely that the compound is a solid. I_2 is the heaviest and is a solid; Cl_2 is the lightest and is a gas.

11-10. If H_2O were linear, the two equal bond dipoles would be exactly opposite and would therefore cancel. Hydrogen bonding can occur only when the molecule is polar.

11-12.

NH_3 molecules interact by hydrogen bonding.

11-14. (a) HBr   (b) SO_2 (f) CO

11-17. (a) HF   (c) H_2NCl (d) H_2O (f) HCOOH

11-19. (a) ion–ion   (b) hydrogen bonding plus London (c) dipole–dipole plus London   (d) London only

11-20. $F_2 < HCl < HF < KF$

11-23. CO_2 is a nonpolar molecular compound. SiO_2 is a network solid.

11-25. $PbCl_2$ is most likely ionic (Pb^{2+}, $2Cl^-$), while the melting point of $PbCl_4$ indicates that it is a molecular compound.

11-26. Motor oil is composed of large molecules that increase viscosity. Motor oil also has a higher surface tension.

11-28. Water evaporates quickly on a hot day and thus lowers the air temperature.

11-30. The comparatively low boiling point of ethyl chloride indicates that it has a high vapor pressure at room temperature (25°C). In fact, it boils rapidly, cooling the skin to below the freezing point of water.

11-32. The ethyl ether. The higher the vapor pressure, the faster the liquid evaporates and the liquid cools.

11-33. Equilibrium refers to a state where opposing forces are balanced. In the case of a liquid in equilibrium with its vapor, it means that a molecule escaping to the vapor is replaced by one condensing to the liquid.

11-35. The substance is a gas at 1 atm and 75°C. It would boil at a temperature below 75 °C.

11-38. Ethyl alcohol boils at about 52°C at that altitude. At 10°C ethyl ether is a gas at that altitude.

11-39. Death Valley is below sea level. The atmospheric pressure is more than 1 atm so water boils above its normal boiling point.

11-41. Both exist as atoms with only London forces, but the higher atomic mass of argon accounts for its higher boiling point.

11-43. Figure 11-15 is not very precise; at 10°C the actual vapor pressure of water is 9.2 torr. Liquid water could theoretically exist but it would rapidly evaporate and be changed to ice by the cooling effect.

11-44. The hexane (75 torr) has the lower boiling point. It also probably has the lower heat of vaporization.

11-46. Molecular O_2 is heavier than Ne atoms and so has stronger intermolecular forces (London). CH_3OH has hydrogen bonding, which would indicate a much higher heat of vaporization than the other two.

11-47. This is a comparatively high heat of fusion, so the melting point is probably also comparatively high.

11-49. This is a comparatively high boiling point, so the compound probably also has a high melting point.

11-50.

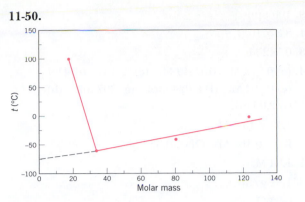

The boiling point of H_2O would be about $-75°C$ without hydrogen bonding.

11-53. 7.07×10^3 J

11-55. 2.54 g H_2O, 1.64 g NaCl, 6.69 g benzene

11-57. 555 cal (ether), 2000 cal (H_2O) Water would be more effective. 1.30×10^4 J (13.0 kJ)

11-59. NH_3: 6.12×10^5 J   Freon: 7.25×10^4 J

On the basis of mass, ammonia is the more effective refrigerant.

11-61. Condensation: 62.2×10^4 J   Cooling: 8.63×10^4 J

Total = $\underline{7.08 \times 10^5 \text{ J} \text{ (708 kJ)}}$

11-63. 2.83×10^4 cal

11-65. heat ice: $132 \text{ g} \times 20.0°C \times \dfrac{0.492 \text{ cal}}{\text{g} \cdot °C} = 1300 \text{ cal}$

melt ice: $132 \text{ g} \times \dfrac{79.8 \text{ cal}}{\text{g}} = 10,500 \text{ cal}$

heat H_2O: $132 \text{ g} \times 100°C \times \dfrac{1.00 \text{ cal}}{\text{g} \cdot °C} = 13,200 \text{ cal}$

vap. H_2O: $132 \text{ g} \times \dfrac{540 \text{ cal}}{\text{g}} = 71,300 \text{ cal}$

Total = $\underline{96,300 \text{ cal} \text{ (96.3 kcal)}}$

11-67. Let $Y =$ the mass of the sample in grams. Then

$\left(\dfrac{2260 \text{ J}}{\text{g}} \times Y\right) + \left(25.0°C \times Y \times \dfrac{4.18 \text{ J}}{\text{g} \cdot °C}\right) = 28,400 \text{ J}$

$Y = \underline{12.0 \text{ g}}$

11-69. 18°C

11-70. (b) melting   (c) boiling

11-71. The average kinetic energy of all molecules of water at the same temperature is the same regardless of the physical state.

11-73. Because H_2O molecules have an attraction for each other, moving them apart increases the potential energy. Since the molecules in a gas at 100°C are farther apart than in a liquid at the same temperature, the potential energy of the gas molecules is greater.

11-74. The water does not become hotter, but it will boil faster.

11-76.

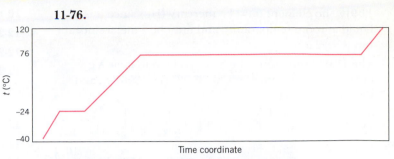

The time of boiling is over ten times longer than the time of melting

11-78. $C_2H_5NH_2$: 17°C, hydrogen bonding
CH_3OCH_3: $-25°C$, dipole–dipole (polar)
CO_2: $-78°C$, London forces only (nonpolar)

11-79. SF_6 is a molecular compound and SnO is an ionic compound (i.e., Sn^{2+}, O^{2-}). All ionic compounds are found as solids at room temperature because of the strong ion–ion forces.

11-81. There is hydrogen bonding in CH_3OH. Hydrogen bonding is a considerably stronger attraction than ordinary dipole–dipole forces. The stronger the attractions the higher the boiling point.

11-83. SiH_4 is nonpolar. PH_3 and H_2S are both polar but H_2S is more polar with more and stronger dipole–dipole attractions.

11-85. Carbon monoxide is polar, whereas nitrogen is not. The added dipole–dipole attraction in CO may account for the slightly higher boiling point.

11-87. $2000 \text{ lb} \times \dfrac{453.6 \text{ g}}{\text{lb}} \times \dfrac{266 \text{ J}}{\text{g}} = 2.41 \times 10^8 \text{ J}$

Heating 1 g of H_2O from 25.0°C to 100.0°C and the vaporizing the water requires

$(75.0°C \times 4.184 \text{ J}/°C) + 2260 \text{ J} = 2.57 \times 10^3 \text{ J/g } H_2O$

$\dfrac{2.41 \times 10^8 \text{ J}}{2.57 \times 10^3 \text{ J/g } H_2O} = 9.38 \times 10^4 \text{ g } H_2O = \underline{93.8 \text{ kg } H_2O}$

$= 9.38 \times 10^4 \text{ g } H_2O = \underline{93.8 \text{ kg } H_2O}$

11-88. V = 100 L, T = 34 + 273 = 307 K,
$P(H_2O) = 0.700 \times 39.0 \text{ torr} = 27.3 \text{ torr}$
Use the ideal gas law to find moles of water.

$n = \dfrac{PV}{RT} = \dfrac{\dfrac{27.3 \text{ torr}}{760 \text{ torr/atm}} \times 100 \text{ L}}{0.0821 \dfrac{\text{L} \cdot \text{atm}}{\text{K} \cdot \text{mol}} \times 307 \text{ K}} = 0.143 \text{ mol}$

$0.143 \text{ mol } H_2O \times \dfrac{18.02 \text{ g } H_2O}{\text{mol } H_2O} = \underline{2.58 \text{ g } H_2O}$

11-90. Calculate the heat required first to heat the water from 25°C to 100°C then to vaporize the water.
3.10×10^5 J (to heat water), 2.260×10^6 J (to vaporize water) = 2570 kJ total

Cr: $\dfrac{2570 \text{ kJ}}{21.0 \text{ kJ/mol}} \times \dfrac{52.00 \text{ g Cr}}{\text{mol Cr}} \times \dfrac{1 \text{ kg}}{10^3 \text{ g}} = \underline{6.36 \text{ kg Cr},}$

$\underline{8.81 \text{ kg Mo}}, \underline{13.5 \text{ kg W}}$

11-91. The element must be mercury (Hg) since it is a liquid at room temperature and must be a metal since it forms a +2 ion.

The 15.0 kJ represents the heat required to melt X g of Hg, heat X g from $-39°C$ to $357°C$ (i.e., $396°C$), and vaporize X g of Hg.

$$(11.5 \text{ J/g} \times X) + \left[396°C \times 0.139 \frac{\text{J}}{\text{g °C}} \times X\right]$$
$$+ (29.5 \text{ J/g} \times X) = 15.00 \text{ kJ} \times \frac{10^3 \text{ J}}{\text{kJ}}$$

$11.5 \text{ J/g } X + 55.0 \text{ J/g } X + 29.5 \text{ J/g } X = 15,000 \text{ J}$

$$X = \underline{156 \text{ g Hg}}$$

11-93. Calculate the mass of water in the vapor state in the room at $-5°C$ (268 K) using the ideal gas law. 2.99 mol H_2O (in vapor) $= 53.9$ g H_2O (capacity of the room) Eventually, all of the 50.0 g should sublime since it is less than the room could contain.

CHAPTER 12

12-1. The ion–ion forces in the crystal hold the crystal together and resist the ion–dipole forces between water and the ions in the crystal. The ion–dipole forces remove the ions from the crystal.

12-3. Calcium bromide is soluble, lead (II) bromide is insoluble, benzene and water are immiscible, and alcohol and water are miscible.

12-4. **(a)** $LiF \longrightarrow Li^+(aq) + F^-(aq)$

  (b) $(NH_4)_3PO_4 \longrightarrow 3NH_4^+(aq) + PO_4^{3-}(aq)$

  (c) $Na_2CO_3 \longrightarrow 2Na^+(aq) + CO_3^{2-}(aq)$

  (d) $Ca(C_2H_3O_2)_2 \longrightarrow Ca^{2+}(aq) + 2C_2H_3O_2^-(aq)$

12-6.

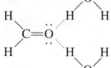

There is hydrogen bonding between solute and solvent.

12-8. At $10°C$, Li_2SO_4 is the most soluble; at $70°C$, KCl is the most soluble.

12-10. **(a)** unsaturated   **(b)** supersaturated

  (c) saturated   **(d)** unsaturated

12-12. A specific amount of Li_2SO_4 will precipitate unless the solution becomes supersaturated.

$$\left[500 \text{ g H}_2\text{O} \times \frac{35 \text{ g}}{100 \text{ g H}_2\text{O}}\right] - \left(500 \text{ g H}_2\text{O} \times \frac{28 \text{ g}}{100 \text{ g H}_2\text{O}}\right)$$
$$= \underline{35 \text{ g Li}_2\text{SO}_4 \text{ (precipitate)}}$$

12-13. 1.49%

12-15. 0.375 mol NaOH

12-17. 30 g KNO_3

12-18. 8.81% NaOH

12-19. 100 ppm

12-21. 8.3×10^7 L

12-23. 0.542 M

12-24. **(a)** 0.873 M   **(b)** 1.40 M   **(c)** 12.4 L   **(d)** 41.4 g

  (e) 0.294 M   **(f)** 0.024 mol   **(g)** 307 mL   **(h)** 49.1 g

  (i) 2.00 mL

12-28. 1.17×10^{-3} M

12-29. Ba^{2+}: 0.166 M OH^-: 0.332 M

12-31. 1.84 M

12-33. 1.32 g/mL

12-34. 0.833 L

12-36. Slowly add 313 mL of the 0.800 M NaOH to about 500 mL of water in a 1-L volumetric flask. Dilute to the 1-L mark with water.

12-38. 0.140 M

12-39. 720 mL

12-41. 2.86 mL

12-42. Find the total moles and the total volume.

  $n = V \times M = 0.150 \text{ L} \times 0.250 \text{ mol/L} = 0.0375$ mol (solution 1)

  $n = V \times M = 0.450 \text{ L} \times 0.375 \text{ mol/L} = 0.169$ mol (solution 2)

  $n_{tot} = 0.206$ mol $V_{tot} = 0.600$ L

  $\dfrac{n}{V} = \dfrac{0.206 \text{ mol}}{0.600 \text{ L}} = \underline{0.343 \text{ M}}$

12-43. $0.500 \text{ L} \times \dfrac{0.250 \text{ mol KOH}}{\text{L}} \times \dfrac{1 \text{ mol Cr(OH)}_3}{3 \text{ mol KOH}} \times$

  $\dfrac{103.0 \text{ g Cr(OH)}_3}{\text{mol Cr(OH)}_3} = \underline{4.29 \text{ g Cr(OH)}_3}$

12-45. 146 g $BaSO_4$

12-46. 4.08 L

12-48. 0.653 L

12-50. 0.864 M

12-52. 0.994 M

12-54. NaOH is the limiting reactant, producing 1.75 g of $Mg(OH)_2$.

12-56. An aqueous solution of AB is a good conductor of electricity. AB is dissociated into ions such as A^+ and B^-. A solution of AC is a weak conductor of electricity, which means that AC is only partially dissociated into ions: i.e., $AC \rightleftharpoons A^+ + C^-$. A solution of AD is a nonconductor of electricity because it is present as undissociated molecules in solution.

12-58. 1 mole of NaBr has a mass of 102.9 g, 1 molar NaBr is a solution containing 102.9 g of NaBr per liter of solution, and 1 molal NaBr is a solution containing 102.9 g of NaBr per kilogram of solvent.

12-59. 2.50 m (The molality is the same in both solvents since the mass of solvent is the same.)

12-61. 15.8 g NaOH

12-63. $-0.37°C$

12-65. The salty water removes water from the cells of the skin by osmosis. After a prolonged period, a person would dehydrate and become thirsty.

12-68. We can concentrate a dilute solution by boiling away some solvent if the solute is not volatile. Reverse osmosis can also be used to concentrate a solution if pressure greater than the osmotic pressure is applied on a concentrated solution separated from the solvent by a semipermeable membrane. As the solution becomes more concentrated, the osmotic pressure becomes greater, and the corresponding pressure that is applied must be increased.

12-69. 1.00 m

12-71. 870 g glycol

12-72. 101.4°C

12-74. 2.9 m

12-77. $-14.5°C$

12-79. HCl is a strong electrolyte that produces two particles (ions) for each mole of HCl that dissolves. HF is a weak electrolyte that is essentially present as nonionized molecules in solution.

12-82. **(a)** $-5.8°C$ **(b)** $-6.4°C$ **(c)** $-5.0°C$

12-83. Dissolve the mixture in 100 g of H_2O and heat to over 45°C. Cool to 0°C where the solution is saturated with about 10 g of KNO_3 and about 25 g of KCl. $50 - 25 = 25$ g of KCl precipitates.

12-85. 90.9% H_2O; 3.03% salt; 6.06% sugar

12-87. 0.10 mol of Ag^+ reacts with 0.10 mol of Cl^-, leaving 0.05 mol of Cl^- in 1.0 L of solution. $[Cl^-] = 0.05$ M

12-89. $0.225 \cancel{L} \times \dfrac{0.196 \cancel{\text{mol HCl}}}{\cancel{L}} \times \dfrac{1 \text{ mol M}}{2 \cancel{\text{mol HCl}}}$

$$= 0.0221 \text{ mol M}$$

$$\dfrac{1.44 \text{ g}}{0.0221 \text{ mol}} = \underline{65.2 \text{ g/mol (Zn)}}$$

12-91. molality $= 0.426$ m molar mass $= 60.1$ g/mol empirical formula $= CH_4N$ molecular formula $= C_2H_8N_2$

12-93. 0.30 m sugar $<$ 0.12 m KCl (0.24 m ions) $<$ 0.05 m $CrCl_3$ (0.20 m ions) $<$ 0.05 m K_2CO_3 (0.15 m ions) $<$ pure water.

12-95. n(hydroxide) $= 0.0584$ mol

n(H_2) $= 0.0292$ mol H_2

$$\dfrac{0.0584}{0.0292} = 2 \,(2 \text{ mol hydroxide: 1 mol } H_2)$$

Balanced Equations:

$$2Na(s) + 2H_2O(l) \longrightarrow 2NaOH(aq) + H_2(g)$$

$$(2 \text{ mol sodium:1 mol } H_2)$$

$$Ca(s) + 2H_2O(l) \longrightarrow Ca(OH)_2(aq) + H_2 \,(g)$$

$$(1 \text{ mol calcium:1 mol } H_2)$$

The answer is sodium since it produces the correct ratio of hydrogen.

12-96. 2.91 L

12-98. 8.60 L

12-100. The formula must be PCl_3. The Lewis structure shows the structure to be trigonal pyramidal.

The reaction is $PCl_3(g) + 3H_2O \longrightarrow H_3PO_3(aq) +3HCl(aq)$

$\underline{0.404 \text{ M HCl}}$

12-102. $(2Na^+S^{2-})$ $10.0 \cancel{\text{g Na}_2\text{S}} \times \dfrac{1 \text{ mol Na}_2\text{S}}{78.05 \cancel{\text{g Na}_2\text{S}}}$

$$= 0.128 \text{ mol Na}_2\text{S}$$

$$\Delta T = 1.86 \times \dfrac{\dfrac{0.128 \text{ mol}}{100 \text{ g}}}{1000 \text{ g/kg}} \times 3 \text{ mol ions/mol Na}_2\text{S}$$

$$= 7.14°C \text{ change}$$

The answer is Na_2S since it would have a freezing point of $-7.14°C$.

12-104. $\Delta T = K_f m$ $1.15°C = 1.86 \times \dfrac{0.806}{1.00 + X}$

Solving for X, $X = 0.30$ kg 0.30 kg $= 300$ g $=$

$$\underline{300 \text{ mL of water added.}}$$

CHAPTER 13

13-1. **(a)** HNO_3, nitric acid   **(b)** HNO_2, nitrous acid

   (c) $HClO_3$, chloric acid   **(d)** H_2SO_3, sulfurous acid

13-3. **(a)** CsOH, cesium hydroxide   **(b)** $Sr(OH)_2$, strontium hydroxide   **(c)** $Al(OH)_3$, aluminum hydroxide

   (d) $Mn(OH)_3$, manganese(III) hydroxide

13-5. **(a)** $HNO_3 + H_2O \longrightarrow H_3O^+ + NO_3^-$

   (b) $CsOH \longrightarrow Cs^+ + OH^-$

   (c) $Ba(OH)_2 \longrightarrow Ba^{2+} + 2OH^-$

   (d) $HBr + H_2O \longrightarrow H_3O^+ + Br^-$

13-7.

Acid	Conjugate base		Acid	Conjugate base
(a) HNO_3	NO_3^-		**(d)** CH_4	CH_3^-
(b) H_2SO_4	HSO_4^-		**(e)** H_2O	OH^-
(c) HPO_4^{2-}	PO_4^{3-}		**(f)** NH_3	NH_2^-

13-9. **(a)** $HClO_4$, ClO_4^- and H_2O, OH^-

   (b) HSO_4^-, SO_4^{2-} and HClO and ClO^-

   (c) H_2O, OH^- and NH_3, NH_2^-

   (d) NH_4^+, NH_3 and H_3O^+, H_2O

13-12. **(a)** $\text{NH}_3 + \text{H}_2\text{O} \longrightarrow \text{NH}_4^+ + \text{OH}^-$

   $B_1 \qquad A_1$
   $A_2 \qquad B_2$

   (b) $\text{N}_2\text{H}_4 + \text{H}_2\text{O} \longrightarrow \text{N}_2\text{H}_5^+ + \text{OH}^-$

   $B_1 \qquad A_1$
   $A_2 \qquad B_2$

   (c) $\text{HS}^- + \text{H}_2\text{O} \longrightarrow \text{H}_2\text{S} + \text{OH}^-$

   $B_1 \qquad A_1$
   $A_2 \qquad B_2$

(d) $H^- + H_2O \longrightarrow H_2 + OH^-$

$\underset{B_1}{\quad} \quad \underset{A_1}{\quad}$

$\underset{A_2}{\quad} \quad \underset{B_2}{\quad}$

(e) $F^- + H_2O \longrightarrow HF + OH^-$

$\underset{B_1}{\quad} \quad \underset{A_1}{\quad}$

$\underset{A_2}{\quad} \quad \underset{B_2}{\quad}$

13-13. base: $HS^- + H_3O^+ \rightleftharpoons H_2S + H_2O$

acid: $HS^- + OH^- \rightleftharpoons S^{2-} + H_2O$

13-16. The information indicates that HBr is a strong acid. The Br^- ion, however, would be a very weak base in water.

13-18. (a) $HNO_3 + H_2O \longrightarrow H_3O^+ + NO_3^-$

(b) $HNO_2 + H_2O \rightleftharpoons H_3O^+ + NO_2^-$

(c) $HClO_3 + H_2O \rightleftharpoons H_3O^+ + ClO_3^-$

(d) $H_2SO_3 + H_2O \rightleftharpoons H_3O^+ + HSO_3^-$

13-20. $(CH_3)_2NH + H_2O \rightleftharpoons (CH_3)_2NH_2^+ + OH^-$

13-22. HX is a weak acid. The concentration of H_3O^+ must equal the concentration of the HX that ionized.

$$\frac{0.010}{0.100} \times 100\% = \underline{10\% \text{ ionized}}$$

13-24. $[H_3O^+] = 0.55$ M

13-25. $[H_3O^+] = 0.030 \times 0.55 = 0.017$ M

13-27. From the first ionization: $[H_3O^+] = 0.354$ M Of that, 25% undergoes further ionization.

$0.25 \times 0.354 = 0.089$ M $[H_3O^+]$

from the second ionization.

The total $[H_3O^+] = 0.354 + 0.089 = \underline{0.443 \text{ M}}$

13-28. $\frac{0.050}{1.0} \times 100\% = \underline{5.0\% \text{ ionized}}$

13-30. The conjugate acid of HSO_4^- is H_2SO_4. This is a strong acid so it completely ionizes in water. Thus it cannot exist in the molecular form in water.

13-31. (a) acid **(b)** salt **(c)** acid **(d)** acid salt **(e)** salt **(f)** base

13-33. (a) $HNO_3 + NaOH \longrightarrow NaNO_3 + H_2O$

(b) $2HI + Ca(OH)_2 \longrightarrow CaI_2 + 2H_2O$

(c) $HClO_2 + KOH \longrightarrow KClO_2 + H_2O$

13-35. (a) $H^+(aq) + NO_3^-(aq) + Na^+(aq) + OH^-(aq) \longrightarrow$
$Na^+(aq) + NO_3^-(aq) + H_2O(l)$

$H^+(aq) + OH^-(aq) \longrightarrow H_2O(l)$ (net ionic)

(b) $2H^+(aq) + 2I^-(aq) + Ca^{2+}(aq) + 2OH^-(aq) \longrightarrow$
$Ca^{2+}(aq) + 2I^-(aq) + 2H_2O(l)$

$H^+(aq) + OH^-(aq) \longrightarrow H_2O(l)$ (net ionic)

(c) $HClO_2(aq) + K^+(aq) + OH^-(aq) \longrightarrow$
$K^+(aq) + ClO_2^-(aq) + H_2O(l)$

$HClO_2(aq) + OH^-(aq) \longrightarrow ClO_2^-(aq) + H_2O(l)$
(net ionic)

13-37. $H_2C_2O_4(aq) + 2NH_3(aq) \longrightarrow (NH_4)_2C_2O_4(aq)$

$H_2C_2O_4(aq) + 2NH_3(aq) \longrightarrow 2NH_4^+(aq) +$
$C_2O_4^{2-}(aq)$

Net ionic equation is the same as the total ionic equation.

13-39. (a) KOH and $HClO_3$ **(b)** $Al(OH)_3$ and H_2SO_3 **(c)** $Ba(OH)_2$ and HNO_2 **(d)** NH_3 and HNO_3

13-41. (a) $2HBr + Ca(OH)_2 \longrightarrow CaBr_2 + 2H_2O$

(b) $2HClO_2 + Sr(OH)_2 \longrightarrow Sr(ClO_2)_2 + 2H_2O$

(c) $2H_2S + Ba(OH)_2 \longrightarrow Ba(HS)_2 + 2H_2O$

(d) $H_2S + 2LiOH \longrightarrow Li_2S + 2H_2O$

13-43. $LiOH + H_2S \longrightarrow LiHS + H_2O$

$LiOH + LiHS \longrightarrow Li_2S + H_2O$

13-44. $H_2S(aq) + OH^-(aq) \longrightarrow HS^-(aq) + H_2O(l)$

$HS^-(aq) + OH^-(aq) \longrightarrow S^{2-}(aq) + H_2O(l)$

13-47. $H_2S + NaOH \longrightarrow NaHS + H_2O$

13-50. The system would not be at equilibrium if $[H_3O^+]$ $= [OH^-] = 10^{-2}$M. Therefore, H_3O^+ reacts with OH^- until the concentration of each is reduced to 10^{-7} M. This is a neutralization reaction,

i.e., $H_3O^+ + OH^- \longrightarrow 2H_2O$

13-51. (a) $[H_3O^+] = \dfrac{K_w}{[OH^-]} = \dfrac{10^{-14}}{10^{-12}} = 10^{-2}$ M

(b) $[H_3O^+] = 10^{-15}$ M

(c) $[OH^-] = 5.0 \times 10^{-10}$ M

13-53. $[H_3O^+] = 0.0250$ M, $[OH^-] = 4.00 \times 10^{-13}$ M

13-55. (a) acidic **(b)** basic **(c)** acidic

13-57. (a) acidic **(b)** basic **(c)** acidic **(d)** basic

13-59. (a) 6.00 **(b)** 9.00 **(c)** 12.00 **(d)** 9.40 **(e)** 10.19

13-61. (a) pH = 4.0, pOH = 10.0 **(b)** pOH = 5.0, pH = 9.0 **(c)** pH = 1.70, pOH = 12.30 **(d)** pOH = 3.495, pH = 10.505

13-63. (a) 1.0×10^{-3} M **(b)** 2.9×10^{-4} M **(c)** 1.0×10^{-6} M **(d)** 2.4×10^{-8} M **(e)** 2.0×10^{-13} M

13-65. For exercise **13-59: (a)** acidic **(b)** basic **(c)** basic **(d)** basic **(e)** basic

For exercise **13-63: (a)** acidic **(b)** acidic **(c)** acidic **(d)** basic **(e)** basic

13-67. pH = 5.0 (less acidic), pH = 2.0 (more acidic)

13-69. pH = 1.12

13-71. pH = 12.56

13-72. $[H_3O^+] = 0.100 \times 0.10 = 0.010 = 1.0 \times 10^{-2}$ M
pH = 2.00

13-74. (a) strongly acidic **(b)** strongly acidic **(c)** weakly acidic **(d)** strongly basic **(e)** neutral **(f)** weakly basic **(g)** weakly acidic **(h)** strongly acidic

13-75. Ammonia (pH = 11.4), eggs (pH = 7.8), rainwater ($[H_3O^+] = 2.0 \times 10^{-6}$ M, pH = 5.70), grape juice ($[OH^-] = 1.0 \times 10^{-10}$ M, pH = 4.0), vinegar ($[H_3O^+] = 2.5 \times 10^{-3}$ M, pH = 2.60), sulfuric acid (pOH = 13.6, pH = 0.4)

13-78. (b) $C_2H_3O_2^- + H_2O \rightleftharpoons HC_2H_3O_2 + OH^-$

(c) $NH_3 + H_2O \rightleftharpoons NH_4^+ + OH^-$

13-80. (a) K^+ The cation of the strong base KOH.

(d) NO_3^- The conjugate base of the strong acid HNO_3.

(e) O_2 dissolves in water without formation of ions.

13-81. (a) HS^- **(b)** H_3O^+ **(c)** OH^- **(d)** $(CH_3)_2NH$

13-83. (a) $F^- + H_2O \rightleftharpoons HF + OH^-$

(b) $SO_3^{2-} + H_2O \rightleftharpoons HSO_3^- + OH^-$

(c) $(CH_3)_2NH_2^+ + H_2O \rightleftharpoons (CH_3)_2NH + H_3O^+$

(d) $HPO_4^{2-} + H_2O \rightleftharpoons H_2PO_4^- + OH^-$

(e) $CN^- + H_2O \rightleftharpoons HCN + OH^-$

(f) no hydrolysis (cation of a strong base)

13-85. $Ca(ClO)_2 \longrightarrow Ca^{2+}(aq) + 2ClO^-(aq)$

$ClO^-(aq) + H_2O \rightleftharpoons HClO(aq) + OH^-(aq)$

13-87. (a) neutral (neither ion hydrolyzes)

(b) acidic (cation hydrolysis)

(c) basic (anion hydrolysis)

(d) neutral (neither ion hydrolyzes)

(e) acidic (cation hydrolysis)

(f) basic (anion hydrolysis)

13-89. $CaC_2(s) + 2H_2O(l) \longrightarrow$
$$C_2H_2(g) + Ca^{2+}(aq) + 2OH^-(aq)$$

13-90. cation: $NH_4^+ + H_2O \rightleftharpoons NH_3 + H_3O^+$

anion: $CN^- + H_2O \rightleftharpoons HCN + OH^-$

Since the solution is basic, the anion hydrolysis reaction must take place to a greater extent than the cation hydrolysis.

13-92. (a), **(b)**, **(e)**, **(h)**, and **(i)** are buffer solutions.

13-93. There is no equilibrium when HCl dissolves in water. A reservoir of nonionized acid must be present to react with any strong base that is added. Likewise, the Cl^- ion does not exhibit base behavior in water, so it cannot react with any H_3O^+ added to the solution.

13-94. $N_2H_4(aq) + H_2O \rightleftharpoons N_2H_5^+(aq) + OH^-(aq)$

added H_3O^+: $H_3O^+ + N_2H_4 \longrightarrow N_2H_5^+ + H_2O$

added OH^-: $OH^- + N_2H_5^+ \longrightarrow N_2H_4 + H_2O$

13-96. $HPO_4^{2-}(aq) + H_2O(l) \rightleftharpoons PO_4^{3-}(aq) + H_3O^+(aq)$

added H_3O^+: $H_3O^+ + PO_4^{3-} \longrightarrow HPO_4^{2-} + H_2O$

added OH^-: $OH^- + HPO_4^{2-} \longrightarrow PO_4^{3-} + H_2O$

13-99. (a) $Sr(OH)_2$ **(b)** H_2SeO_4 **(c)** H_3PO_4

(d) $CsOH$ **(e)** HNO_2 **(f)** $HClO_3$

13-101. $CO_2(g) + LiOH(aq) \longrightarrow LiHCO_3(aq)$

13-102. $2LiNO_3(s)$

13-103. $Fe(s) + 2HI \longrightarrow FeI_2(aq) + H_2(g)$

13-105. $2HNO_3(aq) + Na_2SO_3(aq) \longrightarrow$
$$2NaNO_3(aq) + SO_2(g) + H_2O(l)$$

13-107. (a) $HCN + NH_3 \longrightarrow NH_4^+ + CN^-$

(b) $NH_3 + H^- \longrightarrow H_2 + NH_2^-$

(c) $HCO_3^- + NH_3 \longrightarrow NH_4^+ + CO_3^{2-}$

(d) $NH_4^+Cl^- + Na^+NH_2^- \longrightarrow Na^+Cl^- + 2NH_3$

13-108. (a) $HCl + CH_3OH \longrightarrow CH_3OH_2^+ + Cl^-$

(b) $CH_3OH + NH_2^- \longrightarrow NH_3 + CH_3O^-$

13-110. (a) acidic: $H_2S + H_2O \rightleftharpoons H_3O^+ + HS^-$

(b) basic: $ClO^- + H_2O \rightleftharpoons HClO + OH^-$

(c) neutral

(d) basic: $NH_3 + H_2O \rightleftharpoons NH_4^+ + OH^-$

(e) acidic: $N_2H_5^+ + H_2O \rightleftharpoons N_2H_4 + H_3O^+$

(f) basic: $Ba(OH)_2 \longrightarrow Ba^{2+} + 2OH^-$

(g) neutral

(h) basic: $NO_2^- + H_2O \rightleftharpoons HNO_2 + OH^-$

(i) acidic: $H_2SO_3 + H_2O \rightleftharpoons H_3O^+ + HSO_3^-$

(j) acidic: $Cl_2O_3 + H_2O \longrightarrow 2HClO_2$;
$$HClO_2 + H_2O \rightleftharpoons H_3O^+ + ClO_2^-$$

13-112. LiOH, strongly basic, pH = 13.0; $SrBr_2$, neutral, pH = 7.0; KClO, weakly basic, pH = 10.2; NH_4Cl, weakly acidic, pH = 5.2; HI, strongly acidic, pH = 1.0

When pH = 1.0, $[H_3O^+] = 0.10$ M. If HI is completely ionized, its initial concentration must be 0.10 M.

13-114. 8.2 kg of H_2SO_4

13-115. pH = 0.553 (con) pH = 1.113 (dilute)

13-117. 0.024 mol of HCl remains after neutralization pH = 1.62

13-119. 0.025 mol $Ca(OH)_2$ remaining in 1.00 L

$$[OH^-] = 0.025 \; \overline{\text{mol } Ca(OH)_2} \times \frac{2 \text{ mol } OH^-}{\overline{\text{mol } Ca(OH)_2}}$$

$$= 0.050 \text{ mol/L} \quad pOH = 1.30 \quad \underline{pH = 12.70}$$

CHAPTER 14

14-1. (a) Pb +4, O −2 **(b)** P +5, O −2 **(c)** C −1, H +1

(d) N −2, H +1 **(e)** Li +1, H −1 **(f)** B +3, Cl −1

(g) Rb +1, Se −2 **(h)** Bi +3, S −2

14-3. (a) Li, **(e)** K, and **(g)** Rb

14-5. +3

14-6. (a) P = +5 **(b)** C = +3 **(c)** Cl = +7 **(d)** Cr = +6

(e) S = +6 **(f)** N = +5 **(g)** Mn = +7

14-7. (a) S = +6 **(b)** Co = +3 **(c)** U = +6 **(d)** N = +5

(e) Cr = +6 **(f)** Mn = +6

14-9. (a), **(c)**, and **(f)**

14-10. (a) oxidation **(b)** reduction **(c)** oxidation

(d) reduction **(e)** reduction

14-12.

Reactant Oxidized	Product of Oxidation	Reactant Reduced	Product of Reduction	Oxidizing Agent	Reducing Agent
Br^-	Br_2	MnO_2	Mn^{2+}	MnO_2	Br^-
CH_4	CO_2	O_2	CO_2, H_2O	O_2	CH_4
Fe^{2+}	Fe^{3+}	MnO_4^-	Mn^{2+}	MnO_4^-	Fe^{2+}

14-13.

Reactant Oxidized	Product of Oxidation	Reactant Reduced	Product of Reduction	Oxidizing Agent	Reducing Agent
Al	AlO_2^-	H_2O	H_2	H_2O	Al
Mn^{2+}	MnO_4^-	$Cr_2O_7^{2-}$	Cr^{3+}	$Cr_2O_7^{2-}$	Mn^{2+}

14-16.

$$\overset{-5e^- \times 4 = -20e^-}{\underset{-3 \qquad\qquad +2}{\rule{3cm}{0pt}}}$$

(a) $4NH_3 + 5O_2 \longrightarrow 4NO + 6H_2O$

$$\underset{0 \qquad\qquad -2}{\rule{3cm}{0pt}}$$
$$+4e^- \times 5 = +20e^-$$

$$\overset{-4e^- \times 1 = -4e^-}{\underset{0 \qquad\qquad +4}{\rule{3cm}{0pt}}}$$

(b) $Sn + 4HNO_3 \longrightarrow SnO_2 + 4NO_2 + 2H_2O$

$$\underset{+5 \qquad\qquad +4}{\rule{3cm}{0pt}}$$
$$+1e^- \times 4 = +4e^-$$

(c) Before the number of electrons lost is calculated, notice that a temporary coefficient of "2" is needed for the Na_2CrO_4 in the products since there are 2 Cr's in Cr_2O_3 in the reactants.

$$\overset{+2e^- \times 3 = +6e^-}{\underset{+5 \qquad\qquad +3}{\rule{4cm}{0pt}}}$$

$Cr_2O_3 + 2Na_2CO_3 + 3KNO_3 \longrightarrow 2CO_2 + 2Na_2CrO_4 + 3KNO_2$

$$\underset{+6 \qquad\qquad\qquad +12}{\rule{4cm}{0pt}}$$
$$-6e^- \times 1 = -6e^-$$

$$\overset{-4e^- \times 3 = -12e^-}{\underset{0 \qquad\qquad +4}{\rule{4cm}{0pt}}}$$

(d) $3Se + 2BrO_3^- + 3H_2O \longrightarrow 3H_2SeO_3 + 2Br^-$

$$\underset{+5 \qquad\qquad -1}{\rule{4cm}{0pt}}$$
$$+6e^- \times 2 = +12e^-$$

14-18. (a) $2H_2O + Sn^{2+} \longrightarrow SnO_2 + 4H^+ + 2e^-$

(b) $2H_2O + CH_4 \longrightarrow CO_2 + 8H^+ + 8e^-$

(c) $e^- + Fe^{3+} \longrightarrow Fe^{2+}$

(d) $6H_2O + I_2 \longrightarrow 2IO_3^- + 12H^+ + 10e^-$

(e) $e^- + 2H^+ + NO_3^- \longrightarrow NO_2 + H_2O$

14-20. (a)

$$S^{2-} \longrightarrow S + 2e^- \qquad \times 3$$
$$\underline{3e^- + 4H^+ + NO_3^- \rightarrow NO + 2H_2O \qquad \times 2}$$
$$3S^{2-} + 8H^+ + 2NO_3^- \rightarrow 3S + 2NO + 4H_2O$$

(b) $2S_2O_3^{2-} + I_2 \longrightarrow S_4O_6^{2-} + 2I^-$

(c)
$$H_2O + SO_3^{2-} \longrightarrow SO_4^{2-} + 2H^+ + 2e^- \qquad \times 3$$
$$\underline{6e^- + 6H^+ + ClO_3^- \rightarrow Cl^- + 3H_2O} \qquad \times 1$$
$$3SO_3^{2-} + ClO_3^- \rightarrow Cl^- + 3SO_4^{2-}$$

(d) $2H^+ + 2Fe^{3+} + H_2O_2 \longrightarrow 2Fe^{3+} + 2H_2O$

(e) $AsO_4^{3-} + 2I^- + 2H^+ \longrightarrow I_2 + AsO_3^{3-} + H_2O$

(f) $4Zn + NO_3^- + 10H^+ \longrightarrow 4Zn^{2+} + NH_4^+ + 3H_2O$

14-22. (a) $2OH^- + SnO_2^{2-} \longrightarrow SnO_3^{2-} + H_2O + 2e^-$

(b) $6e^- + 4H_2O + 2ClO_2^- \longrightarrow Cl_2 + 8OH^-$

(c) $6OH^- + Si \longrightarrow SiO_3^{2-} + 3H_2O + 4e^-$

(d) $8e^- + 6H_2O + NO_3^- \longrightarrow NH_3 + 9OH^-$

14-24. (a) In acid solution:
$$4H_2O + S^{2-} + 4I_2 \longrightarrow SO_4^{2-} + 8I^- + 8H^+$$
Add $8OH^-$ to both sides and simplify.
$$8OH^- + S^{2-} + 4I_2 \longrightarrow SO_4^{2-} + 8I^- + 4H_2O$$

(b) $8OH^- + I^- + 8MnO_4^- \longrightarrow$
$$8MnO_4^{2-} + IO_4^- + 4H_2O$$

(c) $2H_2O + SnO_2^{2-} + BiO_3^- \longrightarrow$
$$SnO_3^{2-} + Bi(OH)_3 + OH^-$$

(d) $32OH^- + CrI_3 \longrightarrow$
$$CrO_4^{2-} + 3IO_4^- + 16H_2O + 27e^- \qquad \times 2$$
$$\underline{2e^- + Cl_2 \rightarrow 2Cl^-} \qquad\qquad\qquad \times 27$$
$$2CrI_3 + 64OH^- + 27Cl_2 \longrightarrow$$
$$2CrO_4^{2-} + 6IO_4^- + 32H_2O + 54Cl^-$$

14-26. (a) $2H_2 + O_2 \longrightarrow 2H_2O$

(b)
$$H_2O_2 \longrightarrow O_2 + 2H^+ + 2e^-$$
$$\underline{2e^- + 2H^+ + H_2O_2 \longrightarrow 2H_2O}$$
$$2H_2O_2 \longrightarrow O_2 + 2H_2O$$
$$2OH^- + H_2O_2 \longrightarrow O_2 + 2H_2O + 2e^-$$
$$\underline{2e^- + H_2O_2 \longrightarrow 2OH^-}$$
$$2H_2O_2 \longrightarrow O_2 + 2H_2O$$

14-27. Reactions **(a)**, **(c)**, **(e)**, and **(f)** are predicted to be favorable.

14-29. (a) no reaction

(b) $Br_2(aq) + Sn(s) \longrightarrow SnBr_2(aq)$

(c) no reaction

(d) $O_2(g) + 4H^+(aq) + 4Br^-(aq) \longrightarrow$
$$2Br_2(l) + 2H_2O(l)$$

(e) $3Br_2(l) + 2Al(s) \longrightarrow 2AlBr_3(s)$

14-31. (c) $2F_2(g) + 2H_2O(l) \longrightarrow 4HF(aq) + O_2(g)$

(e) $Mg(s) + 2H_2O(l) \longrightarrow Mg(OH)_2(s) + H_2(g)$

14-33. From Table 14-1:
$$Fe + 2H_2O \longrightarrow \text{no reaction}$$
$$Fe + 2H^+ \longrightarrow Fe^{2+} + H_2$$

Acid rain has a higher $H^+(aq)$ concentration, thus making the second reaction more likely.

14-36. $Zn(s)$ reacts at the anode and $MnO_2(s)$ reacts at the cathode. The total reaction is
$$Zn(s) + 2MnO_2(s) + 2H_2O(l) \longrightarrow$$
$$Zn(OH)_2(s) + 2MnO(OH)(s)$$

14-38. anode: $Cd(s) + 2OH^-(aq) \longrightarrow Cd(OH)_2(s) + 2e^-$

cathode: $NiO(OH)(s) + 2H_2O(l) + e^- \longrightarrow$
$$Ni(OH)_2(s) + OH^-(aq)$$

14-40. The spontaneous reaction is

$$Pb(NO_3)_2(aq) + Fe(s) \longrightarrow Fe(NO_3)_2(aq) + Pb(s)$$

anode: $Fe \longrightarrow Fe^2 + 2e^-$

cathode: $Pb^{2+} + 2e^- \longrightarrow Pb$

14-41. The Daniell cell is more powerful. The greater the separation between oxidizing and reducing agents as shown in Table 14-1, the more powerful is the cell.

14-43. $3Fe(s) + 2Cr^{3+}(aq) \longrightarrow 3Fe^{2+}(aq) + 2Cr(s)$

Zinc would spontaneously form a chromium coating as illustrated by the equation:

$$3Zn(s) + 2Cr^{3+}(aq) \longrightarrow 3Zn^{2+}(aq) + 2Cr(s)$$

14-44. Sodium reacts spontaneously with water since it is an active metal. The actual cathode reaction is

$$2H_2O + 2e^- \longrightarrow H_2 + 2OH^-$$

Elemental sodium is produced by electrolysis of the molten salt, NaCl.

14-47. The strongest oxidizing agent is reduced the easiest. Thus the reduction of Ag^+ to Ag occurs first. This procedure can be used to purify silver.

14-48. K_3N (–3), N_2H_4 (–2), NH_2OH, (–1), N_2 (0),

N_2O (+1), NO (+2), N_2O_3 (+3), N_2O_4 (+4),

$Ca(NO_3)_2$ (+5)

14-49. $Cd(s) + NiCl_2(aq) \longrightarrow Ni(s) + CdCl_2(aq)$

$Zn(s) + CdCl_2(aq) \longrightarrow Cd(s) + ZnCl_2(aq)$

These reactions indicate that Cd^{2+} is a stronger oxidizing agent than Zn^{2+} but weaker than Ni^{2+}. It appears about the same as Fe^{2+}.

14-51. The reactions that occur are

$$4Au^{3+} + 6H_2O \longrightarrow 4Au + 12H^+ + 3O_2$$

$$2Au + 3F_2 \longrightarrow 2AuF_3$$

These reactions and the fact that Au does not react with Cl_2 rank Au^{3+} above H^+ and Cl_2 but below F_2.

14-52. $12H^+(aq) + 5Zn(s) + 2NO_3^-(aq) \longrightarrow$

$$5Zn^{2+}(aq) + N_2(g) + 6H_2O(l)$$

0.658 g N_2 requires 7.68 g Zn.

14-54. $14H^+(aq) + 2NO_3^-(aq) + 3Cu_2O(s) \longrightarrow$

$$6Cu^{2+}(aq) + 2NO(g) + 7H_2O(l)$$

10.0 g of Cu_2O produces 1.04 L of NO measured at STP.

14-56. $7OH^-(aq) + 4Zn(s) + 6H_2O(l) + NO_3^-(aq) \longrightarrow$

$$NH_3(g) + 4Zn(OH)_4^{2-}(aq)$$

6.54 g of Zn produces 0.493 L of NH_3 at 27.0°C and 1.25 atm pressure.

14-57. (a) $MnO_4^- + 8H^+ + 5Fe^{2+} \longrightarrow$

$$5Fe^{3+} + Mn^{2+} + 4H_2O$$

(b) $2MnO_4^- + 16H^+ + 10Br^- \longrightarrow$

$$5Br_2 + 2Mn^{2+} + 8H_2O$$

(c) $2MnO_4^- + 16H^+ + 5C_2O_4^{2-} \longrightarrow$

$$10CO_2 + 2Mn^{2+} + 8H_2O$$

14-58. 179 mL

14-59. 306 mL

CHAPTER 15

15-1. Colliding molecules must have the proper orientation relative to each other at the time of the collision, and the colliding molecules must have the minimum kinetic energy for the particular reaction.

15-3. (a) As the temperature increases, the frequency of collisions between molecules increases, as does the average energy of the collisions. Both contribute to the increased rate of reaction.

(b) The cooking of eggs initiates a chemical reaction that occurs more slowly at lower temperatures.

(c) The average energy of colliding molecules at room temperature is not sufficient to initiate a reaction between H_2 and O_2.

(d) A higher concentration of oxygen increases the rate of combustion.

(e) When a solid is finely divided, a greater surface area is available for collisions with oxygen molecules. Thus it burns faster.

(f) The souring of milk is a chemical reaction that slows as the temperature drops. It takes several days in a refrigerator.

(g) The platinum is a catalyst. Since the activation energy in the presence of a catalyst is lower, the reaction can occur at a lower temperature.

15-6. The rate of the forward reaction was at a maximum at the beginning of the reaction; the rate of the reverse reaction was at a maximum at the point of equilibrium.

15-7. In many cases, reactions do not proceed directly to the right because other products are formed between the same reactants. For example, combustion may produce carbon monoxide as well as carbon dioxide.

15-8. Products are easier to form because the activation energy for the forward reaction is less than for the reverse reaction. This is true of all exothermic reactions. The system should come to equilibrium faster starting with pure reactants.

15-10. (a) right   (b) left   (c) left   (d) right

(e) right   (f) has no effect   (g) yield decreases

(h) yield increases but rate of formation decreases

15-12. (a) increase   (b) increase   (c) decrease

(d) decrease   (e) decrease   (f) no effect

15-14. (a) Since there are the same number of moles of gas on both sides of the equation, pressure (or volume) has no effect on the point of equilibrium.

(b) decrease the amount of NO

(c) decrease the amount of NO

15-15. (a) $K_{eq} = \dfrac{[COCl_2]}{[CO][Cl_2]}$ (b) $K_{eq} = \dfrac{[CO_2][H_2]^4}{[CH_4][H_2O]^2}$

(c) $K_{eq} = \dfrac{[Cl_2]^2[H_2O]^2}{[HCl]^4[O_2]}$ (d) $K_{eq} = \dfrac{[CH_3Cl][HCl]}{[CH_4][Cl_2]}$

15-17. products

15-19. There will be an appreciable concentration of both reactants and products at equilibrium.

15-20. $K_{eq} = 0.34$

15-21. $K_{eq} = 49.2$

15-23.
$$K_{eq} = \frac{[CO_2][H_2]^4}{[CH_4][H_2O]^2} = \frac{(2.20/30.0)(4.00/30.0)^4}{(6.20/30.0)(3.00/30.0)^2} = 0.0112$$

15-25. **(a)** $[H_2] = [I_2] = 0.30$ mol/L

(b) $[H_2] = 0.30$ mol/L; $[I_2] = 0.50$ mol/L

(c) $[HI] = 0.40$ mol/L; $[I_2] = [H_2] = 0.10$ mol/L

(d) $K_{eq} = 0.063$ **(e)** $K_r = 16$.

This is a smaller value than that used in Table 15-1. This indicates that the equilibrium in this problem was established at a different temperature than that of Table 15-1.

15-27. **(a)** $[N_2]_{reacts} = 0.50 - 0.40 = 0.10$ mol/L

$$0.10 \text{ mol } N_2 \times \frac{3 \text{ mol } H_2}{1 \text{ mol } N_2} = 0.30 \text{ mol/L } H_2 \text{(reacts)}$$

$[H_2]_{eq} = 0.50 - 0.30 = \underline{0.20 \text{ mol/L}}$

$$0.10 \text{ mol } N_2 \times \frac{2 \text{ mol } NH_3}{1 \text{ mol } N_2} = 0.20 \text{ mol/L } NH_3 \text{ formed}$$

$[NH_3]_{eq} = 0.50 + 0.20 = \underline{0.70 \text{ mol/L}}$

(b) $K_{eq} = \dfrac{(0.70)^2}{(0.20)^3(0.40)} = \underline{150}$

15-28. **(a)** The concentration of O_2 that reacts is

$$0.25 \text{ mol } NH_3 \times \frac{5 \text{ mol } O_2}{4 \text{ mol } NH_3} = 0.31 \text{ mol/L } O_2$$

(b) $[NH_3] = 0.75$ mol/L; $[O_2] = 0.69$ mol/L O_2;

$[NO] = 0.25$ mol/L NO (formed)

$[H_2O] = 0.38$ mol/L H_2O

(c) $K_{eq} = \dfrac{[NO]^4[H_2O]^6}{[NH_3]^4[O_2]^5} = \dfrac{(0.25)^4(0.38)^6}{(0.75)^4(0.69)^5}$

15-30. $[PCl_5] = 0.28$ mol/L

15-32. $[H_2O] = 0.26$ mol/L

15-33. Let $x = [HCl] = [CH_3Cl]$

$$K_{eq} = \frac{[HCl][CH_3Cl]}{[CH_4][Cl_2]} = \frac{x^2}{(0.20)(0.40)} = 56$$

$x^2 = 4.5$ $x = \underline{2.1 \text{ mol/L}}$

15-35. **(a)** $K_a = \dfrac{[H_3O^+][BrO^-]}{[HBrO]}$

(b) $K_b = \dfrac{[NH_4^+][OH^-]}{[NH_3]}$ **(c)** $K_a = \dfrac{[H_3O^+][HSO_3^-]}{[H_2SO_3]}$

(d) $K_a = \dfrac{[H_3O^+][SO_3^{2-}]}{[HSO_3^-]}$ **(e)** $K_a = \dfrac{[H_3O^+][H_2PO_4^-]}{[H_3PO_4]}$

(f) $K_b = \dfrac{[(CH_3)_2NH_2^+][OH^-]}{[(CH_3)_2NH]}$

15-37. The acid HB is weaker because it produces a smaller hydronium ion concentration (higher pH) at the same initial concentration of acid. The stronger acid HX has the larger value of K_a.

15-39. **(a)** $[HOCN]_{eq} = 0.20 - 0.0062 = 0.19$

(b) $K_a = \dfrac{[H_3O^+][OCN^-]}{[HOCN]} = \dfrac{(6.2 \times 10^{-3})(6.2 \times 10^{-3})}{0.19} = 2.0 \times 10^{-4}$

(c) pH = 2.21

15-40. **(a)** $HX + H_2O \rightleftharpoons H_3O^+ + X^-$ $K_a = \dfrac{[H_3O^+][X^-]}{[HX]}$

(b) From the equation $[H_3O^+] = [X^-] = 0.100 \times 0.58 = 0.058$ $[HX] = 0.58 - 0.058 = 0.52$

(c) $K_a = 6.5 \times 10^{-3}$ **(d)** pH = 1.24

15-43. $Nv + H_2O \rightleftharpoons NvH^+ + OH^-$ $K_b = 6.6 \times 10^{-6}$

15-44. $[H_3O^+] = 0.300 - 0.277 = 0.023$

pH = 1.64 $K_a = 1.9 \times 10^{-3}$

15-45. pH = 4.43

15-47. $[OH^-] = 3.2 \times 10^{-3}$

15-49. pH = 12.43

15-51. pH = 9.40

15-53. $pK_a = 8.68$ $pH = 8.68 + \log \dfrac{0.20/0.850}{0.60/0.850} =$

$8.68 - 0.48 = \underline{8.20}$

15-56. $pK_b = 6.01$ $pOH = 6.01 + \log \dfrac{0.0288 \text{ mol acid}}{0.0469 \text{ mol base}}$

$pOH = 6.01 - 0.21 = 5.80$ $pH = \underline{8.20}$

15-58. When the concentrations of acid and base species are equal, $pH = pK_a$ and $pOH = pK_b$. If a buffer of pH = 7.50 is required, then we look for an acid with $K_a = [H_3O^+] = 3.2 \times 10^{-8}$ or a base with $K_b = [OH^-] = 3.2 \times 10^{-7}$. Since $K_a = 3.2 \times 10^{-8}$ for HClO, an equimolar mixture of HClO and KClO produces the required buffer.

15-60. **(a)** $FeS(s) \rightleftharpoons Fe^{2+}(aq) + S^{2-}(aq)$

$K_{sp} = [Fe^{2+}][S^{2-}]$

(b) $Ag_2S(s) \rightleftharpoons 2Ag^+(aq) + S^{2-}(aq)$

$K_{sp} = [Ag^+]^2[S^{2-}]$

(c) $Zn(OH)_2(s) \rightleftharpoons Zn^{2+}(aq) + 2OH^-(aq)$

$K_{sp} = [Zn^{2+}][OH^-]^2$

15-62. $K_{sp} = 4.8 \times 10^{-12}$

15-64. $Ca(OH)_2(s) \longrightarrow Ca^{2+}(aq) + 2OH^-(aq)$

$K_{sp} = [Ca^{2+}][OH^-]^2$

At equilibrium $[Ca^{2+}] = 1.3 \times 10^{-2}$ mol/L; $[OH^-] = 2 \times (1.3 \times 10^{-2})$ mol/L $= 2.6 \times 10^{-2}$ mol/L

$K_{sp} = [1.3 \times 10^{-2}][2.6 \times 10^{-2}]^2 = 8.8 \times 10^{-6}$

15-66. 9.1×10^{-9} mol/L

15-68. $AgBr(s) \longrightarrow Ag^+(aq) + Br^-(aq)$

$K_{sp} = [Ag^+][Br^-] = 7.7 \times 10^{-13}$

$[7 \times 10^{-6}][8 \times 10^{-7}] = 6 \times 10^{-12}$

This is a larger number than K_{sp}, so a precipitate of AgBr does form.

15-70. (a) $3Cl_2(g) + NH_3(g) \rightleftharpoons NCl_3(g) + 3HCl(g)$

(b) $K_{eq} = \dfrac{[NCl_3][HCl]^3}{[NH_3][Cl_2]^3}$

(c) no effect

(d) decreases $[NH_3]$ **(e)** reactants

(f) $[NH_3] = 0.083$ mol/L

15-72. (a) $2N_2O(g) \rightleftharpoons 2N_2(g) + O_2(g)$

(b) $K_{eq} = \dfrac{[N_2]^2[O_2]}{[N_2O]^2}$

(c) decreases $[N_2]$

(d) decreases $[N_2O]$

(e) $[N_2O]_{eq} = 0.10 - (0.015 \times 0.10) = 0.10$ mol/L;

$[N_2] = 0.015 \times 0.10 = 1.5 \times 10^{-3}$ mol/L;

$[O_2] = \dfrac{1.5 \times 10^{-3}}{2} = 7.5 \times 10^{-4}$ mol/L

$K_{eq} = \underline{1.7 \times 10^{-7}}$

15-74. (a) A solution of $NaHCO_3$ is slightly basic because the hydrolysis reaction occurs to a greater extent than the acid ionization reaction. That is, K_b is larger than K_a.

(b) It is used as an antacid to counteract excess stomach acidity.

(c) $HCl + H_2O \longrightarrow H_3O^+ + Cl^-$ (HCl is a strong acid.)

$H_3O^+(aq) + HCO_3^-(aq) \longrightarrow H_2CO_3(aq) + H_2O(l)$

$H_2CO_3(aq) \longrightarrow H_2O(l) + CO_2(g)$

(d) No. The HSO_4^- does not react with H_3O^+ because it would form H_2SO_4, which is a strong acid. The molecular form of a strong acid does not exist in water.

15-76. $0.265 \, \cancel{L} \times 0.22 \, \text{mol}/\cancel{L} = 0.058$ mol $HC_2H_3O_2$

$0.375 \, \cancel{L} \times \dfrac{0.12 \, \text{mol } \cancel{Ba(C_2H_3O_2)_2}}{\cancel{L}} \times$

$\dfrac{2 \, \text{mol } C_2H_3O_2^-}{\text{mol } \cancel{Ba(C_2H_3O_2)_2}} = 0.090$ mol $C_2H_3O_2^-$ $\qquad$ pH $= 4.93$

15-77. The addition of the NaOH neutralizes part of the acetic acid to produce sodium acetate.

$HC_2H_3O_2 + NaOH \longrightarrow NaC_2H_3O_2 + H_2O$

$1.00 - 0.20 = 0.80$ mol of $HC_2H_3O_2$ remains, and 0.20 mol of $C_2H_3O_2^-$ is formed after 0.20 mol of OH^- is added.

$\underline{\text{pH} = 4.14}$

15-78. $[OH^-] = 3.2 \times 10^{-4}$; $[Mg^{2+}] = [OH^-]/2 = 1.6 \times 10^{-4}$

$K_{sp} = [1.6 \times 10^{-4}][3.2 \times 10^{-4}]^2 = 1.6 \times 10^{-11}$

15-80. 1.0×10^{-5} mol/L, 0.023 g $BaSO_4$/L

CHAPTER 16

16-1. (a) $^{210}_{84}Po$ **(b)** $^{84}_{38}Sr$ **(c)** $^{257}_{100}Fm$ **(d)** $^{206}_{82}Pb$ **(e)** $^{233}_{92}U$

16-3. (a) ^4_2He **(b)** $^0_{-1}e$ **(c)** ^1_0n

16-4. ^2_1H

16-6. (a) $^{206}_{82}Pb$ **(b)** $^{148}_{62}Sm$ **(c)** $^{248}_{98}Cf$ **(d)** $^{262}_{107}Bh$

16-8. (a) ^3_2He **(b)** $^{153}_{65}Tb$ **(c)** $^{59}_{27}Co$ **(d)** $^{24}_{12}Mg$

16-11. $^{51}_{25}Mn \rightarrow \, ^{51}_{24}Cr + \, ^0_{+1}e$

16-14. $^{68}_{32}Ge + \, ^0_{-1}e \rightarrow \, ^{68}_{31}Ga$

16-16. (a) $^0_{-1}e$ **(b)** $^{90}_{38}Sr$ **(c)** $^{26}_{13}Al$ **(d)** $^{231}_{90}Th$

(e) $^{179}_{72}Hf$ **(f)** $^{41}_{20}Ca$ **(g)** $^{210}_{81}Tl$

16-18. (a) $^{230}_{90}Th \longrightarrow \, ^{226}_{88}Ra + \, ^4_2He$

(b) $^{214}_{84}Po \longrightarrow \, ^{210}_{82}Pb + \, ^4_2He$

(c) $^{210}_{84}Po \longrightarrow \, ^{210}_{85}At + \, ^0_{-1}e$

(d) $^{218}_{84}Po \longrightarrow \, ^{214}_{82}Pb + \, ^4_2He$

(e) $^{14}_6C \longrightarrow \, ^{14}_7N + \, ^0_{-1}e$

(f) $^{50}_{25}Mn \longrightarrow \, ^{50}_{24}Cr + \, ^0_{+1}e$

(g) $^{37}_{18}Ar + \, ^0_{-1}e \longrightarrow \, ^{37}_{17}Cl$

16-19. $^{234}_{90}Th, \, ^{234}_{91}Pa, \, ^{234}_{92}U, \, ^{230}_{90}Th, \, ^{226}_{88}Ra, \, ^{222}_{86}Rn, \, ^{218}_{84}Po, \, ^{214}_{82}Pb,$ $^{214}_{83}Bi, \, ^{210}_{81}Tl, \, ^{210}_{82}Pb, \, ^{210}_{83}Bi, \, ^{210}_{84}Po, \, ^{206}_{82}Pb$

16-20. $\dfrac{1}{16}$

16-22. (a) 5 mg **(b)** 1.25 mg **(c)** 1.25 mg

16-23. 2.50 g

16-25. about 11,500 years old

16-27. The energy from the radiation causes an electron in an atom or a molecule to be expelled, leaving behind a positive ion. When a molecule in a cell is ionized, it is damaged and may die or mutate.

16-29. Radiation entering a chamber causes ionization. The electrons formed migrate to the central electrode (positive) and cause a burst of current that can be detected and amplified. In a scintillation counter, the radiation is detected by phosphors that glow when radiation is absorbed.

16-30. (a) $^{35}_{16}S$ **(b)** ^2_1H **(c)** $^{30}_{15}P$ **(d)** $8\,^0_{-1}e$ **(e)** $^{254}_{102}No$

(f) $^{237}_{92}U$ **(g)** $4\,^1_0n$

16-32. one neutron

16-34. $^{208}_{82}Pb$

16-35. $^{262}_{107}Bh$

16-36. (a) $^{87}_{34}Se$ **(b)** $^{97}_{38}Sr$

16-38. $^2_1H + \, ^2_1H \Big\langle \begin{array}{l} \nearrow \, ^3_2He + \, ^1_0n \\ \searrow \, ^3_1H + \, ^1_1H \end{array}$

16-39. (a) $^{106}_{46}Pd + \, ^4_2He \longrightarrow \, ^{109}_{47}Ag + \, ^1_1H$

(b) $^{266}_{109}Mt \longrightarrow \, ^{262}_{107}Bh + \, ^4_2He$

$^{262}_{107}Bh \longrightarrow \, ^{258}_{105}Db + \, ^4_2He$

(c) $^{212}_{83}Bi \longrightarrow \, ^{212}_{84}Po + \, ^0_{-1}e$

(d) $^{60}_{30}Zn \longrightarrow \, ^{60}_{29}Cu + \, ^0_{+1}e$

(e) $^{239}_{94}Pu + \, ^1_0n \longrightarrow \, ^{140}_{55}Cs + \, ^{97}_{39}Y + 3\,^1_0n$

(f) $^{206}_{82}Pb + \, ^{54}_{24}Cr \longrightarrow \, ^{257}_{106}Sg + 3\,^1_0n$

(g) $^{93}_{42}Mo + \, ^0_{-1}e \longrightarrow \, ^{93}_{41}Nb$

16-41. (a) $^{90}_{39}Y$ **(b)** 50 years

16-44. 60 years is five half-lives. $0.42 \, \mu g \times 2^5 = \underline{13.4 \, \mu g}$

16-46. $^{272}_{111}Rg$

16-48. ^{70}Zn

Photo Credits

Index

Prefixes and Numerical Values for SI Units

Prefix	Symbol	Numerical value	Power of 10 equivalent
exa	E	1,000,000,000,000,000,000	10^{18}
peta	P	1,000,000,000,000,000	10^{15}
tera	T	1,000,000,000,000	10^{12}
giga	G	1,000,000,000	10^{9}
mega	M	1,000,000	10^{6}
kilo	k	1,000	10^{3}
hecto	h	100	10^{2}
deka	da	10	10^{1}
—	—	1	10^{0}
deci	d	0.1	10^{-1}
centi	c	0.01	10^{-2}
milli	m	0.001	10^{-3}
micro	μ	0.000001	10^{-6}
nano	n	0.000000001	10^{-9}
pico	p	0.000000000001	10^{-12}
femto	f	0.000000000000001	10^{-15}
atto	a	0.000000000000000001	10^{-18}

SI Units and Conversion Factors

Length

SI unit: meter (m)

1 meter	=	1000 millimeters
1 meter	=	1.0936 yards
1 centimeter	=	0.3937 inch
1 inch	=	2.54 centimeters (exactly)
1 kilometer	=	0.62137 mile
1 mile	=	5280 feet
1 mile	=	1.609 kilometers
1 angstrom	=	10^{-10} meter

Mass

SI unit: kilogram (kg)

1 kilogram	=	1000 grams
1 kilogram	=	2.20 pounds
1 gram	=	1000 milligrams
1 pound	=	453.59 grams
1 pound	=	0.45359 kilogram
1 pound	=	16 ounces
1 ton	=	2000 pounds
1 ton	=	907.185 kilograms
1 ounce	=	28.3 grams
1 atomic mass unit	=	1.6606×10^{-27} kilograms

Volume

SI unit: cubic meter (m³)

1 liter	=	1000 milliliters
1 liter	=	$10^{-3} m^3$
1 liter	=	1 dm³
1 liter	=	1.0567 quarts
1 gallon	=	4 quarts
1 gallon	=	8 pints
1 gallon	=	3.785 liters
1 quart	=	32 fluid ounces
1 quart	=	0.946 liter
1 quart	=	4 cups
1 fluid ounce	=	29.6 mL

Temperature

SI unit: kelvin (K)

$0 \text{ K} = -273.15°\text{C}$

$= -459.67°\text{F}$

$\text{K} = °\text{C} + 273.15$

$°\text{C} = \dfrac{(°\text{F} - 32)}{1.8}$

$°\text{F} = 1.8(°\text{C}) + 32$

Energy

SI unit: joule (J)

1 joule	=	1 kg m²/s²
1 joule	=	0.23901 calorie
1 calorie	=	4.184 joules

Pressure

SI unit: pascal (Pa)

1 pascal	=	1 kg/(ms²)
1 atmosphere	=	101.325 kilopascals
1 atmosphere	=	760 torr
1 atmosphere	=	760 mm Hg
1 atmosphere	=	14.70 pounds per square inch (psi)